"Why do engineers prefer this FE/EIT study guide?"

Some notes from the GLP in-basket....

"This book provided all of the necessary information for passing the FE exam. I was able to review for and *pass* the exam in only 3 weeks!"

—*Joseph Rozza, Orlando, Florida, a recent university graduate*

"I passed the first time I took the test. Potter's book is a great, great review. It's right to-the-point and doesn't include extra stuff you don't need. All I did was work straight through the book—it didn't take that long—and went into the test and passed it. A lot of my friends took review courses and spent all kinds of time studying, didn't use Potter's book, and failed the FE/EIT! The other books have way too much detail." — *Dana, Boston, Massachusetts*

"Your review material is perfect for the non-traditional 'night student.' It refreshes and restores many years of part-time study, without being overly exhaustive."

—*Jerome Bobak, New York, Mechanical. Engineer.*

"Before I bought your book I thought, 'How on earth am I going to pass this exam?!' Then I opened it and everything was so simple, so clear. The problems and examples of what was going to be on the test were easy to find—that's all that's in the book! Now I'm setting aside other materials and focusing on yours because it's all I need." — *Blandey Doll, SWE officer, Ohio Northern University*

"I bought the big yellow book before I found yours to study for the FE/EIT exam. I used your FE book and thought it was far superior to any other. It was 'to-the point' and completely focused on the exam—I took the exam and did great! Now I need your PE review."

—*Tom Roach, Indiana Midwest Steel*

"I bought two books for the exam and I liked yours much better. The other was four times as long and didn't have anything more than your *FE* that I needed to study for the exam. Your book was very informative and more than adequate to pass the exam. I'm calling to see if you have a review for the PE exam." —*Burton Mills, Fruehauf Company, Fort Madison, Iowa*

Professors' comments....

"I was talking to a student a few hours ago and he said the morning problems on the FE/EIT were like the sample problems in your book—verbatim. It's the best review I've ever seen. When we started using your review in 1994, our students' pass rates went up 20%! Now we will continue to use your book because we take professional licensing seriously."

—*Dr. Gary Rogers, Virginia Military Institute*

"The best book I've found after an exhaustive search. Simple, but not shallow. High information density with low alpha-numeric effect. Recommended to all intelligent reviewers. My students will benefit greatly from this book. It's also very cost-effective."

—*Dr. Xueshu Song, NIU, Dekalb, Illinois*

"The Fundamentals of Engineering is very concise; it does a great job with respect to the time that you have to study, getting you through the material for the test. It's right where it should be in terms of length and depth of review. It doesn't overwhelm those studying for the exam; I agree with the approach you took. It's a very useful book."

—*Professor, University of California, Berkeley*

Important Information

FE examination date: ____________________ time: ________________

location: ____

____

Examination Board Address: ____

____

____

phone: ____

Application was requested on this date: ____

Application was received on this date: ____

Application was accepted on this date: ____

This book belongs to: ____

____

____

____

phone: ____

...from the Professors who know it best!...

FUNDAMENTALS OF ENGINEERING

Merle C. Potter PhD, PE — Editor

THE MOST EFFECTIVE FE/EIT REVIEW

Authors:

D. Farnum, PhD	Chemistry
B. Weinberg, PhD, PE	Computer Science
F. Hatfield, PhD, PE	Civil Engineering
K. Mukherjee, PhD	Materials Science
M. Potter, PhD, PE	Mechanical Engineering
J. Soper, PhD	Electrical Engineering
G. Mase, PhD	Engineering Mechanics

Our authors are award-winning full professors. All presently teach at Michigan State University, with the exception of Dr. Soper, who teaches at Michigan Technological University.

published by

GREAT LAKES PRESS

P.O. Box 483

Okemos, MI 48805

International Standard Book Number 1-881018-14-8

Cover and page design by Ross W. Orr.

All comments and inquiries should be addressed to:
Great Lakes Press
PO Box 172
Grover, MO 63040-0172
Phone (314) 273-6016.

Library of Congress
Cataloging-in-Publication Data

Printed in the USA by Braun-Brumfield, Inc. of Ann Arbor, Michigan.

10 9 8 7 6 5 4 3 2 1

Table of Contents

PREFACE

PART I: PASSING THE FE EXAM

PART II: DIAGNOSTIC EXAM

PART III : THE FE SUBJECT REVIEWS

Chapter 2—DIGITAL COMPUTING Bernhard Weinberg

Chapter 3.—MATERIALS SCIENCE Kalinath Mukherjee

Chapter 4—CHEMISTRY Donald G. Farnum

Chapter 5—ENGINEERING ECONOMICS Frank Hatfield

Chapter 6—ELECTRICAL THEORY Jon A. Soper

Chapter 11—FLUID MECHANICS Merle C. Potter

PART IV: PRACTICE EXAM

EQUATION SUMMARY Appendix A

UNITS &CONVERSIONS Appendix B

INDEX

Preface

An Efficient FE Exam Review

This book was written by seven teaching professors to provide the most efficient review possible for the FE exam. The page layout and overall organization of our new fifth edition is easier to use and quicker to study through than ever before. The longer review books available make preparing for the FE—and recalling what you've studied during the exam—twice as difficult as need be. Many users have told us that our review, which is focused solely on the essentials, is much more useful.

Lecturing professors—who we have asked to help create this book—are clearly the people who know how to teach the material best, as the FE is an exam that has a strictly academic basis. The result of their efforts is a succinct, authoritative, thorough review for the Fundamentals of Engineering (FE) Examination—formerly the Engineer-In-Training (EIT) Examination. Credit is also due Jeff Potter, Michelle Gruender and Ross Orr for their help in making this review book the best available.

This volume includes a review of the most important terms and concepts and the major equations in each of the ten major subject areas covered by the FE exam. *The goal of this review is to enable the reader to pass the FE exam.* Therefore, subjects that make up an insignificant portion of the exam have been intentionally omitted for the sake of efficiency. By far the vast majority of question topics will be covered. If you look at other resources available, you'll see that they either include much extraneous material not on the FE exam, or they just present sample problems with no subject-matter review. We feel our approach is the most effective, and the most time-saving solution of the FE test-prep problem.

Sample Problems Like the FE Questions

The practice problems are designed to be as exam-like as possible. We have included theory questions as well as numerical problems, all with five-part multiple-choice answers—just like the exam. In the more extensive chapters selected problems are starred (*) so that choosing to work only those problems gives you the option of a quicker review. (We have not starred any problems in the matematics chapter; you should make sure that you have a solid math capability by working *all* those problems.) It may seem that some of our questions are simplistic. This, too, is intentional. By and large, most of the exam questions will be quite simple, with only one principle involved. After reviewing this material and working the practice problems, you should be ready to pass the FE exam. (Detailed solutions to all practice problems are provided in this book.)

Mix Practice Tests with Study for Best Results

To give you further confidence we have included two full 8-hour practice exams that provide an excellent indicator of whether you are completely ready for the examination. You may never have taken an eight-hour exam before; it is very demanding intellectually, physically and emotionally. Actual simulations of the exam will prepare you for the total experience. The first exam, called a Diagnostic Exam, is intended to be taken when you begin your review—it will expose your weak areas. The review material comes next, and finally the second exam to determine if you're completely ready.

Keep in mind that the objective of your preparation is to pass the Fundamentals of Engineering Examination—your first step toward becoming a licensed engineer. The point is *not* to develop a exhaustive proficiency in all subject areas included on the exam. It is sufficient simply to understand the basic principles of engineering.

Use your time most wisely by reviewing in detail the *afternoon* subject areas, as you will be tested over those areas twice! Get a good night's sleep before the exam, and we here at Great Lakes Press wish you well!

Dr. Merle C. Potter
Okemos, Michigan

P.S. You are a vital link in our unique user-friendly method—and your comments are greatly appreciated! Please fill out and return the comment and errata cards included in the back of this volume!

Passing the FE/EIT Exam

This book presents an efficient review of the subject areas covered on the FE/EIT and also serves as an excellent desk reference.

Certain chapters of this book are also recommended for review in preparation for the second exam one takes toward an engineering license, the PE exam. If you are preparing for the PE exam, select those chapters that are related to the primary subject area of the particular PE exam that you plan to take.

This book contains:

- A full 8-hour Diagnostic Exam with solutions.
- Short, succinct reviews of the theoretical aspects of the subject areas
- Equation Summary Sheets for pertinent subject areas which allow optimum use of the new NCEES Reference Handbook
- Example Problems with detailed solutions illustrating important concepts
- Practice Problems similar to those that will be encountered on the FE/EIT exam
- Solutions to all Practice Problems
- A full 8-hour Practice Exam with solutions—a readiness check
- An NCEES Handbook glued inside the back cover

How Do I Prepare for the FE Exam?

As of October, 1993, the FE exam changed to a closed-book format, although an NCEES Handbook of equations and tables is provided at the exam site. A copy is rubber-cemented to the back cover of this book and another will be sent to you after you register for the exam. Your first impression of this handbook may leave you overwhelmed! It has a tremendous amount of equations, figures, and information that may, at first glance, seem unfamiliar to you. Unfortunately, from the credits given in the handbook it is obvious that considerable material has been derived from obscure or outdated sources. But do not let this discourage you. We suggest that you let your opinion of the material be heard at your state board, however. To meet this new challenge, in this review book we have used the NCEES handbook to compile summaries of the more important equations from the most significant subject areas. In addition, we have adapted our nomenclature throughout to parallel the sometimes unusual nomenclature of the NCEES Handbook.

We strongly urge you to use a highlighting marker to *identify the equations you most often use* in the handbook. Initially, you can use our Equation Summary Sheets and your own best judgment to highlight the key equations. Thus you may familiarize yourself with the location of the most useful equations, and can then easily find them again in the unmarked handbook which will be given you at the test-site. If you do not prepare this way, you might become 'lost' in the handbook during the exam.

Why We Created This FE Review

We at Great Lakes Press developed this reference with the cooperation of a team of colleagues in response to our experience in coordinating review courses. The kind of well-planned material that an engineer naturally desires to use did not exist at the time. The options were either to pick and choose from a large, encyclopedia-like book that covers almost everything in engineering, or to use material that didn't even attempt to cover the material. It was either feast or famine. And both options were often rather expensive! So we recruited popular lecturers from the university campus and prepared this study guide. Its overwhelming acceptance has encouraged us to keep it continuously in tune with the changing tenor of the FE exam. Our goal is to always act as an advocate for the test-taker in our interactions with engineering departments, associations, review courses and registration governing bodies. We are dedicated to making the entire licensing and registration process as reasonable and brief as possible. Our present combination of pre-study diagnostic exam plus concise, focused exam-topic reviews (complete with the sometimes-unusual nomenclature used in the official exam-site NCEES Handbook) followed by a final readiness-check practice test makes for ideal preparation. Not too little, not too much, presented by the teachers who know it best, at a fair price, all in one volume: easily the winningest combination for you.

How To Become A Professional Engineer

To become registered as an engineer, a state may require that you:

1. Graduate from an ABET-accredited engineering program.
2. Pass the *Fundamentals of Engineering* exam.
3. Pass the *Principles and Practice of Engineering* (PE) exam after several years of engineering experience.

Requirements vary from state to state, so you should obtain local guidelines and follow them carefully.

Why Get an Engineering License?

Registration is necessary if an engineer works as a consultant, and is highly recommended in certain industries—especially when one is, or hopes to be, in a management position.

Examination Format

The FE is composed of a four-hour morning session, followed by a one-hour break, then a four-hour afternoon session. The sessions have a total of 210 multiple-choice questions covering the subject areas listed in the following tables:

Morning Session

Subject Area	*Approximate Number*
Chemistry	14
Dynamics	14
Economics	11
Electrical Circuits	14
Fluids	14
Materials Science	14
Mathematics (includes Digital Computing)	20
Mechanics of Materials (Solids)	11
Statics	14
Thermodynamics	14
	Total Questions: 140

Afternoon Session

Subject Area	*Number*
Economics	10
Mathematics	20
Mechanics •Statics •Dynamics •Solids	20
Electrical Theory	10
Thermodynamics/Fluids	10
	Total Questions: 70

How The FE Exam Is Scored

In the morning session, each problem is worth one-half point. Thus a maximum score of 70 is possible.

The afternoon session is *different* from the morning in that each problem is worth one point. With half the number of questions, the maximum possible score in the afternoon session is also 70.

The two-session total is 140 points, with both sessions having equal weight. A predetermined passing percentage is not established, nor is the exam graded on a curve. From recent exams, a score of 70 (50% correct) seems to be close to the passing score.

It is obvious, once you understand the scoring plan—which is not overtly described in NCEES material—that the subjects covered in the afternoon session carry much more scoring-power. Hence, engineers looking to optimize the performance for a given effort (as engineers always should) deduce that the afternoon subjects deserve much more attention!

English vs. SI Units

Questions can be worked using either English units or SI units. The answers from the array of choices will be the same using either set of units. The problems in this book are mostly in SI units. Practice problem sets in the three chapters involving pressure or stress are given in English units as well as SI. A table of conversion factors is presented later in this introductory chapter.

Strategies for Study and Exam-Day

To effectively prepare for the exam you should first develop a strategy. The afternoon session of the test includes questions in eight subject areas. All eight areas are also covered in the morning session. Therefore, these eight subject areas are extremely important to review. Be certain that you are most confident of your skill in those eight areas. Make them your #1 priority.

The following strategies will help you to use this book most effectively to prepare for the FE exam:

1. Focus your review on the eight subjects of the afternoon session.
2. *Quickly* review the remaining three major subject areas, Chemistry, Digital Computing (computing is covered in the math portion of the exam, and Materials Science.
3. Spend the majority of your time reviewing material with which you are <u>most</u> confident.
4. Inside the back cover of this book you'll find a copy of the NCEES Handbook prepared by the organization that writes the FE exam (NCEES). You will not be allowed to bring this handbook (or any other material) with you into the exam. However, a clean copy of this same handbook will be given to you upon entering the exam site. In our book, we have provided most of the same equations that you will find in the NCEES handbook for the major subjects. Go through the handbook and highlight those important equations which we have also included. This will allow you to familiarize yourself with the location of the equations you need most, and you will be able to quickly locate those equations in the handbook during the exam. You will find this particularly helpful since the handbook is filled with extraneous material. If you do not train yourself in this way, you may well find it difficult to quickly find appropriate equations during the exam.
5. The problems in the morning session are, for the most part, unrelated. Consequently, *less than two minutes* (on the average) can be spent on each problem. This makes fast recall *essential* as time does not allow you to contemplate various methods of solution. But each problem is only worth one-half point—so do not fall into the trap of spending too much time on any one problem from the morning session.
6. In the afternoon session, you can spend an average of $3\frac{1}{2}$ minutes per individual question. Many questions in the afternoon session are related to a common problem, divided into 5 sub-problems, thus you may maintain a good pace and still spend up to 20-25 minutes solving a problem set that is more difficult for you.
7. Chemistry and Materials Science make up only 14 points each from the total of 140 points. (Note: some questions on Physics may be included in the domain of Materials.) Do not spend much time reviewing these subject areas if you are short on time. The problems on Economics are usually quite easy so a brief review of that subject should allow you to answer most of those problems correctly.
8. If the time you have available for review is quite limited, study only those subjects that are included in *both* the morning and afternoon sessions.

Process of Elimination

Sometimes the best way to find the right answer is to look for the wrong ones and cross them out. On questions which are difficult for you, wrong answers are often much easier to find than right ones!

Answers are seldom given with more than three significant figures, and may be given with two significant figures. The choice *closest* to your own solution should be selected.

Use the *process of elimination* when *guessing* at an answer. If only one answer is negative and four answers are positive, eliminate the one odd answer from your guess. Also, when in doubt, work backwards and eliminate those answers that you believe are untrue until you are left with only 2 or 3 and then guess. By using a combination of methods, you greatly improve your odds of answering correctly.

Should I Guess?

Leave the last ten minutes of each session for making educated guesses. **Do not leave any answers blank** on your answer key. A guess *cannot* hurt you, only help you. Your score is based on the number of questions you answer correctly. An incorrect answer does not harm your score.

Place a question mark beside choices you are uncertain of, but seem correct. If time prevents you from re-working that problem, you will have at least identified your best guess.

Be sure to make a best guess if you have spent some time on a problem. Then circle the problem number, so you can go back to it later if time permits. But your guess is most "educated"—and most likely to be correct—if you make it *immediately* after reading the problem.

Difficult Problems

If at first glance you know that a certain problem will require much time and is exceptionally difficult for you, make your best guess then *skip right over it*. Be sure to mark the problem in a unique manner (we suggest that you circle the problem number) so that if time permits, you may come back to it. (Note: it is not possible to return to the Morning Session problems in the Afternoon Session.)

When you are working through a problem and decide to move on due to difficulty, be sure to write down in your test booklet your notes and conclusions up to that point in case you have time to return to it. Then make your best guess and circle the problem number to return to if you have time.

If you feel you know how to work a difficult problem and could answer it with more time, identify it by circling the entire problem, not just the problem number—this identifies it as a 'most likely' candidate for your set-aside ten minutes of 'guess' time.

Time-Saving Tips

Once you determine your answer, *always write the letter corresponding to the correct answer in the margin of the test booklet* beside the question. At the end of the page, you can then transfer all the answers from that page to the answer key at once. This will save you considerable time and help you maintain concentration as well!

Cross out choices that you have eliminated in your test booklet on the problems you will return to. Otherwise, you will have to reread them as you make your last-ditch deliberations.

Write Out Your Work

Feel free to write all over the writing space provided for you in your test booklet. Do not hesitate to work out a problem, no matter how simple it may be. Doing as much work as you can on paper will ease your mind and leave it less 'cluttered'—and it will help you if you need to return to a problem. You may think you're saving time, but your exam performance is *not* improved when you work problems in your head. (Note: Room for working out problems is provided in the test booklet.)

Bring a Calculator—Ideally, an HP-48G

You must take a silent calculator (it may be pre-programmed) into the exam. A calculator is essential when solving many problems. In fact, with the exception of a couple states, the premier engineering calculator, the HP-48G, is allowed into the exam (check with your state board). This calculator is a hand computer which has hundreds of basic equations and constants preprogrammed. If you need extra help with exam calculations, the HP-48G will be very useful. We at GLP offer this calculator and its higher-powered relative, the GX, for sale at substantial discounts. (Our 48G price $115; list $165.) We also offer a short manual to guide you through the steps to using this calculator effectively for the FE Exam. (The manufacturer's manual is difficult to use for even basic operations.) Our engineering-oriented manual is available for $9.95. Call us at 1-800-837-0201 for ordering or info.

A Strategic Guide to Students—How to Prepare for the FE

For a senior in an engineering college who has a busy schedule, we suggest 8 weeks as an ideal study period. During those 8 weeks, you must be willing to perform a fairly concentrated study. You should set aside blocks of 3 hours at least 2 days a week. In any case, we recommend that you study no less than 4 weeks for the FE exam. For those of you who have been away from this material for a while, base the length of your study period on the number of years you have been away from school and your own memory capability.

Perform an initial review during the first half of your selected study period (for engineering seniors, 4 weeks). But if you are trying to get through your review quickly, only the Practice Problems that have been starred (*) should be worked and studied.

Halfway through your review, set aside 9 hours to take the Practice Exam. Start at 8 a.m., take a one hour break from 12 to 1 p.m., and finish at 5 p.m. If this is not possible for you, set aside two 4-hour blocks of time within 2 days of each other, one to take the Morning Session and the second to take the Afternoon Session of the Practice Exam. It is best if you simulate the true FE Exam experience and attempt a continuous 8-hour exam session. In any case, try to take the Practice Exam during the morning or afternoon and avoid taking it late at night.

Upon scoring your Practice Exam, you should be able to select no more than 5 key areas out of the 10 subject areas that need the most additional review. If you are weak in more than five areas, you need substantially more study. Be sure <u>not</u> to select subject areas that are tested lightly (i.e., Chemistry and Materials Science) as your key areas. For the next half of your study period (engineering seniors: 4 weeks) you may be able to reduce your study to 2 hours at least 2 days a week. Or, you may realize that you need to study an additional day a week or increase your study session to 4 hours. By now you should have a definite plan of how much time you will commit to each key subject area, and distribute your study time evenly throughout the 4 week period—an even, consistent study schedule builds skills best and minimizes stress and fatigue.

Two days prior to exam day, review all subject areas briefly, using the highlighted handbook of equations and tables that you will be using during the exam. Be sure you can quickly find the equations you will use most often in the *unmarked* booklet you are given on exam day.

The day before the exam, relax and go to bed early. Do *not* cram or perform any panic studying. By now, you are as prepared as you can be for the exam.

The morning of your test, get up early and have a light, healthy breakfast (and maybe some coffee!). Arrive at the exam site at least 20 minutes early. You need to allow time for parking and getting settled—be sure to bring some change to pay for parking! During the one-hour lunch break, it is best to plan to meet with a friend who can help you relax and get refreshed for the Afternoon Session. If there are no restaurants nearby, bring a bag lunch and eat outside on the lawn somewhere. Get some fresh air. Do *not* spend the entire hour reviewing. Try not to talk to other test-takers about how it is going for them. This can easily induce either insecurity or false confidence. If you understand engineering principles and have prepared well, after the dust and sweat of the test day clears away, you'll find you've passed!

State Boards of Registration Addresses and Phone Numbers

All State Boards of Registration administer the National Council of Engineering Examiners and Surveyors (NCEES) uniform examination. The dates of the exams cover a span of three days in mid-April and three days in late October. The specific dates are selected by each State Board. To be accepted to take the FE exam, an applicant must apply well in advance. For information regarding the specific requirements in your state, contact your state board's office. Your state board address can be obtained from the Executive Director of NCEES, P. O. Box 1686, Clemson, SC 29633-1686, phone (803) 654-6824. Any comments relating to the exam or the Reference Handbook should be addressed to NCEES at the above address.

Board Addresses

ALABAMA: State Board of Registration for Professional Engineers and Land Surveyors, P. O. Box 304451, Montgomery 36130-4451. Executive Secretary, Telephone: (205) 242-5568.

ALASKA: State Board of Registration for Architects, Engineers and Land Surveyors, Pouch D, Juneau 99811. Licensing Examiner, Telephone: (907) 465-2540.

ARIZONA: State Board of Technical Registration, 1951 W. Camelback Rd., Suite 250, Phoenix 85015. Executive Director, Telephone: (602) 255-4053.

ARKANSAS: State Board of Registration for Professional Engineers and Land Surveyors, P. O. Box 2541, Little Rock 72203. Secretary-Treasurer, Telephone: (501) 324-9085.

CALIFORNIA: State Board of Registration for Professional Engineers and Land Surveyors, 2535 Capitol Oaks Dr., Sacramento 95853. Executive Secretary, Telephone: (916) 263-2222.

COLORADO: State Board of Registration for Professional Engineers and Professional Land Surveyors, 1860 Broadway, Suite 1370, Denver 80202. Program Administrator, Telephone: (303) 894-7788.

CONNECTICUT: State Board of Registration for Professional Engineers and Land Surveyors, The State Office Building, Room G-3A, 165 Capitol Avenue, Hartford 06106. Administrator, Ph: (203) 566-3290.

DELAWARE: Delaware Association of Professional Engineers, 2005 Concord Pike, Wilmington 19803. Executive Secretary, Telephone: (302) 577-6500.

DISTRICT OF COLUMBIA: Board of Registration for Professional Engineers, 614 H Street, N.W., Room 910, Washington 20001. Executive Secretary, Telephone: (202) 727-7468.

FLORIDA: Department of Business and Professional Regulations, Board of Professional Engineers, Northwood Centre, 1940 N. Monroe Street, Tallahassee 32399-0755. Executive Director, Telephone: (904) 488-9912.

GEORGIA: State Board of Registration for Professional Engineers and Land Surveyors, 166 Pryor Street, SW, Atlanta 30303-3465. Executive Director, Telephone: (404) 656-3926.

GUAM: Territorial Board of Registration for Professional Engineers, Architects and Land Surveyors, Department of Public Works, Government of Guam, P. 0. Box 2950, Agana 96911. Chairman, Telephone: (671) 646-3115.

HAWAII: State Board of Registration for Professional Engineers, Architects, Land Surveyors and Landscape Architects, P. 0. Box 3469, (1010 Richards Street), Honolulu 96801. Executive Secretary, Telephone: (808) 586-2693.

IDAHO: Board of Professional Engineers and Land Surveyors, 600 S. Orchard, Suite A, Boise 83705. Executive Secretary, Telephone: (208) 334-3860.

ILLINOIS: Department of Registration and Education, Professional Engineers' Examining Committee, 320 West Washington, 3rd Floor, Springfield 62786. Unit Manager, Telephone: (217) 782-0458.

INDIANA: Indiana Professional Licensing Agency, 302 W. Washington St., Room E034, Indianapolis 46204. Executive Director, Telephone: (317) 232-2980.

IOWA: State Board of Engineering Examiners, Capitol Complex, 1918 S.E. Hulsizer, Ankeny 50021. Executive Secretary, Telephone: (515) 281-5602.

KANSAS: State Board of Technical Professions, 900 Jackson, Room 507, Topeka 66612. Executive Secretary, Telephone: (913) 296-3053.

KENTUCKY: State Board of Registration for Professional Engineers and Land Surveyors, 160 Democrat Dr., Frankfort 40601. Executive Director, Telephone: (502) 573-2680.

LOUISIANA: State Board of Registration for Professional Engineers and Land Surveyors, 1055 St. Charles Avenue, Suite 415, New Orleans 70130. Executive Secretary, Telephone: (504) 568-8450.

MAINE: State Board of Registration for Professional Engineers, State House, Station 92, Augusta 04333. Secretary, Telephone: (207) 289-3236.

MARYLAND: State Board of Registration for Professional Engineers, 501 St. Paul Place, Room 902, Baltimore 21202. Executive Secretary, Telephone: (410) 659-6322.

MASSACHUSETTS: State Board of Registration of Professional Engineers and of Land Surveyors, Room 1512, Leverett Saltonstall Building, 100 Cambridge Street, Boston 02202. Secretary, Ph: (617) 727-9957.

MICHIGAN: Board of Professional Engineers, P. O. Box 30018, (611 West Ottawa), Lansing 48909. Administrative Secretary, Telephone: (517) 335-1669.

MINNESOTA: State Board of Registration for Architects, Engineers, Land Surveyors and Landscape Architects, 133 7th St. E., St. Paul 55101. Executive Secretary, Ph: (612) 296-2388.

MISSISSIPPI: State Board of Registration for Professional Engineers and Land Surveyors P. 0. Box 3, Jackson 39205. Executive Director, Telephone: (601) 359-6160.

MISSOURI: Board of Architects, Professional Engineers and Land Surveyors, P. O. Box 184, Jefferson City 65102. Executive Director, Telephone: (314) 751-0047.

MONTANA: State Board of Professional Engineers and Land Surveyors, Department of Commerce, 111 N. Jackson, POB 200513, Helena 59620-0513. Administrative Secretary, Telephone: (406) 444-4285.

NEBRASKA: State Board of Examiners for Professional Engineers and Architects, P. 0. Box 94751, Lincoln 68509. Executive Director, Telephone: (402) 471-2021.

NEVADA: State Board of Registered Professional Engineers and Land Surveyors, 1755 East Plum Lane, Ste. 135, Reno 89502. Executive Secretary, Telephone: (702) 688-1231.

NEW HAMPSHIRE: State Board of Professional Engineers, 57 Regional Drive, Concord 03301. Executive Secretary, Telephone: (603) 271-2219.

NEW JERSEY: State Board of Professional Engineers and Land Surveyors, P.O. Box 45015, Newark 07101. Executive Secretary-Director, Telephone: (201) 504-6460.

NEW MEXICO: State Board of Registration for Professional Engineers and Land Surveyors, 1010 Marquez Pl., Santa Fe 87501. Secretary, Telephone: (505) 827-7561.

NEW YORK: State Board for Engineering and Land Surveying, The State Education Department, Cultural Education Center, Madison Avenue, Albany 12230. Executive Secretary, Telephone: (518) 474-3846.

NORTH CAROLINA: State Board of Professional Engineers and Land Surveyors, 3620 Six Forks Rd., Suite 300, Raleigh 27609. Executive Secretary, Telephone (919) 781-9499

NORTH DAKOTA: State Board of Registration for Professional Engineers and Land Surveyors, P. 0. Box 1357, Bismarck 58502. Executive Secretary, Telephone: (701) 258-0786.

OHIO: State Board of Registration for Professional Engineers and Surveyors, 77 S. High St., 16th Fl., Columbus 43266-0314. Executive Secretary, Telephone: (614) 466-8999.

OKLAHOMA: State Board of Registration for Professional Engineers and Land Surveyors, Oklahoma Engineering Center, Room 120, 201 N.E. 27th Street, Oklahoma City, 73105. Executive Secretary, Telephone: (405) 521-2874.

OREGON: State Board of Engineering Examiners, Department of Commerce, 750 Front St., NE, Suite 240, Salem 97310. Executive Secretary, Telephone: (503) 378-4180.

PENNSYLVANIA: State Registration Board for Professional Engineers, P. O. Box 2649, Harrisburg 17120-264g. Administrative Assistant, Telephone: (800) 877-3926.

PUERTO RICO: Board of Examiners of Engineers, Architects, and Surveyors, Box 3271, San Juan 00904. Director, Examining Boards, Telephone: (809) 722-2122.

RHODE ISLAND: State Board of Registration for Professional Engineers and Land Surveyors, 10 Orms St., Suite 324, Providence 02904. Administrative Assistant, Telephone: (401) 277-2565.

SOUTH CAROLINA: State Board of Registration for Professional Engineers and Land Surveyors, P.O. Drawer 50408, Columbia 29250. Agency Director, Ph: (803) 734-9166.

SOUTH DAKOTA: State Commission of Engineering, Architectural Examiners and Land Surveying, 2040 West Main Street, Suite 304, Rapid City 57702-2447. Executive Secretary, Telephone: (605) 394-2510.

TENNESSEE: State Board of Architectural and Engineering Examiners, 3rd Fl. Volunteer Plaza, 500 James Robertson Pkwy, Nashville 37243-1142. Administrator, Telephone: (615) 741-3221.

TEXAS: State Board of Registration for Professional Engineers, P.O. Drawer 18329, Austin 78760-8329. Executive Director, Telephone: (512) 440-7723.

UTAH: Division of Occupational and Professional Licensing, P.O. Box 45805, Salt Lake City 84145-0805. Director, Ph: (801) 530-6628.

VERMONT: State Board of Registration for Professional Engineers, Division of Licensing and Registration, Pavilion Building, Montpelier 05602. Executive Secretary, Telephone: (802) 828-2875.

VIRGINIA: State Board of Architects, Professional Engineers, Land Surveyors and Certified Landscape Architects, 3600 West Broad St., Seaboard Building, 5th Floor, Richmond 23230-4917. Assistant Director, Telephone: (804) 367-8506.

VIRGIN ISLANDS: Board for Architects, Engineers and Land Surveyors, Bldg 1, Sub-Base, Rm 205, St. Thomas 00801. Secretary, Telephone: (809) 774-3130.

WASHINGTON: State Board of Registration for Professional Engineers and Land Surveyors, P.O. Box 9649, Olympia 98504. Executive Secretary, Telephone: (206) 753-3634.

WEST VIRGINIA: State Board of Registration for Professional Engineers, 608 Union Building, Charleston 25301. Executive Director, Telephone: (304) 348-3554.

WISCONSIN: State Examining Board of Professional Engineers, P.O. Box 8935, Madison 53708. Administrator, Telephone: (608) 266-1397.

WYOMING: State Board of Examining Engineers, Herschler Building, Room 4135, Cheyenne 82002. Secretary-Accountant, Telephone: (307) 777-6156.

FE Diagnostic Examination

Until you sit to take it, do not look or even glance at the material in this practice examination. To properly take this exam, spend at most four hours on each part.

You may use only the Reference Handbook supplied by NCEES. One is bound with this book; another will be sent to you after you register for the exam. (You may also order additional copies by calling Great Lakes Press at 1-800-837-0201.) If you are studying without the NCEES Handbook, use the equation summaries provided in this manual to familiarize yourself with the most important equations.

Battery-operated silent calculators are allowed. Indicate your answers on the separate answer sheet. Be sure and answer all questions. Guess at those you do not have time to work through!

You may write on scratch paper now, but remember, scratch paper is not allowed during the actual examination; large margins will provide sufficient work space.

To score yourself, add your correct responses from the afternoon part to one half of your correct responses from the morning part (a maximum score of 140). If you score less than 70 (50%) you may have difficulty with the actual examination. If your score is substantially greater than 70, you should be ready to pass the exam! If your score is marginal, study those areas that have a large numbers of problems in which you scored the lowest.

Morning Session Answer Sheet

Math/Computers

1. A B C D E
2. A B C D E
3. A B C D E
4. A B C D E
5. A B C D E
6. A B C D E
7. A B C D E
8. A B C D E
9. A B C D E
10. A B C D E
11. A B C D E
12. A B C D E
13. A B C D E
14. A B C D E
15. A B C D E
16. A B C D E
17. A B C D E
18. A B C D E
19. A B C D E
20. A B C D E

score________

Materials

21. A B C D E
22. A B C D E
23. A B C D E
24. A B C D E
25. A B C D E
26. A B C D E
27. A B C D E
28. A B C D E
29. A B C D E
30. A B C D E
31. A B C D E
32. A B C D E
33. A B C D E
34. A B C D E

score________

Chemistry

35. A B C D E
36. A B C D E
37. A B C D E
38. A B C D E
39. A B C D E
40. A B C D E
41. A B C D E
42. A B C D E
43. A B C D E
44. A B C D E
45. A B C D E
46. A B C D E
47. A B C D E
48. A B C D E

score________

Economics

49. A B C D E.
50. A B C D E
51. A B C D E
52. A B C D E
53. A B C D E
54. A B C D E
55. A B C D E
56. A B C D E
57. A B C D E
58. A B C D E
59. A B C D E

score________

Electrical Theory

60. A B C D E
61. A B C D E
62. A B C D E
63. A B C D E
64. A B C D E
65. A B C D E
66. A B C D E
67. A B C D E
68. A B C D E
69. A B C D E
70. A B C D E
71. A B C D E
72. A B C D E
73. A B C D E

score________

Thermodynamics

74. A B C D E
75. A B C D E
76. A B C D E
77. A B C D E
78. A B C D E
79. A B C D E
80. A B C D E
81. A B C D E
82. A B C D E
83. A B C D E
84. A B C D E
85. A B C D E
86. A B C D E
87. A B C D E

score________

Statics

88. A B C D E
89. A B C D E
90. A B C D E
91. A B C D E
92. A B C D E
93. A B C D E
94. A B C D E
95. A B C D E
96. A B C D E
97. A B C D E
98. A B C D E
99. A B C D E
100. A B C D E
101. A B C D E

score________

Dynamics

102. A B C D E
103. A B C D E
104. A B C D E
105. A B C D E
106. A B C D E
107. A B C D E
108. A B C D E
109. A B C D E
110. A B C D E
111. A B C D E
112. A B C D E
113. A B C D E
114. A B C D E
115. A B C D E

score________

Mechanics of Materials

116. A B C D E
117. A B C D E
118. A B C D E
119. A B C D E
120. A B C D E
121. A B C D E
122. A B C D E
123. A B C D E
124. A B C D E
125. A B C D E
126. A B C D E

score________

Fluid Mechanics

127. A B C D E
128. A B C D E
129. A B C D E
130. A B C D E
131. A B C D E
132. A B C D E
133. A B C D E
134. A B C D E
135. A B C D E
136. A B C D E
137. A B C D E
138. A B C D E
139. A B C D E
140. A B C D E

score________

tear out answer sheet

Morning Session—Diagnostic Exam

4 hours maximum

Each of the 140 problems is followed by five answers. Select the response that is best and circle the corresponding letter on the answer sheet.

---

1. Determine the equation of a straight line that passes through the point (6, –3) if it is perpendicular to the line $y = 2x - 3$.

(A) $y = -\frac{1}{2}x$ (B) $y = \frac{1}{2}x - 6$ (C) $y = -\frac{1}{2}x - 6$

(D) $y = 2x - 15$ (E) $y = -\frac{2}{3}x + 1$

2. The equation $x^2 - 4xy + y^2 + 4x - 6y + 12 = 0$ represents:

(A) a circle (B) an ellipse (C) a parabola

(D) a hyperbola (E) a sphere

3. The expression $\frac{\cos\theta}{\sin\theta} + \frac{\sin\theta}{\cos\theta}$ can be written as:

(A) $\frac{1}{\sin 2\theta}$ (B) $\frac{\sin 2\theta}{2}$ (C) $\frac{2}{\cos 2\theta}$ (D) $\frac{2\cos 2\theta}{\sin 2\theta}$ (E) $\frac{2}{\sin 2\theta}$

4. $(1-2i)^{-1}$ can be written as:

(A) $\frac{1}{5}+\frac{2}{5}i$ (B) $\frac{1}{5}-\frac{2}{5}i$ (C) $-\frac{1}{3}-\frac{2}{3}i$ (D) $-\frac{1}{3}+\frac{2}{3}i$ (E) $\frac{1}{3}-\frac{2}{3}i$

5. Find the adjoint of $\begin{bmatrix} 2 & 3 \\ -1 & 0 \end{bmatrix}$.

(A) $\begin{bmatrix} \frac{1}{2} & \frac{1}{3} \\ -1 & 0 \end{bmatrix}$ (B) $\begin{bmatrix} 2 & 3 \\ -1 & 0 \end{bmatrix}$ (C) $\begin{bmatrix} 0 & -3 \\ 1 & 2 \end{bmatrix}$ (D) $\begin{bmatrix} 0 & 1 \\ -3 & 2 \end{bmatrix}$ (E) $\begin{bmatrix} 0 & -1 \\ 3 & 2 \end{bmatrix}$

6. Determine y if
$$\begin{aligned} x - y + z &= 2 \\ x \quad\;\; - z &= 0 \\ x + y \quad\;\; &= -3 \end{aligned}.$$

(A) $-\frac{2}{3}$ (B) -8 (C) $-\frac{4}{3}$ (D) $-\frac{8}{3}$ (E) 8

7. Find the eigenvalues associated with the matrix $\begin{bmatrix} 1 & 1 \\ 2 & 0 \end{bmatrix}$.

(A) 0, 1 (B) –2, 1 (C) 0, –1 (D) 2, –1 (E) 2, 0

8. What is the particular solution to the differential equation $\dot{y} + 4y = 3e^{-t}$?

(A) $3e^{-t}$ (B) e^{-4t} (C) c_1e^{-4t} (D) $\frac{3}{5}e^{-t}$ (E) e^{-t}

9. A spring-mass system is modeled by the differential equation $\ddot{y} + 4\dot{y} + 4y = 8\sin 2t$. What is the amplitude of the steady-state solution?

(A) 1 (B) 2 (C) 4 (D) 6 (E) 8

10. The parabola $x = 2y^2$ is rotated about the y-axis. What is the volume between $y = 0$ and $y = 2$?

(A) $\frac{512\pi}{5}$ (B) $\frac{128\pi}{5}$ (C) 8π (D) 32π (E) $\frac{16}{3}\pi$

11. If a vector field is conservative which of the following is zero?

(A) curl (B) cross product (C) divergence
(D) dot product (E) gradient

12. The cross product of $\mathbf{A} = 4\mathbf{i} + 2\mathbf{j}$ with $\mathbf{B}$ is zero. The dot product $\mathbf{A} \cdot \mathbf{B} = 30$. Find $\mathbf{B}$.

(A) $-6\mathbf{i} - 3\mathbf{j}$ (B) $6\mathbf{i} - 3\mathbf{j}$ (C) $6\mathbf{i} + 3\mathbf{j}$ (D) $-6\mathbf{i} + 3\mathbf{j}$ (E) $4\mathbf{i} + 2\mathbf{j}$

13. Determine $\lim_{x \to 0} \frac{\sin^2 x}{x(1-e^x)}$.

(A) ∞ (B) -2 (C) 1 (D) -1 (E) 0

14. Express the base 4 number 101.1 as a base 10 number.

(A) 17.25 (B) 68.25 (C) 21.25 (D) 20.25 (E) 17.4

15. Find the y-coordinate of the centroid of the area bounded by $x = 2y^2$, the x-axis, and the line $x = 2$.

(A) 2/5 (B) 1/4 (C) 1/3 (D) 1/2 (E) 3/8

16. The differential equation $y'' + 3x^2y' + \sin x = 0$ is:

(A) linear, homogeneous, constant coefficient
(B) nonlinear, homogeneous, constant coefficient
(C) linear, nonhomogeneous, variable coefficient
(D) nonlinear, nonhomogeneous, variable coefficient
(E) linear, homogeneous, variable coefficient

17. Select the Taylor series representation of $\cos 2x$.

(A) $2x - \frac{4x^3}{3} + \frac{4x^5}{15} + \cdots$ (B) $1 + 2x^2 + \frac{2x^4}{3} + \cdots$ (C) $1 - \frac{x^2}{2} + \frac{x^4}{24} + \cdots$

(D) $1 - 2x^2 + \frac{2x^4}{3} + \cdots$ (E) $x - \frac{x^3}{6} + \frac{x^5}{120} + \cdots$

18. A chain saw makes a 10-inch straight cut in the side of a 24-inch diameter tree. Calculate the angle made by the two radii connecting the ends of the cut.

(A) 131° (B) 49.2° (C) 45.2° (D) 24.6° (E) 22.6°

19. Find the y-intercept of the line tangent to the parabola $x = 2y^2$ at the point (2, 1).

(A) -7 (B) 7 (C) 3/2 (D) 1/2 (E) $-1/2$

20. What function $f(t)$ has a Laplace transform $\dfrac{s+3}{s^2+4s+8}$?

(A) $\left(\cos 2t + \frac{1}{2}\sin 2t\right)e^{-2t}$ (B) $(\cos 2t + \sin 2t)e^{-2t}$ (C) $e^{-2t}\cos 2t$

(D) $(\cos 4t + \sin 4t)e^{-2t}$ (E) $\left(\cos 8t + \frac{1}{2}\sin 8t\right)e^{-2t}$

21. A solid is good electrical conductor if

(A) the electronic band-gap is > 10 eV
(B) the electronic band-gap is < 5 eV
(C) the conduction band is full
(D) the conduction band is half full
(E) the valance band is full and the conduction band is empty

22. A solid is an intrinsic semiconductor if

(A) there are no electrons in the conduction band, and the band-gap is large
(B) the conduction band is half full with electrons
(C) the conduction band is empty and the band-gap is about ≈ 1 eV or less at ordinary temperature
(D) the impurity atoms provide extra electrons
(E) the electrons in the valance band have intrinsic mobility

23. An n-type semiconductor is a covalent solid in which

(A) the energy gap is very small
(B) an impurity level just above the valance band, is added by the impurity atoms
(C) an impurity level just below the conduction band is added by the impurity atoms
(D) the solid is of ultra high purity
(E) "holes" are present in the valance band which help negative charge flow

24. In a covalently bonded solid, the binding is produced by

(A) sharing of electrons between adjacent atoms
(B) by the coulombic attraction between oppositely charged atoms
(C) by the minimization of kinetic energy of free electrons
(D) by the delocalization of electrons associated with individual atoms
(E) by the absence of homopolar binding

25. A given monovalent metallic crystal has 2×10^{10} atoms. Which of the following statements are true for this crystal?
   (A) There are 2×10^{10} electrons in the valance band.
   (B) There are 2×10^{10} empty states in the valance band.
   (C) There are 4×10^{10} available quantum states in the valance band.
   (D) There are 2×10^{10} unoccupied quantum states in the valance band.
   (E) There are 2×10^{10} available quantum states in the conduction band

26. Which of the following statements about atomic packing in a crystal lattice are true?
   (A) Highest packing fraction is only possible in a FCC crystal.
   (B) Packing fraction for a BCC lattice is higher than that for a HCP lattice.
   (C) Packing fraction for a HCP lattice is the same as for a FCC lattice.
   (D) Packing fraction of a BCC lattice is 0.74.
   (E) Closest packing of atoms is found in the (100) plane of a FCC lattice.

27. The intercepts of a crystallographic plane with the three crystallographic axes, x, y, z, are expressed as fractions of the unit cell dimension along a given direction. Thus, the intercepts of (12) plane are:
   (A) 1, 2, 0 (B) 2, 1, 0 (C) 0, 2, 1 (D) 1, ½, 0 (E) 1, ½, ∞

28. In a cubic lattice, which of the following directions are perpendicular to the (111) plane?
   (A) [110]
   (B) [111]
   (C) [100]
   (D) [010]
   (E) [001]

29. If the hardening of an alloy is induced by diffusion controlled precipitation of second phase particles, then the rate of hardening is
   (A) exponentially dependent on time
   (B) independent of time
   (C) independent of temperature
   (D) exponentially dependent on temperature
   (E) linearly dependent on time and temperature

30. All of the following statements about metallic corrosion are true except:
   (A) Corrosion requires simultaneous oxidation and reduction reactions.
   (B) Iron does not rust if it is kept immersed in water which does not have any dissolved oxygen or air.
   (C) Iron corrodes substantially when in contact with zinc, in an aqueous environment.
   (D) Aluminum, which is more anodic than magnesium, is protected by magnesium in a corrosive environment.
   (E) Oxidation reaction of iron is associated with a release of two electrons.

31. The following statements about toughness of a material are true except:
    (A) Charpy impact test is a measure of toughness.
    (B) High hardness is an indication of high toughness.
    (C) In a tough material extensive necking occurs before tensile failure.
    (D) The area under the tensile stress-strain curve up to fracture is large.
    (E) True strain to fracture is large.

32. Yield point phenomenon in mild steel occurs due to,
    (A) large grain size
    (B) poor tempering
    (C) presence of too much martensite
    (D) interaction of carbon with dislocations
    (E) very slow rate of deformation

33. In this phase diagram, the following are true except:
    (A) an eutectic reaction occurs at a temperature T_3
    (B) an eutectic reaction occurs at a temperature T_5
    (C) an eutectic reaction occurs at a composition C_1
    (D) an eutectic reaction occurs at a composition C_2
    (E) an eutectic reaction occurs at a composition C_3

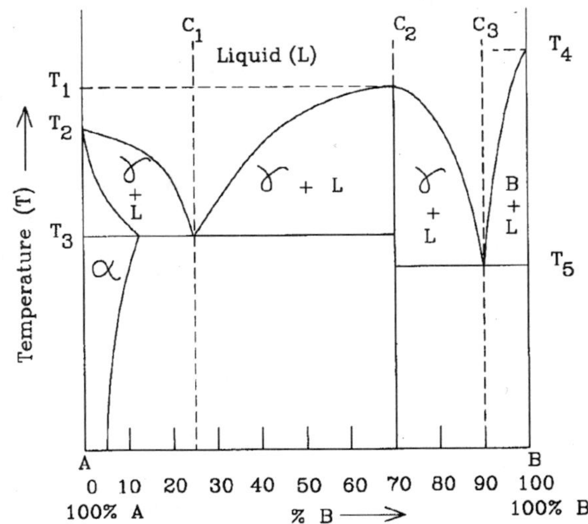

34. In the phase diagram for the problem above, the following are true except:
    (A) an intermetallic phase γ forms at composition C_2
    (B) an eutectic reaction occurs at a composition of 25% A and 75% B and at a temperature T_3
    (C) pure solid B melts at a temperature T_2
    (D) an eutectic reaction occurs at a composition of 90% A and 10% B and at a temperature T_5
    (E) at temperatures less than T_5 and in the composition range greater than 70% B, the solid has a microstructure consisting of pure A and γ phase

35. The nucleus of a neutral atom contains 42 protons and 54 neutrons. The element is

(A) calcium	(B) manganese	(C) gold
(D) molybdenum	(E) mercury	

36. How many protons, neutrons, and electrons are in an ion of the rhodium isotope $^{103}Rh^{3+}$?

	Protons	Neutrons	Electrons
(A)	45	103	45
(B)	45	58	45
(C)	58	45	55
(D)	45	55	45
(E)	45	58	42

37. Which compound does not contain a covalent bond?
(A) HBr (B) $NaNO_3$ (C) KCl (D) CH_4 (E) H_3PO_4

38. Which ion does not have the same electron configuration as a noble gas?
(A) O^{2-} (B) Pb^{2+} (C) Mg^{2+} (D) Br^- (E) N^{3-}

39. In the presence of an iron catalyst, hydrogen gas can be produced from carbon monoxide gas and steam according to the equation:

$$CO(g) + H_2O(g) = CO_2(g) + H_2(g)$$

At a certain temperature, analysis showed the presence of 0.80 mol of CO, 0.25 mol of water vapor, 0.50 mol of H_2, and 2.4 mol of carbon dioxide in a liter of the equilibrium mixture. What is the value of the equilibrium constant at that temperature?
(A) 12.0 (B) 6.0 (C) 4.8 (D) 1.5 (E) 0.17

40. How many valence electrons does an element, such as arsenic, from Group 5A have?
(A) 3 (B) 4 (C) 5 (D) 6 (E) 7

41. What is the ground-state electron configuration of a barium ion, Ba^{2+}?
(A) [Rn] (B) [Xe] $6s^2$ (C) [Xe] (D) [Ar] (E) [Kr] $5p^4$

42. What is the percentage by mass of carbon in the hydrocarbon propane, C_3H_8?
(A) 18 (B) 36 (C) 44 (D) 82 (E) 92

43. When a nonmetallic element in Group 6A reacts to form an ionic compound, the element usually gains an ionic charge of
(A) −2 (B) −1 (C) 0 (D) +1 (E) +2

44. Nitric acid will oxidize sulfur to sulfur dioxide. What is the mole ratio of sulfur to nitric acid in the balanced redox equation?
(A) 1:2 (B) 1:3 (C) 2:3 (D) 3:4 (E) 4:5

45. Which of the following factors does not influence the rate of a chemical reaction?
(A) The nature of the reactants
(B) The amount of a solid reactant present
(C) The reactant concentrations or state of subdivision
(D) The temperature
(E) The presence of a catalyst

46. What is the electron configuration of gallium?
 (A) $1s^2\ 2s^2\ 2p^6\ 2d^{10}\ 3s^2\ 3p^6\ 4s^2\ 4p^1$
 (B) $1s^2\ 2s^2\ 2p^6\ 3s^2\ 3p^6\ 3d^6\ 3p^7$
 (C) $1s^2\ 2s^2\ 2p^6\ 3s^2\ 2p^6\ 4s^2\ 4p^6\ 4d^5$
 (D) $1s^2\ 2s^2\ 2p^6\ 3s^2\ 3p^6\ 4s^2\ 3d^8\ 4p^1$
 (E) $1s^2\ 2s^2\ 2p^6\ 3s^2\ 2p^6\ 4s^2\ 3d^{10}\ 4p^1$

47. In the exothermic reaction

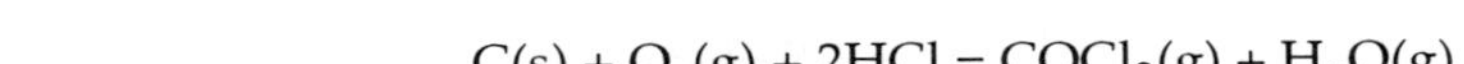

$$C(s) + O_2(g) + 2HCl = COCl_2(g) + H_2O(g)$$

at equilibrium, how is it possible to increase the amount of heat evolved?
 (A) add some carbon (B) add water (C) add a catalyst
 (D) increase the volume (E) increase the pressure

48. Which of the following equilibrium constant expressions is associated with the equation: $BaO(s) + CO_2(g) = Ba\ CO_3(s)$?
 (A) $K_p = 1/p(CO_2)$
 (B) $K_p = p(BaCO_3)/[p(BaO) \times p(CO_2)]$
 (C) $K_p = 1/[p(BaO) \times p(CO_2)]$
 (D) $K_p = 1/p(BaCO_3)$
 (E) $K_p = p(BaCO_3)$

49. At the end of each of twelve months Terry deposited \$100 in a savings account that pays 6% annual interest compounded monthly. Compute the balance in Terry's account immediately after the twelfth payment.
 (A) \$1200 (B) \$1206 (C) \$1234 (D) \$1272 (E) \$1687

50. How many years are required for an amount of money to triple if it is invested at a 4% interest rate?
 (A) 3 (B) 7 (C) 13 (D) 28 (E) 75

51. An alumnus has given Michigan State University ten million dollars to build and operate a laboratory. Annual operating cost is estimated to be one hundred thousand dollars. The endowment will earn 6% interest. Assume an infinite life for the laboratory and determine how much money may be used for its construction.
 (A) $\$5.00 \times 10^6$ (B) $\$8.33 \times 10^6$ (C) $\$8.72 \times 10^6$
 (D) $\$9.90 \times 10^6$ (E) $\$9.98 \times 10^6$

52. An investment pays \$6000 at the end of the first year, \$4000 at the end of the second year, and \$2000 at the end of the third year. Compute the present value of the investment if a 10% rate of return is required.
 (A) \$8333 (B) \$9667 (C) \$10,300 (D) \$12,000 (E) \$14,000

53. An amount F is accumulated by investing a single amount P for n compounding periods with interest rate of i. Select the formula that relates P to F.
 (A) $P = F(1+i)^{n-1}$ (B) $P = F(1+i)^n$ (C) $P = F(1+n)^{-i}$
 (D) $P = F(1+ni)^{-1}$ (E) $P = F(1+i)^{-n}$

54. A set of speakers may be purchased now for $400 or by making a down payment of $35 and additional payments of $45 at the end of each of the next ten months. Compute the nominal annual interest rate for the time payment plan.

(A) 11.2% (B) 21.2% (C) 23.3% (D) 48.0% (E) 60.1%

55. Same as preceding question except find the effective annual interest rate.

(A) 11.2% (B) 21.2% (C) 23.3% (D) 48.0% (E) 60.1%

56. Twenty-one equal end-of-year deposits are made into a savings account that pays 4% interest. Compute the amount of each deposit that will permit withdrawals of $10,000 at the ends of the eighteenth through twenty-first years, leaving the account empty.

(A) $1328 (B) $1667 (C) $1905 (D) $2123 (E) $2587

57. Excelsys Inc. purchased a desk top computer for $4500. Using the general straight line method, compute the depreciation for the first year if the recovery period is three years and the salvage value is $1500.

(A) $500 (B) $750 (C) $1000 (D) $1500 (E) $2250

58. Profits generated by Excelsys Inc. were $10,000 for the first year and increased by $1000 each year. If the profits for five years of operation were invested at an 8% rate of return, how much money had accumulated at the end of the fifth year?

(A) $47,300 (B) $55,000 (C) $60,000 (D) $65,800 (E) $69,500

59. Development costs of a new product are estimated to be $70,000 per year for three years. Annual profits from the sale of the new product, estimated to be $80,000, will begin in the fourth year and continue for ten years. Using a rate of return of 20%, compute the present value of the venture.

(A) $14,300 (B) $46,600 (C) $187,900 (D) $309,200 (E) $590,000

60. The equivalent resistance between terminals *a*–*b* in ohms, is:

(A) 3 (B) 5 (C) 8

(D) 10 (E) 16

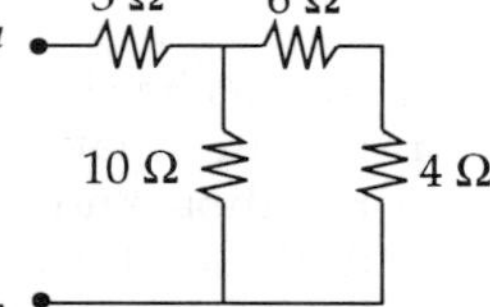

61. The Thevinen Equivalent voltage between terminals *a*–*b* is, in volts:

(A) 3 (B) 4 (C) 6

(D) 8 (E) 9

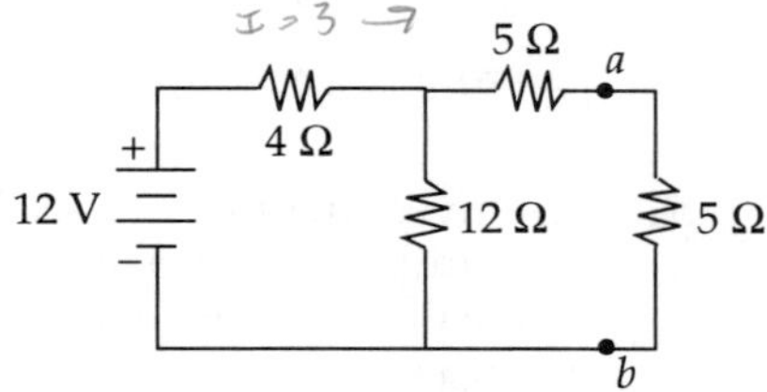

62. The Thevinen Equivalent resistance for the terminals *a*–*b* in the circuit of Problem 61 is, in ohms:

(A) 4 (B) 8 (C) 12 (D) 17 (E) 20

63–64 Given the circuit:

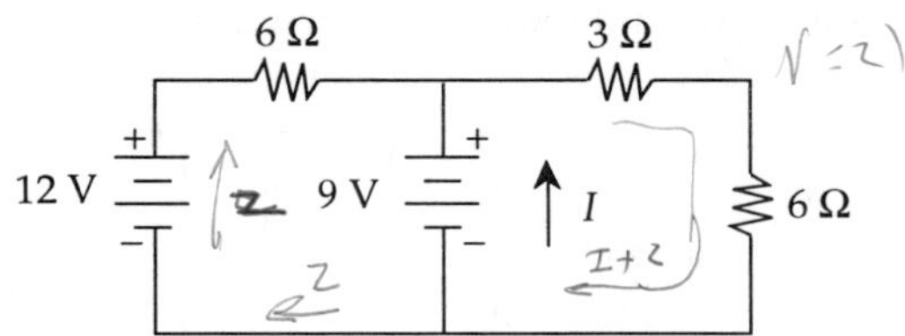

63. The current I, through the 9-volt battery in amperes, is:
(A) 0 (B) 0.5 (C) 1.0 (D) 1.5 (E) 2.0

64. The power delivered by the 12-volt battery in watts, is:
(A) 3 (B) 6 (C) 9 (D) 12 (E) 15

65. The rms current through the 45-ohm resistor in the circuit shown is, in amperes:

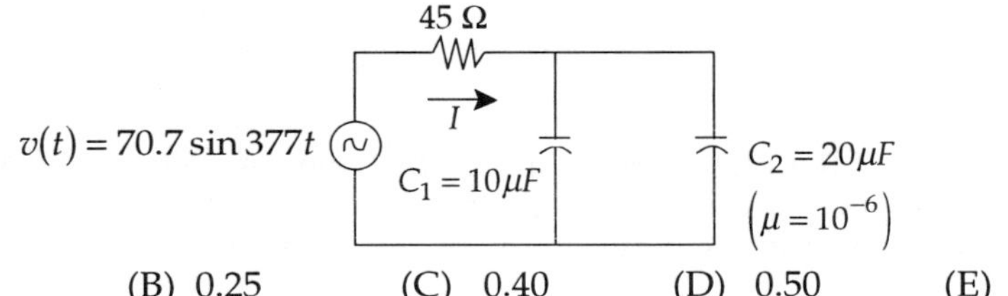

(A) 0.15 (B) 0.25 (C) 0.40 (D) 0.50 (E) 0.75

66. The rms voltage across the 40-ohm resistor is, in volts:

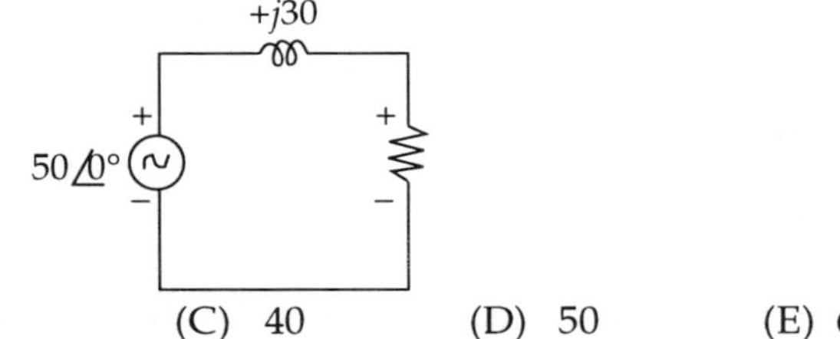

(A) 20 (B) 30 (C) 40 (D) 50 (E) 60

67. The power deliver to the 40-ohm resistor in Problem 66 is, in watts:
(A) 20 (B) 30 (C) 40 (D) 50 (E) 60

68. The instantaneous current through the inductor at $t = 0.001$ sec in amperes is:

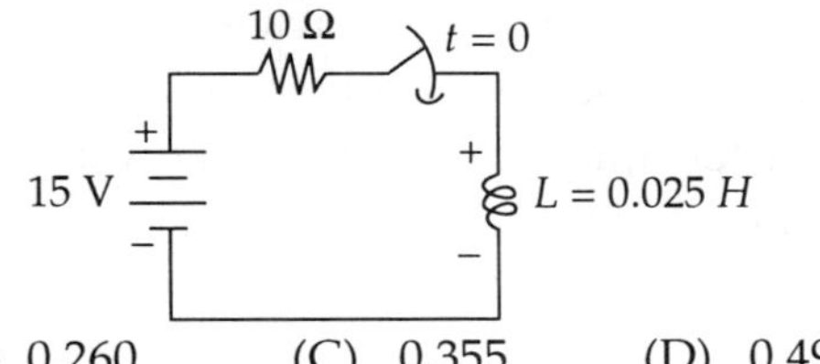

(A) 0.175 (B) 0.260 (C) 0.355 (D) 0.495 (E) 0.600

69. The instantaneous voltage across the inductor at $t = 0.001$ sec in the circuit of Problem 68 is, in volts:
(A) 6.75 (B) 8.25 (C) 9.15 (D) 10.05 (E) 12.55

70. The current in the 100-ohm resistor in amperes, is:

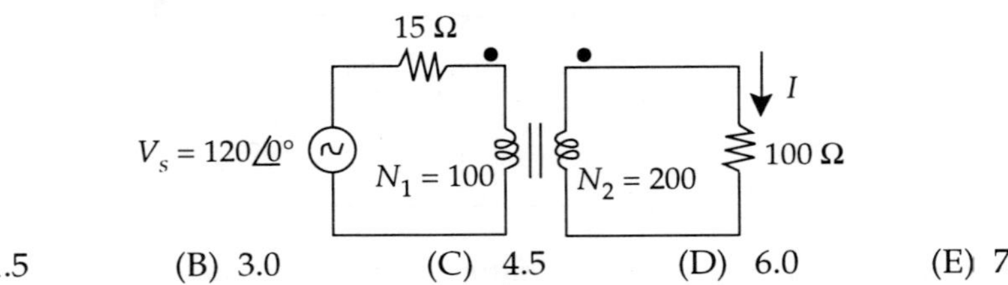

(A) 1.5 (B) 3.0 (C) 4.5 (D) 6.0 (E) 7.5

71. Two parallel conductors carry 50 amperes each in the same direction. The conductors are 5 cm apart. What is the magnetic force on a 1 meter length of one conductor in newtons?

(A) 0.01
(B) 0.02
(C) 0.03
(D) 0.04
(E) 0.05

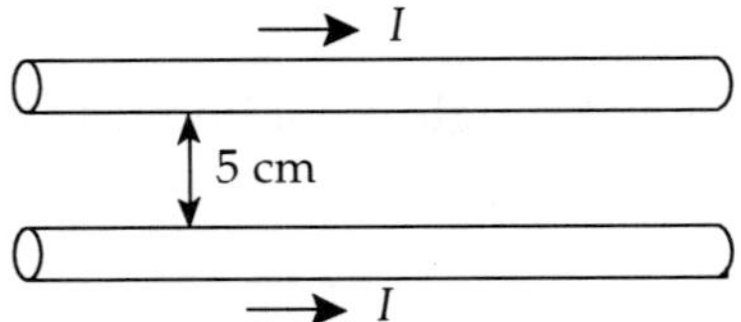

72. Two parallel flat plates are separated by 2 cm. A voltage of 500 V is applied to the parallel plates. What is the maximum electric field force experienced by an electron passing between the plates, in newtons?

(A) 4×10^{-15}

(B) 20×10^{-15}

(C) 30×10^{-15}

(D) 40×10^{-15}

(E) 80×10^{-15}

73. A copper conductor 10 m long and 3 mm in diameter has a resistance of 0.024 ohms at room temperature (20°C). What is the temperature in degrees Celsius if this conductor has a resistance of 0.036 ohms? ($\alpha_c = 0.0039$.)

(A) 87 (B) 118 (C) 128 (D) 148 (E) 172

74. Select the best response for an isolated system.
(A) The entropy of system remains constant.
(B) The heat transfer equals the work done.
(C) The heat transfer equals the internal energy change.
(D) Frictional effects can be neglected.
(E) The heat transfer is zero.

75. When one system interacts with another system which of the following is possible?
(A) Internal energy can be transferred from one system to the other.
(B) Entropy can be transferred from one system to the other.
(C) One system can induce a force on the other system.
(D) One system can cause an irreversibility in the other system.
(E) Temperature can be transferred from one system to the other.

76. Which of the following is not an extensive property?

(A) Momentum (B) Kinetic energy (C) Enthalpy
(D) Density (E) Mass

77. Which of the following is not an acceptable SI unit?

(A) Distance measured in centimeters.
(B) Temperature measured in kelvins.
(C) Pressure measured in newtons per square meter.
(D) Volume measured in cubic centimeters.
(E) Density measured in grams per cubic centimeter.

78. The pressure-temperature diagram for water is shown. The names for points A and B and line ℓ are, respectively:

P
ℓ
B
A
T

(A) triple, critical, fusion
(B) critical, triple, sublimation
(C) triple, critical, sublimation
(D) critical, triple, fusion
(E) triple, critical, vaporization

Select either of the following two problems:

79. (metric) A tank contains 0.02 m^3 of liquid and 1.98 m^3 of vapor. If the density of the liquid is 960 kg/m^3 and that of the vapor is 0.5 kg/m^3, what is the quality of the mixture?

(A) 5.2% (B) 4.9% (C) 2.04% (D) 1.01% (E) 0.052%

79. (English) A tank contains 0.3 ft^3 of liquid and 29.7 ft^3 of vapor. If the density of the liquid is 62 lbm/ft^3 and that of the vapor is 0.503 lbm/ft^3, what is the quality of the mixture?

(A) 5.1% (B) 4.6% (C) 1.96% (D) 0.98% (E) 0.051%

Select either of the following two problems:

80. (metric) Two kilograms of air are contained in a cylinder. If 80 kJ of heat are added to the air, estimate the temperature rise if the pressure is held constant. $c_p = 1.0$, $c_v = 0.716$ kJ/kg·K, $k = 1.4$.

(A) 56°C (B) 40°C (C) 33°C (D) 28°C (E) 21°C

80. (English) Four pounds of air are contained in a cylinder. If 80 BTU of heat are added to the air, estimate the temperature rise if the pressure is held constant. $c_p = 0.24$, $c_v = 0.171$ BTU/lbm·°R, $k = 1.4$.

(A) 117°F (B) 83°F (C) 69°F (D) 52°F (E) 48°F

---

81. A long metal rod can be used as a very stiff spring. The deformation of a rod is given by $\delta = PL/AE$, as defined in Mechanics of Materials. What is the "spring constant" of the metal rod?

(A) AE/L (B) $1/AE$ (C) L/AE (D) AE (E) $AE\delta/L$

82. Clothes are hung out to dry in very cold weather. The water in the clothes freezes, but a day later when the clothes are brought inside they are dry. By what process did the drying occur?
(A) vaporization (B) condensation (C) evaporation
(D) sublimation (E) melting

Select either of the following two problems:

83. (metric) Air is compressed in an ideal, adiabatic compressor from 100 kPa and 20°C to 800 kPa. What is the temperature at the compressor exit? $k = 1.4$.
(A) 1440°C (B) 368°C (C) 258°C (D) 167°C (E) 36°C

83. (English) Air is compressed in an ideal, adiabatic compressor from 15 psia and 70°F to 120 psia. What is the temperature at the compressor exit? $k = 1.4$.
(A) 966°F (B) 687°F (C) 500°F (D) 462°F (E) 127°F

---

84. Vapor refrigerant enters and liquid refrigerant leaves the coils on the back of a refrigerator. These coils are the:
(A) evaporator (B) intercooler (C) reheater
(D) condenser (E) regenerator

85. Steam at high temperature and pressure passes through a half open globe valve. Select the property that remains constant through the valve.
(A) enthalpy (B) temperature (C) pressure
(D) entropy (E) density

86. Air undergoes a three-process cycle that has a constant pressure process $(1)-(2)$, an isothermal process $(2)-(3)$, and a constant volume process $(3)-(1)$. Select the correct response for the piston-cylinder arrangement.
(A) $W_{1-2} = 0$ (B) $Q_{3-1} = 0$ (C) $Q_{2-3} = 0$ (D) $W_{3-1} = 0$ (E) $Q_{1-2} = 0$

Select either of the following two problems:

87. (metric) A 2.5-cm-thick substance has a thermal resistance of 2.0 $\text{hr}\cdot\text{m}\cdot°\text{C/kJ}$. Estimate the heat transferred in 15 minutes through a 3 m by 8 m wall if the inside and outside temperatures are –10°C and 25°C, respectively.
(A) 6300 kJ (B) 5400 kJ (C) 4800 kJ (D) 4200 kJ (E) 2050 kJ

87. (English) A 1-inch-thick substance has a thermal resistance of 10 hr-ft-°F/BTU. Estimate the heat transferred in 15 minutes through a 10 ft by 25 ft wall if the inside and outside temperatures are 10°F and 70°F, respectively.
(A) 6700 BTU (B) 5600 BTU (C) 5000 BTU (D) 4500 BTU (E) 2200 BTU

---

88. Determine the magnitude of the resultant of the following three forces:

$$\mathbf{F}_1 = 100\mathbf{i} - 200\mathbf{k} \quad \text{acting at } (0,\ 0,\ 0)$$
$$\mathbf{F}_2 = 50\mathbf{j} + 50\mathbf{k} \quad \text{acting at } (3,\ 1,\ 2)$$
$$\mathbf{F}_3 = -200\mathbf{i} + 100\mathbf{j} \quad \text{acting at } (0,\ 2,\ 0)$$

(A) 457 (B) 365 (C) 283 (D) 234 (E) 205

89. Find the tension in cable AB.

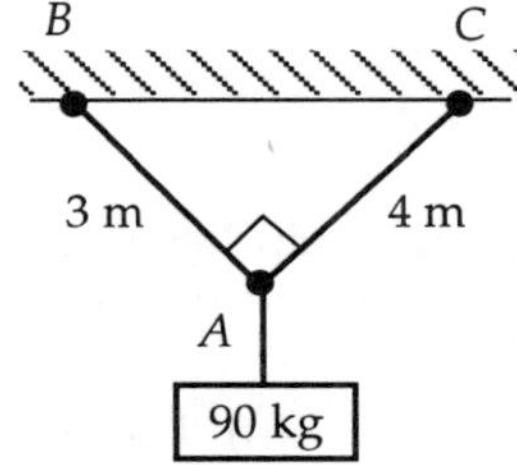

(A) 706 N
(B) 530 N
(C) 264 N
(D) 72 N
(E) 53 N

90. The z-component of the moment of the force $\mathbf{F} = 20\mathbf{i} - 30\mathbf{j} + 40\mathbf{k}$ acting at (1, 2, 3) about the point (0, 2, 0) is

(A) 89 (B) 80 (C) 60 (D) 40 (E) 30

91. If two forces hold a rigid body in equilibrium, they must

(A) form a couple.
(B) be non-concurrent.
(C) be collinear.
(D) act at a right angle.
(E) form an equilateral triangle.

92. A force and a moment are needed at the wall to hold the rigid link in equilibrium. What is the moment?

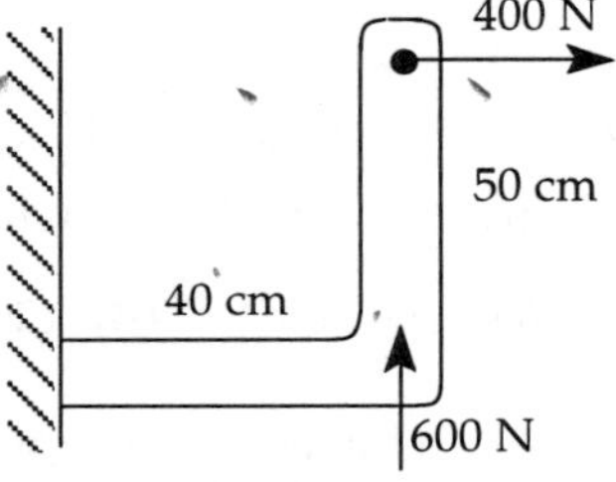

(A) 20 N·m
(B) 40 N·m
(C) 140 N·m
(D) 240 N·m
(E) 480 N·m

93. Find the force in member AB.

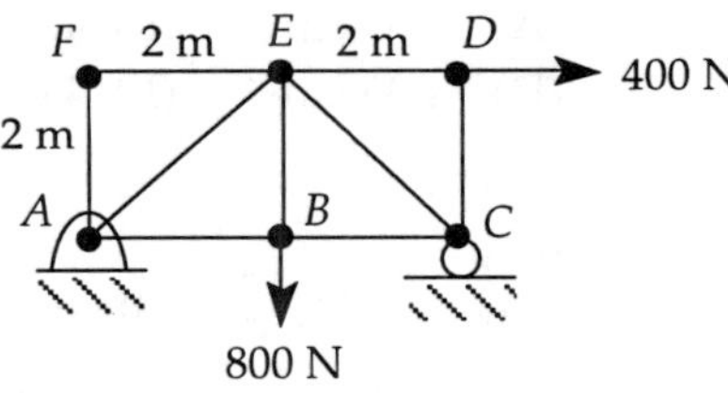

(A) 0 (B) 200 N (C) 400 N
(D) 566 N (E) 600 N

94. What is the force in member AF of Problem 93?

(A) 0 (B) 200 N (C) 400 N
(D) 566 N (E) 600 N

95. What force will cause the 20 kg block to move?

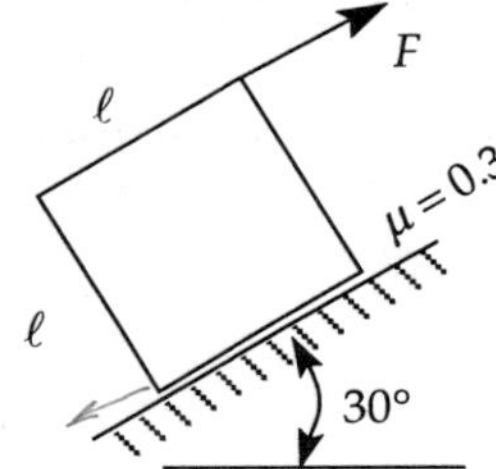

(A) 103 N (B) 121 N (C) 134 N
(D) 149 N (E) 158 N

96. Find the magnitude of the force at support C.

(A) 537 N
(B) 424 N
(C) 387 N
(D) 300 N
(E) 280 N

97. A 4-m-long chain lies in a straight line on a table. How much of the chain can hang over an end of the table without the entire chain slipping off? $\mu = 0.4$.

(A) 1.02 m (B) 1.14 m (C) 1.92 m (D) 2.67 m (E) 2.94 m

98. A wire connects the middle of the two links. What is the tension in the wire?

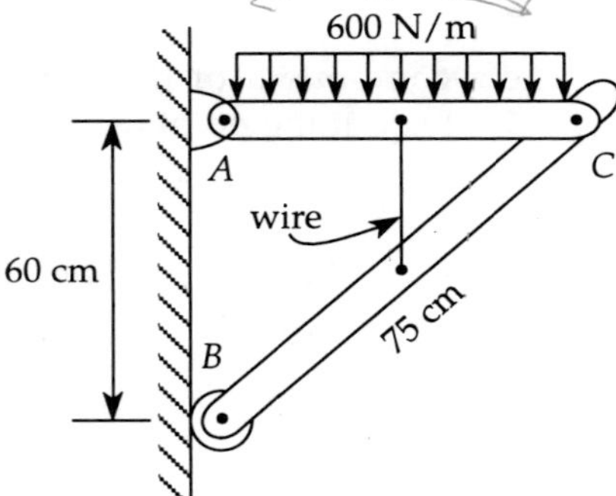

(A) 540 N
(B) 405 N
(C) 270 N
(D) 195 N
(E) 101 N

99. A plane body has its centroid C at (2 cm, 4 cm). If its area is 100 cm^2 and $I_y = 2000\text{ cm}^4$, what is $(I_y)_C$?

(A) 1600 cm^4 (B) 1200 cm^4 (C) 1000 cm^4 (D) 800 cm^4 (E) 400 cm^4

100. The U-beam is composed of two 2 cm by 10 cm plates and a 2 cm by 4 cm plate. What is the y-coordinate of the centroid?

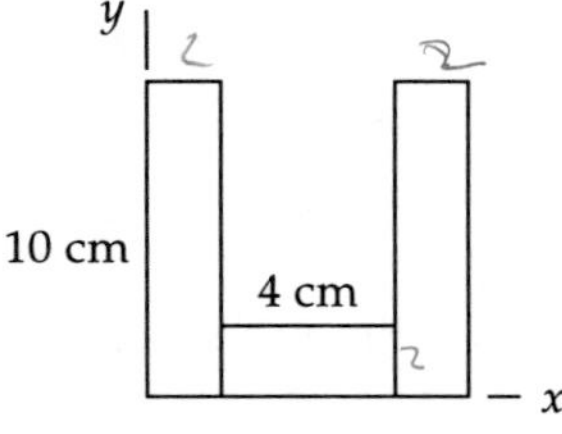

(A) 4.83 cm (B) 4.69 cm (C) 4.52 cm
(D) 4.33 cm (E) 4 cm

101. What is $(I_x)_C$ for the U-beam of Problem 100?

(A) 960 cm^4 (B) 782 cm^4 (C) 664 cm^4 (D) 574 cm^4 (E) 444 cm^4

102. An object experiences rectilinear acceleration $a(t) = 10 - 2t$. How far does it travel in 6 seconds if its initial velocity is 10 m/s?

(A) 182 m (B) 168 m (C) 142 m (D) 126 m (E) 108 m

103. A wheel is rotating at 4000 rpm. If it experiences a deceleration of 20 rad/s^2, through how many revolutions will it rotate before it stops?

(A) 4400 (B) 3200 (C) 2100 (D) 1400 (E) 700

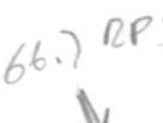

104. An automobile is accelerating at 5 m/s^2 on a straight road on a hill where the radius of curvature of the hill is 200 m. What is the magnitude of the total acceleration when the car's speed is 30 m/s?

(A) 5 m/s^2 (B) 5.46 m/s^2 (C) 6.04 m/s^2 (D) 6.73 m/s^2 (E) 7.93 m/s^2

105. Find an expression for the maximum range of a projectile with initial velocity v_0 at angle θ with the horizontal.

(A) $\frac{v_0^2}{g}\sin 2\theta$ (B) $\frac{v_0^2}{2g}\sin^2\theta$ (C) $\frac{v_0^2}{2g}\sin\theta$

(D) $\frac{v_0^2}{g}\cos\theta$ (E) $\frac{v_0^2}{2g}\sin\theta\cos\theta$

106. An 80-cm-diameter wheel is accelerating at 10 m/s^2 without slipping on a flat surface. What is the magnitude of the acceleration of the very top of the wheel when the velocity of the wheel is 4 m/s?

(A) 20 m/s^2 (B) 25 m/s^2 (C) 40 m/s^2 (D) 45 m/s^2 (E) 64 m/s^2

107. A particle experiences the displacement shown. What is its velocity at $t = 1$ s?

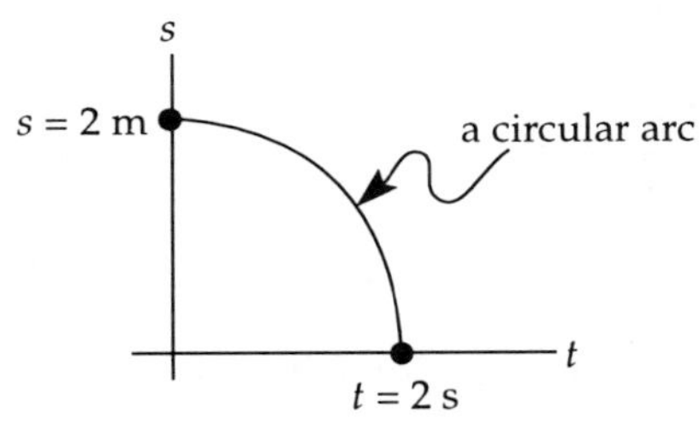

(A) −1.24 m/s (B) − 1.0 m/s (C) −0.92 m/s (D) −0.76 m/s (E) − 0.58 m/s

108. If $\omega = 10$ rad/s, find the velocity of slider A.

(A) 0.5 m/s
(B) 0.577 m/s
(C) 0.6 m/s
(D) 0.866 m/s
(E) 1.0 m/s

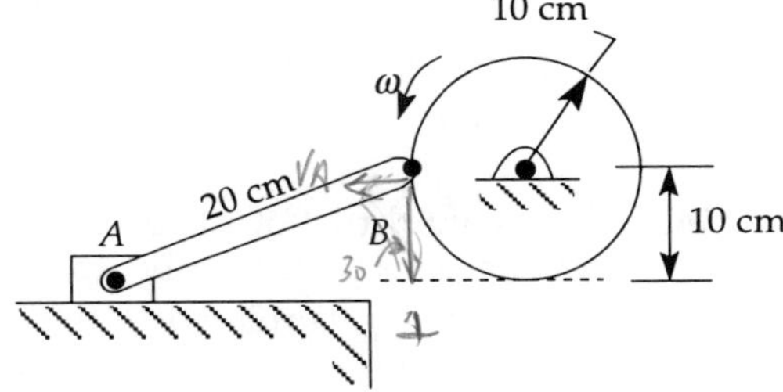

109. If the angular acceleration of the driver wheel in Problem 108 is zero, find the acceleration of slider A.

(A) 12.6 m/s^2 (B) 15.1 m/s^2 (C) 17.7 m/s^2 (D) 20.3 m/s^2 (E) 26.2 m/s^2

110. Find the tension in the string connecting the two masses. The pulley is massless and frictionless.
(A) 240 N
(B) 560 N
(C) 1260 N
(D) 1960 N
(E) 2050 N

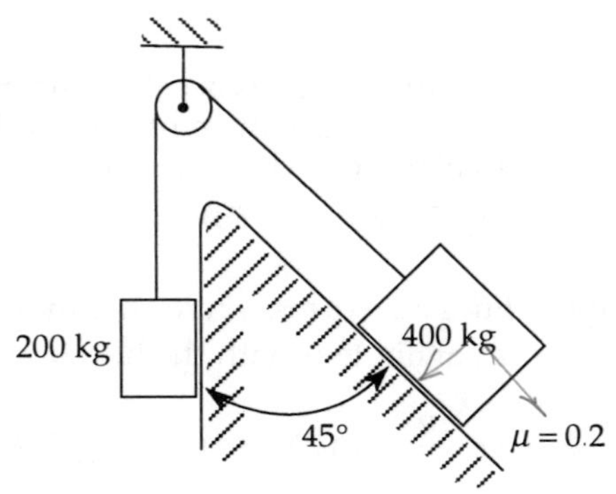

111. If the 200 kg mass in Problem 110 moves 2 m, what is its velocity?
(A) 1.31 m/s (B) 2.33 m/s (C) 2.84 m/s (D) 3.54 m/s (E) 4.02 m/s

112. The force $F(t)$ acts on the mass shown. What is its velocity after 20 s if it starts from rest?

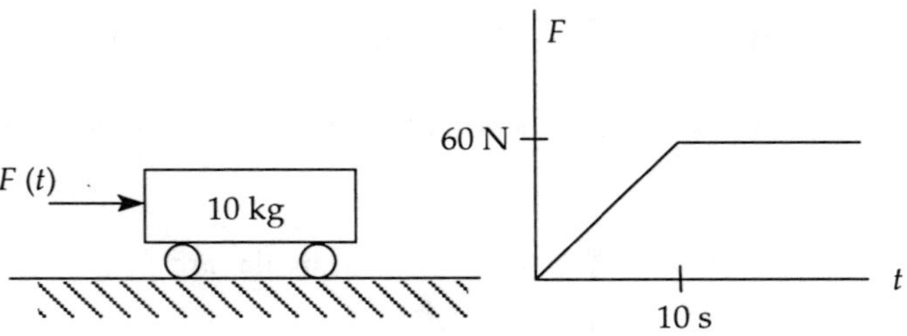

(A) 60 m/s (B) 70 m/s (C) 80 m/s (D) 90 m/s (E) 100 m/s

113. The spring is stretched until the force it exerts on the cylinder is 400 N. It is attached to a rope wrapped around the stationary cylinder. What is the cylinder's speed when the spring force is zero? (No slipping occurs.)
(A) 5.16 m/s (B) 5.98 m/s (C) 6.32 m/s
(D) 7.48 m/s (E) 8.36 m/s

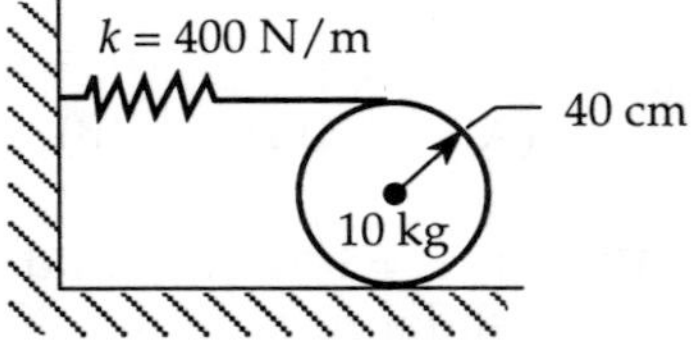

114. A constant thrust of 20 000 N launches on a 1500-kg rocket vertically upward. If the drag force F_D is related to the velocity, as shown, find the velocity of the rocket after 20 seconds.
(A) 38 m/s (B) 42 m/s (C) 58 m/s
(D) 70 m/s (E) 92 m/s

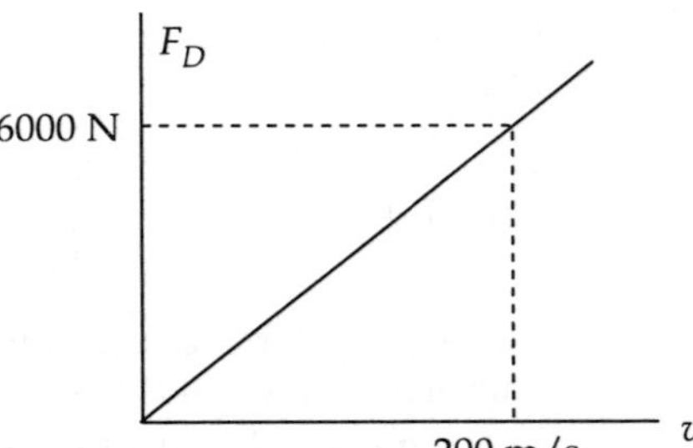

115. The two identical balls collide as shown. What is v_2'?
(A) 5.4 m/s (B) 5.2 m/s (C) 5.0 m/s
(D) 4.8 m/s (E) 4.6 m/s

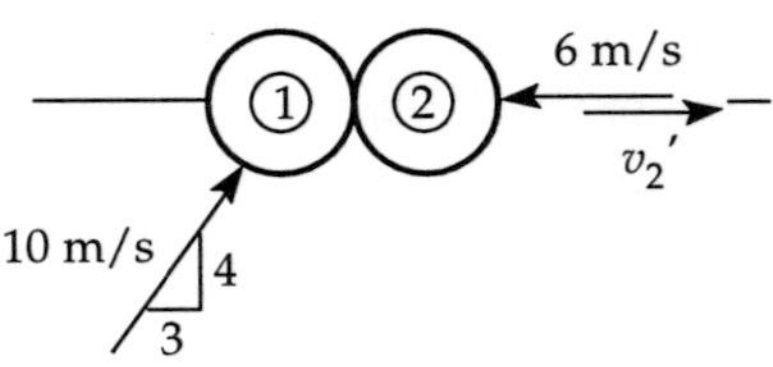

116. Two linear springs, with spring constants k_1 and k_2, are connected in series. What is the over-all spring constant of the two-spring assembly?

(A) $k_1 + k_2$ (B) $\frac{1}{2}(k_1 + k_2)$ (C) $\frac{k_1^2 + k_2^2}{k_1 + k_2}$ (D) $\frac{2k_1k_2}{k_1 + k_2}$ (E) $\frac{k_1k_2}{k_1 + k_2}$

117. A solid steel cylinder of radius r and a hollow brass cylinder of outer radius $3r/2$ support the load. If both cylinders are of the same length, what percentage of the load is carried by the steel cylinder? ($E_{steel} = 2E_{brass}$.)

Load
brass
steel

(A) 72.5% (B) 61.5% (C) 55.5%
(D) 42.5% (E) 37.5%

118. Wood has grain resulting in material properties quite different normal to the grain compared with properties parallel to the grain. Such a material is

(A) nonhomogeneous (B) nonholomorphic (C) nonorthotropic
(D) nonisotropic (E) nonisentropic

119. Steel is used to reinforce concrete because

(A) its density is correctly related to the density of concrete.
(B) it is relatively inexpensive compared to other metals.
(C) its coefficient of thermal expansion is the same as that of concrete.
(D) its Poisson's ratio is the same as that of concrete.
(E) it does not adhere to concrete.

Select either of the following two problems:

120. (metric) Calculate the maximum shearing stress at a point where $\tau_{xy} = 10$ MPa, $\sigma_x = 40$ MPa, and $\sigma_y = 50$ MPa.

(A) 46.1 MPa (B) 36.5 MPa (C) 23.2 MPa (D) 18.3 MPa (E) 11.2 MPa

120. (English) Calculate the maximum shearing stress at a point where $\tau_{xy} = 1000$ psi, $\sigma_x = 4000$ psi, and $\sigma_y = 5000$ psi.

(A) 4610 psi (B) 3650 psi (C) 2320 psi (D) 1830 psi (E) 1120 psi

---

121. A simply-supported beam of length L has a uniform load w over the entire length. What is the ratio of the maximum normal stress to the maximum vertical shearing stress if the cross section is a square $b \times b$?

(A) L/b (B) $2L/b$ (C) $3L/b$ (D) $4L/b$ (E) $5L/b$

122. The bending moment diagram of a simply-supported beam with a uniform load over the entire length has the shape of a

(A) rectangle (B) triangle (C) parabola (D) circle (E) straight line

Select either of the following two problems:

123. (metric) A hollow shaft with an inner radius of 2 cm and an outer radius of 3 cm transmits a torque of 300 N·m. What is the maximum shearing stress?
(A) 17.6 MPa (B) 14.2 MPa (C) 12.6 MPa (D) 9.32 MPa (E) 8.81 MPa

123. (English) A hollow shaft with an inner radius of 1 inch and an outer radius of 1.5 inches transmits a torque of 200 ft-lb. What is the maximum shearing stress?
(A) 1130 psi (B) 960 psi (C) 846 psi (D) 720 psi (E) 564 psi

Select either of the following two problems:

124. (metric) Find the maximum tensile stress if the rectangular cross section is 2 cm × 6 cm (6 cm is vertical).

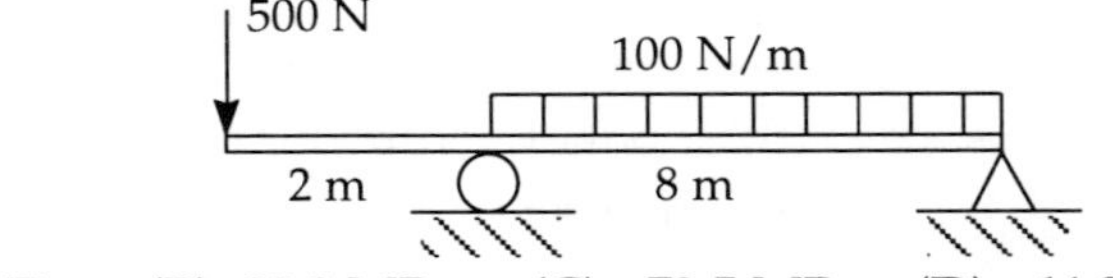

(A) 92.3 MPa (B) 83.3 MPa (C) 72.5 MPa (D) 64.8 MPa (E) 46.5 MPa

124. (English) Find the maximum tensile stress if the rectangular cross section is 1 in × 3 in (3 inches is vertical).

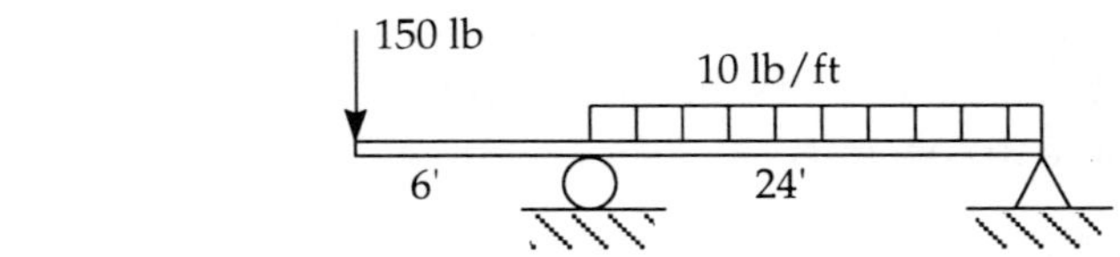

(A) 8400 psi (B) 7200 psi (C) 6000 psi (D) 5400 psi (E) 4800 psi

Select either of the following two problems:

125. (metric) A solid circular shaft, 8 cm in diameter, transmits a torque of 1200 N·m. Calculate the maximum normal stress in the shaft.
(A) 24 MPa (B) 18 MPa (C) 14 MPa (D) 12 MPa (E) 10 MPa

125. (English) A solid circular shaft, 4 inches in diameter, transmits a torque of 1200 ft-lb. Calculate the maximum normal stress in the shaft.
(A) 2300 psi (B) 1800 psi (C) 1380 psi (D) 1150 psi (E) 940 psi

Select either of the following two problems:

126. (metric) A 4-cm-diameter cantilever steel beam is 6 m long. A uniform load of 100 N/m is applied over 4 m, beginning from the wall. The remaining 2 m is unloaded. What is the deflection of the free end? ($E_{\text{steel}} = 210 \times 10^6$ kPa.)
(A) 20.2 cm (B) 16.4 cm (C) 14.7 cm (D) 12.1 cm (E) 8.2 cm

126. (English) A 2-in-diameter cantilever steel beam is 18 ft long. A uniform load of 8 lb/ft is applied over 12 ft, beginning from the wall. The remaining 6 ft is unloaded. What is the deflection of the free end? ($E_{\text{steel}} = 30 \times 10^6$ psi.)
(A) 1.61 in (B) 1.23 in (C) 1.02 in (D) 0.92 in (E) 0.84 in

127. A fluid is a substance that
(A) is compressible with a sufficiently large pressure.
(B) has a viscosity that decreases with increased temperature.
(C) flows if subjected to a shearing stress.
(D) vaporizes at sufficiently low temperature.
(E) moves when subjected to a positive pressure.

128. The viscosity of a gas increases with increased temperature because
(A) internal stickiness of the gas decreases.
(B) internal molecular activity decreases.
(C) internal stickiness of the gas increases.
(D) molecular activity increases.
(E) the shearing stress becomes larger.

129. If a fluid "fails" in a piping system, the "failure" is related to which of the following properties?
(A) surface tension (B) bulk modulus (C) viscosity
(D) speed of sound (E) vapor pressure

Select either of the following two problems:

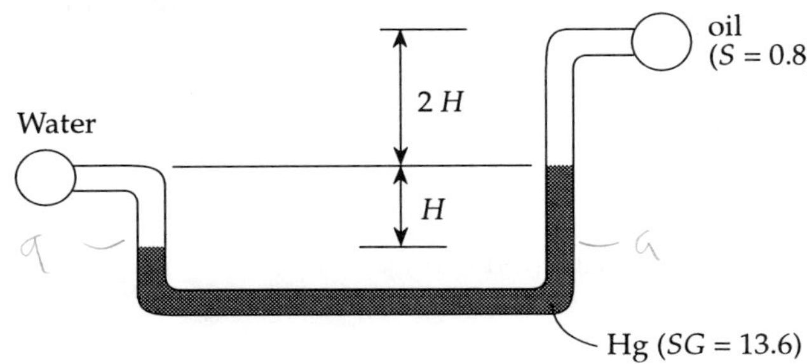

130. (metric) Find the difference in pressure between the water and oil if $H = 25$ cm.
(A) 42.3 kPa (B) 37.2 kPa (C) 34.8 kPa (D) 30.6 kPa (E) 24.6 kPa

130. (English) Find the difference in pressure between the water and oil if $H = 10$ in.
(A) 6.22 psi (B) 5.47 psi (C) 5.13 psi (D) 4.5 psi (E) 3.62 psi

---

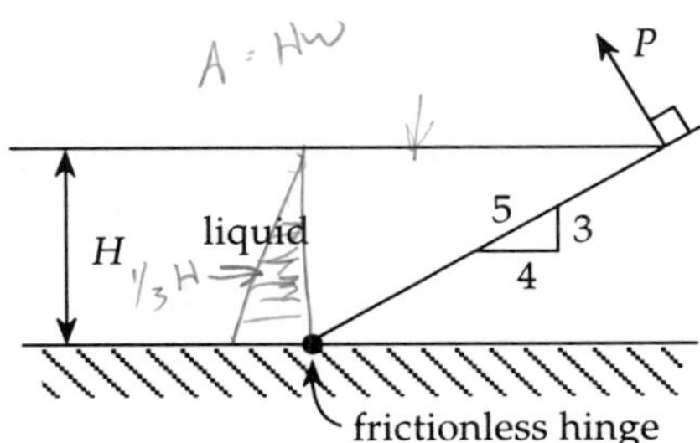

131. Find an expression for the force P needed to hold the gate of width w in the position shown.
(A) $\frac{1}{3}\gamma w H^2$ (B) $\frac{1}{6}\gamma w H^2$ (C) $\frac{2}{9}\gamma w H^2$ (D) $\frac{1}{2}\gamma w H^2$ (E) $\frac{5}{18}\gamma w H^2$

132. Water flows in a pipe of diameter D with a velocity V. It enters at the center of two parallel disks of radius R separated by a distance t. The water flows radially outward between the disks. The velocity with which the water leaves the disks is

(A) $\frac{D^2V}{4R^2}$ (B) $\frac{D^2V}{8Rt}$ (C) $\frac{DtV}{8R^2}$ (D) $\frac{DtV}{4R^2}$ (E) $\frac{DtV}{2R^2}$

133. The Froude number is a ratio of
(A) inertial forces to viscous forces.
(B) body forces to viscous forces.
(C) body forces to pressure forces.
(D) inertial forces to body forces.
(E) inertial forces to pressure forces.

134. Bernoulli's equation cannot be used to approximate the pressure drop for which of the following?
(A) across an orifice through which water flows
(B) across a nozzle through which water flows
(C) from the free stream to the stagnation point on an airfoil of a small aircraft
(D) across a Venturi meter
(E) from upstream to the stagnation region of wind blowing on the window of a house

135. A laminar flow exists in a pipe flow. We know that
(A) the velocity profile is linear.
(B) the Reynolds number is less than 2000.
(C) the shear stress distribution is linear.
(D) the pipe is smooth.
(E) inertial effects dominate the viscous effects.

Select either of the following two problems:

136. (metric) The pressure drop over 15 m of 2-cm-diameter galvanized iron pipe is measured to be 60 kPa. If the pipe is horizontal, estimate the flow rate of water. ($\nu = 10^{-6}\ \text{m}^2/\text{s}$.)

(A) 6.82 L/s (B) 2.18 L/s (C) 0.682 L/s (D) 0.218 L/s (E) 6.82×10^{-4} L/s

136. (English) The pressure drop over 50 ft of 1-in-diameter galvanized iron pipe is measured to be 10 psi. If the pipe is horizontal, estimate the flow rate of water. ($\nu = 10^{-5}\ \text{ft}^2/\text{sec}$.)

(A) 0.48 cfs (B) 0.14 cfs (C) 0.048 cfs (D) 0.014 cfs (E) 4.8×10^{-5} cfs

Select either of the following two problems:

137. (metric) The pressure drop across a valve, through which 0.04 m^3/s of water flows, is measured to be 100 kPa. Estimate the loss coefficient if the nominal diameter of the valve is 8 cm.

(A) 0.32 (B) 0.79 (C) 3.2 (D) 8.7 (E) 31

137. (English) The pressure drop across a valve, through which 1.2 ft^3/sec of water flows, is measured to be 15 psi. Estimate the loss coefficient if the nominal diameter of the valve is 3 inches.

(A) 0.37 (B) 0.85 (C) 3.7 (D) 9.4 (E) 36

Select either of the following two problems:

138. (metric) A large vehicle is slowed by lowering a 2-m-wide scoop into a reservoir of water. What force is exerted on the scoop if the vehicle is traveling 60 m/s and it scoops off 5 cm of water? The scoop diverts the water through 180°.

(A) 1760 kN (B) 720 kN (C) 360 kN (D) 12 kN (E) 7.2 kN

138. (English) A large vehicle is slowed by lowering a 6-ft-wide scoop into a reservoir of water. What force is exerted on the scoop if the vehicle is traveling 180 ft/sec and it scoops off 2 inches of water? The scoop diverts the water through 180°.

(A) 310,000 lb (B) 126,000 lb (C) 63,000 lb (D) 2100 lb (E) 1260 lb

Select either of the following two problems:

139. (metric) What is the energy requirement of an 85% efficient pump that transports 0.04 m^3/s of water if it increases the pressure from 200 kPa to 1200 kPa?

(A) 4.8 kW (B) 14.2 kW (C) 34.0 kW (D) 47.1 kW (E) 56.5 kW

139. (English) What is the energy requirement of an 85% efficient pump that transports 1.2 ft^3/sec of water if it increases the pressure from 30 psi to 180 psi?

(A) 5.4 Hp (B) 19.8 Hp (C) 40.0 Hp (D) 55.4 Hp (E) 66.5 Hp

---

140. The velocity profile between parallel plates is calculated to be Vy/δ, where y is measured from the bottom plate and δ is the distance between plates. We know that:

I. The flow is laminar. II. The flow is turbulent. III. The flow is steady.
IV. One plate is stationary and the other is moving with velocity V.

(A) I, III, and IV only (B) II, III and IV only (C) I and III only
(D) II and IV only (E) I and IV only

STOP!

•The Afternoon Session is Next•

If you finish before 4 hours is up, you may return to any of the problems in the morning session. Do not look at the afternoon session problems. Take a 1 hour break and then continue with the afternoon session.

Afternoon Session Answer Sheet

Mathematics

1. A B C D E
2. A B C D E
3. A B C D E
4. A B C D E
5. A B C D E
6. A B C D E
7. A B C D E
8. A B C D E
9. A B C D E
10. A B C D E
11. A B C D E
12. A B C D E
13. A B C D E
14. A B C D E
15. A B C D E
16. A B C D E
17. A B C D E
18. A B C D E
19. A B C D E
20. A B C D E

score________

Mechanics

21. A B C D E
22. A B C D E
23. A B C D E
24. A B C D E
25. A B C D E
26. A B C D E
27. A B C D E
28. A B C D E
29. A B C D E
30. A B C D E
31. A B C D E
32. A B C D E
33. A B C D E
34. A B C D E
35. A B C D E
36. A B C D E
37. A B C D E
38. A B C D E
39. A B C D E
40. A B C D E

score________

Electrical Circuits

41. A B C D E
42. A B C D E
43. A B C D E
44. A B C D E
45. A B C D E
46. A B C D E
47. A B C D E
48. A B C D E
49. A B C D E
50. A B C D E

score________

Economics

51. A B C D E
52. A B C D E
53. A B C D E
54. A B C D E
55. A B C D E
56. A B C D E
57. A B C D E
58. A B C D E
59. A B C D E
60. A B C D E

score________

Thermodynamics/ Fluid Mechanics

61. A B C D E
62. A B C D E
63. A B C D E
64. A B C D E
65. A B C D E
66. A B C D E
67. A B C D E
68. A B C D E
69. A B C D E
70. A B C D E

score________

Afternoon Session—Diagnostic Exam

4 hours maximum

Each of the 70 problems is followed by five answers. Select the response that is best and circle the corresponding letter on the answer sheet.

Questions 1-3:

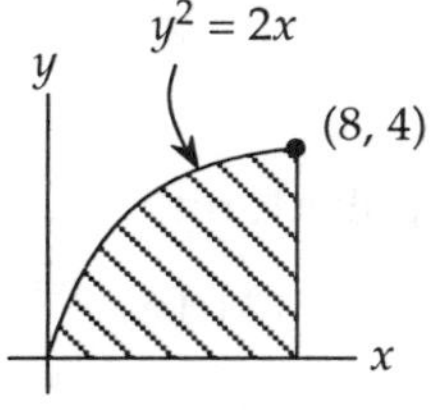

1. Find the cross-hatched area.

   (A) 23 (B) 128/5 (C) 21 (D) 64/5 (E) 64/3

2. Find the y-coordinate of the centroid of the cross-hatched area.

   (A) 5/2 (B) 3/2 (C) 5/3 (D) 4/3 (E) 2/3

3. What volume is generated by rotating the cross-hatched area 60° about the x-axis?

   (A) $32\pi/3$ (B) 16π (C) 32π (D) $64\pi/5$ (E) 64π

Questions 4–6:

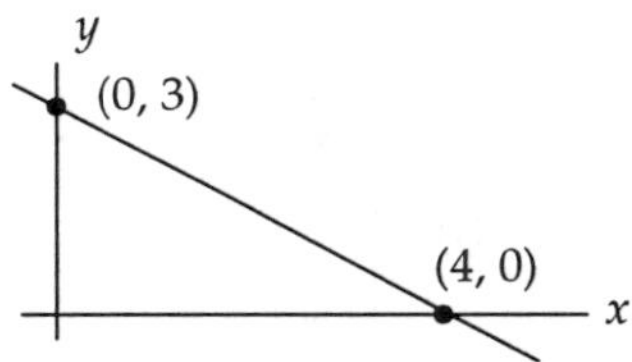

4. What is the equation of the line passing through the two points shown?

   (A) $y = -\frac{4}{3}x + 3$ (B) $4y - 3x = 12$ (C) $3y - 4x = 12$
   (D) $4y + 3x = 12$ (E) $3y + 4x = 9$

5. What is the equation of the line passing through the origin and perpendicular to the angled line?

   (A) $3y = 4x + 4$ (B) $3x + 4y = 0$ (C) $3y - 4x = 0$
   (D) $3x - 4y = 0$ (E) $3y + 4x = 0$

6. Which integral would be used to provide the second moment about the x-axis of the area formed by the straight line, the x-axis, and the y-axis?

   (A) $\int_0^3 y^2 x dy$ (B) $\int_0^4 x^2 y dx$ (C) $\int_0^3 yx^2 dy$ (D) $\int_0^4 y^2 x dy$ (E) $\int_0^3 y^2 x dx$

Questions 7–9:

Three complex numbers are

$$z_1 = 1+2i, \qquad z_2 = 5+3i, \qquad z_3 = i$$

7. Find z_2/z_1.

(A) $\frac{1}{2}(1-11i)$ (B) $\frac{1}{5}(11-7i)$ (C) $\frac{1}{3}(7-6i)$ (D) $\frac{1}{2}(11-7i)$ (E) $\frac{1}{5}(7-6i)$

8. Find $z_3^{1/2}$.

(A) $\sqrt{2i}$ (B) $\sqrt{2}(-1+i)$ (C) $(-1+i)/\sqrt{2}$ (D) $(1+i)/\sqrt{2}$ (E) $\sqrt{2}(1-i)$

9. Express the product z_1z_2 in polar form.

(A) $17.03\,e^{1.494i}$ (B) $13.04\,e^{-1.494i}$ (C) $17.03\,e^{1.648i}$

(D) $17.03\,e^{-1.494i}$ (E) $13.04\,e^{1.648i}$

Questions 10–14:

Given the three matrices:

$$\mathbf{A} = \begin{bmatrix} 2 & 1 & -2 \\ 0 & 1 & 0 \\ 1 & 2 & 0 \end{bmatrix}, \qquad \mathbf{B} = \begin{bmatrix} 1 & -2 & 0 \end{bmatrix}, \qquad \mathbf{C} = \begin{bmatrix} 3 \\ -1 \\ 0 \end{bmatrix}$$

10. Find $|\mathbf{A}|$.

(A) 0 (B) 1 (C) 2 (D) -2 (E) -1

11. Find $\mathbf{BA}$.

(A) $\begin{bmatrix} 2 & -1 & -2 \end{bmatrix}$ (B) $\begin{bmatrix} 2 \\ -1 \\ -2 \end{bmatrix}$ (C) $\begin{bmatrix} 2 \\ -2 \\ 0 \end{bmatrix}$

(D) $\begin{bmatrix} 2 & -2 & 0 \end{bmatrix}$ (E) $\begin{bmatrix} 2 & -2 & 2 \end{bmatrix}$

12. Find $\mathbf{BC}$.

(A) $\mathbf{0}$ (B) $\begin{bmatrix} 3 & -6 & 0 \\ -1 & 2 & 0 \\ 0 & 0 & 0 \end{bmatrix}$ (C) $\begin{bmatrix} 3 \\ 2 \\ 0 \end{bmatrix}$

(D) $\begin{bmatrix} 5 \end{bmatrix}$ (E) $\begin{bmatrix} 3 & 2 & 0 \end{bmatrix}$

13. Which of the following is an eigenvalue of $\mathbf{A}$?

(A) -1 (B) 0 (C) 1 (D) 2 (E) -2

14. If $\mathbf{Ax} = \mathbf{C}$, what is x_1?

(A) 0 (B) 1 (C) 2 (D) -2 (E) -1

Questions 15–18:

A certain phenomenon is modeled with the following differential equation:

$$\frac{d^2x}{dt^2} + C\frac{dx}{dt} + 4x = f(t)$$

with initial conditions $x(0) = 0$ and $x'(0) = 10$.

15. If C = 0, the homogenous solution is

(A) $c_1e^{2t} + c_2e^{-2t}$ (B) $c_1e^{2it} + c_2e^{-2it}$ (C) $c_1 \cos 4t + c_2 \sin 4t$

(D) $c_1e^{2t} + c_2te^{2t}$ (E) $c_1e^{2it} + c_2te^{2it}$

16. If $C = 0$ and $f(t) = 8e^{2t}$, the particular solution is

(A) e^{2t} (B) Ate^{2t} (C) At^2e^{2t} (D) e^{2it} (E) te^{2it}

17. If $C = 5$ and $f(t) = 20 \sin 2t$, the solution of the initial-value problem is

(A) $e^{-t} - e^{-4t} + 20 \sin 2t$

(B) $6 \sin t - 4 \cos 4t + 4 \cos 2t$

(C) $4 \cos t - 6 \sin 4t - 4 \cos 2t$

(D) $6e^{-t} - 4e^{-4t} - 2 \cos 2t$

(E) $6 \cos t - 6 \cos 4t + 6 \sin 2t$

18. If C = 0 and $f(t) = 4 \cos \omega t$, select the value for ω that would make $x(t)$ unbounded as $t \to \infty$.

(A) 0 (B) 1 (C) 2 (D) 3 (E) 4

Questions 19–20:

$$f(x) = \sin 2x, \qquad g(x) = x$$

19. What is $\lim_{x \to 0} \frac{f^2}{g^2}$?

(A) 0 (B) 2 (C) 4 (D) -2 (E) ∞

20. What is $\int_0^{\pi} f(x)g(x)dx$?

(A) $\pi/2$ (B) π (C) 0 (D) $-\pi$ (E) $-\pi/2$

Questions 21–23:

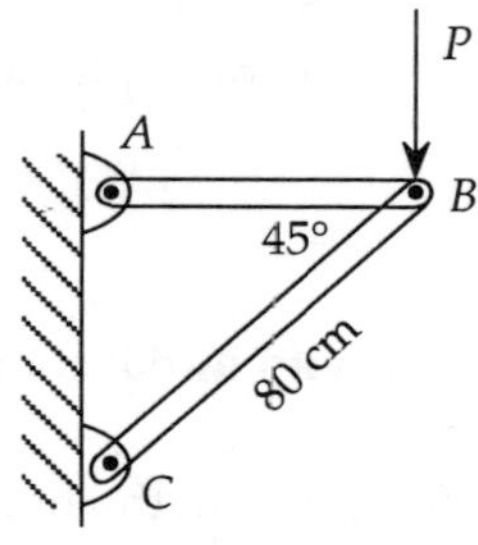

Each link has a circular cross section of radius 8 mm and is made of steel.

$E_{steel} = 210 \times 10^6$ kPa, $\alpha = 11.7 \times 10^{-6}$ °C^{-1}

21. Find the force in member AB if $P = 400$ N.
    (A) 0 (B) 283 N (C) 400 N (D) 476 N (E) 566 N

22. What force P is required to just buckle link BC?
    (A) 9430 N (B) 7370 N (C) 5980 N (D) 4720 N (E) 3260 N

23. If the temperature of link BC were increased 40°C while that of link AB were held constant, how far would point B move up?
    (A) 0.187 mm (B) 0.265 mm (C) 0.529 mm (D) 0.765 mm (E) 0.921 mm

Questions 24–26:

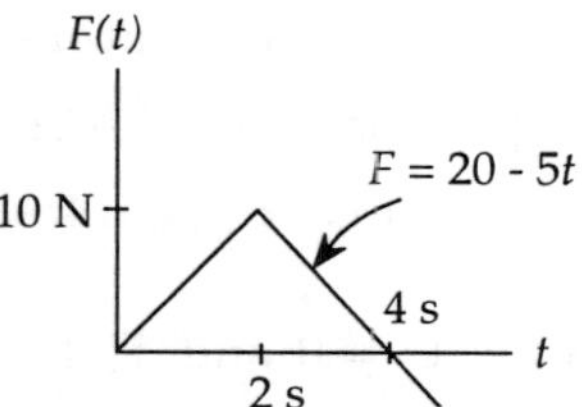

The force $F(t)$ acts on a 2 kg mass that moves in a straight line on a frictionless plane. It starts from rest at $x = 0$ at $t = 0$.

24. What is the displacement x after 2 seconds?
    (A) 5 m (B) 5/3 m (C) 3 m (D) 10/3 m (E) 2 m

25. What is the velocity after 4 seconds?
    (A) 5 m/s (B) 6 m/s (C) 8 m/s (D) 9 m/s (E) 10 m/s

26. How long does it take for the mass to reach its maximum displacement from $x = 0$?
    (A) 2 s (B) 4 s (C) 6.83 s (D) 7.24 s (E) 8.61 s

Questions 27–31:

The 50 kg solid cylinder rolls without slipping. It is initially at rest.

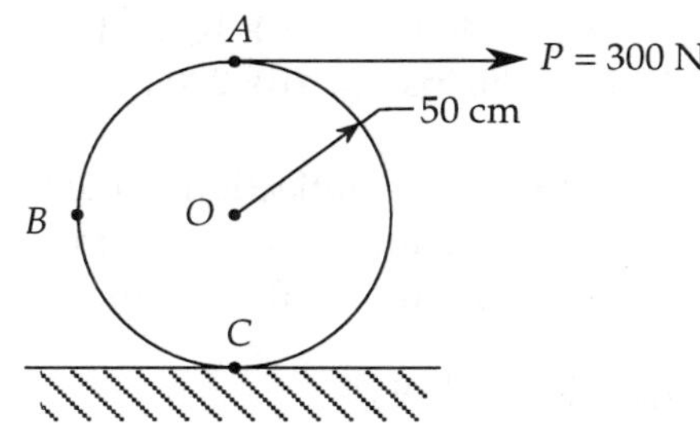

27. Find the speed of point B when the speed of A is 20 m/s.
(A) 20 m/s (B) 14.14 m/s (C) 12.12 m/s (D) 10 m/s (E) 7.07 m/s

28. Find the acceleration of point O.
(A) 20 m/s^2 (B) 16 m/s^2 (C) 12 m/s^2 (D) 8 m/s^2 (E) 4 m/s^2

29. Find the friction force.
(A) 100 N right (B) 100 N left (C) 50 N right (D) 50 N left (E) 0

30. If P acts for 2 seconds, what is the velocity of point O ?
(A) 8 m/s (B) 12 m/s (C) 16 m/s (D) 24 m/s (E) 32 m/s

31. If P acts over a distance of 2 m, what is the velocity of point O?
(A) 4 m/s (B) 6 m/s (C) 8 m/s (D) 10 m/s (E) 12 m/s

Questions 32–36:

	metric	English
$P =$	400 N	100 lb
$\omega =$	200 N/m	40 lb/ft
$\ell =$	3 m	9 feet
$b =$	80 mm	4 inches
$t =$	10 mm	0.5 inches
$y_C =$	27.5 mm	1.375 inches
$I_{N.A.} =$	1.848×10^{-6} m^4	7.771 in^4
$E =$	210×10^9 Pa	30×10^6 psi

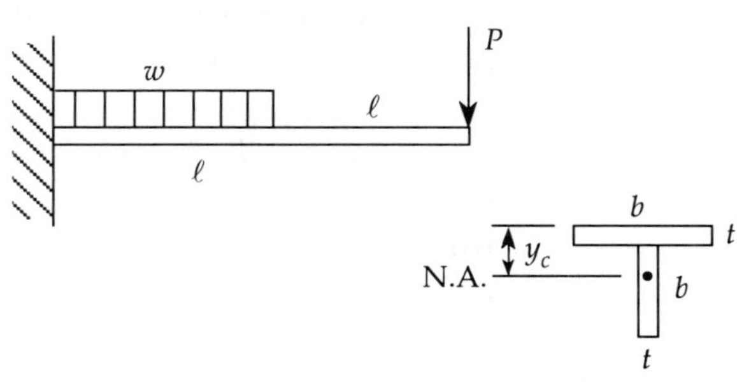

32. The maximum moment acting in the beam is
(metric) (A) 5600 N·m (B) 4200 N·m (C) 3300 N·m (D) 2800 N·m (E) 1800 N·m
(English) (A) 5740 ft-lb (B) 4640 ft-lb (C) 3420 ft-lb (D) 2960 ft-lb (E) 1940 ft-lb

33. The maximum tensile stress in the beam is
(metric) (A) 112 MPa (B) 91.2 MPa (C) 76.5 MPa (D) 60.4 MPa (E) 49.1 MPa
(English) (A) 16,500 psi (B) 13,600 psi (C) 11,500 psi (D) 9000 psi (E) 7262 psi

34. The maximum compressive stress in the beam is
(metric) (A) 112 MPa (B) 91.2 MPa (C) 76.5 MPa (D) 60.4 MPa (E) 49.1 MPa
(English) (A) 16,500 psi (B) 13,600 psi (C) 11,500 psi (D) 9000 psi (E) 7262 psi

35. The maximum vertical shearing stress in the beam is

	(A)	(B)	(C)	(D)	(E)
(metric)	1240 kPa	1060 kPa	840 kPa	720 kPa	560 kPa
(English)	352 psi	290 psi	230 psi	196 psi	152 psi

36. Estimate the deflection of the free end.

	(A)	(B)	(C)	(D)	(E)
(metric)	51.2 mm	63.6 mm	77.9 mm	86.4 mm	98.2 mm
(English)	1.23 in	1.48 in	1.76 in	2.01 in	2.23 in

Questions 37–40:

The stress state at a point in a beam is as shown.

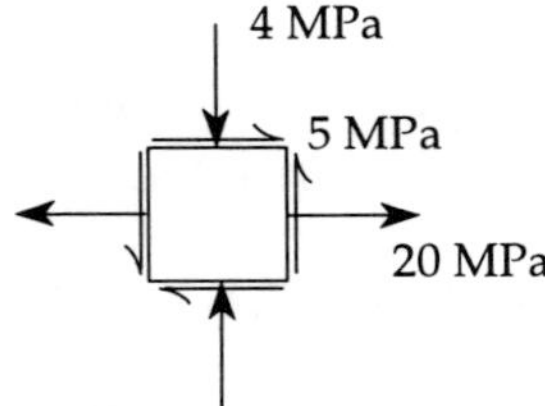

37. What is the maximum tensile stress at the point?
(A) 5 MPa (B) 13 MPa (C) 21 MPa (D) 26 MPa (E) 33 MPa

38. What is the maximum shearing stress at the point?
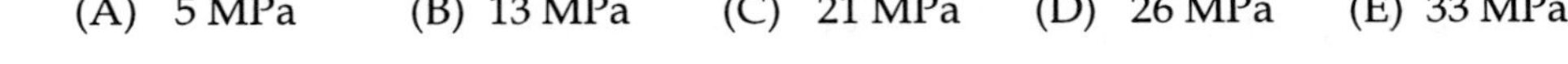
(A) 5 MPa (B) 13 MPa (C) 21 MPa (D) 26 MPa (E) 33 MPa

39. The maximum shearing stress occurs on a plane that makes an angle θ with the 20 MPa stress shown. What is θ?
(A) 67.4° (B) 51.9° (C) 42.8° (D) 33.7° (E) 29.4°

40. The element can be rotated through an angle ϕ such that there is no normal stress on the face with the 4 MPa stress shown. Find ϕ.
(A) 10° (B) 13° (C) 16° (D) 19° (E) 22°

41. What value of R_X in the circuit shown will make $I_G = 0$?
(A) 10
(B) 20
(C) 30
(D) 40
(E) 60

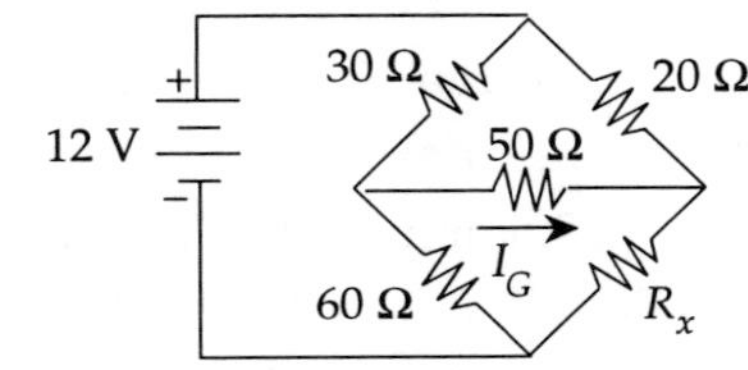

Questions 42–43:

Given the network:

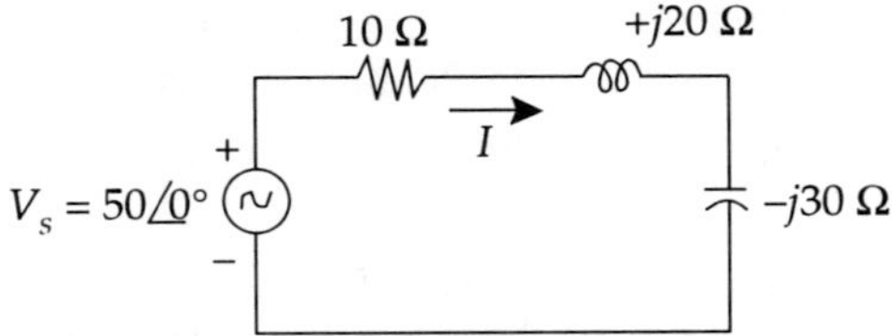

42. The magnitude and phase of the current in the circuit is
(A) 1.72 ∠45° A (B) 3.54 ∠45° A (C) 4.62 ∠−45° A
(D) 5.00 ∠0° A (E) 7.25 ∠53° A

43. The magnitude of the voltage across the capacitor is
(A) 32 V (B) 41 V (C) 50 V (D) 84 V (E) 106 V

Questions 44–45:

Given the circuit:

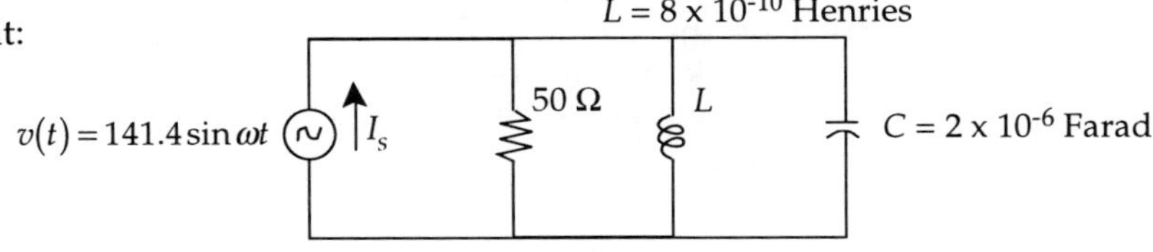

44. What value of ω will make the rms of $|I_S|$ a minimum?

(A) 2×10^6 (B) 8×10^6 (C) 25×10^6 (D) 50×10^6 (E) 100×10^6

45. What is the rms magnitude of the minimum source current?

(A) 1.0 A (B) 1.5 A (C) 2.0 A (D) 2.5 A (E) 3.0 A

46. The power delivered to the load $R_L = 100\ \Omega$ is

(A) 15.6 W
(B) 28.1 W
(C) 32.4 W
(D) 39.1 W
(E) 51.7 W

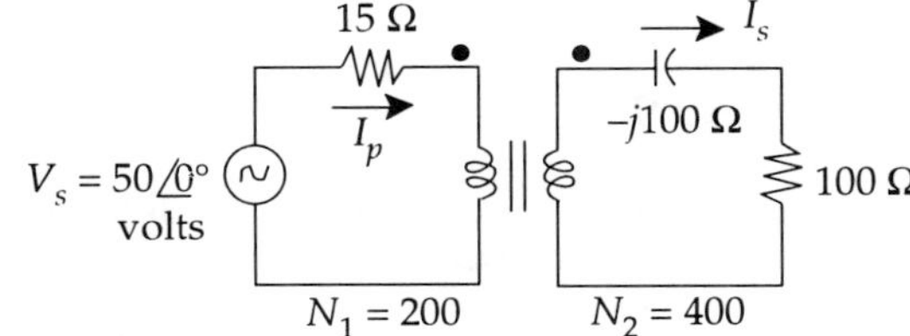

47. The current in a circuit is digital as shown in the scope trace. The rms value of this current is

(A) 0 A
(B) 8.71 A
(C) 16.33 A
(D) 20.0 A
(E) 47.68 A

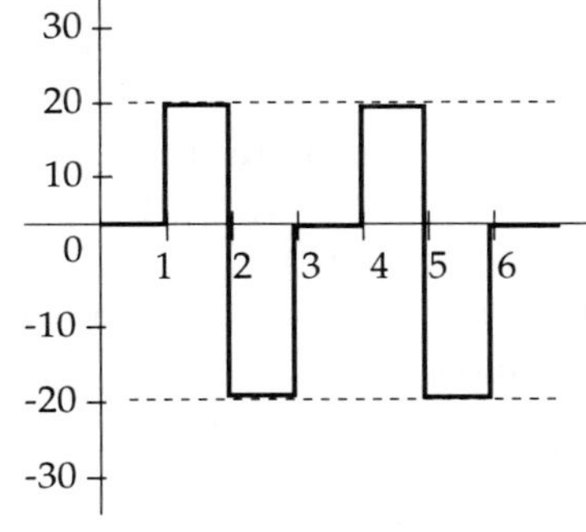

48. Compute the reactive power in volt amperes delivered by the source in the circuit shown.

(A) 741
(B) 892
(C) 1287
(D) 1346
(E) 1623

$+j4\ \Omega$ 25 Ω

$V_s = 120\angle 0°$ volts $-j10\ \Omega$ $+j25\ \Omega$

49. The gain (v_{out}/v_{in}) of the op-amp circuit shown is

(A) +2
(B) −4
(C) +6
(D) −8
(E) +10

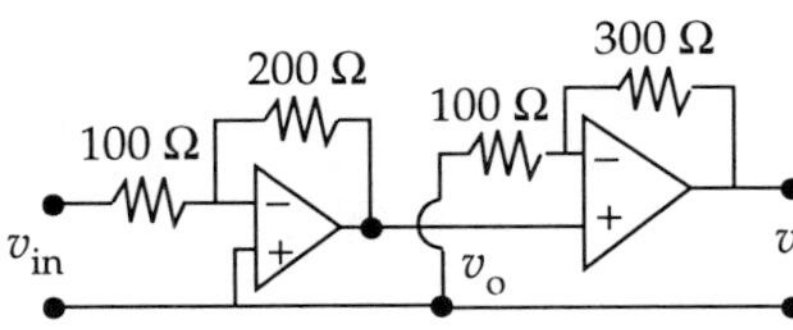

50. An op-amp integrator is used to integrate the square wave shown. What must be the value of R to make the peak value of the triangular wave equal 150 volts?
 (A) 10 000 ohms
 (B) 16 667 ohms
 (C) 25 000 ohms
 (D) 33 333 ohms
 (E) 40 000 ohms

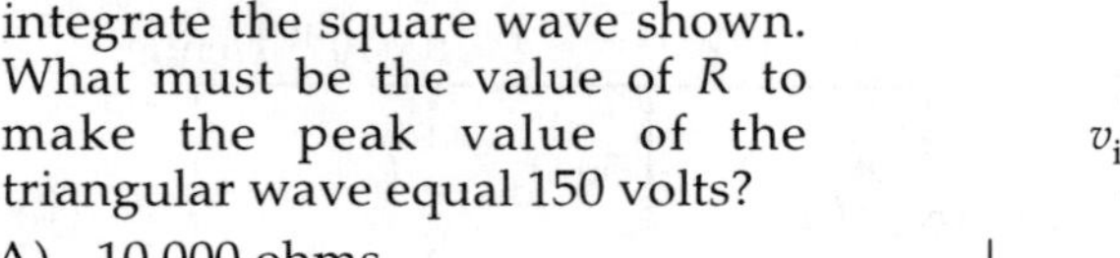

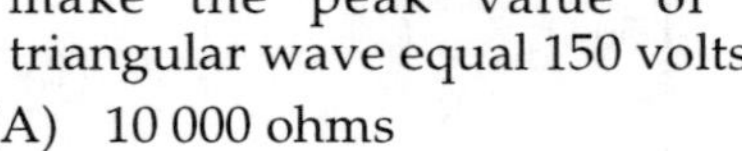

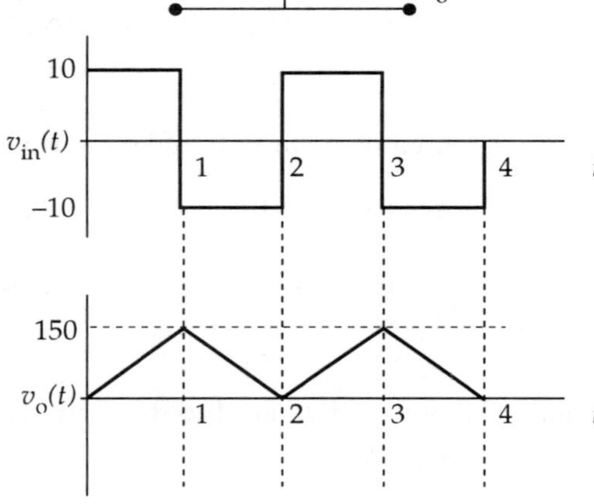

<u>**Questions 51–54:**</u>

The market for laser ignition units is estimated to be 1000 per month if the selling price is $400 per unit. The monthly cost of production is

$$C = 35,000 + 0.5S^2$$

where S is the number of units produced each month. Assume that monthly sales and production rates are equal.

51. Compute the monthly profit if production is 1000 units per month.
 (A) –$135,000 (B) $112 (C) $135 (D) $35,800 (E) $45,000

52. Compute the profitable range of production rates.
 (A) 100 – 400 units/mo (B) 100 – 700 units/mo (C) 265 – 400 units/mo
 (D) 265 – 700 units/mo (E) 400 – 700 units/mo

53. Compute the maximum possible monthly profit.
 (A) –$135,000 (B) $112 (C) $135 (D) $35,800 (E) $45,000

54. Compute the maximum possible profit per unit.
 (A) –$135,000 (B) $112 (C) $135 (D) $35,800 (E) $45,000

Questions 55–57:

Flood damage to businesses and homes in Riverbend City averages five million dollars annually. A levee system costing fifty million dollars would prevent all flood damage. The annual maintenance and interest costs for the levee system would be three million dollars. Riverbend City could finance construction by selling bonds with a 20-year term that would pay 5% interest on par value. The market rate for similar bonds is 6%.

55. Estimate the rate of return provided by the levee system.
    (A) 4% (B) 5% (C) 6% (D) 10% (E) 12%

56. Compute the benefit/cost ratio for the levee system using 6% interest rate.
    (A) 0.10 (B) 0.50 (C) 0.67 (D) 1.00 (C) 1.67

57. Estimate the market price for a bond with $5000 par value.
    (A) $2867 (B) $4426 (C) $5000 (D) $5574 (E) $6000

Questions 58–60:

As a byproduct of its manufacturing process, Technocel Inc. generates 100 kg of highly toxic waste per year. Currently Technocel pays Enviroserv, a waste management service, $15 per kg to remove and treat the waste. Alternatively, Technocel could treat and dispose of the waste for $5 per kg if it installed a $4000 on-site treatment reactor. The estimated life of the reactor is five years, and it would have no salvage value. Technocel's before-tax minimum attractive rate of return is 20%.

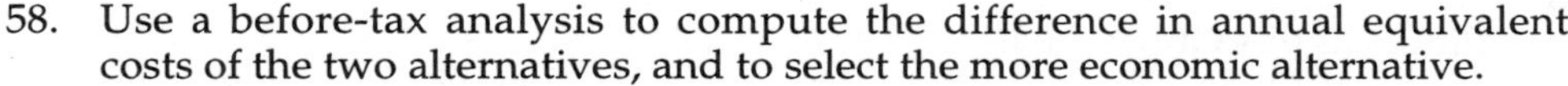

58. Use a before-tax analysis to compute the difference in annual equivalent costs of the two alternatives, and to select the more economic alternative.
    (A) ±$200, prefer Enviroserv
    (B) ±$200, prefer on-site treatment
    (C) ±$338, prefer Enviroserv
    (D) ±$338, prefer on-site treatment
    (E) equal costs, no preference

59. Compute the annual mass of toxic waste above which on-site treatment is the more economic alternative.
    (A) 80 kg/yr (B) 100 kg/yr (C) 123 kg/yr (D) 134 kg/yr (E) 165 kg/yr

60. Use the general straight line method to compute depreciation of the on-site treatment reactor for the second year, and its book value at the end of that year. The recovery period is three years and there is no salvage value.
    (A) $1000, $2000
    (B) $1333, $1333
    (C) $1333, $2000
    (D) $1333, $2667
    (E) $2000, 0

Questions 61–65:

	metric	English
T_{high} =	1200°C	2200°F
T_{low} =	20°C	60°F
P_{low} =	100 kPa	14.7 psia
compression ratio =	8	8
R_{air} =	$0.287\ \frac{\text{kJ}}{\text{kg}\cdot\text{K}}$	$53.3\ \frac{\text{ft-lbf}}{\text{lbm-°R}}$

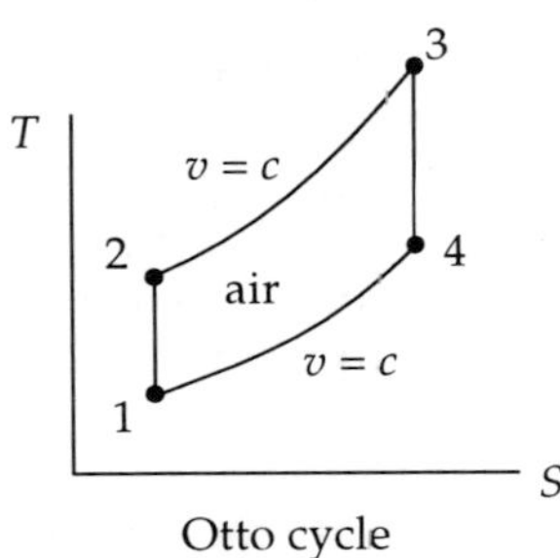

Otto cycle

61. Find the added heat.

(metric) (A) 440 kJ/kg (B) 575 kJ/kg (C) 620 kJ/kg (D) 800 kJ/kg (E) 960 kJ/kg

(English) (A) 200 BTU/lbm (B) 250 BTU/lbm (C) 300 BTU/lbm (D) 350 BTU/lbm (E) 400 BTU/lbm

62. Find the net work per cycle.

(metric) (A) 325 kJ/kg (B) 400 kJ/kg (C) 425 kJ/kg (D) 450 kJ/kg (E) 475 kJ/kg

(English) (A) 140 BTU/lbm (B) 160 BTU/lbm (C) 180 BTU/lbm (D) 200 BTU/lbm (E) 220 BTU/lbm

63. The efficiency of the Otto cycle is

(A) 42% (B) 46% (C) 56% (D) 68% (E) 80%

64. The efficiency of a Carnot cycle operating between the same temperature limits as this Otto cycle is

(A) 42% (B) 46% (C) 56% (D) 68% (E) 80%

65. The work needed to compress the air from BDC to TDC is approximated by which expression?

(A) $RT_2 \ln v_1/v_2$ (B) $c_p(T_2 - T_1)$ (C) $P_2(v_1 - v_2)$

(D) $c_v(T_2 - T_1)$ (E) $P_1(v_1 - v_2)$

Questions 66–70:

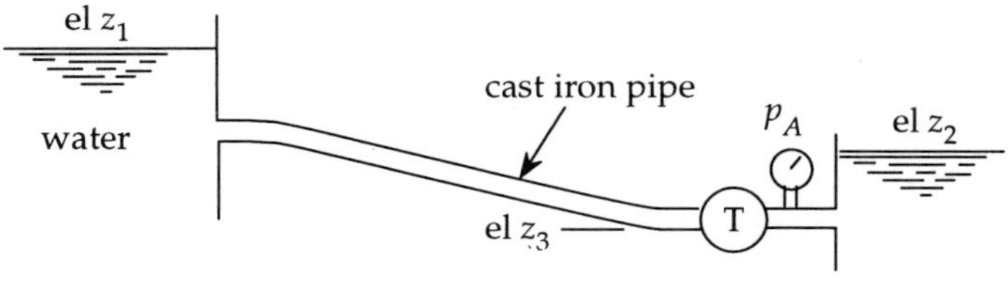

	metric	English
pipe diameter =	1.2 m	4 ft
pipe length =	800 m	2500 ft
water viscosity =	10^{-3} N·s/m^2	2×10^{-5} lb-sec/ft^2
flow rate =	0.8 m^3/s	25 ft^3/sec
z_1 =	40 m	120 ft
z_2 =	10 m	30 ft
z_3 =	0	0

66. Estimate the pressure p_A just upstream of the pipe exit.

(metric) (A) 102 kPa (B) 98 kPa (C) 82 kPa (D) 76 kPa (E) 62 kPa
(English) (A) 14 psi (B) 13 psi (C) 12 psi (D) 11 psi (E) 10 psi

67. Estimate the pressure in the pipe at the turbine inlet.

(metric) (A) 356 kPa (B) 365 kPa (C) 372 kPa (D) 389 kPa (E) 392 kPa
(English) (A) 42.3 psi (B) 48.6 psi (C) 50.2 psi (D) 51.7 psi (E) 52.0 psi

68. If the turbine is 89% efficient, what is the power output?

(metric) (A) 161 kW (B) 186 kW (C) 207 kW (D) 236 kW (E) 258 kW
(English) (A) 175 Hp (B) 196 Hp (C) 225 Hp (D) 250 Hp (E) 271 Hp

69. The turbine has a set of nozzles that direct high speed water into the turbine blades. What is the maximum velocity that the water can have exiting the nozzles?

(metric) (A) 31.3 m/s (B) 27.9 m/s (C) 23.2 m/s (D) 20.1 m/s (E) 18.2 m/s
(English) (A) 99.3 ft/sec (B) 87.6 ft/sec (C) 80.1 ft/sec (D) 71.3 ft/sec (E) 62.4 ft/sec

70. If a model study were used to study the proposed designs of the blades on the turbine, which of the following parameters would be used in the study?

(A) Reynolds number (B) Froude number (C) Weber number
(D) Cauchy number (E) Mach number

STOP!

•You are finished with the Diagnostic Exam!•

If you finish before 4 hours is up, you may return to any of the problems in the afternoon session. You may not return to the morning session.

Answers to the Diagnostic Examination

The Morning Session

1. A
2. D
3. E
4. A
5. C
6. D
7. D
8. A
9. A
10. B
11. A
12. C
13. D
14. A
15. E
16. C
17. D
18. B
19. D
20. A
21. D
22. C
23. A
24. A
25. C
26. C
27. E
28. B
29. D
30. C
31. B
32. D
33. D
34. D
35. D
36. E
37. C
38. B
39. B
40. C
41. C
42. D
43. A
44. D
45. B
46. E
47. E
48. A
49. C
50. D
51. B
52. C
53. E
54. D
55. E
56. A
57. C
58. E
59. B
60. C
61. E
62. B
63. B
64. B
65. D
66. C
67. C
68. D
69. D
70. A
71. A
72. A
73. D
74. E
75. C
76. D
77. E
78. A
79. B
80. B
81. A
82. D
83. C
84. D
85. A
86. D
87. D
88. D
89. A
90. E
91. C
92. B
93. E
94. A
95. C
96. B
97. D
98. C
99. A
100. D
101. E
102. B
103. E
104. D
105. A
106. D
107. E
108. B
109. C
110. E
111. A
112. D
113. A
114. C
115. D
116. E
117. B
118. D
119. C
120. E
121. A
122. C
123. E
124. B
125. D
126. A
127. C
128. D
129. E
130. C
131. E
132. B
133. D
134. A
135. C
136. C
137. C
138. B
139. D
140. A

The Afternoon Session

1. E
2. B
3. A
4. D
5. C
6. A
7. B
8. D
9. E
10. C
11. A
12. D
13. C
14. B
15. B
16. A
17. D
18. C
19. C
20. E
21. C
22. B
23. C
24. D
25. E
26. C
27. B
28. D
29. A
30. C
31. A
32. C
33. E
34. A
35. B
36. D
37. C
38. B
39. D
40. A
41. A
42. B
43. E
44. C
45. C
46. B
47. C
48. E
49. D
50. D
51. A
52. B
53. E
54. C
55. A
56. C
57. B
58. C
59. D
60. B
61. B
62. A
63. C
64. E
65. D
66. B
67. D
68. C
69. A
70. A

Diagnostic Exam Solutions—Morning

1. **A** The slope of the line $y = 2x - 3$ is 2. Its negative reciprocal is $-\frac{1}{2}$.
Hence, $y = -\frac{1}{2}x + b$ is the line. Substitute in:
$-3 = -\frac{1}{2}(6) + b$. $\therefore b = 0$ and $y = -\frac{1}{2}x$.

2. **D** Refer to the general form $Ax^2 + 2Bxy + Cy^2 + \cdots = 0$.
$B^2 - AC = (-2)^2 - (1)(1) = 3$. Since $B^2 - AC > 0$, this is a hyperbola.

3. **E** $\frac{\cos\theta}{\sin\theta} + \frac{\sin\theta}{\cos\theta} = \frac{\cos^2\theta + \sin^2\theta}{\cos\theta\sin\theta} = \frac{1}{\cos\theta\sin\theta} = \frac{2}{2\cos\theta\sin\theta} = \frac{2}{\sin 2\theta}$.

4. **A** $(1-2i)^{-1} = \frac{1}{1-2i} = \frac{1}{1-2i}\frac{1+2i}{1+2i} = \frac{1+2i}{1+4} = \frac{1}{5} + \frac{2}{5}i$.

5. **C** $\text{adj}\begin{bmatrix} 2 & 3 \\ -1 & 0 \end{bmatrix} = \begin{bmatrix} 2 & 3 \\ -1 & 0 \end{bmatrix}^+ = \begin{bmatrix} 0 & 1 \\ -3 & 2 \end{bmatrix}^T = \begin{bmatrix} 0 & -3 \\ 1 & 2 \end{bmatrix}$.
Note: The elements in the transpose are the cofactors of the elements of $\begin{bmatrix} 2 & 3 \\ -1 & 0 \end{bmatrix}$.

6. **D** $y = \frac{\begin{vmatrix} 1 & 2 & 1 \\ 1 & 0 & -1 \\ 1 & -3 & 0 \end{vmatrix}}{\begin{vmatrix} 1 & -1 & 1 \\ 1 & 0 & -1 \\ 1 & 1 & 0 \end{vmatrix}} = \frac{-8}{3} = -\frac{8}{3}$

7. **D** $\begin{vmatrix} 1-\lambda & 1 \\ 2 & -\lambda \end{vmatrix} = (1-\lambda)(-\lambda) - 2 = \lambda^2 - \lambda - 2 = 0.$ $(\lambda - 2)(\lambda + 1) = 0.$ $\therefore \lambda = 2, -1.$

8. **E** Let $y_p = Ae^{-t}$. Substitute in:
$-Ae^{-t} + 4Ae^{-t} = 3e^{-t}$. $\therefore -A + 4A = 3$. $\therefore 3A = 3$. $A = 1$. $\therefore y_p(t) = e^{-t}$.

9. **A** Damping makes $y_h \to 0$ as $t \to \infty$. $\therefore y_{\text{steady-state}} = y_p$. Let
$y_p = A\sin 2t + B\cos 2t$. Substitute in:
$-4A\sin 2t - 4B\cos 2t + 8A\cos 2t - 8B\sin 2t + 4A\sin 2t + 4B\cos 2t = 8\sin 2t$.
$\therefore 8A = 0$ and $-8B = 8$. $\therefore A = 0$ and $B = -1$. $\therefore$ Amplitude = 1.

10. **B** $V = \int_0^2 \pi x^2 dy = \int_0^2 \pi 4y^4 dy = 4\pi\left(\frac{2^5}{5}\right) = \frac{128\pi}{5}$.

11. **A** The curl of a vector field is zero if it is conservative.

12. **C** $\mathbf{A}\times\mathbf{B}=(4\mathbf{i}+2\mathbf{j})\times(B_x\mathbf{i}+B_y\mathbf{j})=0. \quad \therefore 4B_y-2B_x=0.$

$\mathbf{A}\cdot\mathbf{B}=(4\mathbf{i}+2\mathbf{j})\cdot(B_x\mathbf{i}+B_y\mathbf{j})=30. \quad \therefore 4B_x+2B_y=30.$ $\Big\}$ $B_y=3.\ B_x=6.$

$\therefore \mathbf{B}=6\mathbf{i}+3\mathbf{j}.$

13. **D** $$\lim_{x\to 0}\frac{\sin^2 x}{x\left(1-e^x\right)}=\lim_{x\to 0}\frac{2\sin x\cos x}{1-e^x-xe^x}=\lim_{x\to 0}\frac{2\left(\cos^2 x-\sin^2 x\right)}{-e^x-e^x-xe^x}=\frac{2}{-2}=-1.$$

Note: We had to apply L'Hospital's rule twice.

14. **A** $(101.1)_4=1\times 4^2+0\times 4^1+1\times 4^0+1\times 4^{-1}=16+0+1+0.25=17.25.$

15. **E** $$\bar{y}=\frac{\int y\,dA}{\int dA}=\frac{\int_0^1 y\left(2-2y^2\right)dy}{\int_0^1\left(2-2y^2\right)dy}=\frac{1-\frac{2}{4}}{2-\frac{2}{3}}=\frac{1/2}{4/3}=\frac{3}{8}.$$

16. **C** The term sin x makes it nonhomogeneous. (If it were sin y it would be nonlinear.) It has a variable coefficient, x^2. It is linear since the dependent variable y (or its derivatives) appear to the first power.

17. **D** $$\cos 2x=1-\frac{(2x)^2}{2!}+\frac{(2x)^4}{4!}-\cdots=1-2x^2+\frac{2}{3}x^4-\cdots.$$

Note: Since cos 0 = 1, the series must begin with 1. This is a Taylor series with $a=0$:

$$\cos 2x=\cos(0)+\frac{-2\sin(2\cdot 0)}{1!}x+\frac{-4\cos(2\cdot 0)}{2!}x^2+\cdots=1+0-\frac{4}{2}x^2+\cdots.$$

18. **B** Sketch the circle: $\sin\frac{\theta}{2}=\frac{5}{12}. \quad \therefore \frac{\theta}{2}=24.62°$ and $\theta=49.2°.$

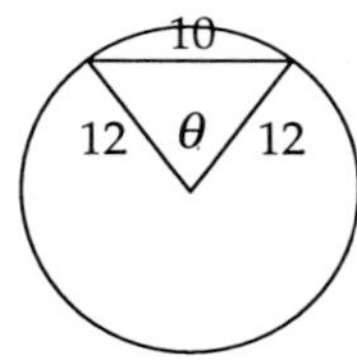

19. **D** At the point (2, 1) the slope of $x=2y^2$ is found by differentiating: $1=2(2y)\frac{dy}{dx}. \quad \therefore \frac{dy}{dx}=\frac{1}{4}.$ Hence the line is $y=\frac{1}{4}x+b.$ If it passes through (2, 1), then $1=\frac{1}{4}(2)+b. \quad \therefore b=\frac{1}{2}.$

20. **A** $$\frac{s+3}{s^2+4s+8}=\frac{(s+2)+1}{(s+2)^2+4}=\frac{s+2}{(s+2)^2+4}+\frac{1}{(s+2)^2+4}.$$

$\therefore f(t)=e^{-2t}\cos 2t+\frac{1}{2}e^{-2t}\sin 2t.$

21. **D** The half-filled conduction band has the available energy levels for Fermi electrons to move up by acquiring extra kinetic energy when an electrical potential is applied.

22. **C** In such a material electrons from the valance band can be thermally activated across the band-gap to the conduction band, and once thus promoted the electrons have mobility under an applied voltage.

23. **A** The energy gap between the impurity level and the bottom of the conduction band is small, and thus electrons from the impurity level can be thermally activated to the conduction band. This is called n-type because the charge carriers are negative (electrons).

24. **A** Covalent bonding is produced by electron sharing between atoms as in the case of diamond (carbon) or Si.

25. **C** If there are N atoms then each atom contributes 2 s-states in s-band. Thus there are $2N$ available states in the s-band. However, if the atom is a monovalent metal then only N electrons are available. That is, half of the states in the s-band are unoccupied.

26. **C** Packing factor of a BCC lattice is less than that for a FCC lattice since the packing factors for BCC and FCC lattice are 0.68 and 0.74, respectively. Packing factor for FCC and HCP are equal, and it is 0.74. Note that (111) plane of a FCC lattice is the most densely packed plane, whereas (001) type planes are the most densely packed in HCP.

27. **E** The three Miller indices are reciprocals of the intercepts. Note that the plane (120) is parallel to the z-axis, i.e., its intercept with this axis is at ∞.

28. **B** In a cubic crystal, Miller indices of a direction perpendicular to a plane are the same as that of the plane. Note that [111] is the body diagonal of a cubic lattice. If you draw the (111) plane, you will note that one of the body diagonals is perpendicular to this plane.

29. **D** Diffusional processes are thermally activated and the rate increases exponentially with increasing temperature.

30. **C** Iron is actually protected by zinc. Zn, being more anodic in Galvanic series than iron, is preferentially corroded while iron is protected. Zinc in this case acts as the sacrificial electrode.

31. **B** High hardness does not mean high toughness. To the contrary, brittle materials (low toughness) often have high hardness.

32. **D** Carbon atoms form a locking of dislocations (called Cottrell atmosphere) thereby a higher stress is necessary to initiate dislocation motion. Once the dislocations are freed, they can move under a lower applied stress, and thus the stress drops giving rise to an yield point.

33. **D** An eutectic point is where two solids coexist with one liquid. Composition C_2 is not such a composition.

34. **D** The second eutectic in the phase diagram is at a composition of 90% B and 10% A, corresponding to the composition C_3 shown in the diagram.

35. **D** The number of protons (42) defines the element. From the Periodic Table it must be Molybdenum (Mo), number 42.

36. **E** The symbol (Rh) identifies the atom as number 45 in the Periodic Table. It must, therefore have 45 protons. The atomic mass (103) tells you it must have 103 – 45 = 58 neutrons. In order to have a charge of +3 it must have 45 – 3 = 42 electrons.

37. **C** Covalent bonds occur between atoms of similar electronegativity. Hydrogen always forms covalent bonds. Ionic bonds form between atoms of very different electronegativity. Only KCl has no possibility of covalent bonding. $NaNO_3$ contains Na^+ ions and NO^{3^-} ions, but the N-O bond is covalent.

38. **B** Find the column or Group for each element in the Periodic Table, then move one column to the right for each negative charge, or one column to the left for each positive charge. If you end up in Group O, the Noble Gases, then the ion has the same electron configuration as a Noble Gas. For example: O^{2-}; Oxygen is in Group 6A. Two columns to the right (because there are 2 negative charges) brings you to Group 0, the Noble Gases. Therefore O^{2-} does have a Noble Gas electron configuration. Pb^{2+}; Lead (82) is in Group 4A. Two columns left brings you to Group 2B, which is not the Noble Gases. Therefore the correct answer is Pb^{2+}.

39. **B** The equilibrium constant (in moles/l) is $[CO_2][H_2]/[CO][H_2O]$ = [2.4][0.50]/[0.80][0.25] = 6

40. **C** As (element number 33) has 33 electrons. The first three shells accommodate 2 + 8 + 18 = 28 electrons. As therefore has 33 – 28 = 5 valence electrons.

41. **C** From the Periodic Table, neutral Barium (number 56) has 56 electrons. Ba^{2+} has 56-2 = 54 electrons, the same as Xe (number 54).

42. **D** The mass of Carbon in a mole of propane equals 3 (the number of atoms of carbon per molecule) times 12 (the atomic weight of carbon) = 36 g. The mass of propane in a mole = 36 + 8 = 44 g (the molecular weight of propane = 44). The %C = 100(36)/44 = 82%.

43. **A** Elements in Group 6A are two electrons short of a Noble Gas configuration. Therefore they will gain a charge of –2.

44. **D** First you must find the change in oxidation numbers for S and N. Sulfur changes from 0 to +4, a change of +4. Nitrogen changes from +5 to +2, a change of –3. Therefore 4 equivalents of nitric acid must react with three equivalents of sulfur to balance the change in oxidation numbers. The S/HNO_3 ratio = 3/4.

45. **B** There are few reaction variables that do not influence the reaction rate. The amount of a solid reactant present is one of them. A change in the total amount of material present without changing concentrations is another (i.e. the rate of conversion of one liter of NO_2 to N_2O_4 at STP is the same as the rate of conversion of 2 liters at STP.

46. E Gallium, element number 31 in the Periodic Table, has 31 electrons. Answer D) has only 29 electrons. Answer A) has 2*d* electrons. There are no *d* orbitals in the 2 shell. Answer B) has 3*f* electrons. There are no *f* orbitals in the 3 shell. Answer C) has empty 3*d* orbitals. Only E) remains and is correct.

47. E In this reaction the pressure decreases as products on the right are formed since there are 3 molecules of gaseous reactants on the left (O_2 + 2HCl) and only 2 on the right. therefore increasing the pressure should drive the reaction to the right in order to decrease the pressure (Le Chatelier Principle)and will increase the amount of heat evolved.

48. A Since solids do not enter into the equilibrium constant only P(CO_2) is important.

49. C $F = 100(F/A)_{12}^{0.5} = 1234.$

50. D $3P = P(F/P)_n^4.\ (F/P)_n^4 = (1.04)^n = 3.\ \therefore n = 28.$

51. B $10\times 10^6 = X + 0.1\times 10^6 (P/A)_\infty^6.\ \therefore X = 8.33\times 10^6.$

52. C $P = 6000(P/A)_3^{10} - 2000(P/G)_3^{10} = 10,300.$

53. E By inspection.

54. D $400 = 35 + 45(P/A)_{10}^i.\ (P/A)_{10}^i = 8.11.\ i = 4\%.\ i_n = 12i.\ \therefore i_n = 48\%.$

55. E From preceding solution $i = 4\%$.
$i_e = (1+0.04)^{12} - 1 = 0.601.\ \therefore i_e = 60.1\%.$

56. A $0 = X(F/A)_{21}^4 - 10,000(F/A)_4^4.\ \therefore X = 1328.$

57. C $D = (4500 - 1500)/3 = 1000.$

58. E $F = 10,000(F/A)_5^8 + 1000(F/A)_5^8(A/G)_5^8 = 69,500.$

59. B $P = -70,000(P/A)_3^{20} + 80,000(P/A)_{10}^{20}(P/F)_3^{20} = 46,600.$

60. C

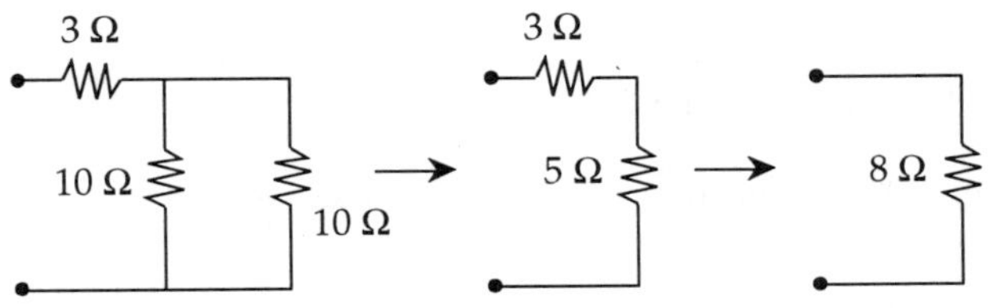

3 Ω 3 Ω
10 Ω 10 Ω → 5 Ω → 8 Ω

61. E With terminal *a–b* open, there is no current in the 5 Ω resistor, therefore no voltage drop in the 5 Ω resistor.

$$V_{ab} = \frac{12\Omega}{(12\Omega + 4\Omega)} \times 12 \text{ volts} = 9 \text{ volts} = V_{\text{Thevinen}}.$$

62. B

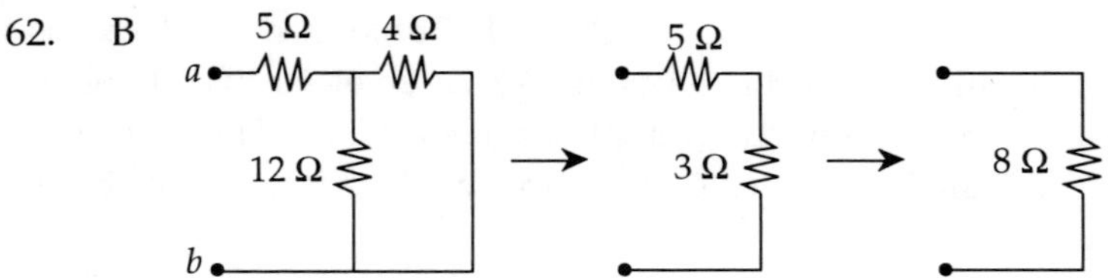

63. **B**

Write KVL $= \sum V_{drop} = 0$

Loop 1 $-12 + 6I_1 + 9 = 0$

Loop 2 $-9 + 9(I_1 + I) = 0$

Solve Loop 1: $I_1 = \frac{12-9}{6} = 0.5$. Solve Loop 2: $I_1 + I = 1,\ I = 1 - 0.5 = 0.5$ A.

64. **B** Power $= VI = 12 \times 0.5 = 6$ W.

65. **D** Combine the parallel capacitors:

$$v(t) = 70.7 \sin 377t$$

$$V_{rms} = \frac{V_{max}}{\sqrt{2}} = \frac{70.7}{1.414} = 50$$

$$X_C = \frac{1}{\omega C} = \frac{1}{377 \times 30 \times 10^{-6}} = 88.42$$

$$I = \frac{50\angle 0}{45 - j88.42} = 0.504\angle 63° \text{ amp}$$

66. **C** $I = \frac{50\angle 0°}{40 + j30} = \frac{50\angle 0°}{50\angle 36.9°} = 1\angle -36.9°,\quad V_{40} = IR = 40$ V.

67. **C** $P = |I|^2 R = (1)^2 40 = 40$ W.

68. **D** $I(t) = I_\infty + (I_0 - I_\infty)e^{-\frac{t}{T}},\ I_0 = 0,\quad I_\infty = \frac{15 \text{ V}}{10\ \Omega} = 1.5$ A,

$T = \frac{L}{R} = \frac{0.025}{10} = 0.0025.$

$I(t) = 1.5 + (0 - 1.5)e^{-\frac{t}{0.0025}},\quad I(0.001) = 1.5e^{-0.4} = 0.495$ A.

69. **D** $V_L = L\frac{dI}{dt} = 0.025\frac{d}{dt}\left[1.5 - 1.5e^{-\frac{t}{0.0025}}\right] = 0.025\left[0 + \frac{1.5}{0.0025}e^{-\frac{t}{0.0025}}\right]$

$= 15e^{-\frac{0.001}{0.0025}} = 15e^{-0.4} = 10.05$ V

70. **A** Reflect the 100 Ω resistance to the primary side:

$$Z_P = \left(\frac{N_1}{N_2}\right)^2 Z_S = \left(\frac{100}{200}\right)^2 100 = 25\ \Omega$$

$$I_P = \frac{120\angle 0°}{40} = 3\angle 0° \text{ A}$$

$$I_S = \left(\frac{N_1}{N_2}\right) I_P = \left(\frac{100}{200}\right) 3 = 1.5 \text{ A}$$

71. A $\overline{\mathbf{F}} = \bar{\mathbf{I}}L \times \overline{\mathbf{B}}$, $B = \frac{\mu I}{2\pi r}$ where r is the distance between the wires.

$B = \frac{4\pi \times 10^{-7} \times 50}{2\pi \times 0.05} = \frac{10^{-5}}{0.05} = 20 \times 10^{-5}$. Therefore,

$F = 50 \times 1 \times 20 \times 10^{-5} = 10^{-2}$ N.

72. A The field between the plates is $E = \frac{V}{d} = \frac{500}{0.02} = 25\,000$ V/m.

The force on the electron: $F = qE = 1.6 \times 10^{-19} \times 25\,000 = 4 \times 10^{-15}$ N.

73. D $R = R_0\left[1 + \alpha(T - T_0)\right]$, $0.036 = 0.024\left[1 + 0.0039(T - 20)\right]$.
Solve for T: $T = 148$ °C.

74. E An isolated system is defined as a system that does not interact with its surroundings, i.e., $Q = W = 0$. Even though $Q = W$, $Q = 0$ is the best response.

75. C Work can occur by one system inducing a force on the other.

76. D Density does not depend on the mass, i.e., when the mass is increased the density does not increase simultaneously.

77. E Distance, area, or volume may be measured in cm, cm^2, or cm^3, respectively, but grams are not acceptable in SI and cm cannot be used in combination with other units.

78. A

79. B (metric) $x = \frac{m_g}{m_g + m_f} = \frac{1.98 \times 0.5}{1.98 \times 0.5 + 0.02 \times 960} = 0.049$ or 4.9%.

(English) $x = \frac{m_g}{m_g + m_f} = \frac{29.7 \times 0.03}{29.7 \times 0.03 + 0.3 \times 62} = 0.046$ or 4.6%.

80. B (metric) $Q = m\Delta h = mc_p\Delta T$. $80 = 2 \times 1.00\Delta T$. $\therefore \Delta T = 40$°C.

(English) $Q = m\Delta h = mc_p\Delta T$. $80 = 4 \times 0.24\Delta T$. $\therefore \Delta T = 83.3$°F.

81. A $F = kx$. $P = \frac{AE}{L}\delta$. $\therefore k = \frac{AE}{L}$.

82. D Ice changes directly to vapor by the process called sublimation.

83. C (metric) $T_2 = T_1\left(\frac{P_2}{P_1}\right)^{\frac{k-1}{k}} = 293\left(\frac{800}{100}\right)^{\frac{0.4}{1.4}} = 531$ K or 258° C.

(English) $T_2 = T_1\left(\frac{P_2}{P_1}\right)^{\frac{k-1}{k}} = 530\left(\frac{120}{15}\right)^{\frac{0.4}{1.4}} = 960$ °R or 500° F.

84. D The refrigerant enters as a vapor and leaves as a liquid: a condenser.

85. **A** The energy equation: $q - w_s = \Delta h$ gives $\Delta h = 0$ with $q = w_s = 0$. Note: w_s is the shaft work. It differs from the total work w by the flow work Pv. For steam $h = h(T, P)$ so $\Delta T \neq 0$, in general.

86. **D** $W = \int P dV = 0$ if $dV = 0$. Note: $W_{2-3} = Q_{2-3} = mRT \ln V_2/V_1$.

87. **D** (metric) $Q = -kA\frac{\Delta T}{L}\Delta t = -\frac{1}{2}\times(3\times 8)\frac{35}{0.025}\times\frac{15}{60} = -4200$ kJ.

Note: Make sure the units all check out: $\frac{\text{kJ}\cdot\text{m}^2\cdot{}^\circ\text{C}\cdot\text{hr}}{\text{hr}\cdot\text{m}\cdot{}^\circ\text{C}\cdot\text{m}} = \text{kJ}$.

(English) $Q = -kA\frac{\Delta T}{L}\Delta t = -\frac{1}{10}\times(10\times 25)\frac{60}{1/12}\times\frac{15}{60} = -4500$ BTU.

88. **D** $\mathbf{F}_1 + \mathbf{F}_2 + \mathbf{F}_3 = 100\mathbf{i} - 200\mathbf{k} + 50\mathbf{j} + 50\mathbf{k} - 200\mathbf{i} + 100\mathbf{j} = -100\mathbf{i} + 150\mathbf{j} - 150\mathbf{k}$.
$F = \sqrt{100^2 + 150^2 + 150^2} = 234$.

89. **A** Sum forces along AB: $90\times 9.8\cos\theta = F_{AB}$.
$\cos\theta = \frac{4}{5}$ so that $F_{AB} = 706$ N. Note: Exaggerate the angle in your sketch so you know that θ is the small angle, i.e., $\cos\theta = 4/5$, and not $3/5$.

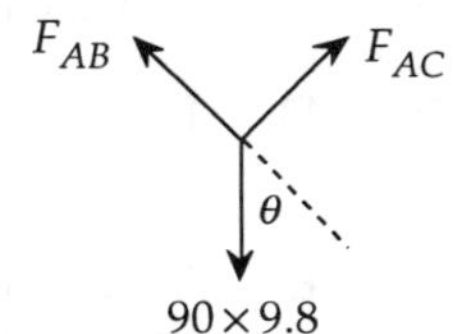

90. **E** $\mathbf{M} = \mathbf{r}\times\mathbf{F} = [(0-1)\mathbf{i} + (2-2)\mathbf{j} + (0-3)\mathbf{k}]\times(20\mathbf{i} - 30\mathbf{j} + 40\mathbf{k})$
$= (-\mathbf{i} - 3\mathbf{k})\times(20\mathbf{i} - 30\mathbf{j} + 40\mathbf{k})$.
The z-component is $-\mathbf{i}\times(-30\mathbf{j}) = 30\mathbf{k}$.

91. **C**

92. **B** $M_{\text{wall}} = 600\times 0.4 - 400\times 0.5 = 240 - 200 = 40$ N·m.

93. **E** $\sum M_A$: $4\times C_y = 2\times 800 + 2\times 400$. $\therefore C_y = 600$ N. $F_{CD} = 0$ from point D.
$\therefore (F_{EC})_y = 600$ N. $(F_{EC})_x = 600\text{ N} = F_{BC} = F_{AB}$.

94. **A** Analyze joint F. $F_{AF} = 0$ and $F_{EF} = 0$.

95. **C** <u>If the block slides</u>:
Friction force $= \mu N = \mu W\cos 30° = 0.3\times(20\times 9.8)\times 0.866 = 50.9$ N.
$\therefore F = W\sin 30° + 50.9 = 20\times 9.8\times 0.5 + 50.9 = 149$ N.
<u>If the block tips</u>:
$F\times\ell = Wg\cos 30°\times\frac{\ell}{2} + Wg\sin 30°\times\frac{\ell}{2}$. $\therefore F = 20\times 9.8\left(\frac{0.866}{2} + \frac{0.5}{2}\right) = 134$ N.

96. **B** $\sum M_A$: $900\times 8 = C_y\times 24$. $\therefore C_y = 300$ N. Since link BC is a two-force member, $\mathbf{F}_{BC}$ must be directed from B to C.
$\therefore C_x = C_y = 300$ N. $\therefore F_C = \sqrt{300^2 + 300^2} = 424$ N.

97. **D** Length = 4. Length over = ℓ. Friction force = $(4-\ell)mg\mu$.
Weight of part hanging over = $mg\ell$.

$\therefore mg\ell = (4-\ell)mg\mu$, or $\ell = (4-\ell)\times 0.4$. $\therefore \ell = 1.14$ m. Note: We let m be the mass per unit length of chain.

98. **C** $\sum M_A$: $B_x \times 60 = (600\times 0.45)\times\frac{45}{2}$. $\therefore B_x = 101$ N.

$\sum M_C$ on link BC: $101\times 60 = F_{\text{wire}}\times\frac{45}{2}$. $\therefore F_{\text{wire}} = 270$ N.

99. **A** $I_y = (I_y)_C + Ad^2$. $2000 = (I_y)_C + 100\times 2^2$. $\therefore (I_y)_C = 1600\ \text{cm}^4$.

100. **D** $y_c = \frac{y_{c1}A_1 + y_{c2}A_2 + y_{c3}A_3}{A_1+A_2+A_3} = \frac{5\times 20 + 1\times 8 + 5\times 20}{20+8+20} = 4.33$ cm.

101. **E** $I_x = \frac{1}{3}b_1h_1^3 + \frac{1}{3}b_2h_2^3 + \frac{1}{3}b_3h_3^3 = \frac{1}{3}\times 2\times 10^3 + \frac{1}{3}\times 4\times 2^3 + \frac{1}{3}\times 2\times 10^3 = 1344\ \text{cm}^4$.

$(I_x)_C = 1344 - Ad^2 = 1344 - 48\times 4.33^2 = 444\ \text{cm}^4$.

102. **B** $a = \frac{dv}{dt} = 10 - 2t$. $\therefore \int_{10}^{v} dv = \int_0^t (10-2t)dt$. $v - 10 = 10t - t^2$.

$\therefore v = \frac{ds}{dt} = 10t - t^2 + 10$. $\therefore \int_0^s ds = \int_0^6 \left(10t - t^2 + 10\right)dt$.

$s = 5\times 6^2 - \frac{6^3}{3} + 10\times 6 = 168$ m. (Make sure $v \neq 0$ for $0 < t < 6$.)

103. **E** $\alpha = \frac{d\omega}{dt}$. $\omega = \frac{d\theta}{dt}$ or $dt = \frac{d\theta}{\omega}$. $\therefore \alpha = \omega\frac{d\omega}{d\theta}$. $\int_0^\theta \alpha d\theta = \int_{\omega_0}^0 \omega d\omega$.

$\therefore \alpha\theta = -\frac{\omega_0^2}{2}$. $-20\theta = -\frac{1}{2}\left(\frac{4000\times 2\pi}{60}\right)^2$. $\therefore \theta = 4386.5$ rad or 698 rev.

104. **D** $a_t = 5$. $a_n = \frac{v^2}{R} = \frac{30^2}{200} = 4.5$. $a = \sqrt{a_t^2 + a_n^2} = \sqrt{5^2 + 4.5^2} = 6.73\ \text{m/s}^2$.

105. **A** Because of symmetry, the time is twice that to reach the maximum height.

Let $v_y = 0$: $gt = v_0\sin\theta$. $\therefore t = 2\left(\frac{v_0}{g}\sin\theta\right)$.

$\therefore x_{\max} = v_0\left(\frac{2v_0}{g}\sin\theta\right)\cos\theta = \frac{v_0^2}{g}\sin 2\theta$.

106. **D** $\mathbf{a}_T = \mathbf{a}_O + \mathbf{a}_{T/O} = 10\mathbf{i} + 0.4\times 25\mathbf{i} - 0.4\times 10^2\mathbf{j} = 20\mathbf{i} - 40\mathbf{j}$.

$a = \sqrt{20^2 + 40^2} = 44.7\ \text{m/s}^2$.

We used $\alpha = \frac{a_O}{r} = \frac{10}{0.4} = 25\ \text{rad/s}^2$ and $\omega = \frac{v}{r} = \frac{4}{0.4} = 10$ rad/s.

107. **E** The equation of the circle is

$s^2 + t^2 = 4$. $2s\frac{ds}{dt} + 2t = 0$. $\therefore \frac{ds}{dt} = -\frac{2t}{2s} = -\frac{t}{s} = -\frac{1}{\sqrt{4-1}} = -0.577$ m/s.

108. **B** $\mathbf{v}_A = \mathbf{v}_B + \mathbf{v}_{A/B}$. $v_B = r\omega = 0.1 \times 10 = 1$ m/s.

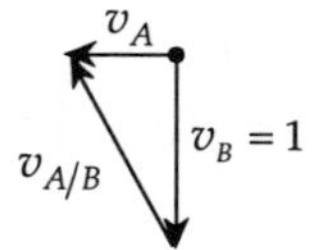

$\therefore v_A = v_B \times \frac{10}{\sqrt{300}} = 0.577$ m/s.

Note: The velocity triangle is similar to the triangle with legs 10 and $\sqrt{300}$.

109. **C** $\mathbf{a}_A = \mathbf{a}_B + \mathbf{a}_{A/B}$. From the velocity diagram of Number 108,

$\mathbf{v}_{A/B} = 0.2\omega_{AB} = 1 \times \frac{20}{\sqrt{300}}$. $\therefore \omega_{AB} = 5.77$ rad/s.

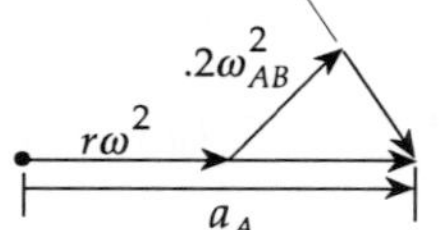

$r\omega^2 = 0.1 \times 10^2 = 10$ rad/s^2 and $.2\omega_{AB}^2 = 6.66$.

Finally, recognizing that a_A is horizontal,

$a_A = 10 + 6.66 / \cos 30° = 17.7$ m/s^2.

110. **E** Assume the 200 kg mass rises: $T - 200 \times 9.8 = 200a$. $\therefore T = 200a + 1960$. The friction force is $0.2(400 \times 9.8 \times 0.707) = 554.3$ N. Then summing forces on the 400-kg mass: $400 \times 9.8 \times 0.707 - 554.3 - T = 400a$. Substituting T from above: $2217 - 200a - 1960 = 400a$. $\therefore a = 0.429$ m/s^2 and $T = 2046$ N.

111. **A** Work-energy: Use friction force = 554.3 N, then
$-200 \times 9.8 \times 2 + 400 \times 9.8 \times 0.707 \times 2$

$$-554.3 \times 2 = \frac{1}{2}200v^2 + \frac{1}{2}400v^2. \quad \therefore v = 1.31 \text{ m/s}.$$

112. **D** Impulse-momentum:

$Fdt = mdv$. $\int_0^{10} 6t\,dt = \int_0^{v_1} 10\,dv$. $300 = 10v_1$. $\therefore v_1 = 30$ m/s.

$\int_{10}^{20} 60\,dt = \int_{30}^{v_2} 10\,dv$. $60(20-10) = 10(v_2 - 30)$. $\therefore v_2 = 90$ m/s.

113. **A** Work-energy:

$\frac{1}{2}kx^2 = \frac{1}{2}mv^2 + \frac{1}{2}I\omega^2$. $\frac{1}{2} \times 400 \times \left(\frac{400}{400}\right)^2 = \frac{1}{2} \times 10v^2 + \frac{1}{2}\left(\frac{1}{2}10 \times 0.4^2\right)\frac{v^2}{0.4^2}$.

$\therefore v = 5.16$ m/s. We used $\omega = v/r$ and $I = \frac{1}{2}mr^2$ for a cylinder.

114. **C** $F_D = 30v$. Sum forces:

$20\,000 - 1500 \times 9.8 - 30v = 1500\frac{dv}{dt}$. $5300 - 30v = 1500\frac{dv}{dt}$, or

$\frac{-30dv}{5300 - 30dv} = \frac{-dt}{50}$. $\therefore \int_0^v \frac{-30dv}{5300 - 30v} = -\int_0^{20} \frac{dt}{50}$. $\ln\frac{5300 - 30v}{5300} = -\frac{20}{50}$.

$\therefore v = 58.2$ m/s.

115. **D** Along the centerline connecting the two balls:
$e = \frac{v_2' - v_1'}{6 - (-6)} = 0.8$. $\therefore v_2' - v_1' = 9.6$. Also, for the normal direction,
$m_1 \times 6 - m_2 \times 6 = m_1v_1' + m_2v_2'$. Since $m_1 = m_2$, $v_1' = -v_2'$ and
$v_2' = 4.8$ m/s.

116. **E** $F = kx$. $x_1 = \frac{F}{k_1}$ and $x_2 = \frac{F}{k_2}$. $x_1 + x_2 = x$.

$\therefore \frac{F}{k_1} + \frac{F}{k_2} = \frac{F}{k}$ or $\frac{1}{k_1} + \frac{1}{k_2} = \frac{1}{k}$. $\therefore k = \frac{k_1 k_2}{k_1 + k_2}$.

117. **B** $\delta_B = \delta_S$ or $\frac{P_B L_B}{A_B E_B} = \frac{P_S L_S}{A_S E_S}$. $\therefore P_S = P_B \frac{A_S}{A_B}\frac{E_S}{E_B} = P_B \frac{\pi r^2}{\frac{9}{4}\pi r^2 - \pi r^2}\frac{2E_B}{E_B} = \frac{8}{5}P_B$.

$P = P_S + P_B = P_S + \frac{5}{8}P_S$. $\therefore P_S = \frac{8}{13}P$ or $P_S = 0.615P$ or 61.5% of P.

118. **D**

119. **C**

120. **E** (metric) The center of Mohr's circle is at 45.

$\therefore \tau_{max} = \text{radius} = \sqrt{5^2 + 10^2} = 11.2$ MPa.

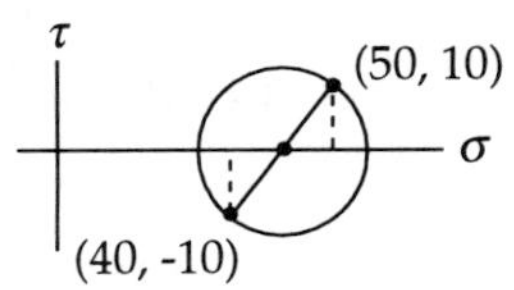

(English) The center of Mohr's circle is at 4500.

$\therefore \tau_{max} = \text{radius} = \sqrt{500^2 + 1000^2} = 1120$ psi.

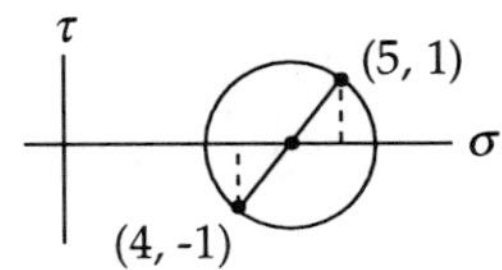

121. **A** $V_{max} = wL/2$. $M_{max} = wL^2/8$. $\sigma_{max} = \frac{My}{I} = \frac{(wL^2/8)(b/2)}{b^4/12} = \frac{3wL^2}{4b^3}$.

$\tau_{max} = \frac{VQ}{Ib} = \frac{(wL/2)(b^3/8)}{(b^4/12)b} = \frac{3wL}{4b^2}$. $\sigma_{max}/\tau_{max} = \frac{3wL^2}{4b^3} \times \frac{4b^2}{3wL} = \frac{L}{b}$.

Note: $Q = b \times \frac{b}{2} \times \frac{b}{4} = b^3/8$.

122. **C** The vertical shear diagram is triangular.

123. **E** (metric) $\tau = \frac{Tr}{J} = \frac{300 \times 0.03}{\pi(.03^4 - .02^4)/2} = 8.81 \times 10^6$ Pa or 8.81 MPa.

(English) $\tau = \frac{Tr}{J} = \frac{(200 \times 12) \times 1.5}{\pi(1.5^4 - 1^4)/2} = 564$ psi.

124. **B** (metric) The force on the left support is found from $800 \times 4 + 500 \times 10 = 8P$. $\therefore P = 1025$ N. The force on the right support is 275 N. The greatest moment is at the left support: $M_{max} = 500 \times 2 = 1000$ N·m.

$\therefore \sigma_{max} = My/I = 1000 \times 0.03/(0.02 \times 0.06^3/12) = 83.3 \times 10^6$ Pa.

(English) The force on the left support is found from $240 \times 12 + 150 \times 30 = 24P$. $\therefore P = 307.5$ lb. The force on the right support is 275 N. The greatest moment is at the left support: $M_{max} = 150 \times 6 = 900$ ft-lb.

$\therefore \sigma_{max} = My/I = (900 \times 12) \times 1.5/(1 \times 3^3/12) = 7200$ psi.

125. **D** (metric) $\tau_{max} = \frac{Tr}{J} = \frac{1200 \times 0.04}{\pi \times 0.04^4/2} = 11.9 \times 10^6$ Pa.

$\therefore \sigma_{max} = 11.9 \times 10^6$ Pa.

τ, τ_{max}, σ, σ_{max}

(English) $\tau_{max} = \frac{Tr}{J} = \frac{(1200 \times 12) \times 2}{\pi \times 2^4/2} = 1146$ psi. $\therefore \sigma_{max} = 1146$ psi.

126. **A** (metric) $\delta = \frac{wL^4}{8EI} + \phi_{max} \times \ell =$

$$\frac{100 \times 4^4}{8 \times 210 \times 10^9 \times \pi(0.02^4)/4} + \frac{100 \times 4^3}{6 \times 210 \times 10^9 \times \pi(0.02^4)/4} \times 2 = 0.202 \text{ m.}$$

(English) $\delta = \frac{wL^4}{8EI} + \phi_{max} \times \ell =$

$$\frac{(8/12) \times (12 \times 12)^4}{8 \times 30 \times 10^6 \times \pi(1^4)/4} + \frac{(8/12) \times (12 \times 12)^3}{6 \times 30 \times 10^6 \times \pi(1^4)/4} \times 6 = 1.61 \text{ in.}$$

127. **C**

128. **D**

129. **E**

130. **C** (metric) $p_w + \gamma H = p_o + 0.8\gamma(2H) + 13.6\gamma H$.

$p_w - p_o = 14.2\gamma H = 14.2 \times 9800 \times 0.25 = 34\ 790$ Pa.

(English) $p_w + \gamma H = p_o + 0.8\gamma(2H) + 13.6\gamma H$.

$p_w - p_o = 14.2\gamma H = 14.2 \times 62.4 \times \frac{10}{12} = 738$ psf or 5.13 psi.

131. **E** $F = \gamma h_c A = \gamma \frac{H}{2}\left(\frac{5}{3} Hw\right) = \frac{5}{6}\gamma H^2 w$. This force acts 1/3 up from hinge.

$\therefore P = F/3$.

132. **B** $\frac{\pi D^2}{4} \times V = 2\pi Rt \times V_2$. $\therefore V_2 = \frac{D^2 V}{8Rt}$.

133. **D**

134. **A** All choices except A involve flows in which the losses are negligible.

135. **C** Laminar flow can occur in a pipe upwards of 40 000 Reynolds number. At low Reynolds number a laminar flow occurs in even a rough pipe. Viscous effects dominate in a laminar flow.

136. **C** (metric) $\frac{e}{D} = \frac{0.15}{20} = 0.0075.\quad \therefore f = 0.034$ (assume fully rough).

$\Delta p = f\frac{L}{D}\frac{V^2}{2g}\gamma.\quad 60\,000 = 0.034\frac{15}{0.02}\frac{V^2}{2\times 9.8}\times 9800.\quad \therefore V = 2.17\ \text{m/s}.$

$Q = AV = \pi\times .01^2 \times 2.17 = 6.82\times 10^{-4}\ \text{m}^3/\text{s}$ or $0.682\ \text{L/s}$.

(English) $\frac{e}{D} = \frac{0.0005}{1/12} = 0.006.\quad \therefore f = 0.032$ (assume fully rough).

$\Delta p = f\frac{L}{D}\frac{V^2}{2g}\gamma.\quad 10\times 144 = 0.032\frac{50}{1/12}\frac{V^2}{2\times 32.2}\times 62.4.\quad \therefore V = 8.80$ fps.

$Q = AV = 0.048$ cfs.

137. **C** (metric) $\frac{\Delta p}{\gamma} = C\frac{V^2}{2g}.\quad V = \frac{Q}{A} = \frac{0.04}{\pi\times .04^2} = 7.96\ \text{m/s}.$

$\therefore C = \frac{100\,000\times 2\times 9.8}{9800\times 7.96^2} = 3.16.$

(English) $\frac{\Delta p}{\gamma} = C\frac{V^2}{2g}.\quad V = \frac{Q}{A} = \frac{1.2}{\pi\times 1.5^2/144} = 24.4$ fps.

$\therefore C = \frac{15\times 144\times 2\times 32.2}{62.4\times 24.4^2} = 3.73.$

138. **B** (metric) $-F_x = \rho Q(V_{2x} - V_{1x}) = 1000\times(2\times 0.05)\times 60[60-(-60)] = 720\,000$ N.

(English) $-F_x = \rho Q(V_{2x} - V_{1x}) = 1.94\times\left(6\times\frac{2}{12}\right)\times 180[180-(-180)] = 125{,}700$ lb.

139. **D** (metric) $\dot{W}_p = Q\gamma h/\eta = 1.2\times 62.4\times\left[\frac{(180-30)144}{62.4}\right]\Big/0.85 = 47.1$ kW.

Note: If pressure is in kPa, $\dot{W}$ will be in kW.

(English) $\dot{W}_p = Q\gamma h/\eta = 1.2\times 62.4\times\left[\frac{(180-30)144}{62.4}\right]\Big/0.85\Big/550 = 55.4$ Hp.

Note: $\dot{W}$ is in ft-lb/sec so the 550 is needed to convert to horsepower.

140. **A** The flow is laminar and steady, and $v = V$ at $y = \delta$ and $v = 0$ at $y = 0$.

Diagnostic Exam Solutions—Afternoon

1. **E** $A = \int_0^8 ydx = \int_0^8 \sqrt{2}x^{1/2}dx = \frac{\sqrt{2}\,8^{3/2}}{3/2} = \frac{2}{3}\sqrt{2}\times 8\times 2\sqrt{2} = 64/3.$

2. **B** $y_c = \frac{\int_0^8 \frac{y}{2}(ydx)}{A} = \frac{3}{64}\frac{1}{2}\int_0^8 2xdx = \frac{3}{64}\times\frac{64}{2} = \frac{3}{2}.$

3. **A** $V = \frac{60}{360}\int_0^8 \pi y^2 dx = \frac{\pi}{6}\int_0^8 2xdx = \frac{\pi}{6}\times 8^2 = 32\pi/3.$
   Note: With a vertical strip, you must use $y/2$, the distance to the centroid of the strip.

4. **D** The slope is $m = \frac{-3}{4}$. $\therefore y = -\frac{3}{4}x + 3$, or $4y + 3x = 12.$

5. **C** The slope of the desired line is $-\frac{1}{m} = \frac{4}{3}$. $\therefore y = \frac{4}{3}x$, or $3y - 4x = 0.$

6. **A** Using a horizontal strip: $I = \int_A y^2 dA = \int_0^3 y^2 x dy.$

7. **B** $\frac{z_2}{z_1} = \frac{5+3i}{1+2i} = \frac{5+3i}{1+2i}\,\frac{1-2i}{1-2i} = \frac{5+6+3i-10i}{1+4} = \frac{11-7i}{5}.$

8. **D** $z_3 = i = e^{\pi i/2}$. $z_3^{1/2} = e^{(\pi i/2)/2} = e^{\pi i/4} = \cos\frac{\pi}{4} + i\sin\frac{\pi}{4} = \frac{1}{\sqrt{2}} + i\frac{1}{\sqrt{2}} = (1+i)/\sqrt{2}.$

9. **E** $z_1 z_2 = (1+2i)(5+3i) = 5 - 6 + 10i + 3i = -1 + 13i = re^{i\theta}$. $r = \sqrt{13^2 + 1^2} = 13.04.$
   $\theta = \tan^{-1}\frac{13}{-1} = -85.6°$. $\therefore \theta = 180 - 85.6 = 94.4° = 1.648$ rad.
   Note: $0 < \theta \le 180°$, so it is in the second quadrant.

10. **C** $|\mathbf{A}| = \begin{vmatrix} 2 & 1 & -2 \\ 0 & 1 & 0 \\ 1 & 2 & 0 \end{vmatrix} = 2.$

$$\begin{array}{ccccc} 2 & 1 & -2 & 2 & 1 \\ 0 & 1 & 0 & 0 & 1 \\ 1 & 2 & 0 & 1 & 2 \end{array} = 0+0+0-(-2)-0-0 = 2\ .$$

− − −
+ + +

11. **A** $[1\ \ -2\ \ 0]\begin{bmatrix} 2 & 1 & -2 \\ 0 & 1 & 0 \\ 1 & 2 & 0 \end{bmatrix} = [2+0+0\quad 1-2+0\quad -2+0+0] = [2\ \ -1\ \ -2].$

12. **D** $\begin{bmatrix}1 & -2 & 0\end{bmatrix}\begin{bmatrix}3\\-1\\0\end{bmatrix}=[3+2+0]=[5].$

13. **C** $\begin{vmatrix}2-\lambda & 1 & -2\\0 & 1-\lambda & 0\\1 & 2 & -\lambda\end{vmatrix}=(1-\lambda)[(2-\lambda)(-\lambda)+2]=(1-\lambda)(\lambda^2-2\lambda+2)=0.$

$\therefore \lambda=1$ is one of the eigenvalues.

14. **B** $\begin{bmatrix}2 & 1 & -2\\0 & 1 & 0\\1 & 2 & 0\end{bmatrix}\begin{bmatrix}x_1\\x_2\\x_3\end{bmatrix}=\begin{bmatrix}3\\-1\\0\end{bmatrix}.$ $\quad x_1=\dfrac{\begin{vmatrix}3 & 1 & -2\\-1 & 1 & 0\\0 & 1 & 0\end{vmatrix}}{\begin{vmatrix}2 & 1 & -2\\0 & 1 & 0\\1 & 2 & 0\end{vmatrix}}=\dfrac{2}{2}=1.$

15. **B** $x''+4x=0$ is the homogeneous equation. $r^2+4=0.$ $r=\pm 2i.$

$\therefore x(t)=C_1e^{2it}+C_2e^{-2it}=A\cos 2t+B\sin 2t.$ Either form is acceptable.

16. **A** Assume $x_p(t)=Ae^{2t}.$ $x_p'=2Ae^{2t}$ and $x_p''=4Ae^{2t}.$

Then $4Ae^{2t}+4Ae^{2t}=8e^{2t}.$ $\therefore 8A=8.$ $\therefore A=1$ and $x_p(t)=e^{2t}.$

17. **D** $x''+5x'+4x=0.$ $r^2+5r+4=0.$ $(r+4)(r+1)=0.$ $r=-4,\ -1.$

$\therefore x_h(t)=C_1e^{-4t}+C_2e^{-t}.$

Assume $x_p(t)=A\sin 2t+B\cos 2t.$ Then

$-4A\sin 2t-4B\cos 2t+10A\cos 2t-10B\sin 2t+4A\sin 2t+4B\cos 2t=20\sin 2t.$

$\therefore A=0$ and $-10B=20$ or $B=-2.$

$\therefore x_p(t)=C_1e^{-4t}+C_2e^{-t}-2\cos 2t.$ $\quad \left.\begin{aligned}x(0)&=0=C_1+C_2-2.\\x'(0)&=10=-4C_1-C_2\end{aligned}\right\}\therefore C_1=-4$ and $C_2=6.$

18. **C** This is resonance: the input frequency ω equals the natural frequency of 2. (See Problem 15 for the natural frequency.)

19. **C** $\lim\limits_{x\to 0}\dfrac{\sin^2 2x}{x^2}=\lim\limits_{x\to 0}\dfrac{2\sin 2x(2\cos 2x)}{2x}=\lim\limits_{x\to 0}\dfrac{2\sin 2x\cos 2x}{x}$

$=\lim\limits_{x\to 0}\dfrac{2\left[2\cos^2 2x-2\sin^2 2x\right]}{1}=4.$

Note: We had to differentiate twice.

20. **E** $\int_0^{\pi}x\sin 2x\,dx\begin{pmatrix}u=x; & dv=\sin 2x\,dx\\du=dx; & v=-\frac{1}{2}\cos 2x\end{pmatrix}=-\frac{x}{2}\cos 2x\Big|_0^{\pi}+\frac{1}{2}\int_0^{\pi}\cos 2x\,dx$

$=-\frac{\pi}{2}(1)+\frac{1}{4}\sin 2x\Big|_0^{\pi}=-\frac{\pi}{2}.$

21. **C** $\sum M_C = 0. \quad \therefore 400 \times 80 \cos 45° = F_{AB} \times 80 \sin 45°. \quad \therefore F_{AB} = 400$ N.

22. **B** Link BC is pinned at both ends.

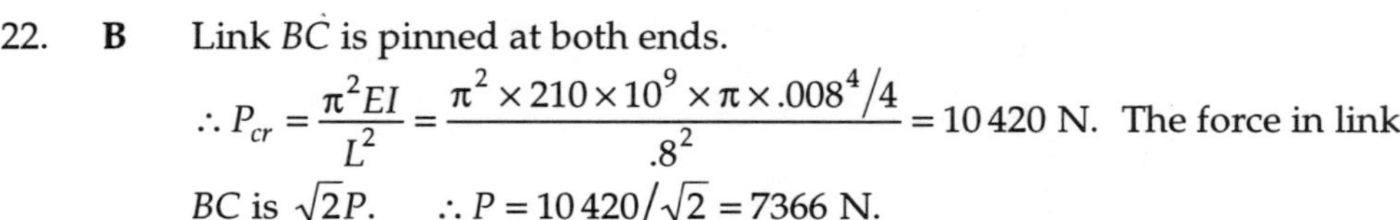

$$\therefore P_{cr} = \frac{\pi^2 EI}{L^2} = \frac{\pi^2 \times 210 \times 10^9 \times \pi \times .008^4/4}{.8^2} = 10\,420 \text{ N.}$$ The force in link BC is $\sqrt{2}P$. $\quad \therefore P = 10\,420/\sqrt{2} = 7366$ N.

23. **C** $\delta_{BC} = \alpha L(T_2 - T_1) = 11.7 \times 10^{-6} \times 800 \times 40 = 0.374$ mm.

$\therefore \overline{BB}' = \sqrt{2} \times .374 = .529$ mm.

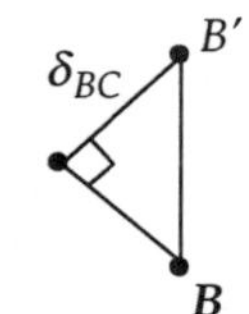

24. **D** $F = 5t$ for $t < 2$ s. $\quad \therefore 5t = 2\frac{dv}{dt}$ or $\int_0^t 5t\,dt = \int_0^v 2\,dv. \quad \therefore v = \frac{5}{4}t^2.$

$v = \frac{dx}{dt} = \frac{5}{4}t^2. \qquad \therefore \int_0^x dx = \frac{5}{4}\int_0^2 t^2 dt$ and $x = \frac{5}{4} \times \frac{2^3}{3} = \frac{10}{3}$ m.

25. **E** $v = \frac{5}{4}t^2 = \frac{5 \times 2^2}{4} = 5$ m/s after 2 seconds. For $t > 2$, $\ 20 - 5t = 2\frac{dv}{dt}.$

$\therefore \int_2^4 (20 - 5t)dt = 2\int_5^v dv. \quad 20 \times (4-2) - \frac{5}{2}(16-4) = 2(v-5). \quad \therefore v = 10$ m/s.

26. **C** From Problem 25,

$\int_2^t (20 - 5t)dt = 2\int_5^v dv$ or $20(t-2) - \frac{5}{2}\left(t^2 - 4\right) = 2(v-5).$

The mass reaches its maximum displacement when $v = 0$. Set $v = 0$ and solve for t:

$20t - 40 - \frac{5}{2}t^2 + 10 = -10. \quad \therefore t^2 - 8t + 8 = 0. \quad \therefore t = \frac{8 \pm \sqrt{64 - 32}}{2} = 6.83$ s.

27. **B** Point C is the instant center for velocity. $\therefore \omega = \frac{v}{r} = \frac{20}{1.00} = 20$ rad/s.

The speed of point B is $v_B = r_{BC}\omega = 0.5\sqrt{2} \times 20 = 14.14$ m/s.

28. **D** There is a friction force acting at C in the x-direction.

$\sum F_x = ma_O \qquad \sum M_O = I\alpha$

$P - F = ma_O \qquad Pr + Fr = \frac{1}{2}mr^2\alpha$

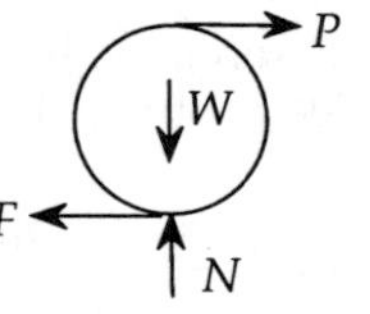

Combine the two equations, letting $r\alpha = a_O$, and

$2P = \frac{3}{2}ma_O. \quad \therefore a_O = \frac{4P}{3m} = \frac{4 \times 300}{3 \times 50} = 8 \text{ m/s}^2.$

29. **A** $F = P - ma_O = 300 - 50 \times 8 = 300 - 400 = -100$ N. $\quad \therefore F$ acts to the right.

30. **C** Impulse-momentum since force and time are specified:

$$\sum F_x \Delta t = m\Delta v.$$

$$(300 + 100)2 = 50(v - 0). \quad \therefore v = 16 \text{ m/s}.$$

Note: The friction force acting to the right must be included.

31. **A** Work-energy since force and distance are specified:

Work $= 300 \times 2 = 600$ N·m.

$$\text{Energy} = \frac{1}{2}mv_0^2 + \frac{1}{2}I\omega^2 = \frac{1}{2}50v_0^2 + \frac{1}{2}\left(\frac{1}{2}50\times.5^2\right)\frac{v_0^2}{.5^2} = \frac{3}{4}\times 50v_0^2.$$

$$\therefore 600 = \frac{3}{4}\times 50v_0^2. \quad \therefore v_0 = 4 \text{ m/s}.$$

Note: The friction force does no work since point C moves vertically.

32. **C** (metric) $M_{max} = P(2\ell) + w\ell(\ell/2) = 400\times 2\times 3 + 200\times 3^2/2 = 3300 \text{ N}\cdot\text{m}.$

(English) $M_{max} = P(2\ell) + w\ell(\ell/2) = 100\times 2\times 9 + 40\times 9^2/2 = 3420 \text{ ft - lb}.$

33. **E** (metric) $\sigma_{max} = \dfrac{My}{I} = \dfrac{3300\times.0275}{1.848\times 10^{-6}} = 49.1\times 10^6 \text{ Pa}.$

(English) $\sigma_{max} = \dfrac{My}{I} = \dfrac{(3420\times 12)\times 1.375}{7.771} = 7262 \text{ psi}.$

34. **A** (metric) $\sigma_{max} = \dfrac{My}{I} = \dfrac{3300\times(.09-.0275)}{1.848\times 10^{-6}} = 111.6\times 10^6 \text{ Pa}.$

(English) $\sigma_{max} = \dfrac{My}{I} = \dfrac{(3420\times 12)\times(4.5-1.375)}{7.771} = 16,500 \text{ psi}.$

35. **B** (metric) $\tau_{max} = \dfrac{VQ}{Ib} = \dfrac{1000\times.01\times(.09-.0275)^2/2}{1.848\times 10^{-6}\times.01} = 1.057\times 10^6 \text{ Pa}.$

(English) $\tau_{max} = \dfrac{VQ}{Ib} = \dfrac{460\times 0.5\times(4.5-1.375)^2/2}{7.771\times.5} = 289 \text{ psi}.$

36. **D** (metric) $\delta = \delta_P + \delta_w + \ell\theta_w = \dfrac{P(2\ell)^3}{3EI} + \dfrac{w_0\ell^4}{8EI} + \ell\dfrac{w_0\ell^3}{6EI}$

$$= \left(\frac{400\times 6^3}{3} + \frac{200\times 3^4}{8} + \frac{200\times 3^4}{6}\right)\Big/ 210\times 10^9 \times 1.848\times 10^{-6} = 0.0864 \text{ m}.$$

(English) $\delta = \delta_P + \delta_w + \ell\theta_w = \dfrac{P(2\ell)^3}{3EI} + \dfrac{w_0\ell^4}{8EI} + \ell\dfrac{w_0\ell^3}{6EI}$

$$= \left(\frac{100\times 18^3}{3} + \frac{40\times 9^4}{8} + \frac{40\times 9^4}{6}\right)\times 12^3 \Big/ 30\times 10^6 \times 7.771 = 2.01 \text{ in}.$$

37. **C**

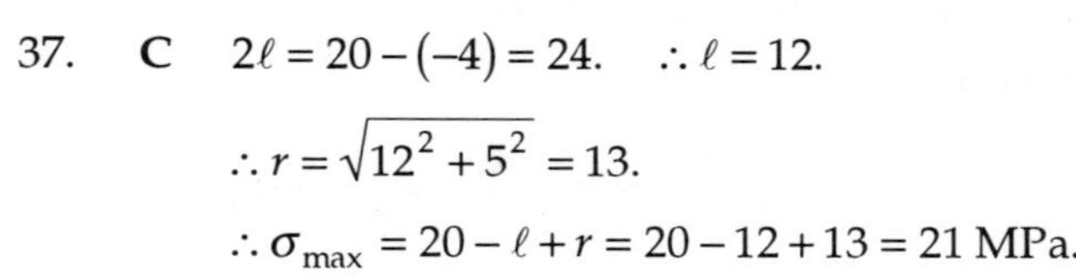

$2\ell = 20-(-4) = 24. \quad \therefore \ell = 12.$

$\therefore r = \sqrt{12^2+5^2} = 13.$

$\therefore \sigma_{max} = 20-\ell+r = 20-12+13 = 21$ MPa.

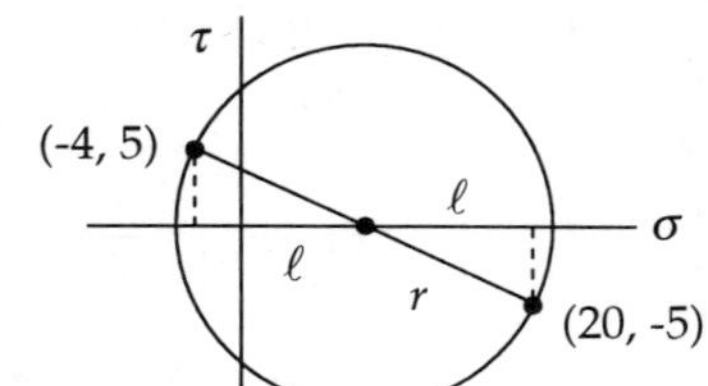

38. **B** $\tau_{max} = r$ in above diagram. $\therefore \tau_{max} = 13$ MPa.

39. **D** $\sin 2\theta = \frac{12}{13}$. $\therefore 2\theta = 67.4°$ and $\theta = 33.7°$.
Note: 2θ on Mohr's circle is θ on the actual element.

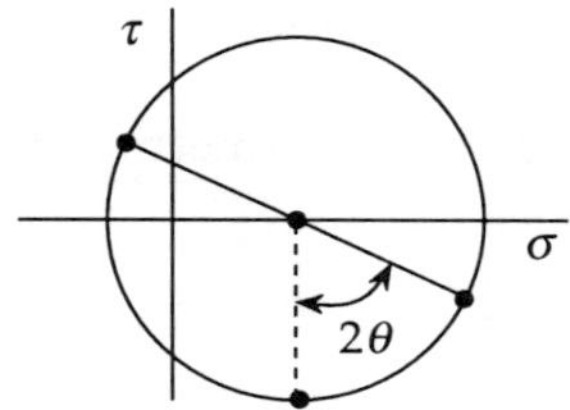

40. **A** $\sin 2\alpha = \frac{4}{13}$. $\therefore 2\alpha = 17.92°$

$\sin(2\alpha+2\phi) = \frac{12-4}{13}$. $\therefore 2\alpha+2\phi = 37.98°$.

$\therefore \phi = 10.03°$.

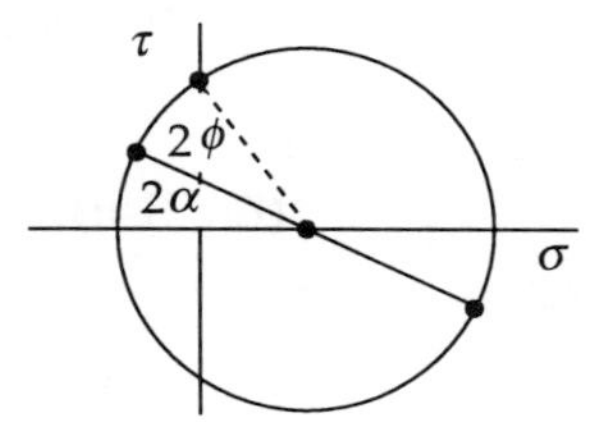

41. **D** If $I_G = 0$ then the current in the $30\,\Omega$ resistor equals the current in the $60\,\Omega$ resistor, likewise in R_X and the $20\,\Omega$ resistor. Therefore

$$\left.\begin{matrix} I_1 30 = I_2 20 \\ I_1 60 = I_2 R_X \end{matrix}\right\} \Rightarrow R_X = \frac{60}{30}\times 20 = 40\ \Omega.$$

42. **B** $V_S - I(10) - I(j20) - I(-j30) = 0.$

$$I = \frac{V_S}{10+j20-j30} = \frac{50\angle 0°}{10-j10} = 2.5+j2.5 = 3.54\angle 45°\ \text{A}.$$

43. **E** $V_C = I(-jX_C) = (3.54\angle 45°)(30\angle -90°) = 106.2\angle -45°.$

44. **C** Recall $I_S = V_S Y_T$ where Y_T is the total parallel admittance of the circuit:

$$Y_T = \frac{1}{Z_T} = \frac{1}{Z_R}+\frac{1}{Z_L}+\frac{1}{Z_C} = \frac{1}{50}+\frac{1}{j\omega L}+\frac{1}{\frac{1}{j\omega C}} = 0.02 - j\frac{1}{\omega L} + j\omega C$$

If Y_T is minimum, the $I_S = V_S Y_T$ is minimum. For Y_T minimum, set

$-j\frac{1}{\omega L}+j\omega C = 0 \Rightarrow \omega^2 = \frac{1}{LC}$. $\therefore$Resonance.

$$\omega = \frac{1}{\sqrt{8\times 10^{-10}\times 2\times 10^{-6}}} = 0.25\times 10^8 = 25\times 10^6\ \text{rad/s}.$$

45. **C** rms $V_S = \frac{V_{max}}{\sqrt{2}} = \frac{141.4}{\sqrt{2}} = 100.$ $\quad I_S \min = 100\times 0.02 = 2$ A.

46. **B** First reflect the 100 Ω resistor to the transformer primary and find I_P.

$$Z_P = \left(\frac{N_1}{N_2}\right)^2 Z_S = \left(\frac{1}{2}\right)^2 (100 - j100) = 25 - j25$$

$$I_P = \frac{50\angle 0°}{40 - j25} = 1.06\angle 32° \text{ A}$$

$$I_S = \left(\frac{N_1}{N_2}\right) I_P = \left(\frac{1}{2}\right) 1.06 = 0.53 \text{ A}$$

$$P = I^2 R = (0.53)^2 100 = 28.1 \text{ W.}$$

47. **C** $I_{\text{rms}} = \sqrt{\frac{1}{T}\int_0^T i^2(t)dt}$, find $\int_0^T i^2(t)dt$ graphically

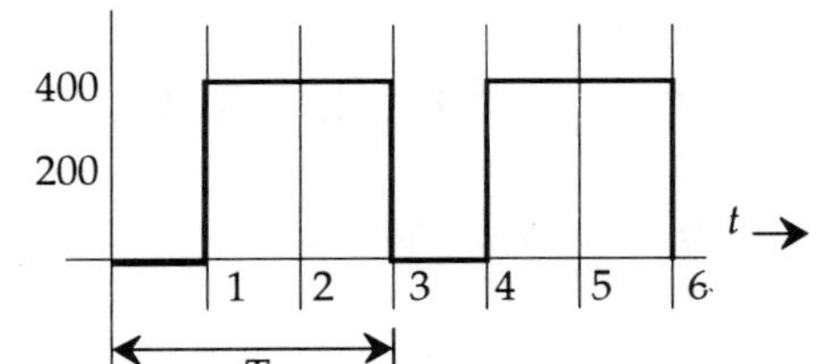

$$\int_0^T i^2 dt = 400 \times 2 = 800 \text{ amp}^2 \cdot \text{sec}$$

$$I_{\text{rms}} = \sqrt{\frac{1}{3} \times 800} = 16.33 \text{ A}$$

48. **E** Find the source current I_S. Reactive power $Q = V_S I_S \sin\theta$. The total impedance seen at the source is:

$$Z_S = +j4 + \frac{(-j10)(25 + j25)}{(-j10) + (25 + j25)} = 8.30\angle -69.25° \text{ ohm}$$

$$I_S = \frac{120\angle 0°}{8.30\angle -69.25°} = 14.46\angle 69.25°$$

$$Q = (120)(14.46)\sin 69.25 = 1623 \text{ VARS.}$$

49. **D** The gain of stage 1: $\frac{v_{01}}{v_{\text{in}}} = -\frac{R_F}{R_i} = -\frac{200}{100} = -2$

The gain of stage 2: $\frac{v_0}{v_{01}} = \left(1 + \frac{R_F}{R_i}\right) = 1 + \frac{300}{100} = 4$

The overall gain: $\frac{v_0}{v_{\text{in}}} = \frac{v_{01}}{v_{\text{in}}} \times \frac{v_0}{v_{01}} = (-2)(4) = -8.$

50. **D** For the integrator $v_0 = \frac{1}{RC}\int v_i dt$. Note from the plot of v_i that $\int v_i dt = 10 \times 1 = 10$.

$$\therefore v_0 = 150 = \frac{1}{RC}(10), \quad R = \frac{10}{150C} = \frac{10}{150 \times 2 \times 10^{-6}} = \frac{10^7}{300} = 33\,333 \ \Omega.$$

51. **A** $P = 400S - (35000 + 0.5S^2)$. $S = 1000$ gives $P = -\$135{,}000$.

52. **B** $P = 0.$ $S = \left(-400 \pm \sqrt{400^2 - 4 \times 0.5 \times 35000}\right) \Big/ (-2 \times 0.5)$.

$\therefore S = 400 \pm 300$ units/mo

53. **E** $dP/dS = 400 - 2\times 0.5S = 0.\ \ S = 400.\ \ P = \$45{,}000$ per month.

54. **C** $U = P/S = 400 - (35000S^{-1} + 0.5S).\ \ dU/dS = 35{,}000S^{-2} - 0.5 = 0.$
$S = 265.\ \ U = \$135/\text{unit}.$

55. **A** $0 = (5-3)\times 10^6 (P/A)^i_\infty - 50\times 10^6.\ \ (P/A)^i_\infty = 1/i = 25.\ \ \therefore i = 4\%.$

56. **C** $B/C = \left[(5-3)\times 10^6\right] / \left[50\times 10^6 (A/P)^6_\infty\right] = 0.67.$

57. **B** $A = 0.05\times 5000 = \$250.\ \ P = 250(P/A)^6_{20} + 5000(P/F)^6_{20} = \$4426.$

58. **C** $\text{ANEV} = \left[-4000(A/P)^{20}_5 - 500\right] - [-1500] = -\$338.$

59. **D** $\text{ANEV} = 0 = \left[-4000(A/P)^{20}_5 - 5X\right] - [-15X].\ \ \therefore X = 134\ \text{kg/yr}.$

60. **B** $D = 4000/3 = \$1333.\ V_2 = 4000 - 2\times 1333 = \$1333.$

61. **B** (metric) $q_{in} = q_{2-3} = u_3 - u_2 + \cancel{w}^{0}_{2-3} = c_v(T_3 - T_2)$
$= 0.718(1473 - 673) = 574\ \text{kJ/kg}.$

We used $T_2 = T_1\left(\dfrac{v_1}{v_2}\right)^{k-1} = 293\times 8^{1.4-1}$
$= 673\ \text{K}.$

(English) $q_{in} = c_v(T_3 - T_2).$
$= 0.171(2660 - 1195) = 251\ \text{Btu/lbm}.$

We used $T_2 = T_1\left(\dfrac{v_1}{v_2}\right)^{k-1} = 520\times 8^{1.4-1}$
$= 1195\ °\text{R}.$

62. **A** (metric) $w_{net} = q_{net} = q_{in} - q_{out}.\qquad q_{out} = -q_{4-1} = c_v(T_4 - T_1).$
$= 574 - 0.718(641 - 293) = 324\ \text{kJ/kg}.$

We used $T_4 = T_3\left(\dfrac{v_4}{v_3}\right)^{k-1} = 1473\left(\dfrac{1}{8}\right)^{0.4}$
$= 641\ \text{K}.$

(English) $w_{net} = q_{net} = q_{in} - q_{out}.\qquad q_{out} = c_v(T_4 - T_1).$
$= 251 - 0.171(1158 - 520) = 142\ \text{Btu/lbm}.$

We used $T_4 = T_3\left(\dfrac{v_4}{v_3}\right)^{k-1} = 2660\left(\dfrac{1}{8}\right)^{0.4}$
$= 1158\ °\text{R}.$

Note: q_{in} was taken from Problem 61. Also, $w_{net} = q_{net}$ is the 1st law for a cycle.

63. **C** (metric) $\eta = \dfrac{w_{\text{net}}}{q_{\text{add}}} = \dfrac{324}{574} = 0.564.$

(English) $\eta = \dfrac{w_{\text{net}}}{q_{\text{add}}} = \dfrac{142}{251} = 0.566.$

64. **E** (metric) $\eta_{\text{Carnot}} = 1 - \dfrac{T_L}{T_H} = 1 - \dfrac{293}{1473} = 0.801.$

(English) $\eta_{\text{Carnot}} = 1 - \dfrac{T_L}{T_H} = 1 - \dfrac{520}{2660} = 0.805.$

65. **D** $-w_{1-2} = (u_2 - u_1) = c_v(T_2 - T_1)$. Note: q_{1-2} is zero since it's an isentropic process. We use Δu since it's a piston-cylinder, i.e., a system (not a control volume).

66. **B** (metric) $p_A = \gamma h = 9800 \times 10 = 98\,000$ Pa.

h

(English) $p_A = \gamma h = 62.4 \times 30 = 1872$ psf or 13 psi.

Note: The pressure inside the exiting stream is the same as that outside the stream. So the pressure is simply due to the static head.

67. **D** (metric) $V = \dfrac{Q}{A} = \dfrac{0.8}{\pi \times 0.6^2} = 0.707 \text{ m/s}. \quad \dfrac{e}{D} = \dfrac{0.25}{1200} = 0.00021.$

$\text{Re} = \dfrac{0.707 \times 1.2}{10^{-6}} = 8.5 \times 10^5. \quad \therefore f = 0.015.$

Energy:

$$40 = \frac{p_i}{9800} + \frac{0.707^2}{2 \times 9.8} + \left(0.5 + 0.015 \times \frac{800}{1.2}\right)\frac{0.707^2}{2 \times 9.8}. \qquad \therefore p_i = 389\,000 \text{ Pa.}$$

(English) $V = \dfrac{Q}{A} = \dfrac{25}{\pi \times 2^2} = 1.99 \text{ fps}. \quad \dfrac{e}{D} = \dfrac{0.00085}{4} = 0.00021.$

$\text{Re} = \dfrac{1.99 \times 4}{10^{-5}} = 8 \times 10^5. \quad \therefore f = 0.015.$

Energy:

$$120 = \frac{p_i}{62.4} + \frac{1.99^2}{2 \times 32.2} + \left(0.5 + 0.015 \times \frac{2500}{4}\right)\frac{1.99^2}{2 \times 32.2}. \qquad \therefore p_i = 7446 \text{ psf (51.7 psi).}$$

68. **C** (metric) $h_T = \dfrac{\Delta p}{\gamma} = \dfrac{389\,000 - 98\,000}{9800} = 29.7 \text{ m.}$

$\therefore \dot{W}_T = Q\gamma h_T \eta = 0.8 \times 9800 \times 29.7 \times 0.89 = 2.07 \times 10^5 \text{ W}$

(English) $h_T = \dfrac{\Delta p}{\gamma} = \dfrac{7446 - 1872}{62.4} = 89.3 \text{ ft.}$

$\therefore \dot{W}_T = Q\gamma h_T \eta = \dfrac{25 \times 62.4 \times 89.3 \times 0.89}{550} = 225 \text{ Hp.}$

69. **A** (metric) $p_{min} = -100$ kPa. $\frac{0.707^2}{2\times 9.8}+\frac{389\,000}{9800}=\frac{V_{max}^2}{2\times 9.8}+\frac{-100\,000}{9800}$.

$\therefore V_{max} = 31.3$ m/s.

(English) $p_{min} = -14.7$ psi. $\frac{1.99^2}{2\times 32.2}+\frac{7446}{62.4}=\frac{V_{max}^2}{2\times 32.2}+\frac{-14.7\times 144}{62.4}$.

$\therefore V_{max} = 99.3$ fps.

Note: The lowest possible pressure at the nozzle exit is absolute zero (vapor pressure is quite close to absolute zero).

70. **A** Flow around blades involves viscous and inertia forces: Reynolds number.

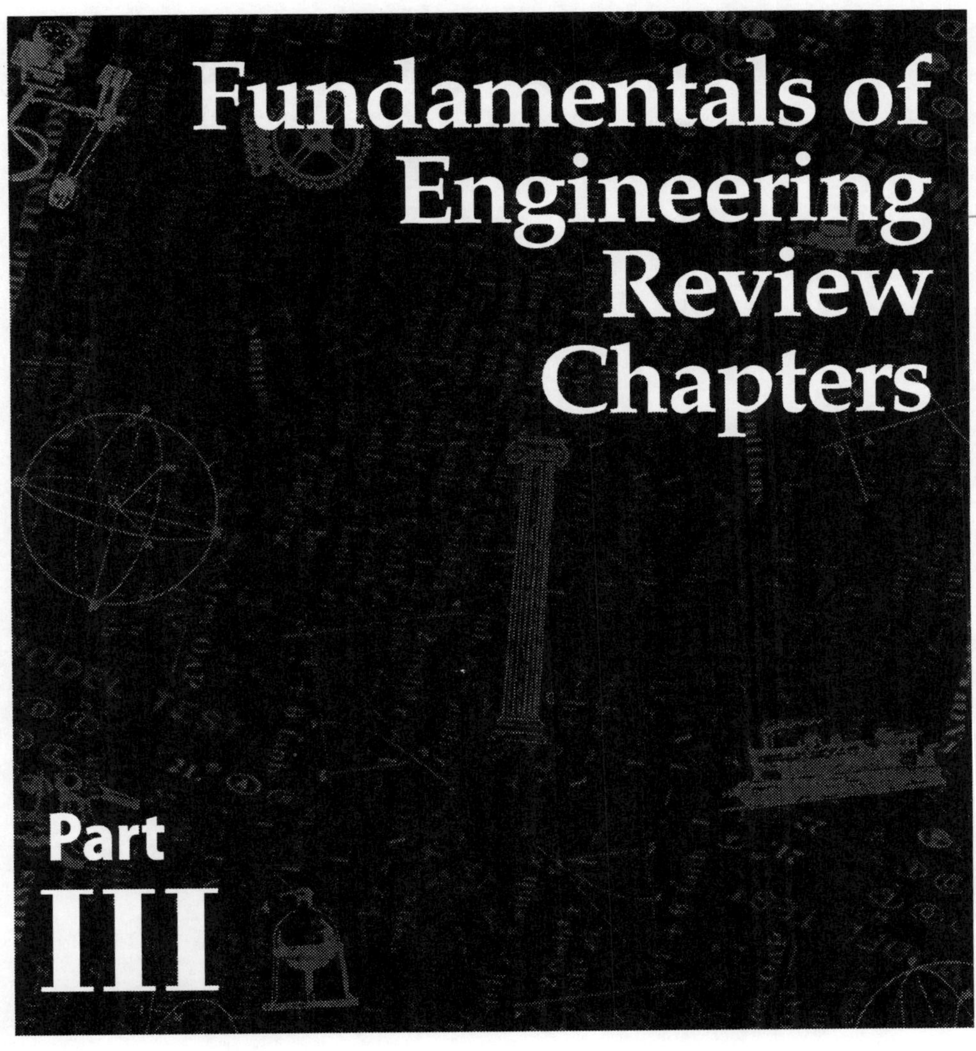

If you seek to reduce the amount of time required for your review, ideally you will have already completed the Diagnostic Exam and identified your weakest subjects. You may then concentrate your review on your weaker areas.

However, if you are reviewing all the material, the chapters are organized so that you may begin with the math chapter and continue in sequence through the book. Note that we present Statics, then Dynamics, then Solid Mechanics, then Fluid Mechanics—the same order in which the courses are offered in most engineering programs. Chemistry, Materials, Economics, Thermodynamics, and Electrical Theory are independent subjects, and can be reviewed in any order.

You'll find that the chapters with numerous Practice Problems will have some of the problems starred. If you are trying to save time in your review, then work only the starred problems. These will cover the most essential information contained in those chapters.

The problem solutions following each chapter outline only the most crucial steps. The Examples in each chapter provide more detailed analyses of the most essential problem-solving methods. By both studying the examples and working the Practice Problems, you should be well prepared for the FE/EIT examination.

Mathematics

by Merle C. Potter

The engineer uses mathematics as a tool to help solve the problems encountered in the analysis and design of physical systems. We will review those parts of mathematics that are used fairly often by the engineer and which may appear on the exam. The topics include: algebra, trigonometry, analytic geometry, linear algebra (matrices), calculus, differential equations, and probability and statistics. The review here is intended to be brief and not exhaustive. The majority of the questions on the exam will be based on the material included in this chapter. There may be a few questions, however, that will require information not covered here; to cover all possible points would not be in the spirit of an efficient review.

1.1 Algebra

It is assumed that the reader is familiar with most of the rules and laws of algebra as applied to both real and complex numbers. We will review some of the more important of these and illustrate several with examples. The three basic rules are:

commutative law:	a + b = b + a	ab = ba	**(1.1.1)**
distributive law:	a(b + c) = ab + ac		**(1.1.2)**
associative law:	a + (b + c) = (a + b) + c	a(bc) = (ab)c	**(1.1.3)**

1.1.1 Exponents

Laws of exponents are used in many manipulations. For positive x and y we use

$$x^{-a} = \frac{1}{x^a}$$

$$x^a x^b = x^{a+b}$$

$$(xy)^a = x^a y^a \qquad (1.1.4)$$
$$x^{ab} = (x^a)^b$$

1.1.2 Logarithms

Logarithms are actually exponents. For example if $b^x = y$ then $x = \log_b y$; that is, the exponent x is equal to the logarithm of y to the base b. Most engineering applications involve common logs which have a base of 10, written as log y, or natural logs which have a base of e (e = 2.7183 ...), written as ln y. If any other base is used it will be so designated, such as $\log_5 y$.

Remember, logarithms of numbers less than one are negative, the logarithm of one is zero, and logarithms of numbers greater that one are positive. The following identities are often useful when manipulating logarithms:

$$\ln x^a = a \ln x$$
$$\ln(xy) = \ln x + \ln y$$
$$\ln(x/y) = \ln x - \ln y$$
$$\ln x = 2.303 \log x$$
$$\log_b b = 1 \qquad (1.1.5)$$
$$\ln 1 = 0$$
$$\ln e^a = a$$

1.1.3 The Quadratic Formula and the Binomial Theorem

We often encounter the quadratic equation $ax^2 + bx + c = 0$ when solving engineering problems. The *quadratic formula* provides its solution; it is

$$x = \frac{-b \pm \sqrt{b^2 - 4ac}}{2a}. \qquad (1.1.6)$$

If $b^2 < 4ac$, the two roots are complex numbers. Cubic and higher order equations are most often solved by trial and error.

The *binomial theorem* is used to expand an algebraic expression of the form $(a + x)^n$. It is

$$(a + x)^n = a^n + na^{n-1}x + \frac{n(n-1)}{2!}a^{n-2}x^2 + \cdots. \qquad (1.1.7)$$

If n is a positive integer, the expansion contains (n + 1) terms. If it is a negative integer or a fraction, an infinite series expansion results.

1.1.4 Partial Fractions

A rational fraction $P(x) / Q(x)$, where $P(x)$ and $Q(x)$ are polynomials, can be resolved into partial fractions for the following cases.

Case 1: $Q(x)$ factors into n different linear terms,

$$Q(x) = (x - a_1)(x - a_2)\ldots(x - a_n).$$

Then

$$\frac{P(x)}{Q(x)} = \sum_{i=1}^{n} \frac{A_i}{x - a_i}. \qquad (1.1.8)$$

Case 2: $Q(x)$ factors into n identical terms,

$$Q(x) = (x - a)^n.$$

Then

$$\frac{P(x)}{Q(x)} = \sum_{i=1}^{n} \frac{A_i}{(x-a)^i}.(1.1.9)$$

Case 3: $Q(x)$ factors into n different quadratic terms,

$$Q(x) = \left(x^2 + a_1 x + b_1\right)\left(x^2 + a_2 x + b_2\right)\ldots\left(x^2 + a_n x + b_n\right).$$

Then

$$\frac{P(x)}{Q(x)} = \sum_{i=1}^{n} \frac{A_i\, x + B_i}{x^2 + a_i\, x + b_i}. \quad \textbf{(1.1.10)}$$

Case 4: $Q(x)$ factors into n identical quadratic terms,

$$Q(x) = \left(x^2 + ax + b\right)^n.$$

Then

$$\frac{P(x)}{Q(x)} = \sum_{i=1}^{n} \frac{A_i\, x + B_i}{\left(x^2 + a\, x + b\right)^i}. \quad \textbf{(1.1.11)}$$

Case 5: $Q(x)$ factors into a combination of the above. The partial fractions are the obvious ones from the appropriate expansions above.

Example 1.1

The temperature at a point in a body is given by $T(t) = 100e^{-0.02t}$. At what value of t does $T = 20$?

Solution. The equation takes the form

$$20 = 100e^{-0.02t}$$

$$0.2 = e^{-0.02t}.$$

Take the natural logarithm of both sides and obtain

$$\ln 0.2 = \ln e^{-0.02t}.$$

Using a calculator, we find

$$-1.6094 = -0.02t.$$

$$\therefore t = 80.47.$$

Example 1.2

Find an expansion for $(9+x)^{1/2}$.

Solution. Using the binomial theorem, Eq. 1.1.7, we have

$$(9+x)^{1/2} = 3\left(1+\frac{x}{9}\right)^{1/2}$$

$$= 3\left[1 + \frac{1}{2}\left(\frac{x}{9}\right) + \frac{1/2(-1/2)}{2}\left(\frac{x}{9}\right)^2 + \frac{1/2(-1/2)(-3/2)}{3\cdot 2}\left(\frac{x}{9}\right)^3 + \cdots\right]$$

$$= 3 + \frac{x}{6} - \frac{x^2}{216} + \frac{x^3}{3888} + \cdots.$$

Note: We factored out $9^{1/2} = 3$ so that in Eq. 1.1.7 $a = 1$; this simplifies the expansion.

Example 1.3

Resolve $\frac{x^2+2}{x^4+4x^3+x^2}$ into partial fractions.

Solution. The denominator is factored into

$$x^4+4x^3+x^2 = x^2\left(x^2+4x+1\right).$$

Using Cases 2 and 3 there results

$$\frac{x^2+2}{x^4+4x^3+x^2} = \frac{A_1}{x}+\frac{A_2}{x^2}+\frac{A_3x+B_3}{x^2+4x+1}.$$

This can be written as

$$\frac{x^2+2}{x^4+4x^3+x^2} = \frac{A_1x\left(x^2+4x+1\right)+A_2\left(x^2+4x+1\right)+\left(A_3x+B_3\right)x^2}{x^2\left(x^2+4x+1\right)}$$

$$= \frac{\left(A_1+A_3\right)x^3+\left(4A_1+A_2+B_3\right)x^2+\left(A_1+4A_2\right)x+A_2}{x^2\left(x^2+4x+1\right)}.$$

The numerators on both sides must be equal. Equating the coefficients of the various powers of *x* provides us with the four equations:

$$A_1+A_3=0$$

$$4A_1+A_2+B_3=1$$

$$A_1+4A_2=0$$

$$A_2=2.$$

These are solved quite easily to give $A_2=2$, $A_1=-8$, $A_3=8$, $B_3=31$. Finally,

$$\frac{x^2+2}{x^4+4x^3+x^2} = \frac{8}{x}+\frac{2}{x^2}+\frac{8x+31}{x^2+4x+1}.$$

1.2 Trigonometry

The primary functions in trigonometry involve the ratios between the sides of a right triangle. Referring to the right triangle in Fig. 1.1, the functions are defined by

$$\sin\theta = \frac{y}{r}, \qquad \cos\theta = \frac{x}{r}, \qquad \tan\theta = \frac{y}{x}. \tag{1.2.1}$$

In addition, there are three other functions that find occasional use, namely,

$$\cot\theta = \frac{x}{y}, \qquad \sec\theta = \frac{r}{x}, \qquad \csc\theta = \frac{r}{y}. \tag{1.2.2}$$

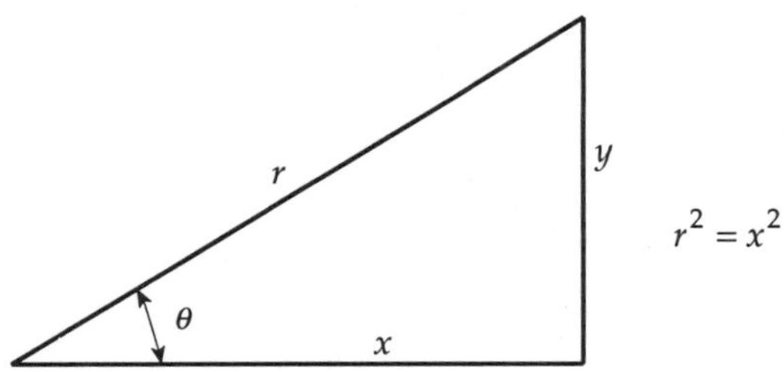

Figure 1.1 A right triangle.

The trig functions $\sin\theta$ and $\cos\theta$ are periodic functions with a period of 2π. Fig. 1.2 shows a plot of the three primary functions.

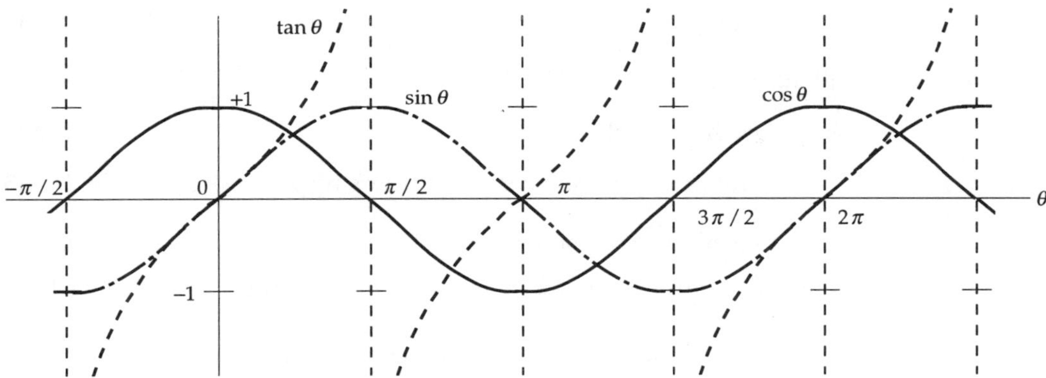

Figure 1.2 The Trig Functions.

In the above relationships, the angle θ is usually given in radians for mathematical equations. It is possible, however, to express the angle in degrees; if that is done it may be necessary to relate degrees to radians. This can be done by remembering that there are 2π radians in 360°. Hence, we multiply radians by $(180/\pi)$ to obtain degrees, or multiply degrees by $(\pi/180)$ to obtain radians. A calculator may use either degrees or radians for an input angle.

Most problems involving trigonometry can be solved using a few fundamental identities. They are

$$\sin^2\theta + \cos^2\theta = 1 \tag{1.2.3}$$

$$\sin 2\theta = 2\sin\theta\cos\theta \tag{1.2.4}$$

$$\cos 2\theta = \cos^2\theta - \sin^2\theta \tag{1.2.5}$$

$$\sin(\alpha \pm \beta) = \sin\alpha\cos\beta \pm \sin\beta\cos\alpha \tag{1.2.6}$$

$$\cos(\alpha \pm \beta) = \cos\alpha\cos\beta \mp \sin\alpha\sin\beta. \tag{1.2.7}$$

A general triangle may be encountered such as that shown in Fig. 1.3. For this triangle we may use the following equations:

$$\text{law of sines:} \quad \frac{\sin\alpha}{a} = \frac{\sin\beta}{b} = \frac{\sin\gamma}{c} \tag{1.2.8}$$

$$\text{law of cosines:} \; a^2 = b^2 + c^2 - 2bc\cos\alpha \tag{1.2.9}$$

Note that if $\gamma = 90°$, the law of cosines becomes the *Pythagorean Theorem*

$$c^2 = a^2 + b^2. \tag{1.2.10}$$

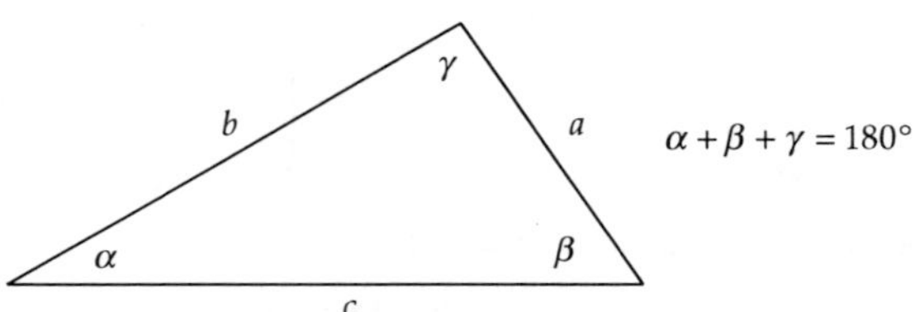

Figure 1.3 A general triangle.

The hyperbolic trig functions also find occasional use. They are defined by

$$\sinh x = \frac{e^x - e^{-x}}{2}, \quad \cosh x = \frac{e^x + e^{-x}}{2}, \quad \tanh x = \frac{\sinh x}{\cosh x}. \tag{1.2.11}$$

Useful identities follow:

$$\cosh^2 x - \sinh^2 x = 1 \tag{1.2.12}$$

$$\sinh(x \pm y) = \sinh x\cosh y \pm \cosh x\sinh y \tag{1.2.13}$$

$$\cosh(x \pm y) = \cosh x\cosh y \pm \sinh x\sinh y \tag{1.2.14}$$

The values of the primary trig functions of certain angles are listed in Table 1.1.

TABLE 1.1 Functions of Certain Angles.

	0	**30°**	**45°**	**60°**	**90°**	**135°**	**180°**	**270°**	**360°**
$\sin\theta$	0	1/2	$\sqrt{2}/2$	$\sqrt{3}/2$	1	$\sqrt{2}/2$	0	–1	0
$\cos\theta$	1	$\sqrt{3}/2$	$\sqrt{2}/2$	1/2	0	$-\sqrt{2}/2$	–1	0	1
$\tan\theta$	0	$1/\sqrt{3}$	1	$\sqrt{3}$	∞	–1	0	$-\infty$	0

Example 1.4

Express $\cos^2\theta$ as a function of $\cos 2\theta$.
Solution. Substitute Eq. 1.2.3 into Eq. 1.2.5 and obtain

$$\begin{aligned}\cos 2\theta &= \cos^2\theta - \left(1-\cos^2\theta\right) \\ &= 2\cos^2\theta - 1.\end{aligned}$$

There results

$$\cos^2\theta = \frac{1}{2}(1+\cos 2\theta).$$

Example 1.5

If $\sin\theta = x$, what is $\tan\theta$?
Solution. Think of $x = x/1$. Thus, the hypotenuse of an imaginary triangle is of length unity and the side opposite θ is of length x. The adjacent side is of length $\sqrt{1-x^2}$. Hence,

$$\tan\theta = \frac{x}{\sqrt{1-x^2}}.$$

Example 1.6
An airplane leaves Lansing flying due southwest at 300 km/hr, and a second leaves Lansing at the same time flying due west at 500 km/hr. How far apart are the airplanes after 2 hours?
Solution. After 2 hours, the respective distances from Lansing are 600 km and 1000 km. A sketch is quite helpful. The distance d that the two airplanes are apart is found using the law of cosines:

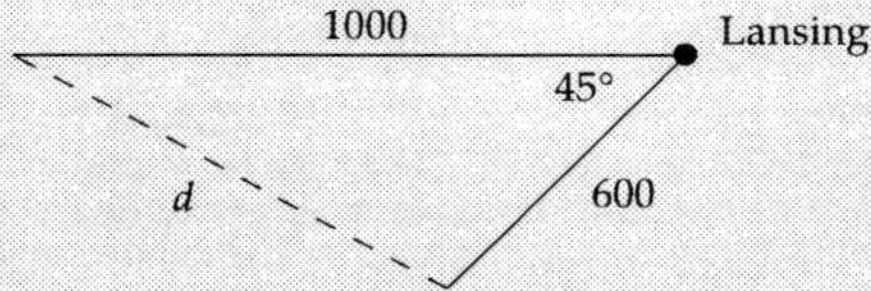

$$\begin{aligned}d^2 &= 1000^2 + 600^2 - 2\times 1000\times 600\cos 45^\circ \\ &= 511470. \\ \therefore d &= 715.2 \text{ km}.\end{aligned}$$

1.3 Geometry

A regular polygon with n sides has a vertex angle (the central angle subtended by one side) of $2\pi/n$. The included angle between two successive sides is given by $\pi(n-2)/n$.

Some common geometric shapes are displayed in Fig. 1.4.

The equation of a straight line can be written in the general form

$$Ax + By + C = 0. \tag{1.3.1}$$

There are three particular forms that this equation can take. They are:

$$\text{Point-slope:} \quad y - y_1 = m(x - x_1) \tag{1.3.2}$$

$$\text{Slope-intercept:} \quad y = mx + b \tag{1.3.3}$$

$$\text{Two-intercept:} \quad \frac{x}{a} + \frac{y}{b} = 1. \tag{1.3.4}$$

In the above equations m is the slope, (x_1, y_1) a point on the line, "a" the x-intercept, and "b" the y-intercept (see Fig. 1.5). The perpendicular distance d from the point (x_3, y_3) to the line $Ax + By + C = 0$ is given by (see Fig. 1.5).

$$d = \frac{|Ax_3 + By_3 + C|}{\sqrt{A^2 + B^2}}. \tag{1.3.5}$$

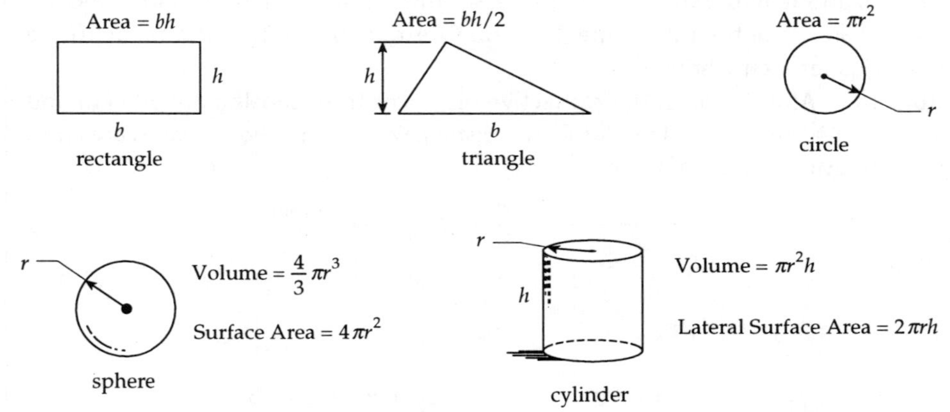

Figure 1.4 Common geometric shapes.

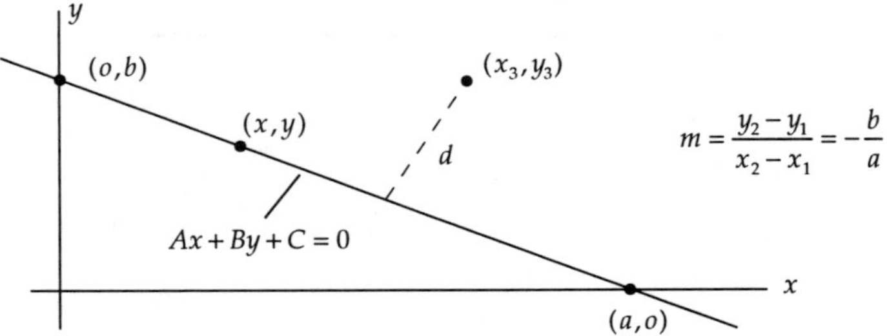

Figure 1.5 A straight line.

The equation of a plane surface is given as

$$Ax + By + Cz + D = 0.(1.3.6)$$

The general equation of second degree

$$Ax^2 + 2Bxy + Cy^2 + 2Dx + 2Ey + F = 0 \qquad \textbf{(1.3.7)}$$

represents a set of geometric shapes called *conic sections.* They are classified as follows:

$$\begin{aligned} &\text{ellipse:} && B^2 - AC < 0 (\text{circle: } B = 0,\ A = C) \\ &\text{parabola:} && B^2 - AC = 0 \\ &\text{hyperbola:} && B^2 - AC > 0 \end{aligned} \qquad \textbf{(1.3.8)}$$

If $A = B = C = 0$, the equation represents a line in the xy-plane, not a parabola. Let's consider each in detail.

Circle: The circle is a special case of an ellipse with $A = C$. Its general form can be expressed as

$$(x-a)^2 + (y-b)^2 = r^2 \qquad \textbf{(1.3.9)}$$

where its center is at (a, b) and r is the radius.

Ellipse: The sum of the distances from the two foci, F, to any point on an ellipse is a constant. For an ellipse centered at the origin

$$\frac{x^2}{a^2} + \frac{y^2}{b^2} = 1 \qquad \textbf{(1.3.10)}$$

where a and b are the semi-major and semi-minor axes. The foci are at $(\pm\, c,0)$ where $c^2 = a^2 - b^2$. See Fig. 1.6a.

Parabola: The locus of points on a parabola are equidistant from the focus and a line (the directrix). If the vertex is at the origin and the parabola opens to the right, it is written as

$$y^2 = 2px \qquad \textbf{(1.3.11)}$$

where the focus is at $(p/2, 0)$ and the directrix is at $x = -p/2$. See Fig. 1.6b. For a parabola opening to the left, simply change the sign of p. For a parabola opening upward or downward, interchange x and y.

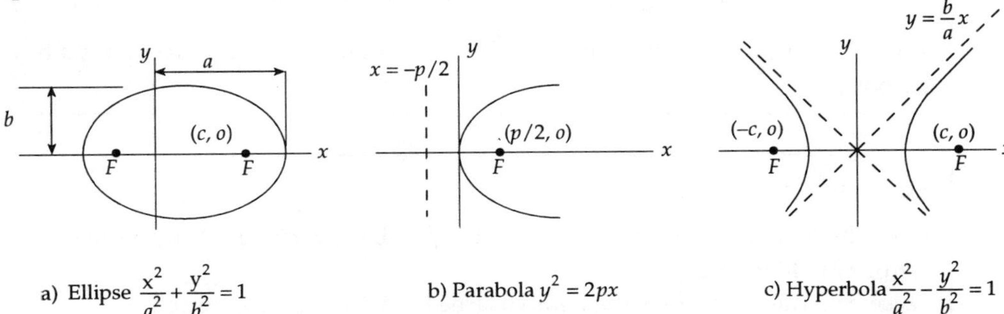

a) Ellipse $\frac{x^2}{a^2} + \frac{y^2}{b^2} = 1$ b) Parabola $y^2 = 2px$ c) Hyperbola $\frac{x^2}{a^2} - \frac{y^2}{b^2} = 1$

Figure 1.6 The three conic sections.

Hyperbola: The difference of the distances from the foci to any point on a hyperbola is a constant. For a hyperbola centered at the origin opening left and right, the equation can be written as

$$\frac{x^2}{a^2} - \frac{y^2}{b^2} = 1. \qquad \textbf{(1.3.12)}$$

The lines to which the hyperbola is asymptotic are asymptotes:

$$y = \pm \frac{b}{a} x . \quad \textbf{(1.3.13)}$$

If the asymptotes are perpendicular, a rectangular hyperbola results. If the asymptotes are the x and y axes, the equation can be written as

$$xy = \pm k^2 . \quad \textbf{(1.3.14)}$$

Finally, in our review of geometry, we will present three other coordinate systems often used in engineering analysis. They are the polar (r, θ) coordinate system, the cylindrical (r, θ, z) coordinate system, and the spherical (r, θ, ϕ) coordinate system. The polar coordinate system is restricted to a plane:

$$x = r\cos\theta, \quad y = r\sin\theta . \quad \textbf{(1.3.15)}$$

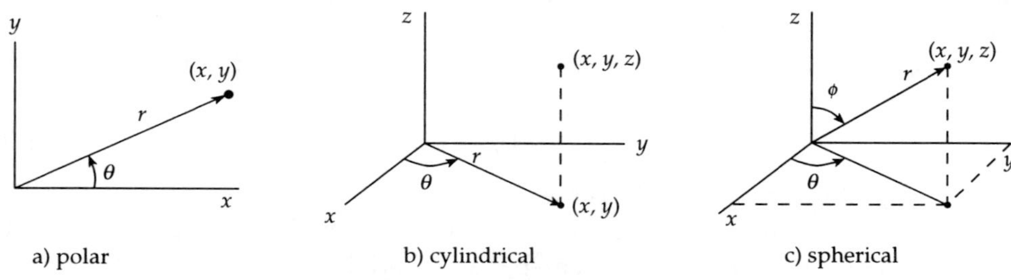

Figure 1.7 The polar, cylindrical and spherical coordinate systems.

For the cylindrical coordinate system

$$x = r\cos\theta, \quad y = r\sin\theta, \quad z = z . \quad \textbf{(1.3.16)}$$

And, for the spherical coordinate system

$$x = r\sin\phi\cos\theta, \quad y = r\sin\phi\sin\theta,$$
$$z = r\cos\phi . \quad \textbf{(1.3.17)}$$

Example 1.7

What conic section is represented by $2x^2 - 4xy + 5x = 10$?

Solution. Comparing this with the general form Eq. 1.3.7, we see that

$$A = 2, \quad B = -2, \quad C = 0.$$

Thus, $B^2 - AC = 4$, which is greater than zero. Hence, the conic section is a hyperbola.

Example 1.8

Write the general form of the equation of a parabola, vertex at (2, 4), opening upward, with directrix at $y = 2$.

Solution. The equation of the parabola (see Eq. 1.3.11) can be written as

$$(x - x_1)^2 = 2p(y - y_1)$$

where we have interchanged x and y so that the parabola opens upward. For this example, $x_1 = 2$, $y_1 = 4$ and $p = 4$ ($p/2$ is the distance from the vertex to the directrix). Hence, the equation is

$$(x-2)^2 = 2(4)(y-4),$$

or, in general form,

$$x^2 - 4x - 8y + 36 = 0.$$

Example 1.9

Express the rectangular coordinates (3, 4, 5) in cylindrical coordinates and spherical coordinates.

Solution. In cylindrical coordinates

$$r = \sqrt{x^2 + y^2} = \sqrt{3^2 + 4^2} = 5,$$

$$\theta = \tan^{-1} y/x = \tan^{-1} 4/3 = 0.927 \text{ rad}.$$

Thus, in cylindrical coordinates, the point is located at (5, 0.927, 5).

In spherical coordinates

$$r = \sqrt{x^2 + y^2 + z^2} = \sqrt{3^2 + 4^2 + 5^2} = 7.071,$$

$$\phi = \cos^{-1} z/r = \cos^{-1} 5/7.071 = 0.785 \text{ rad},$$

$$\theta = \tan^{-1} y/x = \tan^{-1} 4/3 = 0.927 \text{ rad}.$$

Finally, in spherical coordinates, the point is located at (7.071, 0.927, 0.785).

1.4 Complex Numbers

A complex number consists of a real part x and an imaginary part y, written as $x + iy$, where $i = \sqrt{-1}$. (In electrical engineering, however, it is common to let $j = \sqrt{-1}$ since i represents current.) In real number theory, the square root of a negative number does not exist; in complex number theory, we would write $\sqrt{-4} = \sqrt{4(-1)} = 2i$. The complex number may be plotted using the real x-axis and the imaginary y-axis, as shown in Fig. 1.8.

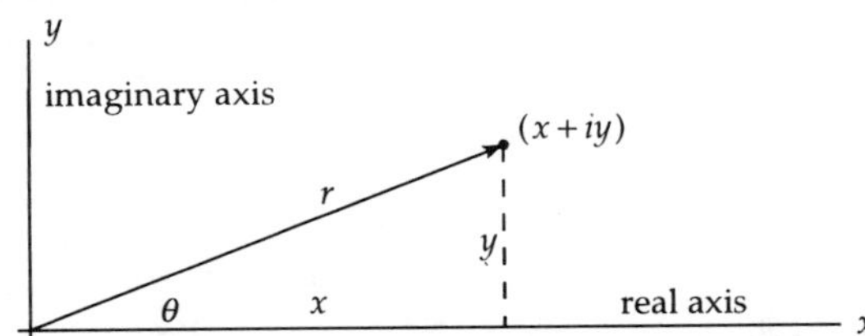

Figure 1.8 The complex number.

It is often useful to express a complex number in polar form as (both $re^{i\theta}$ and $r\angle\theta$ are polar forms)

$$x + iy = re^{i\theta} \quad \text{or} \quad r\angle\theta \tag{1.4.1}$$

where we use *Euler's equation*

$$e^{i\theta} = \cos\theta + i\sin\theta \tag{1.4.2}$$

to verify the relations

$$x = r\cos\theta, \quad y = r\sin\theta. \tag{1.4.3}$$

Note that $e^{i\theta} = e^{i(\theta+2n\pi)}$ where n is an integer. This simply adds 360° (2π radians) to θ and hence in Fig. 1.8 $re^{i\theta}$ and $re^{i(\theta+2n\pi)}$ represent the identical point.

Multiplication and division are accomplished with either form:

$$\begin{aligned}(a+ib)(c+id) &= ac - bd + i(ad+bc) \\ &= r_1 e^{i\theta_1} r_2 e^{i\theta_2} = r_1 r_2 e^{i(\theta_1+\theta_2)} \\ &= r_1 r_2 \angle \theta_1 + \theta_2\end{aligned} \tag{1.4.4}$$

$$\begin{aligned}\frac{a+ib}{c+id} = \frac{a+ib}{c+id}\,\frac{c-id}{c-id} &= \frac{(a+ib)(c-id)}{c^2+d^2} \\ &= \frac{r_1}{r_2} e^{i(\theta_1-\theta_2)} = \frac{r_1}{r_2} \angle \theta_1 - \theta_2\end{aligned} \tag{1.4.5}$$

It is usually easier to find powers and roots of complex numbers using the polar form:

$$(x+iy)^k = r^k e^{ik\theta}, \quad (x+iy)^{1/k} = r^{1/k} e^{i\theta/k} \tag{1.4.6}$$

When finding roots, more than one root results by using $e^{i\theta}$ and $e^{i(\theta+2n\pi)}$. An example illustrates. Remember, in mathematical equations we usually express θ in radians; however, when displaying the angle as $\angle\theta$ it is expressed in degrees.

Using Euler's equation we can show that

$$\sin\theta = \frac{e^{i\theta} - e^{-i\theta}}{2i}, \qquad \cos\theta = \frac{e^{i\theta} + e^{-i\theta}}{2} \tag{1.4.7}$$

Example 1.10

Divide (3 + 4*i*) by (4 + 3*i*).

Solution. We perform the division as follows:

$$\frac{3+4i}{4+3i} = \frac{3+4i}{4+3i}\cdot\frac{4-3i}{4-3i} = \frac{12+16i-9i+12}{16+9} = \frac{24+7i}{25} = 0.96 + 0.28i.$$

Note that we multiplied the numerator and the denominator by the *complex conjugate* of the denominator. A complex conjugate is formed simply by changing the sign of the imaginary part.

Example 1.11

Find $(2+3i)^4$.

Solution. This can be done using the polar form. Hence,

$$r = \sqrt{2^2+3^2} = \sqrt{13}, \quad \theta = \tan^{-1} 3/2 = 0.9828 \text{ rad}.$$

We normally express θ in radians. The complex number, in polar form, is

$$2+3i = \sqrt{13}\, e^{0.9828i}.$$

Thus,

$$(2+3i)^4 = \left(\sqrt{13}\right)^4 e^{4(0.9828i)} = 169\, e^{3.9312i}.$$

Converting back to rectangular form we have

$$169\, e^{3.9312i} = 169(\cos 3.9312 + i \sin 3.9312)$$

$$= 169(-0.7041 - 0.7101i) = -119 - 120i.$$

Alternatively, we recognize that this could be expressed in polar form as 169 $\angle 225.2^\circ$ or $\angle -134.8^\circ$, where a positive angle is measured counterclockwise.

Example 1.12

Find the three roots of $2 + 3i$.

Solution. We express the complex number (see Example 1.11) in polar form as

$$2+3i = \sqrt{13}\, e^{0.9828i}.$$

Since the trig functions are periodic, we know that

$$\sin\theta = \sin(\theta+2\pi) = \sin(\theta+4\pi)$$

$$\cos\theta = \cos(\theta+2\pi) = \cos(\theta+4\pi).$$

Thus, in addition to the first form, we have

$$2+3i = \sqrt{13}\, e^{(0.9828+2\pi)i} = \sqrt{13}\, e^{(0.9828+4\pi)i}.$$

Taking the one-third root of each form, we find the three roots to be

$$(2+3i)^{1/3} = \left(\sqrt{13}\right)^{1/3} e^{0.3276i}$$

$$= 1.533(0.9468 + 0.3218i) = 1.452 + 0.4935i \quad \text{or} \quad 1.533\angle 18.77^\circ.$$

$$(2+3i)^{1/3} = \left(\sqrt{13}\right)^{1/3} e^{2.422i}$$

$$= 1.533(-0.7521 + 0.6591i) = -1.153 + 0.010i \quad \text{or} \quad 1.533\angle 138.77^\circ.$$

$$(2+3i)^{1/3} = \left(\sqrt{13}\right)^{1/3} e^{4.516i}$$

$$= 1.533(-0.1951 - 0.9808i) = -0.2991 - 1.504i \quad \text{or} \quad 1.533\angle 258.77^\circ.$$

If we added 6π to the angle we would be repeating the first root, so obviously this is not done.

1.5 Linear Algebra

The primary objective in linear algebra is to find the solution to a set of n linear algebraic equations for n unknowns. To do this we must learn how to manipulate a matrix, a rectangular array of quantities arranged into rows and columns.

An $m \times n$ matrix has m rows (the horizontal lines) and n columns (the vertical lines). An $m \times n$ matrix multiplied by an $n \times s$ matrix produces an $m \times s$ matrix. When multiplying two matrices the columns of the first matrix must equal the rows of the second. Their product is a third matrix:

$$\left[c_{ij}\right] = \sum_{k=1}^{n}\left[a_{ik}\right]\left[b_{kj}\right] \tag{1.5.1}$$

We are primarily interested in square matrices since we usually have the same number of equations as unknowns, such as

$$\begin{aligned} a_{11}x_1 + a_{12}x_2 + a_{13}x_3 + a_{14}x_4 &= r_1 \\ a_{21}x_1 + a_{22}x_2 + a_{23}x_3 + a_{24}x_4 &= r_2 \\ a_{31}x_1 + a_{32}x_2 + a_{33}x_3 + a_{34}x_4 &= r_3 \\ a_{41}x_1 + a_{42}x_2 + a_{43}x_3 + a_{44}x_4 &= r_4. \end{aligned} \tag{1.5.2}$$

In matrix form this can be written as

$$\left[a_{ij}\right]\left[x_j\right] = \left[r_i\right] \quad \text{or} \quad \mathbf{Ax} = \mathbf{r} \tag{1.5.3}$$

where $\left[x_j\right]$ and $\left[r_i\right]$ are column matrices. (A column matrix is often referred to as a *vector*.) The coefficient matrix $\left[a_{ij}\right]$ and the column matrix $\left[r_i\right]$ are assumed to be known quantities. The solution $\left[x_j\right]$ is expressed as

$$\left[x_j\right] = \left[a_{ij}\right]^{-1}\left[r_i\right] \quad \text{or} \quad \mathbf{x} = \mathbf{A}^{-1}\mathbf{r} \tag{1.5.4}$$

where $[a_{ij}]^{-1}$ is the *inverse* matrix of $\left[a_{ij}\right]$. It is defined as

$$\left[a_{ij}\right]^{-1} = \frac{\left[a_{ij}\right]^{+}}{\left|a_{ij}\right|} \quad \text{or} \quad \mathbf{A}^{-1} = \frac{\mathbf{A}^{+}}{|\mathbf{A}|} \tag{1.5.5}$$

where $[a_{ij}]^{+}$ is the *adjoint* matrix and $\left|a_{ij}\right|$ is the *determinant* of $\left[a_{ij}\right]$. Let us review how the determinant and the adjoint are evaluated.

In general, the determinant may be found using the *cofactor* A_{ij} of the element a_{ij}. The cofactor is defined to be $(-1)^{i+j}$ times the *minor*, the determinant obtained by deleting the ith row and the jth column. The determinant is then

$$\left|a_{ij}\right| = \sum_{j=1}^{n} a_{ij}\,A_{ij}, \tag{1.5.6}$$

where i is any value from 1 to n. Recall that the third-order determinant can be evaluated by writing the first two columns after the determinant and then summing the products of the elements of the diagonals, using negative signs with the diagonals sloping upward.

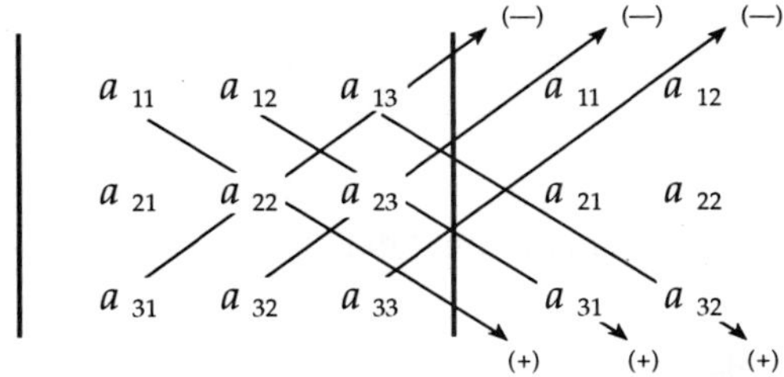

The elements of the adjoint $[a_{ij}]^+$ are the cofactors A_{ij} of the elements a_{ij}; for the matrix $[a_{ij}]$ of Eq. 1.5.2 we have

$$\left[a_{ij}\right]^+ = \begin{bmatrix} A_{11} & A_{21} & A_{31} & A_{41} \\ A_{12} & A_{22} & A_{32} & A_{42} \\ A_{13} & A_{23} & A_{33} & A_{43} \\ A_{14} & A_{24} & A_{34} & A_{44} \end{bmatrix} \qquad \textbf{(1.5.7)}$$

Note that A_{ij} takes the position of a_{ji}.

Finally, the solution $[x_j]$ results if we multiply a square matrix by a column matrix. In general, we multiply the elements in each left-hand matrix row by the elements in each right-hand matrix column, add the products, and place the sum at the location where the row and column intersect. The following examples will illustrate.

Before we work some examples though, we should point out that the above matrix presentation can also be presented as *Cramer's rule,* which states that the solution element x_n can be expressed as

$$x_n = \frac{\left|b_{ij}\right|}{\left|a_{ij}\right|} \qquad \textbf{(1.5.8)}$$

where $\left|b_{ij}\right|$ is formed by replacing the nth column of $\left|a_{ij}\right|$ with the elements of the column matrix $\left[r_i\right]$.

Notes: If the system of equations is homogeneous, i.e., $r_i = 0$, a solution may exist if $\left|a_{ij}\right|$ = 0. If the determinant of a matrix is zero, that matrix is *singular.*

Example 1.13

Calculate the determinants of $\begin{bmatrix} 2 & -3 \\ 1 & 4 \end{bmatrix}$ and $\begin{bmatrix} 2 & 3 & 0 \\ 1 & 4 & -2 \\ 0 & 3 & 5 \end{bmatrix}$.

Solution. For the first matrix we have

$$\begin{vmatrix} 2 & -3 \\ 1 & 4 \end{vmatrix} = 2 \times 4 - 1(-3) = 11.$$

The second matrix is set up as follows:

$$\begin{vmatrix} 2 & 3 & 0 \\ 1 & 4 & -2 \\ 0 & 3 & 5 \end{vmatrix} \begin{matrix} 2 & 3 \\ 1 & 4 \\ 0 & 3 \end{matrix} = 40 + 0 + 0 - 0 - (-12) - 15 = 37.$$

Example 1.14

Expanding with cofactors, evaluate the determinant $D = \begin{vmatrix} 1 & 0 & -2 \\ -1 & 2 & 0 \\ 1 & 2 & 1 \end{vmatrix}$.

Solution. Choosing the first row ($i = 1$ in Eq. 1.5.6) we have

$$D = 1\begin{vmatrix} 2 & 0 \\ 2 & 1 \end{vmatrix}(-1)^2 - 0\begin{vmatrix} -1 & 0 \\ 1 & 1 \end{vmatrix}(-1)^3 + (-2)\begin{vmatrix} -1 & 2 \\ 1 & 2 \end{vmatrix}(-1)^4$$

$$= 2 + 0 - 2(-4) = 10.$$

Note: If any two columns (or rows) are a multiple of each other, a determinant is zero.

Example 1.15

Find the adjoint of the matrix if

$$[a_{ij}] = [b_{in}][c_{nj}] \quad \text{where } [b_{in}] = \begin{bmatrix} 0 & 0 & 1 \\ 4 & 1 & -2 \\ -1 & 1 & 0 \end{bmatrix} \text{ and } [c_{nj}] = \begin{bmatrix} 0 & 0 & -1 \\ 1 & 2 & 0 \\ 1 & 0 & -2 \end{bmatrix}.$$

Solution. First, multiply the two matrices using Eq. 5.1.1. If we desire a_{23} we use the second row of $[b_{in}]$ and the third column of $[c_{nj}]$ so that $a_{23} = b_{21}c_{13} + b_{22}c_{23} + b_{23}c_{33} = 4(-1) + 1(0) + (-2)(-2) = 0$. Doing this for all elements of $[a_{ij}]$ we find

$$[a_{ij}] = \begin{bmatrix} 1 & 0 & -2 \\ -1 & 2 & 0 \\ 1 & 2 & 1 \end{bmatrix}.$$

The cofactor of each element of $[a_{ij}]$ must be determined. The cofactor is found by multiplying $(-1)^{i+j}$ times the determinant formed by deleting the ith row and the jth column. They are found to be

$$A_{11} = 2, \quad A_{12} = 1, \quad A_{13} = -4$$

$$A_{21} = -4, \quad A_{22} = 3, \quad A_{23} = -2$$

$$A_{31} = 4, \quad A_{32} = 2, \quad A_{33} = 2.$$

The adjoint is then

$$[a_{ij}]^+ = [A_{ji}] = \begin{bmatrix} 2 & -4 & 4 \\ 1 & 3 & 2 \\ -4 & -2 & 2 \end{bmatrix}.$$

Note: The matrix $[A_{ji}]$ is called the *transpose* of $[A_{ij}]$, i.e., $[A_{ji}] = [A_{ij}]^T$.

Example 1.16

Find the inverse of the matrix $[a_{ij}] = \begin{bmatrix} 1 & 0 & -2 \\ -1 & 2 & 0 \\ 1 & 2 & 1 \end{bmatrix}$.

Solution. The inverse is defined to be the adjoint matrix divided by the determinant $|a_{ij}|$. Hence, the inverse is (see Examples 1.14 and 1.15)

$$[a_{ij}]^{-1} = \frac{1}{10}\begin{bmatrix} 2 & -4 & 4 \\ 1 & 3 & 2 \\ -4 & -2 & 2 \end{bmatrix} = \begin{bmatrix} 0.2 & -0.4 & 0.4 \\ 0.1 & 0.3 & 0.2 \\ -0.4 & -0.2 & 0.2 \end{bmatrix}.$$

Example 1.17

Find the solution to

$$\begin{aligned} x_1 \quad\quad - 2x_3 &= 2 \\ -x_1 + 2x_2 \quad\quad &= 0 \\ x_1 + 2x_2 + x_3 &= -4 \end{aligned}.$$

Solution. The solution matrix is (see Example 1.14 and 1.15 or use $[a]^{-1}$ fromExample 1.16)

$$[x_j] = [a_{ij}]^{-1}[r_i] = \frac{[a_{ij}]^{+}}{|a_{ij}|}[r_i]$$

$$= \frac{1}{10}\begin{bmatrix} 2 & -4 & 4 \\ 1 & 3 & 2 \\ -4 & -2 & 2 \end{bmatrix}\begin{bmatrix} 2 \\ 0 \\ -4 \end{bmatrix}.$$

First, let's multiply the two matrices; they are multiplied row by column as follows:

$$2 \cdot 2 + (-4) \cdot 0 + 4 \cdot (-4) = -12$$

$$1 \cdot 2 + 3 \cdot 0 + 2 \cdot (-4) = -6$$

$$-4 \cdot 2 - 2 \cdot 0 + 2 \cdot (-4) = -16$$

The solution vector is then

$$[x_i] = \frac{1}{10}\begin{bmatrix} -12 \\ -6 \\ -16 \end{bmatrix} = \begin{bmatrix} -1.2 \\ -0.6 \\ -1.6 \end{bmatrix}.$$

In component form, the solution is

$$x_1 = -1.2, \qquad x_2 = -0.6, \qquad x_3 = -1.6.$$

Example 1.18
Use Cramer's rule and solve

$$\begin{aligned} x_1 \qquad\quad - 2x_3 &= 2 \\ -x_1 + 2x_2 \qquad\quad &= 0 \\ x_1 + 2x_2 + x_3 &= -4 \end{aligned}$$

Solution. The solution is found (see Example 1.14) by evaluating the ratios as follows:

$$x_1 = \frac{\begin{vmatrix} 2 & 0 & -2 \\ 0 & 2 & 0 \\ -4 & 2 & 1 \end{vmatrix}}{D} = \frac{-12}{10} = -1.2 \qquad x_2 = \frac{\begin{vmatrix} 1 & 2 & -2 \\ -1 & 0 & 0 \\ 1 & -4 & 1 \end{vmatrix}}{D} = \frac{-6}{10} = -0.6$$

$$x_3 = \frac{\begin{vmatrix} 1 & 0 & 2 \\ -1 & 2 & 0 \\ 1 & 2 & -4 \end{vmatrix}}{D} = \frac{-16}{10} = -1.6$$

where

$$D = \begin{vmatrix} 1 & 0 & -2 \\ -1 & 2 & 0 \\ 1 & 2 & 1 \end{vmatrix} = 10.$$

Note that the numerator is the determinant formed by replacing the ith column with right-hand side elements r_i when solving for x_i.

1.6 Calculus

1.6.1 Differentiation

The slope of a curve $y = f(x)$ is the ratio of the change in y to the change in x as the change in x becomes infinitesimally small. This is the first derivative, written as

$$\frac{dy}{dx} = \lim_{\Delta x \to 0} \frac{\Delta y}{\Delta x}. \tag{1.6.1}$$

This may be written using abbreviated notation as

$$\frac{dy}{dx} = Dy = y' = \dot{y}. \tag{1.6.2}$$

The second derivative is written as

$$\frac{d^2y}{dx^2} = D^2y = y'' = \ddot{y}, \tag{1.6.3}$$

and is defined by

$$\frac{d^2y}{dx^2} = \lim_{\Delta x \to 0} \frac{\Delta y'}{\Delta x}. \tag{1.6.4}$$

Some derivative formulas, where f and g are functions of x, and k is constant, are given below.

$$\begin{aligned}
\frac{dk}{dx} &= 0 \\
\frac{d\left(k\,x^n\right)}{dx} &= k\,n\,x^{n-1} \\
\frac{d}{dx}(f+g) &= f'+g' \\
\frac{df^n}{dx} &= n\,f^{n-1}\,f' \\
\frac{d}{dx}(fg) &= fg'+gf' \\
\frac{d}{dx}(\ln x) &= \frac{1}{x} \\
\frac{d}{dx}\left(e^{kx}\right) &= ke^{kx} \\
\frac{d}{dx}(\sin x) &= \cos x \\
\frac{d}{dx}(\cos x) &= -\sin x
\end{aligned} \qquad \textbf{(1.6.5)}$$

If a function f depends on more than one variable, partial derivatives are used. If $z = f(x, y)$, then $\partial z / \partial x$ is the derivative of z with respect to x holding y constant. It would represent the slope of a line tangent to the surface in a plane of constant y.

1.6.2 Maxima and Minima

Derivatives are used to locate points of inflection, maxima, and minima. Note the following:

$$\begin{aligned}
f'(x) &= 0 \text{ at a maximum or a minimum.} \\
f''(x) &= 0 \text{ at an inflection point.} \\
f''(x) &> 0 \text{ at a minimum.} \\
f''(x) &< 0 \text{ at a maximum.}
\end{aligned}$$

An inflection point always exists between a maximum and a minimum.

1.6.3 L'Hospital's Rule

Differentiation is also useful in establishing the limit of $f(x)/g(x)$ as $x \to a$ if $f(a)$ and $g(a)$ are both zero or $\pm\infty$. *L'Hospital's rule* is used in such cases and is as follows:

$$\lim_{x\to a}\frac{f(x)}{g(x)} = \lim_{x\to a}\frac{f'(x)}{g'(x)} = \lim_{x\to a}\frac{f''(x)}{g''(x)}. \qquad \textbf{(1.6.6)}$$

1.6.4 Taylor's Series

Derivatives are used to expand a continuous function as a power series around $x = a$. Taylor's series is as follows:

$$f(x) = f(a) + (x-a)f'(a) + (x-a)^2 f''(a)/2! + \cdots. \qquad \textbf{(1.6.7)}$$

This series is often used to express a function as a polynomial near a point $x = a$ providing the series can be truncated after a few terms. Using Taylor's series we can show that (expanding about $a = 0$):

$$\begin{aligned} \sin x &= x - x^3/3! + x^5/5! - \cdots \\ \cos x &= 1 - x^2/2! + x^4/4! - \cdots \\ \ln(1+x) &= x - x^2/2 + x^3/3 - \cdots \\ \frac{1}{1-x} &= 1 + x + x^2 + \cdots \\ e^x &= 1 + x + x^2/2! + x^3/3! + \cdots \end{aligned} \tag{1.6.8}$$

When expanding with $a = 0$, the Taylor series may be called a *Maclaurin series.*

1.6.5 Integration

The inverse of differentiation is the process called integration. If a curve is given by $y = f(x)$, then the area under the curve from $x = a$ to $x = b$ is given by

$$A = \int_a^b y\,dx. \tag{1.6.9}$$

The length of the curve between the two points is expressed as

$$L = \int_a^b \left(1 + y'^2\right)^{1/2} dx. \tag{1.6.10}$$

Volumes of various objects are also found by an appropriate integration.

If the integral has limits, it is a *definite integral;* if it does not have limits, it is an *indefinite integral* and a constant is always added. Some common indefinite integrals follow:

$$\begin{aligned} &\int dx = x + C. \\ &\int c y\,dx = c\int y\,dx \\ &\int x^n dx = \frac{x^{n+1}}{n+1} + C \quad n \neq -1 \\ &\int x^{-1} dx = \ln x + C \\ &\int e^{ax} dx = \frac{1}{a} e^{ax} + C \\ &\int \sin x\,dx = -\cos x + C \\ &\int \cos x\,dx = \sin x + C \\ &\int \cos^2 x\,dx = \frac{x}{2} + \frac{1}{4}\sin 2x + C \\ &\int u\,dv = uv - \int v\,du. \end{aligned} \tag{1.6.11}$$

This last integral is often referred to as "integration by parts." If the integrand (the coefficients of the differential) is not one of the above, then in the last integral, $\int v\,du$ may in fact be integrable. An example will illustrate.

Example 1.19

Find the slope of $y = x^2 + \sin x$ at $x = 0.5$.

Solution. The derivative is the slope:

$$y'(x) = 2x + \cos x.$$

At $x = 0.5$ the slope is

$$y'(0.5) = 2 \cdot 0.5 + \cos 0.5 = 1.878.$$

Example 1.20

Find $\frac{d}{dx}(\tan x)$.

Solution. Writing $\tan x = \sin x / \cos x = f(x) \cdot g(x)$ we find

$$\begin{aligned}\frac{d}{dx}(\tan x) &= \frac{1}{\cos x}\frac{d}{dx}(\sin x) + \sin x \frac{d}{dx}(\cos x)^{-1} \\ &= \frac{\cos x}{\cos x} + \frac{\sin^2 x}{\cos^2 x} = 1 + \tan^2 x \\ &= \frac{\cos^2 x + \sin^2 x}{\cos^2 x} = \frac{1}{\cos^2 x} = \sec^2 x.\end{aligned}$$

Either expression is acceptable although the latter is usually used.

Example 1.21

Locate the maximum and minimum points of the function $y(x) = x^3 - 12x - 9$ and evaluate y at those points.

Solution. The derivative is

$$y'(x) = 3x^2 - 12.$$

The points at which $y'(x) = 0$ are at

$$x = 2, -2.$$

At these two points the extrema are:

$$\begin{aligned}y_{\min} &= (2)^3 - 12 \cdot 2 - 9 = -25 \\ y_{\max} &= (-2)^3 - 12(-2) - 9 = 7.\end{aligned}$$

Let us check the second derivative. At the two points we have:

$$\begin{aligned}y''(2) &= 6 \cdot 2 = 12 \\ y''(-2) &= 6 \cdot (-2) = -12\end{aligned}$$

Obviously, the point $x = 2$ is a minimum since its second derivative is positive there.

Example 1.22

Find the limit as $x \to 0$ of $\sin x / x$.

Solution. If we let $x = 0$ we are faced with the ratio of 0/0, an indeterminate quantity. Hence, we use L'Hospital's rule and differentiate both numerator and denominator to obtain

$$\lim_{x\to 0}\frac{\sin x}{x} = \lim_{x\to 0}\frac{\cos x}{1}.$$

Now, we let $x = 0$ and find

$$\lim_{x\to 0}\frac{\sin x}{x} = \frac{1}{1} = 1.$$

Example 1.23

Verify that $\sin x = x - x^3/3! + x^5/5! - \cdots$.

Solution. We expand in a Taylor's series about $x = 0$:

$$f(x) = f(0) + x f'(0) + x^2 f''(0)/2! + \cdots.$$

Letting $f(x) = \sin x$ so that $f' = \cos x$, $f'' = -\sin x$, etc., there results

$$\begin{aligned}\sin x &= 0 + x(1) + x^2(0/2!) + x^3(-1)/3! + \cdots \\ &= = x - x^3/3! + x^5/5! - \cdots.\end{aligned}$$

Example 1.24

Find the area of the shaded area in the figure.

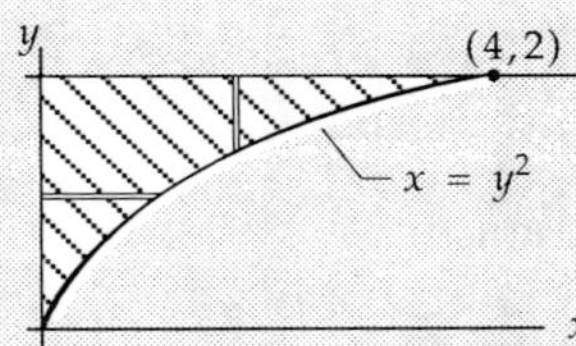

Solution. We can find this area by using either a horizontal strip or a vertical strip. We will use both. First, for a horizontal strip:

$$\begin{aligned}A &= \int_0^2 x\,dy \\ &= \int_0^2 y^2\,dy = \left.\frac{y^3}{3}\right|_0^2 = 8/3.\end{aligned}$$

Using a vertical strip we have

$$A = \int_0^4 (2-y)\,dx,$$

$$= \int_0^4 \left(2 - x^{1/2}\right) dx = \left[2x - \frac{2}{3}x^{3/2}\right]_0^4 = 8 - \frac{2}{3}8 = 8/3.$$

Either technique is acceptable. The first appears to be the simpler one.

Example 1.25

Find the volume enclosed by rotating the shaded area ofExample 1.24 about the *y*-axis.

Solution. If we rotate the horizontal strip about the *y*-axis we will obtain a disc with volume

$$dV = \pi x^2 dy.$$

This can be integrated to give the volume, which is

$$V = \int_0^2 \pi x^2 dy$$

$$= \pi \int_0^2 y^4 dy = \left.\frac{\pi y^5}{5}\right|_0^2 = \frac{32\pi}{5}.$$

Now, let us rotate the vertical strip about the *y*-axis to form a cylinder with volume

$$dV = 2\pi x(2-y)dx.$$

This can be integrated to yield

$$V = \int_0^4 2\pi x(2-y)dx$$

$$= \int_0^4 2\pi x\left(2 - x^{1/2}\right)dx = 2\pi\left[x^2 - \frac{2x^{5/2}}{5}\right]_0^4 = \frac{32\pi}{5}.$$

Again, the horizontal strip is simpler.

Example 1.26

Show that $\int xe^x dx = (x-1)e^x + C$.

Solution. Let's attempt the last integral of (1.6.11). Define the following:

$$u = x, \quad dv = e^x dx$$

Then,

$$du = dx, \quad v = \int e^x dx = e^x$$

and we find that

$$\int xe^x dx = x e^x - \int e^x dx$$

$$= x e^x - e^x + C = (x-1)e^x + C.$$

1.7 Differential Equations

A differential equation is *linear* if no term contains the dependent variable or any of its derivatives to a power other than one (terms that do not contain the dependent variable are not considered in the test of linearity). For example,

$$y'' + 2xy' - y\sin x = 3x^2 \tag{1.7.1}$$

is a linear differential equation. The dependent variable is y and the independent variable is x. If a term contained y'^2, or $y^{1/2}$ or $\sin y$ the equation would be nonlinear.

A differential equation is *homogeneous* if all of its terms contain the dependent variable. Eq. 1.7.1 is *nonhomogeneous* because of the term $3x^2$.

The *order* of a differential equation is established by its highest order derivative. Eq. 1.7.1 is a second order differential equation.

The general solution of a differential equation involves a number of arbitrary constants equal to the order of the equation. If conditions are specified, the arbitrary constants may be calculated.

1.7.1 First Order

A first order differential equation is *separable* if it can be expressed as

$$M(x)\,dx + N(y)\,dy = 0. \tag{1.7.2}$$

The solution follows by integrating each of the terms. This equation is, in general, nonlinear.

If a first order differential equation is not separable, we will consider only those that are linear with a constant coefficient, written as

$$y' + ay = f(x) \tag{1.7.3}$$

For the homogeneous equation, $f(x) = 0$, it has the solution

$$y(x) = Ce^{-ax} \tag{1.7.4}$$

The constant C would be determined from an imposed condition

$$y(0) = C \tag{1.7.5}$$

1.7.2. Second Order

The general form of a second order, linear, homogeneous differential equation with constant coefficients is

$$y'' + 2ay' + by = 0. \tag{1.7.6}$$

To find a solution we assume $y = e^{rx}$ and solve the *characteristic equation*

$$r^2 + 2ar + b = 0. \tag{1.7.7}$$

Let r_1 and r_2 be its roots. If $r_1 \neq r_2$ and both are real, the general solution is

$$y(x) = C_1 e^{r_1 x} + C_2 e^{r_2 x}. \tag{1.7.8}$$

If $r_1 = r_2$, the general solution is

$$y(x) = C_1 e^{r_1 x} + C_2 xe^{r_2 x}. \tag{1.7.9}$$

Finally, if $r_1 = \alpha + i\beta$ and $r_2 = \alpha - i\beta$, the general solution is

$$y(x) = (C_1 \sin \beta x + C_2 \cos \beta x)\, e^{-ax}. \quad \textbf{(1.7.10)}$$

where $\beta = \sqrt{b - a^2}$. and $\alpha = -a$.

1.7.3 Linear, Nonhomogeneous, with Constant Coefficients

If Eq. 1.7.6 were nonhomogeneous, it would be written as

$$y'' + 2ay' + by = f(x). \quad \textbf{(1.7.11)}$$

The general solution is found by finding the solution $y_h(x)$ to the homogeneous equation (simply let the right-hand side be zero and solve the equation as in Section 1.7.2) and adding to it a particular solution $y_p(x)$ found by using Table 1.2. This is expressed as

$$y(x) = y_h(x) + y_p(x). \quad \textbf{(1.7.12)}$$

TABLE 1.2 Particular Solutions

$f(x)$	$y_p(x)$	provisions
A	C	
$Ax + B$	$Cx + D$	
e^{Ax}	Ce^{Ax}	if r_1 or $r_2 \neq a$
	Cxe^{Ax}	if r_1 or $r_2 = a$
$B \sin Ax$	$C \sin Ax + D \cos Ax$	if $r_{1,2} \neq \pm ai$
	$Cx \sin Ax + Dx \cos Ax$	if $r_{1,2} = \pm ai$
$B \cos Ax$	(same as above)	

Example 1.27

Find the general solution to $2xy' + y^2 = 2y$.

Solution. Write the differential equation as

$$2x\frac{dy}{dx} = 2y - y^2.$$

This is separable and written as

$$\frac{2dy}{y(2-y)} = \frac{dx}{x}$$

This is rewritten as

$$\frac{dy}{y} + \frac{dy}{2-y} = \frac{dx}{x}$$

Integrating provides a solution:

$$\ln y - \ln(2-y) = \ln x - \ln C$$

Rewrite this as

$$\ln\frac{Cy}{2-y} = \ln x,$$

Finally, the solution is best written as

$$\frac{Cy}{2-y} = x \quad \text{or} \quad y(x) = \frac{2x}{C+x}$$

Example 1.28

Find the solution to

$$y' + 2y = 4x, \qquad y(0) = 2.$$

Solution. We will use the method outlined in Section 1.7.2. The characteristic equation of the homogeneous equation is

$$r + 2 = 0. \qquad \therefore r = -2.$$

The homogeneous solution is then

$$y_h(x) = Ce^{-2x}.$$

The particular solution is assumed to be of the form

$$y_p(x) = Ax + B.$$

Substituting this into the original differential equation gives

$$A + 2(Ax + B) = 4x. \qquad \therefore A = 2, \quad B = -1.$$

The solution is then

$$\begin{aligned} y(x) &= y_h(x) + y_p(x) \\ &= Ce^{-2x} + 2x - 1. \end{aligned}$$

Using the given condition

$$y(0) = 2 = C - 1. \quad \therefore C = 3.$$

Finally,

$$y(x) = 3e^{-2x} + 2x - 1.$$

Example 1.29

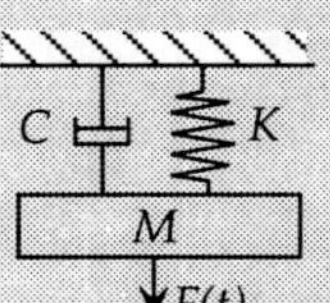

The motion of a simple spring-mass system is represented by

$$M\ddot{y} + C\dot{y} + Ky = F(t)$$

where the mass *M*, the damping coefficient *C*, the spring constant *K*, and the forcing function *F(t)* have the appropriate units. Find the general solution if $M = 2$, $C = 0$, $K = 50$, and $F(t) = 0$.

Solution. The differential equation simplifies to $2\ddot{y} + 50y = 0$.

The characteristic equation is then

$$2r^2 + 50 = 0.$$

$$\therefore r_1 = 5i, \quad r_2 = -5i.$$

The solution is then (see Eq. 1.7.10)

$$y(t) = C_1 \sin 5t + C_2 \cos 5t.$$

This situation with zero damping represents simple *harmonic motion.*
Note that we have used dots to represent the time derivative $dy / dt = \dot{y}$. Also, note that the coefficient of *t* in the undamped system is the *natural frequency* $\omega_n = \sqrt{K/M}$; the frequency in hertz is $f = \omega_n / 2\pi$; and the *period* is $\tau = f^{-1} = 2\pi/\omega_n$. For this problem $\omega_n = 5$ rad/s, $f = 5/2\pi$ Hz, and $\tau = 2\pi/5$ seconds.

Example 1.30

In the equation of Ex. 1.29, let *M* = 2, *C* = 12, *K* = 50, and *F(t)* = 60 sin 5*t*. Find the general solution.

Solution. The differential equation is

$$2\ddot{y} + 12\dot{y} + 50y = 60 \sin 5t.$$

The characteristic equation of the homogeneous differential equation is found by letting $y = e^{rx}$:

$$2r^2 + 12r + 50 = 0.$$

$$\therefore r_1 = -3 + 4i, \quad r_2 = -3 - 4i.$$

The homogeneous solution is (see Eq. 1.7.10)

$$y_h(t) = e^{-3t}\left(C_1 \sin 4t + C_2 \cos 4t\right).$$

The particular solution is found by assuming that

$$y_p(t) = A \sin 5t + B \cos 5t.$$

Substitute this into the original differential equation:

$$2[-25A \sin 5t - 25B \cos 5t] + 12[5A \cos 5t - 5B \sin 5t] + 50[A \sin 5t + B \cos 5t]$$
$$= 60 \sin 5t$$

Equating coefficients of sin terms and then cos terms:

$$-50A - 60B + 50A = 60. \quad \therefore B = -1.$$

$$-50B + 60A + 50B = 0. \quad \therefore A = 0.$$

The two solutions are superposed as in Eq. 1.7.12 to give

$$y(t) = \left(C_1 \sin 4t + C_2 \cos 4t\right) e^{-3t} - \cos 5t.$$

1.7.4 Laplace Transforms

The solution of nonhomogeneous, constant coefficient, linear differential equations can be obtained quite easily using Laplace transforms. The initial conditions must be given and with them we obtain the solution directly; we do not superpose homogeneous and particular solutions.

The *Laplace transform* $F(s)$ of $f(t)$ is defined as

$$F(s) = \int_0^\infty f(t)e^{-st}dt \,. \tag{1.7.13}$$

We also use the notation

$$\mathcal{L}(f) = F(s)$$

where $\mathcal{L}(f)$ is the Laplace transform of $f(t)$.

By applying Eq. 1.7.13 to a variety of functions, including derivatives, we can obtain the Laplace transforms presented in Table 1.3.

Table 1.3 Laplace Transforms

f(t)	*F(s)*	*f(t)*	*F(s)*
(t), Impulse at $t = 0$	1	$\mathcal{L}\left\{\dfrac{d^n f(t)}{dt^n}\right\}$	$s^n F(s) - \sum_{m=0}^{n-1} s^{n-m-1} \dfrac{d^m f(0)}{dt^m}$
$u(t)$, Step at $t = 0$	$1/s$		
$t\,u(t)$, Ramp at $t = 0$	$1/s^2$	$\int_0^t f(\tau)d\tau$	$(1/s)F(s)$
$e^{-\alpha t}$	$1/(s+\alpha)$	$\int_0^t f(t-\tau)h(\tau)d\tau$	$F(s)H(s)$
$te^{-\alpha t}$	$1/(s+\alpha)^2$	$\lim_{t\to\infty} f(t)$	$\lim_{s\to 0} sF(s)$
$e^{-\alpha t}\sin\beta t$	$\beta/\left[(s+\alpha)^2+\beta^2\right]$	$\lim_{t\to 0} f(t)$	$\lim_{s\to\infty} sF(s)$
$e^{-\alpha t}\cos\beta t$	$(s+\alpha)/\left[(s+\alpha)^2+\beta^2\right]$		
$t\sin\beta t$	$2bs/\left(s^2+\beta^2\right)^2$		
$t\cos\beta t$	$\left(s^2-\beta^2\right)/\left(s^2+\beta^2\right)^2$		

We can now use Table 1.3 to solve ordinary differential equations. We can write:

$$\begin{aligned}\mathcal{L}\{y'(t)\} &= sY(s) - y(0)\\ \mathcal{L}\{y''(t)\} &= s^2Y(s) - sy(0) - y'(0)\end{aligned} \tag{1.7.15}$$

If we take the Laplace transform of Eq. 1.7.11, replacing the independent variable x with t, we find the *subsidiary* equation:

$$s^2Y(s) + 2asY(s) + bY(s) - sy(0) - y'(0) - y(0) = F(s) \tag{1.7.16}$$

This can be rearranged as

$$Y(s)\frac{sy(0) + y'(0) + y(0)}{s^2 + 2as + b} + \frac{F(s)}{s^2 + 2as + b} \tag{1.7.17}$$

The solution $y(t)$ is then the inverse Laplace transform

$$y(t) = \mathcal{L}^{-1}\{Y(s)\} \tag{1.7.18}$$

To find the inverse Laplace transform we find the inverse functions of the fractions on the right-hand side of Eq. 1.7.17. Section 1.1.4 on partial fractions will aid in this process.

Example 1.31

Solve the differential equation $2y'' + 12y' + 50y = 60\sin 5t$ if $y(0) = 0$ and $y'(0) = 10$.

Solution. We will find the solution using Laplace transforms. The Laplace transform of our differential equation is (first divide by 2 so that the coefficient of y'' is 1)

$$s^2 Y(s) + 6sY(s) + 25Y(s) - s\,y(0) - 10 - y(0) = \frac{30\,(5)}{s^2 + 5^2}$$

where we have used $\alpha = 0$ and $\beta = 5$ in the Laplace transform of $e^{-\alpha t}\sin \beta t$. Next, we see, with some algebra, that

$$\begin{aligned} Y(s) &= \frac{10}{s^2 + 6s + 25} + \frac{150}{(s^2 + 25)(s^2 + 6s + 25)} \\ &= \frac{10}{s^2 + 6s + 25} + \frac{-s}{s^2 + 25} + \frac{s + 6}{s^2 + 6s + 25} \\ &= \frac{13}{(s+3)^2 + 16} - \frac{s}{s^2 + 25} + \frac{s + 3}{(s+3)^2 + 16}. \end{aligned}$$

The inverse Laplace transform is found using Table 1.3 to be:

$$\begin{aligned} y(t) &= \frac{13}{4} e^{-3t} \sin 4t - \cos 5t + e^{-3t} \cos 4t \\ &= \left(\frac{13}{4} \sin 4t + \cos 4t \right) e^{-3t} - \cos 5t \end{aligned}$$

Compare this solution to that ofExample 1.30. Also, check and make sure that those initial conditions are satisfied.

1.8 Vectors

There are two vector multiplications. The first, the *scalar product*, or *dot product*, is the scalar defined by

$$\mathbf{A} \cdot \mathbf{B} = AB\cos\theta \qquad \textbf{(1.8.1)}$$

where θ is the angle between the two vectors, as shown in Fig. 1.9, and A and B are the magnitudes of the two vectors. In a rectangular coordinate system the dot product becomes

$$\begin{aligned} \mathbf{A} \cdot \mathbf{B} &= (A_x\mathbf{i} + A_y\mathbf{j} + A_z\mathbf{k}) \cdot (B_x\mathbf{i} + B_y\mathbf{j} + B_z\mathbf{k}) \\ &= A_xB_x + A_yB_y + A_zB_z. \end{aligned} \qquad \textbf{(1.8.2)}$$

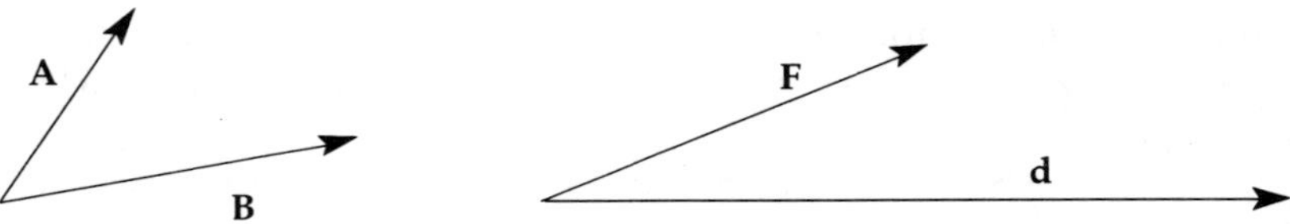

Figure 1.9 Vectors.

The scalar quantity *work* can be defined using the dot product:

$$w = \mathbf{F} \cdot \mathbf{d} \tag{1.8.3}$$

where **d** is the directed distance moved by the force.

The second, the *vector product*, or *cross product*, of the two vectors **A** and **B** is a vector defined by

$$\mathbf{C} = \mathbf{A} \times \mathbf{B} \tag{1.8.4}$$

where the magnitude of **C** is given by

$$C = AB \sin\theta. \tag{1.8.5}$$

The vector **C** acts in a direction perpendicular to the plane of **A** and **B** so that the three vectors form a right-handed set of vectors. (If the fingers curl **A** into **B**, the thumb points in the direction of **C**.) In a rectangular coordinate system the cross product is

$$\mathbf{A} \times \mathbf{B} = \begin{vmatrix} \mathbf{i} & \mathbf{j} & \mathbf{k} \\ A_x & A_y & A_z \\ B_x & B_y & B_z \end{vmatrix} = \left(A_y B_z - A_z B_y\right)\mathbf{i} + \left(A_z B_x - A_x B_z\right)\mathbf{j} + \left(A_x B_y - A_y B_x\right)\mathbf{k}. \tag{1.8.6}$$

The magnitude of **C** is the area of the parallelogram with sides **A** and **B**.

The volume of the parallelepiped with sides **A**, **B**, and **C** is the scalar triple product given by

$$\mathbf{A} \times \mathbf{B} \cdot \mathbf{C} = \begin{vmatrix} A_x & A_y & A_z \\ B_x & B_y & B_z \\ C_x & C_y & C_z \end{vmatrix}. \tag{1.8.7}$$

Using the above definition we can produce the following identities:

$$\begin{aligned} &\mathbf{i} \cdot \mathbf{i} = \mathbf{j} \cdot \mathbf{j} = \mathbf{k} \cdot \mathbf{k} = 1 \\ &\mathbf{i} \times \mathbf{i} = \mathbf{j} \times \mathbf{j} = \mathbf{k} \times \mathbf{k} = 0 \\ &\mathbf{A} \cdot \mathbf{B} = 0 \quad \text{if} \quad \mathbf{A} \perp \mathbf{B} \\ &\mathbf{A} \cdot \mathbf{B} = AB \quad \text{if} \quad \mathbf{A} \parallel \mathbf{B} \\ &\mathbf{i} \cdot \mathbf{j} = \mathbf{j} \cdot \mathbf{k} = \mathbf{k} \cdot \mathbf{i} = 0 \\ &\mathbf{i} \times \mathbf{j} = \mathbf{k}, \quad \mathbf{j} \times \mathbf{k} = \mathbf{i}, \quad \mathbf{k} \times \mathbf{i} = \mathbf{j} \\ &\mathbf{A} \times \mathbf{B} = 0 \quad \text{if} \quad \mathbf{A} \parallel \mathbf{B} \end{aligned} \tag{1.8.8}$$

Example 1.32

Find the projection of **A** on **B** if $\mathbf{A} = 12\mathbf{i} - 18\mathbf{j} + 6\mathbf{k}$ and $\mathbf{B} = 2\mathbf{i} - 4\mathbf{j} + 4\mathbf{k}$.

Solution. Find the unit vector $\mathbf{i}_B$ in the direction of B:

$$\mathbf{i}_B = \frac{\mathbf{B}}{B} = \frac{2\mathbf{i} - 4\mathbf{j} + 4\mathbf{k}}{\sqrt{2^2 + 4^2 + 4^2}} = \frac{1}{6}(2\mathbf{i} - 4\mathbf{j} + 4\mathbf{k}),$$

The projection of **A** on **B** is then

$$\begin{aligned}\mathbf{A}\cdot\mathbf{i}_B &= (12\mathbf{i}-18\mathbf{j}+6\mathbf{k})\cdot\tfrac{1}{6}(2\mathbf{i}-4\mathbf{j}+4\mathbf{k})\\ &= 4+12+4=20.\end{aligned}$$

Example 1.33

Find the area of a parallelogram with two sides identified by vectors from the origin to the points (3,4) and (8,0).

Solution. The two vectors are represented by

$$\mathbf{A}=3\mathbf{i}+4\mathbf{j},\qquad \mathbf{B}=8\mathbf{i}.$$

The area of the parallelogram is then

$$|\mathbf{A}\times\mathbf{B}|=|(3\mathbf{i}+4\mathbf{j})\times 8\mathbf{i}|=32$$

since $\mathbf{i}\times\mathbf{i}=0$ and $\mathbf{j}\times\mathbf{i}=-\mathbf{k}$.

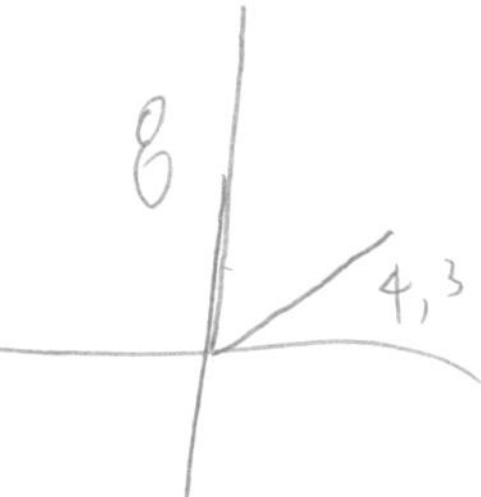

Example 1.34

Find a unit vector perpendicular to the plane that contains both $\mathbf{A}=\mathbf{i}-2\mathbf{j}+3\mathbf{k}$ and $\mathbf{B}=\mathbf{i}+2\mathbf{j}-\mathbf{k}$.

Solution. The vector $\mathbf{C}=\mathbf{A}\times\mathbf{B}$ is perpendicular to the plane of A and B. Using Eq. 1.8.6,

$$\begin{aligned}\mathbf{C} &= [-2(-1)-3\times 2]\mathbf{i}+[3\times 1-1(-1)]\mathbf{j}+[1\times 2-(-2)\times 1]\mathbf{k}\\ &= -4\mathbf{i}+4\mathbf{j}+4\mathbf{k}.\end{aligned}$$

A unit vector in the direction of **C** is the desired unit vector:

$$\mathbf{i}_c=\frac{-4\mathbf{i}+4\mathbf{j}+4\mathbf{k}}{\sqrt{4^2+4^2+4^2}}=\frac{1}{\sqrt{3}}(-\mathbf{i}+\mathbf{j}+\mathbf{k}).$$

1.8.1 The Gradient (may not be tested)

We define a vector differential operator ∇, referred to as *del*, to be

$$\nabla=\frac{\partial}{\partial x}\mathbf{i}+\frac{\partial}{\partial y}\mathbf{j}+\frac{\partial}{\partial z}\mathbf{k} \qquad \textbf{(1.8.9)}$$

The *gradient* of a scalar function ϕ is

$$\nabla\phi=\frac{\partial\phi}{\partial x}\mathbf{i}+\frac{\partial\phi}{\partial y}\mathbf{j}+\frac{\partial\phi}{\partial z}\mathbf{k}. \qquad \textbf{(1.8.10)}$$

It is a vector quantity that points in the direction of the maximum rate of change of ϕ; hence, it is normal to a constant ϕ surface.

The *divergence* of a vector function **u** is the scalar quantity

$$\nabla\cdot\mathbf{u}=\frac{\partial u_x}{\partial x}+\frac{\partial u_y}{\partial y}+\frac{\partial u_z}{\partial z}. \qquad \textbf{(1.8.11)}$$

The *curl* of a vector function **u** is the vector quantity

$$\nabla \times \mathbf{u} = \left(\frac{\partial u_z}{\partial y} - \frac{\partial u_y}{\partial z} \right)\mathbf{i} + \left(\frac{\partial u_x}{\partial z} - \frac{\partial u_z}{\partial x} \right)\mathbf{j} + \left(\frac{\partial u_y}{\partial x} - \frac{\partial u_x}{\partial y} \right)\mathbf{k}. \quad \textbf{(1.8.12)}$$

The *Laplacian* of a scalar function is, using $\nabla \cdot \nabla = \nabla^2$,

$$\nabla^2 \phi = \frac{\partial^2 \phi}{\partial x^2} + \frac{\partial^2 \phi}{\partial y^2} + \frac{\partial^2 \phi}{\partial z^2} \quad \textbf{(1.8.13)}$$

and *Laplace's equation* is

$$\nabla^2 \phi = 0.$$

A vector function **u** is a *conservative* (or *irrotational*) vector function if it is given by the gradient of a scalar function ϕ, i.e.,

$$\mathbf{u} = \nabla \phi.$$

The curl of a conservative vector function is always zero. A vector field is *solenoidal* if its divergence is zero.

Example 1.35

Find a unit vector in the direction of the maximum rate of change of $\phi = x^2 - 2yz + xy$ at the point (2, −1, −1).

Solution. The gradient of ϕ is

$$\begin{aligned} \nabla \phi &= \frac{\partial \phi}{\partial x}\mathbf{i} + \frac{\partial \phi}{\partial y}\mathbf{j} + \frac{\partial \phi}{\partial z}\mathbf{k} \\ &= (2x + y)\mathbf{i} + (-2z + x)\mathbf{j} + (-2y)\mathbf{k} = 3\mathbf{i} + 4\mathbf{j} + 2\mathbf{k}. \end{aligned}$$

A unit vector in the direction of $\nabla \phi$ is

$$\mathbf{i}_n = \frac{\nabla \phi}{|\nabla \phi|} = \frac{3\mathbf{i} + 4\mathbf{j} + 2\mathbf{k}}{\sqrt{3^2 + 4^2 + 2^2}} = \frac{1}{\sqrt{29}}(3\mathbf{i} + 4\mathbf{j} + 2\mathbf{k}).$$

Example 1.36

Given a vector function $\mathbf{u} = x^2\mathbf{i} + 2yz\mathbf{j} + y^2\mathbf{k}$, find a) the divergence, b) the curl, and c) its associated scalar function.

Solution. a) the divergence is the scalar function

$$\begin{aligned} \nabla \cdot \mathbf{u} &= \frac{\partial u_x}{\partial x} + \frac{\partial u_y}{\partial y} + \frac{\partial u_z}{\partial z} \\ &= 2x + 2z + 0 = 2(x + z). \end{aligned}$$

b) The curl is the vector function

$$\begin{aligned} \nabla \times \mathbf{u} &= \left(\frac{\partial u_z}{\partial y} - \frac{\partial u_y}{\partial z} \right)\mathbf{i} + \left(\frac{\partial u_x}{\partial z} - \frac{\partial u_z}{\partial x} \right)\mathbf{j} + \left(\frac{\partial u_y}{\partial x} - \frac{\partial u_x}{\partial y} \right)\mathbf{k} \\ &= (2y - 2y)\mathbf{i} + (0 - 0)\mathbf{j} + (0 - 0)\mathbf{k} = 0, \end{aligned}$$

c)We know that an associated scalar function exists since the curl is everywhere zero. It is found by equating the scalar components of Eq. 1.8.15:

$$u_x = \frac{\partial \phi}{\partial x} \Rightarrow x^2 = \frac{\partial \phi}{\partial x}. \quad \therefore \phi = \frac{x^3}{3} + f(y,z).$$

$$u_y = 2yz = \frac{\partial \phi}{\partial y} = \frac{\partial}{\partial y}\left[\frac{x^3}{3} + f(y,z)\right] = \frac{\partial f}{\partial y}.$$

$$\therefore \frac{\partial f}{\partial y} = 2yz \text{ and } f(y,z) = y^2 z + g(z).$$

$$u_z = y^2 = \frac{\partial \phi}{\partial z} = \frac{\partial}{\partial z}\left[\frac{x^3}{3} + y^2 z + g(z)\right] = y^2 + \frac{dg}{dz}.$$

$$\therefore \frac{dg}{dz} = 0 \quad \text{so that} \quad g(z) = c.$$

Finally,

$$\phi = \frac{x^3}{3} + y^2 z + c.$$

1.9 Probability and Statistics

Events are independent if the probability of occurrence of one event does not influence the probability of occurrence of other events. The number of permutations (a particular sequence) of n things taken r at a time is

$$P(n,r) = \frac{n!}{(n-r)!}. \tag{1.9.1}$$

If the starting point is unknown, as in a ring, the *ring permutation* is

$$P(n,r) = \frac{(n-1)!}{(n-r)!}. \tag{1.9.2}$$

The number of *combinations* (no order-conscious arrangement) of n things taken r at a time is given by

$$C(n,r) = \frac{n!}{r!(n-r)!}. \tag{1.9.3}$$

For independent events of two sample groups A and B the following rules are necessary:

1. The probability of A or B occurring equals the sum of the probability of occurrence of A and the probability of occurrence of B ; that is,

$$P(A \text{ or } B) = P(A) + P(B). \tag{1.9.4}$$

2. The probability of both A and B occurring is given by the product

$$P(A \text{ and } B) = P(A)P(B). \tag{1.9.5}$$

3. The probability of A not occurring is given as

$$P(\text{not } A) = 1 - P(A). \tag{1.9.6}$$

4. The probability of either A or B occurring is given by

$$P(A \text{ or } B) = P(A) + P(B) - P(A)P(B). \tag{1.9.7}$$

The probability of an event occurring is in the range of 0 to 1. An impossible event has a probability of 0 and an event that is certain to occur has a probability of 1.

The data gathered during an experiment can be analyzed using quantities defined by the following:

1. The arithmetic mean $\bar{x}$ is the average of the observations; that is,

$$\bar{x} = \frac{x_1 + x_2 + x_3 + \cdots + x_n}{n}$$

2. The median is the middle observation when all the data is ordered by magnitude; half the values are below the median. The median for an even number of data is the average of the two middle values.
3. The mode is the observed value that occurs most frequently.
4. The standard deviation s of the sample is a measure of variability. It is defined as

$$\sigma = \left[\frac{(x_1 - \bar{x})^2 + (x_2 - \bar{x})^2 + \cdots + (x_n - \bar{x})^2}{n-1}\right]^{1/2} \tag{1.9.8}$$

$$= \left[\frac{x_1^2 + x_2^2 + \cdots + x_n^2 - n\bar{x}^2}{n-1}\right]^{1/2}. \tag{1.9.9}$$

   For large observations (over 50), it is customary to simply use n, rather that $(n - 1)$, in the denominators of the above. In fact, if n is used in the above equations, σ is often referred to as the *standard deviation of the population.*
5. The *variance* is defined to be σ^2.

Example 1.37

How many different ways can seven people be arranged in a lineup? In a circle?

Solution. This is the number of permutations of seven things taken seven at a time. The answer is

$$P(7,7) = \frac{n!}{(n-r)!}$$

$$= \frac{7!}{(7-7)!} = 5040.$$

In a circle we use the ring permutation:

$$P(7,7) = \frac{(n-1)!}{(n-r)!}$$

$$= \frac{(7-1)!}{(7-7)!} = 720.$$

Note that 0! = 1.

Example 1.38

How many different collections of eight people can fit into a six-passenger vehicle? (Only six will fit at a time.)

Solution. The answer does not depend on the seating arrangement. If it did, it would be a permutation. Hence, we use the combination relationship and find

$$C(8,6) = \frac{n!}{(n-r)!r!}$$
$$= \frac{8!}{(8-6)!6!} = 28.$$

Example 1.39

A carnival booth offers $10 if you pick a red ball and then a white ball (the first ball is re-inserted) from a bin containing 60 red balls, 15 white balls, and 25 blue balls. If $1 is charged for an attempt, will the operator make money?

Solution. The probability of drawing a red ball on the first try is 0.6. If it is then re-inserted, the probability of drawing a white ball is 0.15. The probability of accomplishing both is then given by

$$P(\text{red and white}) = P(\text{red})\ P(\text{white})$$
$$= 0.6 \times 0.15 = 0.09,$$

or 9 chances out of 100 attempts. Hence, the entrepreneur will pay out $90 for every $100 taken in and will thus make money.

Example 1.40

If the operator of the bin of balls in Example 1.39 offers a $1.00 prize to contestants who pick either a red ball or a white ball from the bin on the first attempt, and charges $0.75 per attempt, will the operator make money?

Solution. The probability of selecting either a red ball or a white ball on the first attempt is

$$P(\text{red and white}) = P(\text{red}) + P(\text{white})$$
$$= 0.6 + 0.15 = 0.75.$$

Consequently, 75 out of 100 gamblers will win and the operator must pay out $75 for every $75 taken in. The operator would not make any money.

Example 1.41

If the operator of Example 1.40 had two identical bins and offered a $1.00 prize for successfully withdrawing a red ball from the first bin or a white ball from the second bin, will the operator make money if he charges $0.75 per attempt?

Solution. The probability of selecting a red ball from the first bin (sample group A_i) or a white ball from the second bin (sample group B_i) is

$$P(\text{red or white}) = P(\text{red}) + P(\text{white}) - P(\text{red})\ P(\text{white})$$
$$= 0.6 + 0.15 - 0.6 \times 0.15 = 0.66.$$

For this situation the owner must pay \$66 to every 100 gamblers who pay \$75 to participate. The profit is \$9.

Example 1.42

The temperature at a given location in the south at 2 p.m. each August 10 for 25 consecutive years was measured, in degrees Celsius, to be 33, 38, 34, 26, 32, 31, 28, 39, 29, 36, 32, 29, 31, 24, 35, 34, 32, 30, 31, 32, 26, 40, 27, 33, 39. Calculate the arithmetic mean, the median, the mode and the sample standard deviation.

Solution. Using the appropriate equations, we calculate the arithmetic mean to be

$$\overline{T} = \frac{T_1 + T_2 + \cdots + T_{25}}{25}$$

$$= \frac{33 + 38 + 34 + \cdots + 39}{25} = \frac{801}{25} = 32.04°\text{C}$$

The median is found by first arranging the values in order. We have 24, 26, 26, 27, 28, 29, 29, 30, 31, 31, 31, 32, 32, 32, 32, 33, 33, 34, 34, 35, 36, 38, 39, 39, 40. Counting 12 values in from either end, the median is found to be 32°C.

The mode is the observation that occurs most often; it is 32°C.

The sample standard deviation is found to be

$$\sigma = \left[\frac{T_1^2 + T_2^2 + \cdots + T_{25}^2 - n\overline{T}^2}{n-1}\right]^{1/2}$$

$$= \left[\frac{33^2 + 38^2 + \cdots + 39^2 - 25 \times 32.04^2}{25-1}\right]^{1/2} = \sqrt{\frac{26,099 - 25,664}{24}} = 4.26.$$

Practice Problems

ALGEBRA

1.1 A growth curve is given by $A = 10\,e^{2t}$. At what value of t is A = 100?

a) 5.261 b) 3.070 c) 1.151 d) 0.726 e) 0.531

1.2 If $\ln x = 3.2$, what is x?

a) 18.65 b) 24.53 c) 31.83 d) 64.58 e) 126.7

1.3 If $\ln_5 x = -1.8$, find x.

a) 0.00483 b) 0.0169 c) 0.0552 d) 0.0783 e) 0.1786

1.4 One root of the equation $3x^2 - 2x - 2 = 0$ is

a) 1.215 b) 1.064 c) 0.937 d) 0.826 e) 0.549

1.5 $\sqrt{4+x}$ can be written as the series

a) $2 - x/4 + x^2/64 + \cdots$

b) $2 + x/8 - x^2/128 + \cdots$

c) $2 - x^2/4 - x^4/64 + \cdots$

d) $2 + x^2/8 + x^4/128 + \cdots$

e) $2 + x/4 - x^2/64 + \cdots$

1.6 Resolve $\dfrac{2}{x\left(x^2 - 3x + 2\right)}$ into partial fractions.

a) $1/x + 1/(x-2) - 2/(x-1)$

b) $1/x - 2/(x-2) + 1/(x-1)$

c) $2/x - 1/(x-2) - 2/(x-1)$

d) $-1/x + 2/(x-2) + 1/(x-1)$

e) $-1/x - 2/(x-2) + 1/(x-1)$

1.7 Express $\dfrac{4}{x^2\left(x^2 - 4x + 4\right)}$ as the sum of fractions.

a) $1/x + 1/(x-2)^2 - 1/(x-2)$

b) $1/x + 1/x^2 - 1/(x-2) + 1/(x-2)^2$

c) $1/x^2 + 1/(x-2)^2$

d) $1/x + 1/x^2 + 1/(x-2) + 1/(x-2)^2$

e) $1/x^2 - 1/(x-2) + 1/(x-2)^2$

1.8 A germ population has a growth curve of $Ae^{0.4t}$. At what value of t does its original value double?

a) 9.682 b) 7.733 c) 4.672 d) 1.733 e) 0.5641

TRIGONOMETRY

1.9 If $\sin\theta = 0.7$, what is $\tan\theta$?

a) 0.98 b) 0.94 c) 0.88 d) 0.85 e) 0.81

1.10 If the short leg of a right triangle is 5 units long and the long leg is 7 units long, find the angle opposite the short leg, in degrees.

a) 26.3 b) 28.9 c) 31.2 d) 33.8 e) 35.5

1.11 The expression $\tan\theta\sec\theta\left(1-\sin^2\theta\right)/\cos\theta$ simplifies to

a) $\sin\theta$ b) $\cos\theta$ c) $\tan\theta$ d) $\sec\theta$ e) $\csc\theta$

1.12 A triangle has sides of length 2, 3 and 4. What angle, in radians, is opposite the side of length 3?

a) 0.55 b) 0.61 c) 0.76 d) 0.81 e) 0.95

1.13 The length of a lake is to be determined. A distance of 850 m is measured from one end to a point x on the shore. A distance of 732 m is measured from x to the other end. If an angle of 154° is measured between the two lines connecting x, what is the length of the lake?

a) 1542 b) 1421 c) 1368 d) 1261 e) 1050

1.14 Express $2\sin^2\theta$ as a function of $\cos 2\theta$.

a) $\cos 2\theta - 1$ b) $\cos 2\theta + 1$ c) $\cos 2\theta + 2$ d) $2 - \cos 2\theta$ e) $1 - \cos 2\theta$

GEOMETRY

1.15 The included angle between two successive sides of a regular eight-sided polygon is

a) 150° b) 135° c) 120° d) 75° e) 45°

1.16 A large 15-m-dia cylindrical tank that sits on the ground is to be painted. If one liter of paint covers 10 m^2, how many liters are required if it is 10 m high? (Include the top.)

a) 65 b) 53 c) 47 d) 38 e) 29

1.17 The equation of a line that has a slope of –2 and intercepts the x-axis at $x = 2$ is

a) $y + 2x = 4$
b) $y - 2x = 4$
c) $y + 2x = -4$
d) $2y + x = 2$
e) $2y - x = -2$

1.18 The equation of a line that intercepts the x-axis at $x = 4$ and the y-axis at $y = -6$ is

a) $2x - 3y = 12$
b) $3x - 2y = 12$
c) $2x + 3y = 12$
d) $3x + 2y = 12$
e) $3y - 2x = 12$

1.19 The shortest distance from the line $3x - 4y = 3$ to the point (6, 8) is

a) 4.8 b) 4.2 c) 3.8 d) 3.4 e) 2.6

1.20 The equation $x^2 + 4xy + 4y^2 + 2x = 10$ represents which conic section?

a) circle b) ellipse c) parabola d) hyperbola e) plane

1.21 The x- and y-axes are the asymptotes of a hyperbola that passes through the point (2, 2). Its equation is

a) $x^2 - y^2 = 0$ b) $xy = 4$ c) $y^2 - x^2 = 0$ d) $x^2 + y^2 = 4$ e) $x^2y = 8$

1.22 A 100-m-long track is to be built 50 m wide. If it is to be elliptical, what equation could describe it?

a) $50x^2 + 100y^2 = 1000$
b) $2x^2 + y^2 = 250$
c) $4x^2 + y^2 = 2500$
d) $x^2 + 2y^2 = 250$
e) $x^2 + 4y^2 = 10000$

1.23 The cylindrical coordinates (5, 30°, 12) are expressed in spherical coordinates as

a) (13, 30°, 67.4°)
b) (13, 30°, 22.6°)
c) (15, 52.6°, 22.6°)
d) (15, 52.6°, –22.6°)
e) (13, 30°, 67.40°)

1.24 The equation of a 4-m-radius sphere using cylindrical coordinates is

a) $x^2 + y^2 + z^2 = 16$
b) $r^2 = 16$
c) $r^2 + z^2 = 16$
d) $x^2 + y^2 = 16$
e) $r^2 + y^2 = 16$

COMPLEX NUMBERS

1.25 Divide $3 - i$ by $1 + i$.

a) $1 - 2i$ b) $1 + 2i$ c) $2 - i$ d) $2 + i$ e) $2 + 2i$

1.26 Find $(1 + i)^6$.

a) $1 + i$ b) $1 - i$ c) $8i$ d) $-8i$ e) $-1 - i$

1.27 Find the first root of $(1+i)^{1/5}$.

a) $0.168 + 1.06i$
b) $1.06 + 0.168i$
c) $1.06 - 0.168i$
d) $0.168 - 1.06i$
e) $-1.06 - 0.168i$

1.28 Express $(3+2i)\,e^{2it} + (3-2i)\,e^{-2it}$ in terms of trigonometric functions.

a) $3\cos 2t - 4\sin 2t$
b) $3\cos 2t - 2\sin 2t$
c) $6\cos 2t - 4\sin 2t$
d) $3\sin 2t + 2\sin 2t$
e) $6\cos 2t + 4\sin 2t$

1.29 Subtract $5e^{0.2i}$ from $6e^{2.3i}$. Express the answer in polar form.

a) $9.56\angle -21.4°$
b) $3.59\angle 104.5°$
c) $3.59\angle 255.5°$
d) $3.59\angle 201.4°$
e) $9.56\angle 158.6°$

LINEAR ALGEBRA

1.30 Find the value of the determinant $\begin{vmatrix} 3 & 2 & 1 \\ 0 & -1 & -1 \\ 2 & 0 & 2 \end{vmatrix}$.

a) 8 b) 4 c) 0 d) –4 e) –8

1.31 Evaluate the determinant $\begin{vmatrix} 1 & 0 & 1 & 1 \\ 2 & -1 & 0 & 1 \\ 0 & 0 & 2 & 0 \\ 3 & 2 & 1 & 1 \end{vmatrix}$.

a) 8 b) 4 c) 0 d) –4 e) –8

1.32 The cofactor A_{21} of the determinant of Prob. 1.30 is

a) –5 b) –4 c) 3 d) 4 e) 5

1.33 The cofactor A_{34} of the determinant of Prob. 1.31 is

a) 4 b) 6 c) 0 d) –4 e) –6

1.34 Find the adjoint matrix of $\begin{vmatrix} 1 & -4 \\ 0 & 2 \end{vmatrix}$.

a) $\begin{bmatrix} 4 & 2 \\ 0 & 1 \end{bmatrix}$ b) $\begin{bmatrix} 1 & 0 \\ 4 & 2 \end{bmatrix}$ c) $\begin{bmatrix} 2 & 4 \\ 1 & 0 \end{bmatrix}$ d) $\begin{bmatrix} 2 & 4 \\ 0 & 1 \end{bmatrix}$ e) $\begin{bmatrix} 1 & 4 \\ 0 & 2 \end{bmatrix}$

1.35 The inverse matrix of $\begin{bmatrix} 2 & 3 \\ 1 & 1 \end{bmatrix}$ is

a) $\begin{bmatrix} -1 & 3 \\ 1 & -2 \end{bmatrix}$ b) $\begin{bmatrix} 1 & -1 \\ -3 & 2 \end{bmatrix}$ c) $\begin{bmatrix} -1 & 1 \\ -3 & 2 \end{bmatrix}$ d) $\begin{bmatrix} -2 & 3 \\ 1 & -1 \end{bmatrix}$ e) $\begin{bmatrix} 2 & 3 \\ -1 & -1 \end{bmatrix}$

1.36 Calculate $\begin{bmatrix} 2 & -1 \\ 3 & 2 \end{bmatrix} \begin{bmatrix} 2 \\ 1 \end{bmatrix}$.

a) $\begin{bmatrix} 8 \\ 3 \end{bmatrix}$ b) $\begin{bmatrix} 3 \\ 8 \end{bmatrix}$ c) $\begin{bmatrix} -3 \\ -8 \end{bmatrix}$ d) [3,8] e) [8,3]

1.37 Determine $\begin{bmatrix} 1 & 2 \\ 2 & 1 \end{bmatrix} \begin{bmatrix} -1 & 0 \\ 1 & 2 \end{bmatrix}$.

a) $\begin{bmatrix} 1 & 4 \\ -1 & 2 \end{bmatrix}$ b) $\begin{bmatrix} 1 & -1 \\ 4 & 2 \end{bmatrix}$ c) $\begin{bmatrix} 1 \\ -1 \end{bmatrix}$ d) $\begin{bmatrix} 4 \\ 2 \end{bmatrix}$ e) [1,4]

1.38 Solve for $[x_i]$.

$$\begin{aligned} 3x_1 + 2x_2 \quad &= -2 \\ x_1 - \ x_2 + \ x_3 &= \ 0 \\ 4x_1 \qquad + 2x_3 &= \ 4 \end{aligned}$$

a) $\begin{bmatrix} 2 \\ 4 \\ -6 \end{bmatrix}$ b) $\begin{bmatrix} -2 \\ 4 \\ 12 \end{bmatrix}$ c) $\begin{bmatrix} 2 \\ 8 \\ 4 \end{bmatrix}$ d) $\begin{bmatrix} -6 \\ 8 \\ 14 \end{bmatrix}$ e) $\begin{bmatrix} 6 \\ 4 \\ 3 \end{bmatrix}$

CALCULUS

1.39 The slope of the curve $y = 2x^3 - 3x$ at $x = 1$ is

a) 3 b) 5 c) 6 d) 8 e) 9

1.40 If $y = \ln x + e^x \sin x$, find dy/dx at $x = 1$.

a) 1.23 b) 3.68 c) 4.76 d) 6.12 e) 8.35

1.41 At what value of x does a maximum of $y = x^3 - 3x$ occur?

a) 2 b) 1 c) 0 d) –1 e) –2

1.42 Where does an inflection point occur for $y = x^3 - 3x$?

a) 2 b) 1 c) 0 d) –1 e) –2

1.43 Evaluate $\lim\limits_{x \to \infty} \dfrac{2x^2 - x}{x^2 + x}$.

a) 2 b) 1 c) 0 d) –1 e) –2

1.44 If a quantity η and its derivatives η' and η'' are known at a point, its approximate value at a small distance h is

a) $\eta + h^2\eta''/2$ b) $\eta + h\eta/2 + h^2\eta''$ c) $\eta + h\eta' + h^2\eta''/2$

d) $\eta + h\eta' + h^2\eta''$ e) $\eta + h\eta'$

1.45 Find an approximation to $e^x \sin x$ for small x.

a) $x - x^2 + x^3$ b) $x + x^2 + x^3/3$ c) $x - x^2/2 + x^3/6$

d) $x + x^2 - x^3/6$ e) $x + x^2 + x^3/2$

1.46 Find the area between the y-axis and $y = x^2$ from $y = 4$ to $y = 9$.

a) 29/3 b) 32/3 c) 34/3 d) 38/3 e) 43/3

1.47 The area contained between $4x = y^2$ and $4y = x^2$ is

a) 10/3 b) 11/3 c) 13/3 d) 14/3 e) 16/3

1.48 Rotate the shaded area of Example 1.24 about the x-axis. What volume is formed?

a) 4π b) 6π c) 8π d) 10π e) 12π

1.49 Evaluate $\int_0^2 \left(e^x + \sin x\right) dx$.

a) 7.81 b) 6.21 c) 5.92 d) 5.61 e) 4.21

1.50 Evaluate $2\int_0^1 e^x \sin x \, dx$.

a) 1.82 b) 1.94 c) 2.05 d) 2.16 e) 2.22

1.51 Derive an expression $\int x \cos x \, dx$.

a) $x \cos x - \sin x + C$ b) $-x \cos x + \sin x + C$ c) $x \sin x - \cos x + C$
d) $x \cos x + \sin x + C$ e) $x \sin x + \cos x + C$

DIFFERENTIAL EQUATIONS

1.52 The differential equation $y'' + x^2 y' + y + 2 = 0$ is

a) linear and homogeneous.
b) linear and nonhomogeneous.
c) nonlinear and homogeneous.
d) nonlinear and nonhomogeneous.
e) not a differential equation.

1.53 Given: $y' + 2xy = 0$, $y(0) = 2$. Find: $y(2)$.

a) 0.0366 b) 0.127 c) 0.936 d) 2.36 e) 27.3

1.54 Given: $y' + 2x = 0$, $y(0) = 1$. Find: $y(10)$.

a) –100 b) –99 c) –91 d) –86 e) –54

1.55 A spring-mass system is represented by $2\ddot{y} + \dot{y} + 50y = 0$. What frequency, in hertz, is contained in the solution?

a) 0.79 b) 1.56 c) 2.18 d) 3.76 e) 4.99

1.56 Find the solution to $\ddot{y} + 16y = 0$.

a) $C_1 \cos 4t + C_2 \sin 4t$
b) $C_1 e^{4t} + C_2 e^{-4t}$
c) $C_1 e^{4t} + C_2 t e^{4t}$
d) $C_1 \cos 4t + C_2 t \cos 4t$
e) $C_1 \sin 4t + C_2 t \sin 4t$

1.57 Find the solution to $\ddot{y} + 8\dot{y} + 16y = 0$.

a) $C_1 \cos 4t + C_2 \sin 4t$
b) $C_1 t \cos 4t + C_2 \cos 4t$
c) $C_1 \sin 4t + C_2 t \sin 4t$
d) $C_1 e^{4t} + C_2 t e^{4t}$
e) $C_1 e^{-4t} + C_2 t e^{-4t}$

1.58 Solve the equation $\ddot{y} - 5\dot{y} + 6y = 4e^t$.

a) $C_1 e^{2t} + C_2 e^{3t}$
b) $C_1 e^{-2t} + C_2 e^{-3t} + 2e^t$
c) $C_1 e^{2t} + C_2 t e^{3t} + 2e^t$
d) $C_1 e^{2t} + C_2 e^{3t} + 2e^t$
e) $C_1 e^{2t} + C_2 e^{3t} + C_3 e^t$

1.59 Solve $\ddot{y}+16y=8\sin 4t$.

a) $c_1\sin 4t+c_2\cos 4t-t\cos 4t$
b) $c_1\sin 4t+c_2\cos 4t+t\cos 4t$
c) $c_1\sin 4t+c_2\cos 4t-\cos 4t$
d) $c_1\sin 4t+c_2\cos 4t+\cos 4t$
e) $c_1\sin 4t+c_2\cos 4t-\sin 4t$

1.60 Solve $\ddot{y}+5\dot{y}+6y=0$ if $y(0)=2$ and $\dot{y}(0)=0$. Use Laplace transforms.

a) $2e^{-3t}-3e^{-2t}$
b) $2e^{3t}-3e^{2t}$
c) $4e^{-3t}-6e^{-2t}$
d) $-4e^{-3t}+6e^{-2t}$
e) $4e^{3t}-6e^{2t}$

1.61 Solve $\ddot{y}+16y=120te^{-t}$ if $y(0)=0$ and $\dot{y}(0)=12$.

a) $8te^{-t}+\sin 4t$
b) $4te^{t}+2\sin 4t$
c) $6te^{-t}+6\sin t$
d) $6te^{t}+6\sin t$
e) $12te^{-t}$

VECTOR ANALYSIS

1.62 Given: $\mathbf{A}=3\mathbf{i}-6\mathbf{j}+2\mathbf{k}$, $\mathbf{B}=10\mathbf{i}+4\mathbf{j}-6\mathbf{k}$. Find: $\mathbf{A}\cdot\mathbf{B}$.

a) $-6\mathbf{i}$ b) 6 c) −6 d) $30\mathbf{i}-24\mathbf{j}-12\mathbf{k}$ e) $28\mathbf{i}+38\mathbf{j}+72\mathbf{k}$

1.63 Given: $\mathbf{A}=2\mathbf{i}-5\mathbf{k}$, $\mathbf{B}=\mathbf{j}$. Find: $\mathbf{A}\times\mathbf{B}$.

a) 0 b) $5\mathbf{i}+2\mathbf{k}$ c) −3 d) $-3\mathbf{k}$ e) 7

1.64 Find the projection of $\mathbf{A}$ in the direction of $\mathbf{B}$ if $\mathbf{A}=14\mathbf{i}-7\mathbf{j}$ and $\mathbf{B}=6\mathbf{i}+3\mathbf{j}-2\mathbf{k}$.

a) −9 b) $12\mathbf{i}-3\mathbf{j}$ c) 0 d) 15 e) 9

1.65 The equation of a plane perpendicular to and passing through the end of the vector $\mathbf{A}=2\mathbf{i}-4\mathbf{j}+6\mathbf{k}$ is given by $2x-4y+6z=k$ where k is

a) 56 b) 24 c) 0 d) −8 e) −16

1.66 Estimate the area of the parallelogram with sides $\mathbf{A}=2\mathbf{i}+3\mathbf{j}$ and $\mathbf{B}=4\mathbf{i}-6\mathbf{j}+5\mathbf{k}$.

a) 32 b) 30 c) 26 d) 20 e) 10

1.67 Find a unit vector normal to the surface $x^2+3y^2-3z=12$ at the point (1,1,0).

a) $\frac{\mathbf{i}+\mathbf{j}}{\sqrt{2}}$ b) $\frac{2\mathbf{i}-3\mathbf{j}}{\sqrt{13}}$ c) $\frac{\mathbf{i}+3\mathbf{j}-3\mathbf{k}}{\sqrt{19}}$ d) $\frac{\mathbf{i}+3\mathbf{j}}{\sqrt{10}}$ e) $\frac{2\mathbf{i}+6\mathbf{j}-3\mathbf{k}}{7}$

1.68 The divergence of $\mathbf{u}=x^2\mathbf{i}+y^2\mathbf{j}+z^2\mathbf{k}$ at the point (1,1,1) is

a) $2\mathbf{i}+2\mathbf{j}+2\mathbf{k}$ b) $\mathbf{i}+\mathbf{j}+\mathbf{k}$ c) 3 d) 6 e) 0

1.69 The curl of $\mathbf{u} = x^2\mathbf{i} + y^2\mathbf{j} + z^2\mathbf{k}$ at the point (1,1,1) is

a) $2\mathbf{i}+2\mathbf{j}+2\mathbf{k}$ b) $\mathbf{i}+\mathbf{j}+\mathbf{k}$ c) 3 d) 6 e) 0

1.70 The vector function $\mathbf{u} = yz\mathbf{i} + xz\mathbf{j} + xy\mathbf{k}$ is

a) conservative and solenoidal
b) conservative only
c) solenoidal only
d) neither conservative nor solenoidal
e) undefined at the origin

PROBABILITY AND STATISTICS

1.71 You reach into a jelly bean bag and grab one bean. If the bag contains 30 red, 25 orange, 15 pink, 10 green, and 5 black beans, the probability that you will get a black bean or a red bean is nearest to

a) 0.6 b) 0.5 c) 0.4 d) 0.3 e) 0.2

1.72 Two jelly bean bags are identical to that of Prob. 1.69. The probability of selecting a black bean from the first bag and a red bean from the second bag is nearest to

a) 1/100 b) 2/100 c) 3/100 d) 4/100 e) 5/100

1.73 From the bag of Prob. 1.69, the probability of selecting 5 beans, the first three of which are red and the next two of which are orange is nearest to

a) 3/1000 b) 4/1000 c) 5/1000 d) 6/1000 e) 7/1000

1.74 Two bags each contain 2 black balls, 1 white ball, and 1 red ball. What is the probability of selecting the white ball from the first bag or the red ball from the second bag?

a) 1/2 b) 7/16 c) 1/4 d) 3/8 e) 9/16

1.75 A professor gives the following scores to her students. What is the mode?

frequency	1	3	6	11	13	10	2
score	35	45	55	65	75	85	95

a) 65 b) 75 c) 85 d) 11 e) 13

1.76 For the data of Prob. 1.73, what is the arithmetic mean?

a) 68.5 b) 68.9 c) 69.3 d) 70.2 e) 73.1

1.77 Calculate the sample standard deviation for the data of Prob. 1.73.

a) 9.27 b) 10.11 c) 11.56 d) 13.42 e) 13.87

Solutions to Practice Problems

1.1 c) $100 = 10e^{2t}$. $\therefore \ln e^{2t} = \ln 10$. $\therefore 2t = 2.303$ $\therefore t = 1.151$

1.2 b) $\ln x = 3.2$. $\therefore e^{3.2} = x$. $\therefore x = 24.53$

1.3 c) $\ln_5 x = -1.8$. $\therefore 5^{-1.8} = x$. $\therefore x = 0.0552$

1.4 a) $x = \dfrac{-(-2) \pm \sqrt{2^2 - 4(3)(-2)}}{3 \cdot 2} = 1.215$

1.5 e) $(4+x)^{1/2} = 4^{1/2} + \dfrac{1}{2}4^{-1/2}x + \dfrac{\frac{1}{2}\left(1-\frac{1}{2}\right)}{2}4^{-3/2}x^2 + \cdots \quad = 2 + x/4 - x^2/64 + \cdots$

1.6 a) $\dfrac{2}{x\left(x^2-3x+2\right)} = \dfrac{A_1}{x} + \dfrac{A_2}{x-2} + \dfrac{A_3}{x-1} = \dfrac{A_1\left(x^2-3x+2\right) + A_2\left(x^2-x\right) + A_3\left(x^2-2x\right)}{x\left(x^2-3x+2\right)}$

$$\therefore \quad \left.\begin{aligned} A_1 + A_2 + A_3 &= 0 \\ -3A_1 - A_2 - 2A_3 &= 0 \\ 2A_1 &= 2 \end{aligned}\right\} \quad \left.\begin{aligned} A_1 &= 1 \\ A_2 + A_3 &= -1 \\ -A_2 - 2A_3 &= 3 \end{aligned}\right\} \quad \begin{aligned} A_3 &= -2 \\ A_2 &= 1 \end{aligned}$$

1.7 b) $\dfrac{4}{x^2\left(x^2-4x+4\right)} = \dfrac{A_1}{x} + \dfrac{A_2}{x^2} + \dfrac{A_3}{x-2} + \dfrac{A_4}{(x-2)^2}$

$$= \frac{A_1\left(x^3-4x^2+4x\right) + A_2\left(x^2-4x+4\right) + A_3\left(x^3-2x^2\right) + A_4x^2}{x^2(x-2)^2}$$

$$\therefore \quad \left.\begin{aligned} A_1 + A_3 &= 0 \\ -4A_1 + A_2 - 2A_3 + A_4 &= 0 \\ 4A_1 - 4A_2 &= 0 \\ 4A_2 &= 4 \end{aligned}\right\} \quad \begin{aligned} A_2 &= 1 \\ A_1 &= 1 \\ A_3 &= -1 \\ A_4 &= 1 \end{aligned}$$

1.8 d) at $t = 0$, population $= A$. $\therefore 2A = Ae^{0.4t}$. $\ln 2 = 0.4t$. $\therefore t = 1.733$

1.9 a) $\sin\theta = 0.7$ $\therefore \theta = 44.43°$. $\tan 44.43° = 0.980$

1.10 e) $\tan\theta = 5/7$. $\therefore \theta = 35.54°$

1.11 c) $\tan\theta \sec\theta\left(1-\sin^2\theta\right)/\cos\theta = \tan\theta \dfrac{1}{\cos\theta}\cos^2\theta \dfrac{1}{\cos\theta} = \tan\theta$

1.12 d) $3^2 = 4^2 + 2^2 - 2 \cdot 2 \cdot 4\cos\theta$. $\therefore \cos\theta = 0.6875$. $\theta = 46.6°$. $\therefore \text{rad} = 0.813$

1.13 a) $L^2 = 850^2 + 732^2 - 2 \cdot 850 \cdot 732 \cos 154°$. $\therefore L = 1542$ m

1.14 e) $\cos 2\theta = \cos^2\theta - \sin^2\theta = 1 - \sin^2\theta - \sin^2\theta = 1 - 2\sin^2\theta$. $\therefore 2\sin^2\theta = 1 - \cos 2\theta$

1.15 b) $\theta = \pi(n-2)/n = \pi(8-2)/8$ radians. $\frac{6\pi}{8} \times \frac{180}{\pi} = 135°$

1.16 a) $\text{Area} = \text{Area}_{\text{top}} + \text{Area}_{\text{sides}} = \pi R^2 + \pi DL$

$= \pi \times 7.5^2 + \pi \times 15 \times 10 = 648\ \text{m}^2. \quad 648 \div 10 \approx 65$

1.17 a) $y = mx + b. \quad y = -2x + b. \quad 0 = -2(2) + b. \quad \therefore b = 4. \quad \therefore y = -2x + 4$

1.18 b) $y = mx + b. \quad 0 = 4m + b. \quad -6 = b. \quad \therefore m = 3/2. \quad \therefore y = 3x/2 - 6 \quad \text{or} \quad 2y = 3x - 12$

1.19 d) $3x - 4y - 3 = 0. \quad A = 3, B = -4. \quad d = \frac{|3 \times 6 - 4 \times 8 - 3|}{\sqrt{3^2 + (-4)^2}} = 3.4$

1.20 c) $B^2 - AC = 2^2 - 1 \times 4 = 0. \quad \therefore$ parabola

1.21 b) $xy = \pm k^2 = 4$

1.22 c) $\frac{x^2}{25^2} + \frac{y^2}{50^2} = 1. \quad \therefore 4x^2 + y^2 = 2500$

1.23 b) $x = r\cos\theta = 5 \times 0.866 = 4.33. \quad y = r\sin\theta = 5 \times 0.5 = 2.5$
Spherical coordinates:

$$r = \sqrt{4.33^2 + 2.5^2 + 12^2} = 13. \quad \phi = \cos^{-1}\frac{z}{r} = \cos^{-1}\frac{12}{13} = 22.6°.$$

$$\theta = \tan^{-1}\frac{y}{x} = \tan^{-1}\frac{2.5}{4.33}. \quad \therefore \theta = 30°.$$

1.24 c) a) is rectangular coordinates. b) is spherical coordinates.

1.25 a) $\frac{3-i}{1+i} = \frac{3-i}{1+i}\frac{1-i}{1-i} = \frac{3-1-4i}{1-(-1)} = \frac{1}{2}(2-4i) = 1-2i$

1.26 d) $1 + i = re^{i\theta}. \quad r = \sqrt{1^2 + 1^2} = \sqrt{2}. \quad \theta = \tan^{-1}\frac{1}{1} = \pi/4$ rad. $\quad \therefore 1 + i = \sqrt{2}e^{i\pi/4}.$

$\therefore (1+i)^6 = \left(\sqrt{2}\right)^6 e^{i\pi 3/2} = 8\left(\cos\frac{3\pi}{2} + i\sin\frac{3\pi}{2}\right) = -8i$. Note: angles are in radians.

1.27 b) $1 + i = \sqrt{2}e^{i\pi/4}. \quad \therefore (1+i)^{1/5} = 1.414^{1/5}e^{i\pi/20} = 1.072\left(\cos\frac{\pi}{20} + i\sin\frac{\pi}{20}\right)$

$= 1.06 + 0.168i$

1.28 c) $(3+2i)(\cos 2t + i\sin 2t) + (3-2i)(\cos 2t - i\sin 2t) = 6\cos 2t + 4i(i\sin 2t)$

$= 6\cos 2t - 4\sin 2t$

1.29 e) $6(\cos 2.3 + i\sin 2.3) - 5(\cos 0.2 + i\sin 0.2) = 6(-0.666 + 0.746i) - 5(0.98 + 0.199i)$

$= -8.90 + 3.48i$

$= 9.56\angle 158.6°$

1.30 e) $\begin{vmatrix} 3 & 2 & 1 \\ 0 & -1 & -1 \\ 2 & 0 & 2 \end{vmatrix} = -6 - 4 + 2 = -8$

1.31 a) $\begin{vmatrix} 1 & 0 & 1 & 1 \\ 2 & -1 & 0 & 1 \\ 0 & 0 & 2 & 0 \\ 3 & 2 & 1 & 1 \end{vmatrix} = 2\begin{vmatrix} 1 & 0 & 1 \\ 2 & -1 & 1 \\ 3 & 2 & 1 \end{vmatrix} = 2(-1+4+3-2) = 8$

Note: Expand using the third row.

1.32 b) $(-1)^3\begin{vmatrix} 2 & 1 \\ 0 & 2 \end{vmatrix} = -4$

1.33 e) $(-1)^7\begin{vmatrix} 1 & 0 & 1 \\ 2 & -1 & 0 \\ 3 & 2 & 1 \end{vmatrix} = -(-1+4+3) = -6$

1.34 d) $\left[a_{ij}\right]^+ = \begin{bmatrix} A_{11} & A_{21} \\ A_{12} & A_{22} \end{bmatrix} = \begin{bmatrix} 2 & 4 \\ 0 & 1 \end{bmatrix}$

1.35 a) $\left[a_{ij}\right]^{-1} = \dfrac{\left[a_{ij}\right]^+}{\left|a_{ij}\right|} = \dfrac{\begin{bmatrix} 1 & -3 \\ -1 & 2 \end{bmatrix}}{-1} = \begin{bmatrix} -1 & 3 \\ 1 & -2 \end{bmatrix}$

1.36 b) $\begin{bmatrix} 2 & -1 \\ 3 & 2 \end{bmatrix}\begin{bmatrix} 2 \\ 1 \end{bmatrix} = \begin{bmatrix} 4-1 \\ 6+2 \end{bmatrix} = \begin{bmatrix} 3 \\ 8 \end{bmatrix}$

1.37 a) $\begin{bmatrix} 1 & 2 \\ 2 & 1 \end{bmatrix}\begin{bmatrix} -1 & 0 \\ 1 & 2 \end{bmatrix} = \begin{bmatrix} 1 & 4 \\ -1 & 2 \end{bmatrix}$

1.38 d) $\left[a_{ij}\right] = \begin{bmatrix} 3 & 2 & 0 \\ 1 & -1 & 1 \\ 4 & 0 & 2 \end{bmatrix}.\quad \left[a_{ij}\right]^+ = \begin{bmatrix} -2 & -4 & 2 \\ 2 & 6 & -3 \\ 4 & 8 & -5 \end{bmatrix}.\quad \left|a_{ij}\right| = -2.$

$\therefore \left[a_{ij}\right]^{-1} = \dfrac{\left[a_{ij}\right]^+}{\left|a_{ij}\right|} = \begin{bmatrix} 1 & 2 & -1 \\ -1 & -3 & 3/2 \\ -2 & -4 & 5/2 \end{bmatrix}.\quad \left[x_j\right] = \left[a_{ij}\right]^{-1}\left[r_i\right] = \left[a_{ij}\right]^{-1}\begin{bmatrix} -2 \\ 0 \\ 4 \end{bmatrix} = \begin{bmatrix} -6 \\ 8 \\ 14 \end{bmatrix}$

1.39 a) $\dfrac{dy}{dx} = 6x^2 - 3 = 6(1)^2 - 3 = 3$

1.40 c) $\dfrac{dy}{dx} = \dfrac{1}{x} + e^x\cos x + e^x\sin x = 1 + e\cos 1 + e\sin 1 = 4.76.\qquad (\cos 1 = \cos 57.3°)$

1.41 d) $\dfrac{dy}{dx} = 3x^2 - 3 = 0.\quad \therefore x^2 = 1.\quad \therefore x = \pm 1.\qquad \dfrac{d^2y}{dx^2} = 6x.\quad \therefore x = -1$ is a maximum.

1.42 c) $y' = 3x^2 - 3.\quad y'' = 6x.\quad \therefore x = 0$ is inflection.

1.43 a) $\lim\limits_{x\to\infty}\dfrac{2x^2 - x}{x^2 + x} = \lim\limits_{x\to\infty}\dfrac{4x-1}{2x+1} = \lim\limits_{x\to\infty}\dfrac{4}{2} = 2$

1.44 c) $\eta(x+h)=\eta+h\eta'+\frac{h^2}{2}\eta''$

1.45 b) $e^x \sin x=\left(1+x+\frac{x^2}{2}\right)\left(x-\frac{x^3}{6}\right)=x+x^2+\frac{x^3}{2}-\frac{x^3}{6}=x+x^2+x^3/3$

1.46 d) Area $=\int_4^9 x\,dy=\int_4^9 y^{1/2}dy=\frac{2}{3}(27-8)=12\frac{2}{3}$

$y=9$, $y=4$, x

1.47 e) Area $=\int_0^4 (x_2-x_1)dy=\int_0^4\left(2y^{1/2}-\frac{y^2}{4}\right)dy$

$(4, 4)$

$=2\times\frac{2}{3}\times 8-\frac{1}{12}\times 64=16/3$

1.48 c) $V=\int_0^2 2\pi y\,x\,dy=2\pi\int_0^2 y^3 dy=2\pi\times\frac{2^4}{4}=8\pi$

1.49 a) $\int_0^2\left(e^x+\sin x\right)dx=e^x-\cos x\Big|_0^2=e^2-1-\cos 2+1=7.81$

1.50 a) $\int_0^1 e^x \sin x\,dx=e^x \sin x\Big|_0^1-\int_0^1 e^x \cos x\,dx$

$$\underbrace{\begin{matrix} u=\sin x & dv=e^x dx \\ du=\cos x dx & v=e^x \end{matrix}}_{\text{1st integral}}$$

$$\underbrace{\begin{matrix} u=\cos x & dv=e^x dx \\ v=e^x & du=-\sin x dx \end{matrix}}_{\text{2nd integral}}$$

$\therefore \int_0^1 e^x \sin x\,dx=e\sin 1-\left[e^x \cos x\Big|_0^1+\int_0^1 e^x \sin x\,dx\right]$

$\therefore 2\int_0^1 e^x \sin x\,dx=e\sin 1-e\cos 1+1\times 1=1.819.\quad \therefore \int_0^1 e^x \sin x\,dx=0.909$

1.51 e) $\int x\cos x\,dx=x\sin x-\int \sin x\,dx=x\sin x+\cos x+C$

$$\underbrace{\begin{matrix} u=x & dv=\cos x dx \\ du=dx & v=\sin x \end{matrix}}_{\text{1st integral}}$$

1.52 b) linear and nonhomogeneous. The term (+2) makes it nonhomogeneous. x^2y' is linear.

1.53 a) $\frac{dy}{dx}=-2xy.\quad \frac{dy}{y}=-2xdx.\quad \therefore \ln y=-x^2+C.\quad \ln 2=0+C.\quad \therefore C=\ln 2.$

$y(2)=\exp\left(-2^2+\ln 2\right)=0.0366$

1.54 b) $\frac{dy}{dx} = -2x.$ $dy = -2xdx.$ $\therefore y = -x^2 + C.$ $1 = 0 + C.$ $\therefore C = 1.$

$y(10) = -10^2 + 1 = -99$

1.55 a) $2m^2 + m + 50 = 0.$ $\therefore m = \frac{-1 \pm \sqrt{1-400}}{4} = -\frac{1}{4} \pm 4.99i.$

$\therefore y(t) = e^{-t/4}(A\cos 4.99t + B\sin 4.99t).$

$\therefore \omega = 4.99 \text{ rad/s}.$ $\therefore f = \frac{\omega}{2\pi} = \frac{4.99}{2\pi} = 0.794 \text{ hertz}$

1.56 a) $m^2 + 16 = 0.$ $\therefore m = \pm 4i.$ $\therefore y(t) = C_1 \cos 4t + C_2 \sin 4t$

1.57 e) $m^2 + 8m + 16 = 0.$ $(m+4)^2 = 0.$ $m = -4, -4.$ $\therefore y(t) = C_1 e^{-4t} + C_2 t e^{-4t}$

1.58 d) $m^2 - 5m + 6 = 0.$ $m = \frac{5 \pm \sqrt{25-24}}{2} = 3, 2.$ $\therefore y_h = C_1 e^{3t} + C_2 e^{2t}.$

Assume $y_p = Ae^t.$ Then $Ae^t - 5Ae^t + 6Ae^t = 4e^t.$ $\therefore A = 2.$

1.59 a) homogeneous: $m^2 + 16 = 0.$ $\therefore m = \pm 4i.$ $\therefore y_h(t) = C_1 \sin 4t + C_2 \cos 4t$

particular: $y_p = At\cos 4t.$ (This is resonance.) $\dot{y}_p = A\cos 4t - 4At\sin 4t$

$\therefore -4A\sin 4t - 4A\sin 4t - 16At\cos 4t + 16At\cos 4t = 8\sin 4t$

$\therefore -8A = 8.$ $A = -1.$ $\therefore y = y_h + y_p = C_1 \sin 4t + C_2 \cos 4t - t\cos 4t$

1.60 d) $s^2Y - 2s + 5sY - 10 + 6Y = 0.$ $Y = \frac{2s+10}{s^2+5s+6} = \frac{-4}{s+3} + \frac{6}{s+2}.$

$\therefore y(t) = -4e^{-3t} + 6e^{-2t}$

1.61 a) $s^2Y - 12 + 16Y = \frac{120}{(s+1)^2}.$ $Y = \frac{12}{s^2+16} + \frac{120}{(s+1)^2(s^2+16)}$

$$= \frac{12}{s^2+16} + \frac{8}{(s+1)^2} - \frac{8}{s^2+16}.$$

$\therefore y(t) = \sin 4t - 8te^{-t}$

1.62 c) $\mathbf{A} \cdot \mathbf{B} = 3 \cdot 10 + (-6) \cdot 4 + 2(-6) = -6$

1.63 b) $\mathbf{A} \times \mathbf{B} = (2\mathbf{i} - 5\mathbf{k}) \times \mathbf{j} = 2\mathbf{i} \times \mathbf{j} - 5\mathbf{k} \times \mathbf{j} = 2\mathbf{k} - 5(-\mathbf{i}) = 5\mathbf{i} + 2\mathbf{k}$

1.64 e) $\mathbf{i}_\mathbf{B} = (6\mathbf{i} + 3\mathbf{j} - 2\mathbf{k}) \big/ \sqrt{6^2 + 3^2 + 2^2} = \frac{1}{7}(6\mathbf{i} + 3\mathbf{j} - 2\mathbf{k})$

$\mathbf{A} \cdot \mathbf{i}_\mathbf{B} = [14 \cdot 6 - 7(3)]/7 = 12 - 3 = 9$

1.65 a) $[(x\mathbf{i} + y\mathbf{j} + z\mathbf{k}) - (2\mathbf{i} - 4\mathbf{j} + 6\mathbf{k})] \cdot (2\mathbf{i} - 4\mathbf{j} + 6\mathbf{k}) = 0.$

$2(x-2) - 4(y+4) + 6(z-6) = 0.$ $\therefore 2x - 4y + 6z = 56$

1.66 b) $|\mathbf{A} \times \mathbf{B}| = |15\mathbf{i} - 10\mathbf{j} - 24\mathbf{k}| = \sqrt{901} \cong 30$

1.67 e) $\nabla\phi = 2x\mathbf{i} + 6y\mathbf{j} - 3\mathbf{k} = 2\mathbf{i} + 6\mathbf{j} - 3\mathbf{k}$. $|\nabla\phi| = 7$. $\therefore\ \mathbf{i}_n = \nabla\phi/|\nabla\phi| = (2\mathbf{i} + 6\mathbf{j} - 3\mathbf{k})/7$

1.68 d) $\nabla \cdot u = 2x + 2y + 2z = 2 + 2 + 2 = 6$

1.69 e) $\nabla \times \mathbf{u} = (0-0)\mathbf{i} + (0-0)\mathbf{j} + (0-0)\mathbf{k} = 0$

1.70 a) $\nabla \cdot \mathbf{u} = 0 + 0 + 0 = 0$. $\therefore$ solenoidal. $\nabla \times \mathbf{u} = (x-x)\mathbf{i} + (y-y)\mathbf{j} + (z-z)\mathbf{k} = 0$.
$\therefore$ conservative and solenoidal.

1.71 c) $P(A_1 \text{ or } A_2) = P(A_1) + P(A_2) = \frac{5}{85} + \frac{30}{85} = 0.41$

1.72 b) $P(A_1 \text{ and } B_1) = P(A_1)P(B_1) = \frac{5}{85} \times \frac{30}{85} = 0.0208$ or $2/96.3$

1.73 b) $P = \frac{30}{85} \times \frac{29}{84} \times \frac{28}{83} \times \frac{25}{82} \times \frac{24}{81} = 0.0037$ or $4/1077$

1.74 b) $P(A \text{ or } B) = P(A) + P(B) - P(A)P(B) = \frac{1}{4} + \frac{1}{4} - \frac{1}{4} \times \frac{1}{4} = 7/16$

1.75 b) The mode is the observation that occurs most frequently. It is 75.

1.76 d) $\bar{x} = \dfrac{35 + 3\times 45 + 6\times 55 + 11\times 65 + 13\times 75 + 10\times 85 + 2\times 95}{1+3+6+11+13+10+2} = \dfrac{3230}{46} = 70.2$

1.77 e) $\sigma =$

$$\sqrt{\frac{35^2 + 3\times 45^2 + 6\times 55^2 + 11\times 65^2 + 13\times 75^2 + 10\times 85^2 + 2\times 95^2 - 46\times 70.2^2}{45}}$$

$= 13.87$

Digital Computing

by Bernhard Weinberg

Computing may be divided into two categories, *analog* and *digital*. Analog computing deals primarily with the solution and simulation of differential equations and is not reviewed here. Digital computing has much more general application possibilities. Emphasis will be placed on the major aspects only; many additional details would be required for a discussion of computing systems.

2.1 Number Systems

A number may be represented as a polynomial in powers of its *base*, or *radix*. Most digital computers represent numbers internally in base 2 (binary) rather than base 10 (decimal). For clarification, consider the following example.

Example 2.1

Evaluate the base 2 number 1101.01 using base 10.

Solution. $1101.01_2 = 1*2^3 + 1*2^2 + 0*2^1 + 1*2^0 + 0*2^{-1} + 1*2^{-2} = 13.25_{10}$

Of necessity, all numbers in a digital computer are of finite length and hence, numerical precision must always be a consideration. It is convenient when handling long binary numbers to group the digits into groups of three or four. If grouped in threes, the numerical value for each group is between 0 and 7; if grouped in fours, the range is from 0 to 15. Thus, by grouping digits, a binary number may be represented either as a base 8 (octal) or base 16 (hexadecimal) number. In the case of hexadecimal numbers a convenient notation for digits larger than 9 is the use of the letters "A" through "F" for the digits equivalent to 10 through 15. Thus, the integer 28 is 34 base 8 or 1C in base 16; the integer 3AC base 16 is 940 base 10.

Negative integers may be represented as either 1's complement or 2's complement notation. The 1's complement negation of a binary integer requires that each digit be "flipped" independently; that is, a 0 becomes a 1 and vice versa. The left-most digit or bit of the number then indicates its algebraic sign, a positive number by 0 and a negative number by 1. This implies the integer 27 be represented by at least six bits as 011011. If interpreted as a signed integer, the 5-bit pattern 11011 is equivalent to −4 and not +27. The 1's complement representation of 0 is either 0..0 or 1..1 which are called plus zero and minus zero, respectively.

The 2's complement representation may be obtained from the 1's complement representation by adding 0..01 or by subtracting the number from a numeric zero. In 2's complement, zero is only represented by 0..0.

All arithmetic operations with signed numbers must be done with regard to the number of digits in the representation. If 1's complement is used, all carries (borrows) from the left-most position must be brought around and added (subtracted) into the right-most bit position. The end-around carry (borrow) is ignored with 2's complement arithmetic, as the following example illustrates.

Example 2.2

Subtract 5 from 7 using 1's complement and 2's complement.

Solution.

Conventional	1's Complement		2's Complement	
+7	000111		000111	
+ (−5)	+ 111010		+ 111011	
+2	1 000001		1 000010	ignore carry
	+ 1	end-around carry	000010	
	000010			

When negative integers are represented in complement form, the sign bit is treated as an ordinary numerical bit. The end-around carry (borrow) in 1's complement signals the need for the result correction resulting from the dual representation of zero. Numerical overflow during addition is signaled by two numbers of the same sign yielding a sum of the opposite sign. Numerical overflow during subtraction may occur only with oppositely signed numbers. It occurs when the sign of the difference is the opposite of the sign of the minuend (second number).

A collection of 8 bits, without regard to any numerical implications, is referred to as a *byte*. Normally, a character in a text requires one byte. A half-byte, or 4 bits, is referred to as a *nibble*. When dealing with large binary numbers, the base 10 value $1024=2^{10}$ is abbreviated by K. The value $1{,}048{,}576=2^{20}$ is abbreviated by M. Thus, a computer memory that has 64K bytes actually has 65,536 bytes of capacity.

Real numbers are represented by a signed integer exponent and signed mantissa similar to standard scientific notation. The representation format is computer dependent.

2.2 Combinational Logic

Boolean algebra deals with logical values, such as false or true, zero or one. The basic operations are "negation" or "NOT," usually indicated by an overscore; "union" or "OR," usually indicated by a plus sign; and "intersection" or "AND," usually indicated by a multiplication sign, a dot, or implied. These basic operations are supplemented by the "NAND" (Not AND), "NOR" (Not OR), and "XOR" (exclusive OR). These operations are defined in Fig. 2.1 with the standard schematic symbols shown. Other schematic symbols are also in common usage. All the operations may be extended to functions of more than two variables.

A given "black box" may not be classified as to its logical function without a knowledge of how its inputs and output(s) are mapped according to the Boolean values of "0" and "1." Consider a black box whose output is at 0 volts if one, two, or three of its three inputs are at 0 volts, and the output is at 5 volts only if all three inputs are at 5 volts. If 0 and 5 volts are equivalent to the Boolean 0 and 1, respectively, then this black box is a three-input AND gate. If, however, 0 and 5 volts are equivalent to the Boolean 1 and 0, respectively, then this black box is a three-input OR gate.

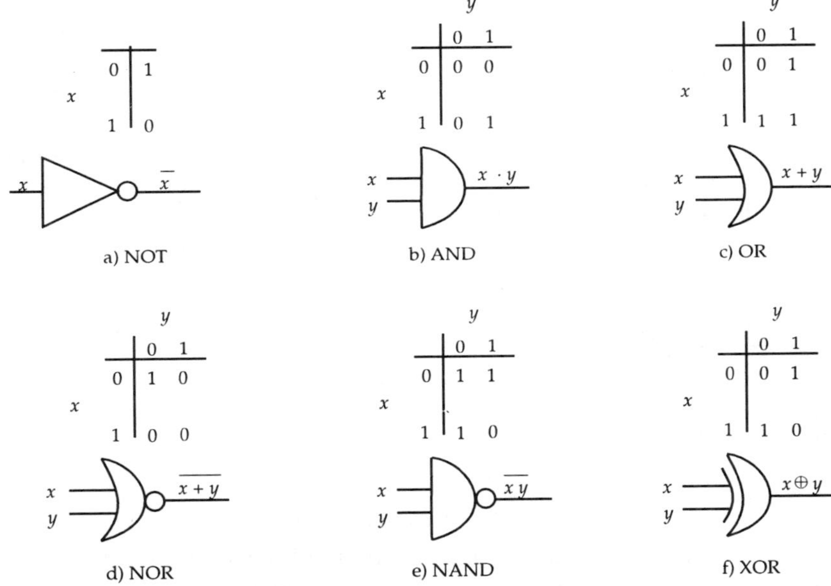

Figure 2.1 Schematic representations of the basic logic operations.

The Karnaugh map is a tool for simplifying the functions of two, three, or four variables. But beyond four variables it rapidly loses its usefulness because of geometric considerations. The map simplifies functions with respect to NOT, OR, and AND operations. Because of input-output loading limitations (fan-in, fan-out) on the electrical circuitry of the gates, the most economical implementation of a Boolean function is not necessarily the simplest. This is particularly true if NAND, NOR, or XOR gates are to be used.

The Karnaugh map displays all 2^N combinations of N variables labeled such that vertically or horizontally adjacent squares differ by only one variable change. The left and right cells are considered adjacent and the top and bottom cells are adjacent. After plotting the function as a '1' for those variable combinations where it is true and a '0' (sometimes omitted) where it is false, simplification may begin. Simplification consists of trying to surround the 1's of the function with the largest possible "circles" consisting of 1, 2, 4, 8, or 16 cells. The minimum number of "circles" are selected such that all the 1's of the function are covered. The larger a "circle" is, the fewer the variables required to describe it. The fewest number of "circles" means the fewest number of terms in the simplification.

There are times when it is known that certain combinations of the variables cannot exist from the physical consideration of the problem. The combinations are called *don't care* functions and may be used in simplifications. They are plotted on the Karnaugh map using an "X" rather than a "1." The name "don't care" arises because the use of an "X" is optional; they may be used as needed to make the largest possible circles and they do not have to be covered. Fig. 2.2 illustrates.

A function may also be simplified algebraically by the method of *prime implicants*. The details of this method will not be shown here, but it follows directly from the map method. The first step is to obtain the prime implicants. They are simply all the possible circles. The second step is to then obtain the minimal covering: select a minimum set of the largest prime implicants (the ones with the fewest variables) in order to cover the original function. Again, "don't care" functions may be used.

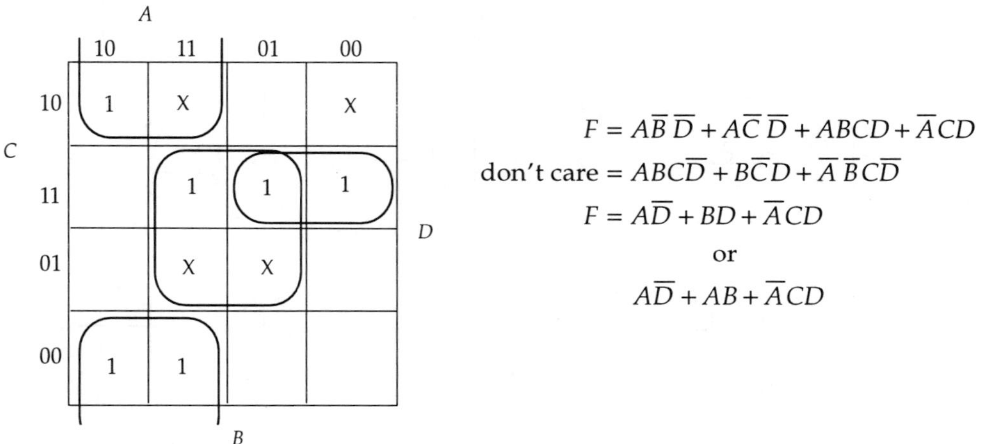

Figure 2.2 A Karnaugh map.

There are two canonical forms for representing Boolean functions. The first is a *sum of products* or *minterm*. The second is a *product of sums* or *maxterm*. Each minterm or maxterm contains all N variables of the function. The function itself may contain a maximum of 2^N min or max terms. Each square on a Karnaugh map represents a minterm.

A function may be negated using a Karnaugh map. The function is plotted as before but the 0's are covered instead. An example is shown below.

Example 2.3
Negate the function $A+\overline{B}C$ using a Karnaugh map.
Solution.

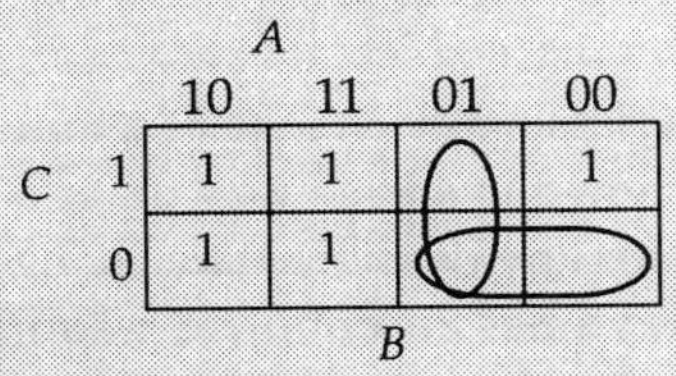

A

C	10	11	01	00
1	1	1		1
0	1	1		

B

$$F = A + \overline{B}C$$

$$\overline{F} = \overline{A}B + \overline{A}\,\overline{C}$$

A function may be negated algebraically with *DeMorgan's Law.* All variables are complemented, with AND and OR interchanged. The trick is to properly parenthesize the expression so the priority of AND over OR is explicit rather than implicit in all cases and then doing the interchanging and complementing. Some parentheses may be removed afterwards.
The dual of a function is obtained by interchanging the AND and OR operators. It is similar to negation but the variables are not complemented. Again, the proper parenthesization should be done beforehand.

Example 2.4
Negate the function $F = A + \overline{B}\,C$ with DeMorgan's Law.
Solution. We may write

$$F = A + \overline{B}\,C = A + \left(\overline{B} * C\right).$$

Upon complementing the variables and interchanging the AND-OR operators, the function becomes

$$\overline{F} = \overline{A} * \left(B + \overline{C}\right) = \overline{A}B + \overline{A}\;\overline{C}.$$

2.3 Sequential Logic

The output in a combinational or memoryless logic circuit is a function of the current input only. If the output is a function of both the current input and past inputs (or outputs), then the circuit is sequential or has memory in it.

The behavior of a sequential circuit is described by a *state table.* The state table describes the output(s) and the next state as functions of the input and present state. The *state* represents the memory. There may exist equivalent *machines* exhibiting identical behavior requiring a lesser number of states. If the outputs of a sequential circuit or machine are a function only of the state, the circuit is classified as a *Moore machine.* In another type called a *Mealy machine,* the outputs are a function of the state and the input. A state is represented by a state variable configuration. N state variables can implement up to 2^N states.

Sequential machines are either synchronous or asynchronous. *Synchronous* machines are defined only at discrete times controlled by an external clock. Their state variables must be implemented with flip-flops in order to hold the state variable values between clock pulses. The *asynchronous* machine is defined for all time and therefore does not need explicit memory for the state variables.

The asynchronous machine, although simpler in concept than a synchronous one, has two implementation restrictions necessary to ensure correct operation due to inherent circuit delays:

1. No more than one input variable may change at a time.
2. State variables must be assigned in such a way that no more than one state variable changes for any possible state changes.

Because of finite signal propagation times, it cannot be assumed that two variables which are supposed to change simultaneously will indeed do so or will always pass through the same intermediate configuration. The actual state transition may depend upon which path wins the "race." For example, the variable change $00 \rightarrow 11$ may actually be $00 \rightarrow 01 \rightarrow 11$ or $00 \rightarrow 10 \rightarrow 11$. Asynchronous machines will not be discussed further. There are no restrictions on the number of variable changes for synchronous machines. Either machine may have "don't care" entries.

Four types of flip-flops are in common usage. If X is the current flip-flop output, these four may be defined in either state table or equation form as shown in Fig. 2.3. The flip-flops also have a "clear" input, which when asserted, forces the output to 0. The general form of the state variable equation is

$$X_{next} = AX + B\overline{X} \tag{2.3.1}$$

where X is the state variable and A, B are not functions of X. For a given state variable application, the flip-flop input(s) must be found. This may be done by equating the desired flip-flop equation (Fig 2.3) to the state variable equation

D	X_{next}
0	0
1	1

$X_{next} = D$

***D* Flip-Flop**

T	X_{next}
0	X
1	$\overline{X}$

$X_{next} = \overline{T}X + T\overline{X}$

***T* Flip-Flop**

J K	X_{next}
0 0	X
0 1	0
1 0	1
1 1	$\overline{X}$

$X_{next} = \overline{K}X + J\overline{X}$

***J-K* Flip-Flop**

S R	X_{next}
0 0	X
0 1	0
1 0	1
1 1	undefined

$X_{next} = S + \overline{R}X$, where $SR = 0$

***S-R* Flip-Flop**

Figure 2.3 The four common flip-flops.

(Eq. 2.3.1) and finding solutions for the input(s). The general solution for each type are:

$D = AX + B\overline{X}$ — *D* Flip-Flop **(2.3.2)**

$T = \overline{A}X + B\overline{X}$ — *T* Flip-Flop **(2.3.3)**

$J = B \qquad K = \overline{A}$ — *J-K* Flip-Flop **(2.3.4)**

$S = B\overline{X},\ = B \qquad R = \overline{A}X,\ = \overline{A}$ if $\overline{A}B = 0$ — *S-R* Flip-Flop **(2.3.5)**

The next example illustrates the above.

Example 2.5

A synchronous Mealy machine with one input, one output, and six states as described below is to be realized. *A* is the initial state.

Present State	Input 0	Input 1	Present State *PQR*	Input $\bar{I}$	Input *I*
A	*C*/0	*C*/0	000	001/0	001/0
B	*A*/0	*A*/0	100	000/0	000/0
C	*D*/0	*E*/0	001	110/0	111/0
D	*B*/0	*F*/0	110	100/0	010/0
E	*F*/0	*F*/0	111	010/0	010/0
F	*A*/0	*A*/1	010	000/0	000/1
			011	xxx/x	xxx/x
			101	xxx/x	xxx/x

next state/output

Solution. This is a minimal state machine requiring at least three state variables. Although state variable assignment is arbitrary, the following rules of thumb seem to yield reduced cost logic:

1. Give adjacent assignments to a state and the state that follows it.
2. If two present states have the same next states, the present states should have adjacent assignments.
3. Use a minimum number of state variables.

These rules of thumb may be contradictory and compromises may have to be made in the assign-ment. Consider, for example, the following assignments:
A = 000, *B* = 100, *C* = 001, *D* = 110, *E* = 111, *F* = 010 a nd "don't care" = 101,011
Note that with these assignments, the initial state may be forced by clearing the state variable flip-flops. By naming the state variables *P*, *Q*, and *R*, and the input variable I, the state and output equations become

$$P_{next} = P\left(Q\overline{R}\,\overline{I}\right) + \overline{P}(R)$$

$$Q_{next} = Q(R + PI) + \overline{Q}(R)$$

$$R_{next} = R\left(\overline{P}\,I\right) + \overline{R}\left(\overline{P}\,\overline{Q}\right)$$

$$\text{output} = \overline{P}\,Q\overline{R}\,I$$

Note that the form of each state equation is the same as that of Eq. 2.3.1. Arbitrarily, let the *P* and *Q* state variables be implemented with *J*-*K* flip-flops and the *R* state variable be implemented by a *T* flip-flop. The flip-flop inputs become

$$J_P = R * \text{clock}, \qquad K_P = \left(\overline{Q} + R + I\right) * \text{clock}$$

$$J_Q = R * \text{clock}, \qquad K_Q = \overline{R}\left(\overline{P} + \overline{I}\right) * \text{clock}$$

$$T_R = \left(PR + \overline{I}R + \overline{P}\,\overline{R}\,\overline{Q}\right) * \text{clock}$$

Some flip-flops have a clock input built in and therefore do not require the explicit AND of the clock with all inputs. The resulting machine is shown as follows:

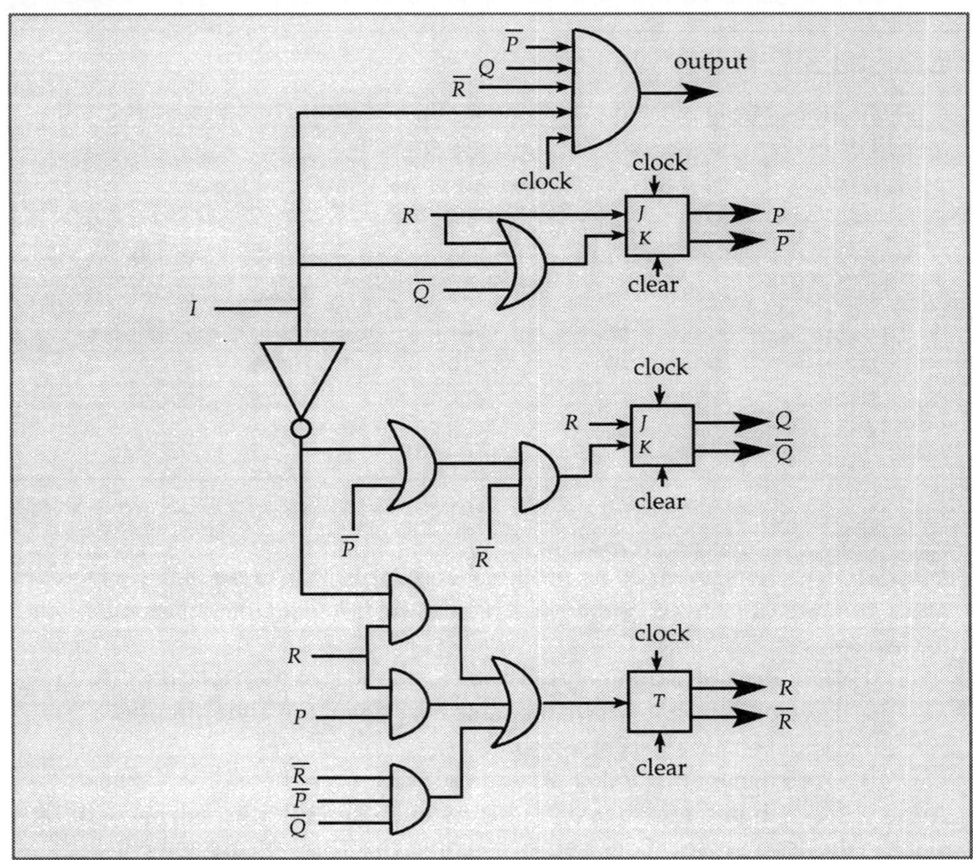

2.4 Operating Systems

Simply stated, an *operating system* is a program for interfacing between user and hardware. Its purpose is to provide an environment in which a user may execute programs conveniently and efficiently. The operating system may serve any number of users concurrently and independently.

A *batch* operating system never interacts between the user and the job once it is submitted. A measure of "goodness" is the average job turnaround time. A *time-sharing* or interactive system does allow a user to interact with the execution of the job. Interaction is accomplished by sharing the processor time and other resources by allocating each user a *time slice* in succession. The average response time to a user directive is the measure of goodness. A *real time* operating system is one that services external processes that have strict time constraints on response, such as for process control. Some interactive or real time operating systems process batch jobs in the *background* when there is no other *on line* or external activity. Most operating systems are *multiprogrammed* wherein several user jobs reside within the computer system at any one time.

A more precise definition of an operating system is that it is a collection of code transparently activated by user application software to provide common tasks such as file storage and character input-output. It is an interface between user application software, not user programs and hardware. The typical application software that comes with an operating system includes a "shell," text editor, language processor, library, and file system. A *shell* is the user interface to the operating system. It permits the user to execute programs and applications. The *text editor* allows the user to enter, modify, and correct programs and data. The *language processor* consists of the assemblers and compilers for translating the user program written in a "source" language into the machine's instructions. The *library,* also called the loader-linker, allows a user program to access utility programs. The *file system* permits the storage of user programs and data for future use. Files are stored on magnetic disks or magnetic tapes.

2.5 Programming

Assembly languages are a means of programming symbolically in machine language. Each line of code writing normally produces one machine instruction. Opposed to this are the high level languages such as FORTRAN or Pascal, wherein a single line of code produces many machine instructions. These so-called "high-level" languages are translated by a process known as compiling.

It is extremely difficult to discuss assembly languages abstractly because they are very dependent upon the architecture of the individual machine. Each line of assembly code has four fields, some of which are optional. The first field is called the *label* or *location.* This field is optional and if the line contains a label, it usually must start in the first column of the line. If the line has no label, column one of the line is left blank. Labels define symbols which usually represent memory location addresses. They are usually required to start with a letter. The second field is required and contains the operation. If the operation is to generate a machine instruction then the field contains the *op-code.* If the operation is a directive to the assembler then this field contains a *pseudo-op.* Pseudo-ops are also used to generate data items. The third field is called the *operand* or *address.* It generally contains symbolic register names or memory locations for op-codes. For data generation pseudo-ops, this field contains the data; there are some operations which do not use this field. The last field is for comments. Fields are separated by one or more blanks or are assumed to start at certain columns in the case of fields following optional or not required fields. Most assemblers treat an entire line as a comment if it starts with a special symbol such as an asterisk or a "C" in column 1.

Assembly languages have a feature that does not appear in most high level languages—the *macro* definition. The macro allows the programmer to name and define parameterized sequences of code that may be referred to later in the program. The facility differs from a subroutine or procedure in that the body of the macro is inserted or assembled into the program at each place the macro name is invoked. If an assembler permits the use of macros it probably permits *conditional assembly.* Conditional assembly allows the testing of parameter attributes and the subsequent conditional generation or assembly of the code depending upon the test outcome. Strictly speaking, a conditional assembly facility is independent of the macro facility of an assembler. It is, however, of very limited utility outside the macro context.

High level languages are either compiled into native code for direct execution on the machine or into a pseudo-code for an interpretive execution. The advantage of an interpreter is that the programmer is allowed a much greater degree of interaction with the executing program than if it were allowed to execute directly on the machine itself. The chief disadvantage of an interpreter is that execution time is much slower than direct execution. Slow-down factors of 5 to 20 are common. FORTRAN is most often compiled into native code while BASIC is mostly interpreted. Pascal seems to be available both ways.

FORTRAN (FORmula TRANslator) was the first commercial high-level language and is the language still used most often by scientists and engineers. It is primarily a numerical computation language and one of the few which directly permits complex number arithmetic. It is easy to use and has a large library. BASIC (Beginner's All-purpose Symbolic Instruction Code) is a variant of FORTRAN. It does not require a differentiation between real and integer variables; nor are any variable declarations required. BASIC is a very easy language to use. Pascal (named after the French mathematician Blaise Pascal) is a "modern" language. It requires that all variables be declared before their use and imposes a rigid structure upon the form of the program code and logic. The use of the "GOTO" is permitted but strongly discouraged by the availability of looping and control constructs. A program written in Pascal requires more preliminary design work before coding than would the same program written in either FORTRAN or BASIC. The net practical effect is that once the compiler-detectable errors are corrected, Pascal programs tend to be more execution-time error-free than either FORTRAN or BASIC programs. Current opinion among software engineers is that FORTRAN and BASIC are suitable for one-off programs, but for commercial quality, error-free, large programs, a modern language such as Pascal is more suitable.

Practice Problems

NUMBER SYSTEMS

Convert the following numbers to their base 10 equivalents.

2.1 101101.11 (base 2)

a) 42.25 b) 45.75 c) 48.68 d) 52.3 e) 55.5

2.2 24.6 (base 8)

a) 12.3 b) 14.75 c) 20.6 d) 20.75 e) 24.6

2.3 3FC.A (base 16)

a) 986.5 b) 995.625 c) 1010.25 d) 1015.75 e) 1020.625

Convert the following numbers to 8-digit binary using 1's complement.

2.4 21 (base 10)

a) 00010001 b) 00010101 c) 10010101 d) 11101110 e) 11101010

2.5 –29 (base 10)

a) 00101001 b) 00011101 c) 11100010 d) 11100000 e) 11100011

2.6 –40 (base 10)

a) 11000000 b) 00101000 c) 11011000 d) 11010110 e) 11010111

Convert the following numbers to 8-digit binary using 2's complement.

2.7 21 (base 10)

a) 0001001 b) 00010101 c) 10101010 d) 11101110 e) 11101010

2.8 –29 (base 10)

a) 11100010 b) 11100011 c) 11100100 d) 00101000 e) 1100100

2.9 –40 (base 10)

a) 11011000 b) 11010111 c) 00100111 d) 00101000 e) 01111110

2.10 How many bytes does a 48K computer memory contain?

a) 47256 b) 48000 c) 48512 d) 49152 e) 51200

COMBINATIONAL LOGIC

2.11 A certain 3-input, 1-output device has the voltage characteristics displayed (numerical values are volts):

A	B	C	out
0	0	—	0
—	—	5	0
0	5	0	5
5	—	0	5

If the logical 0 is equivalent to 5 volts and the logical 1 is equivalent to 0 volts, the logical equation describing this device is

a) $A + C$ b) $A\overline{B}\,C$ c) $AB + \overline{C}$ d) $\overline{A}C + \overline{B}C$ e) $A + \overline{B} + C$

2.12 A minterm is

a) the minimum term in a Boolean function.

b) a prime implicant.

c) always smaller than a maxterm.

d) a circle on a Karnaugh map.

e) a square on a Karnaugh map.

Simplify the following Boolean functions.

2.13 $AB + A\overline{B}C + \overline{A}C$

a) $AB + BC$ b) $AC + B$ c) $AB + C$ d) A e) $A + B$

2.14 $F = \overline{B}\,\overline{C} + \overline{A}C + A\overline{B}C$, don't care $= ABC + \overline{A}B\overline{C}$

a) AC b) $A + C$ c) $AB + \overline{C}$ d) $\overline{A} + B$ e) $\overline{B} + C$

2.15 $F = \overline{A}BC + \overline{A}BD$, don't care $= (AB + \overline{A}\,\overline{B})(C + D)$

a) $BC + BD$ b) $BC + \overline{A}D$ c) $\overline{A}C + A\overline{D}$ d) $\overline{A}C + BD$ e) $\overline{A}\,\overline{B}CD$

Give the simplified complement of the following functions.

2.16 $\overline{C}D + BD$

a) $B + CD$ b) $\overline{C}D + BD$ c) $\overline{D} + \overline{B}C$ d) $\overline{B} + CD$ e) $C + \overline{B}D$

2.17 $A\overline{B}C + A\overline{B}\,\overline{D} + BD$

a) $\overline{A}\,\overline{B} + B\overline{D} + \overline{B}\,\overline{C}D$

b) $AB + \overline{B}\,\overline{D} + \overline{C}D$

c) $A\overline{B}C + D$

d) $AD + BC$

e) $\overline{A}\,\overline{B}\,\overline{C}\,\overline{D}$

SEQUENTIAL LOGIC

Given the transition table of the synchronous Mealy machine and the state variable assignment, answer Problem 2.18–2.20.

PQ	state	MN, input 00	01	11	10
00	a	a/0	a/0	b/0	b/0
01	b	c/0	d/0	c/1	c/1
11	c	a/0	a/0	a/1	–/–
10	d	c/0	a/0	a/0	c/0

2.18 What is the output equation?

a) MQ b) $N\overline{P} + Q$ c) $M\overline{Q}$ d) $\overline{P}\,\overline{N} + PN$ e) $P + MQ$

2.19 If the P state variable is implemented with a J-K flip-flop, what is the "J" equation?

a) Q b) P c) $Q + N$ d) $\overline{Q} + N$ e) $\overline{Q}$

2.20 If the Q state variable is implemented with a S-R flip-flop, what is the "R" equation?

a) $\overline{P}N + \overline{P}M$ b) $(\overline{P}N + \overline{P}M)Q$ c) $(P + \overline{M}N)Q$ d) Q e) P

OPERATING SYSTEMS

2.21 An operating system that allows the user to correct input data has:

a) language processing

b) multiprogramming

c) text editing

d) file storage

e) real time capability

Solutions to Practice Problems

2.1 b) $101101.11_2 = 32+8+4+1+1/2+1/4 = 45.75$

2.2 d) $24.6_8 = 2\times 8+4+6/8 = 20.75$

2.3 e) $3FC.A_{16} = 3\times 256+15\times 16+12+10/16 = 1020.625$

2.4 b) $21_{10} = 16+4+1 = 00010101_2$

2.5 c) $29_{10} = 16+8+4+1 = 00011101_2$
$-29_{10} = 11100010_2$ 1's comp.

2.6 e) $40_{10} = 32+8 = 00101000_2$
$-40_{10} = 11010111_2$ 1's comp.

2.7 b) $21_{10} = 16+4+1 = 00010101_2$

2.8 b) $-29_{10} = 11100010_2$ 1's comp.
$= 11100011_2$ 2's comp.

2.9 a) $-40_{10} = 11010111_2$ 1's comp.
$= 11101000_2$ 2's comp.

2.10 d) $48\times 1024 = 49\ 152$

2.11 c)

A	*B*	*C*	OUT
1	1	–	1
–	–	0	1
1	0	1	0
0	–	1	0

$\text{OUT} = AB + \overline{C}$

2.12 e)

2.13 c)

		A		
C	1	1	1	1
		1		

$A\,B + C$

2.14 e)

		A		
C	1	X	1	1
	1		X	1

$F = \overline{B} + C$

2.15 a), b), d)

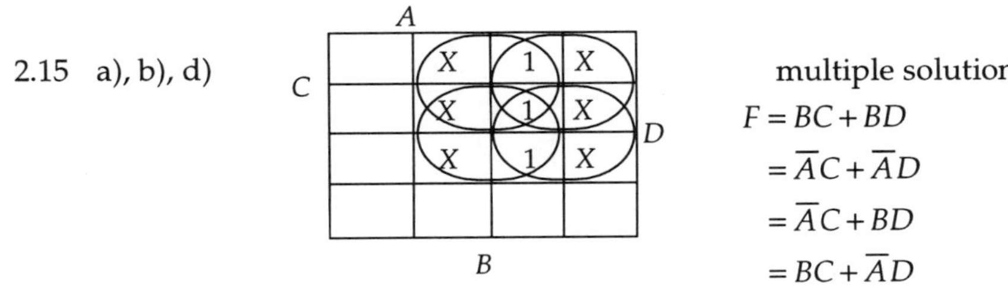

multiple solutions

$F = BC + BD$

$= \overline{A}C + \overline{A}D$

$= \overline{A}C + BD$

$= BC + \overline{A}D$

2.16 c)

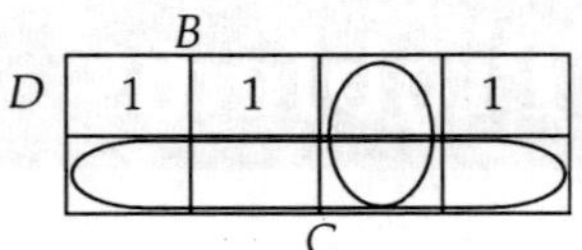

$\overline{F} = \overline{D} + \overline{B}C$

2.17 a)

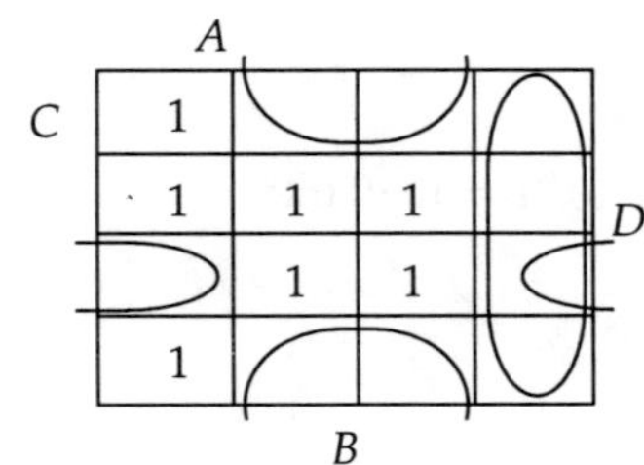

$\overline{F} = \overline{A}\overline{B} + \overline{B}\overline{D} + \overline{B}\overline{C}D$

2.18 a)

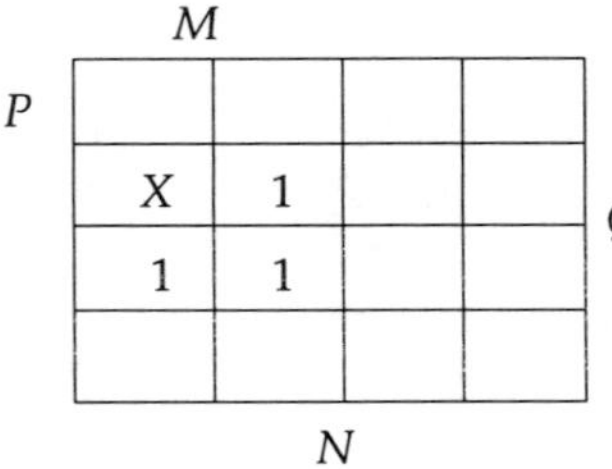

OUTPUT = MQ

2.19 a)

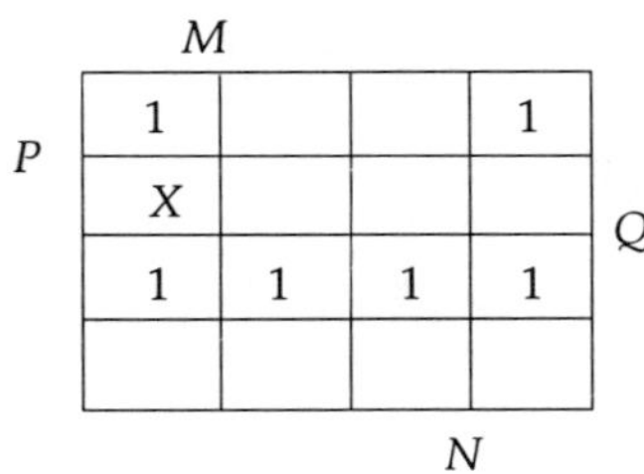

$P_{next} = \overline{Q}MP + Q\overline{P}$

$J_P = Q \; ; \; K_P = \overline{NP}$

2.20 c)

$Q_{next} = (M\overline{P} + N\overline{P})Q + (P\overline{M}\overline{N} + \overline{P}M)\overline{Q}$

$S_Q = (\overline{P}M + P\overline{M}\overline{N})\overline{Q}$

$R_Q = (P + \overline{M}N)Q$

M; P; Q; N

			1
X			
1	1		1
1	1		

2.21 c)

Materials Science

by Kalinath Mukherjee

Materials science is a very broad field of study covering the structures and properties of metallic materials, ceramics, polymers and composites. Hence, it is necessary to discuss the electronic, magnetic, optical, mechanical and chemical properties of various materials in the materials science discipline.

A review of past Fundamentals of Engineering exams indicates that there are two major areas emphasized in materials science. These are related to the fundamentals of 1) strength, deformation and plasticity of crystalline solids, and 2) phase equilibria in metallic systems. Typically the latter is troublesome to engineers who do not have a metallurgy-materials science background. In this review, therefore, these two subject areas are thoroughly treated. (For a more complete understanding of the topic, we recommend that you simply refer to appropriate texts.)

We include a discussion of various solid types, and the nature of atomic bonding which gives rise to different classes of materials. We discuss fundamentals of crystallography, since a majority of engineering solids are crystalline. Thus, it is important to know how to define crystal planes and directions which play an important role in mechanical deformation as we will see. Some discussion on atomic mobility (diffusion) in solids is included, since this mobility is very closely related to phase changes, hardening through heat-treatment etc. A brief discussion of electrical and electrochemical properties is also presented. Please note that corrosion of metallic materials, which can be understood in terms of electrochemical behavior of a solid, is a very important consideration in some branches of engineering. Finally, it may be worth considering that the majority of FE exam materials questions usually hinge on the subjects of our first two articles, with the final eight 'capsule' reviews representing the remainder.

3.1 Mechanical Properties of Metals and Alloys

There are several standard experimental techniques used to determine how a material responds to an applied state of stress. The specific responses and the corresponding tests may be summarized as follows:

- Capacity to withstand static load → Tension or compression test
- Resistance to permanent deformation → Hardness test
- Toughness of a material under shock loading → Impact test
- The useful life of a material under cyclic loading → Fatigue test
- Elevated temperature behavior → Creep and stress rupture tests

In tension testing, there are two distinct stages of deformation:

1) Elastic deformation (reversible)

2) Plastic deformation (irreversible)

The elastic range is characterized by *Hooke's law*, which is the linear equation

$$\sigma = E\varepsilon \qquad \textbf{(3.1.1)}$$

where σ and ε are the stress and strain, respectively, and E (the slope of the line) is the *Young's modulus* of the material, also called the *modulus of elasticity*.

Beyond the elastic range, the material undergoes *plastic deformation*. In this range, the stress-strain relation is nonlinear. It is important to note that, unlike elastic deformation, the volume of the material remains constant during plastic deformation (we will see later why this is so). The stress at which this nonlinearity begins is called the *yield stress* σ_y.

Very frequently it is difficult to obtain a reliable value of σ_y from experimental data because the linear to nonlinear transition of the stress-strain diagram is rather gradual. To avoid this difficulty, an *off-set yield stress* $\sigma_{0.2}$ is defined at a specified percentage of plastic strain (usually 0.2% plastic strain).

The engineering stress-strain curve (Fig. 3.1) shows a maximum value of stress called the *ultimate tensile strength* σ_{uts}. Beyond this value, the engineering stress

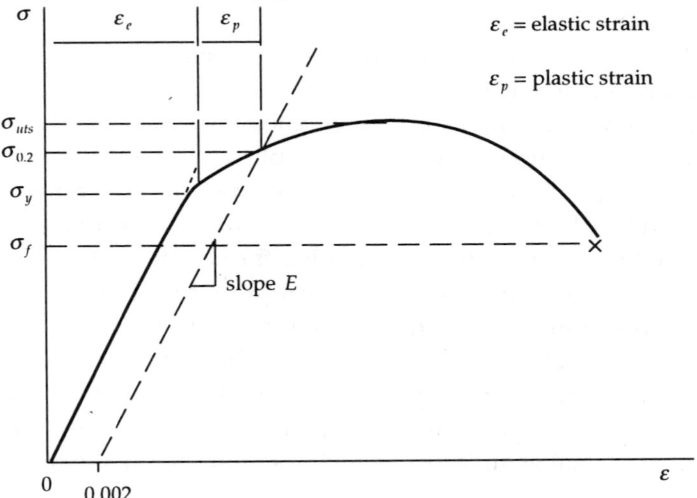

Figure 3.1 Engineering stress-strain diagram.

decreases and the sample fails in tension at a *fracture stress* σ_f, which is (usually) less than σ_{uts}. The diagram in Fig. 3.1 shows various features of the engineering stress-strain diagram.

The reason $\sigma_f < \sigma_{uts}$ is that the engineering stress is obtained by dividing the load by the "original" cross-sectional area of the sample. At stresses $\geq \sigma_{uts}$ the sample starts to neck locally, tri-axial stresses develop and eventually the sample breaks. Since the cross-section is less at the necking, the load bearing capacity goes down. This load divided by the original area is less than the "true" stress beyond this point.

The extent of necking depends on the *ductility* of the material. If the material is very *brittle*, that is, has little or no ductility, then there will be very little necking strain, and in that case $\sigma_f \equiv \sigma_{uts}$. Often it is possible to make a qualitative assessment of ductility from the nature of the tensile fracture. Fig. 3.2 shows three types of necking which might be observed.

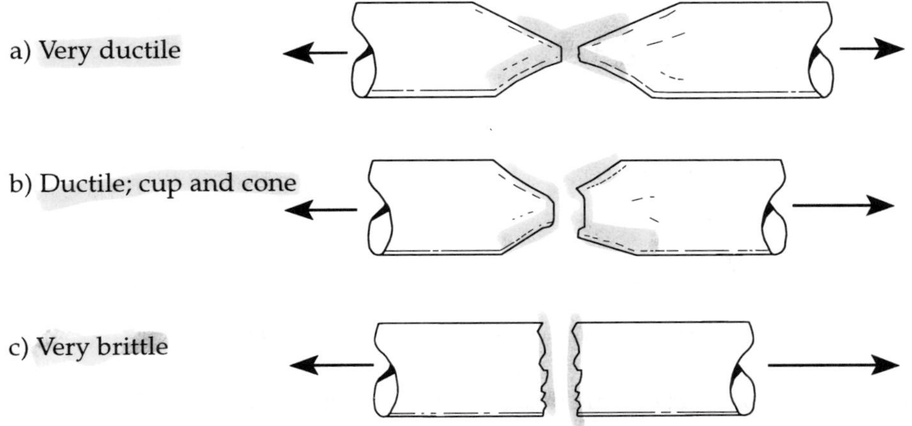

Figure 3.2 Examples of tensile failure.

Plastic deformation at ordinary temperature introduces some additional strength to many metals and alloys. This is called *work-hardening* or *strain-hardening*. The nature of this hardening can be understood by studying Fig. 3.3. If the material is loaded in the plastic range to strain ε_1 and then unloaded, the "apparent" yield stress σ_{y2} on reloading is greater than σ_{y1}. Such a "hardness" is

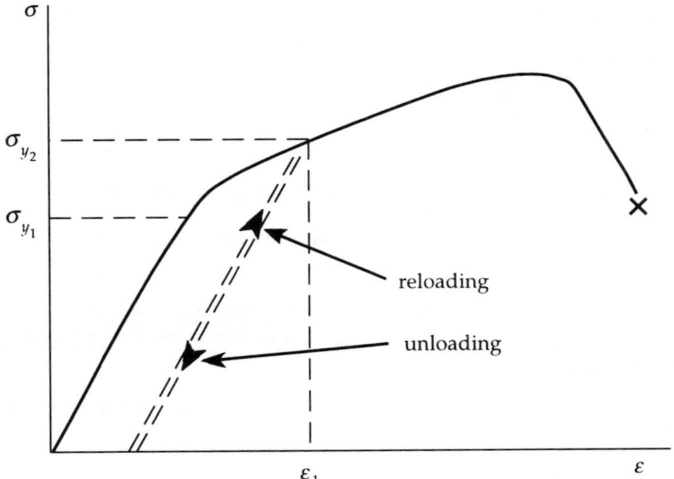

Figure 3.3 Tensile loading-unloading curve.

imparted on cold forged tools, for example. It must be noted that although the yield strength has apparently increased, the material loses some amount of its plastic flow property. The "extra strength" is removed if the material is annealed at an elevated temperature.

In steel and other ferrous alloys, the stress strain curve might look like the one shown in Fig. 3.4 (depending on carbon and other alloy content). The quantities σ_u and σ_l are called *upper* and *lower yield points*. This peculiarity of yielding occurs due to the interaction of carbon atoms with atomic scale defects called *dislocations*.

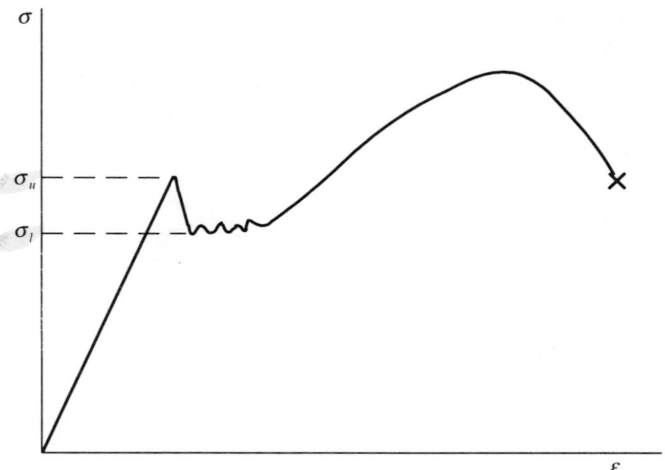

Figure 3.4 Stress-strain diagram of mild steel.

3.1.1 Nature of Plastic Flow

In crystalline materials (all metals and alloys are crystalline) plastic flow or plastic deformation involves the *sliding* of atomic planes, analogous to shearing a deck of cards, as shown in Fig. 3.5. The sliding of atomic planes is called *slip deformation*. Under an applied stress, slip takes place on those crystal planes which have the densest atomic packing. Furthermore, slip directions are restricted to the crystallographic directions along which the atoms are most closely packed.

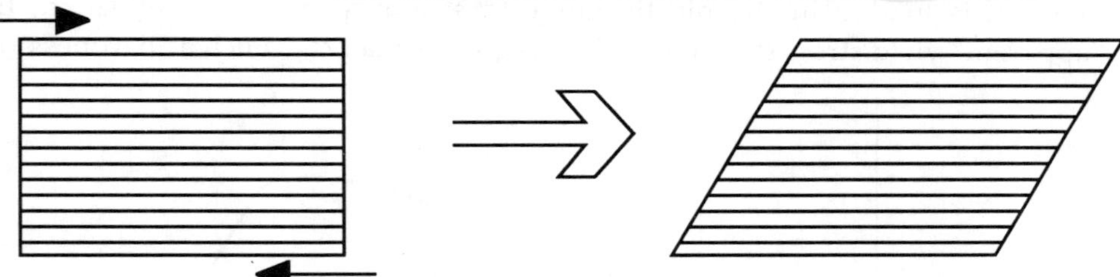

Figure 3.5 Shearing of a deck of cards. (Note shape change, but no volume change.)

The combination of a close-packed plane and a close-packed direction is called a *slip system*. Depending on the crystal structure, some metals and alloys will have more slip systems than others. The higher the number of planes and directions along which slip can take place, the easier it is to produce plastic deformation without brittle fracture.

Slip occurs when the resolved component of shear stress τ_R, given by the expression on the slip plane along the slip direction,

$$\tau_R = \frac{P}{A}\cos\phi\cos\lambda, \tag{3.1.2}$$

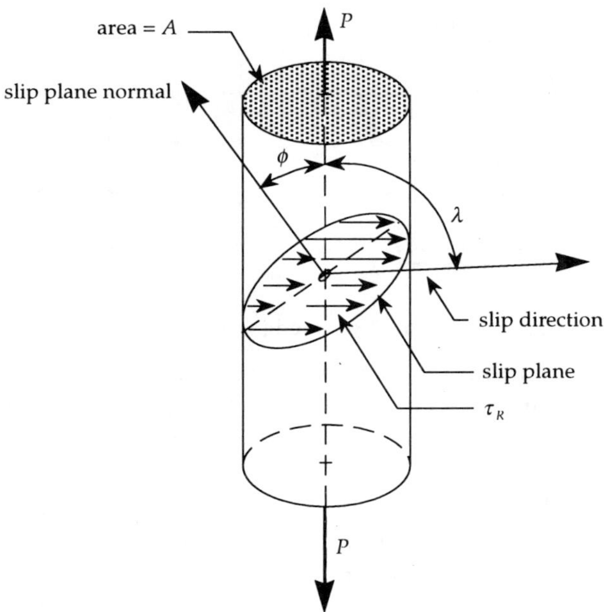

Figure 3.6 Resolved shear stress on the slip plane.

exceeds a critical value called *critical resolved shear stress,* $(\tau_R)_{crit}$ (see Fig. 3.6). This critical value is a property of the material.

The experimental value of $(\tau_R)_{\text{crit}}$ is found to be at least five times smaller than the theoretically calculated value based on the force necessary to slide closely packed planes of atoms. The apparent anomaly here was resolved when dislocations were discovered. Slip becomes easier if dislocations are present. Let us consider an analogy. A long narrow rug on a hallway is to be moved x meters to the right, as shown in Fig. 3.7(a). This could be done by pulling, as shown. It must be noted that considerable friction must be overcome to do this. Alternately, a bulge could be made as shown in Fig. 3.7(b) and then this bulge could be "walked" to the right. When the bulge exits on the right, the rug has moved x meters to the right. Considerably less effort is required in the second procedure.

Analogous to the bulge, there is an extra half-plane of atoms in an *edge dislocation* as shown in Fig. 3.8. The row of atoms at the end of the half-plane is situated between two equilibrium sites. Thus, it takes less force to move this plane.

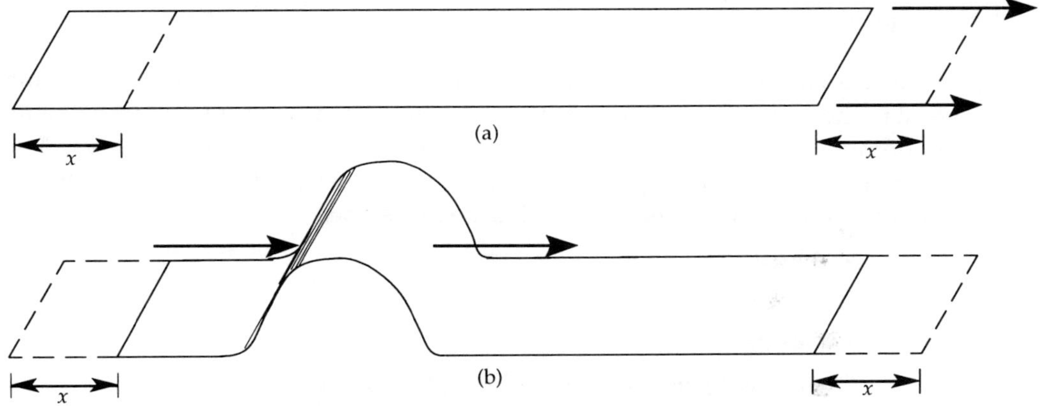

Figure 3.7 Analogy of a dislocation motion.

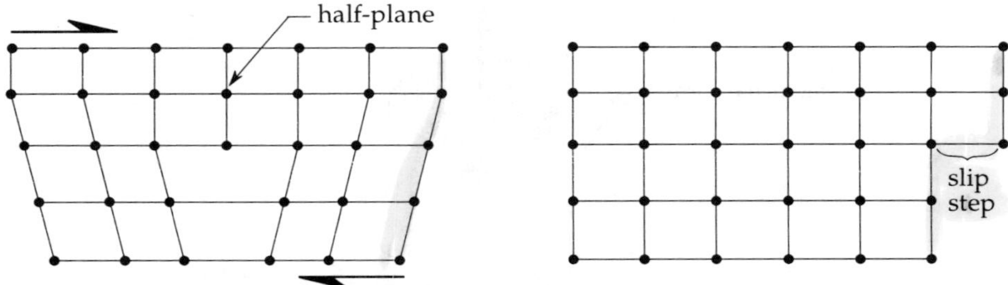

Figure 3.8 An edge dislocation.

When the extra half-plane emerges on the surface, we have one elementary slip step. The row of atoms at the end of the half-plane can be viewed as a line called a *dislocation line*. When a dislocation line moves, plastic deformation occurs.

It has been established that existing dislocation lines in a material can multiply under an applied stress, as shown in Fig. 3.9. This mechanism is called a *Frank-Reed source* and is experimentally verifiable.

Since dislocation motion ≡ plastic deformation, strength can be increased if dislocation motion is blocked. Indeed, the strengthening of alloys can be traced to the interaction of dislocations with dispersed phases, with grain boundaries, and with stress fields of other dislocations (e.g., strain hardening).

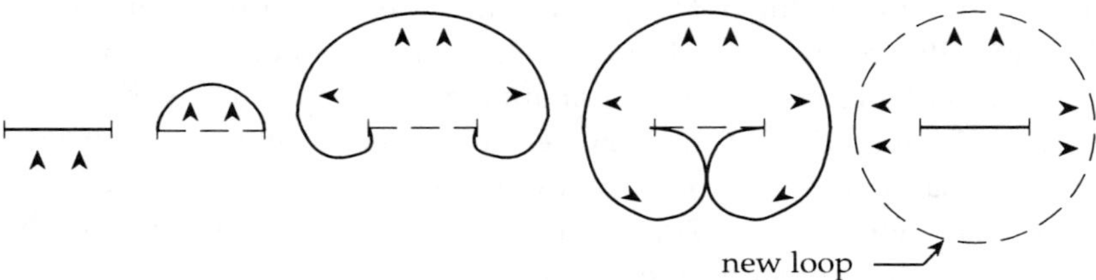

Figure 3.9 Frank-Reed dislocation source.

3.1.2 Compressive Strength

A material's response to a compressive stress is similar to that for a tensile stress, except for the fact that there is no necking involved in pure compression. Compression test data is especially useful for those materials which are quite brittle in tension but have significant compressive load bearing capacities (concrete, cast iron, etc.).

3.1.3 Hardness of Materials

A hardness test is an empirical method which determines the resistance of a material to the penetration of an indenter. Hardness measurement can be useful for obtaining a qualitative estimate of service wear, strength and toughness. Furthermore, for steel, an empirical correlation exists between hardness and tensile strength. The most commonly used hardness tests are:

1) Brinell
2) Rockwell
3) Vickers
4) Microhardness (Vickers or Koop indenters)

Rockwell hardness has scales, A through V (depending on load and type of indenter). Scales A, B, C, and D are commonly used for various steels. Table 3.1

shows a few hardness values along with the tensile strength of steel for comparison.

TABLE 3-1. Hardness and Strength of Steel.*

Brinell	Rockwell			Vickers	Tensile Strength
	C	D	A		MPa (1000 psi)
601	57.3	68.7	79.8	640	2120 (308)
495	51.0	63.8	76.3	528	1740 (253)
401	43.1	57.8	72.0	425	1380 (201)
302	32.1	49.3	66.3	319	1030 (150)
229	20.5	40.5	60.8	241	760 (111)

* A more complete table can be found in "The Testing and Inspection of Engineering Materials": H. F. Danis, G. E. Trowell and C. T. Wiskocil, McGraw Hill Book Co., NY.

3.1.4 Fatigue Test

Life in cyclic loading is important in many applications. Fatigue life is determined from experimental data relating number of cycles (N) to failure with cyclic stress amplitude (S). A schematic S vs. N curve is shown in Fig. 3.10. Note that for steel there is a critical value of stress S_{crit} below which fatigue life is virtually infinity. This limit is called the *endurance limit.* Notice that no such endurance limit exists for Al (or many other nonferrous metals and alloys). Fatigue fractures are progressive, beginning as minute cracks that grow under the action of the fluctuating stress. Fatigue strength is defined as the maximum cyclic stress amplitude for a specified number of cycles until failure.

Fatigue is a surface active failure. Fatigue cracks start at the surface; surface defects such as notches can initiate a crack. A rough surface may reduce fatigue strength by as much as 25%. Cold rolling or shot peening (which introduces surface compressive stress) can increase fatigue strength by as much as 25%. *Corrosion-fatigue* is an important cause of service failure if a corrosive environment and cyclic stresses co-exist. For example, it has been shown that the endurance limit of a steel (tested in air) is altogether eliminated when the sample is tested in pure water. Fatigue life or fatigue strength can be improved by:

1) A highly polished surface
2) Surface hardening (carburizing, nitriding, etc.)
3) Surface compressive stresses (shot peening, cold rolling, etc.)

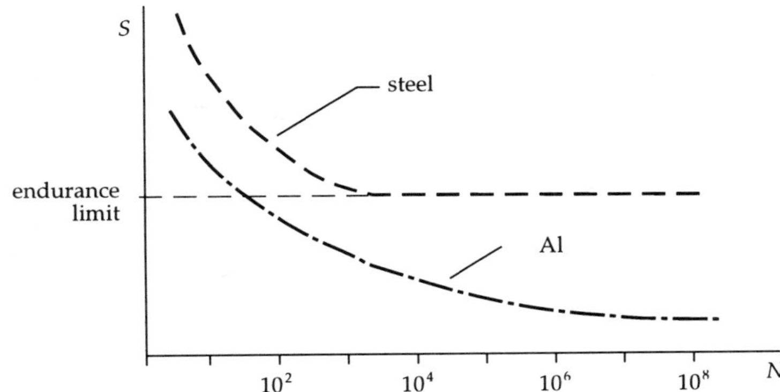

Figure 3.10 *S-N* diagram (schematic) of aluminum and steel.

3.1.5 Toughness and Impact Testing

The *impact value* is a simple evaluation of the notch toughness of the material. And *toughness* is a measure of energy absorbed by the material before fracture. Two types of machines, the Charpy and the Izod, are commonly used to test these qualities, both of which use swinging-pendulum loading with notched-bar samples. In tension, the area under the stress-strain curve is the energy/unit volume to fracture (see Fig. 3.11). The area under the stress-strain diagram of material *A* is less than that of material *B*. The area has the dimension of stress (force/unit area), which if multiplied and divided by length, results in energy/volume. Hence, material *B* can be analyzed as tougher than material *A*.

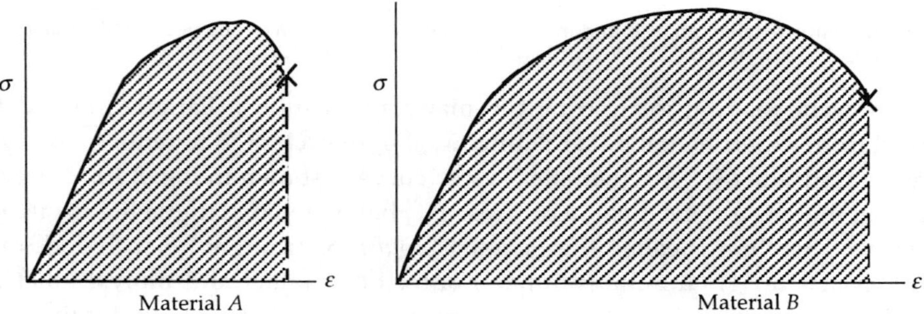

Figure 3.11 Comparison of toughness of two materials.

The presence of a notch introduces tri-axial stresses. Many materials become more brittle under such a state of stress than under uni-axial tension or compression. Temperature has a pronounced effect on the energy absorption and fracture behavior of steel in a notch condition. The sharpness of the transition from tough to brittle fracture depends on the material and also on the notch geometry.

Frequently, Charpy values will change from 35 or 40 to as low as 7 N·m over a temperature interval of 4 to 10°C. Fig. 3.12 shows the qualitative nature of ductile-brittle transition (also the behavior of *Ni*, which does not have a ductile-brittle transition). Service failures testify to the increased hazards of subnormal temperature, and so determination of transition temperature has come to be an important criterion for materials selection. It must be noted that in a large rigid structure, the transition temperature would be considerably higher than that for a standard Charpy sample.

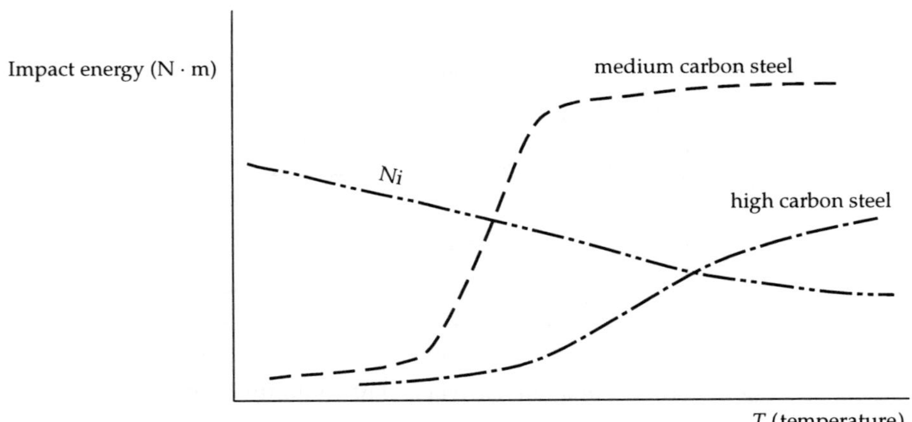

Figure 3.12 Impact energy vs. temperature.

3.1.6 High-Temperature Creep and Stress Rupture

The progressive deformation of a material at constant stress is called *creep*. Below about 40% of the absolute melting temperature, creep strain is negligible for most structural metals and alloys. An idealized shape of a creep curve is shown in Fig. 3.13. Andrade's empirical formula for the creep curve is given by

$$\varepsilon = \varepsilon_0 \left(1 + \beta t^{1/3}\right) e^{kt} \tag{3.1.3}$$

where ε is the strain in time *t*, ε_0 is the initial elastic strain, and β and k are material constants.

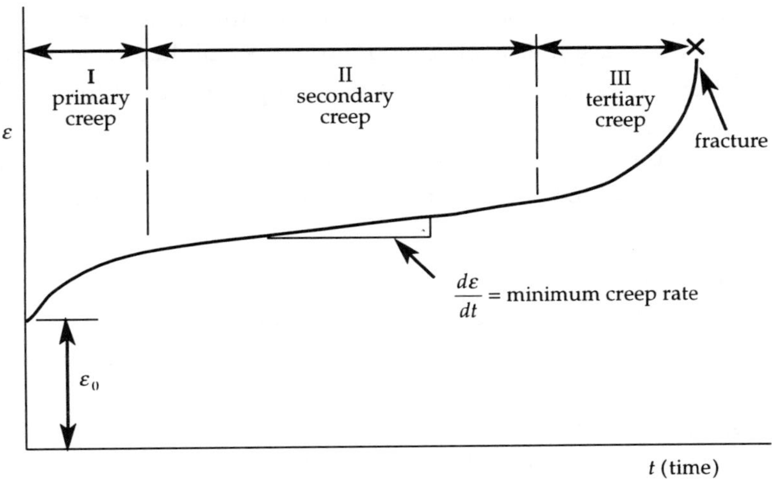

Figure 3.13 A typical creep curve showing three stages of creep.

The stress-rupture test is basically similar to a creep test except that the test is always carried out to failure. Elongation, time to failure, applied load, and temperature are all reported for the purpose of design data.

One particular mode of failure of polycrystalline metals and alloys at elevated temperatures is *grain boundary sliding*. The influence of grain size on creep resistance is not clear-cut. There is some evidence, however, to indicate that creep rate is lower in large grain materials. Because grain boundaries are the nucleation sites for high-temperature fracture, the control (or elimination) of grain boundaries will suppress fracture and increase rupture life. It must be noted that the environment plays an important role in high temperature mechanical properties. The nature of oxides, for example, can influence creep and stress-rupture.

3.1.7 Sound Velocity in Solids and Damping Capacity

Under a rapid cyclic loading of a solid in the elastic range, the elastic stress-strain curve shows a hysteresis. The area inside a hysteresis loop is a measure of energy dissipation per cycle of rapid elastic loading and unloading. This energy loss is said to be due to "internal friction" of the solid. Internal friction is closely related to elastic aftereffect associated with adiabatic temperature rise in the solid. Internal boundaries and defects also contribute to internal friction. Energy dissipation of a vibrating solid through internal friction, damps the vibration amplitude. Thus, higher the internal friction of a solid, higher is its "damping capacity". In many applications, high damping capacity is a desirable property. High damping capacity also retards the sound propagation through the solid. Note that sound propagation through a solid occurs through elastic wave (alternating compression and decompression cycles) propagation. Thus, the velocity of sound wave through an elastic solid is related to the density of the solid and its compressibility.

3.1.8 Metallurgical Variables in Materials Response to Stresses

Microstructural conditions, heat treatment, processing variables, and service conditions can all influence mechanical properties of metals and alloys.

A. Microstructural conditions

1) Grain size effect: At ordinary temperatures, fine grain is better for strength. At high temperatures, perhaps a larger grain size is desirable.
2) Single-phase vs. multiphase alloys: Many times a second phase might have a profound effect on the mechanical properties; for example, the retained austenite may be a problem in fatigue. Deformation behavior of phases may be different, and a simple averaging of properties might not be appropriate.
3) Porosity and inclusions: Poor mechanical properties result from high porosity and inclusions.
4) Directionality of microstructure: Rolling direction vs. transverse direction will have different mechanical properties. These also introduce anisotropy of properties.

B. Effects of heat treatment

1) Annealing: Softening, ductile behavior (depending on alloy).
2) Quenching of steel: Martensite formation, strong but brittle. In high carbon steel, quench-cracks may form.
3) Tempering of martensite: Hardness decreases but toughness increases. Strength is sacrificed to avoid brittle failure.
4) Age hardening: Depending on alloy composition, fine scale ($\approx 10^{-7}\,M$) precipitation may be formed which interacts with dislocations. Increased strength is thus obtained.
5) Case hardening: A hard surface and soft core combination is obtained by carburizing and nitriding. Fatigue strength can be increased by this method. A better wear-resistant surface can be produced.

C. Effects of some processing variables

1) Welding: The heat-affected zone with large grain size will have poorer mechanical properties. Local chemical composition changes can occur, including a loss of carbon in steel. Large parts can have quench cracking due to rapid quenching effects.
2) Flame cutting: Drastic changes of microstructure occur near the flame-cut surface and these changes affect the mechanical properties.
3) Machining, grinding: Cold work results in strain hardening. Excessive cold work may produce surface cracks.

D. Effects of service conditions

1) Extreme low temperature: Ductile-brittle transition occurs in steel.
2) Extreme high temperature: Causes corrosion and oxidation of surface. Surface cracks may form. Results in problems with corrosion fatigue, creep and rupture.
3) Impact loading: Notch sensitivity, surface scratches or corrosion pits can initiate brittle fracture.
4) Corrosive environment: Stress-corrosion, pitting corrosion, and corrosion fatigue result.

These are but a few examples of the service and material variables which can influence the mechanical response of a material. Often, deterioration of material properties with time leads to a service failure. But the most important function of an appropriate materials input in engineering design is to prevent unexpected catastrophic failure.

3.2 Equilibrium Phase Diagrams

For most practical purposes, alloy compositions are listed in weight percentage (wt. %). For example, 70-30 brass means 70 wt. % Cu and 30 wt. % Zn. In this discussion, we will consider a binary alloy (two chemical elements) of elements A and B (e.g., $A = Cu$, $B = Zn$). Sometimes, however, it is convenient to express an alloy composition in atomic percentage (at. %). Weight percent composition can be converted to atomic percent by using the following formulas:

$$\text{at.\%}\ A = \frac{W_A}{W_A + (M_A / M_B)W_B} \times 100, \tag{3.2.1}$$

$$\text{at.\%}\ B = \frac{W_B}{W_B + (M_B / M_A)W_A} \times 100, \tag{3.2.2}$$

where W_A and W_B are the weight percents in the alloy of elements A and B, respectively, and M_A and M_B are the respective atomic weights. Similarly, at. % can be converted to wt. % by using the following formulas:

$$\text{wt.\%}\ A = \frac{P_A M_A}{P_A M_A + P_B M_B} \times 100, \tag{3.2.3}$$

$$\text{wt.\%}\ B = \frac{P_B M_B}{P_A M_A + P_B M_B} \times 100, \tag{3.2.4}$$

where P_A and P_B are atomic percents of A and B, respectively.

There are various methods for determining equilibrium phase diagrams: X-ray diffraction, optical microscopy, calorimetric and thermal analyses. We will consider thermal analysis, i.e., the cooling curve method, here since it is very instructive.

The so-called equilibrium phase diagrams show the existence or coexistence of phases at any given temperature and alloy composition. The term *equilibrium* implies that the alloy is cooled at such a slow rate that thermodynamic equilibrium is attained at each temperature. A *phase* is a volume of material bounded by a distinct boundary within which the chemical composition is uniform. A phase has a fixed crystal structure and thermo-physical properties at a given temperature. The equilibrium between phases is determined by Gibb's phase rule

$$P + F = C + 2 \tag{3.2.5}$$

where P = number of phases, C = number of chemical elements in the alloy and F = the degree of free-dom, i.e., the number of independent variables. Eq. 3.2.5 is the generalized phase rule where both pressure and temperature are independent external variables. If the pressure is kept constant, as is the usual case (usually 1 atmosphere), then we have the *condensed phase rule* given by

$$P + F = C + 1 \tag{3.2.6}$$

We will use Eq. 3.2.6 in our discussion. As an example, consider a binary alloy in which $C = 2$. In this case, if the number of phases P in equilibrium is 3, then Eq. 3.2.6 predicts that $F = 0$, i.e., no degree of freedom. Thus, in such a case, the composition of the phases and the temperature at which the three phases coexist are fixed: no degree of freedom.

Consider the cooling of pure molten metal in a furnace which has the cooling curve depicted in Fig. 3.14(a). If we now place a thermocouple in the molten metal and plot the temperature of the metal as it cools, we will obtain the curve shown in Fig. 3.14(b).

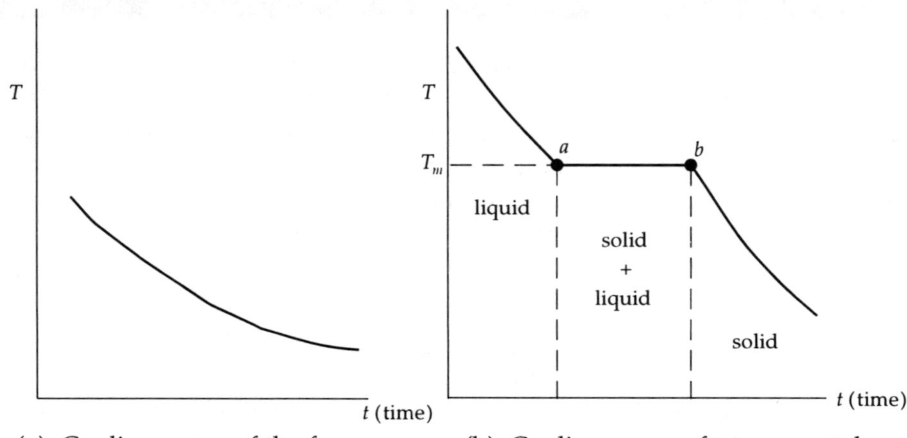

(a) Cooling curve of the furnace (b) Cooling curve of a pure metal

Figure 3.14 The cooling of a pure metal in a furnace.

The horizontal shelf $\overline{ab}$ is the *thermal arrest* at a temperature T_m. In this case T_m is the freezing (or melting) point of the metal where liquid and solid coexist. Until all of the molten metal is solidified, the temperature does not change. An analogy is the equilibrium between ice and water: until the ice cubes melt, the temperature of a glass of water remains at 0°C.

If we now add a small amount of an alloying element in a pure metal and repeat the cooling curve experiment, we will obtain the curve shown in Fig. 3.15. Comparing Fig. 3.15 with Fig. 3.14(b) we note that, unlike a single arrest

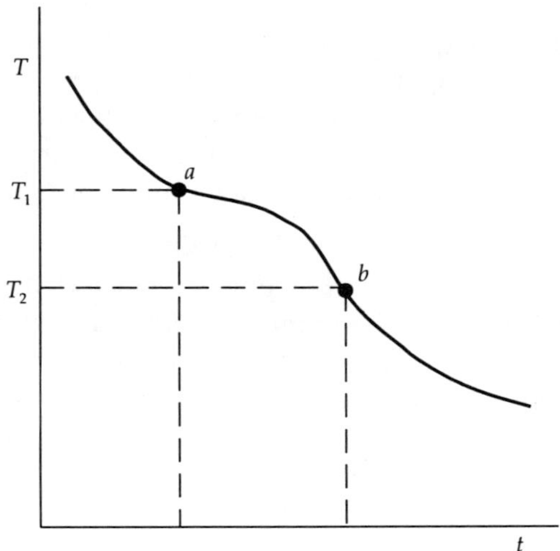

Figure 3.15 Cooling curve of liquid dilute alloy.

temperature, we have a change in slope at point *a*. Actually freezing begins at point *a* and is completed at point *b*. Thus, within the range of temperatures T_1 and T_2, we have a mixture of solid and liquid.

If the alloying element has a higher melting point (or freezing point) than the host pure metal, then at temperature T_1 this element starts to freeze and the remaining liquid has less of the alloying element in solution. Thus, the new liquid has a lower freezing temperature. This process continues until an almost pure host element freezes at temperature T_2.
Consider now a series of alloys of nickel (*Ni*) and copper (*Cu*); we note that the melting point of *Ni* is higher than that of Cu and addition of *Cu* to *Ni* lowers the freezing point (analogous to the lowering of the freezing point of water when salt is added). Several such cooling curves are shown in Fig. 3.16. In all these curves, point *a* corresponds to the beginning of freezing and point *b* the end of freezing. We now plot these temperatures of beginning and end of freezing in a diagram with temperature and alloy composition as coordinates. Such a diagram is shown in Fig. 3.17.

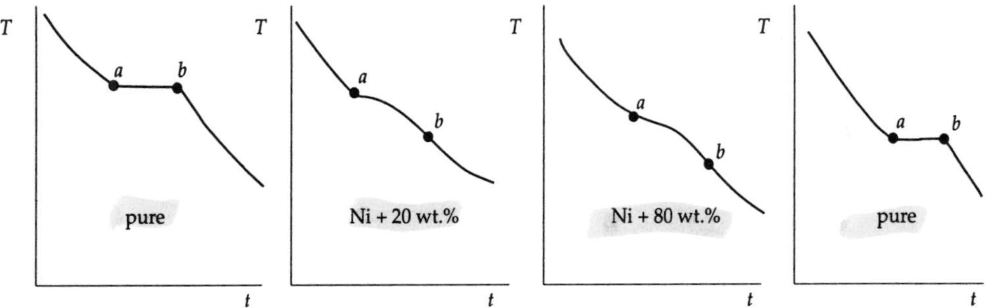

Figure 3.16 Schematic cooling for *Ni, Ni + 20 wt. % Cu, Ni + 80 wt. % Cu, and pure Cu.*

Fig. 3.17 is indeed an *equilibrium phase diagram* of the *Ni-Cu* alloy system. The upper curve defines the temperature above which the alloy is liquid and this curve is called the *liquidus;* the lower curve defines the temperature below which the alloy is solid and it is called the *solidus*. The area bounded by the liquidus and the solidus is the region of two phases: solid + liquid.

Let us now analyze the cooling of an alloy of composition C as shown in Fig. 3.18. At temperature T_1—which is above the liquidus—the alloy of composition C is entirely liquid. At temperature T_2 an extremely minute quantity of solid

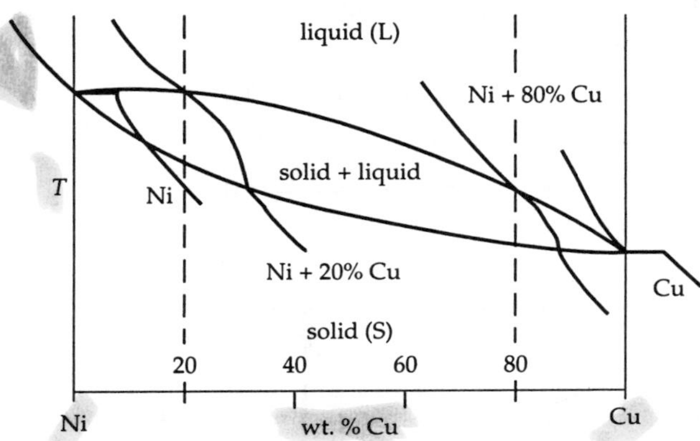

Figure 3.17 Arrest temperatures vs. composition.

forms and at T_3 we are in the two-phase $S + L$ region. At T_3 we have a mixture of solid and liquid, and the composition of the solid is given by the intersection of the temperature-horizontal with the solidus curve. The composition of the liquid phase is given by the intersection of the temperature-horizontal with the liquidus line. The compositions can be obtained by drawing vertical lines through these intersection points—that is, C_L and C_S in the figure are the compositions of the liquid and solid phases, respectively.

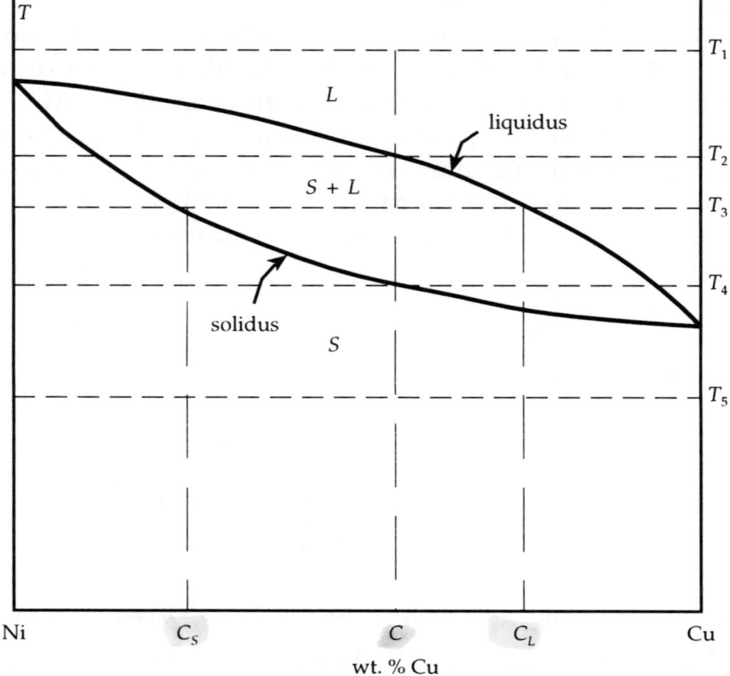

Figure 3.18 Analysis of cooling an alloy of composition *C*; *S* = solid, *L* = liquid.

The next situation we should work through is this: If two phases are present, what are the proportions of these phases? For example, at T_3 in Fig. 3.18 what are the weight percentages of the solid and the liquid phases? These percentages are given by the well-known *Lever rule.* The principle of the Lever rule is based on balancing two weights on a weightless beam across a fulcrum as shown in Fig. 3.19. The alloy composition is the fulcrum, and weights of phases (*S* and *L*) are suspended to balance the beam. The beam will balance if the moments of the weights are equal, i.e.,

$$W_S \times \overline{CC_S} = W_L \times \overline{CC_L}. \tag{3.2.7}$$

It is easy to see from the above that

$$\text{wt. \% } S = \frac{C_L - C}{C_L - C_S} \times 100 \tag{3.2.8}$$

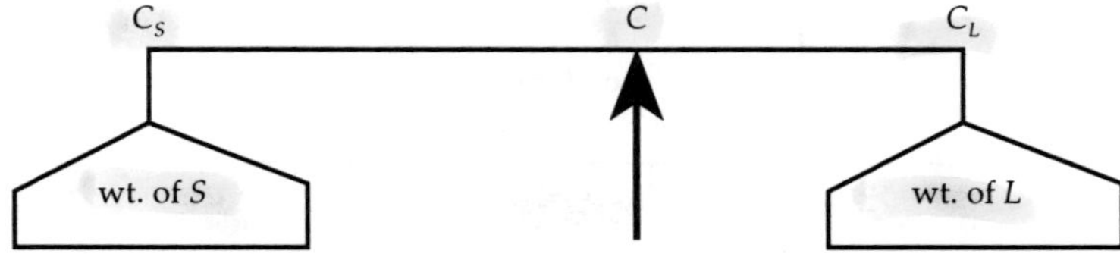

Figure 3.19 Principle of the Lever rule.

$$\text{wt. \% } L = \frac{C - C_S}{C_L - C_S} \times 100. \tag{3.2.9}$$

The *Ni-Cu* equilibrium diagram is a very simple case where *Ni* and *Cu* remain in solution in the solid phase. Such an alloy is called a *solid-solution* alloy. Some elements, however, do not like to remain in solution in the solid state. In some others, the two elements might form a compound at a fixed (or nearly fixed) composition. In others, other phases might form. Thus, in such cases the shapes of the equilibrium diagram are quite different. For example, if the two elements do not mix at all in the solid state, we obtain a phase diagram as shown in Fig. 3.20. This is a *eutectic* diagram.

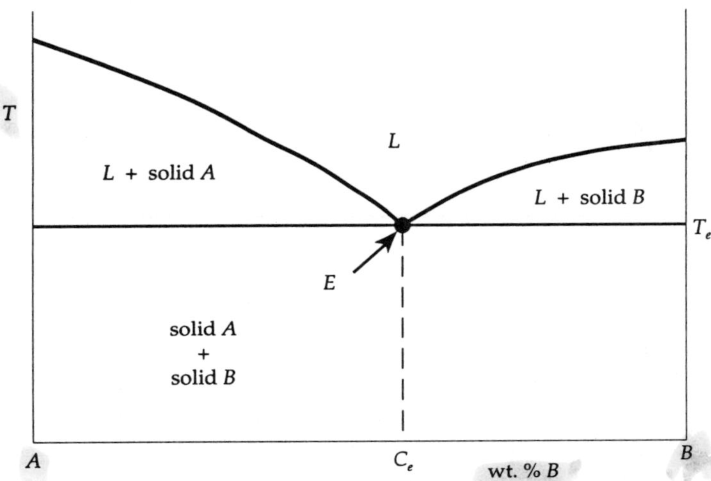

Figure 3.20 A eutectic diagram.

In this diagram (Fig. 3.20), the elements *A* and *B* do not remain in solution below the *eutectic temperature* T_e. If we choose an alloy of composition C_e (as shown in the figure) at T_e, there is a three-phase equilibrium for this composition at point *E*. Point *E* is known as the *eutectic point*. At a temperature infinitesimally below T_e solidification starts. Since *A* and *B* do not mix in the solid state, the solidification process must separate out pure *A* and *B*. In the previous diagram we assumed that the melting point of *A* is higher than that for pure *B*, and thus, a speck of pure *A* solidifies first. This makes the remaining liquid richer in *B* and a speck of pure *B* solidifies. In such a way alternate layers of plates of pure *A* and *B* soli-dify to give rise to a *eutectic microstructure.*

In some cases there might be a limited solid solubility of the two elements. For such a case we obtain a diagram as shown in Fig. 3.21. Below T_e we have two solid phases which we have designated as α and β. Although there are some standard notations—α, β, γ, depending on the crystal structure of the phases—it is important to note that α is, in this case, a solid solution which primarily consists of element *A* with some small amount of *B* dissolved in it. Similarly, β is primarily *B* with some small amount of *A* in it.

Consider the freezing of an alloy of composition *C* as shown in Fig. 3.21. At temperature T_1 the alloy is in the single-phase liquid region. At temperature T_2 the alloy is in the two-phase liquid + α region. Note that the intersection point between the composition-vertical and temperature-horizontal lies in the two-phase region. This region is defined by the liquidus line which delineates the boundary of the liquid, and the solidus line which defines the boundary of the

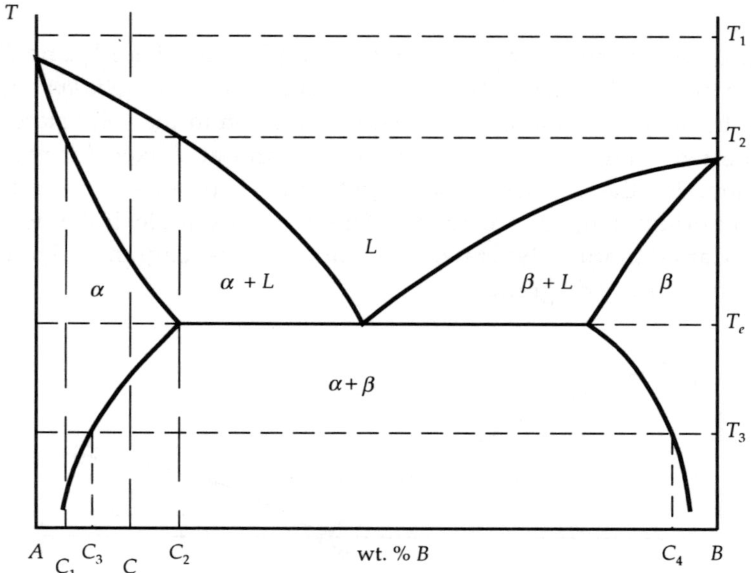

Figure 3.21 Eutectic diagram with limited solid solubility.

solid α-phase. Thus, the region consists of solid α and the liquid alloy. At T_2 the temperature horizontal intersects the solidus and the liquidus. If we drop perpendiculars from these points to the composition axis, we obtain C_1 as the composition of solid α in equilibrium with a liquid of composition C_2 at that temperature. Applying the Lever rule at temperature T_2:

$$\text{wt. \% solid} = \frac{C_2 - C}{C_2 - C_1} \times 100 \tag{3.2.10}$$

$$\text{wt. \% liquid} = \frac{C - C_1}{C_2 - C_1} \times 100 \tag{3.2.11}$$

Similarly, at temperature T_3 the temperature horizontal intersects lines which separate α and β. Thus, at T_3 the two equilibrium phases, for the alloy composition C, are solid α of composition C_3 and solid β of composition C_4, respectively. The proportions of the phases are

$$\text{wt. \% } \alpha = \frac{C_4 - C}{C_4 - C_3} \times 100 \tag{3.2.12}$$

$$\text{wt. \% } \beta = \frac{C - C_3}{C_4 - C_3} \times 100. \tag{3.2.13}$$

Some elements might form a compound-like mixture at a fixed composition such as 75 atomic % A + 25 atomic % B, or 50 atomic % A + 50 atomic % B. These may be correspondingly designated as A_3B (75%:25%), or AB (50%:50%), as if they have a molecular formula like H_2O. When such compounds form, they behave more like pure metal and melt at a fixed single temperature. Formation of a compound splits the phase diagram.

It is easier to depict compound-forming alloys if we plot the phase diagram in at. % composition. Fig. 3.22 (a) and (b) schematically show two cases where the compound formation can be viewed as a separation of the phase diagram into two regions of eutectic type diagrams. Consider alloys of composition C_1 and A + 50% B in Fig. 3.22(b). At temperature T_1 alloy C_1 is liquid, at T_2 it is in two-phase L + (AB) regions (note that (AB) is a compound), and at T_3 it is in two-

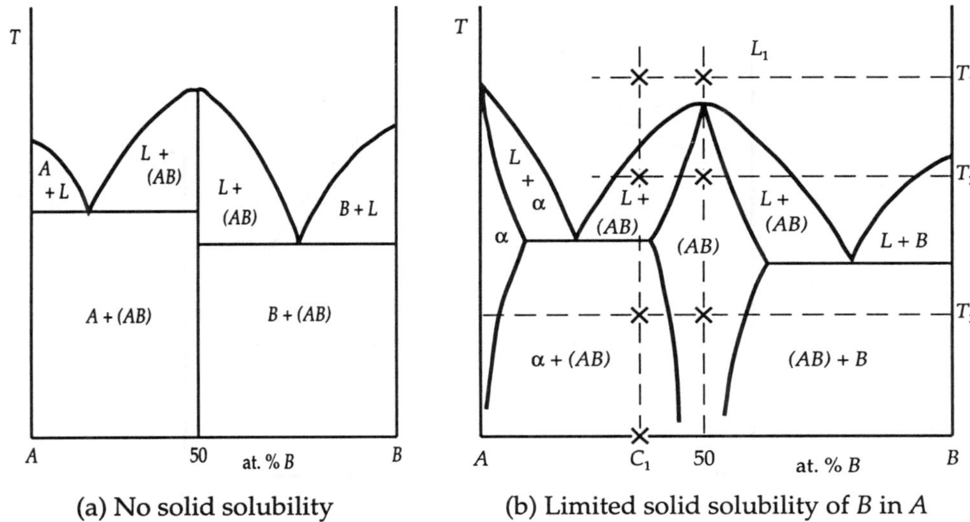

(a) No solid solubility (b) Limited solid solubility of B in A

Figure 3.22 Schematic phase diagrams showing an *AB* compound.

phase α + (AB) regions. Note the intersection points of the temperature-horizontals with the various phase boundaries. For the alloy with A + 50 at. % B, T_1 is the region of liquid phase, but at both T_2 and T_3 the alloy is in a single-phase AB compound.

The eutectic diagram, as discussed earlier, can be viewed as if a chemical reaction occurs at the eutectic point; i.e., in melting a eutectic solid, two solids react to form a liquid of fixed composition. Thus, a eutectic phase separation is sometimes referred to as a *eutectic reaction.*

Other reactions in various binary alloys are similar to the eutectic reaction. For example, a *peritectic reaction* is one in which a liquid L_I reacts with a solid S_I to form a second solid phase S_{II}. Fig. 3.23 shows a schematic diagram with a peri-

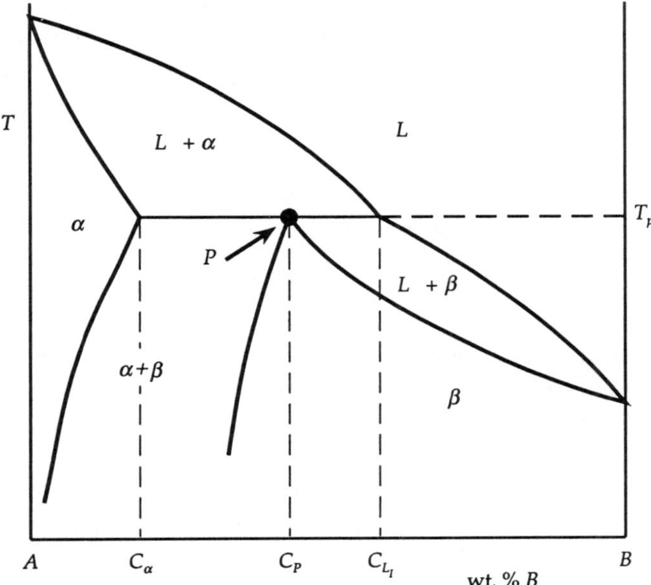

Figure 3.23 Schematic diagram showing a peritectic reaction.

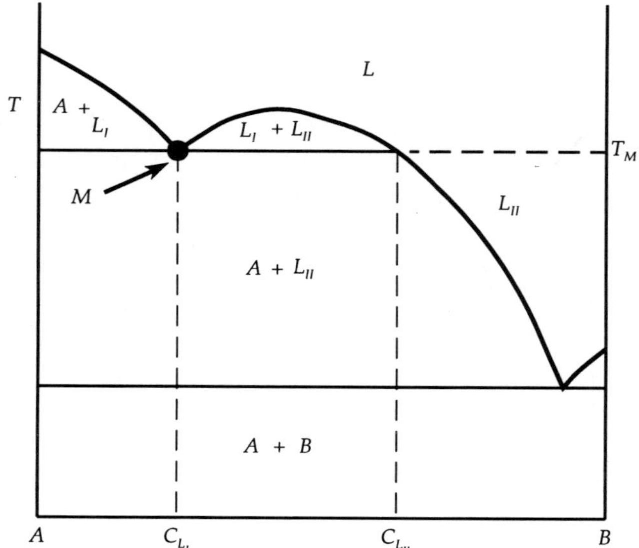

Figure 3.24 Schematic presentation of a monotectic reaction.

tectic reaction. The point P is the peritectic point, and T_p and C_p are the peritectic temperature and composition, respectively. At T_p liquid of composition C_{L_I} reacts with solid α of composition C_α to form a new solid β; for example,

$$L_I + \alpha \rightarrow \beta . \tag{3.2.14}$$

Another type of reaction, known as the *monotectic* reaction, is shown in Fig. 3.24. In this diagram M is the monotectic point. At the monotectic temperature T_M, liquid of composition C_{L_I} reacts with pure A to form another liquid of composition $C_{L_{II}}$. Note that the diagram shows a region where two immissible liquids, L_I and L_{II}, coexist.

Thus, the monotectic reaction in the above diagram could be written as:

$$\text{Solid } A + \text{ liquid } L_I \rightarrow L_{II} \tag{3.2.15}$$

Analogous to the eutectic and peritectic reactions during the freezing of an alloy, we find two other reactions in the solid state diffusion-controlled phase separation. These are called *eutectoid* and *peritectoid* reactions. For a eutectoid reaction, we can write

$$\text{Solid}_I \rightarrow \text{Solid}_{II} + \text{Solid}_{III} \tag{3.2.16}$$

$$\text{e.g.,} \quad \gamma \rightarrow \alpha + \beta$$

and for a peritectoid reaction, we have

$$\text{Solid}_I + \text{Solid}_{II} \rightarrow \text{Solid}_{III} \tag{3.2.17}$$

$$\text{e.g.,} \quad \alpha + \gamma \rightarrow \beta .$$

An important practical example showing a eutectoid reaction is the iron-carbon diagram. Fig. 3.25 (a) and (b) shows some of the important features of the iron-carbon diagram. This diagram is very useful in the determination of heat-treatment procedure for steels.

The eutectoid reaction in the *Fe-C* system produces an alternate plate-like microstructure known as *pearlite.* The alternate plates consist of *ferrite* and *cementite.* Ferrite, denoted by α, is almost pure iron with a small amount of carbon ($\leq$0.02%). Cementite, also known as *carbide,* has a composition Fe + 6.7 wt. % C, and in atomic % has a formula Fe_3C. The phase, designated as γ in this diagram, is known as *austenite*. Since the eutectoid reaction at ≈723°C is controlled by solid state diffusion, this reaction can be suppressed by a rapid quenching of γ. In that case, austenite (γ) can transform to a metastable phase known as *martensite*. Martensite is the phase responsible for the dramatic increase of hardness of steels upon quenching. A steel with < 0.8% C is called a *hypoeutectoid* steel, and one with > 0.8% is called a *hypereutectoid* steel.

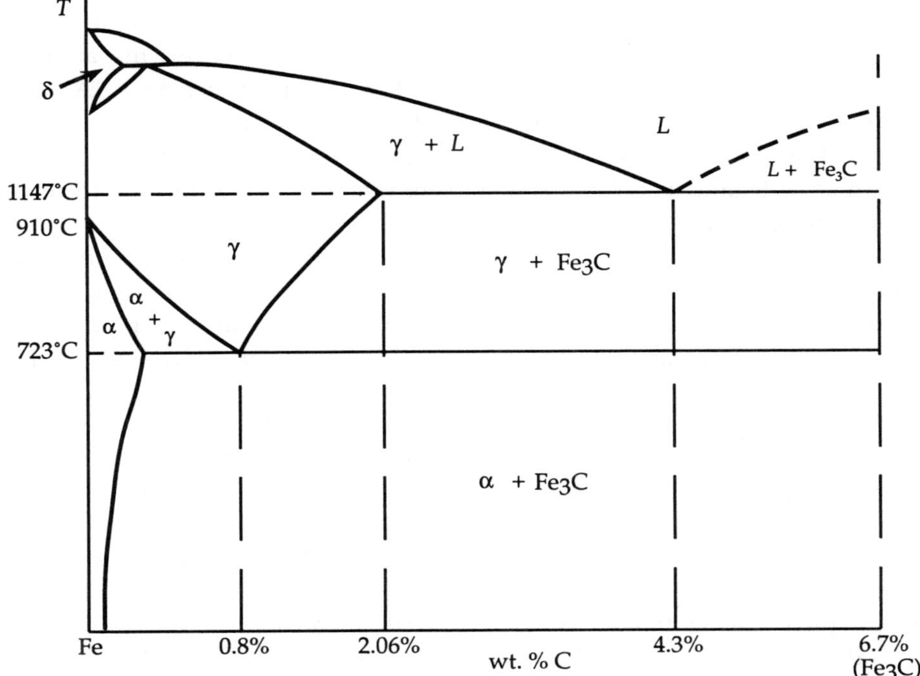

(a) Fe-C phase diagram.

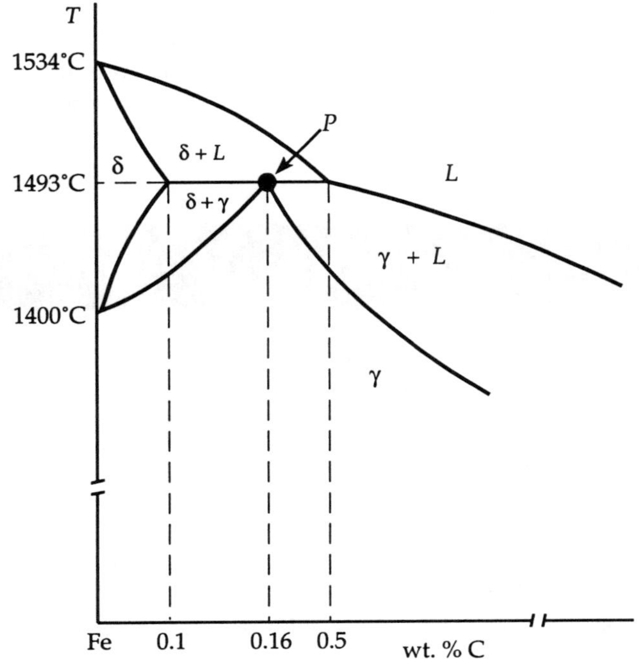

(b) A magnified view of the δ-Fe region. Point P is a peritectic point.

Figure 3.25 The iron-carbon diagram.

3.3 Atomic Bonding and Solid Types

A pure chemical element (or a mixture of elements) can exist in three different forms: (a) gaseous state, (b) liquid state, and (c) solid state. In the gas phase, atoms or molecules are randomly distributed in a given volume, and they are very far apart from each other. They are free to move within the volume of the container that confines the gas. In the liquid form, however, the atoms or molecules are situated at distances comparable to atomic dimensions. A liquid, we know, can not hold its own shape, but takes up the shape of the container. A solid, on the other hand, can maintain its own shape and it has rigidity and strength to withstand a contact pressure. For our present purpose we focus our attention to solids only.

A solid material could be amorphous, crystalline or a mixture of both. In the amorphous form, the atoms or molecules are randomly distributed without any periodicity of arrangement. Glass, for example, is such an amorphous solid. Sometimes we can consider an amorphous solid as a "super cooled liquid." That is, the distribution of the atoms or molecules are similar to that in a liquid, but they are "frozen" in 3-dimensional space by super-cooling bellow its freezing point. In a crystalline solid the atoms or molecules are organized in a distinct 3-dimensional pattern. Like a pattern in a wallpaper, which repeats a basic 'motif' or pattern to fill a large

2-dimensional space, in a crystalline solid, space is filled by the repetition of a 3-dimensional "motif' called an "unit cell" of the solid. Crystallography is a subject which deals with the symbolic presentation and analysis of the unit cell of a crystalline solid. Before we discuss crystallography, it will be instructive to review the nature and type of interatomic force which hold the atoms or molecules in a solid. Basically there are three types of primary bonds between atoms and molecules in an engineering solid. These are: (1) Ionic bonds, (2) Covalent bonds, and (3) Metallic bonds. These bond types are primarily dependent on the electronic structure of the atoms. First, we very briefly review the electronic structure of atoms.

3.4 Electronic Structure of Atoms

Energy of electrons, orbiting around the nucleus of an atom, is quantized. The main quantized energy levels are designated by the primary quantum number n. Where n = 1, 2, 3, etc. Energy levels associated with the quantum number n, is called the nth electronic shell. The number of electrons in a shell is given by $2n^2$. Thus the first shell (n = 1) has 2 electrons, the second shell (n = 2) has 8 electrons, etc.

A second quantum number l, determines the energy sub-levels or sub-shells within a main energy level. The value of the second quantum number can be 0, 1, 2, ...(n-1). A notation of s, p, d, f is used for l = 0, 1, 2, 3 respectively. Thus, when n = 1, l can only be 0 (i.e., n-1), and the first energy level has no sub-level, and the shell is symbolically denoted by 1(s) (1 for n = 1, and s for l being equal to zero). For $n = 2$, l has a value of 0 and 1, i.e., s and p and symbolically, the two sub-shells are denoted as $2(s)$ and $2(p)$ respectively.

Within an energy level, there could be several "quantum states" arising from a third quantum number, called magnetic quantum number. We will use m for this quantum number. This quantum number can have values ranging from $-l$ to $+l$, including 0. Unlike n and l, different value of m, although is a different quantum state, does not give rise to a different value of electronic energy. Different quantum states, which have the same energy are called degenerate states. Note that for $l = 2$ for example, m has values of -2, -1, 0, +1, +2, i.e., there are 5 degenerate quantum states in the 3rd shell ($n = 3$, since $l = n$-1).

A fourth quantum number is the spin quantum number s, which has values of $+1/2$ or $-1/2$. Each quantum state can accommodate two electrons of opposite spins ($+1/2$ or $-1/2$ are conventions of electronic spin up or down, respectively). Pauli exclusion principle states that no more than two electrons (of opposite spin) can occupy a "quantum state." Note that the Pauli exclusion principle applies to quantum states and not to energy levels. Thus, for the above example, n=3 gives $l = 0, 1, 2$, i.e., s, p, and d, respectively. For $l = 0$, that is s state, $m = 0$, and 2 electrons (spin up and spin down) occupy this shell. This is denoted by $3(s)^2$. Note that the first number is for n, and s within the parenthesis stands for the s-state ($l = 0$), and the super script 2 stands for the two electrons as discussed above. In the same example, for $n = 3$, and $l = 1$ that is p-state, m has values of -1, 0, +1; i.e., three quantum states, and each of these states holds two electrons, i.e., a total of 6 electrons. This is denoted by $3(p)^6$. Again, 3 for $n = 3$, p stands for $l = 1$, the superscript 6 is for the six electrons as discussed above. For $n = 3$, and $l = 2$, m = -2, -1, 0, +1, +2, i.e., 5 quantum states. Each of these five quantum states holds two electrons. Thus a total of 10 electrons. This affair is denoted by $3(d)^{10}$. The $n = 3$ electron shell then, consists of 2 electrons in the $3(s)$ state, another 6 electrons in the $3(p)$ state and an additional 10 electrons in $3(d)$ state. Thus a total 18 electrons in the energy level associated with the primary quantum number $n = 3$, and it satisfies the requirement that the total number of electrons is given by $2(3)^2$. In symbolic form , the above mentioned sequence could be written as $3(s)^2 3(p)^6 3(d)^{10}$. It is easy to see that for $n = 1$, we have only $1(s)^2$, for $n = 2$, we have the sequence $2(s)^2 2(p)^6$.

In some elements, the $4(s)$ state might be filled before the $3(d)$ state has all of the ten electrons. This is the case in transition elements such as Fe, Ni, etc. For example, in the above example, if all of the quantum states of an atom up to $3(d)$ were sequentially filled, then there will be a total of 28 electrons in the atom. Since the total number of electrons in an atom is equal to the number of protons in the atom (which defines the atomic number Z). Ni, a transition element, has the atomic number 28. In nickel, the sequence of filed quantum states is given by $1(s)^2 2(s)^2 2(p)^6 3(s)^2 3(p)^6 3(d)^8 4(s)^2$. It is then clear that the $3(d)$ state of Ni is only partially filled.

If we examine the inert gases, He, Ne, Ar, etc., we find that in He the filled electronic shell is $1(s)^2$ (the atomic number of He is 2 and thus a total of 2 electrons), in Ne, the filled shells are sequentially given by $1(s)^2 2(s)^2 2(p)^6$; i.e., 10 electrons, which corresponds to the atomic number of Ne. Similarly for Ar, the sequence is $1(s)^2 2(s)^2 2(p)^6 3(s)^2 3(p)^6$, 18 electrons in total, which corresponds to the atomic number of Ar. In He, there is of course no p-shell, but its s-shell is filled. In Ne, the outer most states are filled; i.e., $2(s)^2 2(p)^6$; in Ar the outer most states $3(s)^2 3(p)^6$ are also filled. Similarly in Kr, and in Xe, the outer most states are given by $4(s)^2 4(p)^6$ and $5(s)^2 5(p)^6$ respectively. Thus, the filled outermost (s) and (p) shells are associated with the inertness or the very high stability of the atom. The electrons which are situated in the outermost s and p states, are called *valence* electrons. In He, the outermost s shell is filled since it has two electrons. Likewise, in Ne, there 8 electrons in the $2(s)2(p)$ states, and in Ar there are 8 electrons in the $3(s)2(p)$ states, and so on. These above mentioned elements have a full count (i.e., 8 electrons except in He) of valence electrons, and we also note that these are chemically inert elements. If we examine the periodic table, we notice that the "metallic elements" have partially filled valence states.

3.5 Ionic Bonding

An electropositive element such as Na and an electronegative element such as Cl can form a strong bond by transferring an electron from Na to Cl. This electron transfer process creates a positively charged Na^+ ion and a negatively charged Cl^- ion. Coulombic force then creates an attraction between these two charges when the pair of ions are not too far from each other. As this attractive force pulls the ions together, and they come in the proximity of each other, a short-range repulsive force then operates due to a repulsion between the negative charge of the electron cloud of the ions. There exists a mutual separation distance at which the attractive and repulsive forces exactly balance each other. At this separation, the potential energy of the system is minimum. The coulomb attractive force is inversely proportion to r^2, where r is the separation distance between the ions. The repulsive force, on the other hand, is inversely proportional to r^n, where $n > 2$ (≈ 6 to about 13). We note that the first derivative of potential energy with respect to separation distance is he force. As a convention, we choose the potential energy resulting from the attractive force as negative, and that resulting from the repulsive force as positive. In Fig. 3.26, we see that the equilibrium position is obtained by the algebraic summation of the attractive and the repulsive potentials.

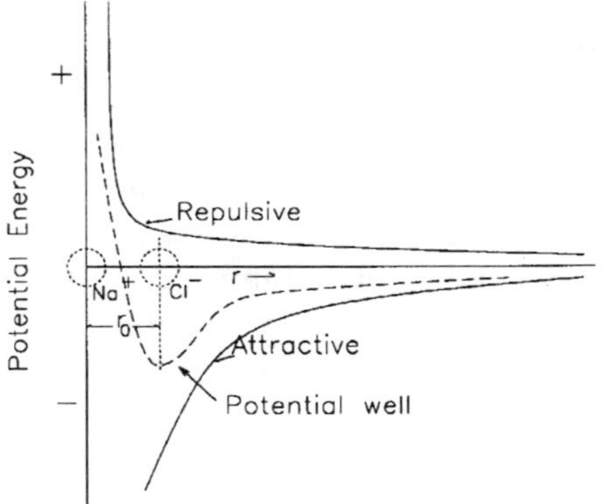

Figure 3.26 Equilibrium positions of atoms as determined by the potential energy minimum. The equilibrium separation distance is r_0.

The resultant potential energy curve has a minimum corresponding to the equilibrium separation. The potential energy minimum, commonly referred to as the "potential well", is a preferred site for an atom in a solid. Even in other types of bonding, potential energy minima associated with equilibrium atomic separation (as shown in Fig. 3.26) exist. What has been said for a pair of ions, can be extended to a group of such ions. The arrangement in such a case is an alternate pairing of + and - ions in the three dimensional space. The situation in a two dimensional plane of a ionic crystal is shown in Fig. 3.27. The ions in this plane are highly organized, and the separation distance a_o is the "interatomic distance" (please note that in the potential energy diagram for a pair of ions, we called this distance r_o).

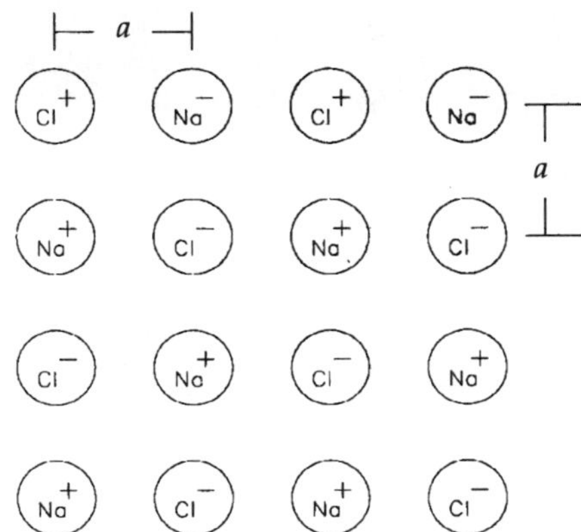

Figure 3.27 A two-dimensional diagram showing a plane of atoms of an ionic solid.

3.6 Covalent Bonding

We saw that in ionic bonding one atom gives up an electron, and another atom gains this electron, thus the donor atom becomes positively charged ion and the other atom becomes a negatively charged ion. Then the coulombic attraction between these two ions is the force that binds a pair of toms. Thus, the genesis of an ionic bond is the opposite polarity of the ions. We could define a bond between two similarly charged ions as a homopolar bonding. In homopolar or covalent bonding, instead of giving up or receiving electrons, the atoms *share* electrons. The bond strength is then derived from this electron sharing. To understand this mechanism, please refer to Section 3.4. Consider two hydrogen atoms adjacent to each other. Hydrogen (atomic number 1) has one election only, and this electron resides in the s-state, i.e., the state is symbolically represented by $1(s)^1$. The "inert" element He (atomic number 2) has two electrons, and its state is represented by $1(s)^2$. Thus, two hydrogen atoms, in close proximity to each other, can share or "pool their two electrons" in a "common orbit" to mimic $1(s)^2$ orbital of the He atom. This bonding in hydrogen is a strong covalent bonding commonly referred to as hydrogen bond. As if the two electrons spend part of the time in the $1(s)$ orbital of each atom. In a similar fashion, in other covalently bonded materials, valence electrons are shared by adjacent atoms in a solid to produce stability, i.e., bonding. The stability is related to the full outermost shell of the type $2(s)^2 2(p)^6$, $3(s)^2 2(p)^6$, etc. The number of nearest neighbors with which an atom can share its electrons, in a homopolar bond, is given by 8-N, where N is the number of valency elections in the atom. Now consider carbon (atomic number 6) which has an electron configuration given by $1(s)^2 2(s)^2 2(p)^2$. That is, the number of valence electrons in carbon is 4, 2 in 2(s) and 2 in 2(p). Hence carbon can make 8 - 4 = 4 bond pairs. Such is the case in diamond, one of the hardest materials, where 4 very strong covalent bonds are made between four nearest carbon atoms in the crystalline solid. Similarly, technologically a very important element Si (extensively used for semiconductors), has 4 valency electrons $3(s)^2 3(p)^2$, and this element crystallizes via a very strong covalent bonding.

3.7 Metallic Bonding

It has been stated earlier that in metallic elements, the number of valance electrons is one or two. The valence electrons are, furthermore, "free" to move around the entire solid piece of metal. The valence elections are no longer "localized" to individual atoms as in the case of ionic and covalent bonds. Instead, these electrons contribute to a common "sea" of electrons, in which the ions are immersed in a three dimensional periodic pattern. The bonding in the solids then is due to an attraction between the positively charged ions and the negatively charged sea of electrons. The motion of these free electrons, through the periodic lattice, however, is governed by quantum mechanical laws. The subject of metallic bonding is much more complicated than that described here. The interested student is advised to read any text book in solid state physics.

3.8 Electrical Properties

3.8.1 Insulators

In a solid in which the bonding between the atoms is either ionic or covalent in nature, valance electrons are localized near the atoms. In ionic solids an electron is actually transferred to the adjacent atom, and in a covalent solid an electron is shared by two adjacent atoms. In either case the electron orbits are tightly bound to the atoms. Thus in these types of solids electrons can not freely move around to produce a current flow when an electrical potential gradient (voltage) is established. These solids are called *insulators*. Insulators have very high electrical resistivity on the order of 10^{16} ohm·cm., compared with a good metallic conductor which has a resistivity on the order of 10^{-6} ohm·cm.

In ionic solids, a current flow can occur if the ions (which carry a positive or a negative charge) can migrate through the solid as charge carriers instead of the electrons. Such an ion mobility, however, requires solid state diffusion. As discussed before, the rate of solid state diffusion increases exponentially with increasing temperature. Thus, the electrical conductivity of ionic solids increases with temperature (i.e., resistivity decreases).

3.8.2 Conductors

In metallic solids, valance electrons are not bound to individual atoms. Instead, these electrons form what may be called a "free electron gas". This electron gas obeys quantum statistical laws rather than the classical statistical mechanics laws which apply to molecular gases. In a crystalline metallic solid, individual atoms are placed in a periodic manner within a 'sea' of free electrons. Since the free electrons are not strongly bound to individual atoms, they can easily move through the solid metal, where a very small electrical potential gradient (voltage) is established. Because of the high mobility of the free electrons, the electrical resistivity of a good metallic conductor like Cu is about 1.7 x 10^{-6} ohm·cm. Electrical resistance, R, of a conductor of length l cm, a cross-sectional area 'a'cm^2 (normal to the direction of current flow) and resistivity ρ, is given by Ohm's law:

$$R = \rho\, l/a \tag{3.8.1}$$

In a metallic conductor, the moving electrons periodically collide with oscillating atoms in the lattice. The average distance traveled by an electron between two successive collisions is called the "mean free path" of an electron. A higher value of the mean free path gives a higher value of conductivity (i.e., lower resistivity). As temperature increases, the amplitude of thermal oscillation of lattice atoms increases, and the mean free path decreases. Thus, in metallic solids resistivity (thus resistance) increases with increasing temperature. Note that this behavior is opposite of that for an ionic conductor as discussed before.

3.8.3 Energy Band

Free atoms have discrete quantum states or levels as has been discussed before. For example Na (a metallic element of atomic number 11) has 11 electrons in energy levels described by: $1(s)^2 2(s)^2 2(p)^6 3(s)^1$. When sodium atoms are brought close together to form a solid, the electronic energy levels of individual atoms overlap. As the atoms come closer together, the energy levels split, and each discrete energy level becomes an energy band with quasi-continuous energy sublevels within each band. If we consider a solid Na block, containing N atoms of Na (for example $N \approx 10^{23}$ for one gram-atom of Na), then $1(S)$ and $2(s)$ bands each contain $2N$ electrons, and $2(p)$ band contains 6N electrons (each N atoms contributing 6 electrons to this band). The outermost band $3(s)$ can accommodate $2N$ electrons, however each Na atom can only contribute 1 electron to this band. Thus the $3(s)$ band is *only half full*. Figure 3.28 shows a schematic picture of energy levels and energy bands of a single Na atom and a crystalline solid block of Na containing N atoms.

Within each energy band Pauli exclusion principle holds, and no more than 2 electrons of opposite spin can stay in each of the states within an energy band as described above. Each band has a quasi-continuous set of energy levels. Electrons fill the bands from the lowest to the highest energy levels. This is a simple manifestation of energy minimization principle. When additional energy is supplied to an electron by either supplying thermal energy or by setting up an electrical potential, the kinetic energy of an electron must increase. Quantum mechanically, then this electron must move up to a higher energy level. Since all of the energy levels from the ground level up are filled, and since Pauli exclusion principle applies, the only electrons which can take up the extra energy and move up to a higher empty level are those near the top of a band.

The probability of finding an electron in a particular energy state E, is given by Fermi-Dirac distribution function $P(E)$,

$$P(E) = 1/\{[e(E - Ef)/kT] - 1\} \tag{3.8.2}$$

where E_f is called the Fermi energy, k is Boltzmann's constant and T is the Kelvin temperature. At $T = 0$, $P(E) = 1$ for $E < E_f$, and $P(E) = 0$ for $E > E_f$. At $T > 0$ K, $P(E) = 1/2$ for $E = E_f$. Figure 3.29 (a) and (b) show the variation of $P(E)$ as a function of energy E at $T = 0$, and at $T > 0$ K, respectively. We note form the above discussion, that at temperatures $T > 0$ K, a small fraction of the energy levels below the Fermi energy are empty, and a small fraction of energy levels above the Fermi level are occupied.

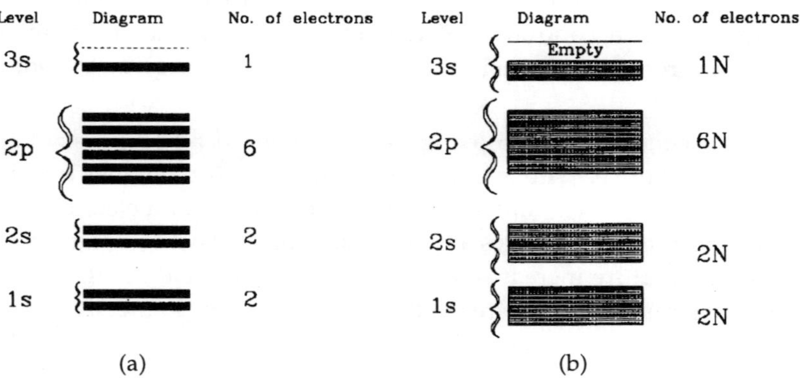

Figure 3.28 Schematic energy level diagram of (a) a single Na atom, and (b) the energy bands of a solid Na containing *N* atoms.

At $T = 0$ K, the Fermi level E_f coincides with the middle line of the half-filled 3(*s*) band in Na. When an electric field is applied at a temperature greater than absolute zero, the electrons very near the Fermi energy can acquire incremental kinetic energy, and thus must move up to a higher empty level. Note that the energy bands are separated by forbidden energy gaps. If a solid is such that the top most band is completely full, which is followed by an energy gap and a completely empty band above that, then the conductivity of this solid depends on the magnitude of the energy gap E_g. At a temperature T, if E_g is less than kT (note that kT has the dimension of energy), then conductivity arises when electrons are raised to the upper empty band. The upper empty band is called "conduction band", and the filled band immediately below it is called the "valence band". In a good insulator, E_g is very large; on the order of 10 electron volts (eV).

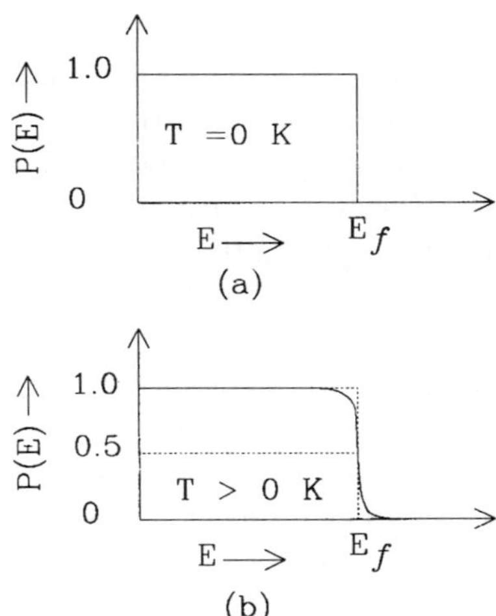

Figure 3.29 *P(E)* vs. *E* plots; (*a*) showing that at $T = 0$ K the probability of finding an electron up to Fermi level is 1; (*b*) at $T > 0$ K, there is a finite probability of finding some electrons above the Fermi level.

3.8.4 Semiconductors

In a material in which the value of E_g is relatively small (on the order of 1 eV), conduction can occur at higher temperatures through electrons which are raised from the full valence band to the empty conduction band across the band gap, via thermal activation. The thermally activated electrons, which are placed in the conduction band, can now move under the influence of an electrical potential. At $T = 0$ K, the conduction band is empty, and has no current flows when an electrical potential is applied. Such a material is called an "intrinsic semiconductor". Si, and Ge are two very common intrinsic semiconductors used in technology today.

Adding impurities, called "doping," to a semiconductor can significantly modify the electrical properties. If an impurity atom is such that it adds an electronic energy level within the E_g and near the bottom of the conduction band, then the electrons form this impurity level can be thermally activated to the empty conduction band, and thus provide a current flow when a voltage is applied. The impurity in this case is called a "donor", since it donates electrons to the conduction band. The energy gap between the donor level and the bottom of the conduction band is typically 0.05 eV. The donor level provides negatively charged electrons for conduction in such a doped material. The semiconductor doped in this fashion is called an "n-type" semiconductor. If however, an added impurity is such that it creates an electron deficient level near the top of the valence band, and this level is separated by an energy gap on the order of 0.05 eV, then electrons can be thermally activated to this donor band, and once again conduction could occur. This impurity level is called "acceptor", since it accepts electrons. When electrons from the filled valence band are promoted to the acceptor level, "holes" are left in the previously filled valence band. Conduction occurs as if by the migration of "holes" in such a semiconductor. Compared with the n-type semiconductors where the charge carrier is negative (electrons), this hole conduction type semiconductors are called p-type (positive carriers) semiconductors. Materials in which semiconducting behavior is ushered in by doping of impurities, are called "extrinsic semiconductors".

3.9 Crystalline State and Crystallography

The glassy or the amorphous state, as discussed earlier, has no long range periodicity of atomic positions. As opposed to an amorphous solid, a crystalline solid is composed of atoms (henceforth "atoms" will be interchangeably used for ions, molecules or atoms) which are arranged in regular periodic patterns. These solid materials where the atoms are arranged in a regular periodic and repetitive manner, following certain geometric and mathematical laws, are called crystalline solids. Many of the engineering solids are crystalline in nature. By long range periodicity we mean periodicity over thousands of atoms in any given direction.

3.9.1 Unit Cell and Lattice Parameters

The fundamental building block of a crystalline solid, in an atomic scale, is called an unit cell or a lattice. A three dimensional stacking of such unit cells produces a macroscopic crystalline solid. An unit cell or a lattice can be defined by three vectors x, y, and z as shown in Fig. 3.30.

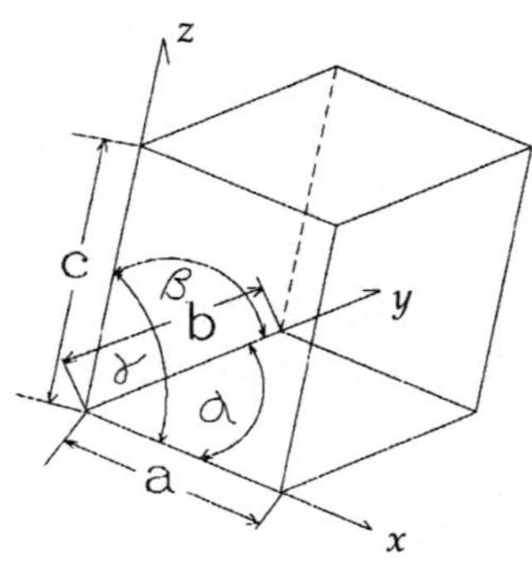

Figure 3.30 Unit cell of a crystal lattice. The dimensions a, b, c and the angles α, β, and γ are the lattice parameters.

In Fig. 3.30, a, b, and c define the dimensions of the three sides of the parallelepiped. The three mutually inclined (non-orthogonal) vectors x, y, and z are chosen to represent a general case. If we now put eight atoms at the eight corners of the parallelepiped shown in Fig. 3.30, then we have, what we may call an unit cell. Note that the size and shape of this unit cell are determined by the three angles, and the three lengths of the three adjacent sides. A special case is when $a = b = c$, and $\alpha = \beta = \gamma = 90^{\circ}$. In that case the shape is a cube, and the unit cell, as shown in Fig. 3.6, is called a simple cubic lattice. The quantities a, b, c, α, β, and γ are called lattice parameters. For an orthogonal axis system, such as the cubic crystal, only one lattice parameter, a, is sufficient to define the unit cell size and shape.

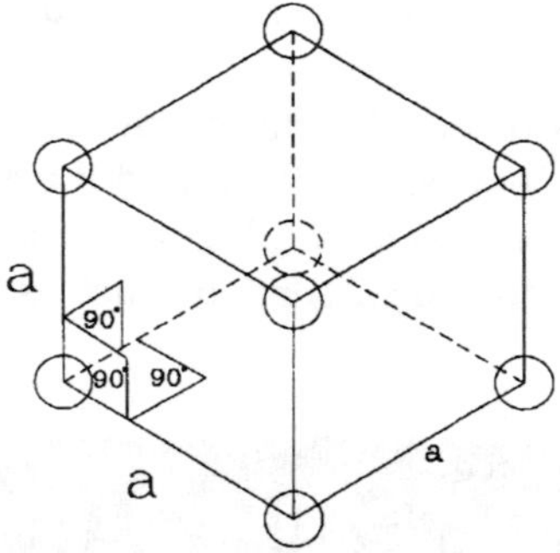

Figure 3.31 A simple cubic unit cell or a simple cubic lattice.

3.9.2 Seven Crystal Systems

Only seven distinct shapes of unit cells can be generated by various combinations of values of a, b, c, α, β, and γ. These are called the seven crystal systems. Note that these crystal unit cells are defined by putting one atom in each corner of the cell. These seven systems are shown in Fig. 3.32.

Note that in the case of hexagonal crystal, only one third of the volume shown in Fig. 3.32 is taken as the unit cell. This volume is a rectangular parallelepiped with two sides equal (shown as a in Fig. 3.32), and the third side is c. The included angle between the two axes, defined as α, is 120°. The various relationships between the dimensions and the angles for the seven crystal systems are shown in Table 3.2.

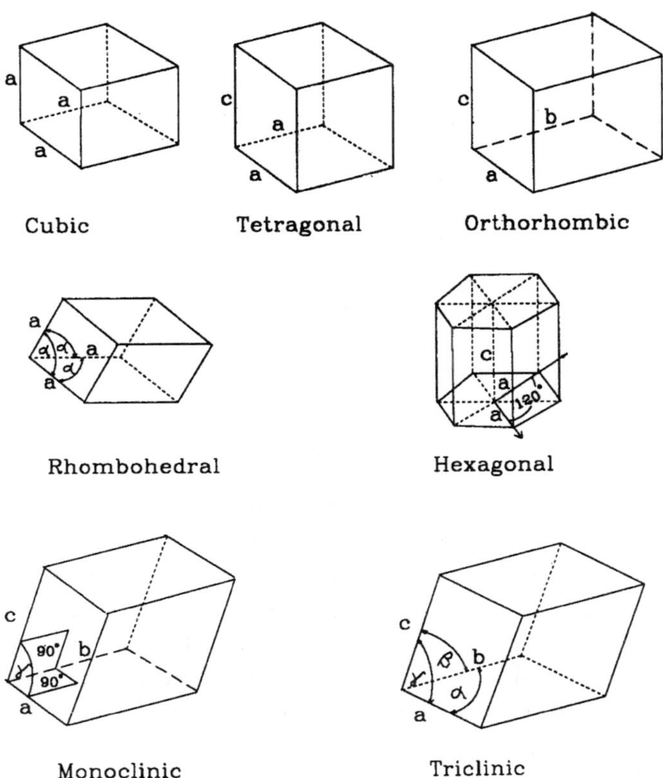

Figure 3.32 Seven crystal systems.

TABLE 3-2. Properties of the seven crystal systems

Crystal system	Dimensions of a, b, and c	Angles between the three axes
Cubic	$a = b = c$	$a = \beta = \gamma = 90^\circ$
Tetragonal	$a = b \neq c$	$\alpha = \beta = \gamma = 90^\circ$
Orthorhombic	$a \neq b \neq c$	$\alpha = \beta = \gamma = 90^\circ$
Rhombohedral	$a = b = c$	$\alpha = \beta = \gamma \neq 90^\circ$
Hexagonal	$a = b \neq c$	$\alpha = 120^\circ, \beta = \gamma = 90^\circ$
Monoclinic	$a \neq b \neq c$	$\alpha = \beta = 90^\circ \neq \gamma$
Triclinic	$a \neq b \neq c$	$\alpha \neq \beta \neq \gamma \neq 90^\circ$

3.9.3 Bravais Lattice and Atom Positions

If we place an atom (remember that we are using "atom" as a generic term for atoms, molecules or ions) at the eight corners of each of the seven cells shown in Fig. 3.32, we obtain seven unit cells. Please note again that for the hexagonal cell, we take 1/3 of the volume shown in (5) of Fig. 3.32. This volume is defined by two axes which have an included angle of 120°, and the third axis is perpendicular to the plane containing the other two axes. With these seven types of elementary unit cells, we can fill up a three dimensional space by stacking them together, much like staking bricks to form a massive structure. If we examine such a large mass, we see hat the adjacent atoms along any direction are obtained by "translations" of vectors along the three defining vectors (called basis vectors) of the unit cell. This is shown in Fig. 3.33. This space filling also produces an identical surrounding for each atom (i.e., number and distance of nearest neighbors to each atom).

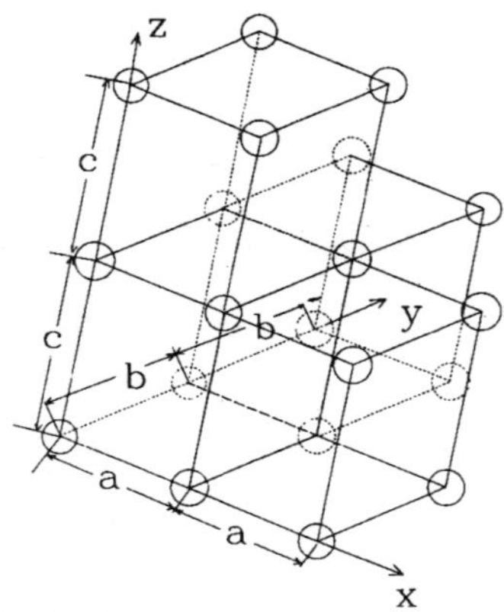

Figure 3.33 Unit translation of atoms along x, y, and z directions for 3-dimensional space filling.

It has been shown mathematically that the above mentioned seven unit cells are not the only possibilities where each atom has an identical surrounding. In fact, altogether fourteen unit cells can be constructed form the above mentioned seven cells where each atom has an identical surroundings. These fourteen are called the Bravais lattices. Because of the brevity of this treatment, we will consider only the cubic and the hexagonal systems, which include a large number of engineering solids.

3.9.4 Simple Cubic Lattice

A simple is very easy to visualize. It consists of an elementary cubic, in which the eight corners of the cube being occupied by eight atoms. This unit cell is already shown in Fig. 3.31.

3.9.5 Body Centered Cubic lattice (BCC)

A body centered unit cell is shown in Fig. 3.34. Aside from the eight corners, there is another atom situated in the interior of the cube. This is at the center of the body, defined by the point of intersections of he body diagonals as shown, and thus the name body centered cubic (BCC) is used. Many metals and alloys crystallize in this form (notably iron and steel).

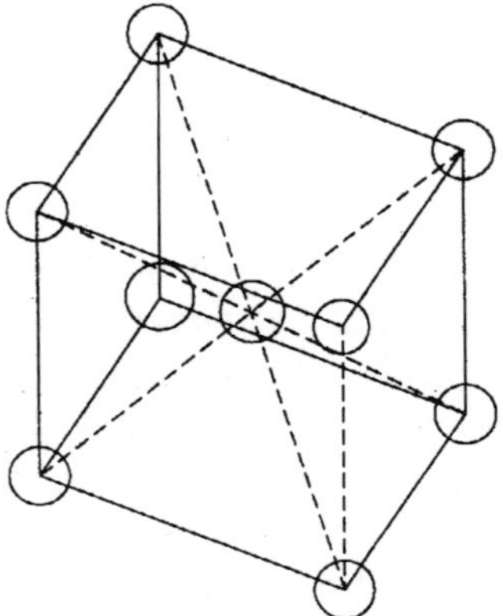

Figure 3.34 Body centered cubic lattice (BCC).

3.9.6 Face Centered Cubic lattice (FCC)

A face centered unit cell (FCC) is shown in Fig. 3.35a. Here, six additional atoms are located at the center of each face of the cube. Hence the name face centered cubic (FCC) is used. The six atom positions are obtained by drawing diagonal lines from he opposite corners of each face as shown in Fig. 3.35b.

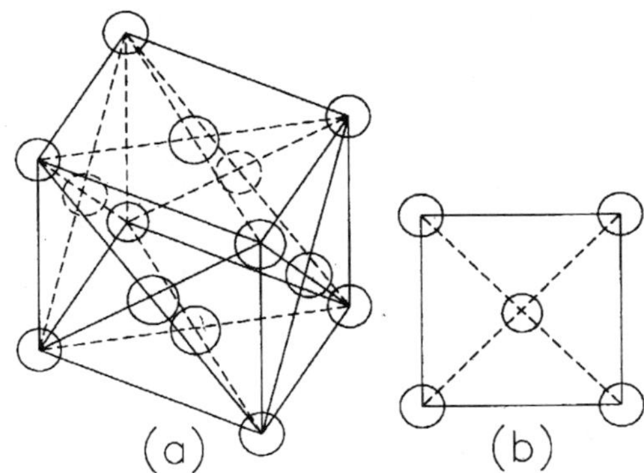

Figure 3.35 (a) A face centered cubic lattice. (b) The location of the face center atom is at the intersection of the face diagonals.

Note that an FCC lattice is a "close-packed" lattice. By this it is meant that, if the atoms are assumed to be like ball bearings of equal size, then this structure provides the maximum packing density. In Fig. 3.36a is shown a plane of atoms in which the atoms are most closely packed. The arrangement of atoms in this plane is shown in Fig. 3.36b. We will return to this subject later.

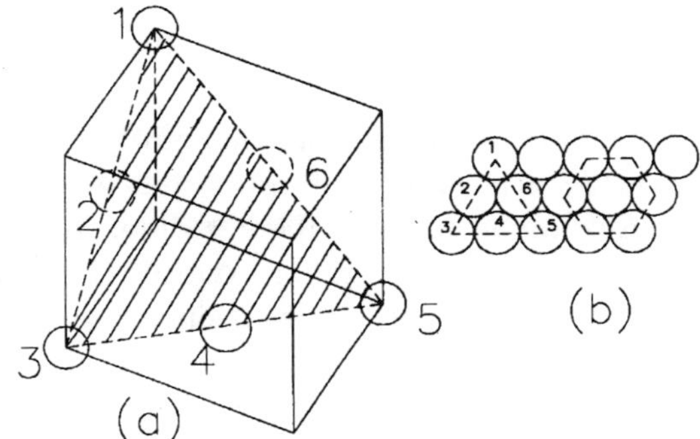

Figure 3.36 (a) Shows a "close packed" plane of the FCC lattice (not all atoms of the cell are shown. (b) Shows the arrangement of atoms in this plane.

3.9.7 Hexagonal Close Packed lattice (HCP)

A hexagonal unit cell is shown in Fig. 3.37a. This lattice is <u>not</u> a close packed cell. Fig. 3.37b is a hexagonal close packed cell (HCP). Note in 3.37b, that there are three additional atoms inside the hexagonal frame, aside from the external corner atoms (not shown in (b) for clarity). The close packed plane is the bottom plane (called the basal plane) and it is shown in Fig. 3.37c. The second layer of atoms (i.e. the interior three atoms) are shown by dotted circles in this figure. Please compare figures 3.36b and 3.37c.

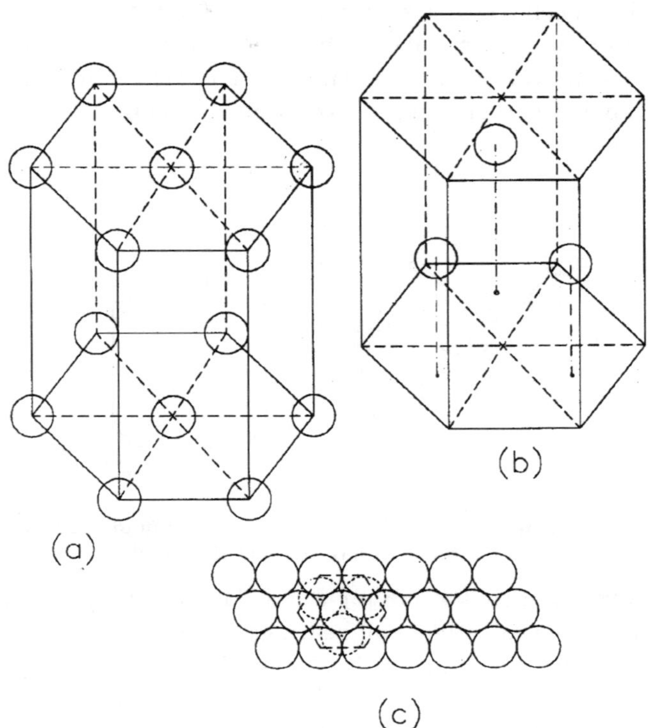

Figure 3.37 Hexagonal lattice. (a) Simple hexagonal, (b) hexagonal close packed HCP, and (c) atomic packing on the basal plane.

3.9.8 Miller Indices of Planes and Directions

Miller indices is a system of notations used for denoting planes and directions in crystals. Letters *h*, *k*, and *l* are used as the indices, and these indices are usually written as (*hkl*) without commas between them. Miller indices are all integers, and they do not contain a common factor. For example (111), (121), (220) etc. are permissible but indices (222), (242) are not used, since taking the common factor out (222) is reduced to (111), and (242) is reduced to (121) etc.

If *a*, *b*, and *c* are the three dimensions of an unit cell as discussed in Section 3.9.1, then a plane with Miller indices (*hkl*) intersects the three coordinate axes *x*, *y*, and *z* as shown in Fig. 3.38a. The intercepts are *a/h*, *b/k*, and *c/l* respectively. If a plane intersects the three axes of the crystal at distances equal to the three dimensions *a*, *b*, and *c* of the unit cell, then the miller indices of the plane is (111). For example the cross-hatched plane shown in Fig. 3.38a, intersects *x*, *y*, and *z* axes at distances equal to the dimension of the side of the cube, and thus this plane is (111). Note therefore, that (111) plane of an FCC crystal is the close packed plane. To determine the Miller indices of a plane we take the intercepts of the plane in terms of the fractions of the lattice dimensions along the three directions. As an example consider an unit cell of dimensions *a*, *b*, and *c*, and a plane which intersects *x*-axis at a distance of 1 *a*, *b*-axis at a distance of 1/2 b, and 1/2 c. Thus the intercepts are 1, 1/2, 1/2. We now take reciprocal of these intercepts, i.e. 1, 2, 2, and write it in the form (122), which is the Miller indices of the plane. Note that this gives the intercepts $a/1$, $b/2$, and $c/2$ as discussed earlier. A plane which intersects the three axes at $a/2$, $b/2$, and $c/2$, has intercepts 1/2, 1/2, 1/2. Reciprocals of which are 2, 2, 2. We eliminate the common factor and write (111) as the Miller indices of that plane. Note that what it means is that (222) plane is parallel to (111) plane. Note that if a plane is parallel to an axis, then the intercept is infinity (the reciprocal is zero). Thus, for example we recognize that (110) plane is parallel to z-axis. A plane which intersects x-axis at unit distance, is parallel to y and z-axes has Miller indices (100), etc.

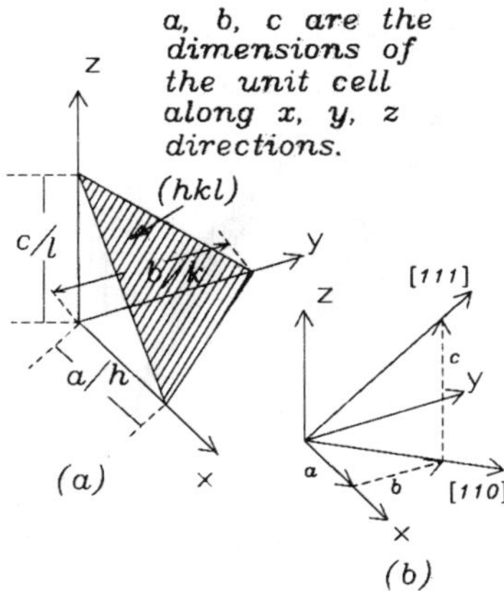

Figure 3.38 Miller indices of planes and directions. (a) Planes, (b) directions.

Miller indices of a direction is obtained by the vector sum of the component vectors along the three axes of the unit cell. Usually Miller indices of a direction is written as [hkl]. for example, [110] is a direction on the x-y plane obtained by adding one lattice vector along x-direction and one lattice vector along y-direction as shown in 3.38b. Fig. 3.38b also shows the [111] direction.

3.9.9 Primitive Cell

An unit cell which only has corner atoms (ions, molecules) is called a primitive cell. Thus, a simple cubic lattice is a primative cell Similarly, a hexagonal lattice (note that one third of the hexagon, as discussed earlier, is the unit cell) only has corner atoms, and thus it is a primitive lattice. However, BCC, FCC, and HCP are not primitive cells.

3.9.10 Number of Atoms per Unit Cell

Let us take a simple cubic lattice. There are eight atoms at the eight corners of a single cell. However, when we pack unit cells to fill up a three-dimensional space, each corner atom is shared by eight other adjacent cells. Thus on the average, only one eighth of a corer atom belongs to one unit cell. Hence, 1/8 x 8 = 1 atom per unit cell. In FCC, each atom at the center of each of the six faces are shared by two adjacent cells during stacking of such cubes. Thus, on the average 1/2 of each face atom belongs to one cell. Thus, in FCC, six faces have each half an atom, and each corner has 1/8 of an atom. Therefore, the total number of atoms per unit cell in FCC is (1/8 x 8 + 1/2 x 6) = 4 atoms. In BCC, the body center atom is not shared by any other cells, and thus BCC has 2 atoms per unit cell.

3.9.11 Interplaner Spacing d_{hkl}

In many problems in materials science, the distance between parallel planes in a given lattice is an important parameter. Inter planer distance d_{hkl} is the perpendicular distance between equivalent planes as shown in Fig. 3.39. The planes shown are of the type are (100), (110) and (111). Note that (100) planes are parallel to y and z-axes, with an intercept on x-axis equal to the unit cell dimension in that direction. Similarly, (110) is parallel to z-axis with unit intercepts along x and y axes. For the (111) plane the intercepts along x, y and z directions are equal to the unit cell dimensions along those directions. Distance between crystallographic planes of an unit cell is expressed in Angstrom unit. One Angstrom is equal to 10^{-8} cm, and it is written as Å.

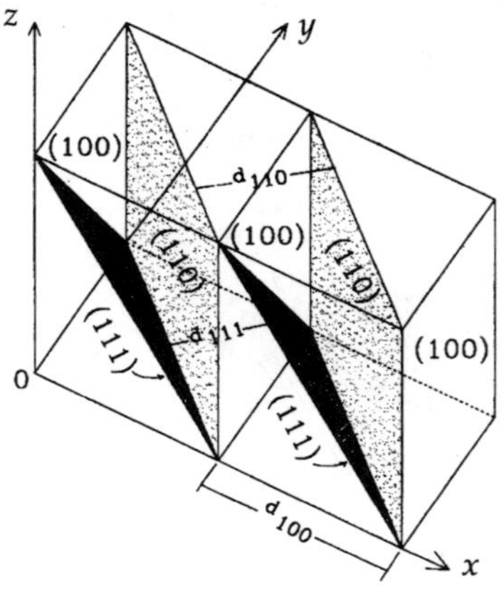

Figure 3.39 Interplaner distance d_{hkl}.

3.9.12 Close Packing of Atoms, and Packing Factor

FCC and HCP are two of the close packed lattices. If we try to arrange a number of dimes or nickels in such a way as to have a most efficient coverage of a two dimensional space, we obtain a configuration similar to the arrangement of atoms on the (111) planes of a FCC lattice or the basal plane of a HCP lattice (see Figs. 3.36a and 3.37c). In examining the close packed arrangement of the coins, we find that there are empty spaces around them. Similarly in close packing of hard spheres, we have empty spaces surrounding the spheres. Thus, the volume bounded by the unit cell consists of the volume occupied by the atoms plus he empty space. We could define an atomic packing factor (APF) as the ratio of the volume of atoms to the volume of the cell.

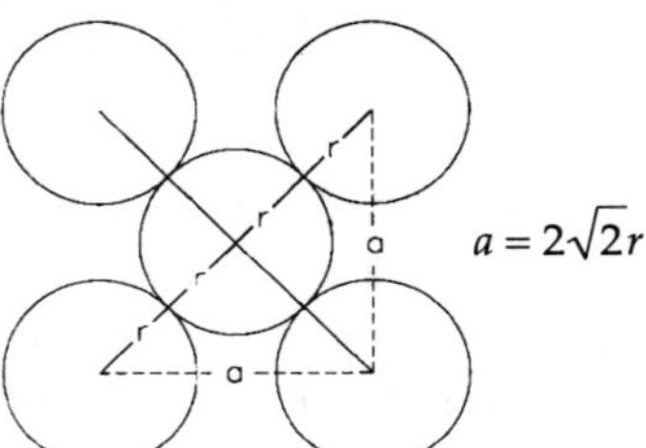

Figure 3.40 Atomic packing on a face plane of a FCC lattice.

Figure 3.40 shows the arrangement of atoms on one of the faces of a FCC unit cell. We note that the face diagonal is equal to four atomic radii, $4r$. The dimension of the side of the square faces is 'a'. Thus from the geometry, we have,

$$a = 2\sqrt{2}r \quad \textbf{(3.9.1)}$$

In a FCC unit cell there are 4 atoms per cell as seen before. The volume of four atoms is given by

$$\text{volume occupied by atoms} = 4 \times (4/3)\,\pi\, r^3 \quad \textbf{(3.9.2)}$$

From Eq. 3.9.1, we can write:

$$\text{volume of the unit cell} = a^3 = (2\sqrt{2})^3 r^3 \quad \textbf{(3.9.3)}$$

Thus, for FCC, the atomic packing factor is given by

$$\text{APF} = [\,4 \times (4/3)\,\pi\, r^3]/[\,(2\sqrt{2})^3 r^3] = 0.74 \quad \textbf{(3.9.4)}$$

In a BCC cell, the atoms touch along the body diagonal as shown in Fig. 3.41a. The geometry of the body diagonal and its relationship with the unit cell dimension 'a' is shown in Fig. 3.41b.

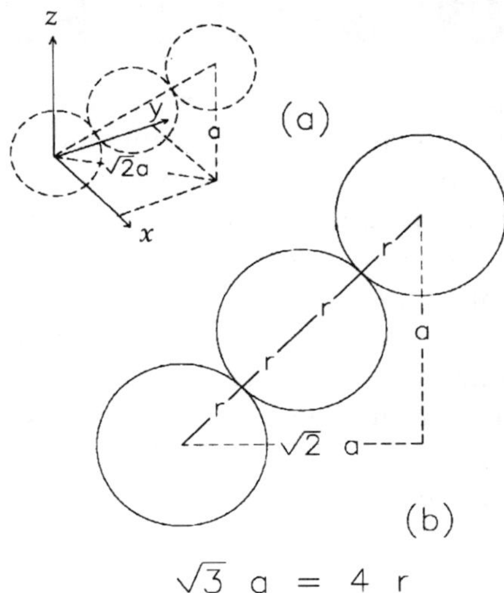

Figure 3.41 Atomic packing along a body diagonal of a BCC cell.

From Fig. 3.41b, we see that

$$\sqrt{3}\, a = 4r \tag{3.9.5}$$

In a BCC unit cell there are 2 atoms per unit cell. The volume occupied by the atoms in the cell is given by

$$\text{volume occupied by atoms} = 2 \times (4/3)\, \pi\, r^3 \tag{3.9.6}$$

From equation (3.75) we can write:

$$\text{volume of the unit cell} = a^3 = (4/\sqrt{3})^3 r^3 \tag{3.9.7}$$

Thus, for BCC, the atomic packing factor is given by

$$\text{APF} = [\, 2 \times (4/3)\, \pi\, r^3]/[\, (4/\sqrt{3})^3 r^3] = 0.68 \tag{3.9.8}$$

We note that the packing efficiency in FCC is higher than in BCC. As the name implies, FCC is a close packed structure, whereas BCC is not a close packed structure.

3.9.13 X-Ray Crystallography

X-ray diffraction is a powerful tool for crystal structure determination. For example, if we determine the spacing of the planes which define the faces of an unit cell, we can determine the size and shape of the unit cell. The well known Bragg's law of X-ray diffraction is given by:

$$2d_{hkl} \sin \theta = n\, \gamma \tag{3.9.9}$$

Where d_{hkl} is the spacing of a set of planes for which an X-ray diffraction maximum is obtained when he angle of incidence of the X-ray beam, measured from the surface of the plane, is θ. γ is the wave length of the incident X-ray, and n is an integer called the order of reflection, and it can have values of 1, 2, 3, ... $N = 1$ is the first order reflection, $n = 2$ is the second order reflection, etc.

3.10 Atomic Mobility, Solid State Diffusion, and Atomic Scale Defects

In the solid state, migration of atoms from one location to another (one lattice site to another) occurs very slowly at ordinary temperatures. However, this atomic migration (called diffusion), at elevated temperatures can cause many important changes, such as recrystallization, precipitation of phases, decomposition of phases etc. Diffusion also plays a role in the high-temperature creep, and creep rupture failure of metals and alloys.

In a crystalline solid, each atomic site is a "potential well" corresponding to the potential energy minima as illustrated in Fig. 3.42. Adjacent atoms in a crystalline solid are situated at the bottom of a periodic well. Separating any two such wells, is a potential energy hill. Thus diffusion requires that an atom has sufficient energy to climb the hill.

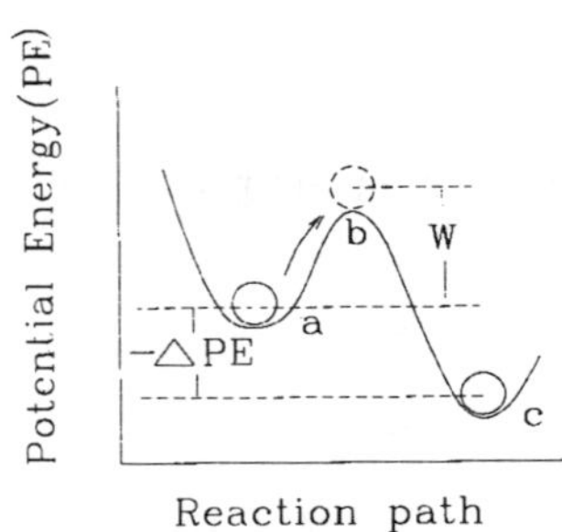

Figure 3.42 Potential wells, potential energy barrier and atomic migration in a solid.

The concept of potential energy barrier, in atomic migration is illustrated in Fig. 3.42. In the solid sate, atoms are constantly oscillating about the lattice sites. As the temperature increases, the amplitude of oscillation of the atoms increases. In the figure, the potential well on the right is deeper (i.e., a site with lower potential energy) than the one on the left of the diagram. A net release of potential energy equal to ΔPE occurs if the atom from site a can move over to site c. However, separating sites a and b is the hill of height W. As the temperature increases, the thermal oscillation of the atom increases, and during its oscillations toward the hill it attempts to surmount the hill. The rate or the kinetics of the atomic migration, in such a case, is then determined by the number of attempts the atom makes per unit time to approach the top of he hill, and the probability that this atom has the additional energy W. This process is called thermally activated process. The number of attempts an atom makes is related to its vibration frequency (number of cycles per second). The rate of such a thermally activated process is given by

$$\text{rate} = f\, e^{-W/kT} \qquad \textbf{(3.10.1)}$$

where, f is the oscillation frequency of the atom, W is the activation barrier, called the activation energy, k is Boltzmann's constant, and T is Kelvin temperature. Note that the rate of any such process increases exponentially with temperature. Note also that an atom can move to another lattice site only if that site is empty. Indeed, such empty sites exists in crystalline solids, and they are called vacancies. Thus, the presence of vacancies is essential to solid state diffusion. Vacancies are also known as point defects.

3.11 Metallic Corrosion

Metallic corrosion is intuitively identified with rusting and perforation of automobile bodies. Billions of dollars are wasted every year due to corrosion damage of engineering structures such as bridges, highways (steel reinforcement bars), storage tanks, automobiles, machineries and many others. The basic phenomena of corrosion can be recognized as "rusting". The fundamental mechanism of rusting is oxidation-reduction reactions. It is recognized that oxidation reactions go hand in hand with reduction reactions. In corrosion, this oxidation-reduction process is electrochemical in nature.

Oxidation is associated with anodic reaction as in an anode of a battery. Reduction reaction, on the other hand is associated with a cathodic reaction. Oxidation of a metal (anodic) takes place when the metal gives up its valance electrons. Reduction, a cathodic reaction, on the other hand occurs through a reaction which accepts the electrons liberated by the anodic reaction. For example, oxidation of two iron atoms could be expressed as follows:

$$2\ Fe \rightarrow 2\ Fe^{+2} + 4e \qquad \textbf{(3.11.1)}$$

where each divalent Fe atom gives up two of its valence electrons to become an Fe^{+2} ions. There must be a corresponding reduction reaction to accept these four liberated electrons. Such a reduction reaction occurs in oxygenated water as follows:

$$O_2 + 2H_2O + 4e \rightarrow 4(OH)^- \qquad \textbf{(3.11.2)}$$

That is, oxygen and two molecules of water combine with four electrons to produce four negatively charged $(OH)^-$ ions. It can be seen that the two Fe^{+2} ions (total positive charge of +4), can be neutralized through the reaction with four $(OH)^-$ ions (total charge -4). This is the basis of oxidation or rusting of iron. Please note that iron will not rust in water which does not have dissolved air or oxygen.

Another form of corrosion, often encountered in improperly designed engineering structures which combine different metals and alloys. This type of corrosion is called "galvanic corrosion." When two dissimilar metals (or alloys) are immersed in a saline solution or in a weak acid solution, and there exists an electrical contact between them, then one of the two metals corrodes away, and an electrical current flows between them (this is how batteries operate. The metal that corrodes, acts as the anode. Which of the two metals corrodes, will depend on their relative positions on the "Galvanic Series". An *emf* or a Galvanic series can be constructed starting from the most corrosion resistance (noble metal or cathodic metal) to the least corrosion (anodic, or base metal) metals and alloys. A partial list of such a galvanic series is shown below.

Noble Anodic →→→ Base
Pt, Au, ...Ag,......Ni, ...Cu, ...Sn, Pb, ..., Fe, ...Al, ...Zn, Mg
←←← Cathodic

It is essential for a design engineer to consult a table of galvanic series before selecting dissimilar metallic materials to be incorporated in a design.

Practice Problems

3.1 Deformation is irreversible if

a) Hooke's law holds.
b) the deformation rate is slow.
c) applied stress is greater than the yield stress.
d) there is no work hardening.
e) modulus of elasticity is low.

3.2 Off-set yield stress is defined as

a) stress at which yielding starts.
b) stress required for inhomogenous deformation.
c) stress corresponding to a specific plastic strain.
d) stress at the proportional limit.
e) a specified percent of fracture stress.

3.3 Ultimate tensile strength is

a) the theoretical strength of metal.
b) the stress to initiate plastic flow.
c) the fracture stress.
d) the maximum engineering stress.
e) the stress at 0.2% plastic strain.

3.4 All of the following statements about fracture are correct, except:

a) extent of necking depends on ductility
b) ductile fracture produces a cup and cone fracture surface
c) in a ductile fracture the reduction in area is zero
d) a low carbon steel can be more brittle at low temperatures
e) high strain rate is more conducive to brittle fracture

3.5 All of the following statements about yield-point phenomenon are true, except:

a) dislocation motion is responsible for yield point
b) presence of carbon in iron is responsible for yield point in steels
c) upper and lower yield points are found in many aluminum alloys
d) yield point indicates inhomogenous deformation
e) high strain rate accentuates yield point

3.6 Plastic deformation is caused by

a) stored elastic energy.
b) dislocation motion.
c) low value of modulus.
d) complex elastic stresses.
e) motion of vacancies.

3.7 All of the following statements regarding slip deformation are correct, except:

a) a volume change is associated with slip
b) slip occurs along close-packed directions
c) some crystals have more slip systems than others
d) dislocation motion is necessary for slip
e) slip is irreversible

3.8 All of the following statements regarding metal fatigue are correct, except:

a) surface roughness decreases fatigue life
b) thermal cycling of a metal can produce fatigue failure
c) carburizing treatment can improve fatigue life
d) a fatigue endurance limit is observed in pure aluminum
e) fatigue failure can occur at stresses below the yield stress

3.9 A solid-solution alloy is

a) produced by melting two metals together which do not mix in the solid state.
b) an alloy of two solid phases.
c) homogeneous and single phase at ordinary temperatures.
d) quickly attacked by an acid.
e) an alloy which does not have any grain boundaries.

3.10 All of the following statements regarding a eutectic alloy are correct, except:

a) eutectic temperature is invariant
b) an alloy of eutectic composition solidifies within a range of temperatures
c) two elements must be partially insoluble in the solid state to form a eutectic
d) eutectic microstructure is easily detectable under a microscope
e) in a eutectic reaction, a liquid decomposes into two solids

3.11 A peritectic reaction is defined as

a) two solids reacting to form a liquid.
b) two liquids reacting to form a solid.
c) a liquid and a solid reacting to form another solid.
d) two solids reacting to form a third solid.
e) a liquid separating into two solids.

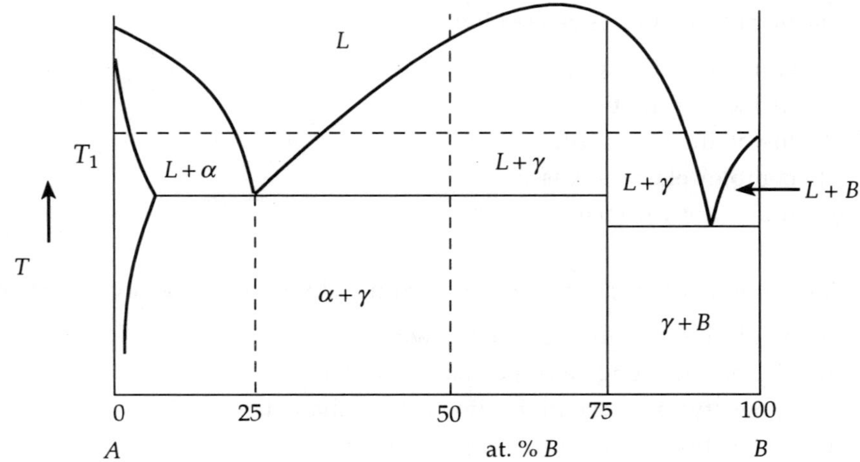

3.12 In the above diagram, which of the following compound or compounds are present?
a) AB b) AB_2 c) A_3B_2 d) AB_3 e) A_2B_3

3.13 For an alloy of A + 50 at. % B as shown in the diagram for Problem 3.12, which phases are present at the temperature T_1?

a) single phase liquid (L)
b) liquid + γ
c) liquid + α
d) $\alpha + \gamma$
e) $\gamma + \beta$

3.14 For the iron-carbon diagram shown in Fig. 3.25, all of the following are true, except:

a) a eutectoid reaction at 1147°C
b) a peritectic reaction at 1493°C
c) maximum carbon content of austenite is 2.06%
d) γ to δ transformation temperature is 1400°C
e) eutectoid carbon composition is 0.8%

3.15 A binary alloy of composition 50 wt. % A and 50 wt. % B has two equilibrium phases, α and β, at a temperature T_1. The composition of α-phase is 75 wt. % A and 25 wt. % B. The composition of β-phase is 25 wt. % A and 75 wt. % B. The proportions of α- and β-phases are

a) 25 wt. % α, 75 wt. % β.
b) 75 wt. % α, 25 wt. % β.
c) 50 wt. % α, 50 wt. % β.
d) 80 wt. % α, 20 wt. % β.
e) 20 wt. % α, 80 wt. % β.

3.16 The most efficient packing of atoms is found in which of the following crystalline lattices:

I) simple cubic
II) face centered cubic
III) body centered tetragonal
IV) rhombohedral
V) hexagonal close packed

a) I only
b) II only
c) III only
d) I and II only
e) II and III only

3.17 Semiconducting materials have which of the following properties:

I) an energy gap on the order of 10 eV between the valance band and the conduction band
II) an energy gap on the order of 0.05 eV between the valance band and the conduction band
III) electrical conductivity increases with increasing temperature

a) I only
b) II only
c) III only
d) I and II only
e) II and III only

3.18 Miller indices of a plane which intersects the three crystal axes at distances equal to $a/2$, $b/3$ and $c/1$ are given by:

a) (1/2 ↖ 1)
b) (312)
c) (2,3,1)
d) (231)
e) (132)

3.19 In a crystalline solid, the rate of a process which is diffusion controlled

I) is a linear function of temperature
II) varies exponentially with temperature
III) is controlled by an activation barrier

a) I only
b) II only
c) III only
d) II and III only
e) I and III only

Solutions to Practice Problems

3.1 c) If applied stress is greater than yield stress a permanent deformation occurs and the strain is irreversible.

3.2 c) Off-set yield stress is defined as the stress corresponding to a specificed plastic strain, usually 0.2%.

3.3 d) When stress is calculated by using the original cross-sectional area, the stress is called the engineering stress. When necking occurs, cross-section decreases and the engineering stress decreases. It reaches a maximum value before the onsset of necking.

3.4 c) Higher the ductility smaller the cross-section at fracture, i.e., greater is the reduction in area. Reduction in area is zero for very brittle material.

3.5 c) Yield point phenomenon is peculiar to carbon steels. Nonferrous alloys, such as aluminum alloys, do not show a yield point.

3.6 b) In a crystalline solid slip is the primary mode of plastic deformation. Dislocation motion is necessary for slip.

3.7 a) Slip produces a change in shape without producing a change in volume, like shearing a deck of cards.

3.8 d) In pure aluminum as the amplitude of alternating stress decreases, the fatigue life monotonically increases. There is no threshold stress and thus there is no endurance limit.

3.9 c) The two elements in a binary alloy must be soluble in a solid state to produce a solid solution alloy.

3.10 b) The eutectic temperature is invariant, i.e., the solidification of the eutectic composition occurs at the single temperature rather than over a range of temperature.

3.11 c) A liquid phase of a fixed composition reacts with a solid phase of another fixed composition to produce a new solid phase at the peritectic temperature.

3.12 d) AB_3 is the only intermetallic compound in this diagram. It consists of 75 at. % B and 25 at. % A.

3.13 b) The horizontal line corresponding to T_1 intersects the vertical, A + 50 at. % B within the $L + \gamma$ region.

3.14 a) According to Fig. 3.25a, 1147°C is the temperature the liquid solidified into γ (solid) and Fe_3C (solid). Thus, this is the eutectic temperature.

3.15 c) The composition of alloy is A + 50 wt. % B. Composition of α is A + 25% B and that of β is A + 75% B. Thus, the lever rule can be applied as follows:

25% B 50% B 75% B
alloy
α β

$$\text{wt. \% } \alpha = \frac{50 - 25}{75 - 25} \times 100 = 50\%$$

$$\text{wt. \% } \beta = \frac{75 - 50}{75 - 25} \times 100 = 50\%$$

3.16 c) 3.17 e) 3.18 d) 3.19 d)

Chemistry

by Donald G. Farnum

Chemistry attempts to describe and predict the macroscopic behavior of matter in terms of the properties of the submicroscopic *atoms* and *molecules* of which it consists. Although the physical properties of matter are of some interest to chemists because of what they reveal about the nature of atoms and mole-cules, the chemical properties or *chemical reactions* of matter are far more important. Thus, an under-standing of the definition, structure, and properties of atoms and molecules, the definition of a chemical reaction, and the effect of variables on a chemical reaction are fundamental to understanding chemistry.

4.1 Atoms

An *atom* is the smallest subdivision or particle of an element that enters into the chemical reactions of that element. An *element* is a pure substance that cannot be decomposed into other pure substances. Each element is composed of only one kind of atom, and is represented by a *chemical symbol*. Thus, gold is an element represented by the symbol *Au*, and is composed entirely of atoms of gold. No chemical or physical process can separate a sample of gold into any other pure substance.

4.1.1 The Periodic Table of Elements

The chemical and physical properties of the elements follow a pattern that is revealed in the *periodic table of the elements* (Table 4.1). In the table, the elements are ordered according to their atomic weights. Note: ordering according to atomic weights—an experimental observation—results in ordering according to *atomic numbers*, an inferred quantity often denoted by Z. The significance of this

(This page is intentionally blank.)

TABLE 4.1. The Periodic Chart of the Elements

I A	II A	III B	IV B	V B	VI B	VII B	VIII B			I B	II B	III A	IV A	V A	VI A	VII A	0
1 H 1.0080																	2 He 4.003
3 Li 6.940	4 Be 9.013											5 B 10.82	6 C 12.011	7 N 14.008	8 O 16.000	9 F 19.00	10 Ne 20.183
11 Na 22.991	12 Mg 24.32											13 Al 26.98	14 Si 28.09	15 P 30.975	16 S 32.066	17 Cl 35.457	18 Ar 39.944
19 K 39.100	20 Ca 40.08	21 Sc 44.96	22 Ti 47.90	23 V 50.95	24 Cr 52.01	25 Mn 54.94	26 Fe 55.85	27 Co 58.94	28 Ni 58.71	29 Cu 63.54	30 Zn 65.38	31 Ga 69.72	32 Ge 72.60	33 As 74.91	34 Se 78.96	35 Br 79.916	36 Kr 83.80
37 Rb 85.48	38 Sr 87.63	39 Y 88.92	40 Zr 91.22	41 Nb 92.91	42 Mo 95.95	43 Tc (99)	44 Ru 101.1	45 Rh 102.91	46 Pd 106.4	47 Ag 107.880	48 Cd 112.41	49 In 114.82	50 Sn 118.70	51 Sb 121.87	52 Te 127.61	53 I 126.91	54 Xe 131.30
55 Cs 132.91	56 Ba 137.36	57 *La 138.92	72 Hf 178.50	73 Ta 180.95	74 W 183.86	75 Re 186.22	76 Os 190.2	77 Ir 192.2	78 Pt 195.09	79 Au 197.0	80 Hg 200.61	81 Tl 204.39	82 Pb 207.21	83 Bi 209.00	84 Po (210)	85 At (210)	86 Rn (222)
87 Fr (223)	88 Ra (226)	89 †Ac (227)	104 Ku (261)	105 Ha (260)													

* Lanthanides	58 Ce 140.13	59 Pr 140.92	60 Nd 144.27	61 Pm (147)	62 Sm 50.35	63 Eu 152.0	64 Gd 157.26	65 Tb 158.93	66 Dy 162.51	67 Ho 164.94	68 Er 167.27	69 Tm 168.94	70 Yb 173.04	71 Lu 174.99
† Actinides	90 Th (232)	91 Pa (231)	92 U 238.07	93 Np (237)	94 Pu (242)	95 Am (243)	96 Cm (247)	97 Bk (249)	98 Cf (251)	99 Es (254)	100 Fm (253)	101 Md (256)	102 No (253)	103 Lw (257)

fact will be discussed when atomic structure is discussed. The *atomic weight* of an element is the weight of an atom of that element in atomic mass units. One *atomic mass unit* (*amu*) is one-twelfth the mass of one normal C atom, very close to that of one *H* atom. Thus, the approximate atomic weight of an element is the weight of one atom (or 10, or 6×10^{23} atoms) of that atomic element relative to the weight of one atom (or 10, or 6×10^{23} atoms) of hydrogen. When ordered in the periodic table according to their atomic weights, the elements fall into families of related physical and chemical properties. A thorough understanding of the periodic table enables one to infer the properties of an unknown element with some reliability from its determined position in the table. Perhaps more important, it is not necessary to memorize the properties of all 106 known elements to have a reasonable grasp of them. It is sufficient to know the properties of a few representative elements and understand the trends within the table. Thus, if you know that the element sodium (*Na*), is a soft, malleable solid with a metallic luster, that can be cut with a knife, conducts heat and electricity, and reacts vigorously with water with bubbling, you can expect that cesium (*Cs*)—in the same column or group—will show the same properties, more or less. The principles discussed here and in Section 4.3 on Chemical Reactions, when combined with practice and experience, will allow the skillful use of the periodic table.

4.1.2 General Trends in the Periodic Table

There is a greater change in the properties of the elements across the table (the rows or *periods*) than there is down the table (the columns, *families*, or *groups*). Note: Hydrogen, the simplest element, is an exception. It exhibits some of the properties of both groups I A and VII A as well as some unique ones. Thus lithium (*Li*, group I A) is much more like francium (*Fr*, group I A) than it is like fluorine (*F*, group VII A). As a result, the elements can be divided into two general classes, the metals and non-metals, by a more or less vertical line. The *metals* are conducting and have low *electron affinities* (tendency to accept an electron). Therefore, they tend to give up electrons easily to form positive ions, and are reducing agents. The *non-metals* are non-conducting and have high electron affinities. They, therefore, tend to accept electrons to form negative ions, and are oxidizing agents. Metallic properties tend to increase for the elements of higher atomic number. As a result, polonium (*Po*, element number 84) is more metallic than tellurium (*Te*, element number 52), and, even though these elements are in the same family (group VI A), polonium is classified as a metal and tellurium as a non-metal. The increase in metallic properties as the atomic number increases thus accounts for the diagonal division separating the metals from the non-metals. The *transition metals* (elements number 21-29, 39-47, 57-79, 89-106) are classified together since their properties are very similar. The changes across the table are dramatically attenuated for the transition metals and they represent a gradual transition from the group II A metals to the group II B metals.

The common names of some of the groups, or families, of elements in the periodic table are as follows:

Group I A: *alkali metals*

Group II A: *alkaline earth metals*

Group VII A: *halogens*

Group 0: *noble gases*

Elements 57-71: *lanthanides* or *rare earths*

Elements 89-106: *actinides*

Groups I B-VII B, VIII: *transition metals*

4.1.3 Atomic Structure and the Properties of the Elements

The periodicity in the properties of the elements can be understood in terms of the structures of their atoms. An atom consists of an extremely dense *nucleus* of *protons* (particles with a unit positive charge and a mass very close to 1 *amu*) and *neutrons* (neutral particles of essentially the same mass as the proton), and a very diffuse surrounding electron cloud containing enough *electrons* (entities with a unit negative charge and a comparatively negligible mass—about 5.5×10^{-4} *amu*) to exactly balance the nuclear charge. Thus, for any atom, the total number of protons equals the total number of electrons. The chemistry of an element is essentially determined by the electron cloud in the atoms of that element. The periodicity of the properties of the elements is thus a reflection of the periodicity of the structure of the electron clouds in atoms. It should now be apparent that the atomic number, which corresponds to the number of protons in the atomic nucleus (hence the number of electrons in the electron cloud), is actually the parameter which determines the position of an element in the periodic table, rather than the atomic weight (which is the sum of the masses of the protons, neutrons, and electrons in the atom).

Electron Orbitals

The theory of wave mechanics applied to atomic structure tells us that electrons (which appear to have properties of both waves and particles) occupy regions in space of specific shape and size called *orbitals*. Note: The older term, *orbit*, derives from a planetary theory of the atom which did not take into account the wave nature of electrons. An orbital can be defined by three *quantum numbers*: $n = 1, 2, 3 \cdots$, which defines the size of the orbital *shell*; $\ell = 0, 1, 2 \cdots n-1$, which defines the shape of the orbital *subshell*; and $m = 0, \pm 1, \pm 2 \cdots \pm \ell$, which defines the orientation of the orbital subshell in space. The larger the *principal quantum*

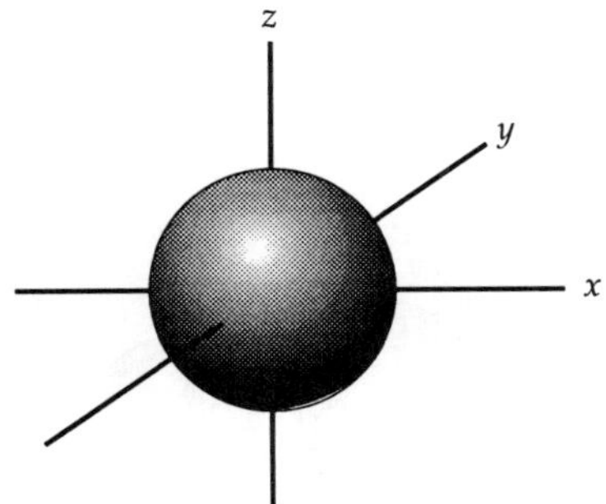

Figure 4.1. Boundary surface representation of an electron in the n = 1 state of the hydrogen atom. (Volume encloses 90% of the electron density. Nucleus is at the origin.)

number, n, the larger the orbital, and the further the average negative charge of the electron from the positive nucleus, hence, the higher the energy. For a principal quantum number $n = 4$, there are different orbital shapes corresponding to the *subsidiary quantum numbers* $\ell = 0, 1, 2, 3$. These shapes are often identified by letter symbols instead of number symbols. An s orbital has $\ell = 0$ and is spherical in shape (Fig. 4.1). A p orbital has $\ell = 1$ and is dumbbell shaped (Fig. 4.2). A d orbital has $\ell = 2$ and is still more complex in shape (Fig. 4.3). An f orbital has $\ell = 3$, and so on. There is only one orientation in space for a spherical shape, and therefore only one value (0) for m, the *magnetic orbital quantum number*,

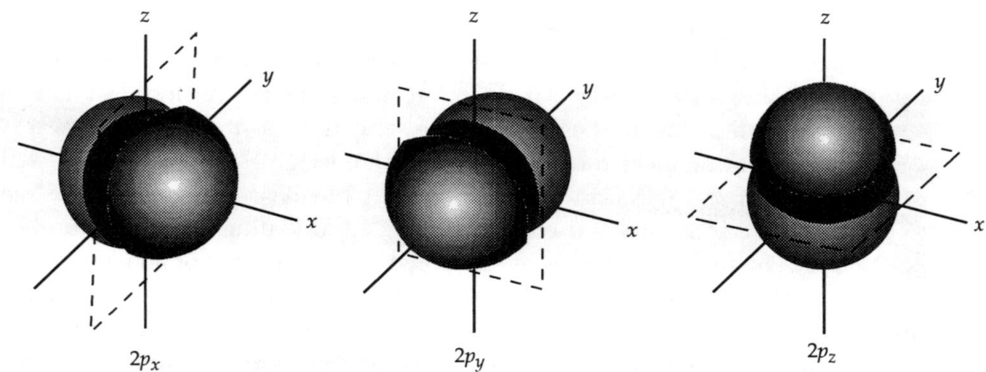

Figure 4.2. Boundary surface diagrams for the 2*p* orbitals.

for an *s* orbital. However, the dumbbell-shaped *p* orbital has three perpendicular orientations in space (see Fig. 4.2), and has three values for *m*, (m = +1, 0, –1). That is, for a given principal quantum number *n* there is only one *s* orbital, but there are three *p* orbitals. For the *d* orbital, there are five values for *m*, (m = +2, +1, 0, –1, –2), hence five different *d* orbitals. The quantum numbers for the thirty atomic orbitals of the first four shells are presented in Table 4.2.

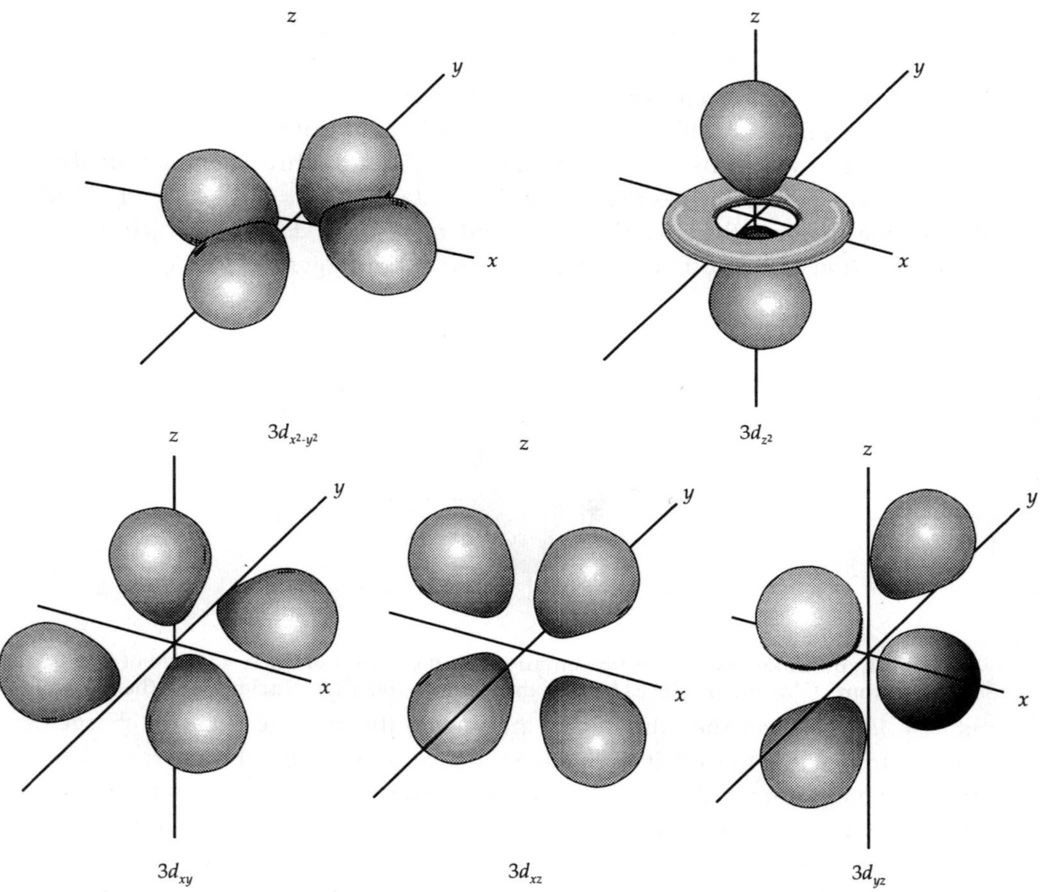

Figure 4.3. Boundary surface diagrams for the 3*d* orbitals.

TABLE 4.2. The Orbitals of the First Four Shells

Shell n	Subshell ℓ	Orbital m	Subshell Notation	Orbitals per Subshell
1	0	0	1*s*	1
2	0	0	2*s*	1
	1	+1, 0, −1	2*p*	3
3	0	0	3*s*	1
	1	+1, 0, −1	3*p*	3
	2	+2, +1, 0, −1, −2	3*d*	5
4	0	0	4*s*	1
	1	+1, 0, −1	4*p*	3
	2	+2, +1, 0, −1, −2	4*d*	5
	3	+3, +2, +1, 0, −1, −2, −3	4*f*	7

Electron Configurations of the Elements

The state of an electron in an atom is completely described by the three quantum numbers defining the electron orbital, and a fourth quantum number m_s, the magnetic spin quantum number, derived from the magnetic field of the spinning electron. The value of m_s can only be $\pm 1/2$. Therefore, a given orbital can represent two and only two electrons with opposite spins. No two electrons can have all four quantum numbers identical. From this *Pauli exclusion principle* and the fact that electrons will assume the lowest values possible for the quantum numbers n and ℓ, the electron configuration for the atoms of the first 18 elements can be described. The higher ones are slightly more complex.

Example 4.1

What is the electron configuration for an aluminum atom?

Solution. From the periodic table we see that aluminum, *Al*, is element No. 13, and therefore has 13 electrons in its atom. From Table 4.2 we can simply fill in up to two electrons in each subshell orbital until we reach 13. Therefore, aluminum has 2 electrons in the 1*s* orbital, 2 electrons in the 2*s* orbital (= 4), 2 electrons in each of the three 2*p* orbitals (total = 10), 2 electrons in the 3*s* orbital, and 1 in the 3*p* orbital (total = 13). This, in shorthand notation, is as follows: 1*s* (2), 2*s* (2), 2*p* (6), 3*s* (2), 3*p* (1).

Example 4.2

What is the electron configuration of a phosphorus atom?

Solution. Phosphorus, *P*, is element 15. Therefore we must arrange 15 electrons as follows: 1*s* (2), 2*s* (2), 2*p* (6), 3*s* (2), 3*p* (3).

In Example 4.2 there is an ambiguity about the state of the 3*p* electrons. We don't know whether to put two of them in one 3*p* orbital or put each of the three in a different 3*p* orbital. This ambiguity is removed by *Hund's rule of maximum multiplicity* which states that, whenever possible, the electrons in a subshell will have the same value for the magnetic spin quantum number, m_s. For phosphorus, this is possible for the 3*p* subshell electrons if each electron is in a different orbital with $m_s = +1/2$ (or $-1/2$). Since the spins of electrons with opposite signs for m_s are said to be paired, an alternate way of stating Hund's rule is that the electrons are distributed among the orbitals of a subshell in a way that gives the maximum number of unpaired electrons with parallel spins. We might, therefore, write the electron configuration of a phosphorus atom in more detail as follows: 1*s* (2), 2*s* (2), 2*p* (6), 3*s* (2), $3p_x$ (1), $3p_y$ (1), $3p_z$ (1). The maximum numbers of electrons that can occupy each subshell for the first four shells are given in Table 4.3.

TABLE 4.3. Maximum Number of Electrons for the Subshells of the First Four Shells

Subshell Notation	Orbitals per Subshell	Electrons per Subshell	Electrons per Shell ($2n^2$)
1*s*	1	2	2
2*s*	1	2	8
2*p*	3	6	
3*s*	1	2	18
3*p*	3	6	
3*d*	5	10	
4*s*	1	2	32
4*p*	3	6	
4*d*	5	10	
4*f*	7	14	

The periodicity of the properties of the elements revealed in the periodic table can now be seen to have its origins in the periodicity of the electronic configurations of the atoms. The chemical properties of an atom are determined by the electrons with the largest value of *n*, that is, the outer shell or *valence electrons*. Thus, the properties of lithium are determined by its lone 2*s* electron, those of sodium by its lone 3*s* electron, those of potassium by its lone 4*s* electron, and those of francium by its lone 7*s* electron. All of the atoms of the elements of Group 0, the noble gases, have *closed-shell configuration*, that is, all of the orbitals of the highest principal quantum number are full of electrons. The closed-shell configuration represents a particularly stable state. Any further electrons would have to occupy higher principal quantum number (and therefore higher energy, less stable) orbital states. All of the atoms of the halogens (Group VII A) are one electron short of closed-shell configurations, hence can accept an electron very readily.

Example 4.3

Determine the electron configuration of the valence electrons of an atom of *N*, and compare it with that for *As*.

Solution. *N* has 7 electrons. The electron configuration is 1*s* (2), 2*s* (2), $2p_x$(1), $2p_y$ (1), $2p_z$(1). The valence electrons are those with $n = 2$: 2*s* (2), $2p_x$(1), $2p_y$(1), $2p_z$ (1).

As has 33 electrons. The first three shells ($n = 1, 2, 3,$) are completely filled to account for $2 + 8 + 18 = 28$ electrons (see Table 4.3). That leaves 5 electrons for the fourth shell, $n = 4$. Therefore, the configuration of the valence electrons of *As* is 4*s* (2), $4p_x$(1), $4p_y$(1), $4p_z$(1). Note the similarity to *N* in the filled *s* orbital and three singly occupied *p* orbitals.

4.1.4 The Properties of the Families of Elements

The similarities in the properties of the elements in each group or family in the periodic table are simply a manifestation of the similarities in the electronic configurations of the valence electrons of the corresponding atoms. Some of those properties will now be examined for the more common families of elements.

Group I A—The Alkali Metals

The alkali metal atoms of Group I A all have one *s* electron beyond a closed-shell configuration. They therefore tend to lose an electron very readily to the environment to give a positively charged species—a *cation*, with a closed-shell configuration—as shown in the following simple chemical equation for sodium:

$$Na \rightarrow Na^{+} + 1e^{-}$$

One sodium atom yields one sodium cation plus one electron.

They are thus said to have a low electron affinity, or desire to accept electrons, and a low *ionization energy*—that is, it doesn't take much energy to make them lose an electron to form a cation. Note: This does not mean that the converse reaction, $Na + 1e^{-} \rightarrow Na^{-}$, is impossible, only that it is extraordinarily difficult. This atomic property of the alkali metals is reflected in their metallic luster and their conductivity (electrons are readily available to form a *conducting band*). It is also reflected in their chemical reactivity as *reducing agents*. Reducing agents give up electrons easily to other substances and are themselves oxidized in the process. A simple example is the chemical reaction of a sodium atom and a chlorine atom:

$$Na + Cl \rightarrow Na^{+}Cl^{-}$$

One sodium atom plus one chlorine atom yields one sodium cation plus one chloride anion.

In the above example, an electron is transferred from sodium to chlorine, and the sodium atom is oxidized and the chlorine atom is reduced. The sodium atom acts as a reducing agent and the chlorine atom acts as an oxidizing agent.

Since the higher atomic number elements in a family have larger electron clouds which screen the nuclear positive charge, electron affinity decreases as the atomic number increases within a family. Thus, potassium (atomic No. 19) is a more powerful reducing agent than lithium (atomic No. 3).

Example 4.4

Complete the following chemical reaction:

$$Rb + F \rightarrow ?$$

Solution. Since rubidium, *Rb*, is in the same family as sodium, it is a reducing agent. Since fluorine, *F*, is in the same family as chlorine, it is an oxidizing agent. Hence, an electron will transfer from rubidium to fluorine:

$$Rb + F \rightarrow Rb^{+} + F^{-}$$

Group II A—The Alkaline Earth Metals

The lightest of the Group II A elements, beryllium (*Be*, atomic No. 4), has the electronic configuration 1s (2), 2s (2). The *s* electrons in the second shell are easily lost to give Be^{+2} with a closed-shell configuration in the first shell. The alkaline earth metals are thus typically reducing agents that give up two electrons. They are said to be *divalent*.

The alkaline earths are less powerful reducing agents than the alkali metals since the higher nuclear charge tends to hold the valence electrons more tightly. In fact, electron affinity increases across the periodic table and decreases down the periodic table, i.e., reducing power decreases across the periodic table and increases down the table.

The alkaline earth metals have a metallic luster and are harder than the alkali metals. They are good conductors of electricity.

Example 4.5

Complete the following chemical reaction:

$$Ca + 2Br \rightarrow ?$$

Solution. Calcium (*Ca*, atomic No. 20) has the following electron configuration: 1*s* (2), 2*s* (2), 2*p* (6), 3*s* (2), 3*p* (6), 4*s* (2). It is an alkaline earth with two valence electrons in the fourth shell. It is, therefore, a reducing agent with a tendency to give up two electrons. Bromine, like chlorine, can accept one electron to give a closed-shell anion. Therefore, two bromine atoms can accept one electron each from one calcium atom:

$$Ca + 2Br \rightarrow Ca^{+2} + 2Br^{-}$$

Example 4.6

Assign the ionization energies in the column on the right to the elements in the column on the left:

Element	Ionization Energy (kJ/mol)
Cs	738
Mg	503
Ca	590
Ba	376

Solution. Magnesium (*Mg*, No. 12), calcium (*Ca*, No. 20), and barium (*Ba*, No. 56) are all alkaline earths. The first ionization energy, the energy required to remove the first valence electron, should decrease with increasing atomic number. Cesium (*Cs*, No. 55) is an alkali metal and should have a lower ionization energy than the next alkaline earth, barium. Cesium, therefore, has the lowest ionization energy of the four, and the correct answer is:

Cs	376
Mg	738
Ca	590
Ba	503

Group VII A—The Halogens

The halogen atoms of Group VII A are all one electron short of a closed-shell configuration. Because they have high nuclear charges which hold their valence electrons very tightly, it is no surprise that they have high electron affinities. They therefore tend to accept electrons readily to give closed-shell configuration anions, and are oxidizing agents. Chlorine (*Cl*, No. 17) is typical with electron configuration 1*s* (2), 2*s* (2), 2*p* (6), 3*s* (2), 3*p* (5):

$$Cl + 1e \rightarrow Cl^-$$

Reactions of halogen atoms as oxidizing agents with metals acting as reducing agents have already been illustrated above. An example is

$$I + K \rightarrow I^- + K^+$$

As you might expect, the oxidizing power of the halogens is greatest for fluorine (*F*, No. 9) and decreases down the table, the inverse of the reducing power of the metals. The halogens tend to be colored. Thus, chlorine is a green gas, bromine is a red volatile liquid, and iodine is a deep purple, volatile solid.

Example 4.7

Which of the following chemical reactions has the greatest tendency to take place?

$$Cl + K \rightarrow Cl^- + K^+$$

$$Br + K \rightarrow Br^- + K^+$$

$$Cl + Li \rightarrow Cl^- + Li^+$$

$$Br + Li \rightarrow Br^- + Li^+$$

Solution. Since oxidizing power decreases down the table, chlorine, *Cl*, is a more powerful oxidizing agent than bromine, *Br*. Since reducing power increases down the table, potassium, *K*, is a more powerful reducing agent than lithium, *Li*. The reaction with the greatest tendency to take place is the one between the best oxidizing and best reducing agents, i.e.,

$$Cl + K \rightarrow Cl^- + K^+$$

Group 0—The Noble Gases

The noble gases of Group 0 all have closed-shell configurations. They therefore have both low electron affinities and high ionization potentials. An added electron would have to go into the next orbital shell of much higher energy, and removal of an electron would have to take place from the closed shell. Because of these properties, the noble gases enter into chemical reactions with extreme reluctance. Some of the properties of the noble gases are given in Table 4.4.

TABLE 4.4. Some Properties of the Noble Gases

Gas	Melting Point (°C)	Boiling Point (°C)	Ionization Energy (kJ/mol)	Abundance in Atmosphere (Volume %)
He	—	–268.9	2.37×10^3	5×10^{-4}
Ne	–248.6	–245.9	2.08×10^3	2×10^{-3}
Ar	–189.3	–185.8	1.52×10^3	0.93
Kr	–157	–152.9	1.35×10^3	1×10^{-4}
Xe	–112	–107.1	1.17×10^3	8×10^{-4}
Rn	–71	–61.8	1.04×10^3	trace

The noble gases are the only non-metallic elements that exist as single atoms in the elemental state. All of the others exist as clusters of atoms bonded together as molecules. Thus, the low melting points and boiling points of the noble gases reflect the stability of the atoms. However, in order to understand the chemical and physical properties of the rest of the elements, it is necessary to understand the molecule and the nature of chemical bonds between atoms.

4.2 Molecular Structure

A *molecule* is the smallest subdivision or particle of a chemical compound that enters into the chemical reactions of that compound. A *chemical compound* is a substance containing more than one element combined in definite proportions. The molecules of a chemical compound are all identical clusters of atoms held together by chemical bonds. Thus, the chemical compound water consists of molecules containing two hydrogen atoms and one oxygen atom joined together by chemical bonds, symbolized by the structure shown in Fig. 4.4.

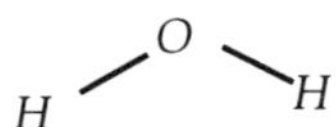

Figure 4.4. The structure of water.

The term *structure* can be used in chemistry on a number of levels, expressed by symbols of different complexity. At the simplest level, the structure of water is defined by its *molecular formula*, H_2O. A molecular formula defines the number of atoms of each element in the molecule of a compound. Thus, water has two hydrogen atoms and one oxygen atom in each molecule. At the next level, the structure of water is further defined by *HOH*, which defines the order in which the atoms are attached to one another. Thus, the two hydrogen atoms are attached to the one oxygen atom. At the next level, the shape of the molecule might be defined as either linear (*H—O—H*), or bent ($\overset{O}{\underset{H\quad H}{/\ \backslash}}$). The water molecule is bent. The shape might then be more precisely defined in terms of the actual bond angles and bond lengths involved. The structure of water (see Fig. 4.5): bond angle 104.5°, *O—H* bond length = 0.96 Å = 0.096 nm = 96 pm.

Molecular distances are often given in Angstrom units (Å): one Å = 10^{-8} centimeters (cm) = 0.1 nanometers (nm) = 100 picometers (pm).

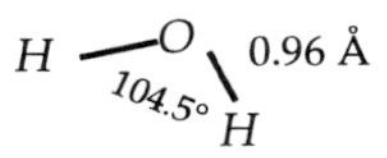

H —O 0.96 Å
104.5° H

Figure 4.5. The detailed structure of water.

For more complex molecules, a structure might include a description of the three-dimensional shape of the molecule. For example, the molecule of methane (the main component of marsh gas and natural gas) has the molecular formula CH_4. The H—C—H bond angles are all exactly 109°28′, and the C—H bond lengths are all 1.09 Å. A qualitative representation of the methane structure might look like that of Fig. 4.6. Thus the carbon atom and the two hydrogens attached to it with solid lines are implied to be in the plane of the paper. The wedged bond (*C* ◀ *H*) is intended to project above the paper, while the dashed bond (*C*- -*H*) is intended to project below the paper.

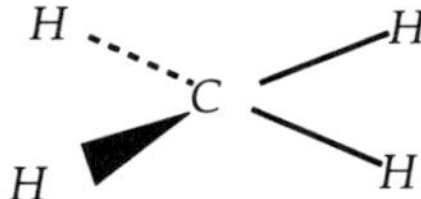

Figure 4.6. The structure of methane.

Perhaps the most complete, though not always the most useful, definition of the structure of a molecule would be a mathematical description of the geometry and motions of the atoms and the states of all the electrons in it. Molecular structures thus require different approaches for each level of complexity.

4.2.1 Determination of the Molecular Formula

If all molecules of a pure compound contain the same elements in the same ratios, then the compound will always contain the same elements combined in the same proportions by mass. This is the *law of definite proportions*. Therefore, if we knew the mass ratio of oxygen to hydrogen in water, and knew the atomic masses of oxygen and hydrogen, we could determine the ratio of oxygen atoms to hydrogen atoms in the water molecule, that is, the *empirical formula*. If we knew the mass of a molecule of water, the *molecular weight*, then we could determine the actual number of oxygen and hydrogen atoms in a water molecule, the *molecular formula*.

The empirical formula can also be determined in some instances by application of *Avogadro's law*, which states: equal volumes of all gases at the same temperature and pressure contain equal numbers of molecules.

Example 4.8

Water can be broken down into elemental hydrogen and oxygen by electrolysis. Every 100 g of water gives 88.8 g of oxygen and 11.2 g of hydrogen. What is the empirical formula of water?

Solution. The atomic weight of oxygen is 16.0, while that of hydrogen is 1.008. The ratio of hydrogen atoms to oxygen atoms is therefore,

$$\frac{11.2 / 1.008}{88.8 / 16.0} = 2.00$$

The empirical formula of water is therefore H_2O.

Example 4.9

Electrolysis of 18 g of water gives 22.4 L of hydrogen gas and 11.2 L of oxygen gas at 22°C and 1 atmosphere pressure. What is the empirical formula of water?
Solution. Since the ratio of the volumes of hydrogen to oxygen is 2:1, Avogadro's law tells us that the ratio of the number of molecules of hydrogen to the number of molecules of oxygen is also 2:1. If hydrogen and oxygen molecules each contain the same number of atoms, then the empirical formula of water is H_2O.

There are several ways of determining the molecular weight of a pure compound. At the present time, an instrument called a mass spectrometer is most often used. However, if the compound is a gas at known temperature and pressure, then the volume of a given mass of the gas can be used to determine the molecular weight. For this purpose, the *gram-atom, gram-mole,* and *mole* are useful and important concepts. A gram-atom of an element is an amount in grams numerically equal to the atomic weight. Thus, 1.000 gram-atom of sodium contains 22.9 grams of sodium. A gram-mole is that amount of a pure compound numerically equivalent in grams to the molecular weight of the compound. Thus 1.000 gram-mole of water contains $2 \times 1.088 + 16.00 = 18.02$ grams of water. Since the weight ratio of a gram-atom of sodium to a gram-atom of hydrogen is the same as the ratio of the atomic weights of the elements (23.0 : 1.0), a gram-atom of sodium contains the same number of atoms of sodium as a gram-atom of hydrogen contains of hydrogen. Similarly, a gram-mole of water contains the same number of water molecules as a gram-mole of methane contains methane molecules. In fact, the number of atoms in a gram-atom of any element and the number of molecules in a gram-mole of any compound is constant. The constant is called *Avogadro's number* and has been experimentally determined to be 6.02205×10^{23}. The amount of any substance that contains Avogadro's number of elementary units is called a *mole* (abbreviated *mol*).*mol*)

With the exception of the noble gases, which are monatomic, all elements which are gases at ordinary temperatures and pressures are diatomic, that is, they contain two atoms per molecule (e.g., N_2, O_2, Cl_2). Note that one mole of any gas will occupy 22.4 L at 22°C and 1 atm—the same as one mole of helium (Avogadro's law). This temperature and pressure is called *Standard Temperature and Pressure*, or STP. Hence, 1 mol of any gas at STP contains 22.4 L.

Example 4.10

A sample of helium gas weighing 4.00 g occupies 22.4 L. The same volume of hydrogen gas weighs 2.02 g. What is the molecular formula of hydrogen gas?
Solution. Since the atomic weight of helium is 4.00, the sample of helium is 1.00 gram-atom. Since the atomic weight of hydrogen is 1.008, the sample of hydrogen is 2.00 gram-atoms and must contain twice as many atoms as the helium sample. Since both occupy the same volume, both must contain the same number of molecules (Avogadro's law). Therefore, a molecule of hydrogen contains twice as many atoms as a molecule of helium. If we assume that a molecule of helium contains only one atom (reasonable, since helium is an unreactive noble gas), then a molecule of hydrogen contains two atoms and has the molecular formula H_2.

Example 4.11

A sample of steam weighing 27.0 g occupies the same volume as a sample of hydrogen weighing 3.02 g at 110°C and 1 atm. What is the molecular formula for water?

Solution. Since the two samples occupy the same volume under the same conditions, they have the same number of molecules. Therefore, the ratio of their weights, 27.0 : 3.02, is equal to the ratio of their molecular weights. If the molecular weight of hydrogen (H_2) is 2.02, the molecular weight of water is 18.0. Thus, the molecular formula for water is H_2O (see Ex. 4.9).

Example 4.12

A sample of ethane is burned in air to give carbon dioxide (molecular formula CO_2) and water. The ratio of the volume of water vapor to that of carbon dioxide at 110° and 1 atm is 3:2. If 1 L of ethane gas at STP gives 2 L of CO_2 gas at STP, how many carbon and hydrogen atoms are there in a molecule of ethane?

Solution. The burning of ethane can be represented symbolically by the following expression:

$$\text{ethane } (g) + O_2(g) \rightarrow CO_2(g) + H_2O(l)$$

Note: The letters in parenthesis simply indicate whether the substance is a gas (*g*), liquid (*l*), or solid (*s*) at STP.

Since equal volumes of CO_2 gas and H_2O gas at the same temperature and pressure will contain equal numbers of molecules, there are twice as many hydrogen atoms per unit volume in the water as there are carbon atoms per unit volume in the carbon dioxide. Since burning ethane gives 3 volumes of water vapor to 2 volumes of carbon dioxide, the ratio of hydrogen atoms to carbon atoms in ethane is $(3 \times 2):(2 \times 1) = 3:1$.

Since 1 L of ethane gives 2 L of carbon dioxide, each molecule of ethane must give two molecules of carbon dioxide. Therefore, there must be two carbon atoms in each molecule of ethane. A molecule of ethane, therefore, contains 2 carbon atoms and 6 hydrogen atoms. Note that we do not yet know the molecular formula of ethane completely, since it may contain other atoms as well (e.g., oxygen).

We can now convert the symbolic expression for the burning or combustion of ethane in Ex. 4.12 into a *balanced chemical equation*. A balanced chemical equation is a symbolic representation of a chemical change in which the number of each kind of atom on the left side of the equation, the reactants, is equal to the number of each kind of atom on the right side of the equation, the products:

$$C_2H_6 + \frac{7}{2}O_2 \rightarrow 2CO_2 + 3H_2O$$

Example 4.13

The density of ethane gas at STP is 1.34 g/L. What is the molecular weight and molecular formula for ethane?

Solution. The molecular weight is (22.4 L × 1.34 g/L) = 30 g. The molecular formula is therefore C_2H_6. Any additional atoms in the molecule would increase the molecular weight.

4.2.2 Determination of Chemical Structure

The determination of the structure of a complex substance of known molecular formula is often a very difficult research task requiring years of effort including the analysis of chemical reactions and the use of sophisticated instruments. Perhaps the most powerful technique is *x-ray crystallography* in which an x-ray picture is taken of the arrangement of the atoms in the regularly spaced molecules of a crystalline substance; pictures taken from several angles allow an accurate three-dimensional structure of the mole-cule to be constructed. For simpler molecules, a knowledge of *chemical bonding* principles can at least limit the number of possible structures for a given molecular formula.

Example 4.14

Given that a carbon atom can bond with no more than four other atoms, and hydrogen with only one, what is the order of connection of atoms in ethane, C_2H_6?

Solution. There is only one logically possible arrangement:

```
    H   H
    |   |
H — C — C — H
    |   |
    H   H
```

4.2.3 Chemical Bonds

There are two limiting types of chemical bonds which hold atoms together—the *ionic bond* and the *covalent bond*. *Metal bonding*, a third and very different type of bonding, is found in the metals where the atoms are bonded together as cations in a "sea" of conducting valence electrons.

The ionic bond, such as that found in sodium chloride, Na^+Cl^-, occurs whenever there is a complete transfer of one or more electrons from one atom to another to give oppositely charged ions. A crystal of sodium chloride, or table salt, might be regarded as a single enormous molecule in which all of the positive sodium ions and neighboring negative chloride ions are bonded together by ionic bonds in the crystal lattice. Ionic bonds are formed between atoms of very different electron affinities and the electron transfer usually brings both atoms to their closed-shell configurations.

The covalent bond, present in the hydrogen molecule (H_2), water (H_2O), methane (CH_4), and ethane (C_2H_6), occurs whenever one or more electrons are shared between two atoms. The force holding the atoms together is then the electrostatic attraction between the shared electrons and the positively charged nuclei. The hydrogen molecule might be regarded as two positively charged protons suspended at equilibrium distance in a diffuse elliptical cloud of two electrons. The covalent bond is formed between atoms of similar electron affinities, the shared electrons generally being sufficiently numerous to bring both atoms to their noble gas configurations. Thus, the two shared electrons in the hydrogen molecule can be considered to fill the 1*s* shell of each hydrogen atom. Although most bonds can be easily classified as ionic or covalent, in practice there is a spectrum of bond types from pure covalent through highly polar covalent to ionic bonds.

Example 4.15

Arrange the following chemical bonds in order of increasing ionic character: *CCl*, *LiCl*, *LiI*, *CC*, *KCl*, *CAl*, *CMg*.

Solution. The most covalent bond is, the *C-C* bond, between identical atoms. The most ionic is the K^+CL^- bond between the atom of lowest electron affinity, *K*, and the one of highest electron affinity, *Cl*. Li^+Cl^- will be less ionic (*Li* has a higher electron affinity than *K*), Li^+I^- still less (*I* has a lower electron affinity than *Cl*), then *CMg*, *CCl*, and *CAl* (reflecting the relative differences in electron affinities). Note that the partial electron transfer is from *C* to *Cl* leaving carbon partially positive for *CCl*, while it is from metal to carbon for *CMg* and *CAl*, leaving carbon partially negative. The correct order of *increasing* ionic character is then: $CC < CAl < CCl < CMg < LiI < LiCl < KCl$.

The number of electrons "transferred" in an ionic bond, and the number of covalent bonds formed to a given atom, is limited by the number of electrons in the valence shell. A convenient, though somewhat arbitrary, way of keeping track of the number of electrons transferred is the *oxidation number*, or oxidation state, or *valence*. (Valence is an older term whose meaning has become confused and is, therefore, rarely used now.) The oxidation number may be defined as the total number of electrons transferred or partially transferred to or from an atom in the bonds it forms with other atoms in the compound. If the electrons are transferred *from* the atoms, its oxidation number is positive. If they are transferred *to* the atom, its oxidation number is negative. For example, in potassium chloride, one electron is transferred from potassium to chlorine. The oxidation number of potassium is therefore 1+, while that of chlorine is 1–. It is obvious that the oxidation numbers of monatomic ions in ionic compounds are equivalent to the charges on the ions. The oxidation number of the atoms in a covalently bonded molecule or ion can be determined by assuming that the electrons are transferred to the atom with the higher electron affinity. For example, in the covalent molecule *H-Cl* we assume that one electron is transferred from hydrogen to chlorine. The oxidation number of hydrogen is therefore 1+ and that of chlorine 1–. For a neutral molecule, the sum of the oxidation numbers of all the atoms must, of course, be zero. For a charged molecule ion or complex ion, the sum of the oxidation numbers of all the atoms must equal the charge on the ion. A covalent bond between two atoms of the same element results in no electron transfer and no change in oxidation number.

The concept of oxidation number is a powerful tool for the analysis of a type of chemical reaction called an *oxidation-reduction* reaction or *Redox reaction*.

Example 4.16

Assign oxidation numbers to all of the atoms in each of the following compounds: *NaBr, $BaCl_2$, CaO, NH_3, CCl_4, OF_2, O_2, C_2H_6, CO, CO_2, Na_2SO_4, $K_2Cr_2O_7$.*

Solution:

Compound	Atom	Oxidation Number	Explanation
NaBr	Na	1+	One electron is transferred from *Na* to *Br*.
	Br	1–	Oxidation number = charge.
$BaCl_2$	Ba	2+	Two electrons transferred from *Ba*. One to
	Cl	1–	each *Cl*.
CaO	Ca	2+	Two electrons transferred from the alkaline
	O	2–	earth, *Ca*, to oxygen.
NH_3	H	1+	*H* is 1+. Therefore, *N* must be 3– for a net = 0
	N	3–	
CCl_4	Cl	1–	One electron considered transferred to each
	C	4+	*Cl* (higher electron affinity (*e.a.*)).
OF_2	F	1–	*F* has the highest *e.a.* of all. It is 1– except in
	O	2+	F_2.
O_2	O	0	Same element. No *e* transfer.
C_2H_6	H	1+	*H* always 1+ with non-metals. *C* must be 3–
	C	3–	for net = 0.
CO	O	2–	*O* generally 2–. *C* must be 2+.
C	2+		
CO_2	O	2–	*C* must be 4+ for net = 0.
	C	4+	
Na_2SO_4	Na	1+	*Na* must be 1+. The anion must be SO_4^{-2}. If
	O	2–	*O* is 2–, then *S* must be 6+.
	S	6+	
$K_2Cr_2O_7$	K	1+	The complex dichromate ion must be $Cr_2O_7^{-2}$.
	O	2–	If *O* is 2–, the two *Cr* atoms must total 12+.
	Cr	6+	Assuming they are in the same oxidation
			state, they are each 6+.

4.3 Chemical Reactions

A *chemical reaction* occurs whenever a chemical compound is formed from the elements or from another chemical compound. Thus, a chemical reaction involves the transformation of the arrangement of atoms in the molecules of the starting materials, or *reactants,* into the arrangement of atoms in the molecules of the *products*. The two main types of variables that affect the course of a reaction are the quantities of reactants and the conditions (e.g., temperature and either pressure for reactions of gases, or concentration for reactions in solution). The two main characteristics of the reaction which affect its course are the *rate* and the *equilibrium.*

4.3.1 Balancing Chemical Equations

In order to understand the effect of the quantities of reactants on the course of a chemical reaction, it is necessary to have a *balanced chemical equation* for the reaction. In a chemical equation, the structures or condensed structures of the reactants and products are written on the left and right, respectively, with an arrow between the two. In a balanced chemical reaction, the numbers of molecules of reactants and products are adjusted such that both the total number of atoms and the total charge of each element on the left is equal to that of the same element on the right. Thus, reaction of magnesium with bromine to give magnesium bromide is symbolized: $Mg + Br_2 \rightarrow MgBr_2$. Reaction of potassium with chlorine to give potassium chloride is symbolized: $2K + Cl_2 \rightarrow 2KCl$. Note that the smallest integers possible are used. We would not write $2Mg + 2Br_2 \rightarrow 2MgBr_2$. However, one might sometimes write $K + \frac{1}{2}Cl_2 \rightarrow KCl$. For the reaction of sodium with water to give sodium hydroxide and hydrogen gas, we might write either $2Na + 2H_2O \rightarrow 2NaOH + H_2$, or $Na + H_2O \rightarrow \frac{1}{2}H_2 + NaOH$.

Slightly more complex reactions can often be balanced by trial and error. For example, rust formation is illustrated by the (unbalanced) equation:

$$Fe + H_2O \rightarrow Fe_2O_3 + H_2$$

In order to balance this equation, we might first locate the molecule that has the largest number of atoms of the elements—the most complex molecule. In this case, it is Fe_2O_3. For a balanced equation we must therefore have at least two atoms of iron in the reactants and three molecules of water:

$$2Fe + 3H_2O \rightarrow Fe_2O_3 + 3H_2$$

Example 4.17

Balance the chemical equation for the complete combustion (with oxygen) of ethane (C_2H_6) to carbon dioxide and water.

Solution. Since ethane has two carbon atoms and six hydrogen atoms, we know that each molecule of ethane must give two molecules of CO_2 and three of H_2O:

$$C_2H_6 + \frac{7}{2}O_2 \rightarrow 2CO_2 + 3H_2O$$

or

$$2C_2H_6 + 7O_2 \rightarrow 4CO_2 + 6H_2O$$

The balanced chemical equation of Ex. 4.17 tells us a number of useful facts. It tells us that two molecules of ethane require seven molecules of oxygen for complete combustion to give four molecules of carbon dioxide and six molecules of water. It tells us that 2 L of ethane requires 7 L of oxygen for complete combustion; less would give incomplete combustion. It tells us that 60 g of ethane (2 mol) will give 108 g of water (6 mol) upon complete combustion.

More complex chemical equations can often be balanced by the use of oxidation numbers. For example, consider the unbalanced reaction:

$$H\overset{5+}{N}O_3 + H_2\overset{2-}{S} \rightarrow \overset{2+}{N}O + \overset{0}{S} + H_2O$$

The oxidation numbers of N and S in reactants and products have been identified. N undergoes a change of –3 (5+ to 2+) from reactants to products. Sulfur undergoes a change of +2 (2– to 0). The total change in oxidation number for the reaction must be zero. Therefore, all we need to do is find the common denominator for 2 and 3, namely 6. Hence we can write

$$2HNO_3 + 3H_2S \rightarrow 2NO + 3S + H_2O$$

The remainder of the equation can be balanced by inspection. Thus, there are eight hydrogen atoms on the left, requiring four water molecules on the right. The balanced equation is then

$$2HNO_3 + 3H_2S \rightarrow 2NO + 3S + 4H_2O.$$

Example 4.18

Balance the following equation:

$$I_2 + H_2O + ClO_3^- \rightarrow IO_3^- + Cl^- + H^+$$

Solution. First identify the oxidation numbers:

$$H_2O + \overset{0}{I_2} + \overset{5+}{Cl}O_3^- \rightarrow \overset{5+}{I}O_3^- + \overset{1-}{Cl^-} + H^+$$

The change for I_2 is +10 (there are 2 *I* atoms in I_2). The change for *Cl* is -6. The common denominator is 30:

$$H_2O + 3I_2 + 5ClO_3^- \rightarrow 6IO_3^- + 5Cl^- + H^+$$

The remainder is balanced by inspection. There are 18 oxygen atoms on the right, requiring 3 from the H_2O on the left, and $6H^+$ on the right:

$$3H_2O + 3I_2 + 5ClO_3^- \rightarrow 6IO_3^- + 5Cl^- + 6H^+$$

Note that the charge also balances.

4.3.2 Chemical Equilibrium

If a sample of the colorless gas N_2O_4 is placed in a container at 25°C and 1 atm pressure, it will slowly turn orange because of the formation of the orange gas NO_2. The balanced chemical reaction can be written as

$$N_2O_4 \leftrightarrow 2\,NO_2$$

Note the arrow in both directions. This implies that the reaction can proceed in either direction. In fact, if a pure sample of the orange gas NO_2 is placed in a container at 25°C and 1 atm, the color will become lighter as N_2O_4 is formed. Whether we start form N_2O_4 or NO_2, the reaction mixture will eventually stabilize at the same color, the same final concentration of N_2O_4 being formed per second from NO_2 as there is dissociating to give NO_2.

The Equilibrium Constant

The position of equilibrium for a given reaction can be defined by the equilibrium constant *K*, a constant for the reaction which is dependent only upon the temperature. The form of the equilibrium constant depends on the *stoichiometry* of the reaction as shown for each of the following reaction types:

Reaction Type	Form of K
$A \leftrightarrow B$	$[B]/[A]$
$A + B \leftrightarrow C$	$[C]/[A][B]$
$A + B \leftrightarrow C + D$	$[C][D]/[A][B]$
$A \leftrightarrow 2B$	$[B]^2/[A]$
$A + 2B \leftrightarrow C + D$	$[C][D]/[A][B]^2$

The numerator in the expression for the equilibrium constant can be obtained by writing the product of the concentrations of each of the substances on the right-hand side of the equation (the products) raised to the power of the number of molecules of that substance in the equation. The denominator is obtained by the same operation for the substances on the left-hand side of the equation (the reactants).

Example 4.19

Write the expression for the equilibrium constant for each of the following chemical reactions (each equation is balanced):

1. $N_2O_4(g) \leftrightarrow 2NO_2(g)$
2. $H_2(g)+I_2(g) \leftrightarrow 2HI(g)$
3. $2CO(g)+O_2(g) \leftrightarrow 2CO_2(g)$
4. $N_2(g)+3H_2(g) \leftrightarrow 2NH_3(g)$

Solution.

1. $K = [NO_2]^2/[N_2O_4]$
2. $K = [HI]^2/[H_2][I_2]$
3. $K = [CO_2]^2/[CO]^2[O_2]$
4. $K = [NH_3]^2/[N_2][H_2]^3$

Example 4.20

At 1 atm and 25 °C, the concentrations of the components of an equilibrium mixture of NO_2 and N_2O_4 are $[N_2O_4] = 4.27 \times 10^{-2}$ mol/L, $[NO_2] = 1.41 \times 10^{-2}$ mol/L. What is the equilibrium constant for the reaction at 25°C (expressed as $N_2O_4 \leftrightarrow 2\ NO_2$)?

Solution. The equilibrium constant is calculated as follows:

$$K_{25^\circ} = [NO_2]^2/[N_2O_4]$$

$$= (1.41 \times 10^{-2})^2 (\text{mol/L})^2 / 4.27 \times 10^{-2} (\text{mol/L})$$

$$= 4.66 \times 10^{-3} \text{mol/L}$$

The chemical reactions in Ex. 4.19 have all reactants and products in the gas phase. For reactions in which one or more of the components are pure solids or liquids, the derivation of the equilibrium constant expression is modified. Since the concentration of pure liquids and solids remains constant, it is not necessary to include them in the expression for the equilibrium constant; their values are absorbed in the constant K. For example, consider the following reaction:

$$CaCO_3 \leftrightarrow CaO + CO_2$$

Since $CaCO_3$ and CaO are pure solids, their concentrations do not appear in the equilibrium constant expression, which is

$$K = [CO_2]$$

That is to say, the concentration of CO_2 over a mixture of $CaCO_3$ and CaO at a given temperature is always the same. It does not depend on the quantities of $CaCO_3$ and CaO present. Similarly, for the reaction of HNO_3 in dilute aqueous solution, the H_2O concentration does not appear in the expression for the equilibrium constant:

$$HNO_3 + H_2O \leftrightarrow H_30^+ + NO_3^-$$

$$K = [H_3O^+][NO_3^-] / [HNO_3]$$

Since water is the solvent, it is present in large excess and its essentially constant concentration is included in the value of K.

Example 4.21

Write the expression for the equilibrium constant for the following reaction:

$$C\,(s) + CO_2\,(g) \leftrightarrow CO\,(g)$$

Solution. First we must balance the equation:

$$C\,(s) + CO_2\,(g) \leftrightarrow 2\,CO\,(g)$$

Then the equilibrium constant may be written as

$$K = [CO]^2 / [CO_2].$$

Note that solid carbon does not appear in the expression.

Since the partial pressure of a gas is a measure of its concentration, equilibrium constants for reactions involving gases may be written in terms of partial pressures. An equilibrium constant expressed this way is designated K_p. For example, the equilibrium constant expressed in partial pressures for the first reaction of Ex. 4.19-1 is

$$K_P = \left(p_{NO_2}\right)^2 / \left(p_{N_2O_4}\right).$$

K_p and K are related by the following expression:

$$K_p = K(RT)^{\Delta n}$$

where Δn is the change in the number of moles of gases between reactants and products. For the reaction under consideration,

$$N_2O_4\,(g) \leftrightarrow 2\,NO_2\,(g).$$

One mole of reactants gives two moles of products, $\Delta n = +1$. The equilibrium constant is then given by

$$K_p = K(RT)^{+1} = KRT.$$

For a reaction with no change in the number of moles of gases, i.e., $\Delta n = 0$, $K_p = K$.

Example 4.22

K_p is 167.5 atm at 1000°C for the following reaction:

$$C\ (s) + CO_2\ (g) \leftrightarrow 2\ CO\ (g)$$

What is the partial pressure of $CO(g)$ in equilibrium when the partial pressure of $CO_2\ (g)$ is 1.0 atm, and when it is 0.1 atm?

Solution. The equilibrium constant is

$$K_P = (p_{CO})^2 / (p_{CO_2}) = 167.5 \text{ atm}.$$

There then follows

$$(p_{CO})^2 / 1 \text{ atm} = 167.5 \text{ atm}.$$

When $p_{CO_2} = 1.0$ atm, the partial pressure of carbon monoxide is

$$p_{CO} = 12.9 \text{ atm}.$$

Likewise, for 0.1 atm

$$(p_{CO})^2 / 0.1 \text{ atm} = 167.5 \text{ atm}$$

$$\therefore p_{CO} = 4.10 \text{ atm}.$$

Factors Affecting Chemical Equilibrium

The value of the equilibrium constant is characteristic of the particular reaction at a particular tem-perature. It depends on the relative stability of reactants and products. If the products are very much more stable than the reactants, the K will be much greater than 1 and the reaction will tend to proceed to complete formation of products, with the generation of a large amount of heat. Such a re-action is said to be *exothermic* as indicated by the following:

$$A + B \leftrightarrow C + D + \text{Heat}$$

In an exothermic reaction, $K \gg 1$.

If the reactants are very much more stable than the products, then K will be much less than 1 and the reaction will tend to favor reactants and will generally proceed with the absorption of heat. Such a reaction, said to be *endothermic*, is indicated by the following:

$$A + B \leftrightarrow C + D - \text{Heat}$$

In an endothermic reaction, $K \ll 1$.

Although external conditions do not generally affect the equilibrium constant significantly, they can affect the relative concentrations of reactants and products dramatically. In Ex. 4.22, for instance, the ratio p_{CO} / p_{CO_2} is 12.9 when p_{CO_2} is 1.0 atm, but changes to 4.1 when p_{CO_2} is 0.10 atm. Note that reducing the pressure on the system increases the relative amount of CO. Concentration changes can have the same effect for certain reactions.

Example 4.23

Calculate the ratio of the concentration of N_2O_4 to NO_2 at equilibrium at 25°C if the concentration of NO_2 is 1 mol/L (see Ex. 4.20). Compare it to the ratio in Ex. 4.20.

Solution. The equilibrium constant is calculated to be

$$K = [NO_2]^2 / [N_2O_4] = 4.66 \times 10^{-3} \text{mol/L}.$$

Hence, there results

$$\frac{1 \text{ mol}^2 / \text{L}^2}{[N_2O_4]} = 4.66 \times 10^{-3} \text{mol/L}.$$

This leads to

$$[N_2O_4] = 2.15 \times 10^2 \text{mol/L}$$

or

$$[N_2O_4]/[NO_2] = 2.15 \times 10^2.$$

From Ex. 4.20

$$[N_2O_4]/[NO_2] = 4.27 \times 10^{-2} / 1.41 \times 10^{-2}$$
$$= 3.03.$$

Note that at the higher concentration of NO_2, the ratio $[N_2O_4]/[NO_2]$ is higher.

Observations of the effect of concentration, pressure, and temperature on the equilibrium com-position of chemical reactions led to the formulation of *LeChatelier's Principle*: a system at equilibrium responds to stress to establish a new equilibrium composition that reduces the stress. For instance, in CO-CO_2 equilibrium, an increase in pressure results in the formation of less CO which reduces the pressure (since there are 2 equivalents of CO formed from each equivalent of CO_2). In the $N_2O_4 — NO_2$ system, an increase in the concentration of NO_2 results in the formation of more N_2O_4, thus decreasing the concentration of NO_2. The extension of LeChatelier's Principle to cover temperature effects suggests that increasing the temperature of an exothermic reaction should cause the equilibrium to shift to the left (more reactants) so that less heat is evolved. Increasing the temperature of an endothermic reaction should cause the equilibrium to shift to the right (more products) so that more heat is absorbed. These predictions are generally true. However, note the important difference that the effect of temperature actually changes the value of the equilibrium constant, while the effect of pressure and concentration changes the composition without changing the value of K.

Example 4.24

For each of the following reactions give the effect of the indicated stress on the stated quantity.

1. $2SO_2(g) + O_2(g) \leftrightarrow 2SO_3(g)$ + Heat
   What is the effect on p_{SO_3} / p_{SO_2} of
   a. Increased total pressure?
   b. Decreased temperature?
2. $CO_2(g) + H_2(g) \leftrightarrow CO(g) + H_2O(g)$ – Heat
   What is the effect on p_{CO} / p_{CO_2} of
   a. Increased total pressure?
   b. Decreased temperature?
3. $2Pb_3O_4(s) \leftrightarrow 6PbO(s) + O_2(g)$ – Heat
   What is the effect on p_{O_2} of
   a. Increased T?
   b. Increased total pressure?
   c. Added Pb_3O_4?
   What is the effect on the ratio of the total quantity of O_2 to that of Pb_3O_4 of
   d. Increased T?
   e. Increased total pressure?
   f. Added Pb_3O_4?

Solutions.

1a. p_{SO_3} / p_{SO_2} will increase. Since there are three equivalents of gases on the left and two on the right, the system will respond to an increase in total pressure by shifting to the right, thereby reducing the number of molecules and the pressure.

1b. p_{SO_3} / p_{SO_2} will increase. The reaction is exothermic. Therefore, decreasing the tem-perature will cause a shift to the right to liberate more heat.

2a. p_{CO} / p_{CO_2} will not change. Since there is no difference in the number of gaseous mole-cules on the two sides of the equation, pressure changes will not change the compo-sition.

2b. p_{CO} / p_{CO_2} will decrease. The reaction is endothermic. A decrease in tempera-ture will result in a shift to the left to liberate more heat (i.e., the reverse reaction is exothermic).

3a. p_{O_2} will increase. The reaction is endothermic. An increase in T will cause a shift to the right with the absorption of more heat.

3b. p_{O_2} will decrease. Since O_2 is the only gas present, an increase in total pres-sure will re-duce the concentration of O_2 in the gas phase in order to reduce the pressure.

3c. p_{O_2} will not change. The concentration of solids remains constant.

3d. O_2/Pb_3O_4 will increase. Although the *concentration* of Pb_3O_4 is constant, more of it will be converted to PbO and O_2 at higher temperatures for this en-dothermic reaction.

3e. O_2/Pb_3O_4 will decrease. At higher total pressure O_2 and PbO will combine to form more Pb_3O_4, thus reducing the pressure.

3f. O_2/Pb_3O_4 will decrease. The added Pb_3O_4 will not affect the total quantity of O_2 present.

4.3.3 Reaction Rate

The fact that a chemical reaction is exothermic and has a large, favorable equilibrium constant does not ensure that products will be formed. For instance, the reaction of the hydrocarbons (compounds containing only carbon and hydrogen, e.g., ethane, C_2H_6) in gasoline with the oxygen in air to give carbon dioxide and water is an exothermic reaction with a large, favorable equilibrium constant. Yet gasoline can stand in air for years without transforming into carbon dioxide and water. However, let someone strike a match or spark, and the favorable equilibrium constant and exothermicity is imme-diately and dramatically revealed. The problem is, of course, in the reaction rate. Unless there is an available chemical path (a mechanism) to get from reactants to products, the rate of conversion of reactants to products will be too slow to be observed. The effect of variables on reaction rate can be understood in terms of reaction rate theories.

Collision Theory of the Reaction Rate

Consider a reaction

$$A + B \rightarrow \text{products}.$$

Since A and B must encounter each other in order to react, the rate will depend upon the number of collisions per second between A and B. This collision frequency or rate will depend upon the concen-trations $[A]$ and $[B]$, that is,

$$\text{Rate} \propto [A][B].$$

The collision frequency will also depend upon the temperature, since increasing temperature increases the velocity of the molecules and therefore the collision frequency:

$$\text{Rate} \propto [A][B] \cdot f(T).$$

Not all collisions will be effective, however, since not all will be energetic enough to result in reaction. We therefore introduce a factor, A (the *Ahrennius factor*), for the fraction of effective collisions. This factor will, of course, also be temperature dependent, since an increase in temperature will increase the fraction of sufficiently energetic collisions. Our rate expression then becomes

$$\text{Rate} = A \cdot f(T) \cdot [A][B]$$

The first part of the expression, $A\,f(T)$, represents a temperature dependent rate constant characteristic of the reaction; the second part, $[A][B]$, represents the effect of the concentration variable. Thus, increasing the temperature of a reaction increases the rate by affecting the rate constant, while increasing the pressure of a gas phase reaction, or the concentration of reactants in a solution reaction, increases the rate by increasing the total collision frequency. For the combustion of gasoline, the match provides the high temperature necessary to obtain an initial high rate of reaction; the heat necessary to maintain a high rate of reaction is provided by the exothermicity of the reaction itself.

Transition State Theory of Reaction Rate

An alternative way of coming to the same qualitative conclusions about reaction rates is to regard the reactants as being in equilibrium with a transition state (TS). The transition state is the most energetic configuration through which the reactants must pass in order to be converted to products. It is represented by

$$A + B \leftrightarrow TS \leftrightarrow \text{products}.$$

The rate of the reaction will depend on the concentration of the transition state, which can be expressed in terms of an equilibrium constant

$$K = [TS]/[A][B]$$

by the equation

$$\text{Rate} \propto [TS]$$

$$= K[A][B]$$

Since the transition state is much higher in heat content than the reactants, its formation is highly endothermic, and increasing the temperature will increase its concentration, expressed as

$$A + B \leftrightarrow TS - \text{Heat.}$$

The rate at which the transition state goes on to products is taken to be a constant. (It is also temperature dependent.) Therefore, the rate expression becomes

$$\text{Rate} = K \cdot f(T) \cdot K \cdot f(T) \cdot [A][B].$$

The form is very similar to the expression derived from collision theory and again implies an increase in rate with increasing temperature, pressure, and concentration.

Practice Problems

ATOMS

4.1 Which statement is incorrect?

a) Solutions may be homogeneous or heterogeneous.
b) Matter may be homogeneous or heterogeneous.
c) Both elements and compounds are composed of atoms.
d) All substances contain atoms.
e) Substances are always homogeneous.

4.2 Which of the following statements is not correct?

a) An element may be separated into atoms.
b) An element may be a gas, a liquid, or a solid.
c) A compound can be separated into its elements by chemical means.
d) An element is always heterogeneous.
e) A compound may be a gas, a liquid, or a solid.

4.3 In relation to the proton, the electron is

a) about the same mass and of opposite charge.
b) about the same mass and of the same charge.
c) about the same mass and with no charge.
d) much lighter and of opposite charge.
e) much heavier and with no charge.

4.4 A negative ion of a certain element can be formed by

a) subtraction of a proton from an atom of that element.
b) subtraction of an electron from an atom of that element.
c) subtraction of a neutron from an atom of that element.
d) addition of an electron to an atom of that element.
e) addition of a neutron to an atom of that element.

4.5 Metallic conduction involves

a) migration of cations toward a positively charged electrode.
b) migration of cations toward a negatively charged electrode.
c) migration of anions toward a positively charged electrode.
d) passage of electrons from one atom of a metal to another.
e) migration of anions toward a negatively charged electrode.

4.6 Which of the following statements is true?

a) Within a group of elements in the periodic table, the largest atom has the highest ionization potential.

b) Within a period of elements in the periodic table, the noble gas has the highest ionization potential.

c) When all valence p orbitals of an atom are half filled, the ionization potential of that atom is lower than the ionization potential of an atom with only two electrons in the valence *p* orbitals.

d) It is easier to form a 2+ ion than a 1+ ion.

e) Ionization potential is the same as electronegativity.

4.7 Which one of the following elements has the largest atomic radius?

a) lithium b) sodium c) beryllium d) magnesium e) phosphorus

4.8 In the series of elements *B*, *Al*, *Ga*, *In*, which of the following is true?

a) Metallic character increases from *B* to *In*.

b) Electronegativity increases from *B* to *In*.

c) Ionization energy increases from *B* to *In*.

d) Nonmetallic character increases from *B* to *In*.

e) None of the above trends is correct.

4.9 Which of the following lists contains only nonmetals?

a) beryllium (*Be*), hydrogen (*H*), osmium (*Os*)

b) germanium (*Ge*), palladium (*Pd*), silicon (*Si*)

c) carbon (C), sulfur (*S*), fluorine (*F*)

d) calcium (*Ca*), chlorine (*Cl*), boron (*B*)

e) zinc (*Zn*), gallium (*Ga*), germanium (*Ge*)

4.10 In an element

a) the atomic number is equal to the number of neutrons in the atom.

b) the number of protons always equals the number of neutrons in the atom.

c) the mass number is equal to the number of electrons in the atom.

d) the atomic number is equal to the number of protons in the atom.

e) the number of electrons can never equal the number of neutrons in the atom.

4.11 What is the ground state electron configuration of aluminum (Al, $Z = 13$)?

a) $1s^2 2s^2 2p^5 3s^2 3p^1$

b) $1s^2 2s^2 2p^6 3s^2 3p^1$

c) $1s^2 2s^2 2p^6 3s^2 4s^1$

d) $1s^2 2s^2 2p^6 3s^2 3p^2$

e) $1s^2 2s^2 2p^6 3s^2 2d^1$

4.12 Which of the following electron configurations is *inconsistent* with Hund's rule (the principle of maximum multiplicity)?

a) $[Kr]\,5s^2\,4d^{10}\,5p_x^2\,5p_y^1\,5p_z^0$
b) $[Kr]\,5s^2\,4d^{10}\,5p_x^1\,5p_y^0\,5p_z^0$
c) $[Kr]\,5s^2\,4d^{10}\,5p_x^1\,5p_y^1\,5p_z^1$
d) $[Kr]\,5s^2\,4d^{10}\,5p_x^2\,5p_y^1\,5p_z^1$
e) $[Kr]\,5s^2\,4d^{10}\,5p_x^1\,5p_y^0\,5p_z^1$

4.13 Which of these electron configurations is found in periodic Group VI?

a) $\cdots ns^2\ np^6$ b) $\cdots np^6$ c) $\cdots ns^6$ d) $\cdots ns^2\ np^4$ e) $\cdots ns^5\ np^{-1}$

4.14 From a consideration of electron configurations, which of the following elements would you expect to be most similar in chemical properties to strontium (Sr, $Z = 38$)?

a) *Rb* b) *V* c) *Sc* d) *Ba* e) *Ti*

4.15 The principal quantum number designates the

a) shape of an orbital.
b) main energy level in which an electron is found.
c) sublevel of energy in which an electron is found.
d) number of electrons allowed in a main energy level.
e) orientation of the orbital in space.

4.16 In any atom, what is the total number of electrons which can have a principal quantum number of 5 and a secondary quantum number (ℓ) of zero?

a) 2 b) 4 c) 5 d) 6 e) 18

4.17 For a neutral atom of an element in its ground state, 35 electrons occupy the energy levels up to and including the $n = 4$ energy level. If all electrons in the valence (outermost) *p*-orbitals are removed by ionization, how many electrons remain in the resulting ion?

a) 18 b) 20 c) 28 d) 30 e) 35

4.18 How many electrons does a phosphorus atom have in its set of valence shell *p* orbitals?

a) 0 b) 1 c) 2 d) 3 e) 10

4.19 An atom of an unknown element *Q* has a mass number of 31 and the nucleus contains 15 protons. The element is:

a) gallium, *Ga*
b) sulfur, *S*
c) phosphorus, *P*
d) palladium, *Pd*
e) scandium, *Sc*

4.20 An ion of an unknown element has an atomic number of 15 and contains 18 electrons. The ion is:

a) P^{3-} b) Ar c) O^{2-} d) Si^{3-} e) S^{+}

MOLECULES

4.21 Consider the following statements about ionic and covalent bonds. Which statement is true?

a) In a covalent molecule, each atom is bonded to only two other atoms.
b) An ionic bond is an electrostatic interaction localized between two definite ions of identical electrical charge.
c) A covalent bond occurs when electrons are completely transferred from one atom to another.
d) When a covalent bond forms between two atoms with different electronegativities, the bond is always polar.
e) A compound never contains both ionic and covalent bonds.

4.22 Which one of the following compounds is classified as an alkane?

a) ethylene b) benzene c) propane d) acetylene e) ethanol

4.23 Which one of the following bonds is most covalent?

a) $MgCl$ b) AlP c) $NaCl$ d) MgS e) NaP

4.24 The sum of the oxidation states of all the atoms in a neutral molecule

a) must be a small positive number.
b) must be a small negative number.
c) must be zero.
d) can be either positive or negative, but not zero.
e) can have any value, including zero.

4.25 The oxidation state of an element bonded only to itself

a) must be a small positive number.
b) must be a small negative number.
c) can be either positive or negative, but not zero.
d) can have any value, including zero.
e) must be zero.

4.26 The oxidation state of sulfur (S) in the ion SO_3^{2-} is

a) 1+ b) 2+ c) 3+ d) 4+ e) 6–

4.27 A mole

a) is a unit of measurement applicable only to molecules.
b) equals the number of atoms in one gram of carbon-12.
c) equals the number of molecules in 20 liters of air.
d) is Avogadro's number of anything.
e) equals the number of atoms in 22.4 liters of a diatomic gas.

4.28 Which statement is incorrect?

a) Avogadro's number equals the number of molecules in one mole of nitrogen molecules.
b) Avogadro's number equals the number of atoms in one mole of nitrogen atoms.
c) Avogadro's number equals the number of atoms in one mole of nitrogen molecules.
d) Avogadro's number equals 6.02×10^{23}.
e) Avogadro's number equals the number of one faraday of electricity (one faraday equals 96,500 coulombs—the charge carried by one mole of electrons).

4.29 An empty aluminum Coke can weighs 50 grams. How many moles of aluminum does one Coke can contain? (Atomic weight of $Al = 27$)

a) 1,350 b) 1.85 c) 1.0×10^{25} d) 3.0×10^{25} e) 27

4.30 A 27 gram sample of oxygen difluoride, OF_2, contains how many molecules? (Atomic weights: $O = 16$, $F = 19$; Avogadro's number: 6.0×10^{23})

a) 3.0×10^{23}
b) $2 \times 6.0 \times 10^{23}$
c) $6.0 \times 10^{23} / 4$
d) $3.0 \times 10^{23} \times 54$
e) 12.0×10^{23}

4.31 How many grams are there in 0.01 mole of Na_2SO_4?

a) 7.1 g b) 14.2 g c) 9.6 g d) 1.42 g e) 0.71 g

4.32 What is the volume at standard temperature and pressure of 16 grams of gaseous sulfur dioxide, SO_2?

a) 22.4 liters b) 11.2 liters c) 5.6 liters d) 16.8 liters e) 64 liters

4.33 What is the percentage by weight of aluminum, Al, in alumina, Al_2O_3? (Atomic weights: $Al = 27$, $O = 16$)

a) 63 b) 37 c) 23 d) 53 e) 64

4.34 A certain compound consists only of sulfur (S) and chlorine (Cl). It contains 47.5 percent by weight of sulfur and has a molecular weight of 135. What is its molecular formula? (Atomic weights: $S = 32$, $Cl = 35.5$)

a) SCl_2 b) SCl c) S_2Cl_2 d) S_2Cl e) S_3Cl

REACTIONS

4.35 An unknown organic compound was analyzed and found to contain 34.6 percent carbon, 3.8 percent hydrogen, and 61.5 percent oxygen. Which one of the following compounds could the unknown be?

a) methanol, CH_3OH
b) oxalic acid, $CO_2H—CO_2H$
c) acetic acid, $CH_3—CO_2H$
d) malonic acid, $CO_2H—CH_2—CO_2H$
e) propionic acid, $CH_3—CH_2—CO_2H$

4.36 What is the expression for the equilibrium constant for the following system?

$$2\ NOCl\ (g) \leftrightarrow 2NO\ (g) + Cl_2\ (g)$$

a) $K = [NO]^2[Cl_2]^2/[NOCl]^2$
b) $K = 2[NO][Cl_2]/2[NOCl]$
c) $K = [NO]^2[Cl_2]/[NOCl]^2$
d) $K = [NO]^2[Cl_2]^2/[NOCl]^2$
e) $K = [NOCl]^2/[NO]^2[Cl_2]$

4.37 For the reaction of solid BaO with carbon dioxide according to the equation $BaO(s) + CO_2(g) \leftrightarrow BaCO_3(s)$, the equilibrium constant may be represented as:

a) $[BaCO_3]/[BaO]$
b) $1/[CO_2]$
c) $[BaO][CO_2]/[BaCO_3]$
d) $[CO_2]$
e) $[BaCO_3]/[BaO][CO_2]$

4.38 Assume excess oxygen reacts with methane to form 14 grams of carbon monoxide according to the equation $2CH_4 + 3O_2 \leftrightarrow 2CO + 4H_2O$. How many moles of methane will be consumed?

a) 2.0 moles methane
b) one-third mole methane
c) 0.25 moles methane
d) 0.5 moles methane
e) 4.0 moles methane

4.39 What coefficient is required for NO_2 in order to balance the equation

$$2Pb(NO_3)_2 \rightarrow 2PbO + NO_2 + O_2$$

a) 0.5 b) 1 c) 1.5 d) 4 e) 10

4.40 What coefficient is required for H_2O in order to balance the equation

$$Be_3N_2 + H_2O \rightarrow 3Be(OH)_2 + 2NH_3$$

a) 1 b) 2 c) 3 d) 6 e) 18

4.41 In the following reaction, determine the change, if any, that occurs in the oxidation number of the underlined element, and whether the element is oxidized, reduced, or unchanged:

$$3Mg + \underline{N}_2 \rightarrow Mg_3\underline{N}_2$$

a) from 0 to +3; oxidized
b) from 0 to –3; oxidized
c) from +3 to +5; oxidized
d) from 0 to –3; reduced
e) from +5 to +3; reduced

4.42 According to the equation $2Al + 6HCl \rightarrow 2AlCl_3 + 3H_2$,

a) production of 1 mole of H_2 requires 3 moles of HCl.
b) production of 1 mole of $AlCl_3$ requires 3 moles of HCl.
c) production of 2 moles of H_2 requires 2 moles of HCl.
d) production of 2 moles of H_2 requires 5 moles of HCl.
e) production of 4 moles of H_2 requires 2 moles of HCl.

4.43 when crystals of sodium sulfate are dissolved in water, the resulting solution feels warmer. The solubility of Na_2SO_4 could be increased by

a) increasing the temperature.
b) increasing the pressure.
c) decreasing the temperature.
d) adding more solute to the solution.
e) stirring the solution.

4.44 Which of the following statements is false?

a) An exothermic reaction always goes faster than an endothermic reaction.
b) A catalyst provides a different route by which the reaction can occur.
c) Some reactions may never reach completion (100% products).
d) The rate of a reaction depends upon the height of the energy barrier (energy of activation).
e) The activation energy is independent of the energy of reaction.

4.45 In which one of the following reactions would an increase in the volume of the container cause an increase in the amount of products at equilibrium? (All substances are gases unless marked otherwise.)

a) $2NO + 5H_2 \leftrightarrow 2NH_3 + 2H_2O$
b) CH_3CHO + heat $\leftrightarrow CH_4 + CO$
c) $SO_2 \leftrightarrow S(s) + O_2$
d) $SO_3 + HF \leftrightarrow HSO_3(l)$
e) $C + H_2O \leftrightarrow CO + H_2$

Solutions to Practice Problems

4.1 a) A solution is defined as homogeneous.

4.2 d) An element is a pure substance and is homogeneous.

4.3 d) By definition.

4.4 d) By definition.

4.5 d) Conduction is a flow of electrons.

4.6 b) The noble gases have closesd shell, stable electronic configurations.

4.7 b) Atomic radius increases down the periodic table, and decreases across (left to right).

4.8 a) Metallic character increases down the table.

4.9 c) See the periodic table.

4.10 d) By definition.

4.11 b) The 13 eletrons must be in the lowest energy orbitals available. Thus, at each quantum level (n = 1, 2, 3) the s-orbitals, then the p-orbitals must be filled in order.

4.12 a) The $5p_x^2$ configuration has two paired electrons which could be unpaired, i.e., $5p'_x$, $5p'_y$, $5p'_z$.

4.13 d) Group VI is two electrons short of a closed shell configuration, therefore must have 2 electrons in a valence s-orbital, and 4 in the p-orbitals.

4.14 d) Ba, z = 56, has a valence electron configuration $6s^2$. Sr, z = 38, has $5s^2$.

4.15 b) By definition.

4.16 a) If $\ell = 0$, the orbital is an s-orbital. There can be only one s-orbital for a given principle quantum number. Therefore there can be only two electrons.

4.17 d) There are 28 electrons in the n = 1, 2, 3 energy levels (see Table 4.3). The configuration for the n = 4 level is, therefore $4s^2$, $4p^3$. If the 5 p-electrons are removed that leaves 30 electrons.

4.18 d) Phosphorus, z = 15 (Table 4.1), has electronic configuration $1s^2, 2s^2, 1p^6, 3s^2, 3p^3$ (see Table 4.3).

4.19 c) See Table 4.1 for atomic number 15.

4.20 a) The element must be P(z = 15). The 18 electrons are 3 in excess of the nuclear charge.

4.21 d) Electrons are polarized toward the more electronegative atom.

4.22 c) Alkane names always end in "ane."

4.23 b) Al and P are the closest in the periodic table.

4.24 c) By definition.

4.25 e) No net electron transfer.

4.26 d) O is 2^-. Three oxygens is 6^-. The net charge is 2^-. $\therefore$ s must be 4^+, i.e., $4-6=-2$.

4.27 d) By definition.

4.28 c) Molecular nitrogen is N_2. There are therefore 2 moles of N atoms in a mole of nitrogen molecules, and twice Avogadro's number of atoms.

4.29 b) $50/27 = 1.85$ mol.

4.30 a) Molecular wt. of $OF_2 = 16 + (2\times 19) = 54$. $27\text{g} = 0.5$ mol. $0.5\times 6\times 10^{23} = 3\times 10^{23}$.

4.31 d) Mol wt. $Na_2SO_4 = (2\times 23) + 32 + (4\times 16) = 142$. $0.01\times 142 = 1.42\,\text{g}$.

4.32 c) Mol wt. $SO_2 = 32 + 2\times 16 = 64$. $16\,\text{g}$ of $SO_2 = 16\div 64 = \frac{1}{4}$ mol. One mol = 22.4 L at STP. $\therefore$ $1/4$ mol $= \frac{1}{4}\times 22.4 = 5.6$ L at STP.

4.33 d) Mol wt. $Al_2O_3 = 2\times 27 + 3\times 16 = 102$. wt % Al = $(54/102)\times 100 = 53\%$.

4.34 c) Compound contains $.475\times 135 = 64.1\text{g}$ S per mol.
$\therefore$ there are $64.1/32 = 2.0$ mol S per mol compound.
The remaining mol wt, $135 - 64 = 71$ is Cl (Atomic wt = 35.5).
There are $\therefore$ 2 mol per mol compound. $\therefore$ formula S_2Cl_2.

4.35 d) Atomic ratio of $C/H = \frac{34.6}{12}\div\frac{3.8}{1} = 0.75$, i.e., 3 : 4 or C_3H_4. Malonic acid is $C_3H_4O_4$. To confirm, the atomic ratio of O/H = $\frac{61.5}{16}\div 3.8 = 1.0$.

4.36 c) See Example 4.18.

4.37 b) Solids do not enter into the K_{eq}.

4.38 d) Mol wt CO = 28. $14\,\text{g}$ CO = $14/28$ = .5 mol. Each mol of CH_4 gives 1 mol CO. $\therefore \frac{1}{2}$ mol $CH_4 \rightarrow \frac{1}{2}$ mol CO.

4.39 d) 2 $Pb\left(NO_3\right)_2$ gives 4 N atoms, requiring 4 NO_2.

4.40 d) There are 12 H atoms on the right requiring 6 H_2O on the left (note that oxygen balances).

4.41 d) An atom bonded only to itself has oxidation number = 0. In Mg_3N_2, Mg has oxidation no. 2^+ (an alkaline earth). $\therefore$ to make a net 0, N must have 3^-. Any reduction in oxidation no. (0 to -3) reduces the element.

4.42 b) The balanced equation shows that 6 moles of HCL gives 2 moles of $AlCl_3$ or 3 moles gives 1 mole.

4.43 c) Le Chatelier principle.

4.44 a) An exothermic reaction has a favorable equibrium constant, but the rate depends on the activation energy.

4.45 b) LeChatelier principle.

Engineering Economics

by Frank Hatfield

Engineering designs are intended to produce good results. In general, the good results are accompanied by undesirable effects including the costs of manufacturing or construction. Selecting the best design requires the engineer to anticipate and compare the good and bad outcomes. If outcomes are evaluated in dollars, and if "good" is defined as positive monetary value, then design decisions may be guided by the techniques known as *engineering economy*. Decisions based solely on engineering economy may be guaranteed to result in maximum goodness only if all outcomes are anticipated and can be monetized (measured in dollars).

5.1 Value and Interest

"Value" is not synonymous with "amount." The value of an amount of money depends on when the amount is received or spent. For example, the promise that you will be given a dollar one year from now is of less value to you than a dollar received today. The difference between the anticipated amount and its current value is called *interest* and is frequently expressed as a time rate. If an interest rate of 10% per year is used, the expectation of receiving $1.00 one year hence has a value now of about $0.91. In engineering economy, interest usually is stated in percent per year. If no time unit is given, "per year" is assumed.

Example 5.1

What amount must be paid in two years to settle a current debt of $1,000 if the interest rate is 6%?

Solution. Value after one year $= 1000 + 1000 \times 0.06$

$$= 1000(1 + 0.06)$$

$$= \$1060$$

Value after two years $= 1060 + 1060 \times 0.06$

$$= 1000(1 + 0.06)^2$$

$$= \$1124$$

Hence, $1,124 must be paid in two years to settle the debt.

5.2 Cash Flow Diagrams

As an aid to analysis and communication, an engineering economy problem may be represented graphically by a horizontal time axis and vertical vectors representing dollar amounts. The cash flow diagram for Example 5.1 is sketched in Fig. 5.1. Income is up and expenditures are down. It is important to pick a point of view and stick with it. For example, the vectors in Fig. 5.1 would have been reversed if the point of view of the lender had been adopted. It is a good idea to draw a cash flow diagram for every engineering economy problem that involves amounts occurring at different times.

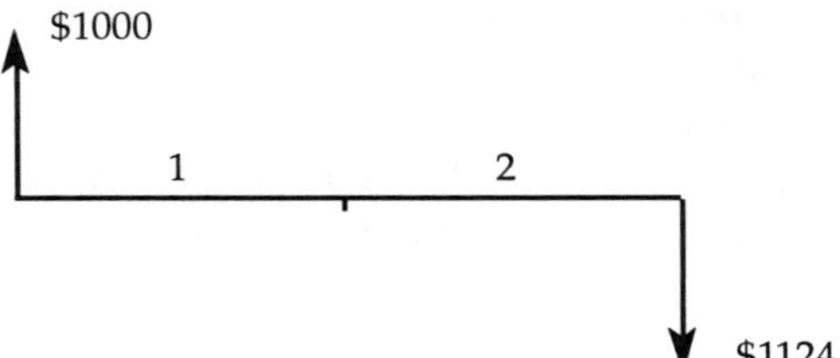

Figure 5.1 Cash flow diagram for Example 5.1.

In engineering economy, amounts are almost always assumed to occur at the ends of years. Consider, for example, the value today of the future operating expenses of a truck. The costs probably will be paid in varied amounts scattered throughout each year of operation, but for computational ease the expenses in each year are represented by their sum (computed without consideration of interest) occurring at the end of the year. The error introduced by neglecting interest for partial years is usually insignificant compared to uncertainties in the estimates of future amounts.

5.3 Cash Flow Patterns

Engineering economy problems involve the following four patterns of cash flow, both separately and in combination:

P-pattern: A single amount P occurring at the beginning of n years. P frequently represents "present" amounts.

F-pattern: A single amount F occurring at the end of n years. F frequently represents "future" amounts.

A-pattern: Equal amounts A occurring at the ends of each of n years. The A-pattern frequently is used to represent "annual" amounts.

G-pattern: End-of-year amounts increasing by an equal annual gradient G. Note that the first amount occurs at the end of the second year. G is the abbreviation of "gradient."

The four cash flow patterns are illustrated in Fig. 5.2.

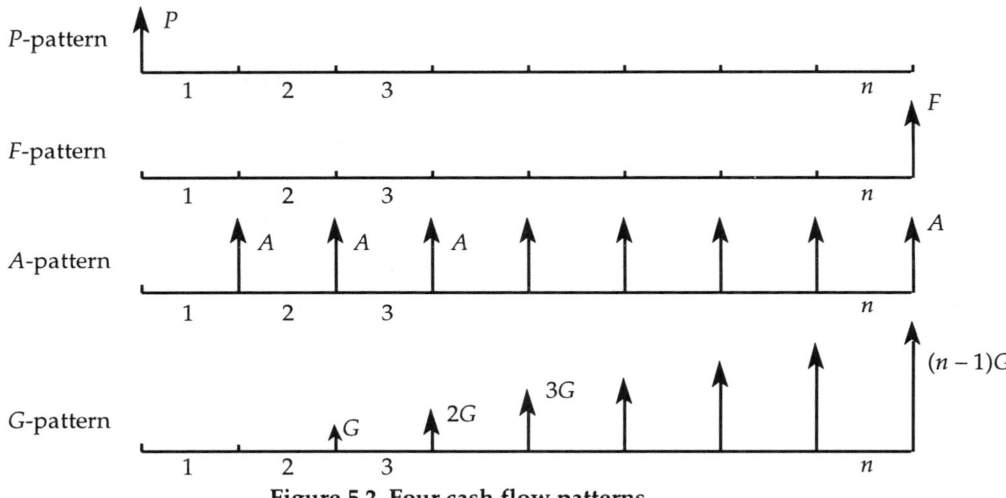

Figure 5.2 Four cash flow patterns.

5.4 Equivalence of Cash Flow Patterns

Two cash flow patterns are said to be equivalent if they have the same value. Most of the computational effort in engineering economy problems is directed at finding a cash flow pattern that is equivalent to a combination of other patterns. Example 5.1 can be thought of as finding the amount in an F-pattern that is equivalent to \$1,000 in a P-pattern. The two amounts are proportional, and the factor of proportionality is a function of interest rate i and number of periods n. There is a different factor of proportionality for each possible pair of the cash flow patterns defined in Section 5.3. To minimize the possibility of selecting the wrong factor, mnemonic symbols are assigned to the factors. For Example 5.1, the proportionality factor is written $(F/P)_n^i$ and solution is achieved by evaluating

$$F = (F/P)_n^i P.$$

To analysts familiar with the canceling operation of algebra, it is apparent that the correct factor has been chosen. However, the letters in the parentheses together with the sub- and super-scripts constitute a single symbol; therefore, the canceling operation is not actually performed. Table 5.1 lists symbols and formulas for commonly used factors. Table 5.2, located at the end of this chapter, presents a convenient way to find numerical values of interest factors. Those values are tabulated for selected interest rates i and number of interest periods n; linear interpolation for intermediate values of i and n is acceptable for most situations.

TABLE 5.1 Formulas for Interest Factors

Symbol	To Find	Given	Formula
$(F/P)_n^i$	F	P	$(1+i)^n$
$(P/F)_n^i$	P	F	$\frac{1}{(1+i)^n}$
$(A/P)_n^i$	A	P	$\frac{i(1+i)^n}{(1+i)^n-1}$
$(P/A)_n^i$	P	A	$\frac{(1+i)^n-1}{i(1+i)^n}$
$(A/F)_n^i$	A	F	$\frac{i}{(1+i)^n-1}$
$(F/A)_n^i$	F	A	$\frac{(1+i)^n-1}{i}$
$(A/G)_n^i$	A	G	$\frac{1}{i}-\frac{n}{(1+i)^n-1}$
$(F/G)_n^i$	F	G	$\frac{1}{i}\left[\frac{(1+i)^n-1}{i}-n\right]$
$(P/G)_n^i$	P	G	$\frac{1}{i}\left[\frac{(1+i)^n-1}{i(1+i)^n}-\frac{n}{(1+i)^n}\right]$

Example 5.2

Derive the formula for $(F/P)_n^i$.

Solution. For $n = 1$,

$$F = (1+i)P$$

that is,

$$(F/P)_1^i = (1+i)^1.$$

For any n,

$$F=(1+i)(F/P)_{n-1}^{i}P$$

that is,

$$(F/P)_n^i=(1+i)(F/P)_{n-1}^i.$$

By induction,

$$(F/P)_n^i=(1+i)^n.$$

Example 5.3

A new widget twister, with a life of six years, would save $2,000 in production costs each year. Using a 12% interest rate, determine the highest price that could be justified for the machine. Although the savings occur continuously throughout each year, follow the usual practice of lumping all amounts at the ends of years.

Solution. First, sketch the cash flow diagram.

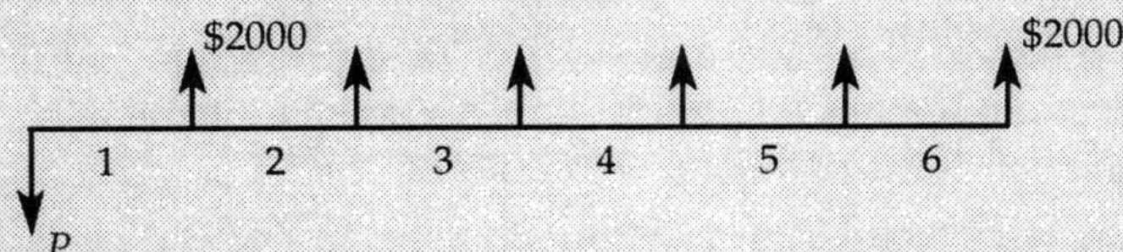

The cash flow diagram indicates that an amount in a *P*-pattern must be found that is equivalent to $2,000 in an *A*-pattern. The corresponding equation is

$$P=(P/A)_n^i A$$

$$=(P/A)_6^{12\%}2000$$

Table 5.2 is used to evaluate the interest factor for $i = 12\%$ and $n = 6$:

$$P = 4.1114 \times 2000$$

$$= \$8223$$

Example 5.4

How soon does money double if it is invested at 8% interest?

Solution. Obviously, this is stated as

$$F = 2P.$$

Therefore,

$$(F/P)_n^{8\%}=2.$$

In the 8% interest table, the tabulated value for (F/P) that is closest to 2 corresponds to $n = 9$ years.

Example 5.5

Find the value in 1987 of a bond described as "Acme 8% of 2000" if the rate of return set by the market for similar bonds is 10%.

Solution. The bond description means that the Acme company has an outstanding debt that it will repay in the year 2000. Until then, the company will pay out interest on that debt at the 8% rate. Unless otherwise stated, the principal amount of a single bond is $1,000. If it is assumed that the debt is due December

31, 2000, interest is paid every December 31, and the bond is purchased January 1, 1987, then the cash flow diagram, with unknown purchase price P, is:

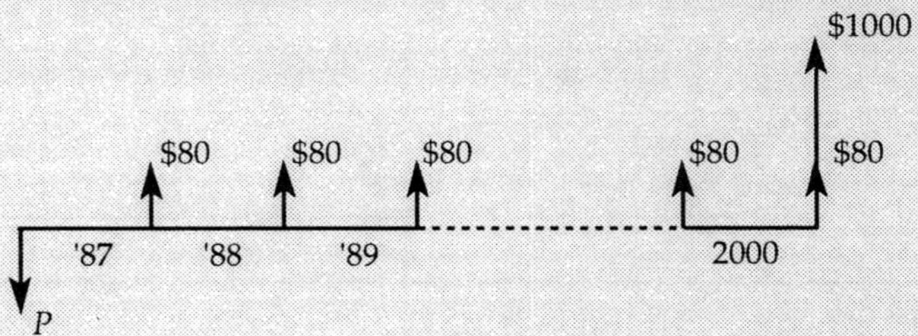

The corresponding equation is

$$P = (P/A)_{14}^{10\%} 80 + (P/F)_{14}^{10\%} 1000$$

$$= 7.3667 \times 80 + 0.2633 \times 1000$$

$$= \$853$$

That is, to earn 10% the investor must buy the 8% bond for \$853, a "discount" of \$147. Conversely, if the market interest rate is less than the nominal rate of the bond, the buyer will pay a "premium" over \$1,000.

The solution is approximate because bonds usually pay interest semiannually, and \$80 at the end of the year is not equivalent to \$40 at the end of each half year. But the error is small and is neglected.

Example 5.6

You are buying a new appliance. From past experience you estimate future repair costs as:

First Year	\$ 5
Second Year	15
Third Year	25
Fourth Year	35

The dealer offers to sell you a four-year warranty for \$60. You require at least a 6% interest rate on your investments. Should you invest in the warranty?

Solution. Sketch the cash flow diagram.

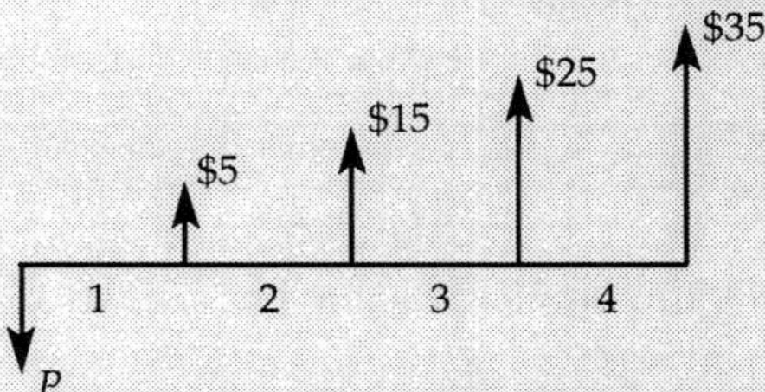

The known cash flows can be represented by superposition of a \$5 A-pattern and a \$10 G-pattern. Verify that statement by drawing the two patterns. Now it is clear why the standard G-pattern is defined to have the first cash flow at the end of the second year. Next, the equivalent amount P is computed:

$$P = (P/A)_4^{6\%} A + (P/G)_4^{6\%} G$$
$$= 3.4651 \times 5 + 4.9455 \times 10$$
$$= \$67$$

Since the warranty can be purchased for less then $67, the investment will earn a rate of return greater than the required 6%. Therefore, you should purchase the warranty .

If the required interest rate had been 12%, the decision would be reversed. This demonstrates the effect of a required interest rate on decision making. Increasing the required rate reduces the number of acceptable investments.

Example 5.7

Compute the annual equivalent maintenance costs over a 5-year life of a laser printer that is warranted for two years and has estimated maintenance costs of $100 annually. Use $i = 10\%$.

Solution. The cash flow diagram appears as:

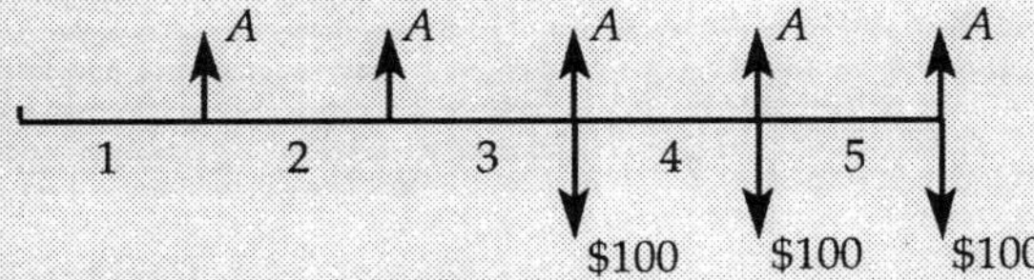

There are several ways to find the 5-year *A*-pattern equivalent to the given cash flow. One of the more efficient methods is to convert the given 3-year *A*-pattern to an *F*-pattern, and then find the 5-year *A*-pattern that is equivalent to that *F*-pattern. That is,

$$A = (A/F)_5^{10\%} (F/A)_3^{10\%} 100$$
$$= \$54.$$

5.5 Unusual Cash Flows and Interest Periods

Occasionally an engineering economy problem will deviate from the year-end cash flow and annual compounding norm. The examples in this section demonstrate how to handle these situations.

Example 5.8

PAYMENTS AT BEGINNINGS OF YEARS

Using a 10% interest rate, find the future equivalent of:

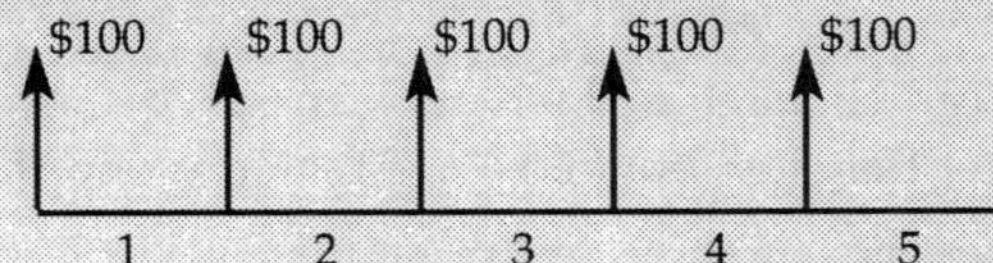

Solution. Shift each payment forward one year. Therefore,

$$A = (F/P)_1^{10\%} 100 = \$110.$$

This converts the series to the equivalent *A*-pattern:

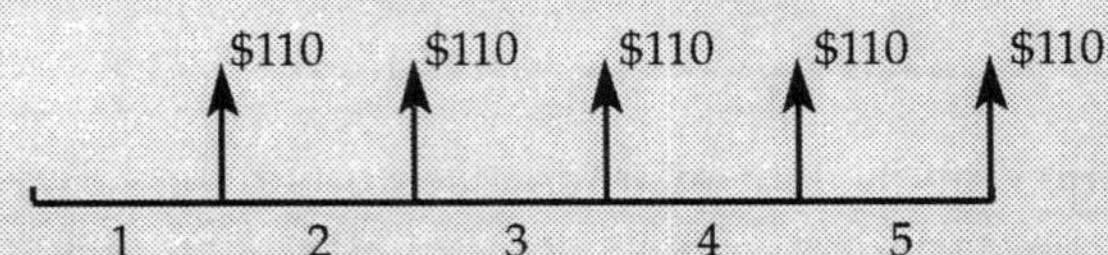

and the future equivalent is found to be

$$F = (F/A)_5^{10\%} 110 = \$672.$$

Alternative Solution. Convert to a six-year series:

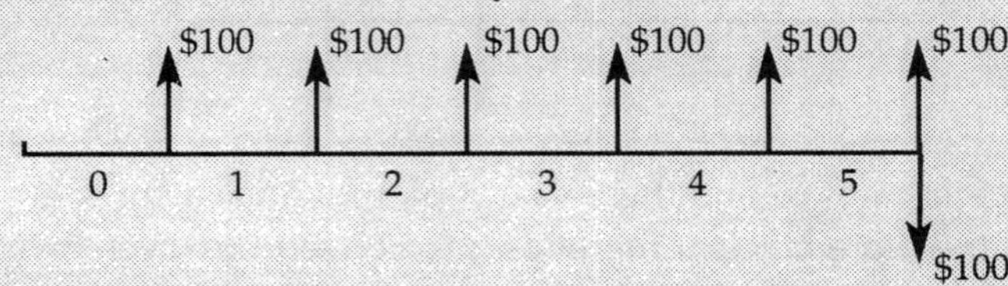

The future equivalent is

$$F = (F/A)_6^{10\%} 100 - 100 = \$672.$$

Example 5.9

SEVERAL INTEREST AND PAYMENT PERIODS PER YEAR

Compute the present value of eighteen monthly payments of $100 each, where interest is 1/2% per month.

Solution. The present value is computed as

$$P = (P/A)_{18}^{1/2\%} 100 = \$1717$$

Example 5.10

ANNUAL PAYMENTS WITH INTEREST COMPOUNDED *m* TIMES PER YEAR

Compute the effective annual interest rate equivalent to 5% nominal annual interest compounded daily. (There are 365 days in a year.)

Solution. The legal definition of nominal annual interest is

$$i_n = mi$$

where *i* is the interest rate per compounding period. For the example,

$$i = i_n/m$$

$$= 0.05/365 = 0.000137 \text{ or } 0.0137\% \text{ per day.}$$

Because of compounding, the effective annual rate is greater than the nominal rate. By equating (*F*/*P*)-factors for one year and *m* periods, the effective annual rate i_e may be computed as follows:

$$(1+i_e)^1 = (1+i)^m$$

$$i_e = (1+i)^m - 1$$

$$= (1.000137)^{365} - 1 = 0.05127 \text{ or } 5.127\%$$

Example 5.11

CONTINUOUS COMPOUNDING

Compute the effective annual interest rate i_e equivalent to 5% nominal annual interest compounded continuously.

Solution. As *m* approaches infinity, the value for i_e is found as follows:

$$i_e = e^{mi} - 1$$

$$= e^{0.05} - 1$$

$$= 0.051271 \text{ or } 5.1271\%$$

Example 5.12

ANNUAL COMPOUNDING WITH *m* PAYMENTS PER YEAR

Compute the year-end amount equivalent to twelve end-of-month payments of $10 each. Annual interest rate is 6%.

Solution. The usual simplification in engineering economy is to assume that all payments occur at the end of the year, giving an answer of $120. This approximation may not be acceptable for a precise analysis of a financial agreement. In such cases, the agreement's policy on interest for partial periods must be investigated.

Example 5.13

ANNUAL COMPOUNDING WITH PAYMENT EVERY *m* YEARS

With interest at 10% compute the present equivalent of

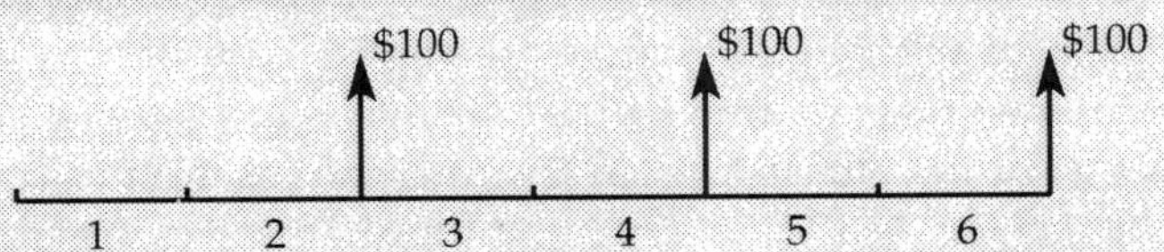

Solution. First convert each payment to an *A*-pattern for the *m* preceding years. That is,

$$A = (A/F)_2^{10\%} 100$$

$$= \$47.62$$

Then, convert the *A*-pattern to a *P*-pattern:

$$P = (P/A)_6^{10\%} 47.62$$

$$= \$207$$

5.6 Evaluating Alternatives

The techniques of engineering economy assume the objective of maximizing net value. For a business, "value" means after-tax cash flow. For a not-for-profit organization, such as a government agency, value may include non-cash benefits, such as clean air, improved public health, or recreation to which dollar amounts have been assigned.

This section concerns strategies for selecting alternatives such that net value is maximized. The logic of these methods will be clear if the following distinctions are made between two different types of interest rates, and between two different types of relationships among alternatives.

Types of Interest Rates

•*Rate of Return* (*ROR*): The estimated interest rate produced by an investment. It may be computed by finding the interest rate such that the estimated income and non-cash benefits (positive value), and the estimated expenditures and non-cash costs (negative value), sum to a net equivalent value of zero.

•*Minimum Attractive Rate of Return (MARR)*: The lowest rate of return that the organization will accept. In engineering economy problems, it is usually a given quantity and may be called, somewhat imprecisely, "interest," "interest rate," "cost of money," or "interest on capital."

Types of Alternative Sets

•*Mutually Exclusive Alternatives*: Exactly one alternative must be selected. Examples: "Shall Main Street be paved with concrete or asphalt?" "In which room will we put the piano?" If a set of alternatives is mutually exclusive, it is important to determine whether the set includes the null (do nothing) alternative. Serious consequences can arise from failure to recognize the null alternative.

•*Independent Alternatives*: It is possible (but not necessarily economical) to select any number of the available alternatives. Examples: "Which streets should be paved this year?" "Which rooms shall we carpet?"

5.6.1 Annual Equivalent Cost Comparisons

The estimated income and benefits (positive) and expenditures and costs (negative) associated with an alternative are converted to the equivalent *A*-pattern using an interest rate equal to *MARR*. The *A*-value is the *annual net equivalent value* (*ANEV*) of the alternative. If the alternatives are mutually exclusive, the one with the largest *ANEV* is selected. If the alternatives are independent, all that have positive *ANEV* are selected.

Example 5.14

A new cap press is needed. Select the better of the two available models described below. *MARR* is 10%.

Model	Price	Annual Maintenance	Salvage Value	Life
Reliable	11,000	1,000	1,000	10 years
Quicky	4,000	1,500	0	5 years

Solution. The *ANEV* is calculated for each model:

Reliable:
$$ANEV = -(A/P)_{10}^{10\%}11000 - 1000 + (A/F)_{10}^{10\%}1000$$
$$= -\$2730$$

Quicky:
$$ANEV = -(A/P)_{5}^{10\%}4000 - 1500$$
$$= -\$2560$$

Negative *ANEV* indicates a rate of return less than *MARR*. However, these alternatives are mutually exclusive and the null is not available. The problem is one of finding the less costly way to perform a necessary function. Therefore, *Quicky* is selected. If *MARR* had been much lower, *Reliable* would have been selected. By setting the *MARR* relatively high, the organization is indicating that funds are not available to invest now in order to achieve savings in the future.

5.6.2 Present Equivalent Cost Comparisons

The estimated income and benefits (positive), and expenditures and costs (negative), associated with an alternative are converted to the equivalent *P*-pattern using an interest rate equal to *MARR*. The *P*-value is the *present net equivalent value* (*PNEV*) of the alternative. If the alternatives are mutually exclusive, the one with the largest *PNEV* is selected. *PNEV* is also called "life cycle cost," "present worth," "capital cost," and "venture worth." If the alternatives are independent, all that have positive *PNEV* are selected.

The present equivalent cost method requires that alternatives be evaluated over the same span of time. If their lives are not equal, the lowest common multiple of the lives is used for the time span, with each alternative repeated to fill the span. A variation, called the *capitalized cost method,* computes the *PNEV* for repeated replacement of the alternatives for an infinite time span. The capitalized cost *P* of an infinite series of equal amounts *A* is given by

$$P = A(P/A)_{\infty}^{i} = A/i\ .$$

Example 5.15

Repeat Example 5.14 using the present equivalent cost method.

Solution. The *PNEV* is calculated for each model:

Reliable:
$$PNEV = -11000 - (P/A)_{10}^{10\%}1000 + (P/F)_{10}^{10\%}1000$$
$$= -\$16,800$$

Quicky:
$$PNEV = -4000 - (P/F)_{5}^{10\%}4000 - (P/A)_{10}^{10\%}1500$$
$$= -\$15,700$$

Note that *Quicky* was replaced in order to fill the ten-year time span. As in Example 5.14, *Quicky* is selected. The two methods will always give the same decision if used correctly. Observe that for both alternatives

$$PNEV = (P/A)_{10}^{10\%} ANEV.$$

5.6.3 Incremental Approach

For a set of mutually exclusive alternatives, only the differences in amounts need to be considered. Compute either the *ANEV* or the *PNEV* and base the decision on the sign of that value.

Example 5.16

Repeat Example 5.14 using an incremental present net equivalent value approach.
Solution. *Reliable* costs $7,000 more than *Quicky* but saves $500 each year in maintenance expenses and eliminates the need for a $4,000 replacement after five years. In addition, *Reliable* has a $1,000 salvage value whereas *Quicky* has none.
Reliable – *Quicky*:

$$PNEV = -7000 + (P/A)_{10}^{10\%} 500 + (P/F)_{5}^{10\%} 4000 + (P/F)_{10}^{10\%} 1000$$

$$= -\$1060$$

The negative result dictates selection of *Quicky*. That is, the additional initial cost required to purchase *Reliable* is not justified.

5.6.4 Rate of Return Comparisons

The expression for *ANEV* or *PNEV* is formulated and then solved for the interest rate that will give a zero *ANEV* or *PNEV*. This interest rate is the rate of return (*ROR*) of the alternative. To apply the rate of return method to mutually exclusive alternatives requires incremental comparison of each possible pair of alternatives; increments of investment are accepted if their rates of return exceed *MARR*. For independent alternatives, all those with *ROR* exceeding *MARR* are accepted. The rate of return method permits conclusions to be stated as functions of *MARR*, which is useful if *MARR* has not been determined precisely.

Example 5.17

A magazine subscription costs $50 for one year or $80 for two years. If you want to receive the magazine for at least two years, which alternative is better?
Solution. The two-year subscription requires an additional initial investment of $30 and eliminates the payment of $50 one year later. The rate of return formulation is:

$$PNEV = 0$$

$$-30 + 50(P/F)_{1}^{i} = 0$$

The solution for *i* is as follows:

$$-30 + 50\frac{1}{(1+i)} = 0$$

$$i = 0.67 \text{ or } 67\%$$

Therefore, if your *MARR* is less than 67%, subscribe for two years.

Example 5.18

Repeat Example 5.14 using the rate of return method.

Solution. Use the incremental expression derived in Example 5.16, but set *PNEV* equal to zero and use the interest rate as the unknown:

$$-7000+(P/A)_{10}^{i}500+(P/F)_{5}^{i}4000+(P/F)_{10}^{i}1000=0$$

By trial and error, the interest rate is found to be 6.6%. Therefore, *Reliable* is preferred if, and only if, *MARR* is less than 6.6%.

5.6.5 Benefit/Cost Comparisons

The benefit/cost ratio is determined from the formula:

$$\frac{B}{C}=\frac{\text{Uniform net annual benefits}}{\text{Annual equivalent of initial cost}}$$

where *MARR* is used in computing the *A*-value in the denominator. As with the rate of return method, mutually exclusive alternatives must be compared incrementally, the incremental investment being accepted if the benefit/cost ratio exceeds unity. For independent alternatives, all those with benefit/cost ratios exceeding unity are accepted.

Note that the only pertinent fact about a benefit/cost ratio is whether it exceeds unity. This is illustrated by the observation that a project with a ratio of 1.1 may provide greater net benefit than a project with a ratio of 10 if the investment in the former project is much larger than the investment in the latter. It is incorrect to rank mutually exclusive alternatives by their benefit/cost ratios as determined by comparing each alternative to the null (do nothing) alternative.

The benefit/cost ratio method will give the same decision as the rate of return method, present equivalent cost method, and annual equivalent cost method if the following conditions are met:

1. Each alternative is comprised of an initial cost and uniform annual benefit.
2. The form of the benefit/cost ratio given above is used without deviation.

Example 5.19

A road resurfacing project costs $200,000, lasts five years, and saves $100,000 annually in patching costs. *MARR* is 10%. Should the road be resurfaced?

Solution. The benefit/cost ratio is

$$\frac{B}{C}=\frac{100,000}{(A/P)_{5}^{10\%}200,000}=1.9$$

Since the ratio exceeds unity, the resurfacing is justified.

5.6.6 A Note on *MARR*

In engineering economy examination problems, *MARR* is a given quantity. However, the following discussion of the determination of *MARR* will help clarify the logic underlying the various comparison methods.

In general, an organization will be able to identify numerous opportunities to spend money now that will result in future returns. For each of these independent investment opportunities, an expected rate of return can be estimated. Similarly, the organization will be able to find numerous sources of funds for investment. Associated with each source of funds is an interest rate. If the source is a loan, the associated interest rate is simply that charged by the lender. Funds generated by operations of the organization, or provided by its owners (if the organization is a business), or extracted from taxpayers (if the organization is a government agency) can be thought of as being borrowed from the owners or taxpayers. Therefore, such funds can be assigned a fictitious interest rate, which should not be less than the maximum rate of return provided by other opportunities in which the owners or taxpayers might invest.

Value will be maximized if the rates of return of all the selected investments exceed the highest interest rate charged for the money borrowed, and if every opportunity has been taken to invest at a rate of return exceeding that for which money can be borrowed. That rate is the Minimum Attractive Rate of Return. No investments should be made that pay rates of return less than *MARR*, and no loans should be taken that charge interest rates exceeding *MARR*. Furthermore, the organization should exploit all opportunities to borrow money at interest rates less than *MARR* and invest it at rates of return exceeding *MARR*.

To estimate *MARR* precisely would require the ability to foresee the future, or at least to predict all future investment and borrowing opportunities and their associated rates. A symptom of *MARR* being set too low is insufficient funds for all the investments that appear to be acceptable. Conversely, if *MARR* has been set too high, some investments will be rejected that would have been profitable.

5.6.7 Replacement Problems

How frequently should a particular machine be replaced? This type of problem can be approached by varying the life n. For each value of n, the annual costs and salvage value are estimated, and then the *ANEV* is computed. The value of n resulting in the smallest annual equivalent cost is the optimum, or economic, life of the machine. This approach is complicated by technological improvements in replacement machinery, which may make it advantageous to replace a machine before the end of its economic life. In practice, technological advances are difficult to anticipate.

Another form of the replacement problem asks if an existing asset should be replaced by a new (and possibly different) one. Again, the annual equivalent cost method is recommended. The *ANEV* of the replacement is computed, using its economic life for n. However, the annual cost of the existing asset is simply the estimated expense for one more year of operation. This strategy is based on the assumption that the annual costs of the existing asset increase monotonically as it ages.

5.6.8 Always Ignore the Past

Engineering economy, and decision making in general, deals with alternatives. But there is only one past and it affects all future alternatives equally. Therefore, past costs and income associated with an existing asset should not be included in computations that address the question of replacing the asset. Only the estimated cash flows of the future are relevant.

The mistake of counting past costs is common in everyday affairs. For example, a student may say, "I paid $90 for this textbook so I will not sell it for $20." A more rational approach would be to compare the highest offered price to the value of retaining the text.

Example 5.20

Yesterday a machine was bought for \$10,000. Estimated life is ten years, with no salvage value at the end of its useful life. Current book value is \$10,000. Today a vastly improved model was announced. It costs \$15,000, has a ten-year life and no salvage value, but reduces operating costs by \$4,000 annually. The current resale value of the older machine has dropped to \$1,000 due to this stunning technological advance. Should the old model be replaced with a new model at this time?

Solution. The purchase price of the old machine, its book value, and the loss on the sale of the old machine are irrelevant to the analysis. The incremental cost of the new machine is \$14,000 and the incremental income is \$4,000 annually. A rate of return comparison is formulated as follows:

$$-14,000+(P/A)_{10}^{i}4000=0$$

Solving for rate of return gives $i = 26\%$, indicating that the older machine should be replaced immediately if *MARR* is less than 26%.

5.6.9 Break-Even Analysis

A break-even point is the value of an independent variable such that two alternatives are equally attractive. For values of the independent variable above the break-even point, one of the alternatives is preferred; for values of the independent variable below the break-even point, the other alternative is preferred. Break-even analysis is particularly useful for dealing with an independent variable that is subject to change or uncertainty since the conclusion of the analysis can be stated as a function of the variable. The rate of return method, as applied to mutually exclusive alternatives, is an example of break-even analysis. The independent variable is *MARR*.

Example 5.21

An item can be manufactured by hand for \$5. Alternatively, the item can be produced by a machine at a fixed annual equivalent cost of \$4,000 plus a variable cost of \$1 per item. Assume that the cost of laying off and hiring workers is zero. For each of the two manufacturing processes, answer the following questions:

a)For what production rate is one method more economical than the other?
b)If the item is sold for \$6, how many must be sold to make a profit?
c)How low must the price fall, in the short term, before production is discontinued?

Solution.

a) Let *P* be production rate in units per year. Production costs for the two processes are equated:

$$\text{Cost by machine} = \text{Cost manually}$$

$$4000+1P=5P$$

$$\therefore P=1000$$

If annual production is expected to be less than 1,000 units, the manual process is more economical. For production rates exceeding 1,000 units per year, the machine process is preferred.

b) Setting profit equal to zero is expressed as:

$$\text{gross income} - \text{cost} = 0$$

$$\text{Manual production: } 6P - 5P = 0$$

$$\therefore P = 0$$

$$\text{Machine production: } 6P - (4000 + 1P) = 0$$

$$\therefore P = 800$$

With price maintained at $6, the mechanized operation will be unprofitable if production rate is less than 800 units per year, but the manual operation is profitable at all production rates.

c) Manual production becomes unprofitable if the price drops below $5, and production will cease at that level. For the machine, the $4,000 cost continues whether or not the machine is running. Incremental income is generated so long as the price stays above the variable (per item) cost. Therefore, production will continue at any price over $1, even though a net loss may be sustained. Of course, if it appears that the price and production rate will not soon increase sufficiently to provide a profit, then the operation will be terminated.

5.7 Income Tax and Depreciation

Business pays to the federal government a tax that is a proportion of taxable income. Taxable income is gross revenue less operating costs (wages, cost of materials, etc.), interest payments on debts, and depreciation. Depreciation is different from the other deductions in that it is not a cash flow.

Depreciation is an accounting technique for charging the initial cost of an asset against two or more years of production. For example, if you buy a $50,000 truck for use in your construction business, deducting its total cost from income during the year of purchase gives an unrealistically low picture of income for that year, and an unrealistically high estimate of income for the succeeding years during which you use the truck. A more level income history would result if you deducted $10,000 per year for five years. In fact, the Internal Revenue Service (IRS) requires that most capital assets used in business be depreciated over a number of years rather than being deducted as expenses during the year of purchase.

An asset is depreciable if it is used to produce income, has a determinable life greater than one year, and decays, wears out, becomes obsolete, or gets used up. Examples include tools, production machinery, computers, office equipment, buildings, patents, contracts, franchises, and livestock raised for wool, eggs, milk or breeding. Non-depreciable assets include personal residence, land, natural resources, annual crops, livestock raised for sale or slaughter, and items intended primarily for resale such as stored grain and the merchandise in a department store.

Since depreciation is not a cash flow, it will not directly enter an engineering economy analysis. However, depreciation must be considered when estimating future income taxes, which are cash flows.

The IRS requires that the Modified Accelerated Cost Recovery System (MACRS) be applied to most tangible property placed in service after 1986. In general, MACRS is based on computing depreciation using a declining balance method or the straight line method treating the property as being placed in service and retired from service at midpoints of tax years, and setting salvage value equal to zero. Older methods, such as the Accelerated Cost Recovery System (ACRS) or the straight line method with a non-zero salvage value, may still show up in engineering economy problems, and therefore are included in this discussion. The following notation will be used in defining methods for computing depreciation:

- B — The installed first cost, or basis.
- n — Recovery period in years.
- D_x — Depreciation in year x. at age n.
- V_x — Undepreciated balance at the end of year x, also called book value.
- V_n — Estimated salvage at age n.

In computing depreciation there is no attempt to equate book value with resale value, productive worth, or any other real figure. A business is not obliged to keep an asset for exactly n years, nor to sell it for exactly its book value or estimated salvage value. These, then, are the depreciation methods:

1. *Declining Balance*: Depreciation is taken as a proportion of book value:

$$D_x = V_{x-1}C/n$$

where n is the recovery period. For values of C equaling 1.25, 1.5 and 2 the method is called, respectively: 125% declining balance, 150% declining balance, and double declining balance.

2. *Straight Line Depreciation*: Depreciation is the same for every full year and is calculated as

$$D_x = (B - V_n)/n.$$

3. *Accelerated Cost Recovery System (ACRS)*: An asset is classed as having a recovery period n of 3, 5, 10, or 15 years using IRS guidelines. For each class, a set of annual rates R_x is specified by the IRS. With 3-year property, for example, $R_1 = 0.25$, $R_2 = 0.38$, $R_3 = 0.37$. Depreciation is calculated as follows:

$$D_x = R_x B.$$

By definition, the salvage value using *ACRS* is zero.

Example 5.22

The purchase price of an over-the-road tractor unit is $100,000, its recovery period is three years, and it can be sold for an estimated $20,000 at that time. Compute the depreciation schedules using each of the methods described.

Solution.

Double Declining Balance (MACRS)

Year	Depreciation	Book Value
		$100,000
1	0.5 × ($100,000 × 2/3) = $33,333	$ 66,667
2	$66,667 × 2/3 = $44,444	$ 22,222
3	$22,222 × 2/3 = $14,815	$ 7,407
4	$7407	$ 0

*The book value must be zero after three years of service, so the formula was not used for the last year.

Straight Line (MACRS)

Year	Depreciation	Book Value
		$100,000
1	$100,000/3 × 0.5 = $16,667	$ 83,333
2	$100,000/3=$33,333	$ 50,000
3	$100,000/3=$33,333	$ 16,667
4	$100,000/3 × 0.5 = $16,667	$ 0

Straight Line (General)

Year	Depreciation	Book Value
		$100,000
1	(100,000 – 20,000)/3 = $26,667	$ 73,333
2	$26,667	$ 46,667
3	$26,667	$ 20,000

Accelerated Cost Recovery System (ACRS]

Year	Depreciation	Book Value
		$100,000
1	0.25 × $100,000 = $25,000	$ 75,000
2	0.38 × $100,000 = $38,000	$ 37,000
3	0.37 × $100,000 = $37,000	$ 0

Note: The factor 0.5 in the MACRS results from the requirement to use the mid-point of a tax year.

5.8 Inflation

The "buying power" of money changes with time. A decline in "buying power" is experienced due to a general increase in prices, called "inflation."

Inflation, if it is anticipated, can be exploited by fixing costs and allowing income to increase. A manufacturing business can fix its costs by entering long-term contracts for materials and wages, by purchasing materials long before they are needed, or by stockpiling its product for later sale. Income is allowed to respond to inflation by avoiding long-term contracts for the product. Borrowing becomes more attractive if inflation is expected since the debt will be paid with the less valuable cash of the future.

MARR may be adjusted for anticipated uniform inflation using the formula

$$d = i + f + if$$

where d is inflation-adjusted *MARR*, i is unadjusted *MARR*, and f is the rate of inflation. This formula facilitates solution of some types of engineering economy problems.

Example 5.23

A machine having a five-year life can replace a worker who is compensated $20,000 per year with 5% annual "cost of living" increases. Operating and maintenance costs for the machine are negligible. *MARR* is 10%. Find the maximum price that can be justified for the machine if:

a) general price inflation is 5%
b) general price inflation is zero

Solution.

a)Although the worker gets a larger amount of money each year, her raises are exactly matched by increased prices, including those of her employer's product. "Buying power" of her annual compensation remains equal to the current value of $20,000. Hence, the maximum justifiable price for the machine is

$$P = (P/A)_5^{10\%} 20,000 = \$75,816$$

b)The maximum justifiable price of the machine is equal to the present equivalent value of the annual amounts of compensation:

$$(P/F)_1^{10\%}(1.05)\ 20,000 = \$19,090$$
$$(P/F)_2^{10\%}(1.05)^2 20,000 = \$18,224$$
$$(P/F)_3^{10\%}(1.05)^3 20,000 = \$17,394$$
$$(P/F)_4^{10\%}(1.05)^4 20,000 = \$16,604$$
$$(P/F)_5^{10\%}(1.05)^5 20,000 = \underline{\$15,850}$$
$$\text{therefore,} \qquad P = \$86,162$$

Example 5.24

Recompute the value, in terms of 1987 "buying power," of the "Acme 8% of 2000" bond discussed in Example 5.5, but assume 6% annual inflation.

Solution. The cash flow for each year must be divided by an inflation factor, as well as multiplied by an interest factor, and then the factored cash flows are added:

$$\begin{aligned}
&(P/F)_1^{10\%} && 80/(1.06) &&= \$\ 69\\
&(P/F)_2^{10\%} && 80/(1.06)^2 &&= \$\ 59\\
&(P/F)_3^{10\%} && 80/(1.06)^3 &&= \$\ 50\\
&&&&& \vdots\\
&(P/F)_{13}^{10\%} && 80/(1.06)^{13} &&= \$\ 11\\
&(P/F)_{14}^{10\%} && 80/(1.06)^{14} &&= \$\ \ \ 9\\
&(P/F)_{14}^{10\%} && 1000/(1.06)^{14} &&= \underline{\$\ 116}\\
&\text{therefore,} && && P = \$\ 541
\end{aligned}$$

Note that investors can account for anticipated inflation simply by using increased values of *MARR*. A *MARR* of 16.6% gives the same conclusions as a *MARR* of 10% with 6% inflation.

Alternative Solution. Using inflation-adjusted *MARR* we have

$$\begin{aligned}
d &= i + f + if\\
&= .10 + .06 + .10 \times .06\\
&= .166 \quad \text{or} \quad 16.6\%
\end{aligned}$$

The value of the bond is

$$\begin{aligned}
P &= (P/A)_{14}^{16.6\%} 80 + (P/F)_{14}^{16.6\%} 1000\\
&= 5.3225 \times 80 + .1165 \times 1000\\
&= \$542
\end{aligned}$$

Formulas from Part 5.4 (a table in the NCEES Handbook) were used to evaluate the interest factors.

Practice Problems

(If you attempt only a few, select those with a star.)

5.1 Which of the following would be most difficult to monetize?

a) maintenance cost b) selling price c) fuel cost
d) prestige e) interest on debt

VALUE AND INTEREST

*5.2 If $1,000 is deposited in a savings account that pays 6% annual interest and all the interest is left in the account, what is the account balance after three years?

a) $840 b) $1,000 c) $1,180 d) $1,191 e) $3,000

*5.3 Your perfectly reliable friend, Merle, asks for a loan and promises to pay back $150 two years from now. If the minimum interest rate you will accept is 8%, what is the maximum amount you will loan him?

a) $119 b) $126 c) $129 d) $139 e) $150

EQUIVALENCE OF CASH FLOW PATTERNS

*5.4 The annual amount of a series of payments to be made at the end of each of the next twelve years is $500. What is the present worth of the payments at 8% interest compounded annually?

a) $500 b) $3,768 c) $6,000 d) $6,480 e) $6,872

*5.5 Consider a prospective investment in a project having a first cost of $300,000, operating and maintenance costs of $35,000 per year, and an estimated net disposal value of $50,000 at the end of thirty years. Assume an interest rate of 8%.

What is the present equivalent cost of the investment if the planning horizon is thirty years?

a) $670,000 b) $689,000 c) $720,000 d) $791,000 e) $950,000

If the project replacement will have the same first cost, life, salvage value, and operating and maintenance costs as the original, what is the capitalized cost of perpetual service?

a) $670,000 b) $689,000 c) $720,000 d) $765,000 e) infinite

*5.6 Maintenance expenditures for a structure with a twenty-year life will come as periodic outlays of $1,000 at the end of the fifth year, $2,000 at the end of the tenth year, and $3,500 at the end of the fifteenth year. With interest at 10%, what is the equivalent uniform annual cost of maintenance for the twenty-year period?

a) $200 b) $262 c) $300 d) $325 e) $342

5.7 After a factory has been built near a stream, it is learned that the stream occasionally overflows its banks. A hydrologic study indicates that the probability of flooding is about 1 in 8 in any one year. A flood would cause about $20,000 in damage to the factory. A levee can be constructed to prevent flood damage. Its cost will be $54,000 and its useful life is thirty years. Money can be borrowed at 8% interest. If the annual equivalent cost of the levee is less than the annual expectation of flood damage, the levee should be built. The annual expectation of flood damage is $(1/8)\times 20{,}000 = \$2{,}500$. Compute the annual equivalent cost of the levee.

a) $1,261 b) $1,800 c) $4,320 d) $4,800 e) $6,750

5.8 If $10,000 is borrowed now at 6% interest, how much will remain to be paid after a $3,000 payment is made four years from now?

a) $7,000 b) $9,400 c) $9,625 d) $9,725 e) $10,700

*5.9 A piece of machinery costs $20,000 and has an estimated life of eight years and a scrap value of $2,000. What uniform annual amount must be set aside at the end of each of the eight years for replacement if the interest rate is 4%?

a) $1,953 b) $2,174 c) $2,250 d) $2,492 e) $2,898

*5.10 The maintenance costs associated with a machine are $2,000 per year for the first ten years, and $1,000 per year thereafter. The machine has an infinite life. If interest is 10%, what is the present worth of the annual disbursements?

a) $16,145 b) $19,678 c) $21,300 d) $92,136 e) $156,600

*5.11 A manufacturing firm entered into a ten-year contract for raw materials which required a payment of $100,000 initially and $20,000 per year beginning at the end of the fifth year. The company made unexpected profits and asked that it be allowed to make a lump sum payment at the end of the third year to pay off the remainder of the contract. What lump sum is necessary if the interest rate is 8%?

a) $85,600 b) $92,700 c) $122,300 d) $196,700 e) $226,000

UNUSUAL CASH FLOWS AND INTEREST PAYMENTS

5.12 A bank currently charges 10% interest compounded annually on business loans. If the bank were to change to continuous compounding, what would be the effective annual interest rate?

a) 10% b) 10.517% c) 12.5% d) 12.649% e) 12.92%

*5.13 Terry bought a CD-ROM drive for $50 down and $30 per month for 24 months. The same drive could have been purchased for $675 cash. What nominal annual interest rate is Terry paying?

a) 7.6% b) 13.9% c) 14.8% d) 15.2% e) 53.3%

5.14 How large a contribution is required to endow perpetually a research laboratory which requires $500,000 for original construction, $200,000 per year for operating expenses, and $100,000 every three years for new and replacement equipment? Interest is 4%.

a) $700,000 b) $6,400,000 c) $7,900,000 d) $10,000,000 e) $12,490,000

ANNUAL EQUIVALENT COST COMPARISONS

*5.15 One of the two production units described below must be purchased. The minimum attractive rate of return is 12%. Compare the two units on the basis of equivalent annual cost.

	Unit A	Unit B
Initial Cost	$16,000	$30,000
Life	8 years	15 years
Salvage value	$ 2,000	$ 5,000
Annual operating cost	$ 2,000	$ 1,000

a) *A*—$5,058; *B*—$5,270
b) *A*—$4,916; *B*—$4,872
c) *A*—$3,750; *B*—$2,667
d) *A*—$1,010; *B*—$1,010
e) *A*—$2,676; *B*—$4,250

5.16 Tanks to hold a corrosive chemical are now being made of material *A*, and have a life of eight years and a first cost of $30,000. When these tanks are four years old, they must be relined at a cost of $10,000. If the tanks could be made of material *B* , their life would be twenty years and no relining would be necessary. If the minimum rate of return is 10%, what must be the first cost of a tank made of material *B* to make it economically equivalent to the present tanks?

a) $38,764 b) $42,631 c) $51,879 d) $58,760 e) $92,361

PRESENT EQUIVALENT COST COMPARISONS

5.17 Compute the life cycle cost of a reciprocating compressor with first cost of $120,000, annual maintenance cost of $9,000, salvage value of $25,000 and life of six years. The minimum attractive rate of return is 10%.

a) $120,000 b) $145,000 c) $149,000 d) $153,280 e) $167,900

5.18 A punch press costs $100,000 initially, requires $10,000 per year in maintenance expenses, and has no salvage value after its useful life of ten years. With interest of 10%, the capitalized cost of the press is:

a) $100,000 b) $161,400 c) $197,300 d) $200,000 e) $262,700

5.19 A utility is considering two alternatives for serving a new customer. Both plans provide twenty years of service, but plan *A* requires one large initial investment, while plan *B* requires additional investment at the end of ten years. Neglect salvage value, assume interest at 8%, and determine the present cost of both plans.

	Plan A	Plan B
Initial Investment	$50,000	$30,000
Investment at end of 10 years	none	$30,000
Annual property tax and maintenance, years 1–10	$ 800	$ 500
Annual property tax and maintenance, years 11–20	$ 800	$ 900

a) *A*—$48,780; *B*—$49,250
b) *A*—$50,000; *B*—$30,000
c) *A*—$50,000; *B*—$60,000
d) *A*—$57,900; *B*—$50,000
e) *A*—$66,000; *B*—$74,000

*5.20 The heat loss of a bare stream pipe costs $206 per year. Insulation *A* will reduce heat loss by 93% and can be installed for $116; insulation *B* will reduce heat loss by 89% and can be installed for $60. The insulations require no additional expenses and will have no salvage value at the end of the pipe's estimated life of eight years. Determine the present net equivalent value of the two insulations if the interest rate is 10%.

a) *A*—$116; *B*—$90

b) *A*—$906; *B*—$918

c) *A*—$1,022; *B*—$978

d) *A*—$1,417; *B*—$1,406

e) *A*—$1,533; *B*—$1,467

INCREMENTAL APPROACH

5.21 A desalinator is needed for six years. Cost estimates for two are:

	The Life of Brine	The Salty Tower
Price	$95,000	$120,000
Annual maintenance	3,000	9,000
Salvage value	12,000	25,000
Life in years	3	6

With interest at 10%, what is the annual cost advantage of the Salty Tower?

a) 0 b) $4,260 c) $5,670 d) $5,834 e) $56,000

5.22 A motor costs $20,000 and has an estimated life of six years. By the addition of certain auxiliary equipment, an annual savings of $300 in operating costs can be obtained, and the estimated life of the motor extended to nine years. Salvage value in either case is $5,000. Interest on capital is 8%. Compute the maximum expenditure justifiable for the auxiliary equipment.

a) $1,149 b) $1,800 c) $2,700 d) $7,140 e) $13,300

*5.23 An existing electrical power line needs to have its capacity increased, and this can be done in either of two ways. The first method is to add a second conductor to each phase wire, using the same poles, insulators and fittings, at a construction cost of $15,000. The second method for increasing capacity is to build a second line parallel to the existing line, using new poles, insulators and fittings, at a construction cost of $23,000. At some time in the future, the line will require another increase in capacity, with the first alternative now requiring a second line at a cost of $32,500, and the second alternative requiring added conductors at a cost of $23,000. If interest rate is 6%, how many years between the initial expenditure and the future expenditure will make the two methods economically equal?

a) 1 b) 3 c) 5 d) 10 e) 25

REPLACEMENT PROBLEMS

5.24 One year ago machine *A* was purchased at a cost of $2,000, to be useful for five years. However, the machine failed to perform properly and costs $200 per month for repairs, adjustments and shut-downs. A new machine *B* designed to perform the same functions is quoted at $3,500, with the cost of repairs and adjustments estimated to be only $50 per month. The expected life of machine *B* is five years. Except for repairs and adjustments, the operating costs of the two machines are substantially equal. Salvage values are insignificant. Using 8% interest rate, compute the incremental annual net equivalent value of machine *B*.

a) – $877 b) $923 c) $1,267 d) $1,800 e) $2,677

BREAK-EVEN ANALYSIS

5.25 Bear Air, an airline serving the Arctic, serves in-flight snacks on some routes. Preparing these snacks costs Bear Air $5000 per month plus $1.50 per snack. Alternatively, prepared snacks may be purchased from a supplier for $4.00 per snack. What is the maximum number of snacks per month for which purchasing from the supplier is justified economically?

a) 769 b) 1250 c) 2000 d) 3333 e) 4000

*5.26 Bear Air has been contracting its overhaul work to Aleutian Aeromotive for $40,000 per plane per year. Bear estimates that by building a $500,000 maintenance facility with a life of 15 years and a salvage value of $100,000, they could handle their own overhauls at a variable cost of only $30,000 per plane per year. The maintenance facility could be financed with a secured loan at 8% interest. What is the minimum number of planes Bear must operate in order to make the maintenance facility economically feasible?

a) 5 b) 6 c) 10 d) 40 e) 50

5.27 It costs Bear Air $1,200 to run a scheduled flight, empty or full, from Coldfoot to Frostbite. Moreover, each passenger generates a cost of $40. The regular ticket costs $90. The plane holds 65 people, but it is running only about 20 per flight. The sales director has suggested selling introductory tickets for $50 to people who have never flown Bear Air.

What is the minimum number of introductory tickets that must be sold in order for a flight to produce a profit?

a) 5 b) 10 c) 15 d) 20 e) 45

What would be the total profit on the flight from Coldfoot to Frostbite if all 65 passengers claimed introductory tickets?

a) – $800 b) – $550 c) 0 d) $400 e) $500

5.28 Two electric motors are being considered for an application in which there is uncertainty concern-ing the hours of usage. Motor *A* costs $4,500 and has an efficiency of 90%. Motor *B* costs $3,000 and has an efficiency of 89%. Each motor has a ten-year life and no salvage value. Electric service costs $18.70 per year per kW of demand and $0.10 per kWh of energy. The output of the motors is to be 75 kW, and interest rate is 8%. At how many hours usage per year would the two motors be equally economical? If the usage is less than this amount, which motor is preferable?

a) 1800, *A* b) 1800, *B* c) 2200, *A* d) 2200, *B* e) 2500, *A*

INCOME TAX AND DEPRECIATION

5.29 A drill press is purchased for $10,000 and has an estimated life of twelve years. The salvage value at the end of twelve years is estimated to be $1,300. Using general straight-line depreciation, compute the book value of the drill press at the end of eight years.

a) $1,300 b) $3,333 c) $3,475 d) $4,200 e) $4,925

5.30 A grading contractor owns earth-moving equipment that costs $300,000 and is classed as 7-year property. After seven years of use, its salvage value will be $50,000. Using the general straight line method, compute the first two depreciation deductions and the book value at the end of four years.

a) $35,714; $35,714; $157,143
b) $85,714; $85,714; $0
c) $21,429; $42,857; $150,000
d) $42,857; $73,469; $93,711
e) $85,714; $61,224; $78,092

5.31 Rework Prob. 5.30 using the MACRS straight line method to compute the first two depreciation deductions and the book value at the end of the fourth tax year.

a) $35,714; $35,714; $157,143
b) $85,714; $85,714; $0
c) $21,429; $42,857; $150,000
d) $42,857; $73,469; $93,711
e) $85,714; $61,224; $78,092

5.32 Rework Prob. 5.30 using the general double-declining balance method to compute the first two depreciation deductions and the book value at the end of the four years.

a) $35,714; $35,714; $157,143
b) $85,714; $85,714; $0
c) $21,429; $42,857; $150,000
d) $42,857; $73,469; $93,711
e) $85,714; $61,224; $78,092

5.33 Rework Prob. 5.30 using the MACRS double-declining balance method to compute the first two depreciation deductions and the book value at the end of the fourth tax year.

a) $35,714; $35,714; $157,143
b) $85,714; $85,714; $0
c) $21,429; $42,857; $150,000
d) $42,857; $73,469; $93,711
e) $85,714; $61,224; $78,092

Solutions to Practice Problems

5.1 d) Prestige.

5.2 d) $1000\times 1.06^3 = \$1191$.

5.3 c) $150/1.08^2 = \$129$.

5.4 b) $500(\mathrm{P/A})_{12}^{8} = 500\times 7.536 = \$3768$.

5.5 b), d) $A:\ 300,000+35,000(P/A)_{30}^{8} - 50,000(P/F)_{30}^{8} = \$689,000$
$B:\ 689,000(A/P)_{30}^{8}(P/A)_{\infty}^{8} = \$765,000$.

5.6 b) $\left[1000(P/F)_{5}^{10} + 2000(P/F)_{10}^{10} + 3500(P/F)_{15}^{10}\right](A/P)_{20}^{10} = \$262$.

5.7 d) $54,000(A/P)_{30}^{8} = \$4800$.

5.8 c) $10,000(F/P)_{4}^{6} - 3000 = \$9625$.

5.9 a) $18,000(A/F)_{8}^{4} = \$1953$.

5.10 a) $1000(P/A)_{\infty}^{10} + 1000(P/A)_{10}^{10} = \$16,145$. Note: for the first 10 years, this accounts for \$2000/yr.

5.11 a) $20,000(P/A)_{6}^{8}(P/F)_{1}^{8} = \$85,600$.

5.12 b) $e^{0.1} - 1 = 0.10517$ or 10.517%.

5.13 b) $675 = 50 + 30(P/A)_{24}^{i}$. $\therefore (P/A)_{24}^{i} = 20.833$.
$\therefore$ by trial and error $i = 0.0116$. $\therefore 12i = 0.139$ or 13.9%.

5.14 b) $500,000 + \left[200,000 + 100,000(A/P)_{3}^{4}\right](P/A)_{\infty}^{4} = \$6,400,000$.

5.15 a) $A:\ -16,000(A/P)_{8}^{12} - 2000 + 2000(A/F)_{8}^{12} = -\$5058$.
$B:\ -30,000(A/P)_{15}^{12} - 1000 + 5000(A/F)_{15}^{12} = -\$5270$

5.16 d) $P(A/P)_{20}^{10} = \left[30,000 + 10,000(P/F)_{4}^{10}\right](A/P)_{8}^{10}$. $\therefore P = \$58,760$.

5.17 b) $120,000 + 9000(P/A)_{6}^{10} - 25,000(P/F)_{6}^{10} = \$145,000$.

5.18 e) $\left[100,000(A/P)_{10}^{10} + 10,000\right](P/A)_{\infty}^{10} = \$262,700$.

5.19 d) $A: \ 50,000 + 800(P/A)_{20}^{8} = \$57,900$

$B: \ 30,000 + 500(A/P)_{20}^{8} + \left[30,000 + 400(P/A)_{10}^{8}\right](P/F)_{10}^{8} = \$50,000.$

5.20 b) $A: \ -116 + 0.93 \times 206(P/A)_{8}^{10} = \$906$

$B: \ -60 + 0.89 \times 206(P/A)_{8}^{10} = \$918$

5.21 b) $A: \ \left[-25,000 + 83,000(P/F)_{3}^{10}\right](A/P)_{6}^{10} - 6000 + 13,000(A/F)_{6}^{10} = \$4260.$

5.22 d) $(20,000 + P)(A/P)_{9}^{8} - 300 - 5000(A/F)_{9}^{8} = 20,000(A/P)_{6}^{8} - 5000(A/F)_{6}^{8}.$

$\therefore \ P = \$7140.$

5.23 b) $(23,000 - 15,000) + (23,000 - 32,500)(P/F)_{N}^{6} = 0. \ \ (1.06)^{-N} = 0.84.$

$\therefore \ N = 3$ yrs.

5.24 b) $-3500(A/P)_{5}^{8} + 12(200 - 50) = \$923.$

5.25 c) $5000 + 1.50n = 4.00n. \ \ \therefore n = 2000.$

5.26 b) $40,000x = 500,000(A/P)_{15}^{8} - 100,000(A/F)_{15}^{8} + 30,000x.$

$\therefore \ x = 5.47.$ Use $x = 6$.

5.27 d) $A) \ \ 1200 + 40(20 + x) = 90(20) + 50x. \ \ \therefore \ x = 20$

$B) \ \ 65(50) - 65(40) - 1200 = -550$

5.28 d) $4500(A/P)_{10}^{8} + (18.7 + .1x)75/.9 = 3000(A/P)_{10}^{8} + (18.7 + .1x)75/.89$.

$\therefore \ x = 2200$ hr., B

5.29 d) $10,000 - (10,000 - 1300)8/12 = 4200$

5.30 a) $(300,000 - 50,000)/7 = 35,714$

$300,000 - 4(35,714) = 157,143$

5.31 c) $(300,000/7)0.5 = 21,419$

$300,000/7 = 42,857$

$300,000 - 3.5(42,857) = 150,000$

5.32 e) $300,000 \times 2/7 = 85,714$

$(300,000 - 85,714)2/7 = 61,224$

$300,000 - 85,714 - 61,224 - 43,732 - 31,237 = 78,092$

5.33 d) $(300,000 \times 2/7)0.5 = 42,857$

$(300,000 - 42,857)2/7 = 73,469$

$300,000 - 42,857 - 73,469 - 52,478 - 37,484 = 93,711$

TABLE 5.2 Compound Interest Factors

i= ½%

n	(P/F)	(P/A)	(P/G)	(F/P)	(F/A)	(A/P)	(A/F)	(A/G)	n
1	.9950	0.995	0.000	1.005	1.000	1.0050	1.0000	0.000	1
2	.9901	1.895	0.990	1.010	2.005	0.5038	0.4988	0.499	2
3	.9851	2.970	2.960	1.015	3.015	0.3367	0.3317	0.997	3
4	.9802	3.950	5.901	1.020	4.030	0.2531	0.2481	1.494	4
5	.9754	4.926	9.803	1.025	5.050	0.2030	0.1980	1.990	5
6	.9705	5.896	14.655	1.030	6.076	0.1696	0.1646	2.485	6
7	.9657	6.862	20.449	1.036	7.106	0.1457	0.1407	2.980	7
8	.9609	7.823	27.176	1.041	8.141	0.1278	0.1228	3.474	8
9	.9561	8.779	34.824	1.046	9.182	0.1139	0.1089	3.967	9
10	.9513	9.730	43.386	1.051	10.228	0.1028	0.0978	4.459	10
11	.9466	10.677	52.853	1.056	11.279	0.0937	0.0887	4.950	11
12	.9419	11.619	63.214	1.062	12.336	0.0861	0.0811	5.441	12
13	.9372	12.556	74.460	1.067	13.397	0.0796	0.0746	5.930	13
14	.9326	13.489	86.583	1.072	14.464	0.0741	0.0691	6.419	14
15	.9279	14.417	99.574	1.078	15.537	0.0694	0.0644	6.907	15
16	.9233	15.340	113.424	1.083	16.614	0.0652	0.0602	7.394	16
17	.9187	16.259	128.123	1.088	17.697	0.0615	0.0565	7.880	17
18	.9141	17.173	143.663	1.094	18.786	0.0582	0.0532	8.366	18
19	.9096	18.082	160.036	1.099	19.880	0.0553	0.0503	8.850	19
20	.9051	18.987	177.232	1.105	20.979	0.0527	0.0477	9.334	20
21	.9006	19.888	195.243	1.110	22.084	0.0503	0.0453	9.817	21
22	.8961	20.784	214.061	1.116	23.194	0.0481	0.0431	10.299	22
23	.8916	21.676	233.677	1.122	24.310	0.0461	0.0411	10.781	23
24	.8872	22.563	254.082	1.127	25.432	0.0443	0.0393	11.261	24
25	.8828	23.446	275.269	1.133	26.559	0.0427	0.0377	ll.N1	25
26	.8784	24.342	297.228	1.138	27.692	0.0411	0.0361	12.220	26
28	.8697	26.068	343.433	1.150	29.975	0.0384	0.0334	13.175	28
30	.8610	27.794	392.632	1.161	32.280	0.0360	0.0310	14.126	30
∞	0	200.000	40000.0	∞	∞	.0050	0	200.00	∞

i=2.00%

n	(P/F)	(P/A)	(P/G)	(F/P)	(F/A)	(A/P)	(A/F)	(A/G)	n
I	.9804	0.9804	0.0000	1.0200	1.0000	1.0200	1.0000	0.0000	1
2	.9612	1.9416	0.9612	1.0404	2.0200	0.5150	0.4950	0.4950	2
3	.9423	2.8839	2.8458	1.0612	3.0604	0.3468	0.3268	0.9868	3
4	.9238	3.8077	5.6173	1.0824	4.1216	0.2626,	0.2426	1.4752	4
5	.9057	4.7135	9.2403	1.1041	5.2040	0.2122	0.1922	1.9604	5
6	.8880	5.6014	13.6801	1.1262	6.3081	0.1785	0.1585	2.4423	6
7	.8706	6.4720	18.905	1.1487	7.4343	0.1545	0.1345	2.9208	7
8	.8535	7.3255	24.8779	1.1717	8.5830	0.1365	0.1165	3.3961	8
9	.8368	8.1622	31.5720	1.1951	9.7546	0.1225	0.1025	3.8681	9
10	.8203	8.9826	38.9551	1.2190	10.9497	0.1113	0.0913	4.3367	10
11	.8043	9.7868	46.9977	1.2434	12.1687	0.1022	0.0822	4.8021	11
12	.7885	10.5753	55.6712	1.2682	13.4121	0.0946	0.0746	5.2642	12
13	.7730	11.3484	64.9475	1.2936	14.6803	0.0881	0.0681	5.7231	13
14	.7579	12.1062	74.7999	1.3195	15.9739	0.0826	0.0626	6.1786	14
15	.7430	12.8493	85.2021	1.3459	17.2934	0.0778	0.0578	6.6309	15
16	.7284	13.5777	96.1288	1.3728	18.6393	0.0737	0.0537	7.0799	16
17	.7142	14.2919	107.5554	1.4002	20.0121	0.0700	0.0500	7.5256	17
18	.7002	14.9920	119.4581	1.4282	21.4123	0.0667	0.0467	7.9681	18
19	.6864	15.6785	131.8139	1.4568	22.8406	0.0638	0.0438	8.4073	19
20	.6730	16.3514	144.6003	1.4859	24.2974	0.0612	0.0412	8.8433	20
21	.6598	17.0112	157.7959	1.5157	25.7833	0.0588	0.0388	9.2760	21
22	.6468	17.6580	171.3795	1.5460	27.2990	0.0566	0.0366	9.7055	22
23	.6342	18.2922	185.3309	1.5769	28.8450	0.0547	0.0347	10.1317	23
24	.6217	18.9139	199.6305	1.6084	30.4219	0.0529	0.0329	10.5547	24
25	.6095	19.5235	214.2592	1.6406	32.0303	0.0512	0.0312	10.9745	25
26	.5976	20.1210	229.1987	1.6734	33.6709	0.0497	0.0297	11.3910	26
28	.5744	21.2813	259.9392	1.7410	37.0512	0.0470	0.0270	12.2145	28
30	.5521	22.3965	291.7164	1.8114	40.5681	0.0446	0.0246	13.0251	30
∞	.0000	50.0000	2500.0000	∞	∞	0.0200	0.0000	50.0000	∞

TABLE 5.2 Compound Interest Factors (continued)

$i = 4.00\%$

n	(P/F)	(P/A)	(P/G)	(F/P)	(F/A)	(A/P)	(A/F)	(A/G)	*n*
1	.9615	0.9615	-0.0000	1.0400	1.0000	1.0400	1.0000	-0.0000	1
2	.9246	1.8861	0.9246	1.0816	2.0400	0.5302	0.4902	0.4902	2
3	.8890	2.7751	2.7025	1.1249	3.1216	0.3603	0.3203	0.9739	3
4	.8548	3.6299	5.2670	1.1699	4.2465	0.2755	0.2355	1.4510	4
5	.8219	4.4518	8.5547	1.2167	5.4163	0.2246	0.1846	1.9216	5
6	.7903	5.2421	12.5062	1.2653	6.6330	0.1908	0.1508	2.3857	6
7	.7599	6.0021	17.0657	1.3159	7.8983	0.1666	0.1266	2.8433	7
8	.7307	6.7327	22.1806	1.3686	9.2142	0.1485	0.1085	3.2944	8
9	.7026	7.4353	27.8013	1.4233	10.5828	0.1345	0.0945	3.7391	9
10	.6756	8.1109	33.8814	1.4802	12.0061	0.1233	0.08333	4.1773	10
11	.6496	8.7605	40.3772	1.5395	13.4864	0.1141	0.0741	4.6090	11
12	.6246	9.3851	47.2477	1.6010	15.0258	0.1066	0.0666	5.0343	12
13	.6006	9.9856	54.4546	1.6651	16.6268	0.1001	0.0601	5.4533	13
14	.5775	10.5631	61.9618	1.7317	18.2919	0.0947	0.0547	5.8659	14
15	.5553	11.1184	69.7355	1.8009	20.0236	0.0899	0.0499	6.2721	15
16	.5339	11.6523	77.7441	1.8730	21.8245	0.0858	0.0458	6.6720	16
17	.5134	12.1657	85.9581	1.9479	23.6975	0.0822	0.0422	7.0656	17
18	.4936	12.6593	94.3498	2.0258	25.6454	0.0790	0.0390	7.4530	18
19	.4746	13.1339	102.8933	2.1068	27.6712	0.0761	0.0361	7.8342	19
20	.4564	13.5903	111.5647	2.1911	29.7781	0.0736	0.0336	8.2091	20
21	.4388	14.0292	120.3414	2.2788	31.9692	0.0713	0.0313	8.5779	21
22	.4220	14.4511	129.2024	2.3699	34.2480	0.0692	0.0292	8.9407	22
23	.4057	14.8568	138.1284	2.4647	36.6179	0.0673	0.0273	9.2973	23
24	.3901	15.2470	147.1012	2.5633	39.0826	0.0656	0.0256	9.6479	24
25	.3751	15.6221	156.1040	2.6658	41.6459	0.0640	0.0240	9.9925	25
26	.3607	15.9828	165.1212	2.7725	44.3117	0.0626	0.0226	10.3312	26
28	.3335	16.6631	183.1424	2.9987	49.9676	0.0600	0.0200	10.9909	28
30	.3083	17.2920	201.0618	3.2434	56.0849	0.0578	0.0178	11.6274	30
∞	.0000	25.000	625.0000	∞	∞	0.0400	0.0000	25.0000	∞

$i = 6.00\%$

n	(P/F)	(P/A)	(P/G)	(F/P)	(F/A)	(A/P)	(A/F)	(A/G)	*n*
1	.9434	0.9434	–0.0000	1.0600	1.0000	1.0600	1.0000	-0.0000	1
2	.8900	1.8334	0.8900	1.1236	2.0600	0.5454	0.4854	0.4854	2
3	.8396	2.6730	2.5692	1.1910	3.1836	0.3741	0.3141	0.9612	3
4	.7921	3.4651	4.9455	1.2625	4.3746	0.2886	0.2286	1.4272	4
5	.7473	4.2124	7.9345	1.3382	5.6371	0.2374	0.1774	1.8836	5
6	.7050	4.9173	11.4594	1.4185	6.9753	0.2034	0.1434	2.3304	6
7	.6651	5.5824	15.4497	1.5036	8.3938	0.1791	0.1191	2.7676	7
8	.6274	6.2098	19.8416	1.5938	9.8975	0.1610	0.1010	3.1952	8
9	.5919	6.8017	24.5768	1.6895	11.4913	0.1470	0.0870	3.6133	9
10	.5584	7.3601	29.6023	1.7908	13.1808	0.1359	0.0759	4.0220	10
11	.5268	7.8869	34.8702	1.8983	14.9716	0.1268	0.0668	4.4213	11
12	.4970	8.3838	40.3369	2.0122	16.8699	0.1193	0.0593	4.8113	12
13	.4688	8.8527	45.9629	2.1329	18.8821	0.1130	0.0530	5.1920	13
14	.4423	9.2950	51.7128	2.2609	21.0151	0.1076	0.0476	5.5635	14
15	.4173	9.7122	57.5546	2.3966	23.2760	0.1030	0.0430	5.9260	15
16	.3936	10.1059	63.4592	2.5404	25.6725	0.0990	0.0390	6.2794	16
17	.3714	10.4773	69.4011	2.6928	28.2129	0.0954	0.0354	6.6240	17
18	.3503	10.8276	75.3569	2.8543	30.9057	0.0924	0.0324	6.9597	18
19	.3305	11.1581	81.3062	3.0256	33.7600	0.0896	0.0296	7.2867	19
20	.3118	11.4699	87.2304	3.2071	36.7856	0.0872	0.0272	7.6051	20
21	.2942	11.7641	93.1136	3.3996	39.9927	0.0850	0.0250	7.9151	21
22	.2775	12.0416	98.9412	3.6035	43.3923	0.0830	0.0230	8.2166	22
23	.2618	12.3034	104.7007	3.8197	46.9958	0.0813	0.0213	8.5099	23
24	.2470	12.5504	110.3812	4.0489	50.8156	0.0797	0.0197	1.87951	24
25	.2330	12.7834	115.9732	4.2919	54.8645	0.0782	0.0182	9.0722	25
26	.2198	13.0032	121.4684	4.5494	59.1564	0.0769	0.0169	9.3414	26
28	.1956	13.4062	132.1420	5.1117	68.5281	0.0746	0.0146	9.8568	28
30	.1741	13.7648	142.3588	5.7435	79.0582	0.0726	0.0126	10.3422	30
∞	.0000	16.6667	277.7778	∞	∞	0.0600	0.0000	16.667	∞

TABLE 5.2 Compound Interest Factors (continued)

$i = 8.00\%$

n	(P/F)	(P/A)	(P/G)	(F/P)	(F/A)	(A/P)	(A/F)	(A/G)	*n*
1	.9259	0.9259	–0.0000	1.0800	1.0000	1.0800	1.0000	–0.0000	1
2	.8573	1.7833	0.8573	1.1664	2.0800	0.5608	0.4808	0.4808	2
3	.7938	2.5771	2.4450	1.2597	3.2464	0.3880	0.3080	0.9487	3
4	.7350	3.3121	4.6501	1.3605	4.5061	0.3019	0.2219	1.4040	4
5	.6806	3.9927	7.3724	1.4693	5.8666	0.2505	0.1705	1.8465	5
6	.6302	4.6229	10.5233	1.5869	7.3359	0.2163	0.1363	2.2763	6
7	.5835	5.2064	14.0242	1.7138	8.9228	0.1921	0.1121	2.6937	7
8	.5403	5.7466	17.8061	1.8509	10.6366	0.1740	0.0940	3.0985	8
9	.5002	6.2469	21.8081	1.9990	12.4876	0.1601	0.0801	3.4910	9
10	.4632	6.7101	25.9768	2.1589	14.4866	0.1490	0.0690	3.8713	10
11	.4289	7.1390	30.2657	2.3316	16.6455	0.1401	0.0601	4.2395	11
12	.3971	7.5361	34.6339	2.5182	18.9771	0.1327	0.0527	4.5957	12
13	.3677	7.9038	39.0463	2.7196	21.4953	0.1265	0.0465	4.9402	13
14	.3405	8.2442	43.4723	2.9372	24.2149	0.1213	0.0413	5.2731	14
15	.3152	8.5595	47.8857	3.1722	27.1521	0.1168	0.0368	5.5945	15
16	.2919	8.8514	52.2640	3.4259	30.3243	0.1130	0.0330	5.9046	16
17	.2703	9.1216	56.5883	3.7000	33.7502	0.1096	0.0296	6.2037	17
18	.2502	9.3719	60.8426	3.9960	37.4502	0.1067	0.0267	6.4920	18
19	.2317	9.6036	65.0134	4.3157	41.4463	0.1041	0.0241	6.7697	19
20	.2145	9.8181	69.0898	4.6610	45.7620	0.1019	0.0219	7.0369	20
21	.1987	10.0168	73.0629	5.0338	50.4229	0.0998	0.0198	7.2940	21
22	.1839	10.2007	76.9257	5.4365	55.4568	0.0980	0.0180	7.5412	22
23	.1703	10.3711	80.6726	5.8715	60.8933	0.0964	0.0164	7.7786	23
24	.1577	10.5288	84.2997	6.3412	66.7648	0.0950	0.0150	8.0066	24
25	.1460	10.6748	87.8041	6.8485	73.1059	0.0937	0.0137	8.2254	25
26	.1352	10.8100	91.1842	7.3964	79.9544	0.0925	0.0125	8.4352	26
28	.1159	11.0511	97.5687	8.6271	95.3388	0.0905	0.0105	8.8289	28
30	.0994	11.2578	103.4558	10.0627	113.2832	0.0888	0.0088	9.1897	30
∞	.0000	12.500	156.2500	∞	∞	0.0800	0.0000	12.5000	∞

$i = 10.00\%$

n	(P/F)	(P/A)	(P/G)	(F/P)	(F/A)	(A/P)	(A/F)	(A/G)	*n*
1	.9091	0.9091	-0.0000	1.1000	1.0000	1.1000	1.0000	-0.0000	1
2	.8264	1.7355	0.8264	1.2100	2.1000	0.5762	0.4762	0.4762	2
3	.7513	2.4869	2.3291	1.3310	3.3100	0.4021	0.3021	0.9366	3
4	.6830	3.1699	4.3781	1.4641	4.6410	0.3155	0.2155	1.3812	4
5	.6209	3.7908	6.8618	1.6105	6.1051	0.2638	0.1638	1.8101	5
6	.5645	4.3553	9.6842	1.7716	7.7156	0.2296	0.1296	2.2236	6
7	.5132	4.8684	12.7631	1.9487	9.4872	0.2054	0.1054	2.6216	7
8	.4665	5.3349	16.0287	2.1436	11.4359	0.1874	0.0874	3.0045	8
9	.4241	5.7590	19.4215	2.3579	13.5795	0.1736	0.0736	3.3724	9
10	.3855	6.1446	22.8913	2.5937	15.9374	0.1627	0.0627	3.7255	10
11	.3505	6.4951	26.3963	2.8531	18.5312	0.1540	0.0540	4.0641	11
12	.3186	6.8137	29.9012	3.1384	21.3843	0.1468	0.0468	4.3884	12
13	.2897	7.1034	33.3772	3.4523	24.5227	0.1408	0.0408	4.6988	13
14	.2633	7.3667	36.8005	3.7975	27.9750	0.1357	0.0357	4.9955	14
15	.2394	7.6061	40.1520	4.1772	31.7725	0.1315	0.0315	5.2789	15
16	.2176	7.8237	43.4164	4.5950	35.9497	0.1278	0.0278	5.5493	16
17	.1978	8.0216	46.5819	5.0545	40.5447	0.1247	0.0247	5.8071	17
18	.1799	8.2014	49.6395	5.5599	45.5992	0.1219	0.0219	6.0526	18
19	.1635	8.3649	52.5827	6.1159	51.1591	0.1195	0.0195	6.2861	19
20	.1486	8.5136	55.4069	6.7275	57.2750	0.1175	0.0175	6.5081	20
21	.1351	8.6487	58.1095	7.4002	64.0025	0.1156	0.0156	6.7189	21
22	.1228	8.7715	60.6893	8.1403	71.4027	0.1140	0.0140	6.9189	22
23	.1117	8.8832	63.1462	8.9543	79.5430	0.1126	0.0126	7.1085	23
24	.1015	8.9847	65.4813	9.8497	88.4973	0.1113	0.0113	7.2881	24
25	.0923	9.0770	67.6964	10.8347	98.3471	0.1102	0.0102	7.4580	25
26	.0839	9.1609	69.7940	11.9182	109.1818	0.1092	0.0092	7.6186	26
28	.0693	9.3066	73.6495	14.4210	134.2099	0.1075	0.0075	7.9137	28
30	.0573	9.4269	77.0766	17.4494	164.4940	0.1061	0.0061	8.1762	30
∞	.0000	10.0000	100.0000	∞	∞	0.1000	0.0000	10.0000	∞

TABLE 5.2 Compound Interest Factors (continued)

i= 12.00%

n	(P/F)	(P/A)	(P/G)	(F/P)	(F/A)	(A/P)	(A/F)	(A/G)	*n*
1	.8929	0.8929	-0.0000	1.1200	1.0000	1.1200	1.0000	–0.0000	1
2	.7972	1.6901	0.7972	1.2544	2.1200	0.5917	0.4717	0.4717	2
3	.7118	2.4018	2.2208	1.4049	3.3744	0.4163	0.2963	0.9246	3
4	.6355	3.073	4.1273	1.5735	4.7793	0.3292	0.2092	1.3589	4
5	.5674	3.6048	6.3970	1.7623	6.3528	0.2774	0.1574	1.7746	5
6	.5066	4.1114	8.9302	1.9738	8.1152	0.2432	0.1232	2.1720	6
7	.4523	4.5638	11.6443	2.2107	10.0890	0.2191	0.0991	2.5515	7
8	.4039	4.9676	14.4714	2.4760	12.2997	0.2013	0.0813	2.9131	8
9	.3606	5.3282	17.3563	2.7731	14.7757	0.1877	0.0677	3.2574	9
10	.3220	5.6502	20.2541	3.1058	17.5487	0.1770	0.0570	3.5847	10
11	.2875	5.9377	23.1288	3.4785	20.6546	0.1684	0.0484	3.8953	11
12	.2567	6.1944	25.9523	3.8960	24.1331	0.1614	0.0414	4.1897	12
13	.2292	6.4235	28.7024	4.3635	28.0291	0.1557	0.0357	4.4683	13
14	.2046	6.6282	31.3624	4.8871	32.3926	0.1509	0.0309	4.7317	14
15	.1827	6.8109	33.9202	5.4736	37.2797	0.1468	0.0268	4.9803	15
16	.1631	6.9740	36.3670	6.1304	42.7533	0.1434	0.0234	5.2147	16
17	.1456	7.1196	38.6973	6.8660	48.8837	0.1405	0.0205	5.4353	17
18	.1300	7.2497	40.9080	7.6900	55.7497	0.1379	0.0179	5.6427	18
19	.1161	7.3658	42.9979	8.6128	63.4397	0.1358	0.0158	5.8375	19
20	.1037	7.4694	44.9676	9.6463	72.0524	0.1339	0.0139	6.0202	20
21	.0926	7.5620	46.8188	10.8038	81.6987	0.1322	0.0122	6.1913	21
22	.0826	7.6446	48.5543	12.1003	92.5026	0.1308	0.0108	6.3514	22
23	.0738	7.7184	50.1776	13.5523	104.6029	0.1296	0.0096	6.5010	23
24	.0659	7.7843	51.6929	15.1786	118.1552	0.1285	0.0085	6.6406	24
25	.0588	7.8431	53.1046	17.0001	133.3339	0.1275	0.0075	6.7708	25
26	.0525	7.8957	54.4177	19.0401	150.3339	0.1267	0.0067	6.8921	26
28	.0419	7.9844	56.7674	23.8839	190.6989	0.1252	0.0052	7.1098	28
30	.0334	8.0552	58.7821	29.9599	241.3327	0.1241	0.0041	7.2974	30
∞	.0000	8.333	69.4444	∞	∞	0.1200	0.0000	8.3333	∞

i = 20.00%

n	(P/F)	(P/A)	(P/G)	(F/P)	(F/A)	(A/P)	(A/F)	(A/G)	*n*
1	.8333	0.8333	-0.0000	1.2000	1.0000	1.2000	1.0000	-0.0000	1
2	.6944	1.5278	0.6944	1.4400	2.2000	0.6545	0.4545	0.4545	2
3	.5787	2.1065	1.8519	1.7280	3.6400	0.4747	0.2747	0.8791	3
4	.4823	2.5887	3.2986	2.0736	5.3680	0.3863	0.1863	1.2742	4
5	.4019	2.9906	4.9061	2.4883	7.4416	0.3344	0.1344	1.6405	5
6	.3349	3.3255	6.5806	2.9860	9.9299	0.3007	0.1007	1.9788	6
7	.2791	3.6046	8.2551	3.5832	12.9159	0.2774	0.0774	2.2902	7
8	.2326	3.8372	9.8831	4.2998	16.4991	0.2606	0.0606	2.5756	8
9	.1938	4.0310	11.4335	5.1598	20.7989	0.2481	0.0481	2.8364	9
10	.1615	4.1925	12.8871	6.1917	25.9587	0.2385	0.0385	3.0739	10
11	.1346	4.3271	14.2330	7.4301	32.1504	0.2311	0.0311	3.2893	11
12	.1122	4.4392	15.4667	8.9161	39.5805	0.2253	0.0253	3.4841	12
13	.0935	4.5327	16.5883	10.6993	48.4966	0.2206	0.0206	3.6597	13
14	.0779	4.6106	17.6008	12.8392	59.1959	0.2169	0.0169	3.8175	14
15	.0649	4.6755	18.5095	15.4070	72.0351	0.2139	0.0139	3.9588	15
16	.0541	4.7296	19.3208	18.4884	87.4421	0.2114	0.0114	4.0851	16
17	.0451	4.7746	20.0419	22.1861	105.9306	0.2094	0.0094	4.1976	17
18	.0376	4.8122	20.6805	26.6233	128.1167	0.2078	0.0078	4.Z975	18
19	.0313	4.8435	21.2439	31.9480	154.7400	0.2065	0.0065	4.3861	19
20	.0261	4.8696	21.7395	38.3376	186.6880	0.2054	0.0054	4.4643	20
21	.0217	4.8913	22.1742	46.0051	225.0256	0.2044	0.0044	4.5334	21
22	.0181	4.9094	22.5546	55.2061	271.0307	0.2037	0.0037	4.5941	22
23	.0151	4.9245	22.8867	66.2474	326.2369	0.2031	0.0031	4.6475	23
24	.0126	4.9371	23.1760	79.4968	392.4842	0.2025	0.0025	4.6943	24
25	.0105	4.9476	23.4276	95.3962	471.9811	0.2021	0.0021	4.7352	25
26	.0087	4.9563	23.6460	114.4755	567.3773	0.2018	0.0018	4.7709	26
28	.0061	4.9697	23.9991	164.8447	819.2233	0.2012	0.0012	4.8291	28
30	.0042	4.9789	24.2628	237.3763	1181.8816	0.2008	0.0008	4.8731	30
∞	.o000	5.0000	25.0000	∞	∞	0.2000	0.0000	5.5000	∞

Electrical Theory

by Jon A. Soper

A review of basic concepts in Electrical Engineering of interest to all engineers is presented in this chapter. The subjects include circuits, electromagnetics, the fundamentals of operational electronics, and electric power distribution and machinery. Each section begins with a brief statement of the concepts to be covered and a listing of pertinent formulas with units.

6.1 Circuits

Electric circuits are an interconnection of electrical components for the purpose of either generating and distributing electrical power, converting electrical power to some other useful form such as light, heat or mechanical torque, or processing information contained in an electrical form (i.e., electrical signals). Most electrical circuits contain a source (or sources) of electrical power, passive components which store or dissipate energy and possibly active components which change the electrical form of the energy/information being processed by the circuit.

Circuits may be classified as *Direct Current* (DC) circuits when the currents and voltages do not vary with time and as *Alternating Current* (AC) circuits when the currents and voltage vary sinusoidally with time. Both DC and AC circuits are said to be operating in the *steady state* when their current/voltage time variation is purely constant or purely sinusoidal with time. A *transient circuit* condition occurs when a switch is thrown turning a source either on or off. This review will cover the DC steady-state and DC transient circuit conditions and the AC steady-state circuit condition. The primary quantities of interest in making circuit calculations are presented in Table 6.1.

TABLE 6.1 Quantities Used in Electric Circuits.

Quantity	Symbol	Unit	Defining Equation	Definition
Charge	Q	coulomb	$Q = \int I dt$	
Current	I	ampere	$I = \frac{dQ}{dt}$	Time rate of flow of charge past a point in the circuit.
Voltage	V	volt	$V = \frac{dW}{dQ}$	Energy per unit charge either gained or lost through a circuit element.
Energy	W	joules	$W = \int V dQ = \int P dt$	
Power	P	watts	$P = \frac{dW}{dt} = IV$	Power is the time rate of energy flow.

CIRCUIT COMPONENTS

The circuits reviewed in this section will contain one or more sources interconnected with passive components. These passive circuit components include resistors, inductors and capacitors:

a) *Resistors* are energy absorbing components and have a resistance value R measured in ohms:

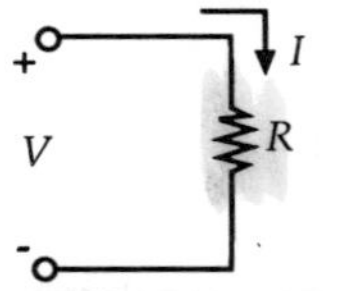

$$I = \frac{V}{R} \tag{6.1.1}$$

$$\text{AMPERES} = \frac{\text{VOLTS}}{\text{OHMS}}$$

b) *Inductors* are energy storage components and have an inductance value L measured in henries:

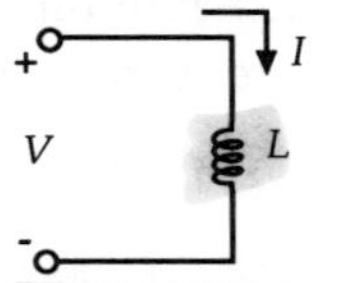

$$I = \frac{1}{L}\int V dt \tag{6.1.2}$$

$$\text{AMPERES} = \frac{\text{VOLT} \cdot \text{SECONDS}}{\text{HENRIES}}$$

c) *Capacitors* are energy storage components and have a capacitance value C measured in farads:

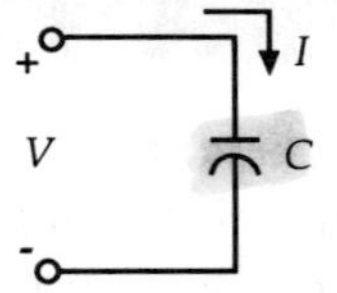

$$I = C\frac{dV}{dt} \tag{6.1.3}$$

$$\text{AMPERES} = \frac{\text{FARAD} \cdot \text{VOLTS}}{\text{SECONDS}}$$

SOURCES OF ELECTRICAL ENERGY

Sources in electric circuits can be either independent of current and/or voltage values elsewhere in the circuit, or they can be dependent upon them. In this section only independent sources will be considered. Fig. 6.1 shows both ideal and linear models for current and voltage sources.

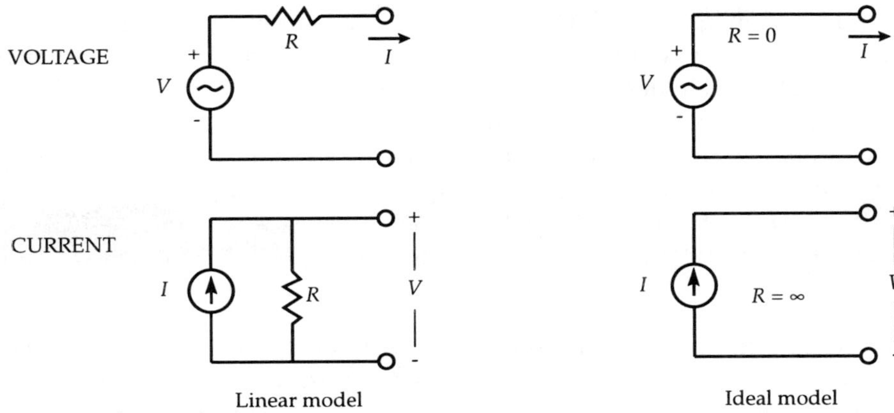

Figure 6.1 Ideal and linear models of current and voltage sources.

KIRCHHOFF'S LAWS

Two laws of conservation govern the behavior of all electrical circuits:

a) *Kirchhoff's voltage law* (KVL), for the conservation of energy, states that the sum of voltage rises or drops around any closed path in an electrical circuit must be zero:

$$\sum V_{DROPS} = 0 \quad \text{around closed path} \qquad \sum V_{RISES} = 0 \quad \text{around closed path} \tag{6.1.4}$$

b) *Kirchhoff's Current Law* (KCL), for the conservation of charge, states that the flow of charges either into (positive) or out of (negative) any node in a circuit must add to zero:

$$\sum I_{IN} = 0 \quad \text{at node} \qquad \sum I_{OUT} = 0 \quad \text{at node} \tag{6.1.5}$$

OHMS LAW

Ohms Law is a statement of the relationship between the voltage across an electrical component and the current through the component. For DC circuits, where the components are resistors, ohms law is

$$V = IR \quad \text{or} \quad I = V/R \tag{6.1.6}$$

For AC circuits, with resistors, capacitors and inductors, ohms law, stated in terms of the component impedance Z (see Table 6.3), is

$$V = IZ \quad \text{or} \quad I = V/Z \tag{6.1.7}$$

REFERENCE VOLTAGE POLARITY AND CURRENT DIRECTION

Circuit analysis requires defining first a reference current direction with an arrow placed next to the circuit component. For each of the components a reference current direction is arbitrarily defined. Once the current reference direction is defined, the voltage reference polarity marks can be placed on each component. The polarity marks on passive components are always placed so that the current flows from the plus (+) mark to the minus (–) mark.

Current values can be either positive or negative. A positive current value shows that the current does in fact flow in the reference direction. A negative current value shows that the current flows opposite to the reference direction. Voltage values can be either positive or negative. A positive voltage value indicates a loss of energy or reduction in voltage when moving through the circuit from the plus polarity mark to the minus polarity mark. A negative voltage value indicates a gain of energy when moving through the circuit from the plus polarity mark to the minus polarity mark.

It is proper electrical jargon to talk about voltage drops and voltage rises. A voltage drop is experienced when moving through the circuit from the plus (+) polarity mark to the minus (–) polarity mark. A voltage rise is experienced when moving through the circuit from the minus (–) polarity mark to the plus (+) polarity mark.

CIRCUIT EQUATIONS

When writing circuit equations, the current is assumed to have a positive value in the reference direction and the voltage is assumed to have a positive value as indicated by the polarity marks. To write the KVL circuit equation one must move around a closed path in the circuit and sum all the voltage rises or all the voltage drops. For example, for the circuit in Fig. 6.2, begin at point a and move to b, then c, then d, and back to a. For $\sum V_{RISES} = 0$ obtain

$$V_s - IR_1 - IR_2 - IR_3 = 0. \tag{6.1.8}$$

For $\sum V_{DROPS} = 0$ obtain

$$-V_s + IR_1 + IR_2 + IR_3 = 0. \tag{6.1.9}$$

Either of these equations can now be solved for the one unknown current

$$I = \frac{V_s}{R_1 + R_2 + R_3} = \frac{V_s}{R_{eq}} \tag{6.1.10}$$

where R_{eq} is the equivalent resistance for the circuit.

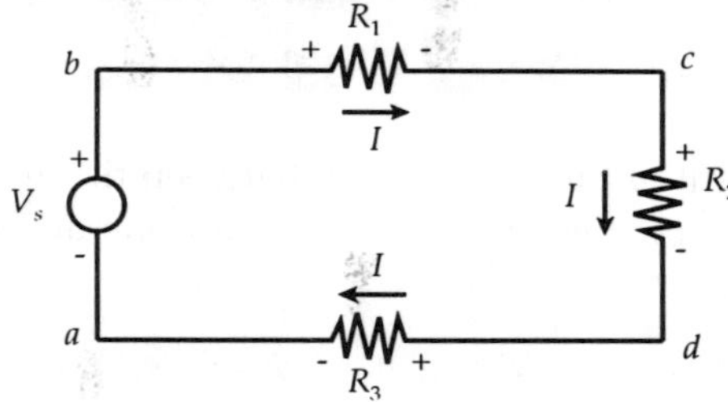

Figure 6.2 A simple circuit.

CIRCUIT EQUATIONS USING BRANCH CURRENTS

The circuit in Fig. 6.3 has meshes around which two voltage equations (KVL) can be written. There are, however, three branches in the circuit. An unknown current with a reference direction is assumed in each branch. The polarity marks are indicated for each resistor so the KVL can be written. Write two KVL equations, one around each mesh. Using $\sum V_{DROPS} = 0$ obtain:

$$-V_s + I_1R_1 + I_3R_2 + I_1R_3 = 0 \tag{6.1.11}$$

$$-I_3R_2 + I_2R_4 + I_2R_5 + I_2R_6 = 0$$

Write one KCL equation at circuit node a. This additional equation is necessary since there are three unknown currents:

$$I_1 - I_2 - I_3 = 0 \tag{6.1.12}$$

These three equations can be solved for I_1, I_2 and I_3. The current I_1 is

$$I_1 = \frac{\begin{vmatrix} V_s & 0 & R_2 \\ 0 & R_4 + R_5 + R_6 & -R_2 \\ 0 & -1 & -1 \end{vmatrix}}{\begin{vmatrix} R_1 + R_3 & 0 & R_2 \\ 0 & R_4 + R_5 + R_6 & -R_2 \\ 1 & -1 & -1 \end{vmatrix}} \tag{6.1.13}$$

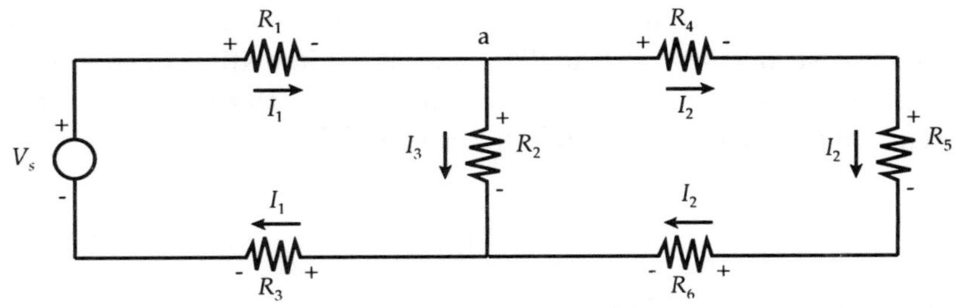

Figure 6.3 A circuit with two meshes and three branches.

CIRCUIT EQUATIONS USING MESH CURRENTS

A simplification in writing the circuit equations for Figure 6.3 occurs if mesh currents are used. Note that

$$I_3 = I_1 - I_2. \tag{6.1.14}$$

Redefine the reference currents in the network of Fig. 6.3 as shown in Fig. 6.4. Now there are only two unknown currents to solve for instead of the three in Fig. 6.3. The current through R_1 and R_3 is I_1. The current through R_4, R_5 and R_6 is I_2. The current through R_2 is $I_1 - I_2$ which is consistent with the KCL equation written for the network of Fig. 6.3. Write two KVL equations, one around each mesh:

$$\begin{aligned} -V_s + I_1(R_1 + R_2 + R_3) - I_2 R_2 &= 0 \\ -I_1 R_2 + I_2(R_2 + R_4 + R_5 + R_6) &= 0 \end{aligned} \tag{6.1.15}$$

These two equations can be solve for I_1 and I_2. The current I_1 is equivalent to that of Eq. 6.1.13:

$$I_1 = \frac{\begin{vmatrix} V_s & -R_2 \\ 0 & R_2 + R_4 + R_5 + R_6 \end{vmatrix}}{\begin{vmatrix} R_1 + R_2 + R_3 & -R_2 \\ -R_2 & R_2 + R_4 + R_5 + R_6 \end{vmatrix}} \tag{6.1.16}$$

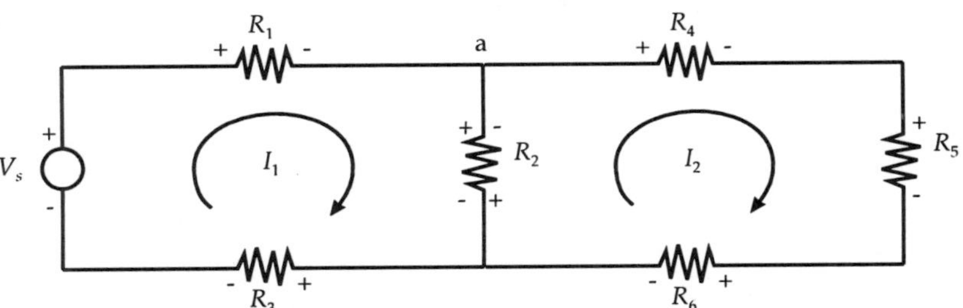

Figure 6.4 A circuit with two meshes.

THÉVENIN EQUIVALENT CIRCUIT

It is often convenient to model a complex circuit which has many electrical components by a single source and single resistor (or impedance) in series. This simplified circuit is called a *Thévenin Equivalent*, as shown in Fig. 6.5. The values of V_{eq} and R_{eq} are found as follows:

- The voltage drop is V_{eq} from a to b in the complex circuit with R_{LOAD} removed. It is evident that V_{eq} is also the open circuit (R_{LOAD} removed) voltage across a–b in the Thévenin equivalent.
- The resistance is R_{eq} looking in at the terminals (a–b) with R_{LOAD} removed and all voltage sources shorted out and all current sources opened (or removed). Any technique for finding equivalent resistance may then be used. A second way to find R_{eq}, once V_{eq} is known, is to short circuit the terminals (a–b) and solve for the current I_{SC} through the short circuit without removing any sources in the complex circuit. Then

$$R_{eq} = \frac{V_{eq}}{I_{SC}} \tag{6.1.17}$$

Example 6.3 illustrates the above.

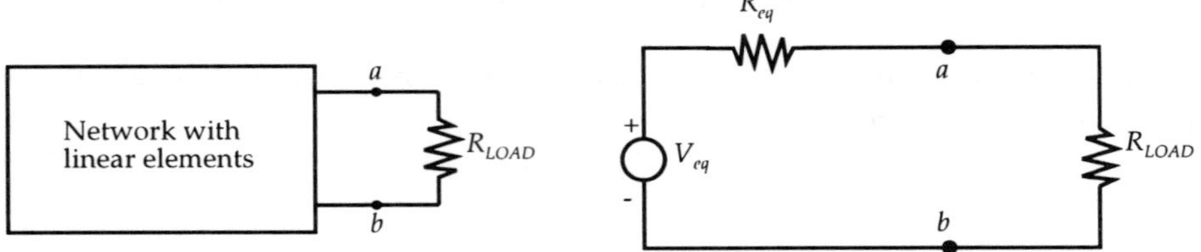

Figure 6.5 A complex circuit and its Thévenin equivalent.

6.1.1 DC Circuits

In a DC circuit the only crucial components are resistors. Another component, the inductor, appears as a zero resistance connection (or short circuit) and a third component, a capacitor, appears as an infinite resistance, or open circuit. The three circuit components are summarized in Table 6.2.

TABLE 6.2 DC Circuit Components

Component	Impedance	Current	Power
Resistor	R	$I = V/R$	$P = I^2R = V^2/R$
Inductor	Zero (Short Circuit)	Unconstrained	None dissipated
Capacitor	Infinite (Open Circuit)	Zero	None dissipated

Example 6.1

Compute the current in the 10 Ω resistor.

Solution. Assume loop currents I_1 and I_2. Write KVL around both meshes:

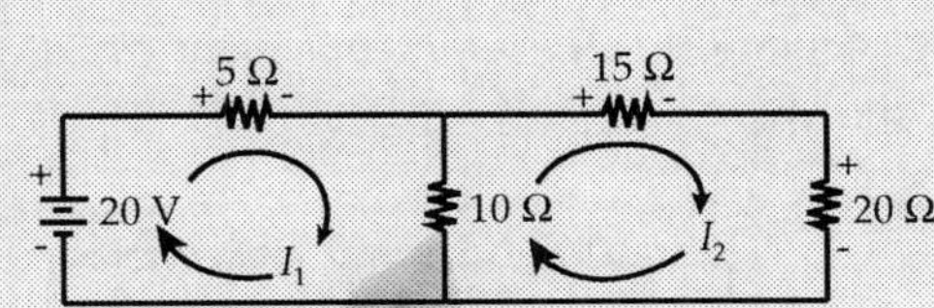

$$\sum V_{DROPS} = 0$$

Mesh 1: $$-20 + 5I_1 + 10I_1 - 10I_2 = 0$$

Mesh 2: $$-10I_1 + 10I_2 + 15I_2 + 20I_2 = 0$$

These are arranged as

$$15I_1 - 10I_2 = 20$$
$$-10I_1 + 45I_2 = 0.$$

The solution is

$$I_1 = 1.57 \text{ A}, \qquad I_2 = 0.35 \text{ A}.$$

The current in the 10 Ω resistor is

$$I = I_1 - I_2$$
$$= 1.57 - 0.35 = 1.22 \text{ A}.$$

Example 6.2

Compute the power delivered to the 6 Ω resistor.

Solution. Only one current path exists since no DC current flows through a capacitor; the voltage drop across the inductor is zero. Write KVL:

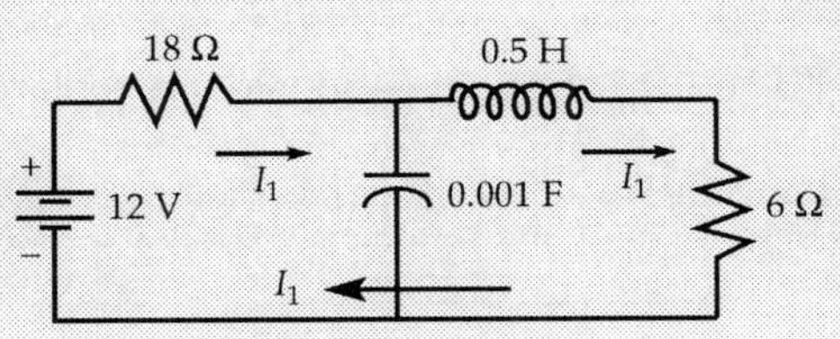

$$\sum V_{DROPS} = 0$$

$$-12 + 18I_1 + 6I_1 = 0$$

$$I_1 = 0.5 \text{ A}$$

The power is then

$$P = I^2R$$
$$= 0.5^2 \times 6 = 1.5 \text{ W}.$$

Example 6.3

Calculate the *Thévenin equivalent circuit* between points *a* and *b* for the circuit shown.

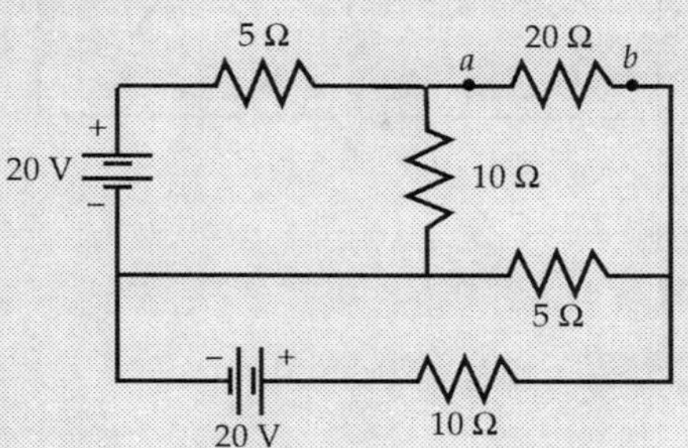

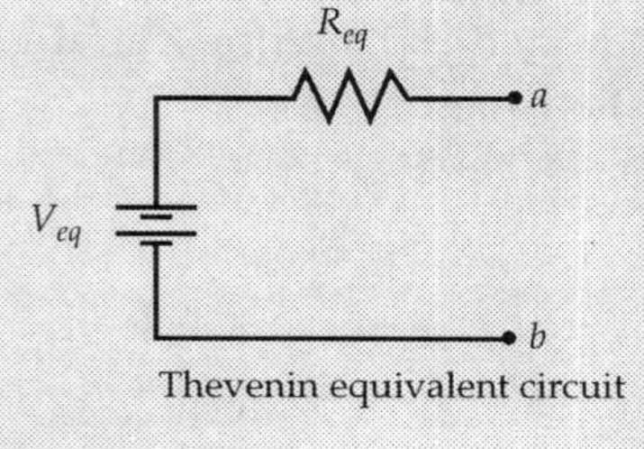

Thevenin equivalent circuit

Solution. V_{eq} is the voltage across *a*–*b* with the 20 Ω resistor out of the circuit. R_{eq} is the resistance across *a*–*b* with the 20 Ω resistor removed and the 20 V sources replaced by short circuits.

The voltage V_{eq} is formed as follows:

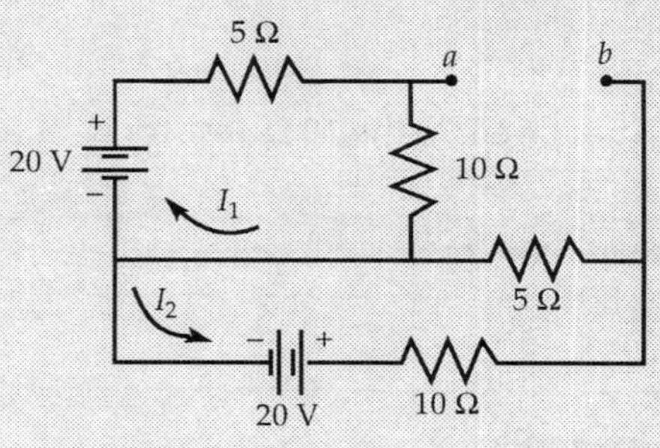

$$I_1 = \frac{20}{15} = \frac{4}{3}\,\text{A}$$

$$I_2 = \frac{20}{15} = \frac{4}{3}\,\text{A}$$

$$-V_{ab} + 10I_1 - 5I_2 = 0$$

$$\therefore V_{ab} = 10 \times \frac{4}{3} - 5 \times \frac{4}{3} = 6.67\ \text{V}$$

$$\therefore V_{eq} = 6.67\ \text{V}$$

To find R_{eq} the voltage sources are replaced by short circuits:

$$R_{eq} = \frac{1}{\frac{1}{5} + \frac{1}{10}} + \frac{1}{\frac{1}{10} + \frac{1}{5}} = 6.67\,\Omega$$

6.1.2 AC Circuits —Single Phase

Single phase AC circuits operate with sinusoidal signal sources and are assumed to be operating steady state, that is, all transients have long since died out. Sinusoidal currents and voltages can be represented either as trigonometric expressions or as complex numbers where the explicit time dependence has been dropped. Consider the following two notations representing a sinusoidal voltage:

Trigonometric	*Complex*
$V(t) = V_m \cos(\omega t + \phi)$	$V = \frac{V_m}{\sqrt{2}} e^{j\phi} = V_{rms} e^{j\phi}$

(6.1.18)

where V_m is the maximum voltage and $j = \sqrt{-1}$. The conversion from trigonometric to complex notation is given by the expression:

$$V(t) = \text{Re}\left[\text{V}_{\text{m}} e^{j\phi} e^{j\omega t}\right] = \text{Re}\left[\sqrt{2} V_{rms} e^{j\phi} e^{j\omega t}\right] \qquad \textbf{(6.1.19)}$$

where "Re" means "the real part of." Once sinusoidal currents and voltages are expressed in complex notation the solution of AC circuits becomes nearly identical to the solution of DC circuits. The difference is that inductors and capacitors can no longer be considered short or open circuit components.

Table 6.3 presents the definitions and relationships between current, voltage, and power for the passive circuit components *R*, *L*, and *C* when used in an AC circuit. All currents and voltages in Table 6.3 and in the rest of this article are rms values.

TABLE 6.3 AC Circuit Components

Component	Impedance (ohms)	Admittance (mhos)	Current (amperes)	Power (watts or vars)
Resistor	R	$G = \frac{1}{R}$	$I = \frac{V}{R}$	$P = I^2R = \frac{V^2}{R}$ real/dissipated
Inductor	$jX_L = j\omega L$	$-jB_L = \frac{-j}{X_L}$	$I = \frac{-jV_L}{\omega L}$	$Q_L = I^2X_L = \frac{{V_L}^2}{X_L}$ reactive/stored
Capacitor	$-jX_C = \frac{-j}{\omega C}$	$jB_C = \frac{j}{X_C}$	$I = jV_C\omega C$	$Q_C = I^2X_C = \frac{{V_C}^2}{X_C}$ reactive/stored
R, L, C	Z	Y	$I = V/Z$	—

For a series connection of components, the impedances sum to give a total impedance for the combination. There results:

$$\text{Series } R\text{–}L\text{:}\quad Z = R + jX_L = \sqrt{R^2 + {X_L}^2}\ \angle\theta_L = |Z|\ \angle\theta_L \qquad \theta_L = \tan^{-1} x_L/R$$

(6.1.20)

$$\text{Series } R\text{–}C\text{:}\quad Z = R - jX_C = \sqrt{R^2 + {X_C}^2}\ \angle\theta_C = |Z|\ \angle\theta_C \qquad \theta_C = -\tan^{-1} x_C/R$$

For a parallel connection of components, the admittances sum to give a total admittance as follows:

$$\text{Parallel } R\text{–}L\text{:}\quad Y = G - jB_L = \sqrt{G^2 + {B_L}^2}\ \angle\theta_L = |Y|\ \angle\theta_L$$

(6.1.21)

$$\text{Parallel } R\text{–}C\text{:}\quad Y = G + jB_C = \sqrt{G^2 + {B_C}^2}\ \angle\theta_C = |Y|\ \angle\theta_C$$

The relationship between Z and Y is always $Z = 1/Y$.

Each circuit parameter in an AC circuit can be represented by a complex number, where a subscript "r" denotes the real part, a subscript "x" denotes the imaginary part, and an asterisk denotes the complex conjugate:

$$\text{Current:}\quad I = I_r + jI_x = |I|\angle\theta$$

$$\text{Voltage:}\quad V = V_r + jV_x = |V|\angle\theta \qquad \textbf{(6.1.22)}$$

$$\text{Power:}\quad S = VI^* = P + jQ$$

where

$$P = \text{real dissipated power (watts)}$$

$$= \text{Re}\left[VI^*\right] = |V||I|\cos\theta \qquad \textbf{(6.1.23)}$$

$$Q = \text{reactive stored power (vars)}$$

$$= \text{Im}\left[VI^*\right] = |V||I|\sin\theta \qquad \textbf{(6.1.24)}$$

and θ in these power equations is as displayed in Fig. 6.6. The quantity $\cos\theta$ is often referred to as the *power factor, pf*. These complex quantities can be visualized using the phasor diagrams and triangles shown in Fig. 6.6.

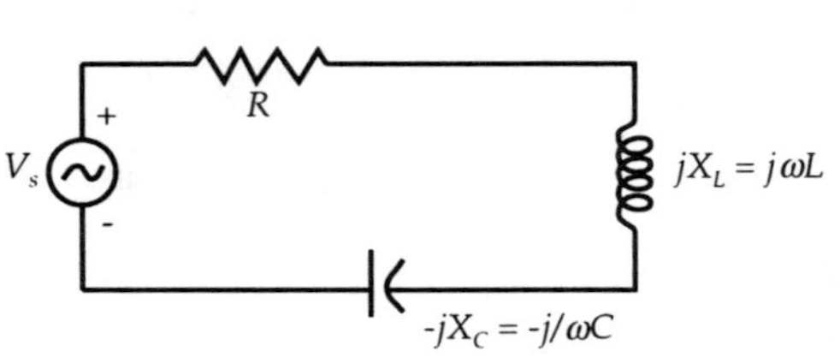

a) A simple AC circuit

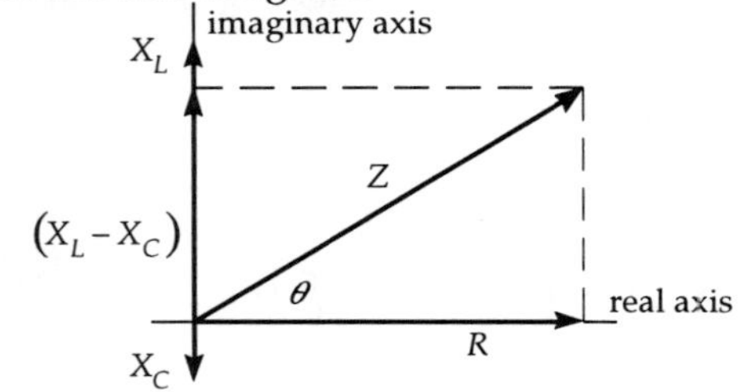

b) Impedance triangle

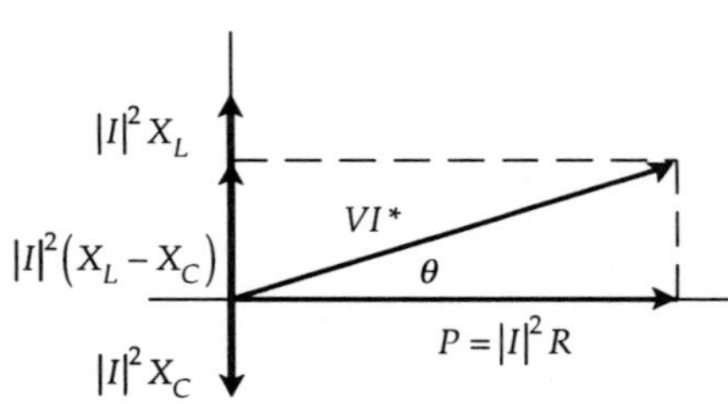

c) The power triangle

d) Current and voltage phasors

Figure 6.6 Diagrams for the simple AC circuit.

RESONANCE

Resonance in an AC circuit occurs at that frequency at which the applied voltage and the input current are in phase, as shown in Fig. 6.7.

Resonance requires that the AC circuit have both capacitive and inductive components, i.e., energy storage components. At resonance the energy stored in the inductor is equal to the energy stored in the capacitor and the circuit timing is such that this energy oscillates back and forth between the inductive and capacitive components. If the circuit did not have any resistance (loss of energy) the oscillations would continue forever.

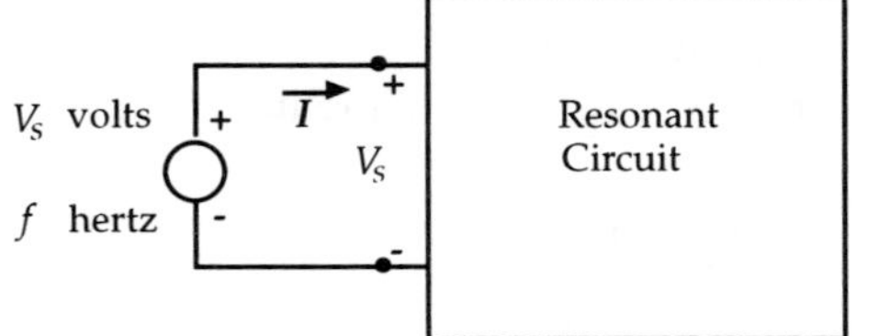

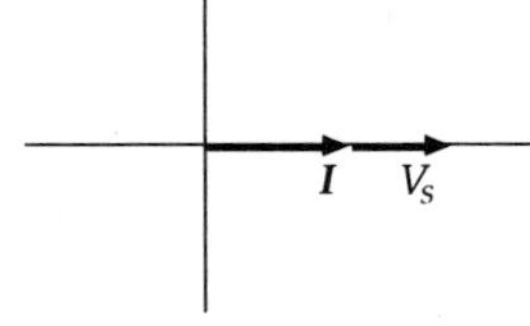

Figure 6.7 Resonance in an AC circuit.

Table 6.4 shows the simple series resonant and parallel resonant circuits along with the input impedance for each circuit. Since the input impedance must be real in order to have the input voltage and current in phase, the reactive part of the input impedance must be zero. This leads to the formulas for the resonant frequencies as shown. See Table 6.3 for the formulas to calculate I, V, and P for each component in a resonant circuit.

TABLE 6.4. Resonant Circuits

	Circuit	Resonant Frequency	Impedance/Admittance
Series		$f = \dfrac{1}{2\pi\sqrt{LC}}$	$Z_{IN} = R + j\left(\omega L - \dfrac{1}{\omega C}\right)$ But $\omega L = \dfrac{1}{\omega C}$ $\therefore\ Z_{IN} = R$
Parallel		$f = \dfrac{1}{2\pi\sqrt{LC}}$	$Y_{IN} = \dfrac{1}{R} + j\left(\omega C - \dfrac{1}{\omega L}\right)$ But $\omega C = \dfrac{1}{\omega L}$ $\therefore\ Y_{IN} = \dfrac{1}{R}$

Example 6.4

Compute the power delivered to the 25 Ω resistor.

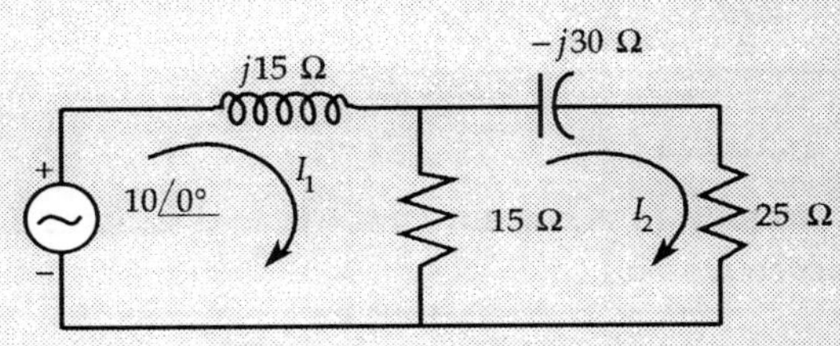

Solution. Use the KVL as follows:

$$\sum V_{DROPS} = 0$$

$$-10 + j15I_1 + 15I_1 - 15I_2 = 0$$

$$-15I_1 + 15I_2 - j30I_2 + 25I_2 = 0$$

These are written as

$$(15 + j15)I_1 - 15I_2 = 10$$

$$-15I_1 + (40 - j30)I_2 = 0$$

A simultaneous solution yields

$$I_2 = -0.179\angle 10.3^\circ\ \text{A}$$

The power delivered is then

$$P = |I_2|^2 R = 0.179^2 \times 25 = 0.800\ \text{W}$$

Example 6.5

Compute the value of X_C so that I is in phase with V_S.

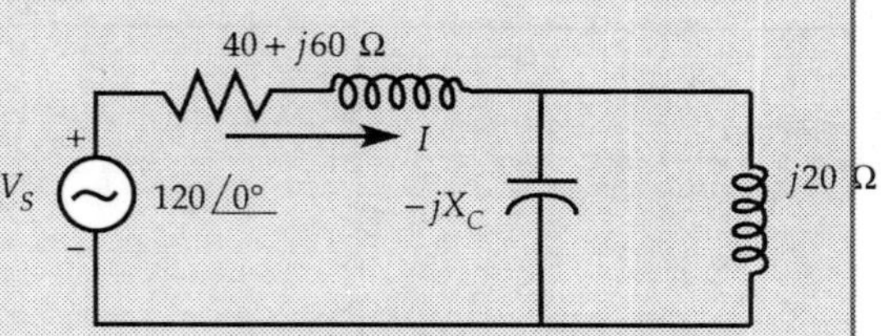

Solution. The impedance of the parallel load is

$$Z_{load} = \frac{-jX_C(j20)}{-jX_C + j20} = \frac{20X_C}{j(20 - X_C)}$$

To make I in phase with V_S, the total impedance seen by the generator must have a zero imaginary part:

$$j60 + \frac{-j20X_C}{20 - X_C} = 0$$

$$\therefore X_C = 15\Omega$$

Example 6.6

Compute the current I if the load dissipates 240 W of real power and −140 vars of reactive power.

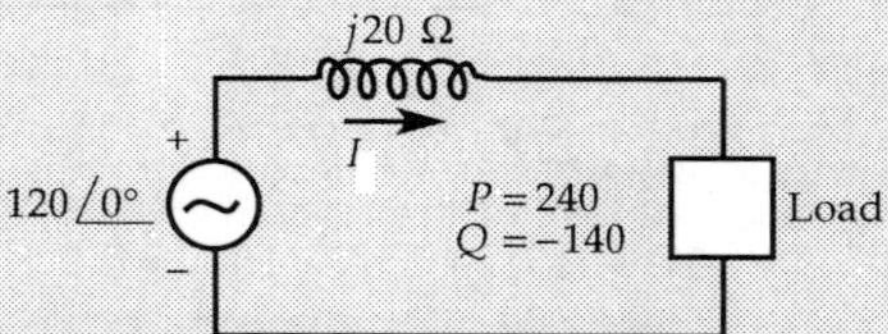

Solution. The complex power consumed by the series combination of the 20 Ω inductor and the load is VI^*. The inductor power is $Q_L = |I|^2 X_L$. The unknown load power is $S = 240 - j140$. Therefore,

$$VI^* = S + jQ_L$$

$$(120\angle 0°)(I_r - jI_x) = j|I|^2 X_L + (240 - j140)$$

This equation can be visualized with the power triangle shown below. Rearranging the equation gives:

$$120(I_r - jI_x) = 240 + j\left[20\left(I_r^{\,2} + I_x^{\,2}\right) - 140\right]$$

Equate real and imaginary parts:

$$\therefore 120I_r = 240 \quad \text{and} \quad -120I_x = 20\left(I_r^{\,2} + I_x^{\,2}\right) - 140$$

The solution is, $I_r = 2, \quad I_x = 0.464, \; -6.464$.

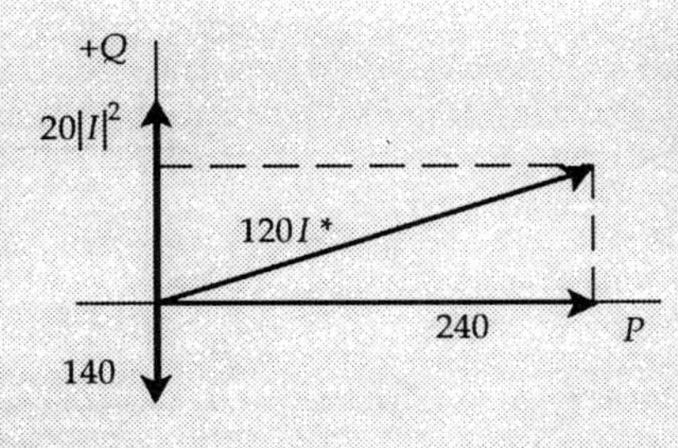

$$\therefore I = 2 + j0.464 \quad \text{or} \quad I = 2 - j6.464.$$

Additional information would be needed to narrow the selection to one of the above answers.

Example 6.7
Compute the resonant frequency and the current through the inductor and capacitor at resonance for the circuit shown.

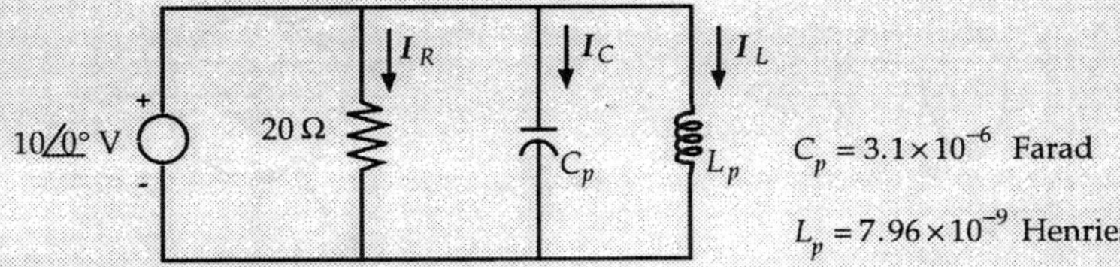

Solution. From Table 6.4 the resonant frequency is

$$f = \frac{1}{2\pi\sqrt{LC}} = \frac{1}{2\pi\sqrt{7.96\times 10^{-9}\times 3.1\times 10^{-6}}} = 10^6 \text{ Hz}$$

The angular frequency is $\omega = 2\pi f$. From Table 6.3 the desired currents are

$$I_C = jV\omega C = j10(2\pi\times 10^6)\times(3.1\times 10^{-6}) = j195 \quad \text{or} \quad 195\angle 90^\circ \text{ A}$$

$$I_L = -\frac{jV}{\omega L} = -\frac{j10}{(2\pi\times 10^6)\times(7.96\times 10^{-9})} = -j200 \quad \text{or} \quad 200\angle -90^\circ \text{ A}$$

6.1.3 AC Circuit —Three Phase

Three-phase circuits are composed of three single-phase circuits where the source voltages for each phase are 120° apart. The load impedances for three-phase circuits can be connected in either a wye (Y) connection or a delta (Δ) connection as shown in Fig. 6.8. Phasor diagrams for a typical set of source voltages are given in Fig. 6.9. One may further specify the source voltage to be either line-to-line or line-to-neutral. Fig. 6.9 shows both the line-to-neutral voltages which can be used for wye-connected load, and the line-to-line voltages, which can be used for either a delta- or a wye-connected load. The advantage of three-phase circuits is that the instantaneous power flow is uniform and not pulsating as it is in a single-phase circuit.

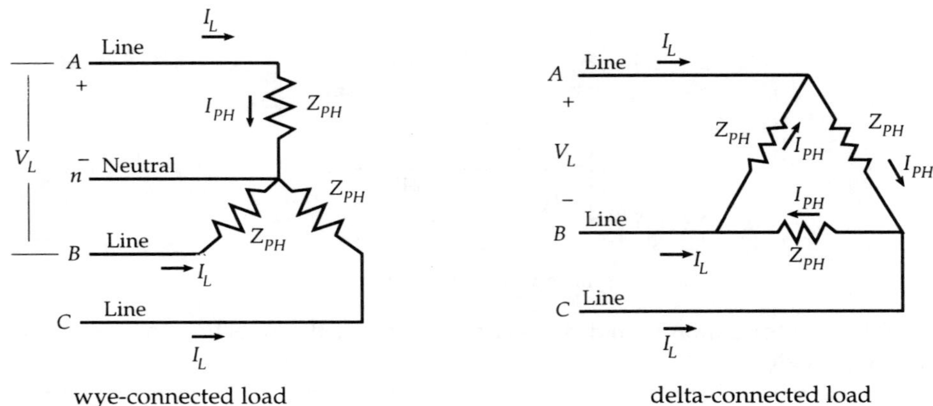

Figure 6.8 Load impedances.

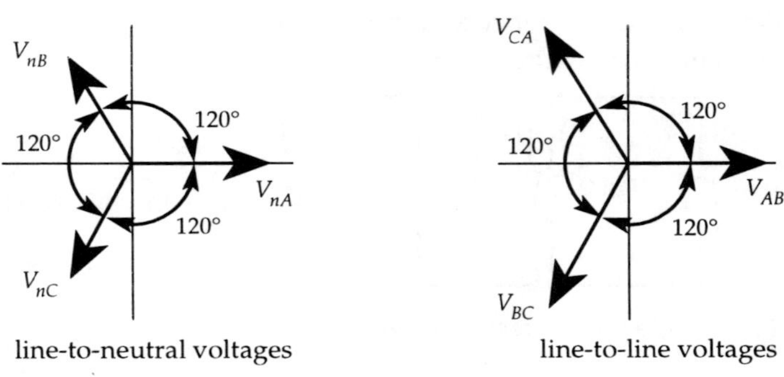

Figure 6.9 Phasor diagrams.

Kirchhoff's circuit laws apply to three-phase circuits just as they apply to all other circuits; however, it is typically not necessary to write complex sets of circuit equations to solve the simple balanced circuits being considered here. Just remember that where all three-phase impedances are equal the circuits are balanced. The expressions used in solving balanced circuits are tabulated in Table 6.5.

TABLE 6.5 AC Circuit Expressions (Three-Phase)

Load	Current	Voltage	Power
wye load	$I_L = I_{PH} = \dfrac{V_{PH}}{Z_{PH}}$	$V_L = \sqrt{3}\,V_{PH}$	$P = 3P_{PH} = 3\lvert V_{PH}\rvert\lvert I_{PH}\rvert\cos\theta$
			$= \sqrt{3}\lvert V_L\rvert\lvert I_L\rvert\cos\theta$
			$Q = 3Q_{PH} = 3\lvert V_{PH}\rvert\lvert I_{PH}\rvert\sin\theta$
			$= \sqrt{3}\lvert V_L\rvert\lvert I_L\rvert\sin\theta$
delta load	$I_L = \sqrt{3}I_{PH}$	$V_L = V_{PH} = I_{PH}Z_{PH}$	same as above

Example 6.8

A balanced delta-connected load is driven from a 208-volt line to a line system as shown. Calculate the line current I_A.

Solution. Sketch the phasor diagram to assist in the calculations. The currents are calculated as follows:

$$I_{AB} = \frac{V_{AB}}{10 - j20} = \frac{208\angle 0^\circ}{10 - j20} = 9.30\angle 63.4^\circ$$

$$\therefore I_{CA} = 9.30\angle 183.4^\circ$$

Kirchhoff's Current Law allows us to find the line current as

$$I_A = I_{AB} - I_{CA}$$

$$= 9.30\angle 63.4^\circ - 9.30\angle 183.4^\circ$$

$$= 4.17 + j8.32 - (-9.28 - j0.55) = 13.45 + j8.87 = 16.11\angle 33.41^\circ \text{ A}$$

We could have found the line current magnitude as follows:

$$I_{PH} = \frac{V_{PH}}{Z_{PH}} = \frac{208}{\sqrt{10^2 + 20^2}} = 9.30 \text{ A}$$

$$\therefore I_L = \sqrt{3} I_{PH} = \sqrt{3} \times 9.30 = 16.11 \text{ A}$$

Example 6.9

A balanced Y-connected load is attached to a 3 ϕ four-wire line with a 120 V line to neutral voltages as shown. Sketch a phasor diagram showing all line and phase currents and voltages. Assume V_{nA} to be at zero phase angle.

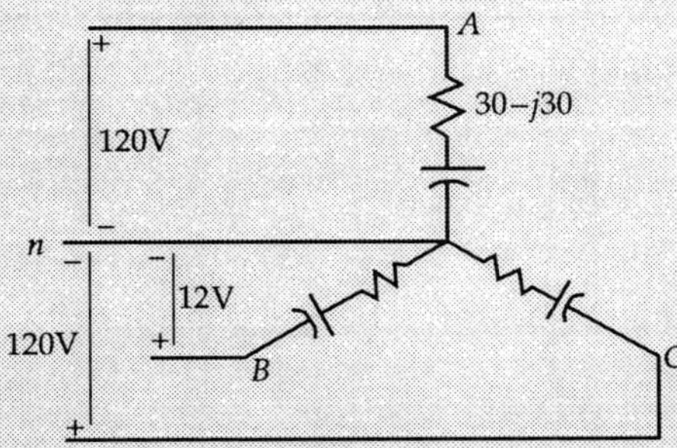

Solution. The phase voltages are expressed as

$$V_{nA} = 120\angle 0^\circ \quad V_{nB} = 120\angle -120^\circ \quad V_{nC} = 120\angle 120^\circ$$

The line voltages are then,

$$V_{AB} = V_{An} + V_{nB} = 208\angle{-150^\circ}$$

$$V_{BC} = V_{Bn} + V_{nC} = 208\angle{90^\circ}$$

$$V_{CA} = V_{Cn} + V_{nA} = 208\angle{-30^\circ}$$

The line currents and phase currents are equal, so that

$$I_A = \frac{V_{nA}}{Z_{PH}} = \frac{120\angle{0^\circ}}{30 - j30} = 2.83\angle{45^\circ}$$

$$I_B = 2.83\angle{-75^\circ} \qquad I_C = 2.83\angle{165^\circ}$$

The phasor diagram is sketched.

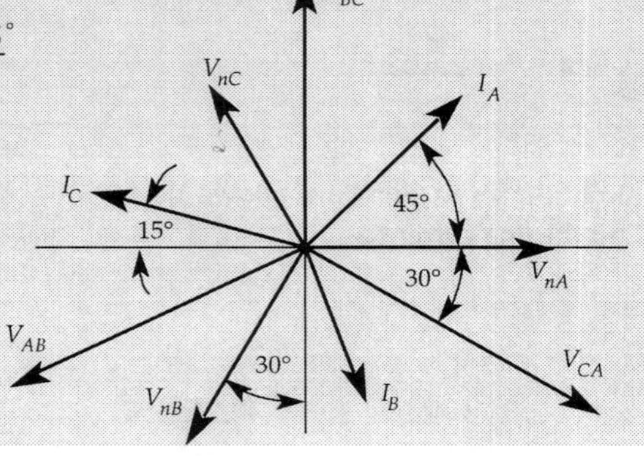

Example 6.10

A balanced Y-connected load is connected to a three-phase source as shown. Determine the power delivered to the load.

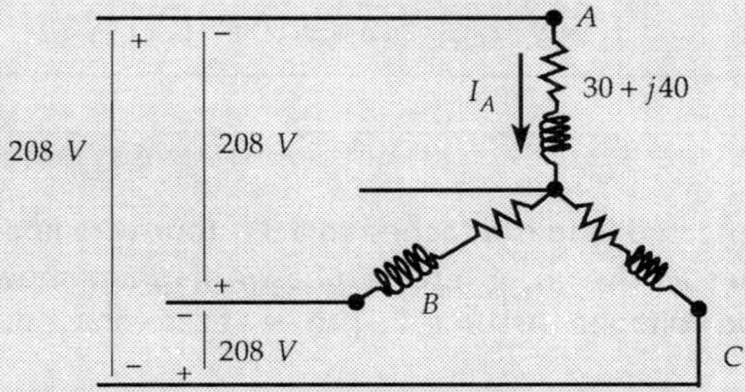

Solution. First calculate the phase voltage V_{PH}:

$$V_{PH} = \frac{V_L}{\sqrt{3}} = V_{nA} = \frac{208\angle{150^\circ}}{\sqrt{3}} = 120\angle{150}\ \text{V}$$

Note: If V_{AB} is assumed to be at 0°, then V_{nA} is at −150°; see Example 6.9. The phase current is

$$I_{PH} = \frac{V_{PH}}{Z_{PH}} = \frac{120\angle{150^\circ}}{50\angle{53.1^\circ}} = 2.4\angle{96.9^\circ}\ \text{A}$$

The power dissipated in phase *A* is,

$$P_{PH} = |V_{PH}||I_{PH}|\cos\theta$$

$$= 120 \times 2.40 \times \cos 53.1^\circ = 172.9\ \text{W}$$

$$P_{\text{total}} = 3 \times 172.9 = 519\ \text{W}$$

This could also have been computed as

$$P_{\text{total}} = \sqrt{3}|V_L||I_L|\cos\theta$$

$$= \sqrt{3} \times 208 \times 2.40 \cos 53.1^\circ = 519\ \text{W}$$

6.1.4 DC Transients

Transients in a DC circuit can exist any time a sudden change is made in the sources applied to the circuit or in the circuit configuration.

Unlike the steady state solution of DC circuits, transient solutions require the inclusion of capacitive and inductive effects as functions of time. If only a single inductor or a single capacitor is contained in a circuit, the current will either grow or decay exponentially. If both an inductor and a capacitor are included in the circuit, oscillations can occur. Only circuits with a single capacitor and/or single inductor are considered here.

For exponentially changing functions a time constant (T) can be defined. The time constant is the time it takes to complete 63.2% of the change that will ultimately be made; see Fig. 6.10. This is a practical number, since in a time corresponding to four or five time constants most transients have disappeared from a circuit.

Several general expressions can be written for the voltage and current in a DC transient circuit. Two of these are

$$I(t) = I_\infty + (I_0 - I_\infty)e^{-t/T}$$
$$V(t) = V_\infty + (V_0 - V_\infty)e^{-t/T} \tag{6.1.25}$$

where

I_0 and V_0 are at the instant of a sudden change, i.e., initial values,

I_∞ and V_∞ will exist in the circuit after the transient has died away (let $t = \infty$),

$T = L/R$ the time constant for series R–L circuit,

$T = RC$ the time constant for series R–C circuit.

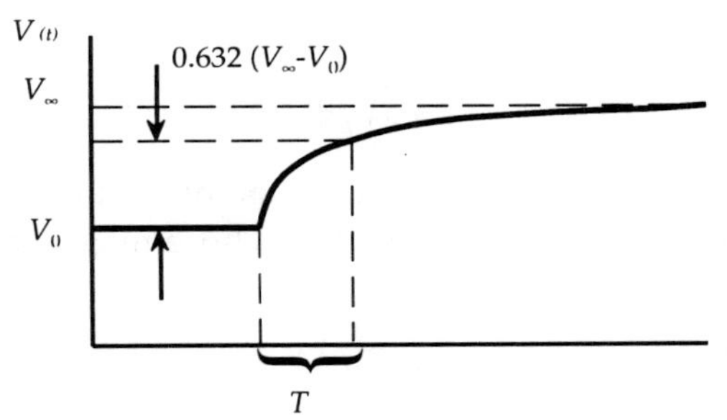

Figure 6.10 A DC transient.

Example 6.11

The switch is closed at $t = 0$. Determine the current in the circuit at $t = 2T$, two time constants after the switch is closed.

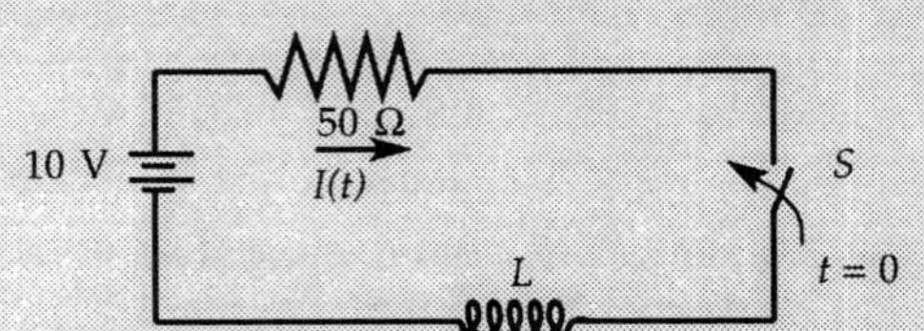

Solution. The general expression for the transient current in a series *R-L* or *R-C* circuit is

$$I(t) = I_\infty + (I_0 - I_\infty)e^{-t/T}$$

For the case being considered

$$I_0 = 0, \quad I_\infty = \frac{V}{R} = \frac{10}{50} = 0.2$$

Thus

$$I(2T) = 0.2 - 0.2e^{-2} = 0.173 \text{ A}$$

6.2 Static Electric Fields

Electric fields exist any time an electric charge is present in the region. The presence of an electric field causes many things to happen. It produces forces on stationary charges, forces on charges moving in conductors and, therefore, forces on the conductors themselves. The forces on the charges in a conductor also cause the motion of those charges to produce current. Forces on the charges in an atom can cause the atom to come apart, creating a material breakdown. In addition to their usefulness in calculating forces, electric field quantities can be used to calculate stored energy and capacitance for any shape of electrodes on which the original charge may exist, as well as the potential difference between those electrodes. Here we will examine only the most basic concepts of electric field theory and attempt to do so with minimal use of mathematics.

The two most basic field quantities are *electric field intensity* **E** (newtons/coulomb or volts/meter) and *electric flux density* **D** (coulombs/meter^2). These are defined as follows:

$$\mathbf{E} = \mathbf{F}/Q \qquad \textbf{(6.2.1)}$$
$$E = dV/dL$$

For the derivative definition, **E** is in the direction of the maximum rate of change. The electric flux density is a vector quantity whose magnitude is the maximum value of the derivative

$$D = dQ/dA \qquad \textbf{(6.2.2)}$$

and **D** is perpendicular to the surface dA.

The two basic field quantities are related to each other by

$$E = D/\varepsilon \qquad \textbf{(6.2.3)}$$

where $\varepsilon = \varepsilon_0 \varepsilon_r$, $\varepsilon_0 = 8.85 \times 10^{-12}$ F/m (the permittivity of free space). The *relative permittivity* ε_r depends on the material and will be specified in a particular problem. The field quantities for some simple geometries are especially useful. They follow:

Point of charge Q (C): $D = \dfrac{Q}{4\pi r^2}$ $E = \dfrac{Q}{4\pi\varepsilon r^2}$ **(6.2.4)**

with vectors **D** and **E** directed radially away from the point charge.

Uniform line of charge ρ_L(C/m): $D = \dfrac{\rho_L}{2\pi r}$ $E = \dfrac{\rho_L}{2\pi\varepsilon r}$ **(6.2.5)**

with vectors **D** and **E** normal to the line of charge.

Uniform sheet of charge $\rho_s(\mathrm{C/m^2})$: $D = \dfrac{\rho_s}{2}$ $E = \dfrac{\rho_s}{2\varepsilon}$ **(6.2.6)**

with vectors **D** and **E** directed normally away from the plane.

If D and E are not known they must be calculated (from the known voltages or charges) using Eqs. 6.2.1 and 6.2.2. Once D and E are known they can be used to calculate the following quantities:

Voltage	$V = \int \mathbf{E} \cdot d\mathbf{L}$	**(6.2.7)**
Charge:	$Q = \oiint \mathbf{D} \cdot d\mathbf{A}$	**(6.2.8)**
Stored electrical energy:	$W_E = \frac{1}{2} \iiint \mathbf{D} \cdot \mathbf{E} d\forall = \frac{1}{2} CV^2$	**(6.2.9)**
Capacitance:	$C = Q/V = 2W_E/V^2$	**(6.2.10)**
Current:	$I = \iint \sigma \mathbf{E} \cdot d\mathbf{A}$	**(6.2.11)**
Force on a charge Q:	$\mathbf{F} = Q\mathbf{E}$	**(6.2.12)**

The following examples illustrate the use of several of the above relationships.

Example 6.12

A point charge of $Q = 5 \times 10^{-9}$ coulombs is located in rectangular coordinates at (1, 3, 5) and a line charge $\rho_L = 1.5 \times 10^{-9}$ coulomb/m is located parallel to the x-axis at $y = 3$ and $z = 0$. Find the point where the net electric field is zero.

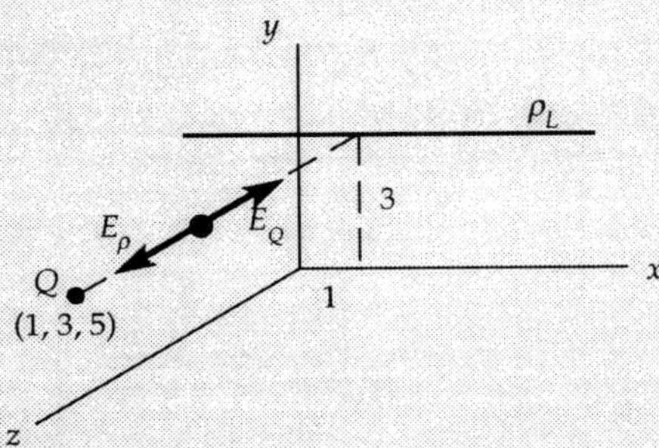

Solution. The field intensity from the line charge is

$$E_\rho = \frac{\rho_L}{2\pi\varepsilon_0 r}$$

The field intensity from the point charge is

$$E_Q = \frac{Q}{4\pi\varepsilon_0 r^2}$$

Since the E fields from the sources are in opposite directions, we simply equate the above two expressions, letting $r = z$ in the expression for E_ρ:

$$\frac{1.5\times10^{-9}}{2\pi z\varepsilon_0} = \frac{5\times10^{-9}}{4\pi\varepsilon_0(5-z)^2}$$

The quadratic $3z^2 - 35z + 75 = 0$ results. The appropriate root is $z = 2.829$ m.

Example 6.13

Concentric coaxial cylinders 50 cm long are separated by air with permittivity ε_0. If the radius of the inner conductor is 0.2 cm, and the inner radius of the outer conductor is 1.0 cm, compute the capacitance per meter between the cylinders.

Solution. Assume a charge ρ_L on the inner conductor, $-\rho_L$ on the outer conductor, and compute the potential difference V. Combining Eqs. 6.2.5 and 6.2.7 we have the voltage on the outer conductor:

$$V = \int_{0.2}^{1.0} \mathbf{E}\cdot d\mathbf{L} = \int_{0.2}^{1.0} \frac{\rho_L}{2\pi r\varepsilon_0} dr$$

$$= \frac{\rho_L}{2\pi\varepsilon_0} \ln\frac{1.0}{0.2}$$

To find the capacitance per meter between the cylinders we use Eq. 6.2.10, replace Q with ρ_L, and obtain

$$C = \frac{\rho_L}{V} = \frac{2\pi\varepsilon_0}{\ln 5}$$

$$= \frac{2\pi\times8.85\times10^{-12}}{\ln 5} = 34.55\times10^{-12}\ \text{F/m}$$

Example 6.14

The voltage distribution between two coaxial cylinders is given by

$$V = 50\ln\frac{2.5}{r}\ \text{volts}$$

where r varies from 0.5 cm to 2.5 cm. At what radius r between the cylinders does the largest magnitude of the E field exist?

Solution. The E field is given by

$$E = \frac{dV}{dr}$$

$$= 50\frac{1}{2.5/r}\left(-\frac{2.5}{r^2}\right) = -\frac{50}{r}$$

By observation the largest magnitude of the E field occurs at the minimum value of r, i.e., $r = 0.5$ cm.

6.3 Static Magnetic Fields

Magnetic fields exist any time there is a moving charge or electric current in the region. Just as electric fields interact with charges, magnetic fields interact with moving charges or currents. A current-carrying conductor in a magnetic field will have a force exerted on it due to its presence in that magnetic field. In addition to their usefulness in calculating forces, magnetic field quantities can be used to calculate energy stored in a magnetic field and inductance for any configuration of current-carrying conductors.

The two most basic magnetic field quantities are *magnetic field intensity* **H** (amperes/meter) and *magnetic flux density* **B** (webers/meter^2 or tesla). Magnetic field intensity is defined by

$$H = dI/dL \tag{6.3.1}$$

where **H** is in the direction of the maximum rate of change. Magnetic flux density is defined by

$$B = F/IL \tag{6.3.2}$$

where **B** is perpendicular to both **F** and $I\mathbf{L}$. **F** is the force on the current element $I\mathbf{L}$ residing in a field with flux density **B**. The relationship between B and H depends on the material in which the field exists through a material parameter called the *permeability*, μ. The relationship is

$$H = B/\mu \tag{6.3.3}$$

where $\mu = \mu_0\mu_r$, $\mu_0 = 4\pi \times 10^{-7}$ H/m (the permeability of free space). The *relative permeability* μ_r depends on the material and will be specified in a particular problem.

The expression for the magnetic field produced by a straight conductor carrying a current I is especially useful. It is

$$B = \frac{\mu I}{4\pi r}(\sin\theta_1 + \sin\theta_2) \tag{6.3.4}$$

where **B** is in a direction perpendicular to a plane containing the point and the straight wire, and where r is the perpendicular distance shown (see fig. 6.11). A general expression, called *Amperes Rule*, for computing the field B given a known current distribution is

$$\mathbf{B} = \frac{\mu}{4\pi}\int \frac{Id\mathbf{L} \times \mathbf{a}_R}{R^2} \tag{6.3.5}$$

The parameters for Eq. 6.3.5 are defined in Fig. 6.12; $\mathbf{a}_R$ is a unit vector. Once B is known, the energy stored in the magnetic field of an inductor can be found using the following:

$$\begin{aligned} W &= \frac{1}{2}\int \mathbf{B} \cdot \mathbf{H} d\forall \\ &= \frac{1}{2}LI^2 \end{aligned} \tag{6.3.6}$$

where I is the current producing the field $\mathbf{B}$. Knowing $\mathbf{B}$, the total force on a current-carrying conductor in the field can be computed using

$$\mathbf{F} = I\mathbf{L} \times \mathbf{B} \tag{6.3.7}$$

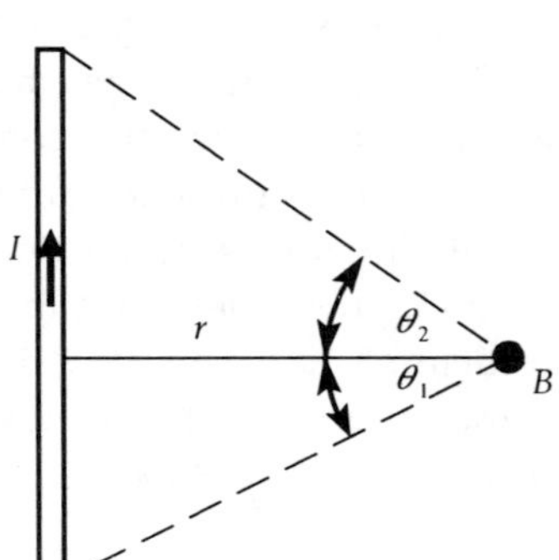

Figure 6.11 Magnetic field due to a straight conductor.

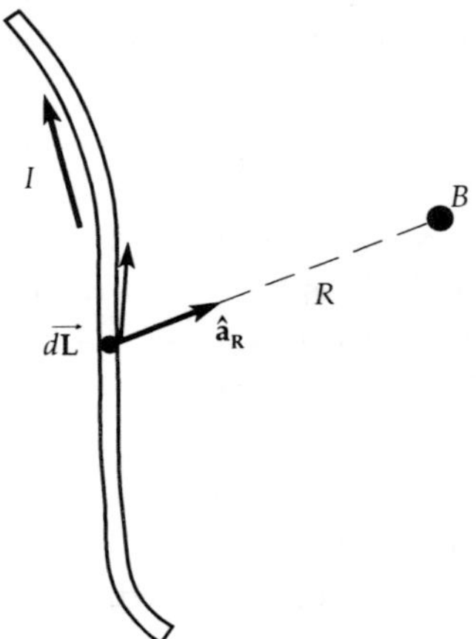

Figure 6.12 Magnetic field due to a curved conductor.

A magnetic circuit is formed when a magnetic field is confined inside a closed circuit of magnetic material. If one recognizes the analogies, the solution of a magnetic circuit is similar to the solution of an electric circuit. Fig. 6.13 shows a simple magnetic circuit along with an analogous electric circuit. The equation similar to Kirchhoff's voltage law is

$$NI = HL \tag{6.3.8}$$

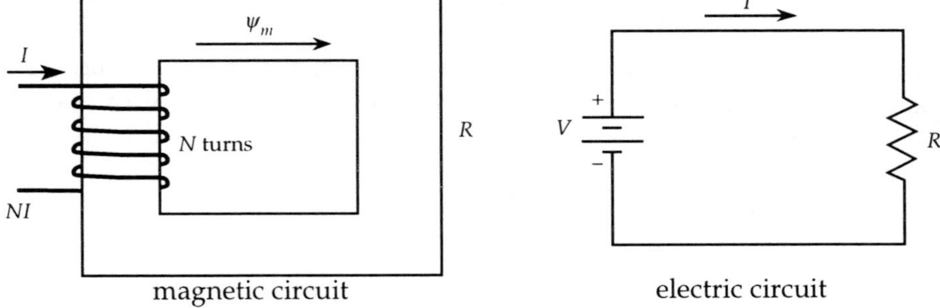

Figure 6.13 A magnetic circuit with its analogous electric circuit.

NI is the *mmf* due to the current (I) in the coil and is similar to the *emf* of the battery in the electric circuit. HL is the magnetic potential drop in the magnetic circuit and is similar to the IR voltage drop in the electric circuit. The magnetic flux ψ_m in the magnetic circuit is similar to the current I in the electric circuit and is related to the flux density B by

$$\psi_m = BA \tag{6.3.9}$$

where A is the cross sectional area. Using Eqs. 6.3.3 and 6.3.8 this can be put in the form

$$\psi_m = \frac{\mu NIA}{L} = \frac{NI}{\Re} \tag{6.3.10}$$

where we have used

$$\Re = \frac{L}{\mu A} \tag{6.3.11}$$

The quantity $\Re$ is called the *reluctance* of the magnetic circuit and is similar to the resistance of the electric circuit. The primary difference between the electric circuit and the magnetic circuit is that R is linear whereas $\Re$ is nonlinear. This means that the solution of a magnetic circuit is best done graphically. An electric current analogy can only be used when $\Re$ is assumed to be linear over the range of flux through which the circuit operates.

Example 6.15
Compute the value of the magnetic flux density at the point (1, –1) in the figure shown.

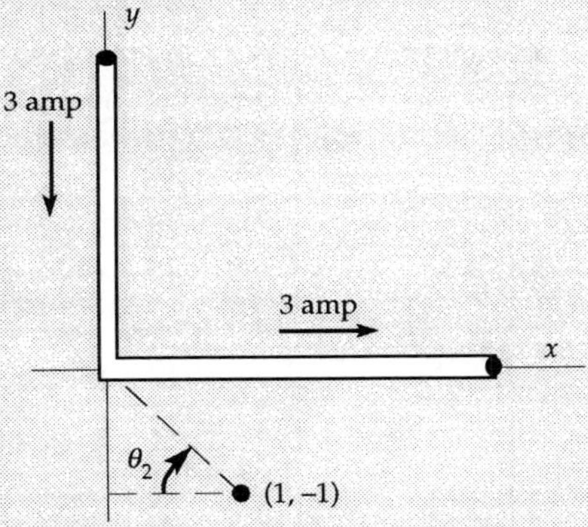

Solution. The field from a straight wire is given by

$$\mathbf{B} = \frac{\mu I}{4\pi r}(\sin\theta_1 + \sin\theta_2)\mathbf{a_R}$$

For the length along the y-axis

$$\mathbf{B} = \frac{3\mu}{4\pi \times 1}(\sin 90° - \sin 45°)\mathbf{z}$$

$$= \frac{3\mu}{4\pi}\left(1 - 1/\sqrt{2}\right)\mathbf{z}$$

For the length along the *x*-axis,

$$\mathbf{B} = \frac{3\mu}{4\pi \times 1}(\sin 90° + \sin 45°)(-\mathbf{z})$$
$$= -\frac{3\mu}{4\pi}\left(1 + 1/\sqrt{2}\right)\mathbf{z}$$

Adding:

$$\mathbf{B}_{total} = -\frac{3\mu}{2\pi\sqrt{2}}\mathbf{z} \quad \text{webers/m}^2$$

Example 6.16

Two parallel conductors each carry 50 amperes, but in opposite directions. If the conductors are one meter apart, compute the force in Newton's per meter on either conductor.

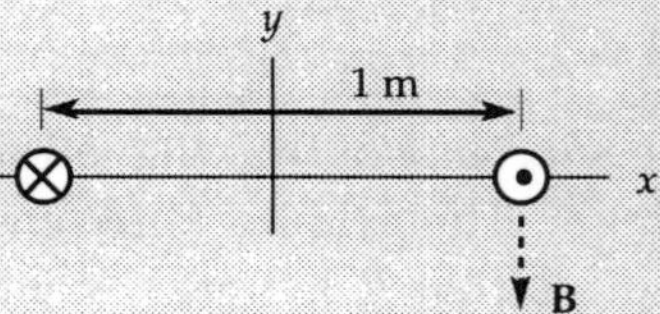

Solution. First, it is necessary to compute the field B that one conductor produces at the location of the other conductor; it is, using μ_0 for air,

$$B = \frac{\mu_0 I}{2\pi r}$$
$$= \frac{4\pi \times 10^{-7} \times 50}{2\pi \times 1} = 10^{-5} \quad \text{Wb/m}$$

If the wires are aligned parallel to the z-axis as shown, then $\mathbf{B} = -10^{-5}\mathbf{y}$ where $\mathbf{y}$ is a unit vector in the y-direction. This is the field at the right conductor due to the current in the left conductor. Then, using Eq. 6.3.7 there results

$$\mathbf{F} = \mathbf{I}L \times \mathbf{B}$$
$$50(1)\mathbf{z} \times \left(-10^{-5}\mathbf{y}\right) = 50 \times 10^{-5}\mathbf{x} \ \ \text{N}$$

Note that the conductors repel one another.

Example 6.17

A magnetization curve for the cast steel of an inductor core is shown. The magnetic flux in the steel is $\psi_m = 8\times10^{-5}$ Wb. Find the current I.

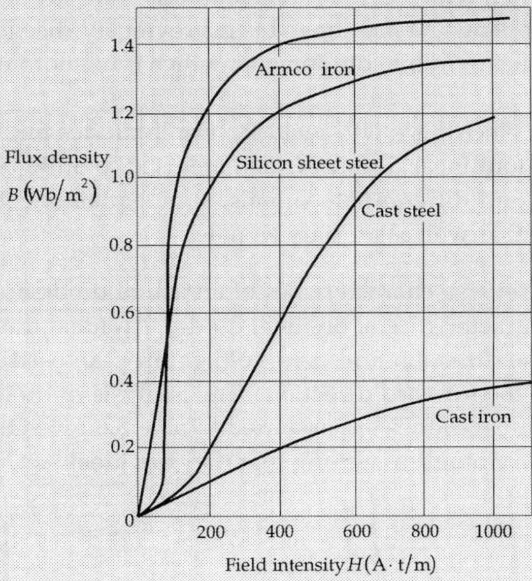

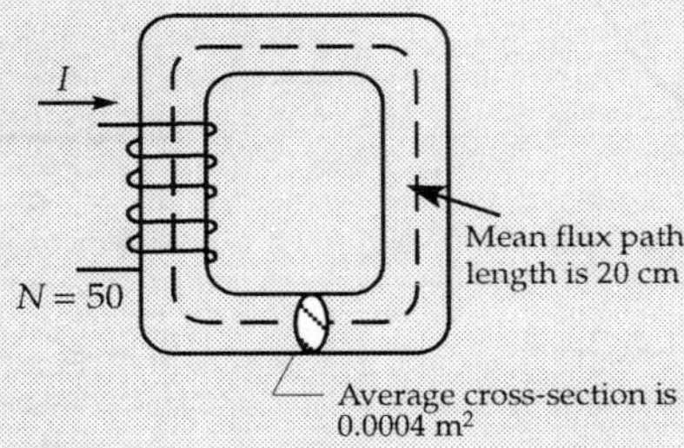

Solution. The flux density B is

$$B = \frac{\psi_m}{A}$$

$$= \frac{8\times10^{-5}}{0.0004} = 0.2 \text{ Wb/m}^2$$

Using the appropriate curve we find H to be

$$H = 200 \text{ A}\cdot\text{turns/m}$$

Using Equation 6.3.8, the current is found to be

$$I = \frac{HL}{N}$$

$$= \frac{200\times0.2}{50} = 0.8 \text{ A}$$

6.4 Electronics

Electronics is the study of circuits which are used to process electrical signals (voltages or currents) which contain information. It involves the use of both passive circuit components—which have been previously discussed—as well as active and/or nonlinear circuit components, which have not yet been discussed in the review.

This review of electronics will include simple diodes and operational amplifiers. Diodes are non-linear devices which are used to build circuits which can be used to integrate and differentiate signals, sum signals together, and filter portions of signals out, as well as amplify signals.

6.4.1 Diodes

The forward and reverse characteristics of an actual diode are shown in Fig. 6.14 along with the characteristics of an ideal diode. An ideal diode has zero current flow in the reverse direction and zero voltage drop across the diode when current is flowing in the forward direction. The analysis of diode circuits is greatly simplified when ideal diodes are assumed. Table 6.6 gives the characteristics of some simple diode circuits, assuming that they are ideal.

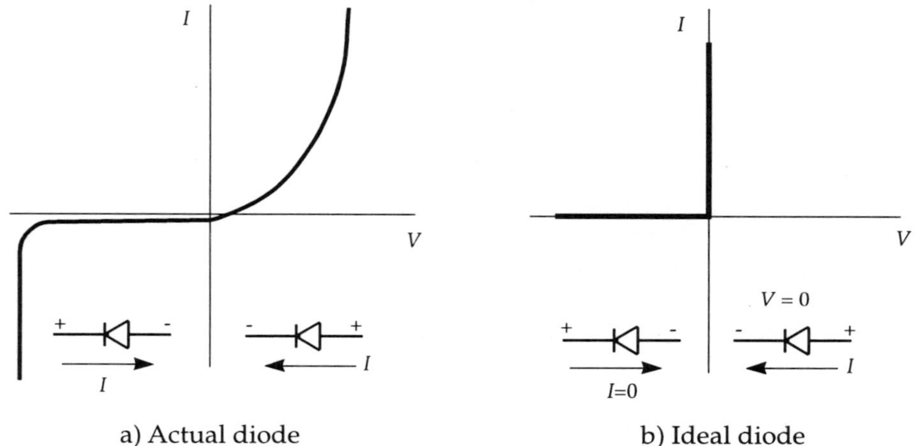

Figure 6.14 Characteristics of a diode.

6.4.2 Operational Amplifiers

The symbolic representation of an operational amplifier is shown in Fig. 6.15. The OP-AMP, as it is often called, has two inputs. One is marked with a "+" sign and is called the non-inverting input, which means that the output is of the same polarity as the input. If the input is positive with respect to ground, then the output is also positive. The other input is marked with a "–" and is called the inverting input, which means that the output is opposite in polarity to the input with respect to ground.

The most important feature of the OP-AMP is the extremely high gain A that it possesses. *Gain* is defined as the ratio of the output voltage to the input voltage. With an extremely high gain it only requires a very, very small input signal to realize a finite output signal. The input current is, in fact, so small that it may

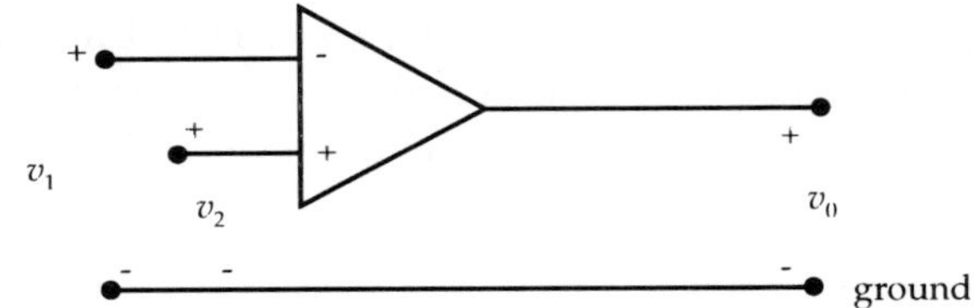

Figure 6.15 An operational amplifier.

safely be assumed to be zero. In the analysis of OP-AMP circuits the input current to either of the input ports is assumed to be zero and the two inputs are assumed to be at the same voltage. Normally, the non-inverting input is tied to ground, so the inverting input is assumed to be at ground potential also.

TABLE 6.6 Some Simple Diode Circuits

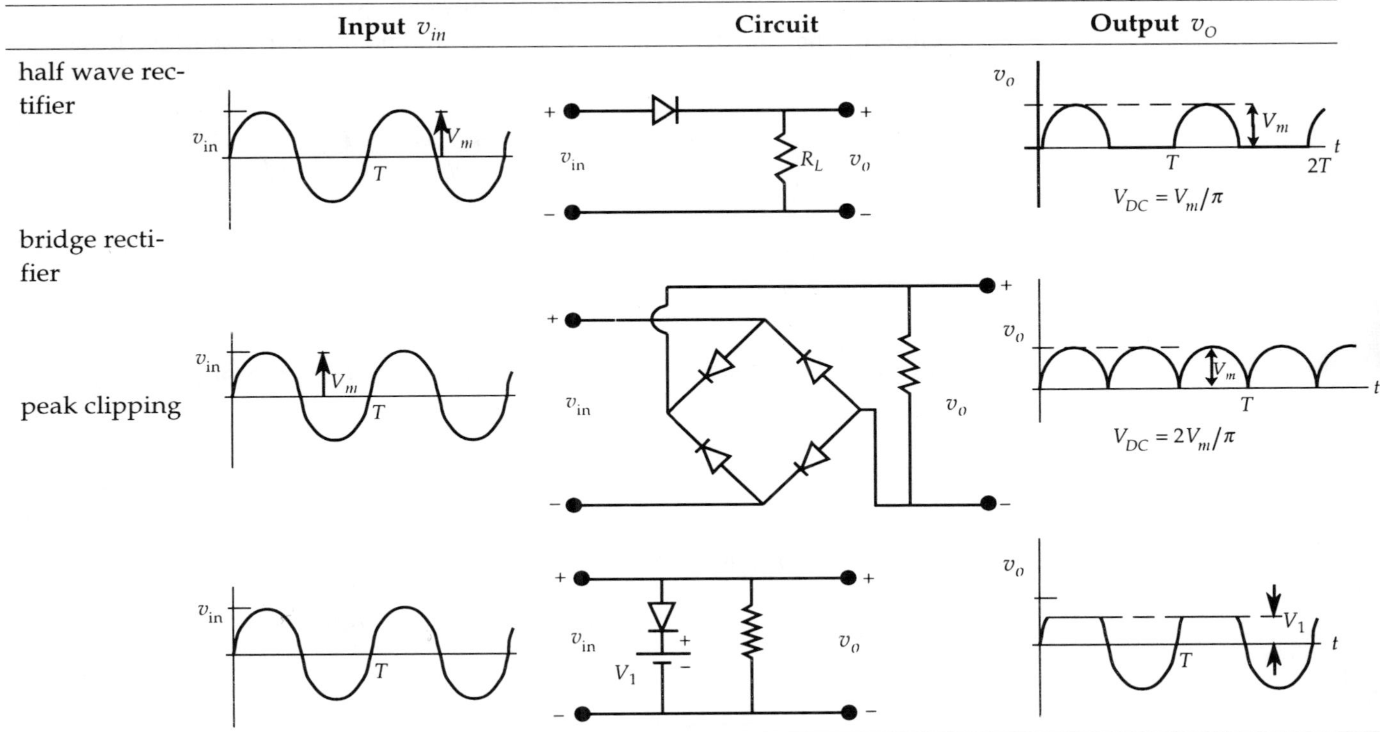

	Input v_{in}	Circuit	Output v_o
half wave rectifier			$V_{DC} = V_m/\pi$
bridge rectifier			$V_{DC} = 2V_m/\pi$
peak clipping			

The typical circuit for an OP-AMP is shown in Fig. 6.16. It has an input impedance Z_i connected in series with the inverting input and a feedback impedance Z_f connected from the output back to the inverting input. The ratio of the output voltage to the input voltage for this circuit configuration is also given in the figure.

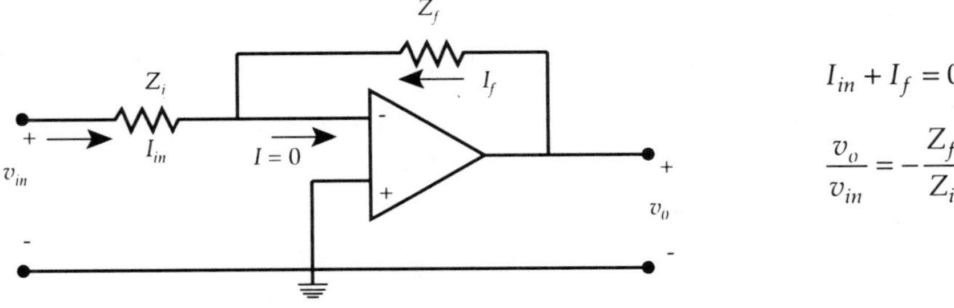

Figure 6.16 A typical OP-AMP circuit.

The impedances Z_i and Z_f can be varied, as indicated in Table 6.7, to cause the OP-AMP to perform many different functions. In this table, the lower case v's represent arbitrary functions of time and the upper case V's represent *rms* values of sinusoidal voltages.

Example 6.18

A half-wave rectifier circuit uses an ideal diode. If the desired DC load voltage is 9 V, what is the *rms* value of the source?

Solution. The time-dependent voltages v_{AC} and v_{DC} appear as sketched below.

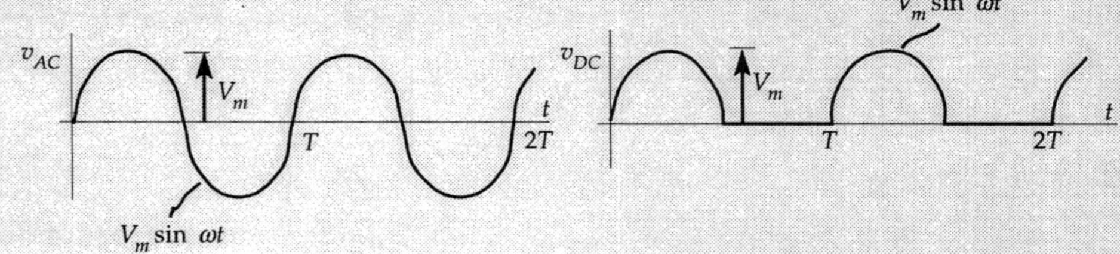

The average, or DC value, of the rectified voltage is given by the integral

$$V_{DC} = \frac{1}{T}\int_0^T v_{DC}(t)dt = \frac{V_m}{\pi}$$

$$\therefore V_m = \pi V_{DC} = 9\pi \text{ volts}$$

The required *rms* value of the source is therefore

$$V_{rms} = \frac{V_m}{\sqrt{2}} = \frac{9\pi}{\sqrt{2}} = 19.99 \text{ volts}$$

Example 6.19

Determine the gain of the OP-AMP circuit shown.

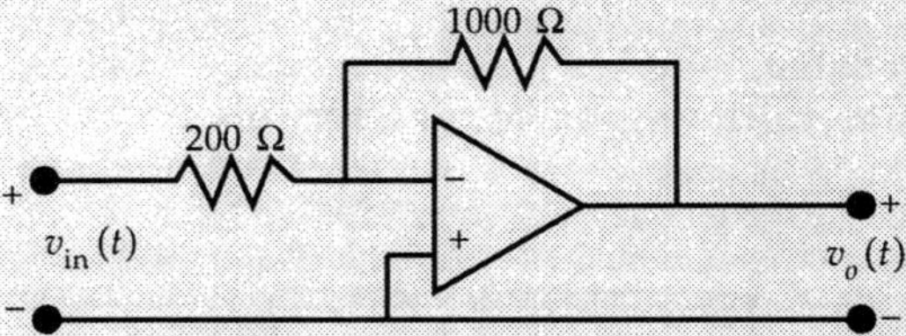

Solution. The gain expression for an OP-AMP circuit is

$$A = -Z_f/Z_i = -R_f/R_i$$

so the OP-AMP gain is

$$A = -1000/200 = -5$$

Example 6.20

A square wave with extrema ± 1 is input to the OP-AMP circuit shown. Sketch $v_0(t)$.

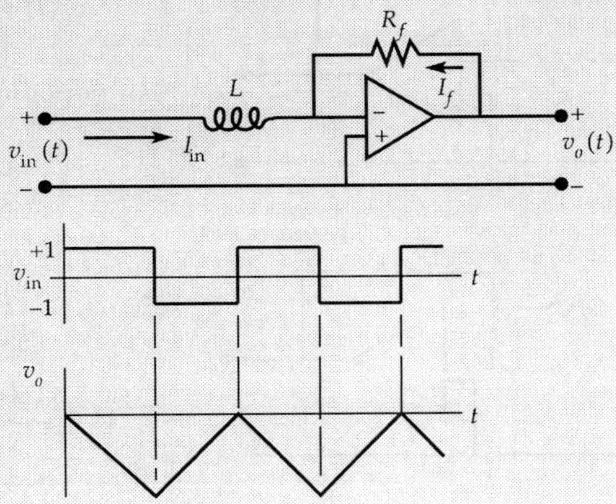

Solution. The KCL allows us to write

$$I_{in} + I_f = 0$$

But we know that

$$I_{in} = \frac{1}{L}\int v_{in} dt \qquad I_f = \frac{v_0}{R_f}$$

Hence,

$$\frac{1}{L}\int v_{in} dt + \frac{v_0}{R_f} = 0$$

or

$$v_0(t) = -\frac{R_f}{L}\int v_{in} dt$$

The circuit shown is an integrator so the output voltage is proportional to the integral of the input voltage.

TABLE 6.7 Operational Amplifier Functions

Name	Circuit	Output
Amplifier		For arbitrary time functions $v_o(t) = -\frac{R_f}{R_i} v_i(t)$ For periodic functions $V_o = -\frac{R_f}{R_i} V_i$
Summer		$v_o = -\frac{R_f}{R_1} v_{i1} - \frac{R_f}{R_2} v_{i2}$ $V_o = -\frac{R_f}{R_1} V_{i1} - \frac{R_f}{R_2} V_{i2}$
Integrator		$v_o = -\frac{1}{R_i C} \int v_i dt$ $V_o = \frac{j}{\omega R_i C} V_i$
Differentiator		$v_o = -\frac{L}{R_i} \frac{dv_i}{dt}$ $V_o = -j \frac{\omega L}{R_i} V_i$
Low Pass Filter		For sinusoidal functions $V_o = -\frac{R_f}{R_i(1 + j\omega R_f C)} V_i$

Practice Problems

(If you attempt only a few, select those with a star.)

DC CIRCUITS

*6.1 For the circuit below, with voltages' polarities as shown, KVL in equation form is

a) $v_1 + v_2 + v_3 - v_4 + v_5 = 0$

b) $-v_1 + v_2 + v_3 - v_4 + v_5 = 0$

c) $v_1 + v_2 - v_3 - v_4 + v_5 = 0$

d) $-v_1 - v_2 - v_3 + v_4 + v_5 = 0$

e) $v_1 - v_2 + v_3 + v_4 - v_5 = 0$

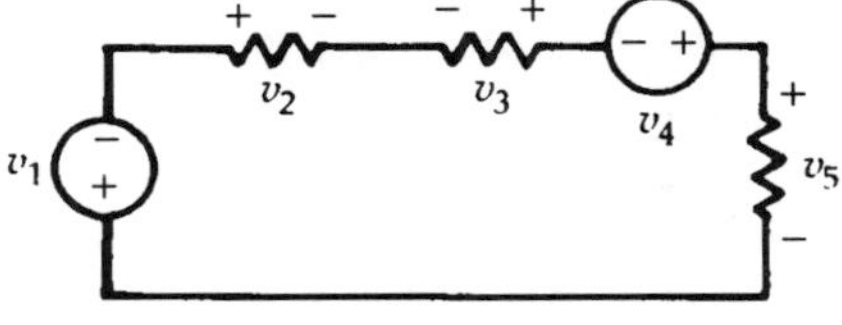

6.2 Find I_1 in amps.

a) 12 d) 21
b) 15 e) 27
c) 18

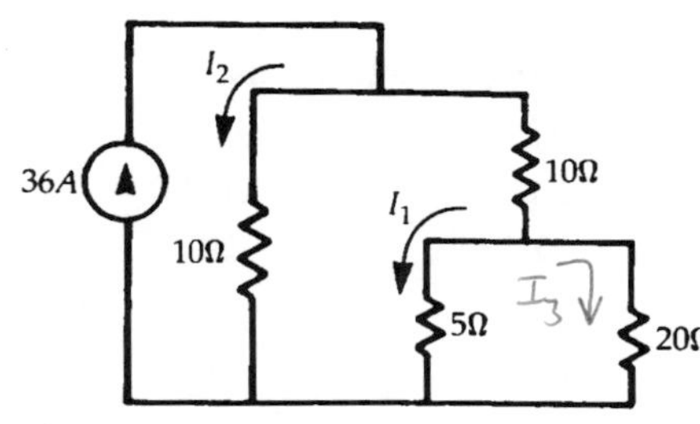

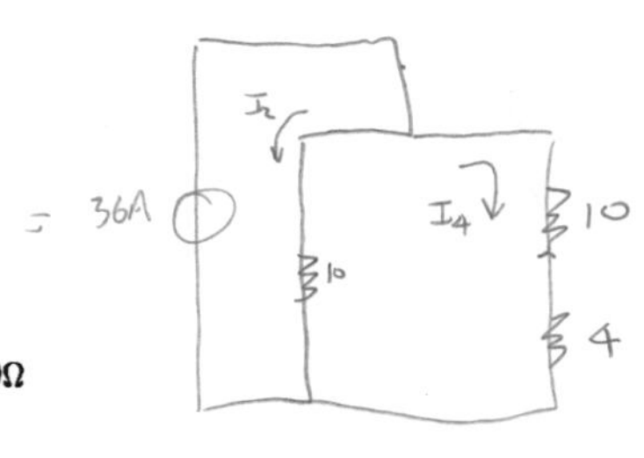

*6.3 Find the magnitude and sign of the power, in watts, absorbed by the circuit element in the box.

a) –20 d) 12
b) –8 e) 20
c) 8

*6.4 For the circuit shown, the voltage across the 4 ohm resistor is, with v=1 V

a) 1/4 d) 2
b) 1/2 e) 4
c) 2/3

6.5 The total conductance, in mhos, in the circuit shown below is

a) 1/5 d) 5
b) 1/2 e) 10
c) 2

6.6 The power, in watts, absorbed by the 6 mho conductance in the circuit below is

a) –0.24 d) 0.48
b) 0.2 e) 0.54
c) 0.24

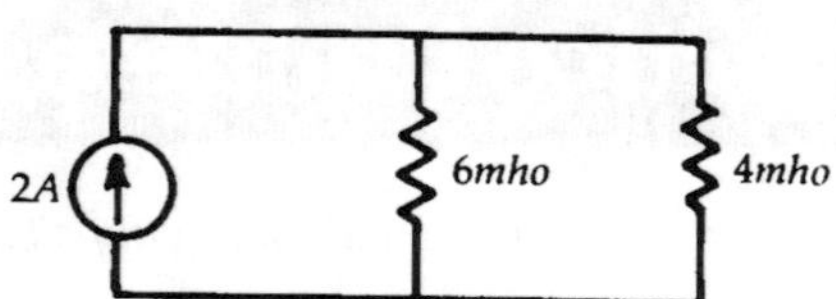

*6.7 The equivalent resistance, in ohms, between points *a* and *b* in the circuit below is

a) 3 d) 8
b) 5 e) 10
c) 7

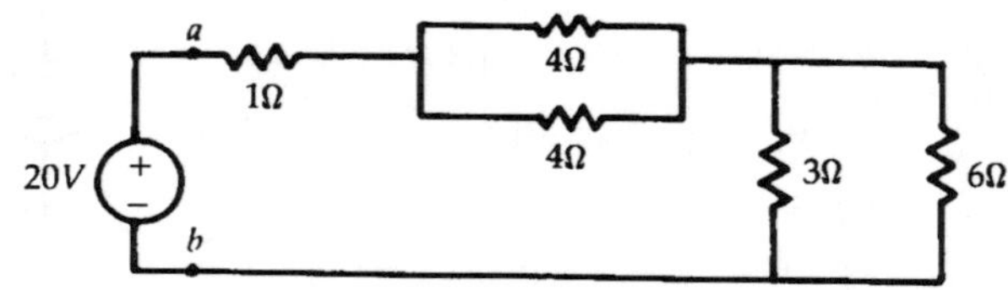

6.8 The voltage V_2 is

a) 6.4 d) 5.6
b) 4.0 e) 3.0
c) 2.0

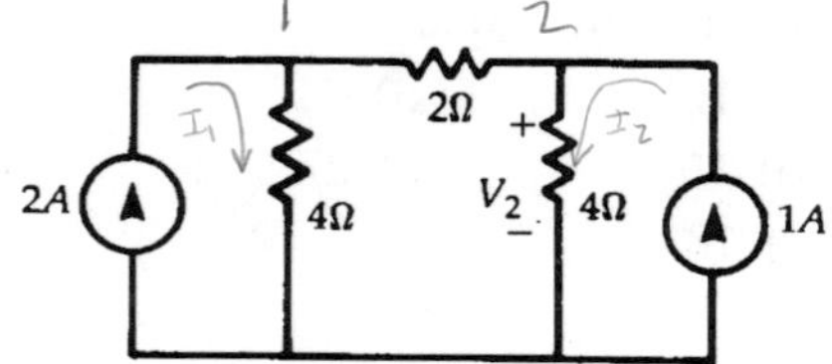

6.9 Find I_1 in amperes.

a) 4.0 d) 2.11
b) 2.0 e) 3.0
c) 4.11

AC CIRCUITS—SINGLE PHASE

*6.10 $(2 + j2)(3 - j4)$ is most nearly

a) $6.0\angle{-21.8°}$ b) $14.1\angle{-21.8°}$ c) $14.1\angle{-8.1°}$ d) $28.0\angle{-8.1°}$ e) $46.0\angle{-8.1°}$

*6.11 The following sinusoid is displayed on an oscilloscope. The RMS voltage and the radian frequency are most nearly

a) 1, 8.33
b) 0.7071, 52.36
c) 1.4142, 52.36
d) 2, 8.33
e) 2, 52.36

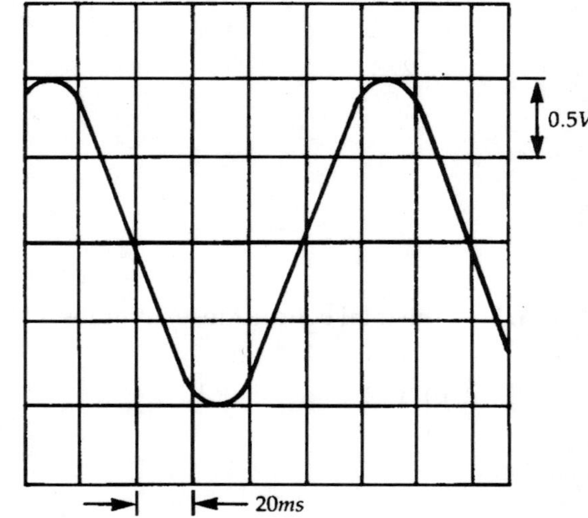

6.12 Find I_2 in amperes.

a) 0.29+j0.68 d) 1–j2
b) –0.12+j0.69 e) –3.33–j4.50
c) –0.82–j0.37

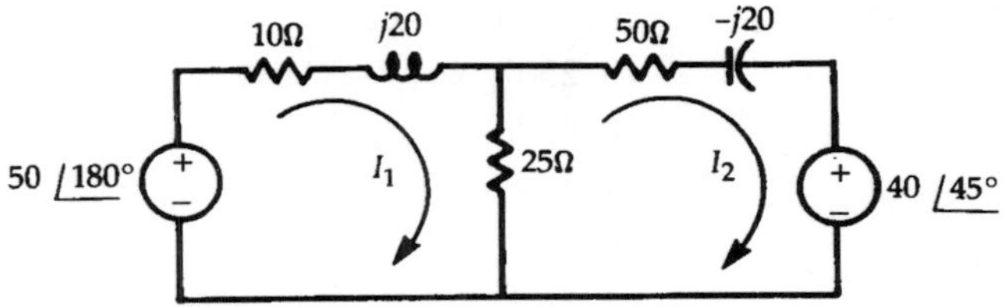

*6.13 Calculate the magnitude of the node voltage V_{AB}.

a) 85.1 d) 92.2
b) 77.2 e) 102.2
c) 68.8

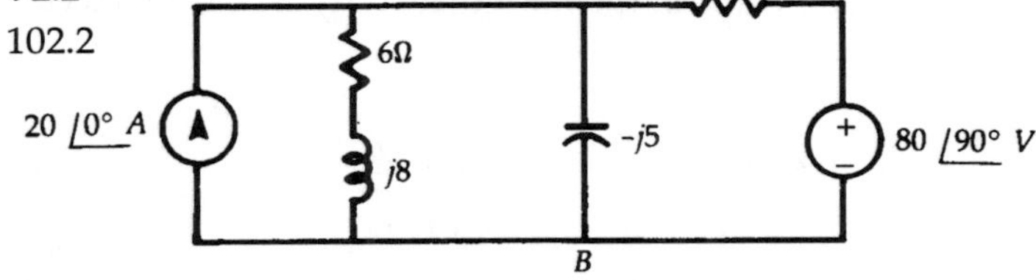

6.14 The peak value of $V(t)$ in the circuit shown is approximately

a) 2.0
b) 3.68
c) 25.9
d) 50.0
e) 71.6

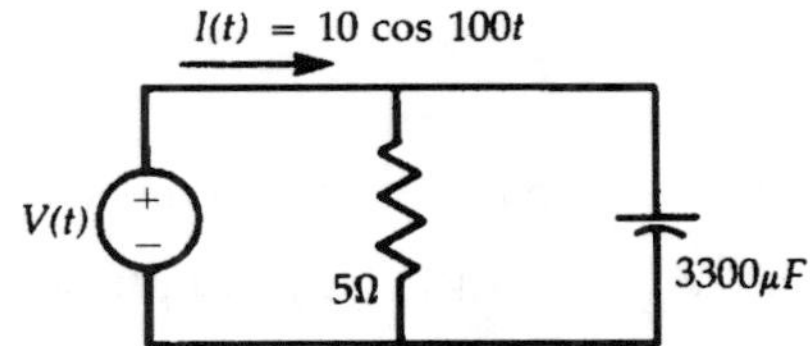

*6.15 The power factor of the circuit shown is most nearly

a) 0.5 d) 0.8
b) 0.6 e) 0.9
c) 0.7

4Ω
j3

6.16 For the circuit shown, the value of capacitance C that will give a power factor of 1.0 is most nearly

a) 0.0173
b) 0.0519
c) 0.0938
d) 0.0393
e) 0.0732

I = 10 cos 50t
C
0.01H
10Ω

6.17 For maximum power dissipation in the load of the circuit shown, R (in ohms) and L (in milli-henries) should be chosen as

a) 26, 50
b) 20, 100
c) 20, 50
d) 25, 100
e) 10, 25

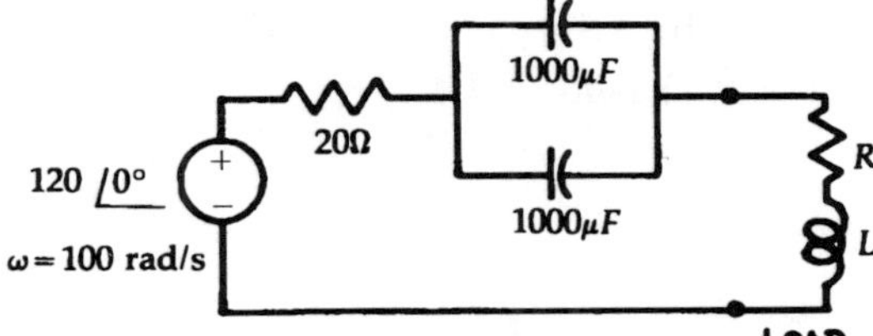

AC CIRCUITS—THREE PHASE

*6.18 Calculate the total average power, in watts, dissipated in the balanced three phase load.

a) 2507
b) 5276
c) 3456
d) 978
e) 1728

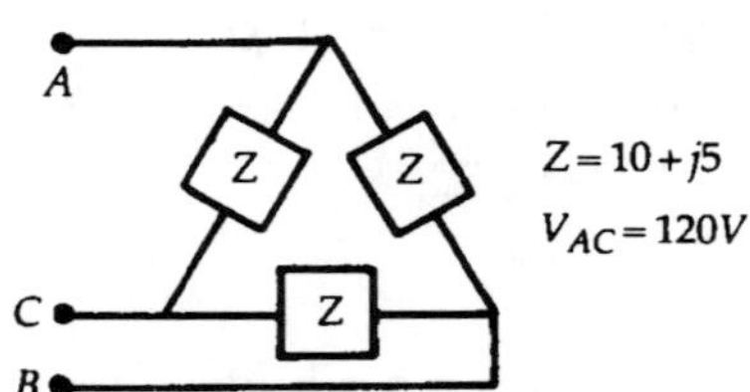

6.19 The value of the line current I_{aA} in the balanced Y-connected system shown is most nearly

a) $20.6\angle 30^\circ$
b) $35.3\angle -45^\circ$
c) $35.3\angle 45^\circ$
d) $15.1\angle -30^\circ$
e) $15.1\angle 30^\circ$

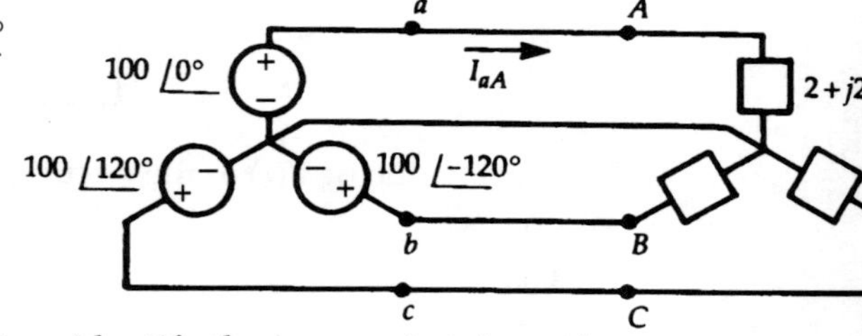

*6.20 For a balanced Y-connected system identify the incorrect statement.

a) $V_{PH} = \sqrt{3}\, V_L$
b) $I_L = I_{PH}$
c) $P_{total} = 3P_{PH}$
d) All phase impedances are equal.
e) The neutral current is zero.

6.21 A 100 μF capacitor has $I_C(t)$. The capacitor voltage $V_c(t)$ at t = 2.5 seconds $\left(V(0) = 1.0 \text{ V}\right)$ is most nearly

a) –24
b) –25
c) 25
d) 26
e) 0.0025

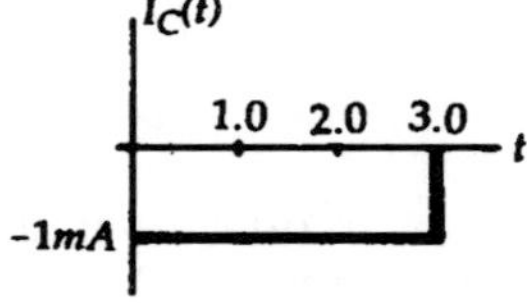

6.22 The voltage across a 10 μF capacitor is $50t^2$ V. The time, in seconds, it will take to store 200 J of energy is most nearly

a) 0.15 b) 0.21 c) 1.38 d) 2.275 e) 11.25

*6.23 The value of the voltage across C at $t = 30 \times 10^{-6}$ s, if the switch is closed at $t = 0$, is

a) 3.51
b) 4.51
c) 5.46
d) 6.32
e) 7.43

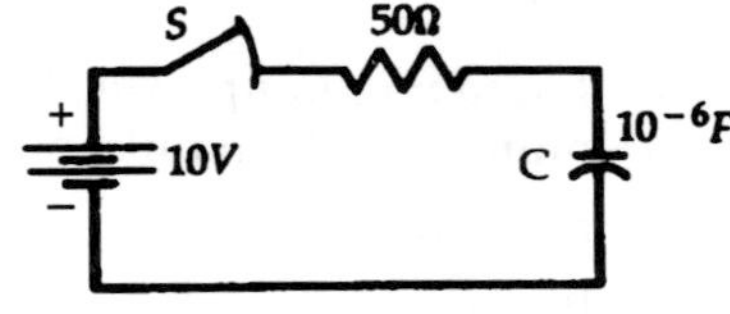

6.24 How long, in microseconds, does it take for the current to reach half its final value, if the switch is closed at $t = 0$?

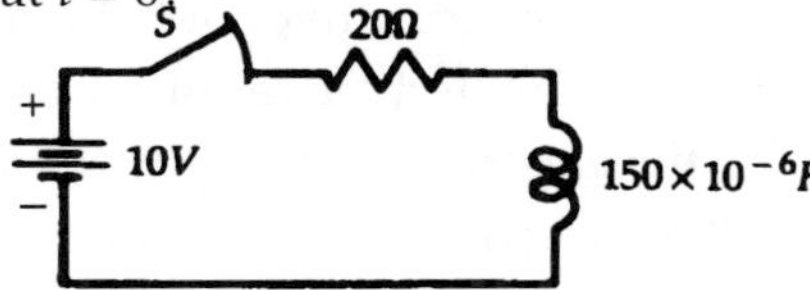

a) 3.1
b) 4.7
c) 5.2
d) 7.3
e) 8.4

ELECTRIC FIELDS

*6.25 The electric flux passing out through a closed surface is equal to

a) the line integral of the current around the surface.
b) zero.
c) the flux density at the surface.
d) the total charge enclosed by the surface.
e) the net surface integral of the current.

*6.26 The direction of the force acting on a moving charge placed in a magnetic field is

a) perpendicular to the magnetic field.
b) opposite to the direction of motion of the charge.
c) along the direction of the magnetic field.
d) along the direction of motion of the charge.
e) at an oblique angle to the magnetic field.

6.27 Two infinitely long lines of charge are parallel to the z-axis and located as shown. The force on an electron at (1,0,0) will be in the direction

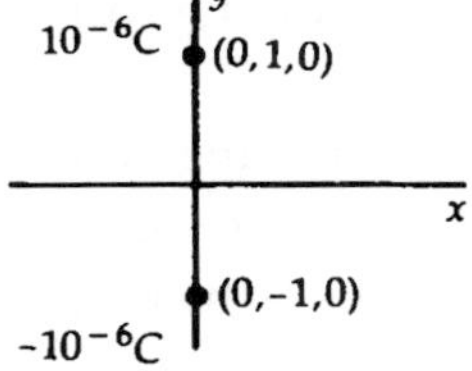

a) +x
b) −x
c) +y
d) −y
e) x + y

6.28 A point charge of 2×10^{-7} C is located at the origin of coordinates. A spherical shell with center at the origin and radius of 20 cm has a surface charge density 1×10^{-7} C/m^2. The electric flux density at $r = 50$ cm, in C/m^2, is

a) 3.18×10^{-8}
b) 7.96×10^{-8}
c) 9.55×10^{-8}
d) 11.14×10^{-8}
e) 14.22×10^{-8}

*6.29 A uniform line charge of $\rho_L = 30$ nC/m lies along the z-axis. The flux density D at (3,–4,5) is:

a) 1.91×10^{-10}
b) 9.55×10^{-10}
c) 11.94×10^{-10}
d) 15.92×10^{-10}
e) 30×10^{-10}

6.30 An electric field in rectangular coordinates is given by $\mathbf{E} = 4y\mathbf{x} + 4x\mathbf{y}$ V/m. The voltage drop from (1,1,1) to (5,1,1) is

a) +12
b) −12
c) +16
d) −16
e) +25

6.31 Static electric field distributions refer to cases where

a) all time derivatives of field quantities are zero.
b) the time derivatives of the displacement current are not zero.
c) the electric fields vary with time.
d) the electric scalar potential is two-dimensional.
e) the electric fields are uniform in space.

*6.32 A point charge of 50×10^{-9} C is placed 10 cm above a perfectly conducting infinitely large flat ground plane. What is the voltage 5 cm above the ground with respect to zero volts on the ground?

a) 3000 b) 4000 c) 5000 d) 6000 e) 7000

MAGNETIC FIELDS

*6.33 Two long, straight conductors located at (0,3,z) and (0,–3,z) each carry 5 amperes in the same direction (distances are in meters). The magnitude of magnetic field intensity at (4,0,0) is

a) $1/\pi$ b) $2/5\pi$ c) $3/5\pi$ d) $4/5\pi$ e) π

6.34 A solenoid has 1000 turns and carries a current of 5 amperes. If $L = 50$ cm and $r_c = 2.5$ cm, what is the magnetic field intensity on the solenoid axis at the center of the solenoid?

a) 10^4 b) 2×10^5 c) 5×10^4 d) 10^5 e) 10^6

*6.35 The inductor shown has an inductance of 4 mH. In order to increase the inductance to 40 mH,

a) increase the current by 10.
b) increase the mean flux path length by 10.
c) increase the number of turns to 10 N.
d) increase the cross sectional area of the iron by 10.
e) increase the number of turns to 100 N.

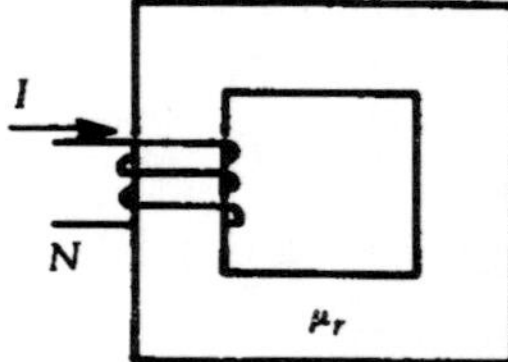

6.36 An iron ring with a mean diameter of 20 cm is wound with a coil of 200 turns. The permeability of the iron is $4\pi \times 10^{-4}$ H/m. A current of 0.05 A is passed through the coil. The magnetic flux density in the iron, in W/m^2, is

a) 0.02 b) $0.01\pi^2$ c) $100/\pi$ d) π e) $10/\pi$

*6.37 An iron core is shown. The relative permeability of the iron is 4000. The reluctance, in H^{-1}, of the magnetic circuit shown is

a) $1/(\pi \times 10^{-8})$

b) $1/(4\pi \times 10^{-5})$

c) $1/(4\pi \times 10^{-8})$

d) $1/(16\pi \times 10^{-8})$

e) $1/(32\pi \times 10^{-8})$

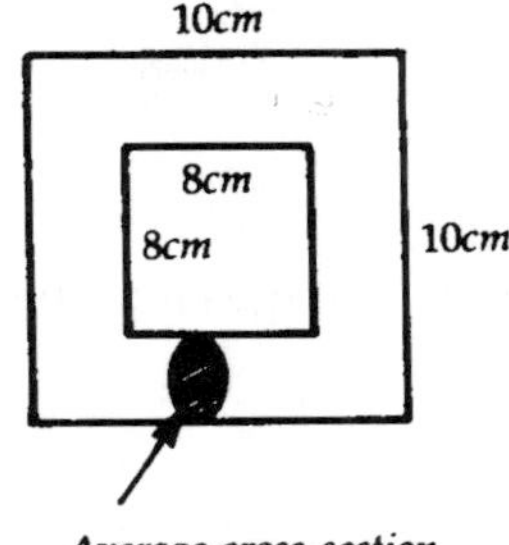

Average cross-section is 0.009 m^2

DIODES

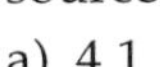

*6.38 If the desired DC load voltage is 9 volts, what is the rms value of the source?

a) 4.1
b) 12.7
c) 20.0
d) 28.3
e) 32.5

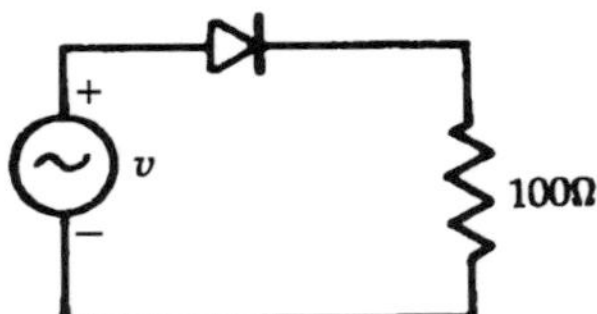

6.39 If the source voltage in the circuit of Prob. 6.38 is $v = 100 \sin 377t$, the peak reverse voltage applied to the diode would be

a) 2.5 b) 100 c) 141.4 d) 31.8 e) 87.2

6.40 If $R_L = 600\ \Omega$, what must be the rms value of the sinusoidal voltage v if $I = 150$ mA?

a) 60
b) 80
c) 100
d) 120
e) 140

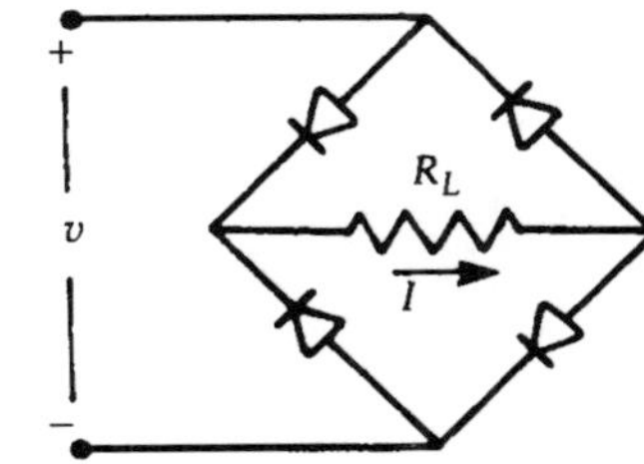

OPERATIONAL AMPLIFIERS

*6.41 Calculate R, in kΩ, so that $v_o/v_{in} = -200$.

a) 50 kΩ
b) 100
c) 200
d) 300
e) 400

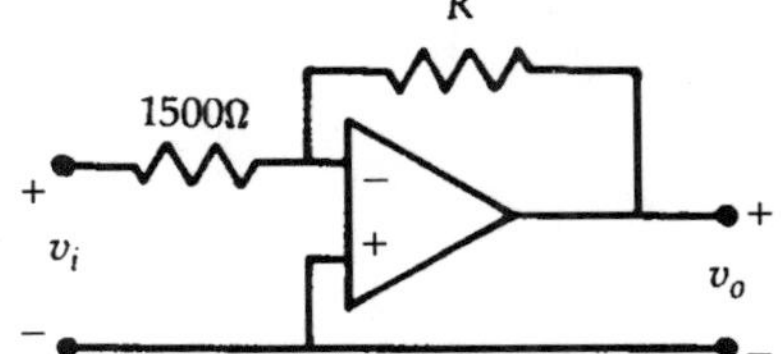

6.42 The gain of the following OP-AMP circuit is

a) –0.2
b) –1.2
c) –4
d) –5
e) –8

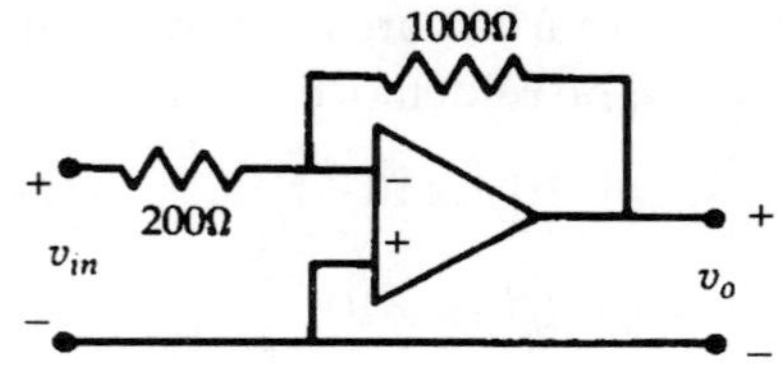

6.43 Given the voltages into the following OP-AMP network, the output voltage is

a) –2
b) –4
c) –7
d) –10
e) –15

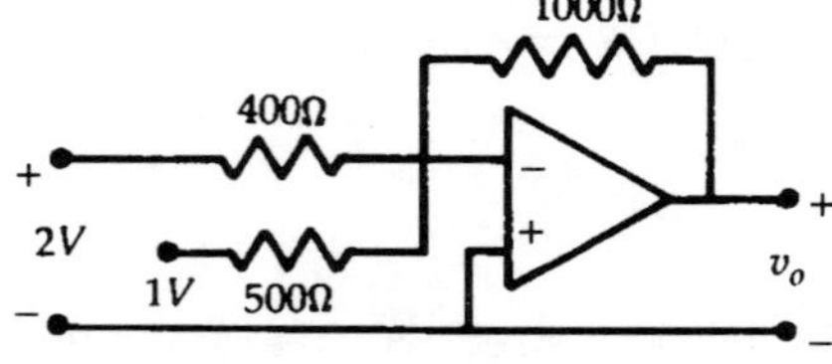

*6.44 The OP-AMP circuit below performs the function of

a) amplification.
b) integration.
c) differentiation.
d) summing.
e) attenuation.

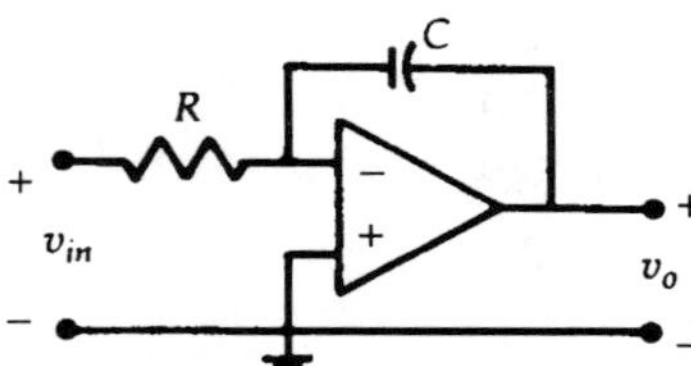

Solutions to Practice Problems

6.1 c) Beginning at lower left, sum the voltage drops around the closed path:

$$v_1 + v_2 - v_3 - v_4 + v_5 = 0$$

6.2 a) An equivalent circuit is:
Use current divider:

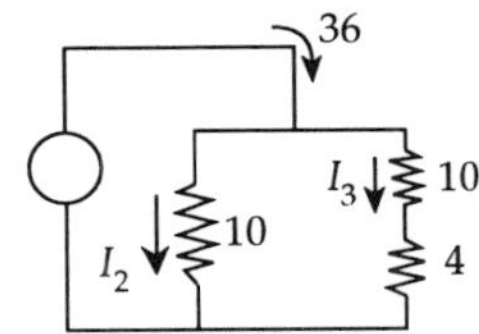

$$I_2 = \frac{14}{24} \times 36 = 21 \text{ A}$$

$$I_1 = \frac{20}{25} I_3 = \frac{20}{25}(36 - 21) = 12 \text{ A}$$

6.3 c) $P_{source} = VI = 10 \times 2 = 20$ W

$P_{resistor} = I^2 R = 2^2 \times 3 = 12$ W.

$\therefore P_{box} = 20 - 12 = 8$ W

6.4 b) A simplified circuit is shown.

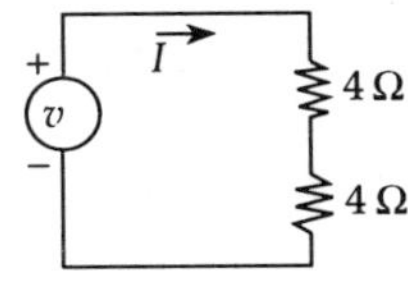

$v - 4I - 4I = 0. \quad \therefore I = v/8.$

$v_4 = IR = \frac{v}{8} \times 4 = v/2.$

$\therefore$ Voltage across 4 Ω resistor if $v = 1V$ is $1/2\ V$.

6.5 a) Total Resistance $= 1 + 2 + 2 = 5\ \Omega$

Conductance $= 1/R = 1/5$ mho

6.6 c) An equivalent circuit is shown.

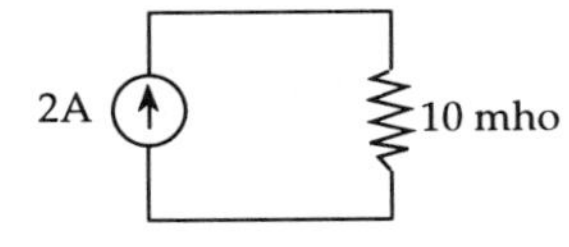

Voltage across 10 mho:

$V = I/G = 2/10 = 0.2$ Volts

Current in 6 mho $= I = VG = 0.2 \times 6 = 1.2$ A

$P = I^2/G = 1.2^2/6 = 0.24$ W.

6.7 b) An equivalent circuit is shown.

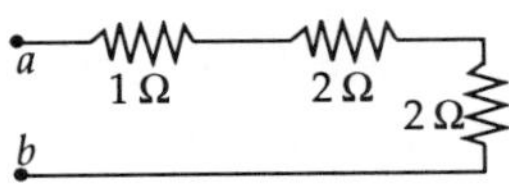

$R_{eq} = 1 + 2 + 2 = 5\ \Omega$

6.8 d) At pt. 1 sum currents:

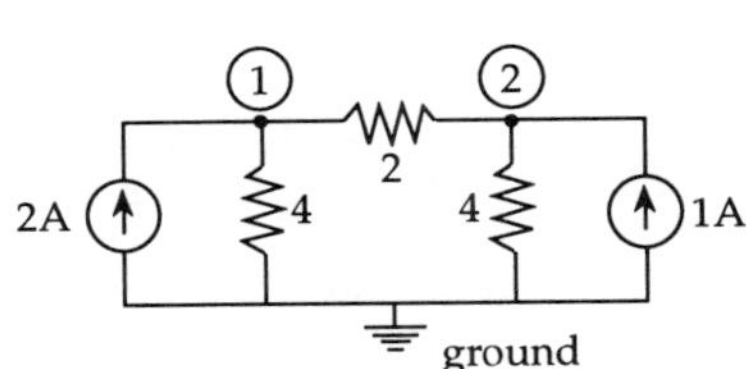

$$2 - \frac{V_1}{4} - \frac{V_1 - V_2}{2} = 0.$$

At pt. 2 sum currents:

$$1 - \frac{V_2}{4} - \frac{V_2 - V_1}{2} = 0.$$

The above equations are: $\left.\begin{array}{r} 3V_1 - 2V_2 = 8 \\ -2_1 + 3V_2 = 4 \end{array}\right\} \quad \therefore V_2 = 5.6$

6.9 d) Convert the 2 A current source with parallel resistor to a voltage source with series resistor.

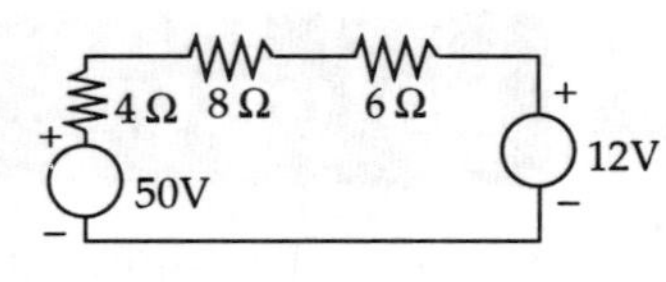

Summing voltages:

$50 - 4I - 8I - 6I - 12 = 0. \quad \therefore I = 38/18 = 2.11 \text{ A}.$

6.10 c) $(2+j2)(3-j4) = (2.82\angle 45^\circ)(5\angle -53.13^\circ) = 14.1\angle -8.13^\circ$.

6.11 b) From graph $V_{max} = 1.0$ V. $\therefore V_{rms} = \dfrac{V_{max}}{\sqrt{2}} = 0.707$ V.

The period $T = 120$ ms. $\therefore f = \dfrac{1}{T} = \dfrac{1}{120 \times 10^{-3}} = 8.33$ Hz

$\therefore$ radian frequency $= \omega = 2\pi f = 52.36$ rad/s.

6.12 c) Write the mesh current equations:

$$50\angle 180^\circ - (10+j20)I_1 - 25I_1 + 25I_2 = 0$$

$$25I_1 - 25I_2 - (50-j2)I_2 - 40\angle 45^\circ = 0$$

$$(35+j20)I_1 - 25I_2 = 50\angle 180^\circ$$

$$-25I_1 + (75-j20)I_2 = -40\angle 45^\circ$$

$$\therefore I_2 = \frac{2285\angle 233^\circ}{2530\angle 18.4^\circ} = 0.90\angle 214.6 = -.82 - j.37$$

6.13 a) Convert voltage source to current source:

Write KCL at A:

$$20\angle 0^\circ - \frac{V_A}{6+j8} - \frac{V_A}{-j5} - \frac{V_A}{4} + 20\angle 90^\circ = 0.$$

$$\therefore V_A = \frac{20\angle 0^\circ + 20\angle 90^\circ}{\left(\dfrac{1}{6+j8} - \dfrac{1}{j5} + \dfrac{1}{4}\right)} = \frac{20+j20}{(.06 - j.08 + j.2 + .25)} = \frac{20+j20}{.31+j.12}$$

$$= \frac{28.28\angle 45^\circ}{0.33\angle 21.16^\circ} = 85.1\angle 23.84^\circ$$

6.14 c) Convert to phasors:

The current $i = 10 \cos 100\,t$ becomes

$I = \dfrac{10}{\sqrt{2}}\angle 0^\circ$ A, $\omega = 100 \dfrac{\text{rad}}{\text{sec}}$

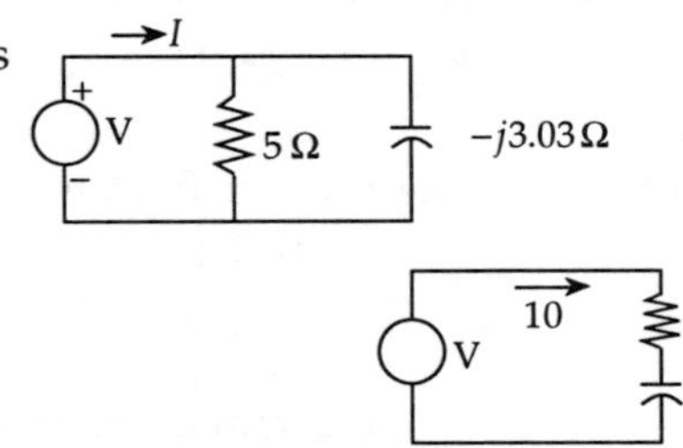

$$-j\frac{1}{\omega c} = -j\frac{1}{100 \times 3300 \times 10^{-6}} = -j3.03$$

Find parallel equivalent impedance:

$$Z = \frac{-j3.03 \times 5}{5 - j3.03} = \frac{15.15\angle -90^\circ}{5.85\angle -31.2^\circ} = 2.59\angle -58.8^\circ$$

$$V = IZ = \left(\frac{10}{\sqrt{2}}\angle 0^\circ\right)(2.59\angle -58.8^\circ) = 18.32\angle -58.8^\circ$$

$v(t) = 25.9\cos(100t - 58.8^\circ)$

6.15 d) $Z = 4 + j3 = 5\angle 36.9°$ $\quad \therefore \theta = 36.9°. \quad pf = \cos\theta = \cos 36.9° = 0.8$

6.16 d) For $pf = 1$: $Y = G + j(\omega C - 1/\omega L) = G + j0$ or $\omega C = \dfrac{1}{\omega L}$.

$$\therefore \; C = \frac{1}{\omega^2 L} = \frac{1}{50^2 \times .01} = 0.04$$

6.17 c) For max. power dissipation choose $z_L = \tilde{z}_{gen}$.

$$Z_{gen} = 20 - j\frac{1}{100 \times 2000 \times 10^{-6}} = 20 - j5.$$

$\therefore Z_L = 20 + j5. \quad \therefore R = 20$ and $\omega L = 5$.

$$\therefore \; L = \frac{5}{100} = .05$$

6.18 c) $\theta_{ph} = \tan^{-1}\dfrac{5}{10} = 26.6°. \quad V_{ph} = 120. \quad I_{ph} = \dfrac{120}{|10 + j5|} = 10.73.$

$$\therefore \; P_L = 3V_{ph}I_{ph}\cos\theta_{ph} = 3 \times 120 \times 10.73 \cos 26.6° = 3456.$$

6.19 b) $I_L = I_{ph} = \dfrac{V_{ph}}{Z_{ph}} = \dfrac{100\angle 0°}{2 + j2} = \dfrac{100\angle 0°}{2.83\angle 45°} = 35.36\angle{-45°}$

6.20 a) (*a*) should be $V_L = \sqrt{3}\, V_{ph}$

6.21 a) $v = \dfrac{1}{C}\int i\,dt + v_o = \dfrac{1}{100 \times 10^{-6}}\left(-10^{-3} \times 2.5\right) + 1.0 = -24$

6.22 e) $w_C(t) = \dfrac{1}{2}Cv^2 = \dfrac{1}{2}\left(10 \times 10^{-6}\right)\left(50t^2\right)^2 = 200. \quad \therefore \; t^4 = 16000. \quad \therefore t = 11.25\,s$

6.23 b) $T = RC = 50 \times 10^{-6}$ sec.

$$v_C(t) = V_\infty + \left(V_o - V_\infty\right)e^{-t/T} = 10 + (0 - 10)e^{-\frac{30\times10^{-6}}{50\times10^{-6}}} = 10 - 10e^{-.6} = 4.51.$$

6.24 c) $I_\infty = 10/20 = 0.5\,\text{A}. \quad T = L/R = 150 \times 10^{-6}/20 = 7.5 \times 10^{-6}$

$$i(t) = I_\infty + \left(I_o - I_\infty\right)e^{-t/T}$$

$$\therefore \; \frac{0.5}{2} = 0.5 + (0 - 0.5)e^{-t/7.5\times10^{-6}}$$

$$\therefore \; t = 5.20 \times 10^{-6} \text{ sec.}$$

6.25 d) This is Gauss' Law.

6.26 a) $\mathbf{F} = q\mathbf{V} \times \mathbf{B}. \quad \therefore \; \mathbf{F} \perp \mathbf{B}$

6.27 c)

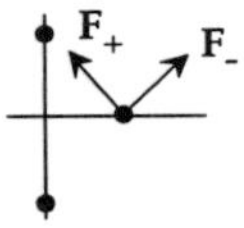

The x-components of $\mathbf{F}_- + \mathbf{F}_+$ cancel and $\therefore$ $\mathbf{F}_{total}$ is in y-direction.

6.28 b)

For the surface charge: $D_s = \rho_s \dfrac{4\pi R^2}{4\pi r_s^2}$.

For the point charge: $D_q = \dfrac{q}{4\pi r_s^2}$.

$\therefore$ The total is $D_t = \dfrac{10^{-7}\times 0.2^2}{0.5^2} + \dfrac{2\times 10^{-7}}{4\pi \times 0.5^2} = 7.96\times 10^{-8}\ \text{C/m}^2$

6.29 b) From the $z-$axis to (3, -4, 5) the distance is

$$r = \sqrt{3^2 + 4^2} = 5$$

$$|D| = \frac{\rho_\ell}{2\pi r} = \frac{30\times 10^{-9}}{2\pi \times 5} = 9.55\times 10^{-10}\ \text{C/m}^2$$

6.30 c) $V = \int_a^b \mathbf{E}\cdot d\mathbf{L} = \int_a^b E_x dx + E_y \overset{0}{\cancel{dy}} + E_z \overset{0}{\cancel{dz}}$

$= \int_1^5 E_x dx = \int_1^5 4y dx = \int_1^5 4\times 1 dx = 4\times 4 = 16$ volts

6.31 a) "Static" implies no time variation in any quantity.

6.32 d) Use image theory and place a second charge at (0, -10 cm). The second charge must be negative.

$$\therefore V = V_+ + V_- = \frac{q}{4\pi\varepsilon r_+} + \frac{-q}{4\pi\varepsilon r_-} = \frac{q}{4\pi\varepsilon}\left(\frac{1}{r_+} - \frac{1}{r_-}\right)$$

$$= \left(50\times 10^{-9}\right)\left(9\times 10^{9}\right)\left(\frac{1}{.05} - \frac{1}{.15}\right) = 6000 \text{ volts}.$$

We used $\dfrac{1}{4\pi\varepsilon} = 9\times 10^9$.

6.33 d)

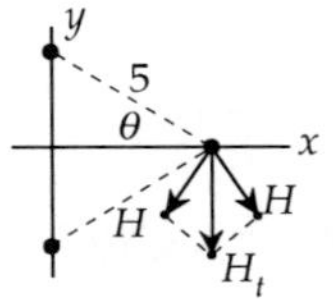

Each H is $\perp$ plane containing wire and point.

$$\theta = \tan^{-1}\frac{3}{4} = 36.9°.$$

$$H = \frac{I}{2\pi r} = \frac{5}{2\pi 5} = \frac{1}{2\pi}$$

$$\therefore\ H_t = 2H\cos\theta = 2\times\frac{1}{2\pi}\times 0.8 = 4/5\pi$$

6.34 a) $H_{\text{center}} = \dfrac{NI}{\ell} = \dfrac{1000\times 5}{0.5} = 10^4\ \text{A/m}.$

6.35 d) $L = \dfrac{N}{I}\Psi = \dfrac{N}{I}\left(\dfrac{NI}{\Re}\right) = N^2/\Re = N^2\Big/\left(\dfrac{\ell}{\mu A}\right) = N^2\mu A/\ell.$

To increase L from 4 to 40 increase A by a factor of 10 .

6.36 a) $B = \dfrac{\Psi}{A} = \dfrac{1}{A}\dfrac{NI}{\Re} = \dfrac{1}{A} NI\left(\dfrac{\mu A}{\ell}\right) = \dfrac{NI\mu}{\ell}$

$$= \frac{200\times 0.05\times 4\pi\times 10^{-4}}{\pi\times 0.2} = 0.02\ \text{Wb/m}^2.$$

6.37 b) $\Re = \dfrac{\ell}{\mu A} = \dfrac{\ell}{\mu_o \mu_r A} = \dfrac{4\times 0.09}{4\pi\times 10^{-7}\times 4000\times 0.009} = \dfrac{1}{4\pi\times 10^{-5}}\ \mathrm{H}^{-1}$

6.38 c) $V_{DC} = V_{max}/\pi$

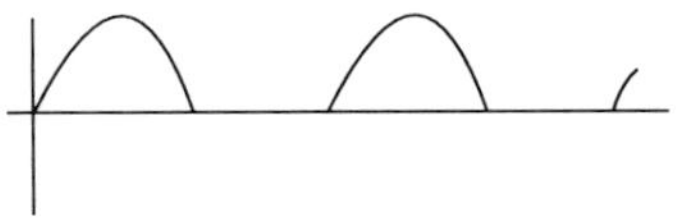

$9 = V_{max}/\pi$.

$\therefore V_{max} = 9\pi$.

$V_{rms} = V_{max}/\sqrt{2} = 9\pi/\sqrt{2} = 20$

6.39 b) The peak reverse voltage occurs when the diode is reverse biased and does not conduct. $\therefore PRV = V_{max} = 100$ volts.

6.40 c) For a bridge rectifier

$V_{DC} = V_{max}/\pi$.

$V_{DC} = I_L R_L = 0.150\times 600 = 90$.

$\therefore V_{max} = \dfrac{\pi}{2}\times 90 = 45\pi$. $\therefore V_{rms} = \dfrac{45\pi}{\sqrt{2}} = 100$ volts.

6.41 d) The gain of an OP - AMP is

$\dfrac{v_o}{v_{in}} = -\dfrac{R_f}{R_i} = -\dfrac{R_2}{1500} = -200$. $\therefore R_2 = 300\ 000\ \Omega$

or 300 kΩ

6.42 d) $A = -\dfrac{R_f}{R_i} = -\dfrac{1000}{200} = -5$

6.43 c) $v_o = -\dfrac{R_f}{R_1}v_1 - \dfrac{R_f}{R_2}v_2$

$= -\dfrac{1000}{400}\times 2 - \dfrac{1000}{500}\times 1$

$= -5 - 2 = -7$ volts

6.44 b) The OP-AMP circuit with capacitor feedback performs integration.

Thermodynamics

by Merle C. Potter

Overview, Definitions and Laws

Thermodynamics involves the storage, transformation and transfer of energy. It is stored as internal energy, kinetic energy and potential energy; it is transformed between these various forms; and, it is transferred as work or heat transfer.

The *macroscopic* approach is used in this presentation, that is, we assume matter occupies all points in a region of interest. This is acceptable providing the density is sufficiently large, which it is in most engineering situations.

Both a *system*, a fixed quantity of matter, and a *control volume*, a volume into which and/or from which a substance flows, can be used in thermodynamics. (A control volume may also be referred to as an *open system*.) A system and its surroundings make up the *universe*. Some useful definitions follow:

phase — matter that has the same composition throughout; it is homogeneous

mixture — a quantity of matter that has more than one phase

property — a quantity which serves to describe a system

simple system — a system composed of a single phase, free of magnetic, electrical, and surface effects. Only two properties are needed to fix a simple system

state — the condition of a system described by giving values to its properties at the given instant

intensive property — a property that does not depend on the mass

extensive property — a property that depends on the mass of the system

specific property — an extensive property divided by the mass

thermodynamic equilibrium — when the properties do not vary from point to point in a system and there is no tendency for additional change

process — the path of successive states through which a system passes

quasi-equilibrium — if, in passing from one state to the next, the deviation from equilibrium is infinitesimal. It is also called a quasistatic process

reversible process — a process which, when reversed, leaves no change in either the system or surroundings

isothermal — temperature is constant

isobaric — pressure is constant

isometric — volume is constant

isentropic — entropy is constant

adiabatic — no heat transfer

Experimental observations are organized into mathematical statements or *laws*. Some of those used in thermodynamics follow:

zeroith law of thermodynamics— If two bodies are equal in temperature to a third, they are equal in temperature to each other.

first law of thermodynamics — During a given process, the net heat transfer minus the net work output equals the change in energy.

second law of thermodynamics — A device cannot operate in a cycle and produce work output while exchanging heat at a single temperature.

Boyle's law — The volume varies inversely with pressure for an ideal gas.

Charles' law — The volume varies directly with temperature for an ideal gas.

Avogadro's law — Equal volumes of different ideal gases with the same temperature and pressure contain the equal number of molecules.

7.1 Density, Pressure, and Temperature

The density ρ is the mass divided by the volume,

$$\rho = \frac{m}{V}. \quad \textbf{(7.1.1)}$$

The specific volume is the reciprocal of the density,

$$v = \frac{1}{\rho} = \frac{V}{m}. \tag{7.1.2}$$

The pressure P is the normal force divided by the area upon which it acts. In thermodynamics, it is important to use *absolute pressure,* defined by

$$P_{\text{abs}} = P_{\text{gauge}} + P_{\text{atmospheric}} \tag{7.1.3}$$

where the atmospheric pressure is taken as 100 kPa (14.7 psi), unless otherwise stated. If the gauge pressure is negative, it is a *vacuum.*

The temperature scale is established by choosing a specified number of divisions, called degrees, between the ice point and the steam point, each at 101 kPa absolute. In the Celsius scale, the ice point is set at 0°C and the steam point at 100°C. In thermodynamics, pressures are always assumed to be given as absolute pressures. Expressions for the absolute temperature in kelvins, and degrees Rankine are, respectively,

$$T = T_{\text{celsius}} + 273; \quad T = T_{\text{fahrenheit}} + 460. \tag{7.1.4}$$

The temperature, pressure and specific volume for an ideal (perfect) gas are related by the *ideal gas law*

$$Pv = RT, \qquad P = \rho RT, \qquad \text{or } PV = mRT \qquad R = \overline{R}/M \tag{7.1.5}$$

where the *universal gas constant* $\overline{R} = 8.134$ kJ/kmol·K (1545 ft-lbf/lbmol-°R) and R is the gas con-stant; for air it is 0.287 kJ/kg·K (53.3 ft-lbf/lbm-°R). (Note: the NCEES Handbook does not list R_{air}. It is a good idea to memorize R_{air} since it is used so often.) Below moderate pressure this equation can be used if the temperature exceeds twice the critical temperature, the temperature above which the liquid and vapor phases do not coexist. Thus, if the temperature of the steam is greater than 800°C (1500°F), Equation 7.1.5 can be used with $R = 0.462$ kJ/kg·K (85.7 ft-lbf/lbm-°R) if the pressure is not excessive; for high pressure ($P > 10$ MPa) the temperature should be higher than 1000°C if Equation 7.1.5 is to be used.

If the temperature is below twice the critical temperature, tables relating the three variables P, v and T must be used. For water such tables are called the "steam tables" and are presented in Table 7.3.1. Data is presented in Table 7.3.1 and 7.3.2 that is used to relate P, v and T when water exists in the liquid phase, as a liquid/vapor mixture, or in the saturated vapor phase. These situations are best described referring to a T-v diagram, shown in Figure 7.1. Suppose we start with a constant pressure container (a cylinder with a floating piston) containing water

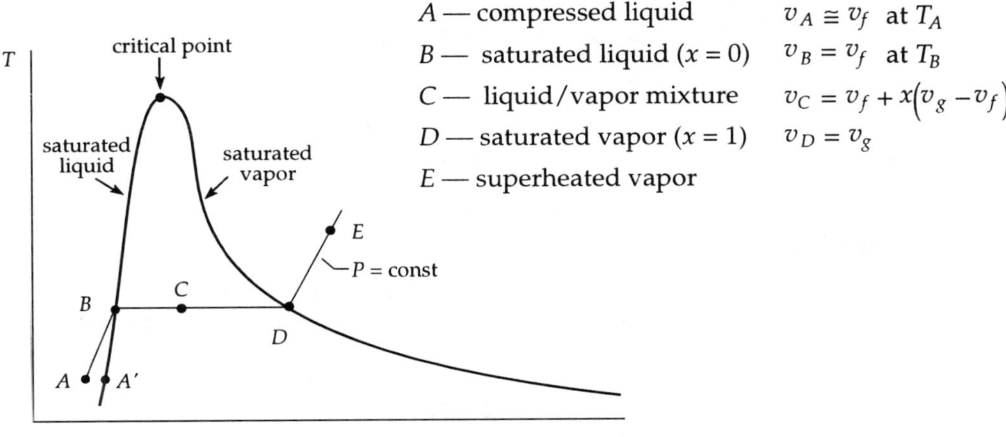

Figure 7.1 A T-v diagram.

at the room temperature; it would undergo the following changes if the heat is transferred to the volume:

- The temperature would rise above T_A in the *compressed liquid* and $v_A \cong v_{A'}$.
- At state *B* *saturated liquid* phase results and vaporization (boiling) begins; $v_B = v_f$. A subscript *f* will always denote a saturated liquid.
- At state *C* the liquid phase and the vapor phase are in equilibrium; a *liquid/vapor mixture* occurs and

$$v = v_f + x(v_g - v_f) \quad \text{or} \quad v = xv_g + (1-x)v_f \tag{7.1.6}$$

  where x is the *quality*. Sometimes we let $v_{fg} = v_g - v_f$.
- At state *D* a *saturated vapor* exists and vaporization is complete; a subscript *g* will always denote a saturated vapor: $v_D = v_g$.
- At state *E* the vapor is *superheated* and v is found in Table 7.3.3.

Equation 7.1.6 above results from the definition of *quality*, the ratio of the vapor mass to the total mass:

$$x = \frac{m_{vapor}}{m_{total}} = \frac{m_g}{m_g + m_f}. \tag{7.1.7}$$

Note that the entries in the tables at the end of this chapter are in absolute pressure.

An alternative to the tabulated properties is the *P-h* diagram which includes the properties P, h, T, v and s. If internal energy is required we use the relation $u = h - Pv$ (see Eq. 7.2.10). We will illustrate its use in Article 7.3.

Even though the NCEES Reference Handbook does not contain the steam tables, it is important that you become familiar with them. Excerpts from the tables will be provided if FE exam problems require their use.

Example 7.1

What mass of air is contained in a room 20 m × 40 m × 3 m at standard conditions?

Solution. Standard conditions are $T = 25°C$ and $P = 100$ kPa abs. Hence,

$$\rho = \frac{P}{RT}$$
$$= \frac{100}{0.287 \times 298} = 1.17 \text{ kg/m}^3, \quad \text{where } R = \frac{8.314}{28.97} = 0.287.$$
$$\therefore m = \rho V$$
$$= 1.17 \times 20 \times 40 \times 3 = 2810 \text{ kg}.$$

Example 7.2

What mass of water contained in 2m^3 at 2000 kPa abs and 200°C?

Solution. The specific volume of compressed water is insensitive to pressure so we use v_f at 200°C. It is, using Table 7.3.1,

$$v_f = 0.001156 \text{ m}^3/\text{kg},$$

The mass is then

$$m = V/v$$
$$= 2/0.001156 = 1730 \text{ kg}.$$

Note: Liquid water has a mass density of about 1000 kg/m^3 or $v = 0.001 \text{ m}^3/\text{kg}$. This is good if the temperature is relatively low, say below 100°C.

Example 7.3

The volume occupied by 20 kg of a water-vapor mixture at 200°C is 2m^3. Calculate the quality.

Solution. The specific volume is

$$v = V/m$$
$$= 2/20 = 0.1 \text{ m}^3/\text{kg}.$$

Using Equation 7.1.6, and Table 7.3.1, we have

$$x = \frac{v - v_f}{v_g - v_f}$$

$$= \frac{0.1 - 0.001156}{0.1274 - 0.001156} = 0.783 \text{ or } 78.3\%.$$

Example 7.4

Find the volume occupied by 20 kg of steam at 4 MPa and 400°C.

Solution. The specific volume is found in Table 7.3.3 to be

$$v = 0.07341 \text{ m}^3/\text{kg}$$

The volume is then

$$V = mv$$

$$= 20 \times 0.07341 = 1.468 \text{ m}^3.$$

7.2 The First Law of Thermodynamics for a System

The first law of thermodynamics, referred to as the "first law" or the "energy equation," is expressed for a cycle as

$$Q_{net} = W_{net} \tag{7.2.1}$$

and for a process as

$$Q - W = \Delta E \tag{7.2.2}$$

where Q is the heat transfer, W is the work, and E represents the energy (kinetic, potential, and internal) of the system.* In thermodynamics attention is focused on internal energy with kinetic and potential energy changes neglected (unless otherwise stated) so that we have

$$Q - W = \Delta U \quad \text{or} \quad q - w = \Delta u \tag{7.2.3}$$

where the specific internal energy is

$$q = \frac{Q}{m}, \quad w = \frac{W}{m}, \quad u = \frac{U}{m}. \tag{7.2.4}$$

Heat transfer may occur during any of the three following modes :

• ***Conduction***—heat transfer due to molecular activity. For steady-state heat transfer through a constant wall area *Fourier's law* states†

$$\dot{Q} = kA\Delta T/L = A\Delta T/R \tag{7.2.5}$$

where k is the *conductivity* (dependent on the material), R is the *resistance factor*, and the length, L, is normal to the heat flow.

• ***Convection***—heat transfer due to fluid motion. The mathematical expression used is

$$\dot{Q} = hA\Delta T = A\Delta T/R \tag{7.2.6}$$

where h is the *convective heat transfer coefficient* (dependent on the surface geometry, the fluid velocity, the fluid viscosity and density, and the temperature difference). The *Nusselt number* ($Nu = hd / k$) is the dimensionless convective heat transfer coefficient.

• ***Radiation***—heat transfer due to the transmission of waves. The heat transfer from body 1 is

$$\dot{Q} = \sigma\varepsilon A\left(T_1^{\,4} - T_2^{\,4}\right)F_{1-2} \tag{7.2.7}$$

in which the *Stefan-Boltzmann constant* is $\sigma = 5.67 \times 10^{-11} \text{kJ/s} \cdot \text{m}^2 \cdot \text{K}^4$ (1.714×10^{-9} Btu / hr - ft^2-°R^4, ε is the *emissivity* ($\varepsilon = 1$ for a black body), and $F_{1\text{-}2}$ is the *shape factor* ($F_{1-2} = 1$ if body 2 encloses body 1).

For a two layer composite wall with an inner and outer convection layer we use the resistance factors and obtain

$$\dot{Q} = A\Delta T/(R_i + R_1 + R_2 + R_0) = UA\Delta T \tag{7.2.8}$$

where U is the *overall heat transfer coefficient*. It is not to be confused with internal energy of Eq. 7.2.3.

In thermodynamics, the heat transfer is usually specified or calculated using Equation 7.2.3; it is usually not calculated with the previous four equations. The equations in the one dimensional form above and in two-dimensions are the focus of attention in a course in Heat Transfer. Thermal properties are presented in Tables 7.5 and 7.6.

---

*In this book heat transferred to the system is positive and work done by the system is positive. It is also conventional to define work done on the system as positive so that $Q + W = \Delta U$.

†A dot signifies a rate, so that $\dot{Q}$ has units of J/s.

Work can be accomplished mechanically by moving a boundary, resulting in a quasi-equilibrium work mode

$$W = \int P dV \qquad (7.2.9)$$

It can also be accomplished in non-quasi-equilibrium modes such as with paddle wheel or by electrical resistance. But then Equation 7.2.9 cannot be used.

We introduce *enthalpy* for convenience, and define it to be

$$H = U + PV \qquad (7.2.10)$$

$$h = u + Pv.$$

For substances such as steam, the specific internal energy and specific enthalpy are either found in the steam tables or from a P-h diagram. For a compressed liquid, u and h are insensitive to pressure and are found in the Table 7.3.1 with the "f" subscripts under the specified temperature. For the liquid/vapor mixture we use

$$u = u_f + x\left(u_g - u_f\right)$$

$$h = h_f + xh_{fg} \qquad (7.2.11)$$

where $h_{fg} = h_g - h_f$.

For ideal gases we assume constant specific heats and use

$$\Delta u = c_v \Delta T \qquad (7.2.12)$$

$$\Delta h = c_p \Delta T \qquad (7.2.13)$$

where c_v is the *constant volume specific heat*, and c_v is the *constant pressure specific heat*. From the differential forms of the above we can find

$$c_p = c_v + R. \qquad (7.2.14)$$

We also define the *ratio of specific heats* k to be

$$k = c_p / c_v. \qquad (7.2.15)$$

For air $c_v = 0.716$ kJ/kg · K (0.171 BTU/lbm - °R), $c_p = 1.00$ kJ/kg · K (0.24 BTU/lbm - °R), $k = 1.4$. For most solids and liquids we can find the heat transfer using

$$Q = mC_p \Delta T. \qquad (7.2.16)$$

For water $C_p = 4.18$ kJ / kg·°K (1.00 BTU / lbm-°R), and for ice $C_p \cong 2.1$ kJ/kg · °K (0.50 BTU/lbm - °R).

When a substance changes phase, *latent heat* is involved. The energy necessary to melt a unit mass of a solid is the *heat of fusion*; the energy necessary to vaporize a unit mass of liquid is the *heat of vaporization*, equal to $(h_g - h_f)$; the energy necessary to vaporize a unit mass of solid is the *heat of sublimation*. For ice, the heat of fusion is approximately 320 kJ/kg (140 BTU/lbm) and the heat of sublimation is about 2040 kJ/kg (877 BTU/lbm); the heat of vaporization varies from 2050 kJ/kg at 0°C (1075 BTU/lbm at 32°F) to zero at the critical point.

For specific processes, we consider the preceding paragraphs and summarize as follows:

Isothermal Process

$$\text{1st law:}\quad Q-W=m\Delta u \quad \text{or} \quad q-w=\Delta u \tag{7.2.17}$$

$$\text{ideal gas:}\quad Q=W=mRT\ln\frac{v_2}{v_1}=mRT\ln\frac{P_1}{P_2} \tag{7.2.18}$$

$$P_2=P_1v_1/v_2 \tag{7.2.19}$$

Constant Pressure (Isobaric)

$$\text{1st law:}\quad Q=m\Delta h \quad \text{or} \quad q=\Delta h \tag{7.2.20}$$

$$W=mP\Delta v \tag{7.2.21}$$

$$\text{ideal gas:}\quad Q=mc_p\Delta T \tag{7.2.22}$$

$$T_2=T_1\,v_2/v_1 \tag{7.2.23}$$

Constant Volume

$$\text{1st law:}\quad Q=m\Delta u \quad \text{or} \quad q=\Delta u \tag{7.2.24}$$

$$W=0 \tag{7.2.25}$$

$$\text{ideal gas:}\quad Q=mc_v\Delta T \tag{7.2.26}$$

$$T_2=T_1P_2/P_1 \tag{7.2.27}$$

Adiabatic Process (Isentropic)

$$\text{1st law:}\quad -W=m\Delta u \quad \text{or} \quad -w=\Delta u \tag{7.2.28}$$

$$Q=0 \tag{7.2.29}$$

$$\text{ideal gas:}\quad -W=mc_v\Delta T \tag{7.2.30}$$

$$T_2=T_1(v_1/v_2)^{k-1}=T_1(P_2/P_1)^{(k-1)/k} \tag{7.2.31}$$

$$P_2=P_1(v_1/v_2)^k \tag{7.2.32}$$

A *polytropic* process results if k in Equations 7.2.31 and 7.2.32 is replaced with n. Then n must be specified. Note that the adiabatic, quasi-equilibrium process is often referred to as an *isentropic process.*

Example 7.5

How much heat must be added to 2 kg of steam contained in a rigid volume, if the initial pressure of 2 MPa abs is increased to 5 MPa abs? $T_1 = 300°C$.

Solution. The first law with $\Delta KE = \Delta PE = 0$ is

$$Q-W=\Delta U.$$

For a rigid container $W=0$ so that

$$Q = m(u_2 - u_1).$$

From the steam Table 7.3.3, we find $u_1 = 2772.6$ kJ/kg and $v_1 = 0.1255\ \text{m}^3/\text{kg}$. We can locate state 2 because the container is rigid, so that

$$v_2 = v_1 \cong 0.126\ \text{m}^3/\text{kg}.$$

The temperature T_2 that has $P_2 = 5$ MPa and $v_2 = 0.126\ \text{m}^3/\text{kg}$ is 1100°C. At that state $u_2 = 4246$ kJ/kg. Thus,

$$Q = m(u_2 - u_1)$$
$$= 2(4246 - 2773) = 2946\ \text{kJ}.$$

Example 7.6

Calculate the heat transfer necessary to raise the temperature of 2 kg of saturated water vapor to 600°C if the pressure is maintained constant at 2000 kPa abs?
Solution. The first law, for a constant pressure process, is

$$Q = m(h_2 - h_1).$$

Using Tables 7.3.2 and 7.3.3, we find $h_1 = 2799.5$ and $h_2 = 3690.1$ kJ/kg. Hence, we have

$$Q = 2(3690.1 - 2799.5) = 1781\ \text{kJ}.$$

Example 7.7

How much heat is needed to completely vaporize 100 kg of ice at $T_1 = -10$°C if the pressure is held constant at 200 kPa abs?
Solution. The heat transfer is related to the enthalpy by

$$Q = m\Delta h$$
$$= m(c\Delta T_{ice} + \text{heat of fusion} + c\Delta T_{water} + \text{heat of vaporization}).$$

Using the values given in Article 7.2, Table 7.2, and Table 7.3.2,

$$Q = 100(2.1 \times 10 + 320 + 4.18 \times 120.2 + 2201.9)$$
$$= 304\ 500\ \text{kJ or } 304.5\ \text{MJ}.$$

Example 7.8

Estimate the heat transfer necessary to increase the pressure of 50% quality steam from 200 kPa abs to 800 kPa abs if the volume is kept constant.
Solution. The first law, with $W = 0$ for a constant volume process, is

$$q = u_2 - u_1.$$

To find state 2 we must use $v_1 = v_2$. At state 1 we have

$$v_1 = v_f + x(v_g - v_f)$$
$$= 0.00106 + 0.5(0.8857 - 0.00106) = 0.4434 \text{ m}^3/\text{kg},$$

$$u_1 = u_f + x(u_g - u_f)$$
$$= 504.5 + 0.5(2529.5 - 504.5) = 1517 \text{ kJ/kg}.$$

At state 2, $P_2 = 0.8$ MPa and $v_2 = 0.4434$ m^3/kg (note: state 2 is superheat) so that at this superheated state $u_2 = 3126$ kJ/kg. Hence,

$$q = 3126 - 1517 = 1609 \text{ kJ/kg}.$$

Note that we have used q rather than Q since mass was not specified.

Example 7.9

Calculate the work done by a piston if the 2 m^3 volume of air is tripled while the temperature is maintained at 40°C. The initial pressure is 400 kPa abs.

Solution. The mass is needed in order to use Equation 7.2.18 to find the work; it is, using $R = \overline{R}/M = 8.314/28.97 = 0.287 \text{ kJ/kg}\cdot\text{K}$,

$$m = \frac{PV}{RT}$$
$$= \frac{400 \times 2}{0.287 \times 313} = 8.91 \text{ kg}, \qquad R = \frac{8.314}{28.97} = 0.287.$$

The work is then found to be

$$W = mRT \ln v_2/v_1$$
$$= 8.91 \times 0.287 \times 313 \ln 3 = 879 \text{ kJ}.$$

Note: The temperature is expressed as 40 + 273 = 313 K.

Example 7.10

How much work is necessary to compress air in an insulated cylinder from 0.2 m^3 to 0.01 m^3? Use $T_1 = 20°\text{C}$ and $P_1 = 100$ kPa abs.

Solution. For an adiabatic process $Q = 0$ so that the first law is

$$-W = m(u_2 - u_1)$$
$$= mc_v(T_2 - T_1).$$

To find the mass m we use the ideal gas equation with

$$R = \overline{R}/M = 8.314/28.97 = 0.287 \text{ kJ/kg}\cdot\text{K},:$$

$$m = \frac{PV}{RT}$$
$$= \frac{100 \times 0.2}{0.287 \times 293} = 0.2378 \text{ kg}.$$

The temperature T_2 is found to be

$$T_2 = T_1(v_1 / v_2)^{k-1}$$
$$= 293(0.2 / 0.01)^{0.4} = 971.1 \text{ K}.$$

The work is then

$$W = 0.2378 \times 0.716(971.1 - 293) = 115.5 \text{ kJ}.$$

Example 7.11

A 10-cm-thick wall made of pine wood is 3 m high and 10 m long. Calculate the heat transfer rate if the temperature is 25°C on the inside and –20°C on the outside. Neglect convection.

Solution. The heat transfer occurs due to conduction. Using *k* from Table 7.5, Eq. 7.2.5 provides

$$\dot{Q} = KA\Delta T/L$$
$$= 0.15 \times (3 \times 10) \times [25 - (-20)]/0.1 = 2025 \text{ J / s}.$$

Example 7.12

The surface of the glass in a $1.2 \text{ m} \times 0.8 \text{ m}$ skylight is maintained at 20°C. If the air temperature is –20°C, estimate the rate of heat loss from the window. Use $h = 12 \text{ J / s} \cdot \text{m}^2 \cdot {}^\circ\text{C}$.

Solution. The convective heat transfer coefficient depends on several parameters so, as usual, it is specified. Using Equation 7.2.6 the rate of heat loss is

$$\dot{Q} = hA\Delta T$$
$$= 12 \times (1.2 \times 0.8) \times [20 - (-20)] = 461 \text{ J / s}.$$

Example 7.13

A 2-cm-diameter heating oven is maintained at 1000°C and the oven walls are at 500°C. If the emissivity of the element is 0.85, estimate the rate of heat loss from the 2-m-long element.

Solution. The heat loss will be primarily due to radiation. Neglecting any convection loss, using $F_{1-2} = 1$ since the oven encloses the element, and Eq. 7.2.7 provides us with

$$\dot{Q} = \sigma \varepsilon A(T_1^{\ 4} - T_2^{\ 4})$$
$$= 5.67 \times 10^{-11} \times 0.85 \times (\pi \times 0.02 \times 2)(1273^4 - 773^4) = 13.7 \text{ kJ / s}.$$

Note that absolute temperature must be used. Also, the area *A* is the surface area of the cylinder.

Example 7.14

Estimate the rate of heat loss from a $3 \text{ m} \times 10 \text{ m}$ wall if it is composed of 2.5 cm of pine wood, 8 cm of wool insulation, and 1.5 cm of plaster. The room is at 22°C. and the outside air is at –20°C. Also, calculate the overall heat transfer coefficient.

Solution. The resistance coefficient is *L/k* for a material layer and 1/h for an air layer. Using Equation 7.2.8, and values from Table 7.5, we have

$$\dot{Q} = A\Delta T/(R_i + R_1 + R_2 + R_3 + R_o)$$

$$= A\Delta T \Big/ \left(\frac{1}{h_i} + \frac{L_1}{k_1} + \frac{L_2}{k_2} + \frac{L_3}{k_3} + \frac{1}{h_o}\right)$$

$$= \frac{(3 \times 10)[22 - (-20)]}{\frac{1}{12} + \frac{0.025}{0.15} + \frac{0.08}{0.038} + \frac{0.015}{0.81} + \frac{1}{12}} = 513 \text{ J/s}.$$

The overall heat transfer coefficient is found as follows:

$$\dot{Q} = UA\Delta T = U \times 30 \times 42 = 513.$$

$$\therefore U = 0.407 \text{ J/s} \cdot \text{m}^2 \cdot {}^\circ\text{C}.$$

7.3 The First Law of Thermodynamics for a Control Volume

The continuity equation, which accounts for the conservation of mass, may be used in certain situations involving control volumes. It is stated as

$$\dot{m} = \rho_1 A_1 V_1 = \rho_2 A_2 V_2 \tag{7.3.1}$$

where, in control volume formulations, V is the velocity; $\dot{m}$ is called the *mass flux*. In the above continuity equation, we assume *steady flow*, that is, the variables are independent of time. For such steady state flow situations, the first law takes the form:

$$\frac{\dot{Q} - \dot{W}_S}{\dot{m}} = \frac{V_2^2 - V_1^2}{2} + h_2 - h_1 + g(z_2 - z_1) \tag{7.3.2}$$

where the dot signifies a rate so that $\dot{Q}$ and $\dot{W}_S$ have the units of kJ/s. In most devices the potential energy is negligible. Also, the kinetic energy change can often be ignored (but if sufficient information is given, it should be included) so that the first law is most often used in the simplified form

$$\dot{Q} - \dot{W}_S = \dot{m}(h_2 - h_1) \qquad \text{or} \qquad q - w_S = h_2 - h_1. \tag{7.3.3}$$

Particular devices are of special interest. The energy equation for a *valve* or a *throttle plate* is simply

$$h_2 = h_1 \tag{7.3.4}$$

providing kinetic energy can be neglected.

For a *turbine* expanding a gas, the heat transfer is negligible so that

$$\dot{W}_T = \dot{m}(h_1 - h_2) \quad \text{or} \quad w_T = h_1 - h_2. \tag{7.3.5}$$

The work input to a *gas compressor* with negligible heat transfer is

$$\dot{W}_C = \dot{m}(h_2 - h_1) \quad \text{or} \quad w_C = h_2 - h_1. \tag{7.3.6}$$

A *boiler* and a *condenser* are simply heat transfer devises. The first law then simplifies to

$$\dot{Q} = \dot{m}(h_2 - h_1) \quad \text{or} \quad q = h_2 - h_1. \tag{7.3.7}$$

For a *nozzle* or a *diffuser* there is no work or heat transfer; we must include, however, the kinetic energy change resulting in

$$0 = \frac{V_2^2 - V_1^2}{2} + h_2 - h_1. \tag{7.3.8}$$

For a *pump* or a *hydroturbine* we take a slightly different approach. We return to Equation 7.3.2 and write it using $v = 1/\rho$, and with Equation 7.2.10, as

$$\frac{\dot{Q} - \dot{W}_S}{\dot{m}} = \frac{V_2^2 - V_1^2}{2} + u_2 - u_1 + \frac{P_2 - P_1}{\rho} + g(z_2 - z_1). \tag{7.3.9}$$

For an ideal situation we do not transfer heat. Neglecting kinetic and potential energy changes we find that

$$-\dot{W}_S = \dot{m}\frac{P_2 - P_1}{\rho}. \tag{7.3.10}$$

This would provide the minimum pump power requirement or the maximum turbine power output. The inclusion of an efficiency would increase the pump power requirement or decrease the turbine output.

A gas turbine or compressor efficiency is based on an isentropic process ($s_2 = s_1$) as the ideal process. For a gas turbine or a compressor we have

$$\eta_T = \frac{\dot{W}_a}{\dot{W}_s} \qquad \eta_C = \frac{\dot{W}_s}{\dot{W}_a} \tag{7.3.11}$$

where $\dot{W}_a$ is the actual power and $\dot{W}_s$ is the power assuming an isentropic process.

Example 7.15

Refrigerant-134a expands through a valve from a state of saturated liquid at 800 kPa abs to a pressure of 100 kPa abs. What is the final quality?

Solution. The first law states that

$$h_1 = h_2.$$

Using Fig. 7.6 (at the end of the chapter) for Refrigerant-134a we find, using $h_1 = h_f$,

$$h_1 = 240 \text{ kJ/kg}.$$

There follows at $P_2 = 100$ kPa, using $h_2 = 240$,

$$x = 0.35 \quad \text{or} \quad 35\%.$$

Example 7.16

Steam expands through a turbine from 6 MPa abs, at 600°C to 2 kPa abs with $x_2 = 1.0$. Find the work output.

Solution. The first law gives

$$\dot{w}_T = h_1 - h_2$$
$$= 3658.4 - 2533.5 = 1125 \text{ kJ/kg},$$

where h_2 is h_g at $P_2 = 0.002$ MPa, as given in Table 7.3.2.

Example 7.17

What is the turbine efficiency in Example 7.16?

Solution. The turbine efficiency is based on an isentropic process. Let state 2′ be at 2 kPa abs with

$$s_{2'} = s_1 = 7.1685 \text{ kJ / kg} \cdot \text{K}.$$

At 0.002 MPa abs we find, from Table 7.3.2

$$s_{2'} = s_f + x_{2'} s_{fg}.$$
$$7.1685 = 0.2606 + x_{2'}(8.4639).$$
$$\therefore x_{2'} = 0.816.$$

At this ideal state we find

$$h_{2'} = h_f + x_{2'} h_{fg}$$
$$= 73.5 + 0.816 \times 2460 = 2080 \text{ kJ / kg}.$$

Finally, we use $\dot{W}_a$ as the answer in Example 7.16 and obtain

$$\eta_T = \frac{\dot{W}_a}{\dot{W}_s} = \frac{\dot{W}_a}{\dot{m}(h_1 - h_{2'})}$$
$$= \frac{11\,250}{10(3658.4 - 2080)} = 0.713 \text{ or } 71.3\%.$$

Example 7.18

What is the minimum power requirement of a pump that is to increase the pressure from 2 kPa abs to 6 MPa abs for a mass flux of 10 kg/s of water?

Solution. With a liquid we let $h_2 - h_1 = u_2 - u_1 + (P_2 - P_1)v$ since $v = \text{const.}$ We let $u_2 - u_1 = 0$ so that the first law simplifies to

$$\dot{W}_P = \dot{m}\frac{P_2 - P_1}{\rho}$$
$$= 10\frac{6000 - 2}{1000} = 59.98 \text{ kW},$$

Note how small this is relative to the power output of the turbine of example 7.16 operating between the same pressures. Because this is less than 1% of the turbine output, the pump work may usually be neglected in the analysis of a cycle involving a turbine and a pump.

Example 7.19

A nozzle accelerates air from 100 m/s, 400°C and 400 kPa abs to a receiver where P = 20 kPa abs. Assuming an isentropic process, find V_2.

Solution. The energy equation takes the form

$$0=\frac{V_2^{\,2}-V_1^{\,2}}{2}+h_2-h_1.$$

Assuming air to be an ideal gas with constant c_p we have

$$c_p(T_1-T_2)=\frac{V_2^{\,2}-V_1^{\,2}}{2}.$$

We can find T_2 from Equation 7.2.31 to be

$$T_2=T_1(P_2\,/\,P_1)^{k-1/k}$$

$$=673(20\,/\,400)^{0.4/1.4}=286\text{ K}.$$

The exiting velocity is found as follows:

$$1000(673-286)=\frac{V_2^{\,2}-100^2}{2}.$$

$$\therefore\; V_2=874\text{ m/s}\,.$$

Note: c_p must be used as $1000\text{ J}/\text{kg}\cdot\text{K}$ so that the units are consistent.

7.4 The Second Law of Thermodynamics

The two scientific statements of the second law of thermodynamics can be shown to be equivalent. They are stated and shown schematically in Figure 7.2.

Clausius Statement—A device which operates in a cycle cannot transfer heat from a cooler body to a hotter body without a work input.

Kelvin-Planck Statement—A device which operates in a cycle cannot produce work while exchanging heat at a single temperature.

Figure 7.3 shows an engine (it produces work) and a refrigerator (it transfers heat from a body at low temperature) that satisfy the second law. The devices of Figure 7.2 do not violate the first law—energy is conserved—however, they represent impossibilities, violations of the second law.

To write a mathematical statement of the second law, we use entropy, defined by $dS=\delta Q/T$. The net entropy change during any process is given by

$$\Delta S_{universe}=\Delta S_{system}+\Delta S_{surroundings}\geq 0 \tag{7.4.1}$$

The equal sign applies to a reversible process, the greater-than sign applies to an irreversible (or real) process. Losses, friction, unrestrained expansion, and heat transfer across a finite temperature difference all lead to irreversibilities. Entropy changes can be found using tables, or for an ideal gas with constant specific heats we can use

$$\Delta s = c_p \ln\frac{T_2}{T_1} - R\ln\frac{P_2}{P_1} \tag{7.4.2}$$

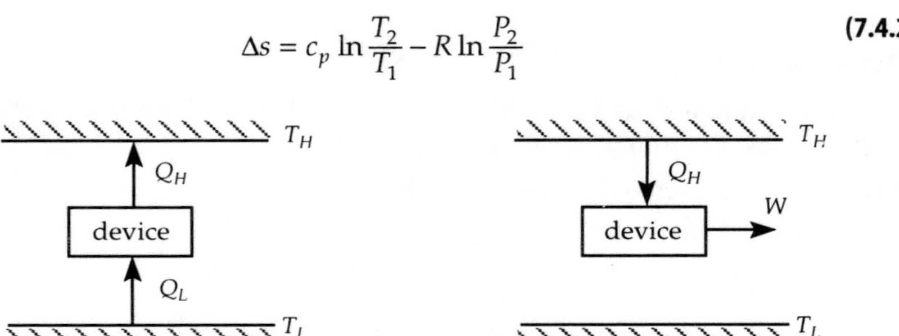

Figure 7.2 Violations of the second law.

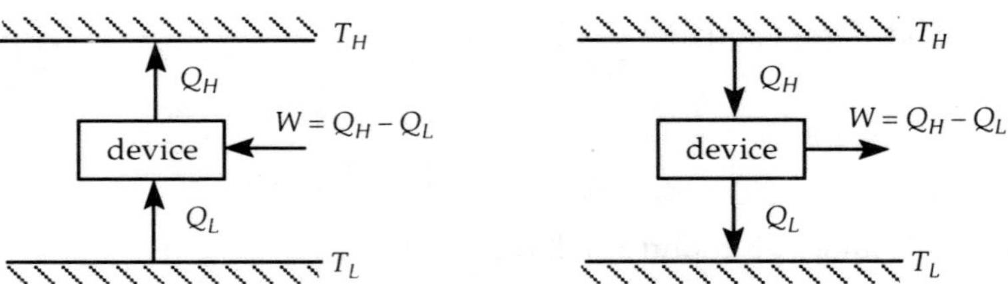

Figure 7.3 Devices that satisfy the second law.

For constant temperature processes, such as heat transfer to a reservoir, we use

$$\Delta S = \frac{Q}{T} \tag{7.4.3}$$

For a solid or a liquid we use

$$\Delta S = mc\ln\frac{T_2}{T_1} \tag{7.4.4}$$

where $c = c_p$ found in Table 7.2.

In general, for a reversible process, we have, from the definition of entropy,

$$Q = \int TdS \tag{7.4.5}$$

which is analogous to $W = \int PdV$ for such a process.

A Carnot engine or a refrigerator is a fictitious device that operates with reversible processes. It provides us with the maximum possible efficiency of an engine in terms of temperatures:

$$\eta = \frac{W_{out}}{Q_{in}}. \qquad \eta_{\max} = 1 - \frac{T_L}{T_H} \tag{7.4.6}$$

It provides the maximum possible *coefficient of performance*, COP, of a refrigerator,

$$\text{COP} = \frac{Q_L}{W_{in}}. \qquad \text{COP}_{\max} = \frac{1}{T_H/T_L - 1} \tag{7.4.7}$$

or, the upper limit for the COP of a heat pump,

$$\text{COP} = \frac{Q_H}{W_{in}}. \qquad \text{COP}_{\max} = \frac{1}{1 - T_L/T_H} \tag{7.4.8}$$

T_H is the temperature of the high temperature reservoir and T_L is the temperature of the low temperature reservoir.

The vapor power cycle is the basic cycle for most power plants; it is sketched in Figure 7.4a. The vapor refrigeration cycle is shown in Figure 7.4b.

The Carnot cycle is sketched in Figure 7.5 along with some other common gas cycles. The efficiency of each of the other cycles is less than that of Carnot cycle, primarily due to the transfer of heat across a finite temperature difference.

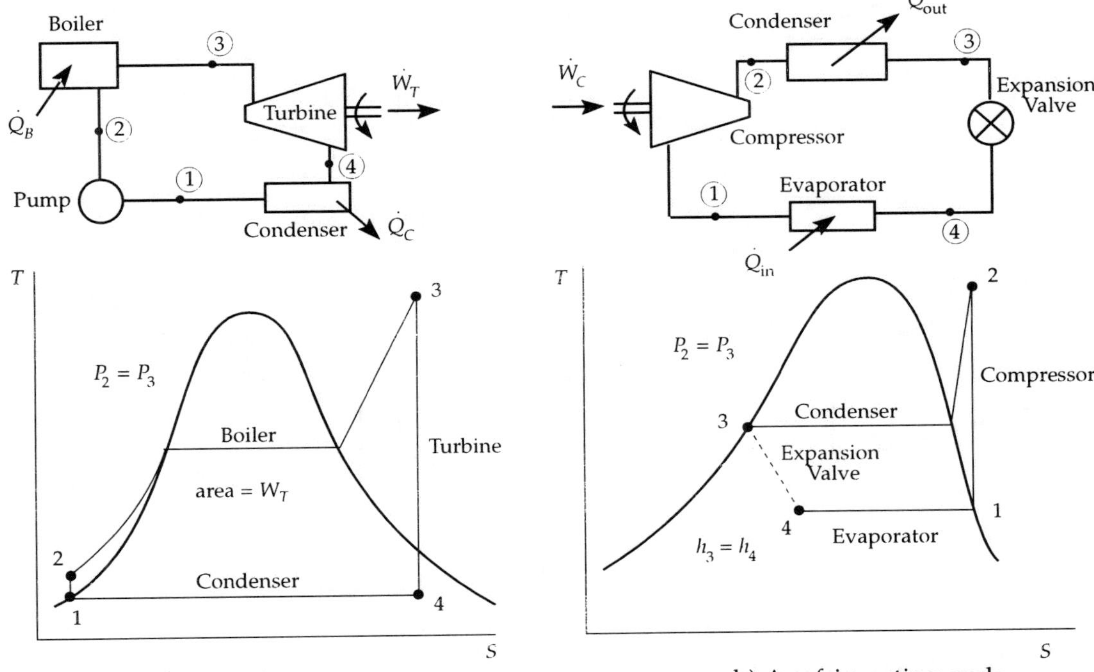

a) A Rankine cycle. b) A refrigeration cycle.

Figure 7.4 Vapor cycles.

Note that all of the above cycles are ideal cycles; the entropy is assumed constant in two processes in each cycle (except for the refrigeration cycle). Actual processes deviate from these ideal processes resulting in lower cycle efficiencies than predicted by the above.

To determine an expression for the maximum work output of a steady flow device, use the first law, neglecting kinetic and potential energy changes, in the form:

$$\dot{Q} - \dot{W}_s = \dot{m}(h_2 - h_1) \quad \textbf{(7.4.9)}$$

If we assume heat is transferred to the surroundings at atmospheric temperature T_a we can relate

$$\dot{m}(s_2 - s_1) = \dot{Q}/T_o. \quad \textbf{(7.4.10)}$$

Substituting into the Equation 7.4.9 we have

$$\begin{aligned} \dot{W}_{max} &= \dot{m}(h_1 - T_o s_1) - \dot{m}(h_2 - T_o s_2) \\ &= \dot{m}(\phi_1 - \phi_2) \end{aligned} \quad \textbf{(7.4.11)}$$

where ϕ is the *availability*. Hence, the maximum work output is the change in the availability.

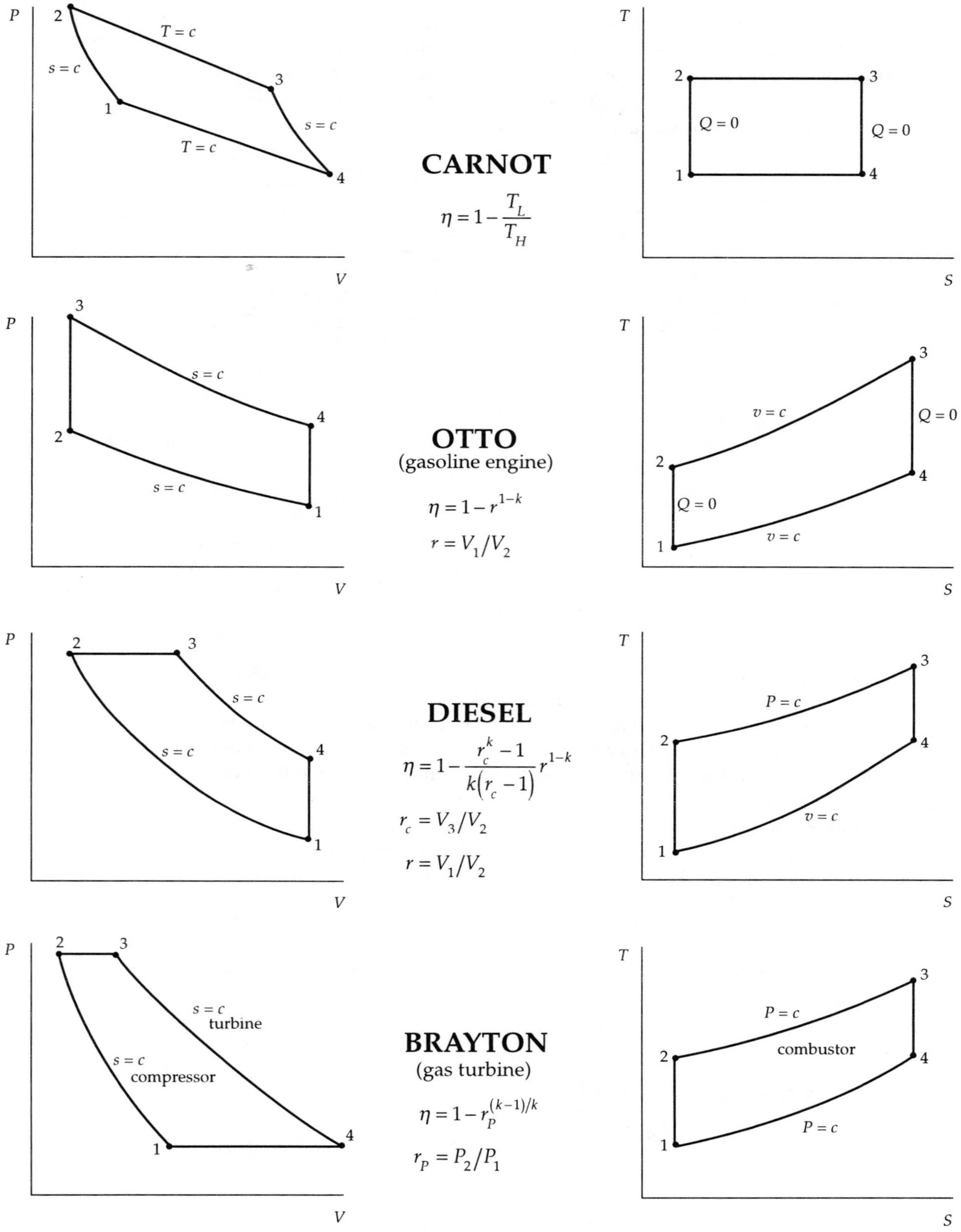

CARNOT

$$\eta = 1 - \frac{T_L}{T_H}$$

OTTO
(gasoline engine)

$$\eta = 1 - r^{1-k}$$

$$r = V_1/V_2$$

DIESEL

$$\eta = 1 - \frac{r_c^k - 1}{k(r_c - 1)} r^{1-k}$$

$$r_c = V_3/V_2$$

$$r = V_1/V_2$$

BRAYTON
(gas turbine)

$$\eta = 1 - r_P^{(k-1)/k}$$

$$r_P = P_2/P_1$$

Figure 7.5 Common thermodynamic gas cycles.

Example 7.20

Ten kilograms of ice at 0°C are melted in 100 kg of water initially at 25°C. Calculate the final temperature and the entropy change. Assume no heat transfer to the surroundings.

Solution. The first law is applied to the ice-water system:

$$Q_{gain} = Q_{lost}$$

$$10 \times 320 + 10(T - 0) \times 4.18 = 100(25 - T) \times 4.18.$$

$$\therefore T = 15.8^\circ \text{C}.$$

The entropy change is found as follows:

$$\text{ice}: \quad \Delta S = \frac{Q}{T_1} + mc \ln \frac{T_2}{T_1}$$

$$= \frac{10 \times 320}{273} + 10 \times 4.18 \ln \frac{288.8}{273} = 14.073 \text{ kJ/K}.$$

$$\text{water}: \quad \Delta S = mc \ln \frac{T_2}{T_1}$$

$$= 100 \times 4.18 \ln \frac{288.8}{298} = -13.107 \text{ kJ/K}.$$

The net entropy change is

$$\Delta S_{net} = 14.073 - 13.107 = 0.966 \text{ kJ / K}.$$

This is positive, as required by the second law.

Example 7.21

An inventor claims to have invented an engine, using a 160°C geothermal heat source, which operates with an efficiency of 30%. If it exhausts to the 20°C atmosphere, is the invention possible?

Solution. The maximum possible efficiency, as limited by the second law, is given by

$$\eta_{\max} = 1 - \frac{T_L}{T_H}$$

$$= 1 - \frac{293}{433} = 0.323 \text{ or } 32.3\%.$$

The proposal is a possibility. However, the proposed efficiency is quite close to the maximum efficiency. It would be extremely difficult to obtain the 30% because of the losses due to heat transfer across a finite temperature difference and friction.

Example 7.22

A heat pump delivers 20 000 kJ /hr of heat with a 1.39 kW input. Calculate the COP.

Solution. Using the definition of the COP we find

$$\text{COP} = \frac{\dot{Q}_H}{\dot{W}_{in}}$$

$$= \frac{20\ 000 / 3600}{1.39} = 4.00.$$

Note, the factor 3600 converts hours into seconds.

Example 7.23

Compare the efficiency of an Otto cycle operating on an 8 to 1 compression ratio ($r = 8$) with a Diesel cycle that has a 20 to 1 compression ratio and a cut-off ratio of 2 to 1 ($r_c = 2$). Use air.

Solution. The efficiency of an Otto cycle (see Figure 7.5) is

$$\eta = 1 - r^{1-k}$$

$$= 1 - 8^{-0.4} = 0.565 \quad \text{or} \quad 56.5\%$$

where air is assumed to be the working fluid.

The efficiency of the Diesel cycle is

$$\eta = 1 - \frac{r_c{}^k - 1}{k(r_c - 1)} r^{1-k}$$

$$= 1 - \frac{2^{1.4} - 1}{1.4(2-1)} 20^{-0.4} = 0.647 \quad \text{or} \quad 64.7\%.$$

The efficiency of the Diesel cycle is higher than that of the Otto cycle because it operates at a higher compression ratio. If the Otto cycle could operate at $r = 20$, its efficiency would be greater than that of the Diesel.

Example 7.24

If a power plant operates on simple Rankine cycle using water between 600°C, 6 MPa abs and a low pressure of 10 kPa, calculate $\eta_{\max}$.

Solution. Referring to Figure 7.4a, we define the efficiency to be

$$\eta_{\max} = \frac{\dot{W}_T}{\dot{Q}_B}.$$

The pump work is neglected (see Example 7.18). To find $\dot{W}_T$ we must find h_4. This is accomplished using $P_3 = 6$ MPa and $T_3 = 600\,°\text{C}$, as follows:

$$s_4 = s_3 = 7.168 \text{ kJ/kg} \cdot \text{K}.$$

$$\text{At } P = 10 \text{ kPa}: \quad 7.168 = 0.6491 + x_4(7.5019).$$

$$\therefore \; x_4 = 0.869.$$

$$\therefore \; h_4 = 191.8 + 0.869 \times 2392.8 = 2271 \text{ kJ/kg}.$$

The turbine output is then (assuming $\dot{m} = 1 \text{ kg/s}$ since it is not given; alternatively we could le t $w_T = h_3 - h_4$)

$$\dot{W}_T = \dot{m}(h_3 - h_4)$$
$$= 1 \times (3658.4 - 2271) = 1387 \text{ kW}.$$

The energy input occurs in the boiler. It is

$$\dot{Q}_B = \dot{m}(h_3 - h_2)$$
$$= 1 \times (3658.4 - 191.8) = 3467 \text{ kW}.$$

Note: Be careful finding h_2. Remember, in a liquid use h_f at the temperature of the liquid, ignoring the pressure; hence, we use h_f at the $T_1 \cong T_2 = 45.8°\text{C}$ (find it in Table 7.3.2).

Finally, the efficiency of this idealized cycle is

$$\eta_{\max} = \frac{1387}{3467} = 0.400 \text{ or } 40.0\%.$$

Example 7.25

A refrigeration system, using Refrigerant-134a, operates between –20°C and 40°C. What is the maximum possible COP?

Solution. The refrigeration effect takes place in the evaporator. Hence, referring to Figure 7.4, the COP is defined as

$$\text{COP} = \frac{\dot{Q}_{in}}{\dot{W}_C}.$$

To find $\dot{W}_C$ we must locate state 2. This is done by following the constant entropy line $(s_1 = s_2 \cong 1.74)$ in Fig. 7.6 at the end of this chapter. Locating states 1 and 2 we find

$$h_1 \cong 385 \text{ kJ/kg}$$
$$h_2 \cong 425 \text{ kJ/kg}$$

The compressor work is, assuming $\dot{m} = 1 \text{ kg/s}$,

$$\dot{W}_C = \dot{m}(h_2 - h_1)$$
$$= 1 \times (425 - 385) = 40 \text{ kW}.$$

To find $\dot{Q}_{in}$ we recognize that $h_4 = h_3 = 255$ kJ/kg, using Fig. 7.6. Thus we find,

$$\dot{Q}_{in} = \dot{m}(h_1 - h_4)$$
$$= 1 \times (385 - 255) = 130 \text{ kW}.$$

The maximum COP for this idealized cycle is

$$\text{COP} = \frac{130}{40} = 3.25.$$

Example 7.26

The maximum pressure in an Otto cycle is 8 MPa. Air is compressed from 85 kPa and 22°C with a compression ratio of 8. What is the required heat addition?
Solution. Referring to the diagrams of the Otto cycle

$$P_1 = 85 \text{ kPa} \;, \quad T_1 = 22°\text{C}. \qquad \therefore v_1 = \frac{RT_1}{P_1} = \frac{0.287 \times 295}{85} = 0.9961 \text{ m}^3/\text{kg}.$$

Since the compression ratio is a volume ratio we have

$$v_3 = v_2 = v_1/8 = 0.9961/8 = 0.1245 \text{ m}^3/\text{kg}.$$

We know that $P_3 = 8$ MPa. Thus

$$T_3 = \frac{P_3 v_3}{R} = \frac{8000 \times 0.1245}{0.287} = 3471 \text{ K}$$

The isentropic process from 1 to 2 allows us to find T_2:

$$T_2 = T_1 \left(\frac{v_1}{v_2}\right)^{k-1} = 295 \times 8^{0.4} = 677.7 \text{ K}.$$

Finally, since the process from 2 to 3 is at constant volume with a system,

$$q_{2-3} = c_v(T_3 - T_2) = 0.717(3471 - 677.7) = 2003 \text{ kJ/kg}.$$

Example 7.27

Steam at 200°C and 200 kPa abs is available to produce work by expanding it to the atmosphere at 20°C and 100 kPa abs. What is $\dot{W}_{max}$ if $\dot{m} = 2$ kg / s?
Solution. We will use the equation

$$\dot{W}_{max} = \dot{m}(\phi_1 - \phi_2).$$

The availabilities are found to be (using Table 7.3)

$$\phi_1 = h_1 - T_0 s_1$$
$$= 2870.5 - 293 \times 7.5074 = 670.8 \text{ kJ / kg}.$$
$$\phi_2 = h_2 - T_0 s_2$$
$$= 83.9 - 293 \times 0.2965 = -3.0 \text{ kJ / kg}.$$

Note: State 2 is at 20°C and 100 kPa abs. This is the liquid state so we use $h_2 = h_f$ and $s_2 = s_f$ at 20°C (see Table 7.3.1), and simply ignore the pressure. Finally,

$$\dot{W}_{max} = 2[670.8 - (-3.0)] = 1348 \text{ kW}.$$

7.5 Psychrometrics

Air is primarily a mixture of nitrogen, oxygen, argon and water vapor. When water vapor is absent, it is *dry air*. In most problems water vapor does not influence calculations, e.g., the problems considered earlier in this chapter. It must, however, be accounted for in problems involving combustion and air conditioning. Here we will present the primary quantities when considering *atmospheric air*, air that contains water vapor. We will make our calculations using the *psychrometric chart* , found in the NCEES Handbook.

The atmospheric pressure P is the sum of the partial pressure P_a of dry air and the partial pressure P_v of the water vapor:

$$P = P_a + P_v . \tag{7.5.1}$$

Relative humidity ϕ is

$$\phi = \frac{m_v}{m_g} = \frac{P_v}{P_g} \tag{7.5.2}$$

where P_g is the saturation pressure at temperature of the air (found in Table 7.3.1, and m_g is the maximum mass of water vapor that the air can hold at air temperature, usually called the *dry-bulb temperature* T_{db}. The *specific humidity* (or *humidity ratio*) ω is

$$\omega = \frac{m_v}{m_a} = 0.622\frac{P_v}{P_a} \tag{7.5.3}$$

The enthalpy of the atmospheric air is

$$h = h_a + \omega h_v \tag{7.5.4}$$

The *dew-point temperature* T_{dp} is the temperature at which condensation begins. The *wet-bulb temperature* T_{wb} is the temperature that a thermometer would read if a saturated wick were attached to its bulb and whirled around through the air. On the T-s diagram shown here, $T_1 = T_2 = T_{db}$, $T_3 = T_{dp}$, and T_{wb} would be between T_2 and T_3.

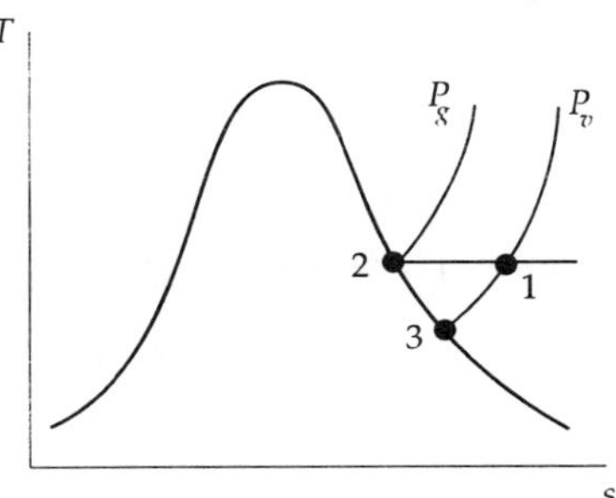

The quantities T_{db}, T_{dp}, T_{wb}, ω, ϕ, and h can be read directly from the psychrometric chart so that the above equations are not needed in most air conditioning problems. Consider a state A, on the psychrometric chart shown, located by specifying the dry-bulb temperature T_{db} and the relative humidity ϕ. We would read T_{wb} at 1, T_{dp} at 2, h at 3, and ω at 4.

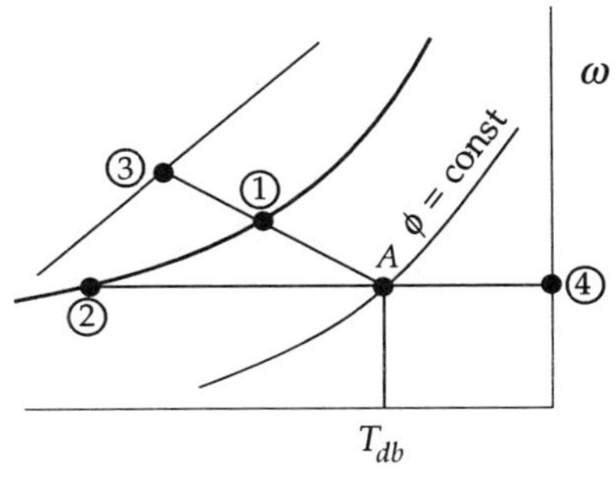

Example 7.28

A thermometer with a wet cloth attached to its bulb reads 20°C when air is blown around it. If the atmospheric air is 33°C, what is the relative humidity and dew-point temperature? How much water could be condensed out of a 100-m^3 volume?

Solution. From the psychrometric chart we read

$$\phi = 30\% \quad \text{and} \quad T_{dp} = 13.2°\text{C}. \quad \text{Also,} \quad \omega = 0.0093 \text{ kg water/kg air}.$$

The amount of water that could be condensed out is

$$\omega \rho_{\text{air}} V_{\text{air}} = 0.0093 \times \frac{100}{0.287 \times 3.6} \times 100 = 1.06 \text{ kg} \quad \text{or} \quad 1.06 \text{ Liters}.$$

Example 7.29

It is desired to condition 35°C, 80% humidity air to 24°C and 50% humidity. If 100 m^3/min of air is to be conditioned, how much energy is required in the cooling process, and how much in the heating process?

Solution. To decrease the moisture content in the air we first cool the air from 1 to 2 at constant ω and then remove water by cooling it further to 3 (along the 100% humidity line). In this process

$$\dot{Q}_{\text{cool}} = \dot{m}\Delta h = \frac{100}{60}\rho_a(h_3 - h_1)$$
$$= \frac{100}{60}\frac{1}{0.913}(105 - 37) = 124 \text{ kJ/s}.$$

Heating takes place from 3 to 4 at constant ω:

$$\dot{Q}_{\text{heat}} = \dot{m}(h_4 - h_3)$$
$$= \frac{100}{60}\frac{1}{0.823}(48 - 37) = 18.3 \text{ kJ/s}.$$

We obtained $v = 1/\rho$ and the above properties from the psychrometric chart found in the NCEES Handbook.

Practice Problems (Metric Units)

(If you choose to work only a few problems, select those with a star.)

GENERAL

*7.1 Which of the following would be considered a system rather than a control volume?

a) a pump b) a tire c) a pressure cooker
d) a turbine e) a jet nozzle

*7.2 Which of the following is an extensive property?

a) Temperature b) Velocity c) Pressure d) Mass e) Stress

7.3 An automobile heats up while sitting in a parking lot on a sunny day. The process can be assumed to be:

a) isothermal b) isobaric c) isometric d) isentropic e) adiabatic

*7.4 If a quasi-equilibrium process exists, we have assumed

a) the pressure at any instant to be everywhere constant.
b) an isothermal process.
c) the heat transfer to be small.
d) the boundary motion to be infinitesimally small.
e) that no friction exists.

*7.5 A scientific law is a statement that

a) we postulate to be true.
b) is generally observed to be true.
c) is derived from a mathematical theorem.
d) is agreed upon by scientific community.
e) is a summary of experimental observation.

DENSITY, PRESSURE, AND TEMPERATURE

*7.6 The density of air at vacuum of 40 kPa and –40°C is, in kg/m^3,

a) 0.598 b) 0.638 c) 0.697 d) 0.753 e) 0.897

7.7 The specific volume of water is 0.5 m^3/kg at a pressure of 200 kPa abs. Find the quality.

a) 0.623 b) 0.564 c) 0.478 d) 0.423 e) 0.356

7.8 The specific volume of steam at 4 MPa abs and 1200°C, in m^3/kg, is

a) 0.20 b) 0.19 c) 0.18 d) 0.17 e) 0.16

7.9 There are 20 kg of steam contained in 2 m^3 at 4 MPa. What is the temperature in °C?

a) 600 b) 610 c) 620 d) 630 e) 640

7.10 The pressure in a cylinder containing water is 200 kPa and the temperature is 115°C. What state is it in?

a) compressed liquid b) saturated liquid c) liquid/vapor mixture
d) saturated vapor e) superheated vapor

*7.11 A cold tire has a volume of 0.03 m^3 at –10°C and 180 kPa gage. If the pressure and temperature increase to 210 kPa gage and 30°C, find the final volume in m^3.

a) 0.0304 b) 0.0308 c) 0.0312 d) 0.0316 e) 0.0320

THE FIRST LAW FOR A SYSTEM

*7.12 A 300-watt light bulb provides energy in a 10-m-dia spherical space. If the outside temperature is 20°C, find the inside steady-state temperature, in °C, if $R = 1.5\ \text{hr} \cdot \text{m}^2 \cdot °\text{C} / \text{kJ}$ for the wall.

a) 40.5 b) 35.5 c) 32.6 d) 29.4 e) 25.2

7.13 A 10 kg mass, which is attached to a pulley and a paddle wheel submerged in water, drops 3 m. Find the subsequent heat transfer, in joules, needed to return the temperature of the water to its original value.

a) –30 b) –95 c) –126 d) –195 e) –294

*7.14 A cycle undergoes the following processes. All units are kJ. Find E_{after} for the process $1 \rightarrow 2$.

	Q	W	ΔE	E_{before}	E_{after}
$1 \rightarrow 2$	20	5		10	
$2 \rightarrow 3$		–5	5		
$3 \rightarrow 1$	30			30	

a) 10 b) 15 c) 20 d) 25 e) 30

7.15 For the cycle of Problem 7.14 find W_{3-1}.

a) 50 b) 60 c) 70 d) 80 e) 90

7.16 A 2000 kg automobile traveling at 25 m/s strikes a plunger in 10 000 cm^3 of water, bringing the auto to a stop. What is the maximum temperature rise, in °C, in the water?

a) 5 b) 10 c) 15 d) 20 e) 25

7.17 Ten kilograms of –10°C ice is added to 100 kg of 20°C water. What is the eventual temperature, in °C, of the water? Assume an insulated container.

a) 9.2 b) 10.8 c) 11.4 d) 12.6 e) 13.9

*7.18 Of the following first law statements, choose the one that is wrong:

a) the net heat transfer equals the net work for a cycle.
b) the heat transfer cannot exceed the work done.
c) the heat transfer equals the work plus the energy change.
d) the heat transfer equals the energy change if no work is done.
e) the energy of an isolated system remains constant.

ISOTHERMAL PROCESS

*7.19 Determine the work, in kJ, necessary to compress 2 kg of air from 100 kPa abs to 4000 kPa abs if the temperature is held constant 300°C.

a) –1210 b) –1105 c) –932 d) –812 e) –733

*7.20 Steam is compressed from 100 kPa abs to 4000 kPa abs holding the temperature constant at 300°C. What is the internal energy change, in kJ, for m = 2 kg?

a) –180 b) –170 c) –160 d) –150 e) –140

7.21 How much heat transfer, in kJ, is needed to convert 2 kg of saturated liquid water to saturated vapor if the temperature is held constant at 200°C?

a) 2380 b) 2980 c) 3210 d) 3520 e) 3880

CONSTANT PRESSURE PROCESS

7.22 There are 200 people in a 2000 m^2 room, lighted with 30 W/m^2. Estimate the maximum temperature increase, in °C, if the ventilation system fails for 20 min. Each person generates 400 kJ/h. The room is 3 m high.

a) 5.6 b) 6.8 c) 8.6 d) 11.4 e) 13.4

*7.23 Estimate the average c_p value, in kJ/kg·K, of a gas if 522 kJ of heat are necessary to raise the temperature from 300 K to 800 K holding the pressure constant.

a) 1.000 b) 1.026 c) 1.038 d) 1.044 e) 1.052

*7.24 How much heat, in kJ, must be transferred to 10 kg of air to increase the temperature from 10°C to 230°C if the pressure is maintained constant?

a) 1780 b) 1620 c) 1890 d) 2090 e) 2200

7.25 Ten kilograms of water, initially at 10°C, are heated until T = 300°C. Estimate the heat transfer, in MJ, if the pressure is held constant at 200 kPa abs.

a) 29.2 b) 29.7 c) 30.3 d) 30.9 e) 31.3

7.26 Calculate the work done, in MJ, in Problem 7.25.

a) 2.63 b) 4.72 c) 8.96 d) 11.4 e) 15.6

CONSTANT VOLUME PROCESS

7.27 How much heat, in kJ, must be added to a rigid volume, containing 2 kg of water/vapor mixture with x = 0.5, to increase the temperature from 200°C to 500°C?

a) 2730 b) 2620 c) 2510 d) 2390 e) 2250

*7.28 A tire is pressurized to 100 kPa gauge in Michigan where T = 0°C. In Arizona the tire is at 70°C. Assuming a rigid tire, estimate the pressure in kPa gauge.

a) 120 b) 130 c) 140 d) 150 e) 160

7.29 A sealed, rigid, 10 m^3 air tank is heated by the sun from 20°C to 80°C. How much energy, in kJ, is transferred to the tank? Assume P_1 = 100 kPa abs.

a) 720 b) 680 c) 620 d) 560 e) 510

7.30 Ten kilograms of water are heated in a rigid container from 10°C to 200°C. What is the final quality if Q = 9000 kJ?

a) 0.258 b) 0.162 c) 0.093 d) 0.078 e) 0.052

ISENTROPIC PROCESS

*7.31 Air expands from an insulated cylinder from 200°C and 400 kPa abs to 20 kPa abs. Find T_2 in °C.

a) −24 b) −28 c) −51 d) −72 e) −93

*7.32 During an isentropic expansion of air, the volume triples. If the initial temperature is 200°C, find T_2 in °C.

a) 32 b) 28 c) 16 d) 8 e) −12

7.33 Superheated steam expands isentropically from 600°C and 6 MPa abs to 10 kPa abs. Find the final quality.

a) 0.79 b) 0.83 c) 0.87 d) 0.91 e) 0.95

7.34 Superheated steam expands isentropically from 600°C and 6 MPa abs to 400 kPa abs. Find T_2 in °C.

a) 220 b) 200 c) 190 d) 160 e) 140

7.35 Find the work, in kJ/kg, needed to compress air isentropically from 20°C and 100 kPa abs to 6 MPa abs.

a) –523 b) –466 c) –423 d) –392 e) –376

*7.36 During an isentropic process, which one of the following is true?

a) The temperature increases as the pressure decreases.
b) The temperature increases as the volume increases.
c) The heat transfer equals the enthalpy change.
d) The heat transfer is zero.
e) The volume decreases as the pressure decreases.

POLYTROPIC PROCESS

*7.37 Find T_2, in °C, if the pressure triples and T_1 = 10°C. Let n = 1.2.

a) 179 b) 113 c) 93 d) 79 e) 67

THE FIRST LAW FOR A CONTROL VOLUME

7.38 Steam enters a turbine in a 20-cm-dia pipe at 600°C and 6 MPa abs. It exits from a 5-cm-dia pipe at 20 kPa abs with x = 1. What is the velocity ratio V_{out}/V_{in}?

a) 1880 b) 1640 c) 1210 d) 820 e) 420

*7.39 Water enters a boiler at 60°C and 4 MPa abs. How much energy, in kJ/kg, must be added to obtain 600°C at the exit if the pressure remains constant?

a) 2340 b) 2630 c) 2970 d) 3280 e) 3420

7.40 A condenser is cooled by heating water from 20°C to 30°C. If the condenser inlets 10 kg/s of saturated water vapor at 20 kPa abs and exits saturated liquid, what is the mass flux, in kg/s, of the cooling water?

a) 640 b) 560 c) 500 d) 410 e) 350

7.41 Steam at 400°C and 4 MPa abs expands isentropically through a turbine to 10 kPa abs. Estimate the maximum work output in kJ/kg.

a) 1030 b) 1050 c) 1070 d) 1090 e) 1110

*7.42 What is the energy requirement, in kW, for a pump that is 75% efficient if it increases the pressure of 10 kg/s of water from 10 kPa to 6 MPa abs?

a) 40 b) 50 c) 60 d) 70 e) 80

7.43 A river 60 m wide and 2 m deep flows at 2 m/s. A hydro plant develops a pressure of 300 kPa gage just before the turbine. What maximum power, in MW, is possible?

a) 40 b) 48 c) 56 d) 64 e) 72

*7.44 A nozzle expands air isentropically from 400°C and 2 MPa abs to the atmosphere at 80 kPa abs. If the inlet velocity is small, what exit velocity, in m/s, can be expected?

a) 500 b) 600 c) 700 d) 800 e) 900

7.45 If the efficiency of a turbine that expands steam at 400°C and 6 MPa abs to 20 kPa abs is 87%, find the work output in kJ/kg.

a) 723 b) 891 c) 933 d) 996 e) 1123

THE SECOND LAW

*7.46 An inventor proposes to take 10 kg/s of geothermal water at 120°C and generate 4180 kW of energy by exhausting very near to the ambient temperature of 20°C. This proposal should not be supported because

a) it violates the first law.
b) it violates the second law.
c) it would be too expensive.
d) friction must be accounted for.
e) the geothermal water is not hot enough.

*7.47 The net entropy change in the universe during any real process

a) is equal to zero.
b) is positive.
c) is negative.
d) must be calculated to determine its sense.
e) is positive if $T_{system} > T_{surroundings}$.

*7.48 A Carnot engine

a) provides a fictitious model which is of little use.
b) can be experimentally modeled.
c) supplies us with the lower limit for engine efficiency.
d) operates between two constant temperature reservoirs.
e) has two reversible and two irreversible processes.

7.49 Ninety kilograms of ice at 0°C are completely melted. Find the entropy change, in kJ/K, if $T_2 = 0$°C.

a) 0 b) 45 c) 85 d) 105 e) 145

7.50 Forty kilograms of ice at 0°C are mixed with 100 kg of water at 20°C. What is the entropy change, in kJ/K?

a) 2.36 b) 2.15 c) 1.96 d) 1.53 e) 1.04

*7.51 An inventor proposes to have developed a small power plant that operates at 70% efficiency. It operates between temperature extremes of 600°C and 50°C. Your analysis shows that the maximum possible efficiency is

a) 56 b) 63 c) 67 d) 72 e) 81

POWER CYCLES

*7.52 A Carnot cycle operates on air such that $P_4 = 160$ kPa abs and $v_4 = 0.5\ \text{m}^3/\text{kg}$ (see Fig. 7.5). If 30 kJ/kg of heat are added at $T_H = 200$°C, find the work produced in kJ/kg.

a) 20.3 b) 18.2 c) 16.8 d) 14.2 e) 12.3

7.53 An Otto cycle operates with volumes of 40 cm^3 and 400 cm^3 at top dead center (TDC) and bottom dead center (BDC), respectively. If the power output is 100 kW, what is the heat input in kJ/s? Assume $k = 1.4$.

a) 166 b) 145 c) 110 d) 93 e) 60

7.54 The volumes of states 1, 2, and 3 in Fig. 7.5 of the Diesel cycle are 450 cm^3, 25 cm^3 and 45 cm^3, respectively. If the power produced is 120 kW, what is the required heat input in kJ/s? Assume that $k = 1.4$.

a) 187 b) 172 c) 157 d) 146 e) 132

7.55 A simple Rankine cycle operates between superheated steam at 600°C and 6 MPa abs entering the turbine, and 10 kPa abs entering the pump. What is the maximum possible efficiency, as a percent?

a) 30 b) 35 c) 40 d) 45 e) 50

7.56 The water vapor that expands from 600°C and 6 MPa abs in a turbine of a Rankine cycle is intercepted at 200 kPa abs and reheated at constant pressure to 600°C in the boiler, after which it is re-injected in the turbine and expanded to 10 kPa abs. (This is the *reheat cycle.*) Calculate the maximum possible efficiency, as a percent?

a) 32 b) 37 c) 41 d) 45 e) 49

7.57 The primary effect of reheating, as illustrated in Problem 7.56 (compare with the result in Problem 7.55), is to

a) increase the efficiency.
b) decrease the pump work.
c) decrease the heat requirement.
d) eliminate unnecessary piping.
e) decrease or eliminate moisture condensation in the turbine.

REFRIGERATION CYCLES

7.58 A refrigeration cycle operates with Refrigerant-134a between 100 kPa and 1000 kPa. What is the maximum possible COP?

a) 1.5 b) 2 c) 2.5 d) 3 e) 3.5

7.59 The quality of the refrigerant immediately after the expansion valve of the cycle of Problem 7.58 is

a) 0.58 b) 0.54 c) 0.50 d) 0.46 e) 0.42

AVAILABILITY

7.60 Determine the maximum possible power output, in kW, if 10 kg/s of air is expanded in a turbine from 100°C and 6 MPa abs to the surroundings at 20°C and 80 kPa abs.

a) 3720 b) 2610 c) 2030 d) 1890 e) 1620

HEAT TRANSFER (See Article 7.2)

*7.61 Heat transfer decreases with increased

a) conductivity b) *R*-factor c) emissivity
d) *U*-factor e) molecular activity

*7.62 Heat transfer due to convection can be determined if we know the temperature difference, fluid properties, the geometry, and the

a) Reynolds No. b) Nusselt No. c) Prandtl No.
d) Froude No. e) Weber No.

*7.63 Frozen soil at –15°C and water at 10°C are separated by a 50 cm-thick concrete wall. Estimate the rate of heat transfer, in J/s, if the wall is 2 m × 8 m.

a) 180 b) 140 c) 100 d) 50 e) 20

7.64 The rate of heat loss, due to 150 °C air blowing over a 4-cm-dia cylinder, is measured to be 1500 kJ over a 10 minute time period. If the cylinder is 1.0-m-long and is maintained at 40°C, estimate the convection heat transfer coefficient in $J/s \cdot m^2 \cdot °C$.

a) 180 b) 140 c) 100 d) 50 e) 20

7.65 A filament with a surface area of 0.8 cm^2 is positioned in a spherical space whose surface is at 80°C. Estimate the temperature of the filament , in °C, if 100 W of power is dissipated. Assume a black body filament.

a) 3000 b) 2600 c) 2300 d) 1900 e) 1600

7.66 Calculate the overall *R*-factor, in $m^2 \cdot °C \cdot s/J$, for a wall composed of 2.5 cm of polystyrene, 8 cm of wool insulation, and 2 cm of pine wood. Assume $h = 10\ J/s \cdot m^2 \cdot °C$.

a) 3.2 b) 3.0 c) 2.8 d) 2.6 e) 2.4

PSYCHROMETRICS (metric only)

7.67 How much liquid water, in liters, is contained in a $3m \times 10m \times 20m$ room if $T = 25°C$ and the humidity is 60%?

a) 5.1 b) 6.6 c) 7.2 d) 7.8 e) 8.2

7.68 A "swamp" cooler operates by blowing relatively dry air past a network of wet wicks. What maximum exiting temperature can be realized if the air is at 35°C with a humidity of 20%?

a) 28°C b) 25°C c) 22°C d) 19°C e) 16°C

Practice Problems (English Units)

(If you choose to work only a few problems, select those with a star.)

GENERAL

*7.1 Which of the following would be considered a system rather than a control volume?

a) a pump b) a tire c) a pressure cooker
d) a turbine e) a jet nozzle

*7.2 Which of the following is an extensive property?

a) Temperature b) Velocity c) Pressure d) Mass e) Stress

7.3 An automobile heats up while sitting in a parking lot on a sunny day. The process can be assumed to be:

a) isothermal b) isobaric c) isometric d) isentropic e) adiabatic

*7.4 If a quasi–equilibrium process exists, we have assumed

a) the pressure at any instant to be everywhere constant.
b) an isothermal process.
c) the heat transfer to be small.
d) the boundary motion to be infinitesimally small.
e) that no friction exists.

*7.5 A scientific law is a statement that
a) we postulate to be true.
b) is generally observed to be true.
c) is derived from a mathematical theorem.
d) is agreed upon by scientific community.
e) is a summary of experimental observation.

DENSITY, PRESSURE, AND TEMPERATURE

*7.6 The density of air at vacuum of 6 psi and –40°F is, in lbm/ft^3,
a) 0.076 b) 0.072 c) 0.068 d) 0.062 e) 0.056

7.7 The specific volume of water is 2 ft^3/lbm at a pressure of 30 psia. Find the quality.
a) 0.183 b) 0.144 c) 0.112 d) 0.092 e) 0.061

7.8 The specific volume of steam, in ft^3/lbm, at 600 psia and 2200°F, is
a) 4.26 b) 3.90 c) 3.21 d) 2.64 e) 2.11

7.9 There are 40 lbm of steam contained in 60 ft³ at 600 psia. What is the temperature in °F?
a) 1121 b) 1084 c) 936 d) 902 e) 876

7.10 The pressure in a cylinder containing water is 30 psia and the temperature is 230°F. What state is it in?
a) compressed liquid b) saturated liquid c) liquid/vapor mixture
d) saturated vapor e) superheated vapor

*7.11 A cold tire has a volume of 0.9 ft³ at 14°F and 26 psia. If the pressure and temperature increases to 30 psia and 90°F, find the final volume in ft³.
a) 0.907 b) 0.906 c) 0.905 d) 0.904 e) 0.903

THE FIRST LAW FOR A SYSTEM

*7.12 A 300-watt light bulb provides energy in a 30-ft-dia spherical space. If the outside temperature is 70°F, find the inside steady-state temperature, in °F, if R = 30 hr-ft²-°F/BTU for the wall.
a) 121.2 b) 115.6 c) 106.2 d) 90.9 e) 80.9

7.13 A 20 lb mass, which is attached to a pulley and a paddle wheel submerged in water, drops 10 ft. Find the subsequent heat transfer, in BTU's, needed to return the temperature of the water to its original value.
a) –0.667 b) –0.576 c) –0.415 d) –0.311 e) –0.257

*7.14 A cycle undergoes the following processes. All units are BTU. Find E_{after} for the process $1 \rightarrow 2$.

	Q	W	ΔE	E_{before}	E_{after}
$1 \rightarrow 2$	20	5		10	
$2 \rightarrow 3$		–5	5		
$3 \rightarrow 1$	30			30	

a) 10 b) 15 c) 20 d) 25 e) 30

7.15 For the cycle of Problem 7.14 find W_{3-1}.
a) 50 b) 45 c) 40 d) 35 e) 30

7.16 A 4000 lb automobile traveling at 90 fps strikes a plunger in 600 in^3 of water, bringing the auto to a stop. What is the maximum temperature rise in °F?

a) 50 b) 40 c) 30 d) 20 e) 10

7.17 Twenty pounds of 15°F ice is added to 200 lbm of 70°F water. What is the eventual temperature in °F, of the water? Assume an insulated container.

a) 58 b) 53 c) 49 d) 44 e) 40

*7.18 Of the following first law statements, choose the one that is wrong:

a) The heat transfer equals the net work for a cycle.
b) The heat transfer cannot exceed the work done.
c) The heat transfer equals the work plus the energy change.
d) The heat transfer equals the energy change if no work is done.
e) The energy of an isolated system remains constant.

ISOTHERMAL PROCESS

*7.19 Determine the work, in ft-lb, necessary to compress 4 lbm of air from 15 psia to 600 psia if the temperature is held constant 600°F.

a) –834,000 b) –726,000 c) –592,000 d) –421,000 e) –302,000

*7.20 Steam is compressed from 20 psia to 600 psia holding the temperature constant at 600°F. What is the internal energy change, in BTU, for $m = 4$ lbs?

a) –144 b) –136 c) –127 d) –112 e) –103

7.21 How much heat transfer, in BTU, is needed to convert 4 lbm of saturated liquid water to saturated vapor if $T = 400$°F?

a) 2320 b) 2560 c) 2810 d) 3020 e) 3310

CONSTANT PRESSURE PROCESS

7.22 There are 200 people in a 20,000 ft^2 room, lighted with 3 W/ft^2. Estimate the maximum temperature increase, in °F, if the ventilation system fails for 20 min. Each person generates 400 BTU/hr. The room is 10 ft high.

a) 18 b) 20 c) 22 d) 24 e) 26

*7.23 Estimate the average c_p value in BTU/lbm-°R of a gas if 520 BTU of heat are necessary to raise the temperature of two pounds from 100°F to 1000°F, holding the pressure constant.

a) 0.202 b) 0.243 c) 0.261 d) 0.289 e) 0.310

*7.24 How much heat, in BTU, must be transferred to 20 lbm of air to increase the temperature from 50°F to 400°F if the pressure remains constant?

a) 2050 b) 1930 c) 1860 d) 1720 e) 1680

7.25 Twenty pounds of water, initially at 40°F, are heated until $T = 600$°F. Estimate the heat transfer, in BTU, if the pressure is held constant at 60 psia.

a) 36,500 b) 31,500 c) 26,500 d) 21,500 e) 16,500

7.26 Calculate the work done, in ft-lb, in Problem 7.25.

a) 1.8×10^6 b) 1.8×10^5 c) 1.8×10^4 d) 1.8×10^3 e) 1.8×10^2

CONSTANT VOLUME PROCESS

7.27 How much heat, in BTU, must be added to a rigid volume containing 4 lbm of water/vapor mixture with $x = 0.5$, to increase the temperature from 500°F to 900°F?

a) 1900 b) 2260 c) 2680 d) 2810 e) 2990

*7.28 A tire is pressurized to 28 psia gauge in Michigan where $T = 30°F$. In Arizona the tire is at 160°F. Assuming a rigid tire, estimate the pressure in psi gauge.

a) 33 b) 35 c) 37 d) 39 e) 41

7.29 A sealed, rigid, 300 ft^3 air tank is heated by the sun from 70°F to 150°F. How much energy, in BTU, is transferred to the tank if $P_1 = 14.7$ psia?

a) 401 b) 382 c) 356 d) 331 e) 308

7.30 Twenty pounds of water are heated in a rigid container from 40°F to 380°F. What is the final quality if $Q = 9000$ BTU?

a) 0.431 b) 0.374 c) 0.246 d) 0.193 e) 0.138

ISENTROPIC PROCESS

*7.31 Air expands from an insulated cylinder from 400°F and 60 psia to 2 psia. Find T_2 in °F.

a) –63 b) –96 c) –110 d) –134 e) –178

*7.32 During an isentropic expansion of air, the volume triples. If the initial temperature is 400°F find T_2 in °F.

a) 94 b) 106 c) 112 d) 126 e) 137

7.33 Superheated steam expands isentropically from 1000°F and 400 psia to 2 psia. Find the initial quality.

a) 0.87 b) 0.89 c) 0.91 d) 0.93 e) 0.95

7.34 Superheated steam expands isentropically from 1000°F and 400 psia to 60 psia. Find T_2 in °F.

a) 500 b) 490 c) 480 d) 470 e) 460

7.35 Find the work, in ft-lb/lbm, needed to compress air isentropically in a cylinder from 70°F and 14.7 psia to 400 psia.

a) 121×10^3 b) 111×10^3 c) 101×10^3 d) 92.1×10^3 e) 85.1×10^3

*7.36 During an isentropic process, which of one the following is true?

a) The temperature increases as the pressure decreases.
b) The temperature increases as the volume increases.
c) The heat transfer equals the enthalpy change.
d) The heat transfer is zero.
e) The volume decreases as the pressure decreases.

POLY TROPIC PROCESS

7.37 Find T_2, in °F, if the pressure triples and $T_1 = 50°F$. Let $n = 1.2$.

a) 194 b) 186 c) 179 d) 161 e) 152

THE FIRST LAW FOR A CONTROL VOLUME

7.38 Steam enters a turbine in a 8-in-dia pipe at 1000°F and 400 psia. It exits from a 2-in-dia pipe at 2 psia with $x = 1$. What is the velocity ratio V_{out} / V_{in}?

a) 1300 b) 1100 c) 900 d) 700 e) 500

7.39 Water enters a boiler at 120°F and 300 psia. How much energy, in BTU/lbm, must be added to obtain 1000°F at the exit if the pressure remains constant?

a) 1080 b) 1110 c) 1260 d) 1390 e) 1440

7.40 A condenser is cooled by heating water from 70°F to 90°F. If the condenser inlets 20 lbm/sec of saturated water vapor at 2 psia and exits saturated liquid, what is the mass flux, in lbm/sec, of the cooling water?

a) 1130 b) 1022 c) 961 d) 902 e) 851

7.41 Steam at 700°F and 300 psia expands isentropically through a turbine to 2 psia. Estimate the maximum work output in ft-lb/lbm.

a) 5.21×10^5 b) 4.12×10^5 c) 3.07×10^5 d) 2.13×10^5 e) 1.09×10^5

7.42 What is the energy requirement, in horsepower, for a pump that is 75% efficient if it increases the pressure of 20 lbm/sec of water from 2 psia to 10 psia?

a) 22.4 b) 28.6 c) 31.6 d) 38.2 e) 44.5

7.43 A river 200 ft wide and 6 ft deep flows at 6 fps. A hydro plant develops a pressure of 50 psi gauge just before the turbine. What maximum power, in horsepower, is possible?

a) 21,300 b) 41,900 c) 65,200 d) 81,100 e) 94,300

*7.44 A nozzle expands air isentropically from 700°F and 100 psia to the atmosphere at 12 psia. If the inlet velocity is small, what exit velocity, in fps, can be expected?

a) 500 b) 1000 c) 1500 d) 2000 e) 2500

7.45 If the efficiency of a turbine that expands steam at 700°F and 400 psia to 2 psia is 87%, find the work output in ft-lb/lbm.

a) 302,000 b) 278,000 c) 233,000 d) 201,000 e) 187,000

THE SECOND LAW

*7.46 An inventor proposes to take 20 lbm/sec of geothermal water at 240°F and generate 2600 hp of energy by exhausting very near to ambient temperature of 70°F. This proposal should not be supported because

a) it violates the first law.
b) it violates the second law.
c) it would be too expensive.
d) friction must be accounted for.
e) the geothermal water is not hot enough.

*7.47 The net entropy change in the universe during any real process

a) is equal to zero.
b) is positive.
c) is negative.
d) must be calculated to determine its sense.
e) is positive if $T_{system} > T_{surroundings}$.

*7.48 A Carnot engine

a) provides a fictitious model which is of little use.
b) can be experimentally modeled.
c) supplies us with the lower limit for engine efficiency.
d) operates between two constant temperature reservoirs.
e) has two reversible and two irreversible processes.

7.49 Two hundred pounds of ice at 32°F are completely melted. Find the entropy change, in BTU/°R, if $T_2 = 32$°F.

a) 87 b) 77 c) 67 d) 57 e) 47

7.50 Eight pounds of ice at 32°F are mixed with 200 lbm of water at 60°F. What is the entropy change in BTU/°F?

a) 1.32 b) 0.926 c) 0.711 d) 0.526 e) 0.312

*7.51 An inventor proposes to have developed a small power plant that operates at 70% efficiency. It operates between temperature extremes of 1000°F and 100°F. Your analysis shows that the maximum possible efficiency is

a) 65 b) 62 c) 57 d) 51 e) 46

POWER CYCLES

*7.52 A Carnot cycle operates on air such that $P_4 = 24$ psia and $v_4 = 8\ \text{ft}^3/\text{lbm}$ (see Figure 7.5). If 30 BTU/lbm of heat are added at $T_H = 400°\text{F}$, find the work produced in ft-lb/lbm.

a) 23,400 b) 14,200 c) 11,600 d) 7820 e) 4630

7.53 An Otto cycle operates with volumes of 4 in^3 and 40 in^3 at top dead center (TDC) and bottom dead center (BDC), respectively. If the power output is 130 hp, what is the heat input in BTU/sec? Assume $k = 1.4$.

a) 153 b) 172 c) 196 d) 210 e) 236

7.54 The volumes of states 1, 2, and 3 in Figure 7.5 of the Diesel cycle are 45 in^3, 3 in^3 and 5 in^3, respectively. If the power produced is 160 hp, what is the required heat input in BTU/sec? Assume $k = 1.4$.

a) 182 b) 171 c) 159 d) 142 e) 136

7.55 A simple Rankine cycle operates between superheated steam at 1000°F and 400 psia entering the turbine, and 2 psia entering the pump. What is the maximum possible efficiency, as a percent?

a) 39 b) 37 c) 35 d) 33 e) 31

7.56 The water vapor that expands from 1000°F and 300 psia in a turbine of a Rankine cycle is intercepted at 20 psia and reheated at constant pressure to 1000°F in the boiler, after which it is re-injected in the turbine and expanded to 1 psia. (This is the *reheat cycle*.) Calculate the maximum possible efficiency, as a percent.

a) 41 b) 39 c) 37 d) 35 e) 33

7.57 The primary effect of reheating, as illustrated in Problem 7.56 (compare with the result in Problem 7.55), is to

a) increase the efficiency.
b) decrease the pump work.
c) decrease the heat requirement.
d) eliminate unnecessary piping.
e) decrease or eliminate moisture condensation in the turbine.

REFRIGERATION CYCLES

7.58 A refrigeration cycle operates with Refrigerant-134a between 10 psia and 100 psia. What is the maximum possible COP?

a) 3.2 b) 2.6 c) 2.1 d) 1.9 e) 1.4

7.59 The quality of the refrigerant immediately after the expansion valve of the cycle of Prob. 7.58 is

a) 0.57 b) 0.53 c) 0.51 d) 0.47 e) 0.43

AVAILABILITY

7.60 Determine the maximum possible power output, in horsepower, if 20 lbm/sec of air is expanded in a turbine from 200°F and 400 psia to the surroundings at 70°F and 12 psia.

a) 3760 b) 3500 c) 3300 d) 3100 e) 2900

HEAT TRANSFER (See Article 7.2)

*7.61 Heat transfer decreases with increased

a) conductivity b) *R*-factor c) emissivity
d) *U*-factor e) molecular activity

*7.62 Heat transfer due to convection can be determined if we know the temperature difference, fluid properties, the geometry, and the

a) Reynolds No. b) Nusselt No. c) Prandtl No.
d) Froude No. e) Weber No.

*7.63 Frozen soil at 0°F and water at 45°F are separated by a 20″-thick concrete wall. Estimate the rate of heat transfer, in J/s, if the wall is 6′ × 25′.

a) 400 b) 350 c) 300 d) 250 e) 200

7.64 The rate of heat loss due to 300°F air blowing over a 2″-dia cylinder is measured to be 1500 BTU over a 10 minute time period. If the cylinder is 3-ft-long and is maintained at 100°F, estimate the convection heat transfer coefficient in Btu/hr-ft^2-°F.

a) 29 b) 26 c) 24 d) 19 e) 14

7.65 A filament with a surface area of 0.12 in^2 is positioned in a spherical space whose surface is at 175°F. Estimate the temperature of the filament if 100 W of power is dissipated. Assume a black body filament.

a) 5000°F b) 4600°F c) 4100°F d) 3500°F e) 3000°F

7.66 Calculate the overall *R*-factor, in ft^2-°F-hr/BTU, for a wall composed of 1″ of polystyrene, 3″ of wool insulation, and ¾″ of pine wood. Assume h = 1.2 BTU/hr-ft^2-°F.

a) 19 b) 17 c) 15 d) 13 e) 11

Solutions to Practice Problems

(Metric Units)

7.1 b) A tire. All other devices have fluid entering and/or leaving.

7.2 d) Mass. All other quantities do not depend on the mass.

7.3 b) Isobaric. The pressure will remain constant due to the inlets and outlets for air.

7.4 a) All properties are uniform throughout the volume.

7.5 e) A scientific law results from experimental observations.

7.6 e) $\rho = 1/v = P/RT = (-40+100)/.287 \times (-40+273) = 0.897 \ \text{kg/m}^3$.
$R = 8.314/28.97 = .287$.

7.7 b) $v = v_f + x(v_g - v_f)$ $P = 0.2$ MPa

$0.5 = .001 + x(.8857 - .001)$ TABLE 7.3.2

$\therefore x = 0.564$

7.8 d) If the state is beyond Table 7.3.3 use the equation of state with

$R = \frac{8.314}{18} = 0.462.$

$$v = \frac{RT}{P} = \frac{0.462 \times 1473}{4000} = 0.17 \text{ m}^3/\text{kg}.$$

Use $T = 1200 + 273 = 1473$ K, $P = 4000$ kPa

7.9 b) $v = \frac{V}{m} = \frac{2}{20} = 0.1,\quad P = 4$ MPa.

From Table 7.3.3 interpolation gives

$T = \frac{0.1 - 0.09885}{0.1109 - 0.09885} \times 100 + 600 = 610° \text{ C}.$

7.10 a) Using Table 7.3.2 we see that 115°C < 120.2°C. $\therefore$ compressed.

7.11 c) $m = \frac{P_1V_1}{RT_1} = \frac{280 \times .03}{.287 \times 263} = 0.1113$ kg, where $R = \frac{8.3.4}{28.97} = 0.287.$

$\therefore V_2 = \frac{mRT_2}{P_2} = \frac{.1113 \times .287 \times 303}{310} = 0.0312 \text{ m}^3$

7.12 e) $\dot{Q} = \frac{1}{R} A \Delta T$ $A = 4\pi r^2$

$\frac{300}{1000} = \frac{1}{1.5} 4\pi \times 5^2 (T - 20) \frac{1}{3600}.$ $\therefore T = 25.2°\text{C}.$

The 1000 converts J/s to kJ/s.

The 3600 converts hr to seconds.

7.13 e) $W = F \times d = (10 \times 9.8) \times 3 = 294$ N·m input. $\therefore$ W = -294 N·m .

$\therefore Q = W = -294$ J. (Heat output is negative.)

7.14 d) $Q - W = E_a - E_b.$

$20 - 5 = E_a - 10.$ $\therefore E_a = 25$ kJ.

7.15 a) For $2 \to 3$ $Q - (-5) = 5.$ $\therefore Q_{2-3} = 0$

$\therefore Q_{net} = 20 + 0 + 30 = 50$

$Q_{net} = W_{net} = 5 - 5 + W_{3-1} = 50.$

$\therefore W_{3-1} = 50$ kJ.

7.16 c) $Q = mc\Delta T.$ Here $Q = \frac{1}{2} mV^2,$ $m = \forall\rho$

$\frac{1}{2} \times 2000 \times 25^2 = 10\ 000 \times 10^{-6} \times 1000 \times 4180\ \Delta T$

$\therefore \Delta T = 14.95°\text{C}$

7.17 b) $(mc\Delta T)_{\text{ice}} + (mc\Delta T)_{\substack{\text{melted}\\\text{ice}}} + mh_{\text{fusion}} = (mc\Delta T)_{\text{water}}$

$10\times 2.1\times 10 + 10\times 4.18(T-0) + 10\times 320 = 100\times 4.18(20-T)$. $\quad \therefore T = 10.77°\text{C}$.

7.18 b) The heat transfer exceeds the work done if $\Delta E > 0$.

7.19 a) $W = mRT\ln P_1/P_2$

$= 2\times .287\times 573\ln 100/4000 = -1213$ kJ

7.20 b) $\Delta U = m(u_2 - u_1)$ Table 7.3.3

$= 2(2725.3 - 2810.4) = -170.2$ kJ

7.21 e) $Q = m\Delta h = mh_{fg} = 2\times 1940.8 = 3881.6$ kJ

7.22 e) Assume P = const since doors would be opened. Cracks abound.

$$Q = mc_p\Delta T\ .$$

$$\frac{400\times 200}{3600} + \frac{30\times 2000}{1000} = \frac{6000\times 1.23\times 1.00\Delta T}{20\times 60}. \qquad \therefore \Delta T = 13.37°\text{ C}$$

7.23 d) $Q = m\Delta h = mc_p\Delta T$

$522 = 1\times c_p\times(800-300)$. $\quad \therefore\ c_p = 1.044\dfrac{\text{kJ}}{\text{kg}\cdot\text{K}}$

7.24 e) $Q = m\Delta h = mc_p\Delta T = 10\times 1.00\times(230-10) = 2200$ kJ

7.25 c) $Q = m\Delta h = 10[3072 - 42] = 30\,300$ kJ

7.26 a) $W = P\Delta V = mP\Delta v = 10(1.316 - .001)\times 200 = 2630$ kJ

7.27 a) $Q = m\Delta u \qquad u_1 = 851 + .5(2595 - 851) = 1723$ kJ/kg.

$T_2 = 500°\text{C}$

$v_2 = .0642 \qquad v_1 = .001 + .5(.1274 - .001) = .0642\ \text{m}^3/\text{kg}$.

Use Table 7.3.3

$u_2 = \dfrac{.0642 - .0566}{.0686 - .0566}(3091 - 3082) + 3082 = 3088$

$\therefore Q = 2(3088 - 1723) = 2730$ kJ

7.28 d) $\dfrac{P_1}{T_1} = \dfrac{P_2}{T_2}$. $\quad (p_2 + 100)273 = (100 + 100)343$

$\therefore P_2 = 151$ kPa gage

7.29 e) $Q = m\Delta u = mc_v\Delta T \qquad m = \dfrac{PV}{RT} = \dfrac{100\times 10}{.287\times 293} = 11.89$ kg

$= 11.89\times .716\times 60 = 511$ kJ

7.30 e) $Q = m\Delta u$. $\quad 9000 = 10(u_2 - 42) \qquad \therefore u_2 = 942$ kJ/kg

$942 = 851 + x_2(2595 - 851) \qquad \therefore x_2 = .0522$

7.31 d) $T_2 = T_1\left(\frac{P_2}{P_1}\right)^{k-1/k} = 473\left(\frac{20}{400}\right)^{.4/1.4} = 201\text{K} \text{ or } -72°\text{C}$

7.32 a) $T_2 = T_1\left(\frac{v_1}{v_2}\right)^{k-1} = 473\left(\frac{1}{3}\right)^{0.4} = 305\text{K} \text{ or } 31.8°\text{C}$

7.33 c) $s_2 = s_1 = 7.1685$. from Table 7.3.2, $7.1685 = .649 + 7.502x_2$. $\therefore x_2 = .869$

7.34 b) $s_2 = s_1 = 7.1685$ from Table 7.3.3. This is slightly less than $s = 7.171$ at $T = 200°\text{C}$ and $P = 0.4\text{MPa}$. $\therefore T_2 = 199°\text{C}$.

7.35 b) $T_2 = T_1\left(\frac{P_2}{P_1}\right)^{k-1/k} = 293\left(\frac{6000}{100}\right)^{.4/1.4} = 944 \text{ K}.$

$W = -m\Delta u = -mc_v\Delta T$ (let $m = 1$)

$= -.716(944 - 293) = -466 \text{ kJ/kg}$.

7.36 d) The heat transfer must be zero.

7.37 e) $T_2 = T_1\left(\frac{P_1}{P_2}\right)^{n-1/n} = 283 \times 3^{.2/1.2} = 340 \text{ K} \text{ or } 66.9°\text{C}$

7.38 a) $\rho_1 = \frac{1}{v_1} = \frac{1}{.06525} = 15.33$ Table 7.3.3

$\rho_2 = \frac{1}{v_2} = \frac{1}{7.649} = 0.1307$ Table 7.3.2

$\rho_1 A_1 V_1 = \rho_2 A_2 V_2$. $15.33\pi \times \frac{.2^2}{4} V_1 = .1307\pi \times \frac{.05^2}{4} V_2$. $\therefore \frac{V_2}{V_1} = 1877$

7.39 e) $Q = h_2 - h_1$. $h_1 = 251$ from Table 7.3.1.

$= 3674 - 251 = 3423 \text{ kJ/kg}$

7.40 b) $\dot{Q} = \dot{m}_w c\Delta T = \dot{m}_s \Delta h$. Table 7.3.1, $h_{fg} = 2358 \text{ kJ/kg}$

$\dot{m} \times 4.18(30 - 20) = 10 \times 2358$ $\therefore \dot{m} = 564 \text{ kg/s}$

7.41 c) $s_2 = s_1 = 6.7698 = .649 + 7.502x_2$. $\therefore x_2 = .816$

$\therefore h_2 = 192 + .816 \times 2393 = 2144$. $W = h_1 - h_2 = 3213.5 - 2144 = 1069 \text{ kJ/kg}$

7.42 e) $\dot{W}_P = \dot{m}\frac{\Delta P}{\rho\eta} = 10\frac{6000 - 10}{1000 \times .75} = 79.9 \text{ kW}$

7.43 e) $\dot{W}_T = \dot{m}\frac{\Delta P}{\rho} = 60 \times 2 \times 2 \times 1000\frac{300}{1000} = 72\ 000 \text{ kW}$

7.44 e) $T_2 = T_1\left(\frac{P_2}{P_1}\right)^{k-1/k} = 673\left(\frac{80}{2000}\right)^{.4/1.4} = 268 \text{ K}.$

$V_2^2/2 = h_1 - h_2 = c_p(T_1 - T_2) = 1000(673 - 268)$. $\therefore V_2 = 900 \text{ m/s}$.

7.45 b) $h_1 = 3177 \quad s_{2'} = s_1 = 6.5415 = .832 + x_{2'} 7.0774. \quad \therefore x_{2'} = .807.$
$\therefore h_{2'} = 251 + .807 \times 2358 = 2153. \quad w = 0.87 \times (3177 - 2153) = 891 \text{ kJ/kg}.$

7.46 b) $\eta_{max} = 1 - \dfrac{T_L}{T_H} = 1 - \dfrac{293}{393} = 0.254$
$\eta = \dfrac{4180}{10 \times 4.18(120 - 20)} = 1.00. \quad \therefore \text{impossible}$

7.47 b) $\Delta s_{net} > 0$ for all processes.

7.48 d) Operates between const. temp. reservoirs.

7.49 d) $\Delta S > Q/T = 90 \times 320/273 = 105.5 \text{ kJ/K}$

7.50 e) $40(320 + 4.18T_2) = 100 \times 4.18(20 - T_2). \quad \therefore T_2 = -7.6°\text{C}.$
Impossible. $\therefore T_2 = 0°\text{C}. \quad 320\, m = 100 \times 4.8 \times 20. \quad \therefore m = 26.1 \text{ kg}.$
$\Delta s = 26.1 \times 320/273 + 100 \times 4.18 \ln 273/293 = 1.04$

7.51 b) $\eta_{max} = 1 - \dfrac{T_L}{T_H} = 1 - 323/873 = 0.63$

7.52 e) $T_4 = P_4 v_4 / R = 160 \times .5/.287 = 279 \text{ K}.$
$w = q\eta = 30(1 - 279/473) = 12.3 \text{ kJ/kg}$

7.53 a) $\eta = 1 - (400/40)^{-.4} = 0.602. \quad \therefore \dot{Q} = \dot{W}/\eta = 100/.602 = 166 \text{ kW}.$

7.54 a) $r_c = 45/25 = 1.8. \qquad r = 450/25 = 18.$
$\dot{Q} = \dot{W}/\eta = 120 \Big/ \left[1 - \dfrac{1.8^{1.4} - 1}{1.4(1.8 - 1)} 18^{-.4}\right] = 187 \text{ kJ/s}.$

7.55 c) $s_4 = s_3 = 7.1685 = .649 + 7.502x_4. \quad \therefore x_4 = .869$
$h_4 = 192 + .869 \times 2393 = 2272. \quad h_3 = 3658, \quad h_1 = 192$
$\eta = \dfrac{w}{q} = (3658 - 2272)/(3658 - 192) = 0.40$

7.56 c) $s_4 = s_3 = 7.168. \quad \therefore$ ④ superheat $\therefore h_4 = 2722. \quad h_3 = 3658.$
After reheat $h_5 = 3704. \quad s_6 = s_5 = 8.778.$
$\therefore h_6 = \dfrac{8.778 - 8.689}{8.9046 - 8.689}(2879.5 - 2783) + 2783 = 2823$
$\eta = \dfrac{w_{3-4} + w_{5-6}}{q_{2-3} + q_{4-5}} = \dfrac{3658 - 2722 + 3704 - 2823}{3658 - 192 + 3704 - 2722} = 0.408$

7.57 e) Decrease moisture content in the turbine.

7.58 b) $s_2 = s_1 = 1.75.$ Fig. 7.6 provides $h_1 = 380$, $h_2 = 430$, $h_3 = 255 = h_4$.
Refer to Fig. 7.4. $COP = \dfrac{q_{4-1}}{w_{1-2}} = \dfrac{380 - 255}{430 - 380} = 2.5.$

7.59 e) $h_4 = h_3 = 255 = h_f + x_4\left(h_g - h_f\right)$

$= 165 + x_4(380 - 165). \quad \therefore x_4 = 0.42 .$

7.60 a) $\dot{W}_{max} = \dot{m}\left(h_1 - T_o s_1\right) - \dot{m}\left(h_2 - T_o s_2\right)$

$= \dot{m}\left[\left(h_1 - h_2\right) + T_o\left(s_2 - s_1\right)\right]$

$= 10\left[1.00(100 - 20) + 293\left(1.00 \ln \frac{293}{373} - .287 \ln \frac{80}{6000}\right)\right] = 3723 \text{ kW}$

7.61 b)

7.62 b)

7.63 c) $\dot{Q} = .13 \times 2 \times 8 \times \frac{25}{.5} = 104 \text{ J/s}$

7.64 a) $\frac{1\,500\,000}{10 \times 60} = h \times \pi \times .04 \times (150 - 40). \quad \therefore h = 181 \text{ J/s} \cdot \text{m}^2 \cdot {}^\circ\text{C}$

7.65 d) $100 = 5.67 \times 10^{-8} \times \left(.8 \times 10^{-4}\right)\left(T^4 - 353^4\right).$

$\therefore T = 2167 \text{ K} \quad \text{or} \quad 1894^\circ \text{ C}$

7.66 b) $R = \frac{1}{h} + \frac{L_1}{k_1} + \frac{L_2}{k_2} + \frac{L_3}{k_3} + \frac{1}{h}$

$= \frac{1}{10} + \frac{.025}{.04} + \frac{.08}{.038} + \frac{.02}{.15} + \frac{1}{10} = 3.06 \ \frac{\text{s} \cdot \text{m}^2 \cdot {}^\circ\text{C}}{\text{J}}$

7.67 e) $\omega = 0.0118. \quad \therefore m_v = \omega m_a = \omega V/v$

$= .0118 \times 600/.861$

$= 8.2 \text{ kg} \quad \text{or } 8.2 \text{ L}$

7.68 d) The enthalpy remains constant since $\dot{W} = \dot{Q} = 0$. States 1 and 2 lie on a constant enthalpy line on the psychrometric chart with $\phi_2 = 100\%$. $\therefore T_2 = 19^\circ\text{C}$.

Solutions to Practice Problems

(English Units)

7.1 b) A tire. All other devices have fluid entering and/or leaving.

7.2 d) Mass. All other quantities do not depend on the mass.

7.3 b) Isobaric. The pressure will remain constant due to the inlets and outlets for air.

7.4 a) All properties are uniform throughout the volume.

7.5 e) A scientific law results from experimental observations.

7.6 e) $\rho = 1/v = P/RT = (-6+14.7)\times 144/(53.3\times 420) = 0.056 \text{ lbm/ft}^3$.

$T = -40+460 = 420°\text{R}$ and

$R = 1545/28.97 = 53.3$

7.7 b) $v = v_f + x(v_g - v_f)$ $\quad P = 30$ psia

$2 = .017 + x(13.75 - .017)$. Table 7.3.2E

$\therefore x = 0.144$

7.8 d) If the state is beyond the Table 7.3.3E use the equation of state with

$R = 1545/18 = 85.8$.

$$v = \frac{RT}{P} = \frac{85.8\times 2660}{600\times 144} = 2.64 \text{ ft}^3/\text{lbm}.$$

7.9 b) $v = \frac{V}{m} = \frac{60}{40} = 1.5$, $P = 600$ psia. From Table 7.3.3E interpolation gives

$$T = \frac{1.5-1.411}{1.622-1.411}\times 200 + 1000 = 1084° \text{ F}.$$

7.10 a) Using Table 7.3.2E we see that $230° \text{ F} < 233.9° \text{ F}$. $\therefore$ compressed liquid.

7.11 c) $m = \frac{P_1V_1}{RT_1} = \frac{26\times 144\times 0.9}{53.3\times 474} = 0.1334$ lbm where $R = 1545/28.97 = 53.3$.

$$\therefore V_2 = \frac{mRT_2}{P_2} = \frac{.1334\times 53.3\times 550}{30\times 144} = 0.905 \text{ ft}^3.$$

7.12 e) $\dot{Q} = \frac{1}{R}A\Delta T$. $\quad A = 4\pi r^2$. $\quad \frac{300}{0.293} = \frac{1}{30}4\pi\times 15^2(T-70)$. $\quad \therefore T = 80.9°\text{F}$.

7.13 e) $W = F\times d = 20\times 10 = 200$ ft – lb input. $Q = W = -\frac{200}{778} = -0.257$ BTU.

(Heat output is neg.)

7.14 d) $Q - W = E_a - E_b$. $\quad 20-5 = E_a - 10$. $\quad \therefore E_a = 25$ BTU.

7.15 a) For $2\rightarrow 3$ $\quad Q-(-5) = 5$. $\quad \therefore Q_{2-3} = 0$.

$\therefore Q_{net} = 20+0+30 = 50$. $\quad Q_{net} = W_{net} = 5-5+W_{3-1} = 50$.

$\therefore W_{3-1} = 50$ BTU.

7.16 c) $Q = mc\Delta T$. Here $Q = \frac{1}{2}mV^2$, $\quad m = \forall\rho$

$$\frac{1}{2}\times\frac{4000\times 90^2}{32.2\times 778} = \frac{600}{1728}\times 62.4\times 1.00\times\Delta T. \quad \therefore \Delta T = 29.9°\text{F}$$

7.17 b) $(mc\Delta T)_{ice} + (mc\Delta T)_{\substack{melted\\ice}} + mh_{melt} = (mc\Delta T)_{water}$

$20\times 0.5\times(32-15) + 20\times 1.00(T-32) + 20\times 140 = 200\times 1.00(70-T)$

$\therefore T = 53.05°$ F.

7.18 b) The heat transfer exceeds the work done if $\Delta E > 0$.

7.19 a) $W = mRT \ln P_1/P_2 = 4 \times 53.3 \times 1060 \ln 15/600 = -834{,}000$ ft - lb.

7.20 b) $\Delta U = m(u_2 - u_1) = 4(1184.5 - 1218.4) = -135.6$ BTU using Table 7.3.3E.

7.21 e) $Q = m\Delta h = mh_{fg} = 4 \times 826.3 = 3305$ BTU

7.22 e) Assume P = const since doors would be opened. Cracks abound.

$$Q = mc_p \Delta T \,.$$

$$200 \times 400 + 3 \times 20{,}000 / 0.293 = \underset{\uparrow \atop V}{200{,}000} \times \frac{\overset{\rho}{\downarrow}\; .076}{\underset{\uparrow \atop \Delta t}{0.333}} \times \underset{\uparrow \atop c_p}{.24} \Delta T. \quad \therefore \Delta T = 26° \text{F}.$$

7.23 d) $Q = m\Delta h = mc_p \Delta T. \quad 520 = 2 \times c_p \times (1000 - 100) \quad \therefore c_p = 0.289$ BTU/lb$_m$ – °F.

7.24 e) $Q = m\Delta h = mc_p \Delta T = 20 \times 0.24 \times (400 - 50) = 1680$ BTU.

7.25 c) $Q = m\Delta h = 20[1332.1 - 8.02] = 26{,}500$ BTU

7.26 a) $W = p\Delta V = mP\Delta v = 20 \times 60 \times 144(10.425 - 0.017) = 1680$ BTU.

7.27 a) $Q = m\Delta u \qquad u_1 = 485 + .5(1117 - 485) = 801$

$v_1 = .020 + .5(.676 - .02) = .348$

$T_2 = 900°$F and $v_2 = 0.348. \quad \therefore$ Using Table 7.3.3E:

$$u_2 = \frac{0.348 - 0.216}{0.353 - 0.216}(1277 - 1242) + 1242 = 1276$$

$\therefore Q = 4(1276 - 801) = 1899$ BTU.

7.28 d) $\frac{P_1}{T_1} = \frac{P_2}{T_2}. \quad (P_2 + 14.7) \times 490 = (28 + 14.7) \times 620. \quad \therefore P_2 = 39.3$ psi gage

7.29 e) $Q = m\Delta u = mc_v \Delta T \qquad m = \frac{pV}{RT} = \frac{14.7 \times 144 \times 300}{53.3 \times 530} = 22.48$ lbm

$= 22.48 \times .171(150 - 70) = 307.5$ BTU

7.30 e) $Q = m\Delta u. \quad 9000 = 20(u_2 - 80). \quad \therefore u_2 = 458$ BTU/lbm.

$458 = 353 + x_2(1114 - 353). \quad \therefore x_2 = 0.138$

7.31 d) $T_2 = T_1\left(\frac{P_2}{P_1}\right)^{k-1/k} = 860\left(\frac{2}{60}\right)^{0.4/1.4} = 325°\text{R}$ or $-134°$F

7.32 a) $T_2 = T_1\left(\frac{v_1}{v_2}\right)^{k-1} = 860\left(\frac{1}{3}\right)^{0.4} = 554°\text{R}$ or $-94°$F

7.33 c) $s_2 = s_1 = 1.7632$, Table 7.3.3E.

Using Table 7.3.2E, $1.7632 = .175 + x_2(1.7448). \quad \therefore \quad x_2 = 0.910$

7.34 b) $s_2 = s_1 = 1.7632$. At 60 psia this is just less than s at 500° F.

Interpolate and find $T_2 = \dfrac{1.763 - 1.713}{1.768 - 1.713} \times 100 + 400 = 491°$ F.

7.35 b) $T_2 = T_1\left(\dfrac{P_2}{P_1}\right)^{k-1/k} = 530\left(\dfrac{400}{14.7}\right)^{0.2857}$

$= 1362°$ R. $W = -m\Delta u = -mc_v\Delta T$. let $m = 1$ lbm

$= 0.171(1362 - 530) = 142.3$ BTU/lbm

or $142.3 \times 778 = 110,700\ \dfrac{\text{ft - lb}}{\text{lbm}}$.

7.36 d) The heat transfer must be zero.

7.37 e) $T_2 = T_1\left(\dfrac{P_2}{P_1}\right)^{n-1/n} = 510 \times 3^{.2/1.2} = 612°$ R or $152°$F.

7.38 a) $\rho_1 = \dfrac{1}{v_1} = \dfrac{1}{2.136} = 0.4682$ Table 7.3.3E

$\rho_2 = \dfrac{1}{v_2} = \dfrac{1}{173.75} = 0.005755$ Table 7.3.2E

$\rho_1 A_1 V_1 = \rho_2 A_2 V_2$

$.4682\pi \times 4^2 \times V_{in} = .005755\pi \times 1^2 \times V_{out}$. $\therefore \dfrac{V_{out}}{V_{in}} = 1302$

7.39 e) $Q = h_2 - h_1$ $h_1 = 88$ from Table 7.3.1E.

$= 1526.5 - 88 = 1438$ BTU/lbm.

7.40 b) $\dot{Q} = \dot{m}_w c\Delta T = \dot{m}_s \Delta h$ Table 7.3.2E gives $h_{fg} = 1022.1$.

$\dot{m}_w \times 1.00 \times (90 - 70) = 20 \times 1022.1$. $\therefore \dot{m}_w = 1022$ lbm/sec

7.41 c) $s_2 = s_1 = 1.6751 = .175 + 1.7448x_2$. $\therefore x_2 = 0.860$.

$\therefore h_2 = 94 + .86 \times 1022 = 973$

$w = h_1 - h_2 = 1368 - 973 = 395$ BTU/lbm. $395 \times 778 = 3.07 \times 10^5$ ft - lb/lbm

7.42 e) $\dot{W}_P = \dot{m}\dfrac{\Delta P}{\rho\eta} = 20\dfrac{398 \times 144}{62.4 \times 0.75} = 24,500$ ft - lb/sec or 44.5 Hp.

7.43 e) $\dot{W}_T = \dot{m}\dfrac{\Delta P}{\rho} = (200 \times 6 \times 6 \times 62.4)\dfrac{50 \times 144}{62.4} = 5.18 \times 10^7\ \dfrac{\text{ft - lb}}{\text{sec}}$ or 94,300 Hp.

7.44 e) $T_2 = T_1\left(\dfrac{P_2}{P_1}\right)^{k-1/k} = 1160\left(\dfrac{12}{100}\right)^{.2857} = 633°$R.

$V_2^2/2 = h_1 - h_2 = c_p(T_1 - T_2) = .24 \times 778(1160 - 633) \times 32.2$. $\therefore V_2 = 2520$ ft/sec.

7.45 b) $h_1 = 1362$. $s_{2'} = s_1 = 1.6397 = .175 + x_{2'}1.7448$. $\therefore x_{2'} = 0.840$.

$\therefore h_{2'} = 94 + .84 \times 1022 = 952$.

$w = 0.87 \times (1362 - 952) = 357$ BTU/lbm or 2.78×10^5 ft - lb/lbm.

7.46 b) $\eta_{max} = 1 - \frac{T_L}{T_H} = 1 - \frac{530}{700} = 0.243.$ $\eta = \frac{2600 \times 550 / 778}{20 \times 1.00(240 - 70)} = 0.54.$ $\therefore$ impossible

7.47 b) $\Delta s_{net} > 0$ for all real processes.

7.48 d) Operates between constant temperature reservoirs.

7.49 d) $\Delta S > Q / T = 200 \times 140 / 492 = 56.9$ BTU/°R.

7.50 e) $80[140 + 1.00(T_2 - 32)] = 200 \times 1.00(60 - T_2).$ $\therefore T_2 = 2.9°$ F.

Impossible. $\therefore$ all the ice doesn't melt and $\therefore T_2 = 32°$F.

$140\ m = 200 \times 1.00 \times (60 - 32).$ $\therefore m = 40$ lbm.

$$\Delta S = \frac{40 \times 140}{492} + 200 \times 1.00 \ln \frac{492}{520} = 0.312 \text{ BTU/°R}.$$

7.51 b) $\eta_{max} = 1 - \frac{T_L}{T_H} = 1 - \frac{560}{1460} = 0.616.$

7.52 e) $T_4 = P_4 v_4 / R = 24 \times 144 \times 8 / 53.3 = 518.7°$R

$w = q\eta = 15 \times (1 - 519/860) \times 778 = 4627$ ft-lb/lbm

7.53 a) $\eta = 1 - (40/4)^{-.4} = 0.602.$ $\therefore \dot{Q} = \dot{W} / \eta = 130 \times \frac{0.746}{1.055} / 0.602 = 153$ BTU/sec .

7.54 a) $r_c = 5/3 = 1.67$ $\quad r = 45/3 = 15$

$$\dot{Q} = \dot{W} / \eta = \frac{160 \times 550}{778} \Big/ \left[1 - \frac{1.67^{1.4} - 1}{1.4(1.67 - 1)} 15^{-.4}\right] = 182 \text{ BTU/sec}.$$

7.55 c) $s_4 = s_3 = 1.7632 = .175 + 1.7448 x_4.$ $\therefore x_4 = 0.910$

$h_4 = 94 + .91 \times 1022 = 1024.$ $\quad h_3 = 1524.$ $\quad h_1 = 94$

$$\eta = \frac{w}{q} = (1524 - 1024)/(1524 - 94) = 0.3495$$

7.56 c) $s_4 = s_3 = 1.7963.$ $\therefore$ superheat $\therefore h_4 = 1208.$ $\quad h_3 = 1526$

after reheat $h_5 = 1534.$ $\quad s_6 = s_5 = 2.10$ (interpolate)

and $h_6 = \frac{2.10 - 2.051}{2.115 - 2.051}(1196 - 1150) + 1150 = 1185$

$$\eta = \frac{W_{3-4} + W_{5-6}}{Q_{2-3} + Q_{4-5}} = \frac{1526 - 1208 + 1534 - 1185}{1526 - 70 + 1534 - 1208} = 0.374$$

7.57 e) Decrease moisture content in the turbine.

7.58 c) $s_2 = s_1 = 0.23$.

Fig. 7.6E provides $h_1 = 98,$ $h_2 = 123,$ $h_3 = 45 = h_4.$ Refer to Fig. 7.4.

$$COP = \frac{Q_{4-1}}{W_{1-2}} = \frac{98 - 45}{123 - 98} = 2.12 .$$

7.59 e) $h_4 = h_3 = 45 = h_f + x_4(h_g - h_f)$

$$= 4 + x_4(100 - 4). \quad \therefore x_4 = 0.427.$$

7.60 a) $\dot{W}_{max} = \dot{m}(h_1 - T_o s_1) - \dot{m}(h_2 - T_o s_2)$

$$= \dot{m}[(h_1 - h_2) + T_o(s_2 - s_1)]$$

$$= 20\left[.24(200 - 70) + 530\left(.24 \ln \frac{530}{660} - \frac{53.3}{778} \ln \frac{12}{400}\right)\right]$$

$$= 2660 \text{ BTU/sec} \quad \text{or} \quad 3760 \text{ Hp}$$

7.61 b)

7.62 b)

7.63 c) $\dot{Q} = .075 \times 6 \times 25 \times 45 \Big/ \frac{20}{12} = 304$

7.64 a) $\frac{1500}{10/60} = h \times \pi \times \frac{2}{12} \times 3(300 - 100). \quad \therefore h = 28.6 \text{ BTU/hr - ft}^2 \text{ - °F}$

7.65 d) $100 \times 3.414 = 1.713 \times 10^{-9} \times \frac{.12}{144}(T^4 - 635^4).$

$\therefore T = 3930° \text{ R} \quad \text{or} \quad 3470° \text{ F}.$

7.66 b) $R = \frac{1}{h} + \frac{L_1}{k_1} + \frac{L_2}{k_2} + \frac{L_3}{k_3} + \frac{1}{h}$

$$= \frac{1}{1.2} + \frac{1/12}{.023} + \frac{3/12}{.022} + \frac{3/12}{.087} + \frac{1}{1.2} = 17.4 \frac{\text{hr - ft}^2 \text{ - °F}}{\text{BTU}}$$

(This page is intentionally blank.)

Tables for Thermodynamics

TABLE 7.1 Properties of Ideal Gases — Metric Units

Gas	Chemical Formula	Molar Mass	$R\ \frac{kJ}{kg \cdot K}$	$c_p\ \frac{kJ}{kg \cdot K}$	$c_v\ \frac{kJ}{kg \cdot K}$	k
Air	—	28.97	0.287 00	1.0035	0.7165	1.400
Argon	Ar	39.948	0.208 13	0.5203	0.3122	1.667
Butane	C_4H_{10}	58.124	0.143 04	1.7164	1.5734	1.091
Carbon Dioxide	CO_2	44.01	0.188 92	0.8418	0.6529	1.289
Carbon Monoxide	CO	28.01	0.296 83	1.0413	0.7445	1.400
Ethane	C_2H_6	30.07	0.276 50	1.7662	1.4897	1.186
Ethylene	C_2H_4	28.054	0.296 37	1.5482	1.2518	1.237
Helium	He	4.003	2.077 03	5.1926	3.1156	1.667
Hydrogen	H_2	2.016	4.124 18	14.2091	10.0849	1.409
Methane	CH_4	16.04	0.518 35	2.2537	1.7354	1.299
Neon	Ne	20.183	0.411 95	1.0299	0.6179	1.667
Nitrogen	N_2	28.013	0.296 80	1.0416	0.7448	1.400
Octane	C_8H_{18}	114.23	0.072 79	1.7113	1.6385	1.044
Oxygen	O_2	31.999	0.259 83	0.9216	0.6618	1.393
Propane	C_3H_8	44.097	0.188 55	1.6794	1.4909	1.126
Steam	H_2O	18.015	0.461 52	1.8723	1.4108	1.327

TABLE 7.2 Specific Heats of Liquids and Solids — Metric Units

c_p kJ/(kg·°C)

A. LIQUIDS

Substance	State	c_p	Substance	State	c_p
Water	1 atm, 25°C	4.177	Glycerin	1 atm, 10°C	2.32
Ammonia	sat., —20°C	4.52	Bismuth	1 atm, 425°C	0.144
	sat., 50°C	5.10	Mercury	1 atm, 10°C	0.138
Refrigerant 12	sat., —20°C	0.908	Sodium	1 atm, 95°C	1.38
	sat., 50°C	1.02	Propane	1 atm, 0°C	2.41
Benzene	1 atm, 15°C	1.80	Ethyl Alcohol	1 atm, 25°C	2.43

B. SOLIDS

Substance	T, °C	c_p	Substance	T, °C	c_p
Ice	–11	2.033	Lead	–100	0.118
	–2.2	2.10		0	0.124
Aluminum	–100	0.699		100	0.134
	0	0.870	Copper	–100	0.328
	100	0.941		0	0.381
Iron	20	0.448		100	0.393
Silver	20	0.233			

TABLE 7.3 Thermodynamic Properties of Water (Steam Tables) — Metric Units

7.3.1 Saturated H_2O — Temperature Table

		Volume, m^3/kg		Energy, kJ/kg		Enthalpy, kJ/kg			Entropy, kJ/(kg·K)		
T, °C	P, MPa	v_f	v_g	u_f	u_g	h_f	h_{fg}	h_g	s_f	s_{fg}	s_g
0.010	0.0006113	0.001000	206.1	0.0	2375.3	0.0	2501.3	2501.3	0.0000	9.1571	9.1571
5	0.0008721	0.001000	147.1	21.0	2382.2	21.0	2489.5	2510.5	0.0761	8.9505	9.0266
10	0.001228	0.001000	106.4	42.0	2389.2	42.0	2477.7	2519.7	0.1510	8.7506	8.9016
20	0.002338	0.001002	57.79	83.9	2402.9	83.9	2454.2	2538.1	0.2965	8.3715	8.6680
30	0.004246	0.001004	32.90	125.8	2416.6	125.8	2430.4	2556.2	0.4367	8.0174	8.4541
40	0.007383	0.001008	19.52	167.5	2430.1	167.5	2406.8	2574.3	0.5723	7.6855	8.2578
50	0.01235	0.001012	12.03	209.3	2443.5	209.3	2382.8	2592.1	0.7036	7.3735	8.0771
60	0.01994	0.001017	7.671	251.1	2456.6	251.1	2358.5	2609.6	0.8310	7.0794	7.9104
70	0.03119	0.001023	5.042	292.9	2469.5	293.0	2333.8	2626.8	0.9549	6.8012	7.7561
80	0.04739	0.001029	3.407	334.8	2482.2	334.9	2308.8	2643.7	1.0754	6.5376	7.6130
90	0.07013	0.001036	2.361	376.8	2494.5	376.9	2283.2	2660.1	1.1927	6.2872	7.4799
100(atm)	0.1013	0.001044	1.673	418.9	2506.5	419.0	2257.0	2676.0	1.3071	6.0486	7.3557
120	0.1985	0.001060	0.8919	503.5	2529.2	503.7	2202.6	2706.3	1.5280	5.6024	7.1304
140	0.3613	0.001080	0.5089	588.7	2550.0	589.1	2144.8	2733.9	1.7395	5.1912	6.9307
160	0.6178	0.001102	0.3071	674.9	2568.4	675.5	2082.6	2758.1	1.9431	4.8079	6.7510
180	1.002	0.001127	0.1941	762.1	2583.7	763.2	2015.0	2778.2	2.1400	4.4466	6.5866
200	1.554	0.001156	0.1274	850.6	2595.3	852.4	1940.8	2793.2	2.3313	4.1018	6.4331
220	2.318	0.001190	0.08620	940.9	2602.4	943.6	1858.5	2802.1	2.5183	3.7686	6.2869
240	3.344	0.001229	0.5977	1033.2	2604.0	1037.3	1766.5	2803.8	2.7021	3.4425	6.1446
260	4.688	0.001276	0.04221	1128.4	2599.0	1134.4	1662.5	2796.9	2.8844	3.1184	6.0028
280	6.411	0.001332	0.03017	1227.4	2586.1	1236.0	1543.6	2779.6	3.0674	2.7905	5.8579
300	8.580	0.001404	0.02168	1332.0	2563.0	1344.0	1405.0	2749.0	3.2540	2.4513	5.7053
320	11.27	0.001499	0.01549	1444.6	2525.5	1461.4	1238.7	2700.1	3.4487	2.0883	5.5370
340	14.59	0.001638	0.01080	1570.3	2464.6	1594.2	1027.9	2622.1	3.6601	1.6765	5.3366
360	18.65	0.001892	0.006947	1725.2	2351.6	1760.5	720.7	2481.2	3.9154	1.1382	5.0536
374.136	22.088	0.003155	0.003155	2029.6	2029.6	2099.3	0.0	2099.3	4.4305	0.0000	4.4305

7.3.2 Saturated H_2O — Pressure Table

		Volume, m^3/kg		Energy, kJ/kg		Enthalpy, kJ/kg			Entropy, kJ/(kg·K)		
P, MPa	T, °C	v_f	v_g	u_f	u_g	h_f	h_{fg}	h_g	s_f	s_{fg}	s_g
0.001	7.0	0.001000	129.2	29.3	2385.0	29.3	2484.9	2514.2	0.1059	8.8706	8.9765
0.002	17.5	0.001001	67.00	73.5	2399.5	73.5	2460.0	2533.5	0.2606	8.4639	8.7245
0.01	45.8	0.001010	14.67	191.8	2437.9	191.8	2392.8	2584.6	0.6491	7.5019	8.1510
0.02	60.1	0.001017	7.649	251.4	2456.7	251.4	2358.3	2609.7	0.8319	7.0774	7.9093
0.04	75.9	0.001026	3.993	317.5	2477.0	317.6	2319.1	2636.7	1.0260	6.6449	7.6709
0.06	85.9	0.001033	2.732	359.8	2489.6	359.8	2293.7	2653.5	1.1455	6.3873	7.5328
0.08	93.5	0.001039	2.087	391.6	2498.8	391.6	2274.1	2665.7	1.2331	6.2023	7.4354
0.1	99.6	0.001043	1.694	417.3	2506.1	417.4	2258.1	2675.5	1.3029	6.0573	7.3602
0.12	104.8	0.001047	1.428	439.2	2512.1	439.3	2244.2	2683.5	1.3611	5.9378	7.2989
0.16	113.3	0.001054	1.091	475.2	2521.8	475.3	2221.2	2696.5	1.4553	5.7472	7.2025
0.2	120.2	0.001061	0.8857	504.5	2529.5	504.7	2201.9	2706.6	1.5305	5.5975	7.1280
0.4	143.6	0.001084	0.4625	604.3	2553.6	604.7	2133.8	2738.5	1.7770	5.1197	6.8967
0.6	158.9	0.001101	0.3157	669.9	2567.4	670.6	2086.2	2756.8	1.9316	4.8293	6.7609
0.8	170.4	0.001115	0.2404	720.2	2576.8	721.1	2048.0	2769.1	2.0466	4.6170	6.6636
1	179.9	0.001127	0.1944	761.7	2583.6	762.8	2015.3	2778.1	2.1391	4.4482	6.5873
1.2	188.0	0.001139	0.1633	797.3	2588.8	798.6	1986.2	2784.8	2.2170	4.3072	6.5242
1.6	201.4	0.001159	0.1238	856.9	2596.0	858.8	1935.2	2794.0	2.3446	4.0780	6.4226
2	212.4	0.001177	0.09963	906.4	2600.3	908.8	1890.7	2799.5	2.4478	3.8939	6.3417
4	250.4	0.001252	0.04978	1082.3	2602.3	1087.3	1714.1	2801.4	2.7970	3.2739	6.0709
6	275.6	0.001319	0.03244	1205.4	2589.7	1213.3	1571.0	2784.3	3.0273	2.8627	5.8900
8	295.1	0.001384	0.02352	1305.6	2569.8	1316.6	1441.4	2758.0	3.2075	2.5365	5.7440
12	324.8	0.001527	0.01426	1472.9	2513.7	1491.3	1193.6	2684.9	3.4970	1.9963	5.4933
16	347.4	0.001711	0.009307	1622.7	2431.8	1650.0	930.7	2580.7	3.7468	1.4996	5.2464
20	365.8	0.002036	0.005836	1785.6	2293.2	1826.3	583.7	2410.0	4.0146	0.9135	4.9281
22.088	374.136	0.003155	0.003155	2029.6	2029.6	2099.3	0.0	2099.3	4.4305	0.0000	4.4305

7.3.3 Superheated Steam

°C	v	u	h	s	v	u	h	s	v	u	h	s
	P = .010 MPa				P = .050 MPa				P = .10 MPa			
100	17.196	2515.5	2687.5	8.4479	3.418	2511.6	2682.5	7.6947	1.6958	2506.7	2676.2	7.3614
200	21.825	2661.3	2879.5	8.9038	4.356	2659.9	2877.7	8.1580	2.172	2658.1	2875.3	7.8343
300	26.445	2812.1	3076.5	9.2813	5.284	2811.3	3075.5	8.5373	2.639	2810.4	3074.3	8.2158
400	31.063	2968.9	3279.6	9.6077	6.209	2968.5	3278.9	8.8642	3.103	2967.9	3278.2	8.5435
500	35.679	3132.3	3489.1	9.8978	7.134	3132.0	3488.7	9.1546	3.565	3131.6	3488.1	8.8342
600	40.295	3302.5	3705.4	10.1608	8.057	3302.2	3705.1	9.4178	4.028	3301.9	3704.7	9.0976
700	44.911	3479.6	3928.7	10.4028	8.981	3479.4	3928.5	9.6599	4.490	3479.2	3928.2	9.3398
800	49.526	3663.8	4159.0	10.6281	9.904	3663.6	4158.9	9.8852	4.952	3663.5	4158.6	9.5652
900	54.141	3855.0	4396.4	10.8396	10.828	3854.9	4396.3	10.0967	5.414	3854.8	4396.1	9.7767
1000	58.757	4053.0	4640.6	11.0393	11.751	4052.9	4640.5	10.2964	5.875	4052.8	4640.3	9.9764
	P = .20 MPa				P = .40 MPa				P = .60 MPa			
200	1.080	2654.4	2870.5	7.5066	.5342	2646.8	2860.5	7.1706	.3520	2638.9	2850.1	6.9665
300	1.316	2808.6	3071.8	7.8926	.6548	2804.8	3066.8	7.5662	.4344	2801.0	3061.6	7.3724
400	1.549	2966.7	3276.6	8.2218	.7726	2964.4	3273.4	7.8985	.5137	2962.1	3270.3	7.7079
500	1.781	3130.8	3487.1	8.5133	.8893	3129.2	3484.9	8.1913	.5920	3127.6	3482.8	8.0021
600	2.013	3301.4	3704.0	8.7770	1.0055	3300.2	3702.4	8.4558	.6697	3299.1	3700.9	8.2674
700	2.244	3478.8	3927.6	9.0194	1.1215	3477.9	3926.5	8.6987	.7472	3477.0	3925.3	8.5107
800	2.475	3663.1	4158.2	9.2449	1.2372	3662.4	4157.3	8.9244	.8245	3661.8	4156.5	8.7367
900	2.706	3854.5	4395.8	9.4566	1.3529	3853.9	4395.1	9.1362	.9017	3853.4	4394.4	8.9486
1000	2.937	4052.5	4640.0	9.6563	1.4685	4052.0	4639.4	9.3360	.9788	4051.5	4638.8	9.1485
1100	3.168	4257.0	4890.7	9.8458	1.5840	4256.5	4890.2	9.5256	1.0559	4256.1	4889.6	9.3381
	P = .80 MPa				P = 1.00 MPa				P = 2.00 MPa			
200	.2608	2630.6	2839.3	6.8158	.2060	2621.9	2827.9	6.6940				
300	.3241	2797.2	3056.5	7.2328	.2579	2793.2	3051.2	7.1229	.1255	2772.6	3023.5	6.7664
400	.3843	2959.7	3267.1	7.5716	.3066	2957.3	3263.9	7.4651	.1512	2945.2	3247.6	7.1271
500	.4433	3126.0	3480.6	7.8673	.3541	3124.4	3478.5	7.7622	.1757	3116.2	3467.6	7.4317
600	.5018	3297.9	3699.4	8.1333	.4011	3296.8	3697.9	8.0290	.1996	3290.9	3690.1	7.7024
700	.5601	3476.2	3924.2	8.3770	.4478	3475.3	3923.1	8.2731	.2232	3470.9	3917.4	7.9487
800	.6181	3661.1	4155.6	8.6033	.4943	3660.4	4154.7	8.4996	.2467	3657.0	4150.3	8.1765
900	.6761	3852.8	4393.7	8.8153	.5407	3852.2	4392.9	8.7118	.2700	3849.3	4389.4	8.3895
1000	.7340	4051.0	4638.2	9.0153	.5871	4050.5	4637.6	8.9119	.2933	4048.0	4634.6	8.5901
1100	.7919	4255.6	4889.1	9.2050	.6335	4255.1	4888.6	9.1017	.3166	4252.7	4885.9	8.7800
	P = 3.0 MPa				P = 4.0 MPa				P = 5.0 MPa			
300	.08114	2750.1	2993.5	6.5390	.05884	2725.3	2960.7	6.3615	.04532	2698.0	2924.5	6.2084
400	.09936	2932.8	3230.9	6.9212	.07341	2919.9	3213.6	6.7690	.05781	2906.6	3195.7	6.6459
500	.11619	3108.0	3456.5	7.2338	.08643	3099.5	3445.3	7.0901	.06857	3091.0	3433.8	6.9759
600	.13243	3285.0	3682.3	7.5085	.09885	3279.1	3674.4	7.3688	.07869	3273.0	3666.5	7.2589
700	.14838	3466.5	3911.7	7.7571	.11095	3462.1	3905.9	7.6198	.08849	3457.6	3900.1	7.5122
800	.16414	3653.5	4145.9	7.9862	.12287	3650.0	4141.5	7.8502	.09811	3646.6	4137.1	7.7440
900	.17980	3846.5	4385.9	8.1999	.13469	3843.6	4382.3	8.0647	.10762	3840.7	4378.8	7.9593
1000	.19541	4045.4	4631.6	8.4009	.14645	4042.9	4628.7	8.2662	.11707	4040.4	4625.7	8.1612
1100	.21098	4250.3	4883.3	8.5912	.15817	4248.0	4880.6	8.4567	.12648	4245.6	4878.0	8.3520
1200	.22652	4460.9	5140.5	8.7720	.16987	4458.6	5138.1	8.6376	.13587	4456.3	5135.7	8.5331
	P = 6.0 MPa				P = 8.0 MPa				P = 10.0 MPa			
300	.03616	2667.2	2884.2	6.0674	.02426	2590.9	2785.0	5.7906				
400	.04739	2892.9	3177.2	6.5408	.03432	2863.8	3138.3	6.3634	.02641	2832.4	3096.5	6.2120
500	.05665	3082.2	3422.2	6.8803	.04175	3064.3	3398.3	6.7240	.03279	3045.8	3373.7	6.5966
600	.06525	3266.9	3658.4	7.1677	.04845	3254.4	3642.0	7.0206	.03837	3241.7	3625.3	6.9029
700	.07352	3453.1	3894.2	7.4234	.05481	3443.9	3882.4	7.2812	.04358	3434.7	3870.5	7.1687
800	.08160	3643.1	4132.7	7.6566	.06097	3636.0	4123.8	7.5173	.04859	3628.9	4114.8	7.4077
900	.08958	3837.8	4375.3	7.8727	.06702	3832.1	4368.3	7.7351	.05349	3826.3	4361.2	7.6272
1000	.09749	4037.8	4622.7	8.0751	.07301	4032.8	4616.9	7.9384	.05832	4027.8	4611.0	7.8315
1100	.10536	4243.3	4875.4	8.2661	.07896	4238.6	4870.3	8.1300	.06312	4234.0	4865.1	8.0237
1200	.11321	4454.0	5133.3	8.4474	.08489	4449.5	5128.5	8.3115	.06789	4444.9	5123.8	8.2055
	P = 15.0 MPa				P = 20.0 MPa				P = 40.0 MPa			
400	.01564	2740.7	2975.5	5.8811	.00994	2619.3	2818.1	5.5540	.001908	1854.6	1930.9	4.1135
500	.02080	2996.6	3308.6	6.3443	.01477	2942.9	3238.2	6.1401	.005622	2678.4	2903.3	5.4700
600	.02491	3208.6	3582.3	6.6776	.01818	3174.0	3537.6	6.5048	.008094	3022.6	3346.4	6.0114
700	.02861	3410.9	3840.1	6.9572	.02113	3386.4	3809.0	6.7993	.009941	3283.6	3681.2	6.3750
800	.03210	3610.9	4092.4	7.2040	.()2385	3592.7	4069.7	7.0544	.011523	3517.8	3978.7	6.6662
900	.03546	3811.9	4343.8	7.4279	.02645	3797.5	4326.4	7.2830	.012962	3739.4	4257.9	6.9150
1000	.03875	4015.4	4596.6	7.6348	.02897	4003.1	4582.5	7.4925	.041324	3954.6	4527.6	7.1356
1100	.04200	4222.6	4852.6	7.8283	.03145	4211.3	4840.2	7.6874	.015642	4167.4	4793.1	7.3364
1200	.04523	4433.8	5112.3	8.0108	.03391	4422.8	5101.0	7.8707	.016940	4380.1	5057.7	7.5224

Figure 7.6 *P-h* Diagram for Refrigerant 134a — Metric Units

(Reproduced by permission of the DuPont Company)

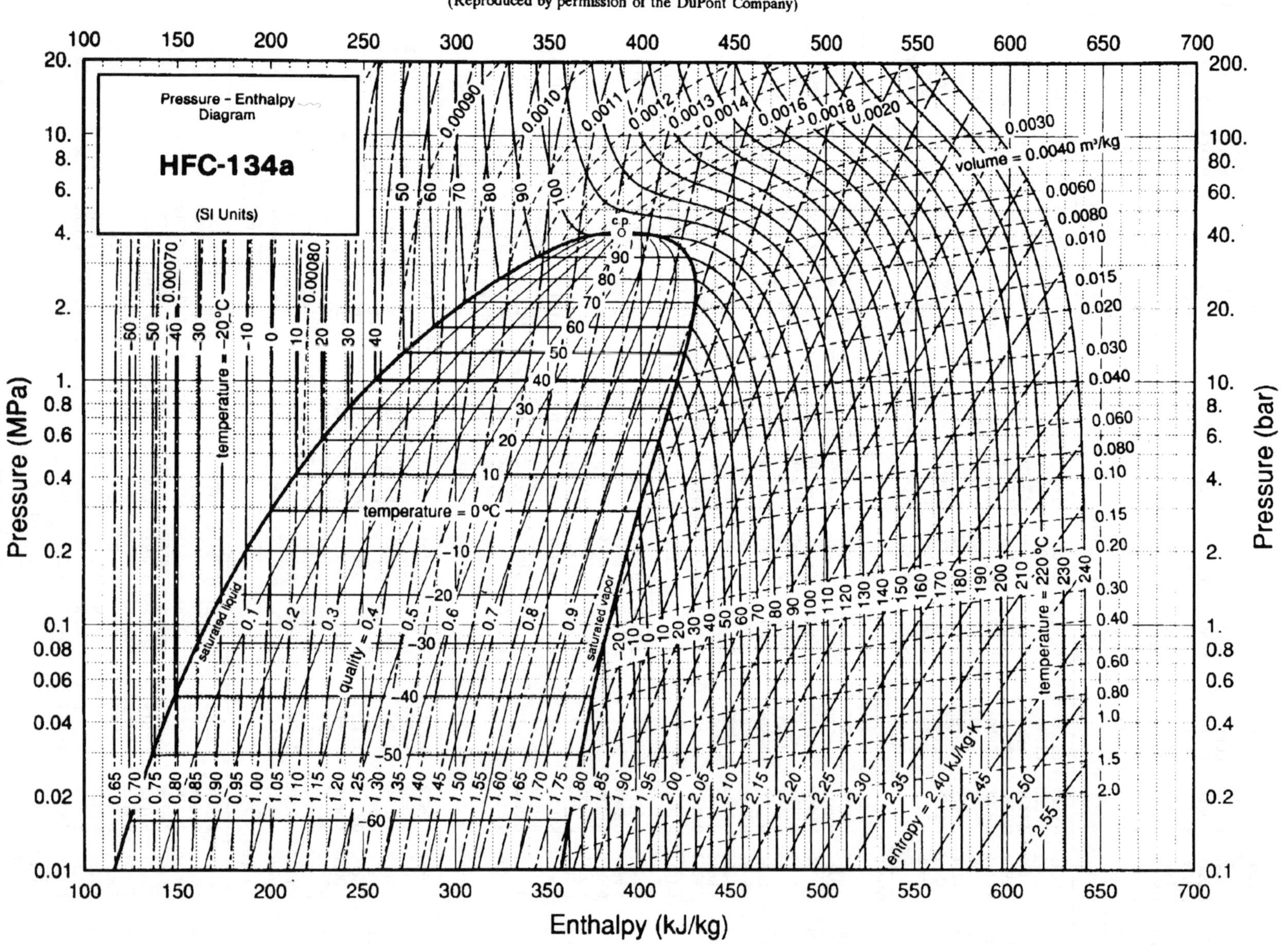

TABLE 7.5 Thermal Properties of Various Materials at 20°C or 68°F.

	Density – ρ		Specific Heat – C_p		Conductivity – k	
Material	**kg/m³**	**lb/ft³**	**J/kg·K**	**BTU/lbm-°R**	**J/s·m·°C**	**BTU/hr-ft-°F**
Aluminum	2700	167	896	0.214	237	136
Bronze	8670	541	343	0.0819	26	15
Concrete	500	30	840	0.2	0.13	0.075
Copper	8930	557	383	0.0915	400	230
Glass	2800	175	800	0.2	0.81	0.47
Ice	910	57	1830	0.44	2.2	1.3
Plaster	1800	112			0.81	0.47
Polystyrene	1210	75			.04	.023
Steel (1% C)	7800	487	473	0.113	43	25
Wood (pine)	420	26	2700	0.64	0.15	0.087
Wool (insulation)	200	12			0.038	0.022

TABLE 7.6 Typical Convective Coefficients *h*.

Fluid	J/s · m² · °C	BTU/hr-ft²-°F
Air, free convection	6 to 30	1 to 5
Air, forced convection	30 to 300	5 to 50
Superheated steam, forced convection	30 to 300	5 to 50
Water, forced convection	300 to 12 000	50 to 2000
Water, boiling	3000 to 60 000	500 to 20,000
Steam, condensing	6000 to 120 000	1000 to 20,000

TABLE 7.1 Properties of Ideal Gases—English Units

Gas	Chemical Formula	Molecular Weight	$R \frac{\text{ft-lb}}{\text{lbm}\cdot°\text{R}}$	$c_p \frac{\text{BTU}}{\text{lbm}\cdot°\text{R}}$	$c_v \frac{\text{BTU}}{\text{lbm}\cdot°\text{R}}$	k
Air	...	28.97	53.34	0.240	0.171	1.400
Argon	Ar	39.94	38.68	0.1253	0.0756	1.667
Butane	C_4H_{10}	58.124	26.58	0.415	0.381	1.09
Carbon Dioxide	CO_2	44.01	35.10	0.203	0.158	1.285
Carbon Monoxide	CO	28.01	55.16	0.249	0.178	1.399
Ethane	C_2H_6	30.07	51.38	0.427	0.361	1.183
Ethylene	C_2H_4	28.054	55.07	0.411	0.340	1.208
Helium	He	4.003	386.0	1.25	0.753	1.667
Hydrogen	H_2	2.016	766.4	3.43	2.44	1.404
Methane	CH_4	16.04	96.35	0.532	0.403	1.32
Neon	Ne	20.183	76.55	0.246	0.1477	1.667
Nitrogen	N_2	28.016	55.15	0.248	0.177	1.400
Octane	C_8H_{18}	114.22	13.53	0.409	0.392	1.044
Oxygen	O_2	32.000	48.28	0.219	0.157	1.395
Propane	C_3H_8,	44.097	35.04	0.407	0.362	1.124
Steam	H_2O	18.016	85.76	0.445	0.335	1.329

TABLE 7.2 Specific Heats of Liquids and Solids—English Units
c_p BTU/(lbm-°F)

A. LIQUIDS

Substance	State	c_p	Substance	State	c_p
Water	1 atm, 77°F	1.00	Glycerin	1 atm, 50°F	0.555
Ammonia	sat., –4°F	1.08	Bismuth	1 atm, 800°F	0.0344
	sat., 120°F	1.22	Mercury	1 atm, 50°F	0.0330
Refrigerant-12	sat.,–4°F	0.217	Sodium	1 atm, 200°F	0.330
	sat., 120°F	0.244	Propane	1 atm, 32°F	0.577
Benzene	1 atm, 60°F	0.431	Ethyl Alcohol	1 atm, 77°F	0.581

B. SOLIDS

Substance	T, °F	c_p	Substance	T, °F	c_p
Ice	–76	0.392	Silver	–4	0.0557
	12	0.486	Lead	–150	0.0282
	28	0.402		30	0.0297
Aluminum	–150	0.167		210	0.0321
	30	0.208	Copper	–150	0.0785
	210	0.225		30	0.0911
Iron	–4	0.107		210	0.0940

TABLE 7.3 Thermodynamic Properties of Water (Steam Tables)—English Units

7.3.1 Saturated H_20—Temperature Table

		Volume, ft^3/lbm		Energy, BTU/lbm		Enthalpy, BTU/lbm			Entropy, BTU/lbm-°R		
T, °F	*P*, psia	v_f	v_g	u_f	u_g	h_f	h_{fg}	h_g	s_f	s_{fg}	s_g
32.018	0.08866	0.016022	3302	0.00	1021.2	0.01	1075.4	1075.4	0.00000	2.1869	2.1869
40	0.12166	0.016020	2445	8.02	1023.9	8.02	1070.9	1078.9	0.01617	2.1430	2.1592
60	0.2563	0.016035	1206.9	28.08	1030.4	28.08	1059.6	1087.7	0.05555	2.0388	2.0943
80	0.5073	0.016073	632.8	48.08	1037.0	48.09	1048.3	1096.4	0.09332	1.9423	2.0356
100	0.9503	0.016130	350.0	68.04	1043.5	68.05	1037.0	1105.0	0.12963	1.8526	1.9822
120	1.6945	0.016205	203.0	87.99	1049.9	88.00	1025.5	1113.5	0.16465	1.7690	1.9336
140	2.892	0.016293	122.88	107.95	1056.2	107.96	1014.0	1121.9	0.19851	1.6907	1.8892
160	4.745	0.016395	77.23	127.94	1062.3	127.96	1002.2	1130.1	0.23130	1.6171	1.8484
180	7.515	0.016509	50.20	147.97	1068.3	147.99	990.2	1138.2	0.26311	1.5478	1.8109
200	11.529	0.016634	33.63	168.04	1074.2	168.07	977.9	1145.9	0.29400	1.4822	1.7762
212	14.698	0.016716	26.80	180.11	1077.6	180.16	970.3	1150.5	0.31213	1.4446	1.7567
220	17.188	0.016772	23.15	188.17	1079.8	188.22	965.3	1153.5	0.32406	1.4201	1.7441
240	24.97	0.016922	16.327	208.36	1085.3	208.44	952.3	1160.7	0.35335	1.3609	1.7143
260	35.42	0.017084	11.768	228.64	1090.5	228.76	938.8	1167.6	0.38193	1.3044	1.6864
280	49.18	0.017259	8.650	249.02	1095.4	249.18	924.9	1174.1	0.40986	1.2504	1.6602
300	66.98	0.017448	6.472	269.52	1100.0	269.73	910.4	1180.2	0.43720	1.1984	1.6356
340	117.93	0.017872	3.792	310.91	1108.0	311.30	879.5	1190.8	0.49031	1.0997	1.5901
380	195.60	0.018363	2.339	352.95	1114.3	353.62	845.4	1199.0	0.54163	1.0067	1.5483
420	308.5	0.018936	1.5024	395.81	1118.3	396.89	807.2	1204.1	0.59152	0.9175	1.5091
460	466.3	0.019614	0.9961	439.7	1119.6	441.4	764.1	1205.5	0.6404	0.8308	1.4712
500	680.0	0.02043	0.6761	485.1	1117.4	487.7	714.8	1202.5	0.6888	0.7448	1.4335
540	961.5	0.02145	0.4658	532.6	1111.0	536.4	657.5	1193.8	0.7374	0.6576	1.3950
580	1324.3	0.02278	0.3225	583.1	1098.9	588.6	589.3	1178.0	0.7872	0.5668	1.3540
620	1784.4	0.02465	0.2209	638.3	1078.5	646.4	505.0	1151.4	0.8398	0.4677	1.3075
660	2362	0.02767	0.14459	702.3	1042.3	714.4	391.1	1105.5	0.8990	0.3493	1.2483
700	3090	0.03666	0.07438	801.7	947.7	822.7	167.5	990.2	0.9902	0.1444	1.1346
705.44	3204	0.05053	0.05053	872.6	872.6	902.5	0	902.5	1.0580	0	1.0580

7.3.2 Saturated H_20—Pressure Table

		Volume, ft^3/lbm		Energy, BTU/lbm		Enthalpy, BTU/lbm			Entropy, BTU/lbm-°R		
P, psia	*T*, °F	v_f	v_g	u_f	u_g	h_f	h_{fg}	h_g	s_f	s_{fg}	s_g
1.0	101.70	0.016136	333.6	69.74	1044.0	69.74	1036.0	1105.8	0.13266	1.8453	1.9779
2.0	126.04	0.016230	173.75	94.02	1051.8	94.02	1022.1	1116.1	0.17499	1.7448	1.9198
4.0	152.93	0.016358	90.64	120.88	1060.2	120.89	1006.4	1127.3	0.21983	1.6426	1.8624
6.0	170.03	0.016451	61.98	137.98	1065.4	138.00	996.2	1134.2	0.24736	1.5819	1.8292
10	193.19	0.016590	38.42	161.20	1072.2	161.23	982.1	1143.3	0.28358	1.5041	1.7877
14.696	211.99	0.016715	26.80	180.10	1077.6	180.15	970.4	1150.5	0.31212	1.4446	1.7567
20	227.96	0.016830	20.09	196.19	1082.0	196.26	960.1	1156.4	0.33580	1.3962	1.7320
30	250.34	0.017004	13.748	218.84	1088.0	218.93	945.4	1164.3	0.36821	1.3314	1.6996
40	267.26	0.017146	10.501	236.03	1092.3	236.16	933.8	1170.0	0.39214	1.2845	1.6767
50	281.03	0.017269	8.518	250.08	1095.6	250.24	924.2	1174.4	0.41129	1.2476	1.6589
60	292.73	0.017378	7.177	262.06	1098.3	262.25	915.8	1178.0	0.42733	1.2170	1.6444
70	302.96	0.017478	6.209	272.56	1100.6	272.79	908.3	1181.0	0.44120	1.1909	1.6321
80	312.07	0.017570	5.474	281.95	1102.6	282.21	901.4	1183.6	0.45344	1.1679	1.6214
90	320.31	0.017655	4.898	290.46	1104.3	290.76	895.1	1185.9	0.46442	1.1475	1.6119
100	327.86	0.017736	4.434	298.28	1105.8	298.61	889.2	1187.8	0.47439	1.1290	1.6034
120	341.30	0.017886	3.730	312.27	1108.3	312.67	878.5	1191.1	0.49201	1.0966	1.5886
140	353.08	0.018024	3.221	324.58	1110.3	325.05	868.7	1193.8	0.50727	1.0688	1.5761
160	363.60	0.018152	2.836	335.63	1112.0	336.16	859.8	1196.0	0.52078	1.0443	1.5651
180	373.13	0.018273	2.533	345.68	1113.4	346.29	851.5	1197.8	0.53292	1.0223	1.5553
200	381.86	0.018387	2.289	354.9	1114.6	355.6	843.7	1199.3	0.5440	1.0025	1.5464
300	417.43	0.018896	1.5442	393.0	1118.2	394.1	809.8	1203.9	0.5883	0.9232	1.5115
400	444.70	0.019340	1.1620	422.8	1119.5	424.2	781.2	1205.5	0.6218	0.8638	1.4856
500	467.13	0.019748	0.9283	447.7	1119.4	449.5	755.8	1205.3	0.6490	0.8154	1.4645
600	486.33	0.02013	0.7702	469.4	1118.6	471.7	732.4	1204.1	0.6723	0.7742	1.4464
800	518.36	0.02087	0.5691	506.6	1115.0	509.7	689.6	1199.3	0.7110	0.7050	1.4160
1000	544.75	0.02159	0.4459	538.4	Il 09.9	542.4	650.0	1192.4	0.7432	0.6471	1.3903
1400	587.25	0.02307	0.3016	592.7	1 096.0	598.6	575.5	1174.1	0.7964	0.5497	1.3461
2000	636.00	0.02565	0.18813	662.4	1066.6	671.9	464.4	1136.3	0.8623	0.4238	1.2861
3000	695.52	0.03431	0.08404	783.4	968.8	802.5	213.0	1015.5	0.9732	0.1843	1.1575
3203.6	705.44	0.05053	0.05053	872.6	872.6	902.5	0	902.5	1.0580	0	1.0580

7.3.3 Superheated Vapor

°F	*v*	*u*	*h*	*s*	*v*	*u*	*h*	*s*	*v*	*u*	*h*	*s*
	1 psia				10 psia				14.7 psia			
200	392.5	1077.5	1150.1	2.0508	38.85	1074.7	1146.6	1.7927	...	...	...	...
300	452.3	1112.1	1195.8	2.1153	45.00	1110.6	1193.9	1.8595	30.53	1109.7	1192.8	1.8160
400	511.9	1147.0	1241.8	2.1720	51.03	1146.1	1240.5	1.9171	34.67	1145.6	1239.9	1.8741
500	571.5	1182.8	1288.5	2.2235	57.04	1182.2	1287.7	1.9690	38.77	1181.8	1287.3	1.9263
600	631.1	1219.3	1336.1	2.2706	63.03	1218.9	1335.5	2.0164	42.86	1218.6	1335.2	1.9737
700	690.7	1256.7	1384.5	2.3142	69.01	1256.3	1384.0	2.0601	46.93	1256.1	1383.8	2.0175
800	750.3	1294.9	1433.7	2.3550	74.98	1294.6	1433.3	2.1009	51.00	1294.4	1433.1	2.0584
1000	869.5	1373.9	1534.8	2.4294	86.91	1373.8	1534.6	2.1755	59.13	1373.7	1534.5	2.1330
1200	988.6	1456.7	1639.6	2.4967	98.84	1456.5	1639.4	2.2428	67.25	1456.5	1639.3	2.2003
1400	1107.7	1543.1	1748.1	2.5584	110.76	1543.0	1748.0	2.3045	75.36	1543.0	1747.9	2.2621
	20 psia				60 psia				100 psia			
300	22.36	1108.8	1191.6	1.7808	7.259	1101.0	1181.6	1.6492	...	...	...	...
400	25.43	1145.1	1239.2	1.8395	8.353	1140.8	1233.5	1.7134	4.934	1136.2	1227.5	1.6517
500	28.46	1181.5	1286.8	1.8919	9.399	1178.6	1283.0	1.7678	5.587	1175.7	1279.1	1.7085
600	31.47	1218.4	1334.8	1.9395	10.425	1216.3	1332.1	1.8165	6.216	1214.2	1329.3	1.7582
700	34.47	1255.9	1383.5	1.9834	11.440	1254.4	1381.4	1.8609	6.834	1252.8	1379.2	1.8033
800	37.46	1294.3	1432.9	2.0243	12.448	1293.0	1431.2	1.9022	7.445	1291.8	1429.6	1.8449
1000	43.44	1373.5	1534.3	2.0989	14.454	1372.7	1533.2	1.9773	8.647	1371.9	1532.1	1.9204
1200	49.41	1456.4	1639.2	2.1663	16.452	1455.8	1638.5	2.0448	9.861	1455.2	1637.7	1.9882
1400	55.37	1542.9	1747.9	2.2281	18.445	1542.5	1747.3	2.1067	11.060	1542.0	1746.7	2.0502
1600	61.33	1633.2	1860.1	2.2854	20.44	1632.8	1859.7	2.1641	12.257	1632.4	1859.3	2.1076
	120 psia				160 psia				200 psia			
400	4.079	1133.8	1224.4	1.6288	3.007	1128.8	1217.8	1.5911	2.361	1123.5	1210.8	1.5600
500	4.633	1174.2	1277.1	1.6868	3.440	1171.2	1273.0	1.6518	2.724	1168.0	1268.8	1.6239
600	5.164	1213.2	1327.8	1.7371	3.848	1211.1	1325.0	1.7034	3.058	1208.9	1322.1	1.6767
700	5.682	1252.0	1378.2	1.7825	4.243	1250.4	1376.0	1.7494	3.379	1248.8	1373.8	1.7234
800	6.195	1291.2	1428.7	1.8243	4.631	1289.9	1427.0	1.7916	3.693	1288.6	1425.3	1.7660
1000	7.208	1371.5	1531.5	1.9000	5.397	1370.6	1530.4	1.8677	4.310	1369.8	1529.3	1.8425
1200	8.213	1454.9	1637.3	1.9679	6.154	1454.3	1636.5	1.9358	4.918	1453.7	1635.7	1.9109
1400	9.214	1541.8	1746.4	2.0300	6.906	1541.4	1745.9	1.9980	5.521	1540.9	1745.3	1.9732
1600	10.212	1632.3	1859.0	2.0875	7.656	1631.9	1858.6	2.0556	6.123	1631.6	1858.2	2.0308
1800	11.209	1726.2	1975.1	2.1413	8.405	1725.9	1974.8	2.1094	6.722	1725.6	1974.4	2.0847
	250 psia				300 psia				400 psia			
500	2.150	1163.8	1263.3	1.5948	1.7662	1159.5	1257.5	1.5701	1.2843	1150.1	1245.2	1.5282
600	2.426	1206.1	1318.3	1.6494	2.004	1203.2	1314.5	1.6266	1.4760	1197.3	1306.6	1.5892
700	2.688	1246.7	1371.1	1.6970	2.227	1244.6	1368.3	1.6751	1.6503	1240.4	1362.5	1.6397
800	2.943	1287.0	1423.2	1.7401	2.442	1285.4	1421.0	1.7187	1.8163	1282.1	1416.6	1.6844
900	3.193	1327.6	1475.3	1.7799	2.653	1326.3	1473.6	1.7589	1.9776	1323.7	1470.1	1.7252
1000	3.440	1368.7	1527.9	1.8172	2.860	1367.7	1526.5	1.7964	2.136	1365.5	1523.6	1.7632
1200	3.929	1453.0	1634.8	1.8858	3.270	1452.2	1633.8	1.8653	2.446	1450.7	1631.8	1.8327
1400	4.414	1540.4	1744.6	1.9483	3.675	1539.8	1743.8	1.9279	2.752	1538.7	1742.4	1.8956
1600	4.896	1631.1	1857.6	2.0060	4.078	1630.7	1857.0	1.9857	3.055	1629.8	1855.9	1.9535
1800	5.376	1725.2	1974.0	2.0599	4.479	1724.9	1973.5	2.0396	3.357	1724.1	1972.6	2.0076
	600 psia				800 psia				1000 psia			
500	0.7947	1128.0	1216.2	1.4592	...	...	...	...	...	...	...	...
600	0.9456	1184.5	1289.5	1.5320	0.6776	1170.1	1270.4	1.4861	0.5140	1153.7	1248.8	1.4450
700	1.0727	1231.5	1350.6	1.5872	0.7829	1222.1	1338.0	1.5471	0.6080	1212.0	1324.6	1.5135
800	1.1900	1275.4	1407.6	1.6343	0.8764	1268.5	1398.2	1.5969	0.6878	1261.2	1388.5	1.5664
900	1.3021	1318.4	1462.9	1.6766	0.9640	1312.9	1455.6	1.6408	0.7610	1307.3	1448.1	1.6120
1000	1.4108	1361.2	1517.8	1.7155	1.0482	1356.7	1511.9	1.6807	0.8305	1352.2	1505.9	1.6530
1200	1.6222	1447.7	1627.8	1.7861	1.2102	1444.6	1623.8	1.7526	0.9630	1441.5	1619.7	1.7261
1400	1.8289	1536.5	1739.5	1.8497	1.3674	1534.2	1736.6	1.8167	1.0905	1531.9	1733.7	1.7909
1600	2.033	1628.0	1853.7	1.9080	1.5218	1626.2	1851.5	1.8754	1.2152	1624.4	1849.3	1.8499
1800	2.236	1722.6	1970.8	1.9622	1.6749	1721.0	1969.0	1.9298	1.3384	1719.5	1967.2	1.9046
	2000 psia				3000 psia				4000 psia			
700	0.2487	1147.7	1239.8	1.3782	0.09771	1003.9	1058.1	1.1944	0.02867	742.1	763.4	0.9345
800	0.3071	1220.1	1333.8	1.4562	0.17572	1167.6	1265.2	1.3675	0.10522	1095.0	1172.9	1.2740
900	0.3534	1276.8	1407.6	1.5126	0.2160	1241.8	1361.7	1.4414	0.14622	1201.5	1309.7	1.3789
1000	0.3945	1328.1	1474.1	1.5598	0.2485	1301.7	1439.6	1.4967	0.17520	1272.9	1402.6	1.4449
1200	0.4685	1425.2	1598.6	1.6398	0.3036	1408.0	1576.6	1.5848	0.2213	1390.1	1553.9	1.5423
1400	0.5368	1520.2	1718.8	1.7082	0.3524	1508.1	1703.7	1.6571	0.2603	1495.7	1688.4	1.6188
1600	0.6020	1615.4	1838.2	1.7692	0.3978	1606.3	1827.1	1.7201	0.2959	1597.1	1816.1	1.6841
1800	0.6656	1712.0	1958.3	1.8249	0.4416	1704.5	1949.6	1.7769	0.3296	1697.1	1941.1	1.7420
2000	0.7284	1810.6	2080.2	1.8765	0.4844	1803.9	2072.8	1.8291	0.3625	1797.3	2065.6	1.7948

Figure 7.6 *P-h* Diagram for Refrigerant 134a — English Units

(Reproduced by permission of the DuPont Company)

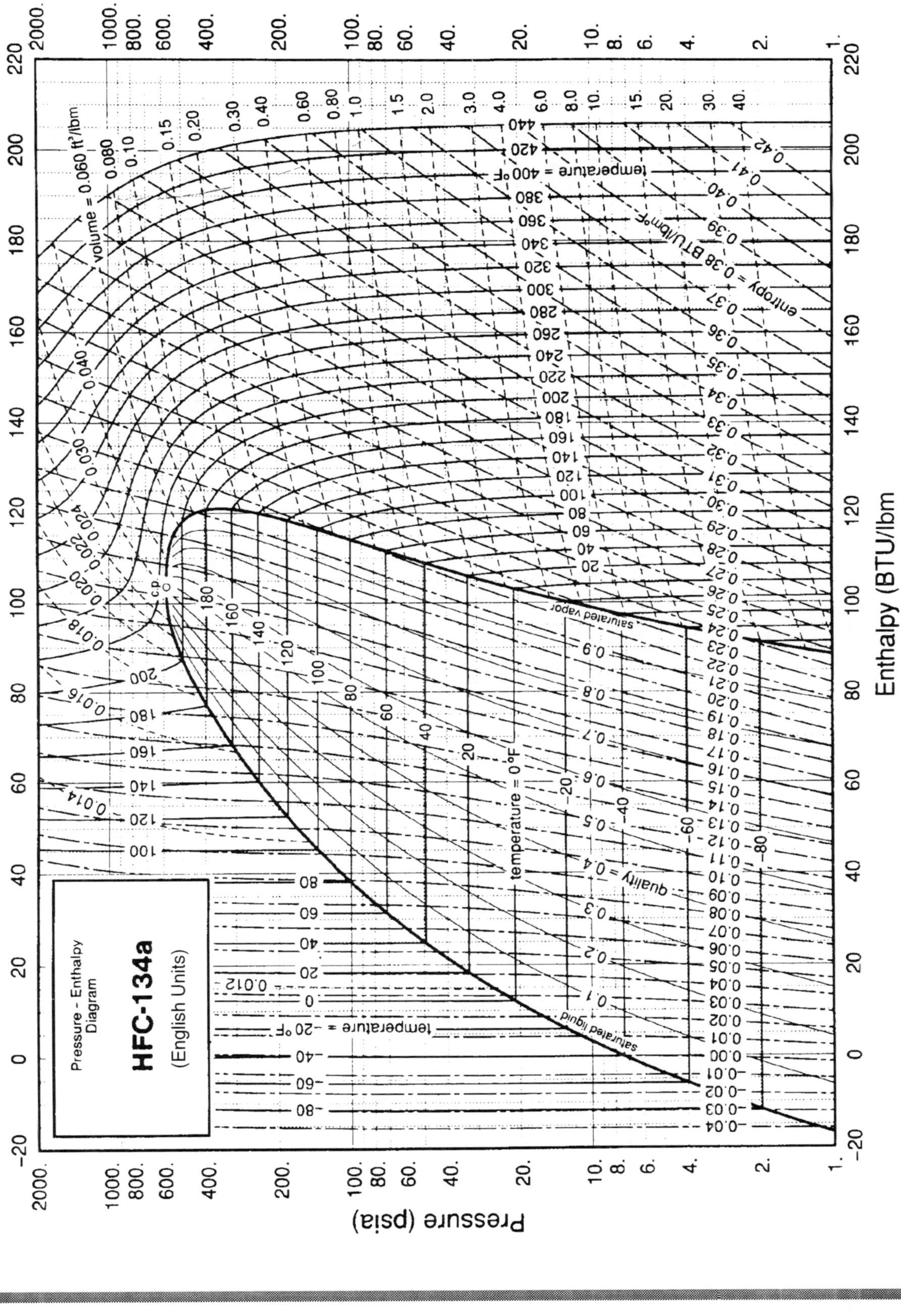

Statics

by George E. Mase

Statics is concerned primarily with the equilibrium of bodies subjected to force systems. Also, traditionally in engineering statics we consider centroids, center of gravity and moments of inertia.

8.1 Forces, Moments and Resultants

A force is the manifestation of the action of one body upon another. Forces arise from the direct action of two bodies in contact with one another, or from the "action at a distance" of one body upon another as occurs with gravitational and magnetic forces. We classify forces as either *body forces* which act (and are distributed) throughout the volume of the body, or as *surface forces*, which act over a surface portion (either external or internal) of the body. If the surface over which the force system acts is very small we usually assume localization at a specific point in the surface and speak of a *concentrated force* at that point.

Mathematically, forces are represented by *vectors*. Geometrically, a vector is a directed line segment having a head and a tail, i.e., an arrow. The length of the arrow corresponds to the magnitude of the force, its orientation defines the line of action, and the direction of the arrow (tail to head) gives the sense of the force.

Systems of concentrated forces are listed as *concurrent* when all of the forces act, or could act, at a single point; otherwise they are termed *non-concurrent* systems. Parallel force systems are in this second group. Also, force systems are often described as two-dimensional (acting in a single plane), or three-dimensional (spatial systems).

In addition to the "push or pull" effect on the point at which it acts, a force creates a *moment* about other points of the body. Conceptually, a moment may be thought of as a torque. Mathematically, the moment of the force **F** with respect to point A when the force acts at point B is defined by the vector cross product

$$\mathbf{M}_A = \mathbf{r}_{B/A} \times \mathbf{F} \tag{8.1.1}$$

where $\mathbf{r}_{B/A}$ is the position vector of B relative to A as shown by Fig. 8.1. Actually, it is easily shown that

$$\mathbf{M}_A = \mathbf{r}_{Q/A} \times \mathbf{F} = \mathbf{r}_{B/A} \times \mathbf{F} \tag{8.1.2}$$

where Q is any point on the line of action of **F**, as shown in Fig. 8.1.

A moment must always be designated with respect to a specific point, and is represented by a vector perpendicular to the plane of **r** and **F**. Moment vectors are denoted by double-headed arrows to distinguish them from force vectors. The component of the moment vector in the direction of any axis (line) passing through A is said to be the moment of the force about that axis. If the direction of the axis is defined by the unit vector $\boldsymbol{\lambda}$, the moment component, a scalar along that axis, is given by

$$M_\lambda = \mathbf{r} \times \mathbf{F} \cdot \boldsymbol{\lambda}\,. \tag{8.1.3}$$

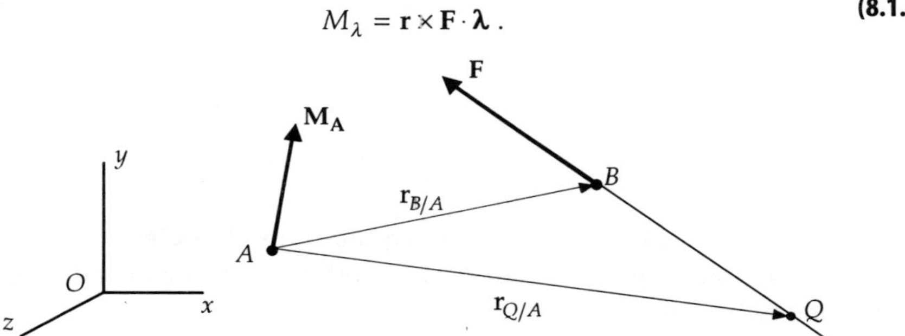

Figure 8.1 Moment of F about A.

The "turning effect" about the hinge axis of a door due to the application of a force to the doorknob is an example. If $\boldsymbol{\lambda} = \mathbf{k}$ in Eq. 8.1.3, the moment component is labeled M_z, the scalar moment about the z-axis, and we find that

$$M_z = xF_y - yF_x \tag{8.1.4}$$

where x and y are the components of $\mathbf{r}_{B/A}$ of Fig. 8.1. Similarly, for $\boldsymbol{\lambda} = \mathbf{j}$ and $\boldsymbol{\lambda} = \mathbf{i}$, respectively,

$$M_y = zF_x - xF_z \tag{8.1.5}$$

$$M_x = yF_z - zF_y. \tag{8.1.6}$$

Furthermore, it may be shown from Eq. 8.1.4 that the scalar moment about any point A in the plane of **r** and **F** is given by

$$M_A = Fd \tag{8.1.7}$$

where d is the perpendicular distance from A to the line of action of **F**.

The *resultant* of a system of forces is the equivalent force and moment of the total system at any point. Thus, the resultant of a concurrent force system is a single force acting at the point of concurrency, and being the vector sum of the individual forces. In contrast, the resultant of a non-concurrent system depends upon the point at which it is determined, and in general consists of a resultant force and resultant moment. Actually, the resultant force (the vector sum of the individual forces) is the same at all points, but the resultant moment will vary (in magnitude and direction) from point to point.

The resultant of a pair of equal, but oppositely directed parallel forces, known as a *couple*, is simply a moment having a magnitude Fd where F is the magnitude of the forces and d the perpendicular distance between their lines of action. Figure 8.2 illustrates equivalent couples and the curly symbol often used.

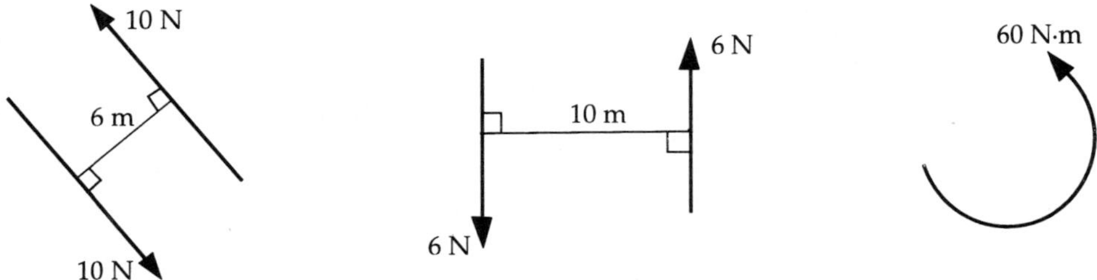

Figure 8.2 Equivalent couples.

Example 8.1

Determine the resultant force for the (a) plane, and (b) space concurrent systems shown.

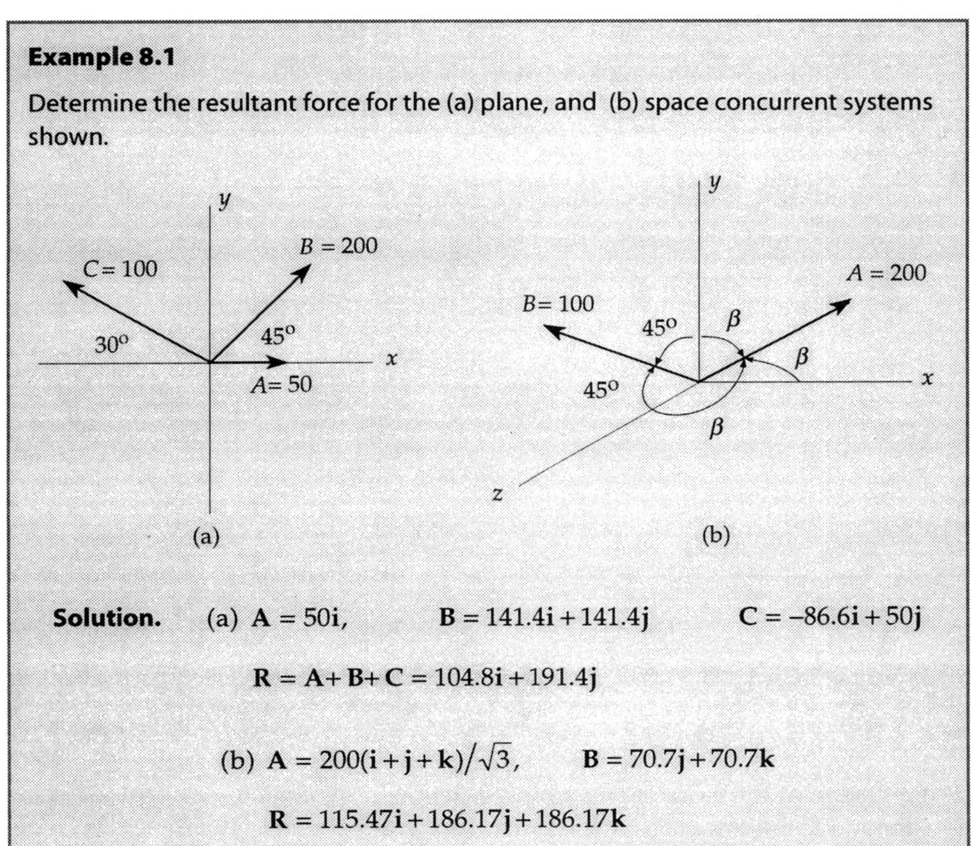

Solution. (a) $\mathbf{A} = 50\mathbf{i}$, $\mathbf{B} = 141.4\mathbf{i} + 141.4\mathbf{j}$ $\mathbf{C} = -86.6\mathbf{i} + 50\mathbf{j}$

$$\mathbf{R} = \mathbf{A} + \mathbf{B} + \mathbf{C} = 104.8\mathbf{i} + 191.4\mathbf{j}$$

(b) $\mathbf{A} = 200(\mathbf{i} + \mathbf{j} + \mathbf{k})/\sqrt{3}$, $\mathbf{B} = 70.7\mathbf{j} + 70.7\mathbf{k}$

$$\mathbf{R} = 115.47\mathbf{i} + 186.17\mathbf{j} + 186.17\mathbf{k}$$

Example 8.2

Determine the moment of force **F**

(a) with respect to the origin. (b) with respect to point *Q*. (c) about the axis *OQ*.
(d) about the *x*-axis. (e) about the *y*-axis.

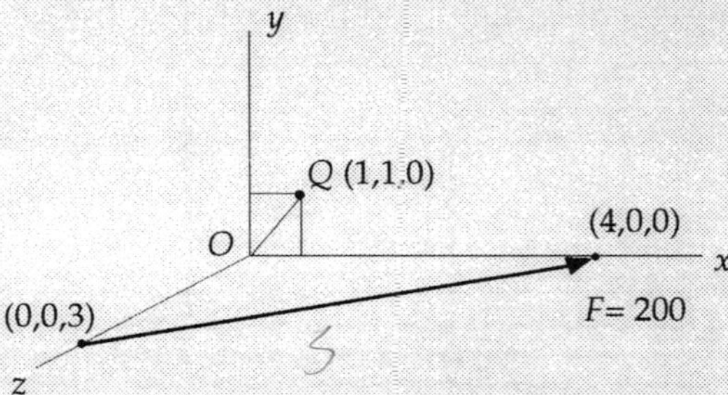

Solution. $\mathbf{F} = 0.8(200)\mathbf{i} - 0.6(200)\mathbf{k} = 160\mathbf{i} - 120\mathbf{k}$.

(a) $\mathrm{M}_O = 4\mathrm{i} \times (160\mathrm{i} - 120\mathrm{k}) = 3\mathrm{k} \times (160\mathrm{i} - 120\mathrm{k}) = 480\mathrm{j}$.

(b) $\mathrm{M}_Q = (3\mathrm{i} - \mathrm{j}) \times (160\mathrm{i} - 120\mathrm{k}) = (-\mathrm{i} - \mathrm{j} + 3\mathrm{k}) \times (160\mathrm{i} - 120\mathrm{k}) = 120\mathrm{i} + 360\mathrm{j} + 160\mathrm{k}$.

(c) $M_{OQ} = (120\mathrm{i} + 360\mathrm{j} + 160\mathrm{k}) \cdot (\mathrm{i} + \mathrm{j})/\sqrt{2}$

$= (120 + 360)/\sqrt{2} = 480/\sqrt{2} = 339.5$.

(d) $M_x = 0$ (force F intersects the x-axis).

(e) $M_y = Fd = 200(3\cos\beta) = 200(2.4) = 480\ \mathrm{N \cdot m}$.

(4, 0, 0)
x
d
β
(0, 0, 3)
z

8.2 Equilibrium

If the system of forces acting on a body is one whose resultant is absolutely zero (vector sum of all forces is zero, and the resultant moment of the forces about every point is zero) the body is in *equilibrium*. Mathematically, equilibrium requires the equations

$$\sum \mathbf{F} = 0, \qquad \sum \mathbf{M}_A = 0 \tag{8.2.1}$$

to be simultaneously satisfied, with A arbitrary. These two vector equations are equivalent to the six scalar equations

$$\sum F_x = 0 \qquad \sum M_x = 0 \qquad \textbf{(8.2.2)}$$

$$\sum F_y = 0 \qquad \sum M_y = 0$$

$$\sum F_z = 0 \qquad \sum M_z = 0$$

which must hold at every point A for any orientation of the xyz-axes. The moment components in Eq. 8.2.2 are the coordinate axes components of $\mathbf{M}_A$. If the forces are concurrent and their vector sum is zero, the sum of moments about every point will be satisfied automatically.

If all the forces act in a single plane, say the xy-plane, one of the above force equations, and two of the moment equations are satisfied identically, so that equilibrium requires only

$$\sum F_x = 0, \qquad \sum F_y = 0, \qquad \sum M_z = 0. \qquad \textbf{(8.2.3)}$$

In this case we can solve for only three unknowns instead of six as when Eqs. 8.2.2 are required.

The solution for unknown forces and moments in equilibrium problems rests firmly upon the construction of a good *free body diagram*, abbreviated FBD, from which the detailed equations 8.2.2 or 8.2.3 may be obtained. A free body diagram is a neat sketch of the body (or of any appropriate portion of it) showing all forces and moments acting on the body, together with all important linear and angular dimensions.

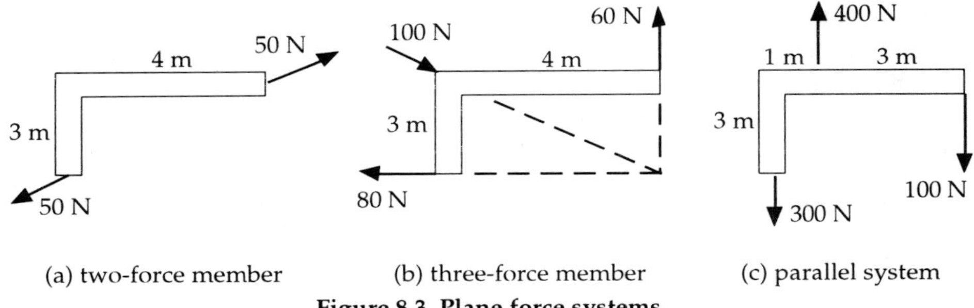

Figure 8.3 Plane force systems.

A body in equilibrium under the action of two forces only is called a *two-force member*, and the two forces must be equal in magnitude and oppositely directed along the line joining their points of application. If a body is in equilibrium under the action of three forces (a *three-force member*) those forces must be coplanar, and concurrent (unless they form a parallel system). Examples are shown in Fig. 8.3.

A knowledge of the possible reaction forces and moments at various supports is essential in preparing a correct free body diagram. Several of the basic reactions are illustrated in Fig. 8.4 showing a block of concrete subjected to a horizontal pull P, Fig. 8.4a, and a cantilever beam carrying both a distributed and concentrated load, Fig. 8.4b. The correct FBDs are to the right of the sketches.

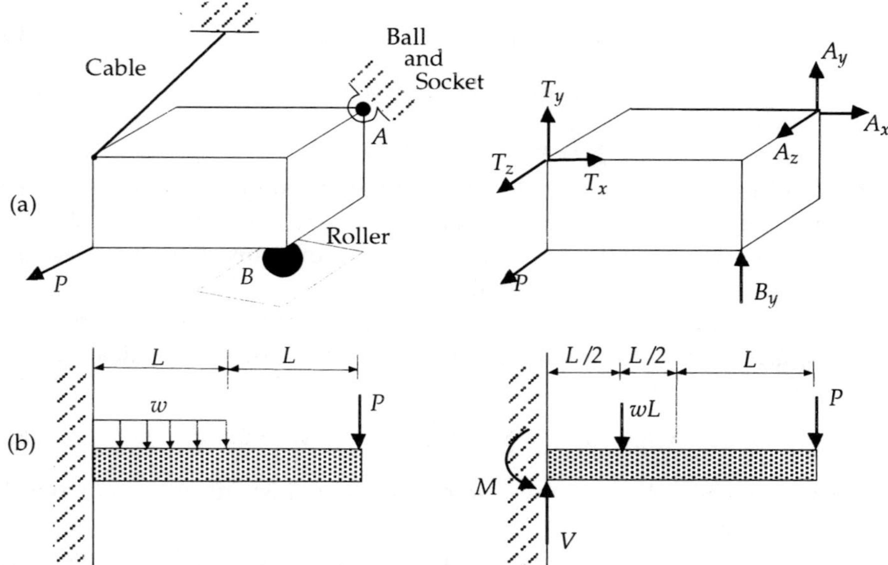

Figure 8.4 Free body diagrams.

Example 8.3

Determine the tension in the two cables supporting the 700 N block.
Solution. Construct the FBD of junction *A* of the cables. Sum forces in *x* and *y* directions:

$$\sum F_x = -0.707T_{AB} + 0.8T_{AC} = 0$$

$$\sum F_y = 0.707T_{AB} + 0.6T_{AC} - 700 = 0.$$

Solve, simultaneously, and find

$$T_{AC} = 700/1.4 = 500 \text{ N}$$

$$T_{AB} = 0.8(500)/0.707 = 565.8 \text{ N}.$$

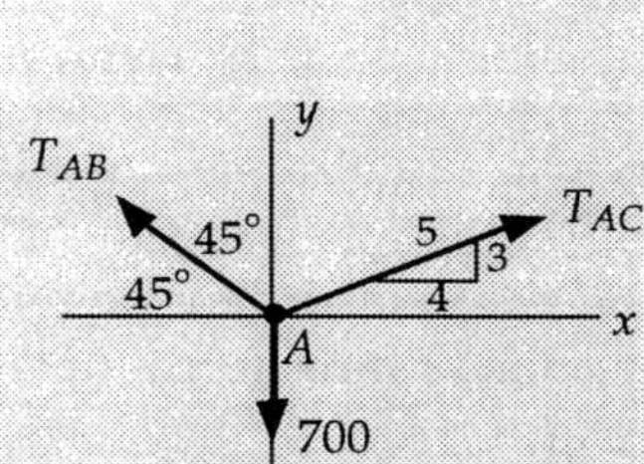

Alternative solution. Draw the force polygon (vector sum of forces) which must close for equilibrium. Determine angles, and use law of sines:

$$\frac{700}{\sin 81.87^\circ} = \frac{T_{AB}}{\sin 53.1^\circ} = \frac{T_{AC}}{\sin 45^\circ}$$

$$T_{AC} = \frac{700(0.707)}{(0.99)} = 500 \text{ N}$$

$$T_{AB} = \frac{700(0.8)}{(0.99)} = 565.8 \text{ N}.$$

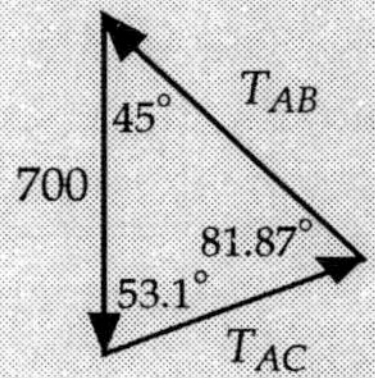

Example 8.4

A 12 m bar weighing 140 N is hinged to a vertical wall at *A*, and supported by the cable *BC*. Determine the tension in the cable together with the horizontal and vertical components of the force reaction at *A*.

Solution. Construct the FBD showing force components at *A* and *B*. Write the equilibrium equations and solve:

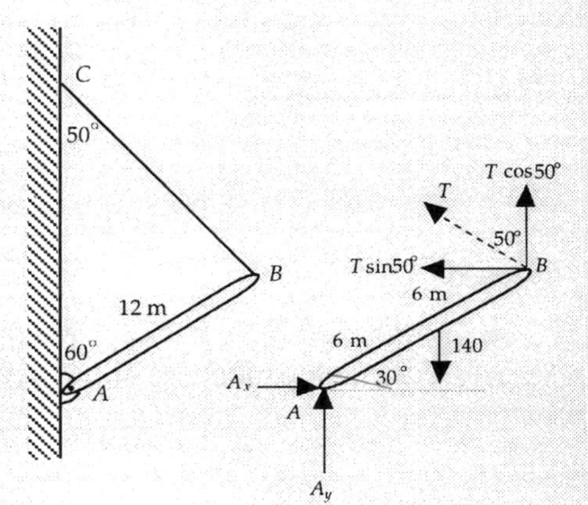

$$\sum M_A = 6T\sin 50^\circ + 6\sqrt{3}T\cos 50^\circ - 140(6)\sqrt{3}/2 = 0$$

$$\therefore T = 64.5 \text{ N}$$

$$\sum F_x = A_x - T\sin 50^\circ = A_x - 64.5(0.766) = 0$$

$$\therefore A_x = 49.4 \text{ N}$$

$$\sum F_y = A_y + T\cos 50^\circ - 140 = A_y + 64.5(0.643) - 140 = 0$$

$$\therefore A_y = 98.5 \text{ N.}$$

8.3 Trusses and Frames

Simple pin-connected trusses and plane frames provide us with elementary examples of structures that may be solved by the equilibrium concepts of statics.

The classic truss problem resembles the one-lane country bridge, as shown schematically in Fig. 8.5a. All members are assumed to be two-force members and are therefore in simple (axial) tension or compression. All loads are assumed to act at the joints (labeled *A*, *B*, *C*, etc.) where the members are pinned together. External reactions such as A_x, A_y and E_y may be determined as a non-concurrent force problem from a FBD of the entire truss. Following that, the internal forces in the members themselves may be determined from a FBD of each joint in turn

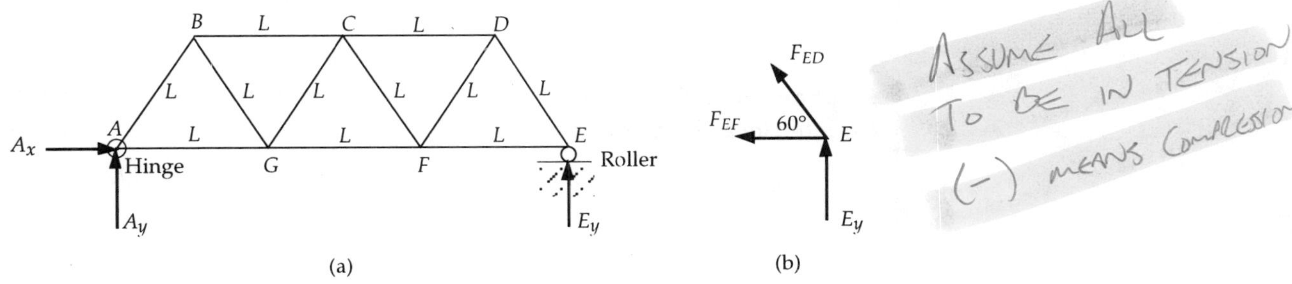

Figure 8.5 Simple truss.

(method of joints) starting, for example, with joint E of the truss as shown in Fig. 8.5b. Thus, we solve a sequence of concurrent force problems at successive joints having only two unknowns. As noted by Fig. 8.5b we may assume the unknown internal forces such as F_{ED}, F_{EF}, etc., to be tension. A negative result indicates compression.

Example 8.5

Using the right sub-truss of Fig 8.6 determine the forces in members *CD*,*CF*and*FG*.
Solution. Summing moments about pin *F* of the FBD of the right sub-truss,

$$\sum M_F = F_{DC}(\sqrt{3}L/2) + 10P(L) - 6P(L/2) = 0.$$

$$\therefore F_{DC} = -8.085P \text{ (comp).}$$

Summing vertical forces on the sub-truss,

$$\sum F_y = 10P - 3P - 6P + 0.866F_{FC} = 0.$$

$$\therefore F_{FC} = -1.156P \text{ (comp).}$$

Summing horizontal forces,

$$\sum F_x = -F_{FG} + 8.085P + 1.156P/2 = 0.$$

$$\therefore F_{FG} = 8.663P \text{ (tens).}$$

Note: These results agree with those determined by the method ofjpints in Example 8.6

Example 8.6

Determine the forces in the members of the pin-connected truss loaded as shown below. All members have length *L*.

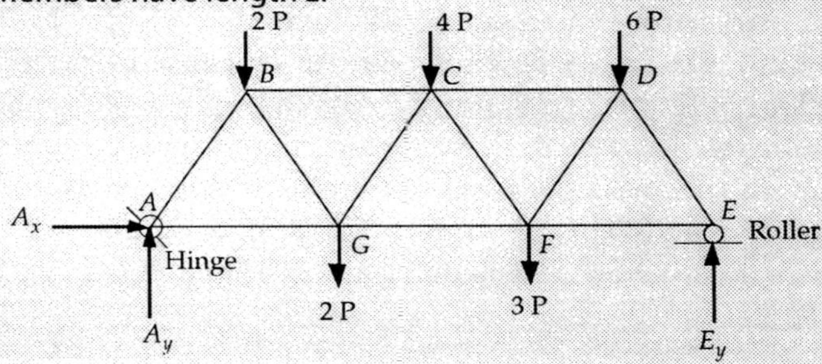

Solution. Summing moments about pin *A* we solve for the reaction at roller *E* (counterclockwise moments are positive):

$$\sum M_A = 3LE_y - 2PL - 3P(2L) - 2P(L/2) - 4P(3L/2) - 6P(5L/2) = 0.$$

$$\therefore E_y = 10P.$$

Also,

$$\sum M_E = -3LA_y + 3P(L) + 2P(2L) + 6P(L/2) + 4P(3L/2) + 2P(5L/2) = 0.$$

$$\therefore A_y = 7P.$$

Summing Horizontal forces,

$$\sum F_x = A_x = 0.$$

Consider joint *E* (see FBD at right),

$$\sum F_y = T_{ED}\sin 60^\circ + 10P = 0$$

$$\therefore T_{ED} = -11.55P \text{ (comp)}.$$

$$\sum F_x = -T_{EF} - T_{ED}\cos 60^\circ = 0$$

$$\therefore T_{EF} = 5.775P \text{ (tens)}.$$

Consider next joint *D* (see FBD at right),

$$\sum F_y = -6P - 0.866F_{DF} + 0.866(11.55P) = 0.$$

$$\therefore F_{DF} = 4.62P \text{ (tens)}.$$

$$\sum F_x = -F_{DC} - 0.5(11.55P) - 0.5(4.62P) = 0$$

$$\therefore F_{DC} = -8.085P \text{ (comp)}.$$

Consider next joint *F* (see FBD at right),

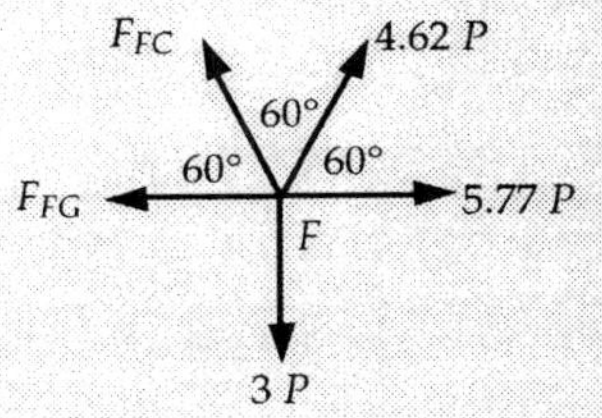

$$\sum F_y = -3P + 0.866(4.62P) + 0.866F_{FC} = 0.$$

$$\therefore F_{FC} = -1.156P \text{ (comp)}.$$

$$\sum F_x = -F_{FG} - 0.5(-1.156P) + 0.5(4.62P) + 5.775P = 0$$

$$\therefore F_{FG} = 8.663P \text{ (tens)}.$$

Note: This example should be completed by considering joint *C* next, then joint *G*, and so on.

If the internal forces in only a few selected members are required, the *method of sections* may be used. For example, to obtain only the forces in members *CD*, *CF* and *GF* of the above truss, we "section" it into two portions by cutting across those members as shown in Fig. 8.6. Each portion becomes a sub-truss, and the internal forces of the sectioned members become external reactions of the two sub-trusses. Both force $\left(\sum F = 0\right)$ and moment $\left(\sum M_A = 0\right)$ equations are useful in this method.

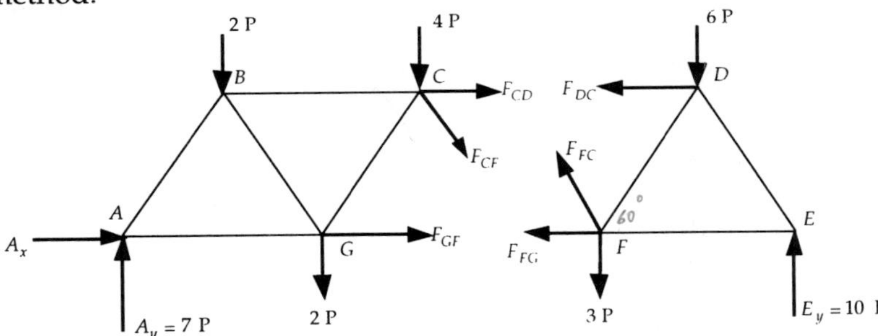

Figure 8.6 Sectioned truss.

A plane frame is a structure that consists of both two-force and three-force members, or even four-force members, etc. Loads may act at any location on the frame. The problem is to determine the components of the reactions at all pins of the frame. This usually requires not only a FBD of the entire frame, but also a FBD of each member. We illustrate by the following example.

Example 8.7

For the frame shown, determine the horizontal and vertical components of the reactions at all pins.

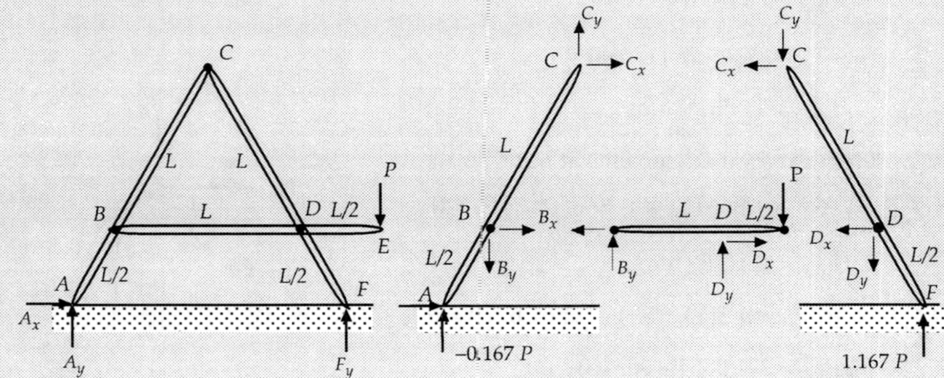

Solution. From a FBD of entire frame,

$$\sum F_x = A_x = 0. \qquad \therefore A_x = 0.$$
$$\sum M_A = (1.5L)F_y - (1.75L)P = 0. \qquad \therefore F_y = 1.167P$$
$$\sum F_y = A_y + 1.167P - P = 0. \qquad \therefore A_y = -0.167P$$

From a FBD of a member BDE,

$$\sum M_D = -(L)B_y - (L/2)P = 0. \qquad \therefore B_y = -0.5P.$$
$$\sum F_y = -0.5P + D_y - P = 0. \qquad \therefore D_y = 1.5P.$$

Transfer the vertical components at *B* and *D* of members *BDE* to members *ABC* and *CDF* by changing directions (action and reaction principle). From a *FBD* of member *ABC*,

$$\sum F_y = -0.167P + 0.5P + C_y = 0. \qquad \therefore C_y = -0.333P$$
$$\sum M_B = -(\sqrt{3}L/2)C_x - (L/2)0.333P + (L/4)0.167P = 0. \qquad \therefore C_x = -0.144P$$
$$\sum F_x = B_x - 0.144P = 0. \qquad \therefore B_x = 0.144P$$

Transfer horizontal components B_x and C_x to members *BDE* and *CDF*. From a FBD of member *BDE*,

$$\sum F_x = -0.144P + D_x = 0. \qquad \therefore D_x = 0.144P.$$

From a FBD of member *CDF*,

$$\sum F_y = 0.333P + D_y + 1.167P = 0. \qquad \therefore D_y = -1.5P.$$

Completed FBDs of all members are shown on the next page.

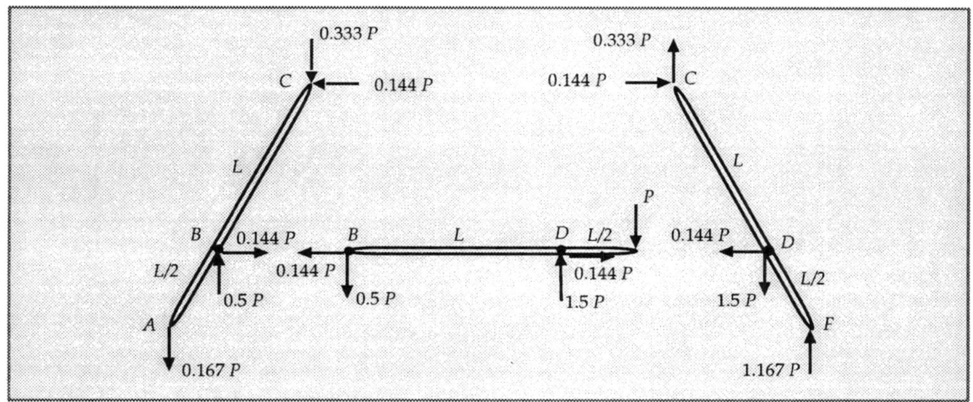

8.4 Friction

Consider a block of weight W at rest on a dry, rough, horizontal plane, Fig. 8.7a. Let a horizontal force P act to the right on the block, Fig. 8.7b. A *friction force* F is developed at the surface of contact between the block and plane, and maintains equilibrium ($F = P$) as long as $P < \mu_s N = \mu_s W$, where μ_s is the *static coefficient of friction*. If P is increased until $P = P^* = \mu_s N = \mu_s W$ (Fig. 8.7c), the equilibrium state is on the verge of collapse, and motion to the right is impending. In summary, dry *Coulomb friction* described here is governed by

$$F \le \mu_s N. \tag{8.4.1}$$

If relative motion between the block and the plane occurs (Fig. 8.7d), the friction force is given by

$$F = \mu_k N \tag{8.4.2}$$

where μ_k is the *kinetic coefficient of friction*, slightly less in value than μ_s, with both coefficients having a range $0 \le \mu \le 1$.

At impending motion, the equality holds in Eq. 8.4.1 and the resultant R of F and N makes an angle ϕ with the normal N, called the *friction angle*. From Fig. 8.8a it is clear that

$$\phi = \tan^{-1} \mu_s. \tag{8.4.3}$$

Also, it may be readily shown that a block will remain at rest (in equilibrium) on a rough inclined plane, Fig. 8.8b, as long as $\beta < \phi$ for the surfaces of contact.

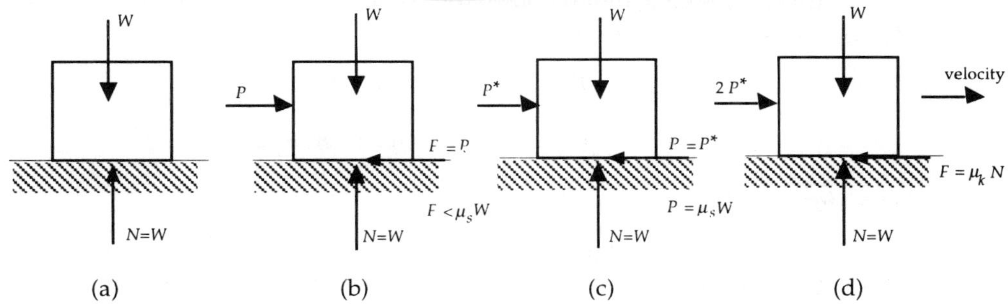

Figure 8.7 Friction forces.

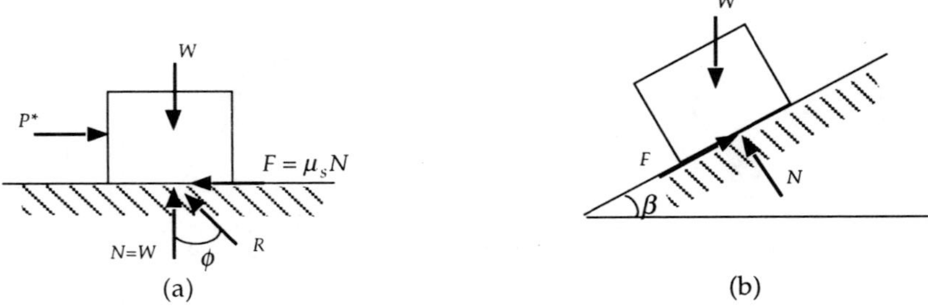

Figure 8.8 Friction angle.

Example 8.8

Determine the horizontal force P required to cause impending motion of the 50 kg block, (a) up the plane, (b) down the plane, if $\mu_s = 0.6$ between the block and plane.

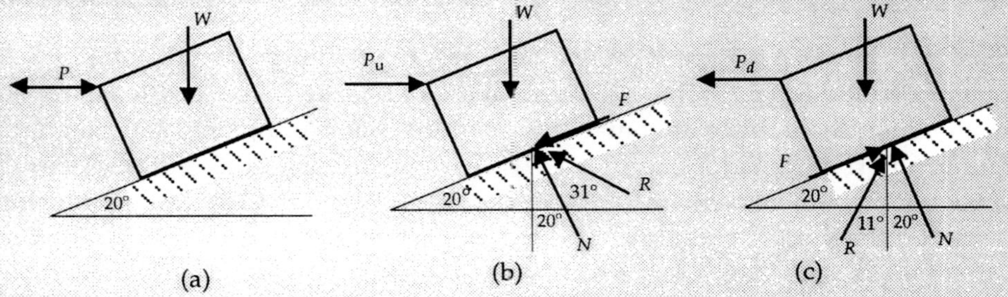

Solution. Note that, $\phi = \tan^{-1}(0.6) = 31°$ so that the block would remain in place if undisturbed.

(a) At impending motion up the plane, Fig. b, the resultant R of the friction force F and normal force N makes an angle of 51° to the right of the vertical and from the FBD of the block,

$$\sum F_x = P_u - R\sin 51^\circ = 0$$

$$\sum F_y = R\cos 51^\circ - 50\times 9.81 = 0.$$

Solving these equations,

$$R = 779 \text{ N}, \qquad P_u = 606 \text{ N}.$$

(b) At impending motion down the plane, R makes an angle of 11° to the left of the vertical, Fig. c, so that now

$$\sum F_x = -P_d + R\sin 11° = 0$$

$$\sum F_y = R\cos 11^\circ - 50\times 9.81 = 0.$$

Therefore

$$R = 499 \text{ N}, \qquad P_d = 95.2 \text{ N}.$$

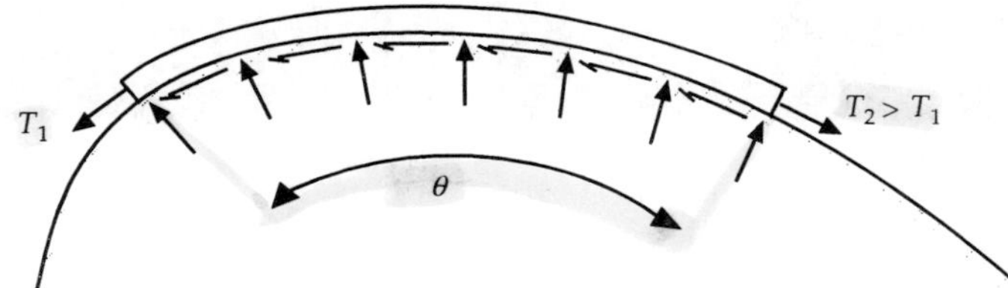

Figure 8.9 A belt with friction.

If a belt, or rope, is pressed firmly against some portion of a rough stationary curved surface, and pulled in one direction or the other, the tension in the belt will increase in the direction of pull due to the frictional resistance between the belt and surface, as shown in Fig. 8.9. It may be shown that

$$T_2 = T_1 e^{\mu_s \theta} \qquad T_2 > T_1 \tag{8.4.4}$$

where θ is the angle of contact, in radians, and μ_s is the static coefficient of friction.

Example 8.9

A 100 kg block rests on a 30° rough inclined plane ($\mu_s = 0.4$) and is attached by a rope to a mass *m* in the arrangement shown. If the static coefficient of friction between the rope and the circular support is 0.25, determine the maximum *m* that can be supported without slipping.

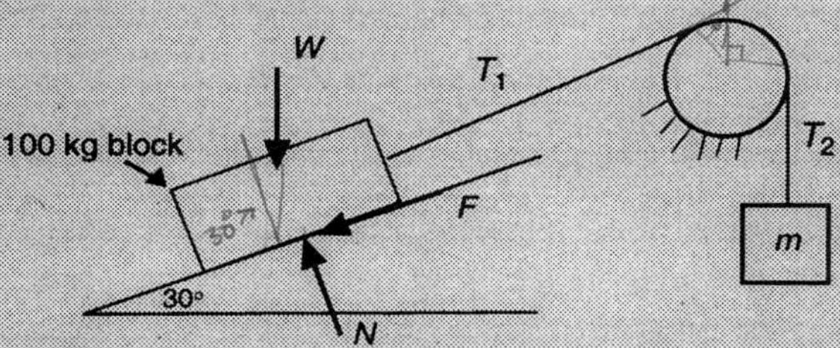

Solution. Summing forces perpendicular to the plane we determine

$$N = 981\cos 30° = 849.5\text{ N}.$$

Thus,

$$F = 0.4(849.5) = 339.8\text{ N}$$

at impending motion, and by summing forces along the plane

$$T_1 = 339.8 + 981\sin 30° = 830.3\text{ N}.$$

For the circular support we have,

$$T_2 = T_1 e^{(0.25)(120\times\pi/180)} = 830.3(1.69) = 1403\text{ N}$$

and

$$m = 1403/9.81 = 143\text{ kg}.$$

8.5 Properties of Plane Areas

Associated with every plane area A (in the xy-plane, for example) there is a point C, known as the *centroid*, whose coordinates $\bar{x}$ and $\bar{y}$ are defined by the integrals

$$\bar{x} = \frac{\int_A x dA}{\int_A dA} \qquad \bar{y} = \frac{\int_A y dA}{\int_A dA} \tag{8.5.1}$$

where dA is differential element of area having coordinates x and y as shown in Fig. 8.10a.

Although the integrals in Eq. 8.5.1 may be evaluated by a double integration using $dA = dxdy$, in practice it is often advantageous to calculate centroidal coordinates by a single integration using either a horizontal or vertical strip for dA (Fig. 8.10b and Fig. 8.10c) for which Eq. 8.5.1 may be expressed in the form

$$\bar{x} = \frac{\int_A x_e dA}{\int_A dA} \qquad \bar{y} = \frac{\int_A y_e dA}{\int_A dA} \tag{8.5.2}$$

where x_e and y_e are the coordinates of the centroids of the strip elements as shown.

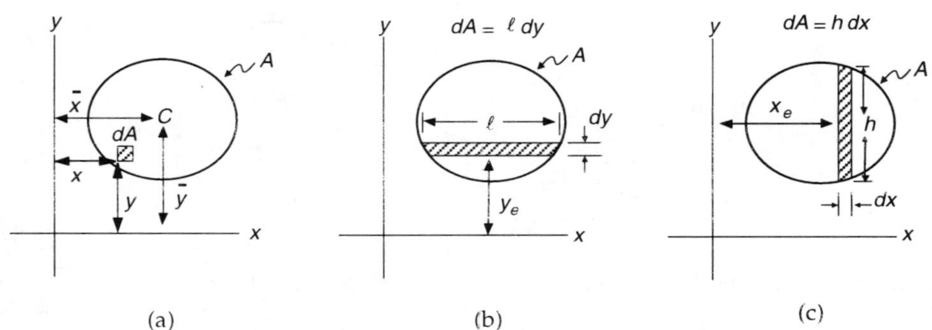

Figure 8.10 Plane area centroid.

The *plane moments of inertia* I_x and I_y of A with respect to the x and y axes, respectively, are defined by

$$I_x = \int_A y^2 dA, \qquad I_y = \int_A x^2 dA \tag{8.5.3}$$

where in integrating for I_x we use the horizontal strip for dA, and for I_y the vertical strip. The *polar moment of inertia* with respect to the origin O is defined by

$$J_O = \int_A r^2 dA = \int_A (x^2 + y^2) dA = I_x + I_y. \tag{8.5.4}$$

All moments of inertia may be expressed in terms of their respective *radii of gyration*. Thus,

$$I_x = r_x^2 A, \qquad I_y = r_y^2 A, \qquad J_O = r_O^2 A. \tag{8.5.5}$$

The *product of inertia* of A is defined with respect to a pair of perpendicular axes. For the coordinate axes we have

$$I_{xy} = \int_A xy\,dA \tag{8.5.6}$$

which normally must be evaluated by a double integration over A. If either one (or both) of the reference axes is an axis of symmetry the product of inertia is zero relative to that pair of axes.

The *transfer theorem* establishes a relationship between the moment of inertia about an arbitrary axis and the moment of inertia about a parallel axis passing through the centroid C. Thus

$$I_P = I_C + Ad^2, \qquad J_P = J_C + Ad^2 \tag{8.5.7}$$

where the subscript C indicates the centroidal moment of inertia, subscript P indicates the moment of inertia about the parallel axis, A is the area, and d the distance separating the two axes. Similarly, for products of inertia,

$$I_{x_P y_P} = I_{x_C y_C} + Ax_1 y_1 \tag{8.5.8}$$

where x_1 and y_1 are the distances between axes x and x_C, and y and y_C, respectively.

Example 8.10

For the shaded area shown by the sketch, determine $\bar{x}$, $\bar{y}$, I_x, I_y, J_O, I_{xy}, $(I_x)_C$, $(I_y)_C$, J_C and $I_{x_C y_C}$. Units are mm.

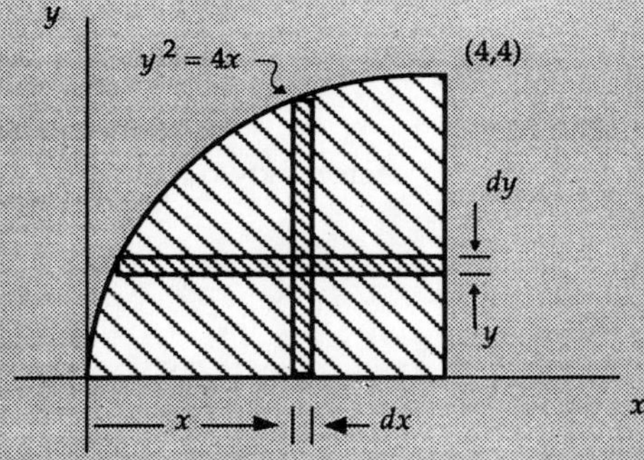

Solution. First we calculate A using the vertical strip, $dA = ydx = 2\sqrt{x}dx$:

$$A = \int_0^4 2\sqrt{x}dx = 32/3 \text{ mm}^2.$$

Thus, from Eq. 8.5.1, with $dA = 2\sqrt{x}dx$,

$$\bar{x} = \frac{3}{32}\int_0^4 x(2\sqrt{x})dx = 2.4 \text{ mm}$$

and, using the horizontal strip $dA = (4 - y^2/4)dy$,

$$\bar{y} = \frac{3}{32}\int_0^4 y(4 - y^2/4)dy = 1.5 \text{ mm}.$$

From Eq. 8.5.3, with $dA = (4 - y^2/4)dy$,

$$I_x = \int_0^4 y^2(4 - y^2/4)dy = 34.13 \text{ mm}^4$$

and, using $dA = 2\sqrt{x}dx$,

$$I_y = \int_0^4 x^2(2\sqrt{x})dx = 73.14 \text{ mm}^4.$$

From Eq. 8.5.4

$$J_O = I_x + I_y = 34.13 + 73.14 = 107.27 \text{ mm}^4.$$

From Eq. 8.5.6, with $dA = dxdy$,

$$I_{xy} = \int_0^4 \int_0^{2\sqrt{x}} xy\, dydx = \int_0^4 \left[\frac{xy^2}{2}\right]_0^{2\sqrt{x}} dx = \int_0^4 2x^2 dx = 42.67 \text{ mm}^4.$$

From Eq. 8.5.7,

$$(I_x)_C = I_x - Ad^2 = 34.13 - 10.67(1.5)^2 = 10.13 \text{ mm}^4$$

$$(I_y)_C = I_y - Ad^2 = 73.14 - 10.67(2.4)^2 = 11.68 \text{ mm}^4$$

$$J_C = J_O - Ad^2 = 107.27 - 10.67\left[(1.5)^2 + (2.4)^2\right] = 21.80 \text{ mm}^4.$$

Note that moments of inertia are always minimum about a centroidal axis. Finally, from Eq. 8.5.8,

$$I_{x_C y_C} = I_{xy} - A(-1.5)(-2.4) = 42.67 - 38.40 = 4.27 \text{ mm}^4$$

Example 8.11

Determine I_x and J_O for the circular sector shown below.

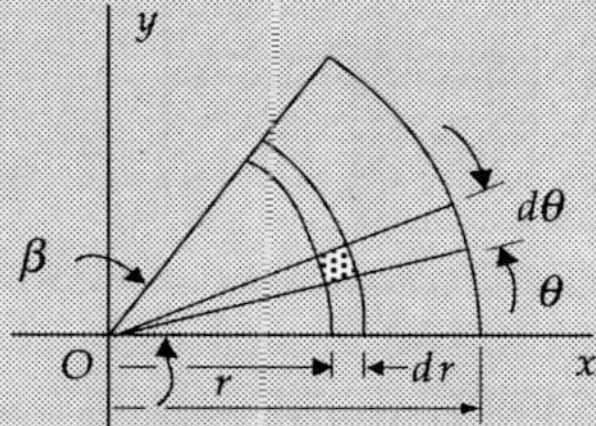

Solution. Using polar coordinates with $dA = rdrd\theta$ and $y = r\sin\theta$ in Eq. 8.5.3:

$$I_x = \int_0^\beta \int_0^a (r\sin\theta)^2 rdrd\theta = \frac{a^4}{4}\left[\frac{\beta}{2} - \frac{\sin 2\beta}{4}\right].$$

When $\beta = \pi/2$,

$$I_x = \pi a^4/16.$$

From Eq. 8.5.4,

$$J_O = \int_0^\beta \int_0^a r^3 drd\theta = \frac{a^4\beta}{4}.$$

When $\beta = \pi/2$,

$$J_O = \pi a^4/8.$$

TABLE 8.1. Properties of Areas.

Shape	Dimensions	Centroid	Inertia
Rectangle		$\bar{x} = b/2$ $\bar{y} = h/2$	$I_C = bh^3/12$ $I_x = bh^3/3$ $I_y = hb^3/3$
Triangle		$\bar{y} = h/3$	$I_C = bh^3/36$ $I_x = bh^3/12$
Circle		$\bar{x} = 0$ $\bar{y} = 0$	$I_x = \pi r^4/4$ $J = \pi d^4/32$
Quarter Circle		$\bar{y} = 4r/3\pi$	$I_x = \pi r^4/16$
Half Circle		$\bar{y} = 4r/3\pi$	$I_x = \pi r^4/8$

The properties of common areas may be determined by integration. A brief list is given in Table 8.1. Using data from Table 8.1, we may calculate centroids and moments of inertia of *composite areas* made up of combinations of two or more (including cutouts) of the common areas. Thus,

$$\bar{x} = \frac{\sum_{i=1}^{N} x_i A_i}{\sum_{i=1}^{N} A_i}, \qquad \bar{y} = \frac{\sum_{i=1}^{N} y_i A_i}{\sum_{i=1}^{N} A_i} \tag{8.5.9}$$

where N is the number of areas, and x_i is the centroidal distance for area A_i. Likewise, for moments of inertia

$$I = \sum_{i=1}^{N} I_i = I_1 + I_2 + \ldots + I_N. \tag{8.5.10}$$

An example illustrates.

Example 8.12

Determine the centroidal coordinates, and I_x and I_y for the composite area shown.

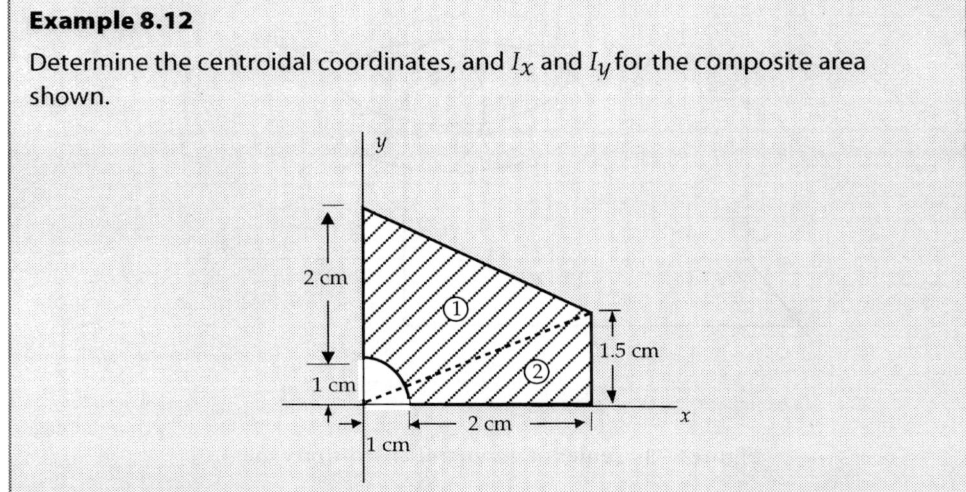

Solution. Decompose the composite into two triangular areas 1 and 2, and the negative quarter circular area 3:

$$A = A_1 + A_2 + A_3 = 4.5 + 2.25 - \pi/4 = 5.97 \text{ cm}^2.$$

$$A\bar{x} = x_1 A_1 + x_2 A_2 + x_3 A_3$$
$$= (1)(4.5) + (2)(2.25) + (4/3\pi)(-\pi/4) = 8.67.$$
$$\therefore \bar{x} = 8.67/5.97 = 1.45 \text{ cm}.$$

$$A\bar{y} = y_1 A_1 + y_2 A_2 + y_3 A_3$$
$$= (1.5)(4.5) + (0.5)(2.25) + (4/3\pi)(-\pi/4) = 7.54.$$
$$\therefore \bar{y} = 7.54/5.97 = 1.26 \text{ cm}.$$

$$I_x = I_{1x} + I_{2x} + I_{3x}$$
$$= \left\{2\left[3(1.5)^3/36\right] + (4.5)(1.5)^2\right\} + 3(1.5)^3/12 - \pi/16 = 11.33 \text{ cm}^4,$$

where we have used the parallel-axis theorem to obtain I_{1x}. Finally,

$$I_y = I_{1y} + I_{2y} + I_{3y}$$
$$= 3(3)^3/12 + 1.5(3)^3/4 - \pi/16 = 16.68 \text{ cm}^4.$$

8.6 Properties of Masses and Volumes

The coordinates of the *center of gravity* G of an arbitrary mass m occupying a volume V of space are defined by

$$x_G = \frac{\int_V x\rho dV}{\int_V \rho dV}, \qquad y_G = \frac{\int_V y\rho dV}{\int_V \rho dV}, \qquad z_G = \frac{\int_V z\rho dV}{\int_V \rho dV} \tag{8.6.1}$$

where ρ is the mass density, dV is the differential element of volume, and x, y and z are the coordinates of dV, as shown in Fig. 8.11.

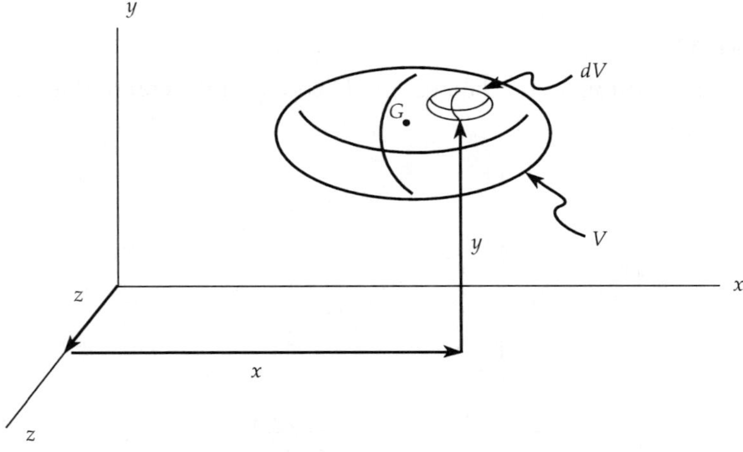

Figure 8.11 Center of gravity of an arbitrary mass.

The density may be a function of the space variables, $\rho = \rho(x,y,z)$. If the density is constant throughout the volume, the integrals in Eq. 8.6.1 reduce to

$$x_C = \frac{\int_V x dV}{V}, \qquad y_C = \frac{\int_V y dV}{V}, \qquad z_C = \frac{\int_V z dV}{V} \qquad \textbf{(8.6.2)}$$

which defines the coordinates of the centroid C of volume V. If ρ is constant, G and C coincide for a given body. As with areas, if an axis of symmetry exists for the volume, C is on that axis. The coordinates of G and C are readily calculated for geometries having an axis of revolution.

Example 8.13

Let the area of Example 8.10 be rotated about the x-axis to form a solid (volume) of revolution. Determine (a) G for the solid if $\rho = \rho_o x$ where ρ_o is a constant, (b) C for the volume.

Solution. Since there is symmetry about the x-axis, and since ρ is at most a function of x, we have

$$y_G = z_G = y_C = z_C = 0$$

(a) Let the element of mass be a thin disk for which $\rho dV = \rho_o x \pi y^2 dx$. Therefore,

$$x_G = \frac{\int_0^4 \rho_o \pi x^2 (4x) dx}{\int_0^4 \rho_o \pi x (4x) dx} = \frac{4 \times 4^4/4}{4 \times 4^3/3} = 3.00 \text{ mm.}$$

(b) For the same element with $\rho = \rho_o$

$$x_C = \frac{\int_0^4 \rho_o \pi x (4x) dx}{\int_0^4 \rho_o \pi (4x) dx} = \frac{4 \times 4^3/3}{4 \times 4^2/2} = 2.67 \text{ mm.}$$

Because the topic of *mass moments of inertia* properly belongs in the subject of dynamics, and is not a factor in statics, we present only a brief comment or two here for comparison with area moments of inertia. The mass moment of inertia of the three-dimensional body in Fig. 8.11 about any axis is defined by the integral

$$I = \int_V r^2 \rho dV = \int_V r^2 dm \qquad \textbf{(8.6.3)}$$

where r is the perpendicular distance of the mass element dm from the axis. With respect to the coordinate axes, Eq. 8.6.3 may be specialized to yield

$$I_x = \int_V (y^2 + z^2) dm \qquad \textbf{(8.6.4)}$$

$$I_y = \int_V (z^2 + x^2) dm \qquad \textbf{(8.6.5)}$$

$$I_z = \int_V (x^2 + y^2) dm. \qquad \textbf{(8.6.6)}$$

There is also a parallel-axis theorem for mass moments of inertia, namely,

$$I_P = I_G + md^2 \qquad \textbf{(8.6.7)}$$

with d being the distance between the center of gravity axis and the parallel axis of interest. Several mass moments of inertia are presented in Table 8.2.

TABLE 8.2 Mass Moments of Inertia.

Shape	Dimensions	Moment of Inertia
Slender rod	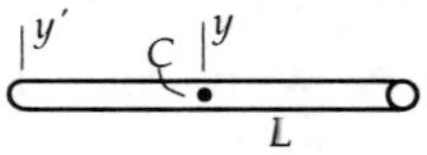	$I_y = mL^2/12$ $I_{y'} = mL^2/3$
Circular cylinder	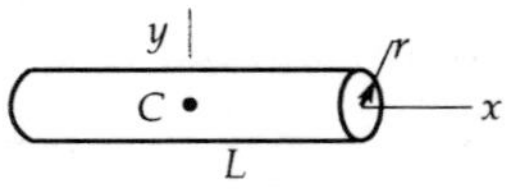	$I_x = mr^2/2$ $I_y = m(L^2 + 3r^2)/12$
Disk	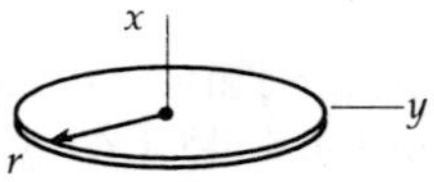	$I_x = mr^2/2$ $I_y = mr^2/4$
Rectangular parallelpiped	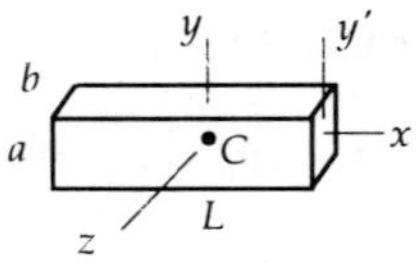	$I_x = m(a^2 + b^2)/12$ $I_y = m(L^2 + b^2)/12$ $I_z = m(L^2 + a^2)/12$ $I_{y'} = m(4L^2 + b^2)/12$
Sphere	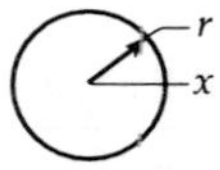	$I_x = 2mr^2/5$

Practice Problems

(If you choose to work only a few problems, select those with a star.)

GENERAL

*8.1 Find the component of the vector $\mathbf{A} = 15\mathbf{i} - 9\mathbf{j} + 15\mathbf{k}$ in the direction of $\mathbf{B} = \mathbf{i} - 2\mathbf{j} - 2\mathbf{k}$.

a) 1 b) 3 c) 5 d) 7 e) 9

8.2 Find the magnitude of the resultant of $\mathbf{A} = 2\mathbf{i} + 5\mathbf{j}$, $\mathbf{B} = 6\mathbf{i} - 7\mathbf{k}$, and $\mathbf{C} = 2\mathbf{i} - 6\mathbf{j} + 10\mathbf{k}$.

a) 8.2 b) 9.3 c) 10.5 d) 11.7 e) 12.8

*8.3 Determine the moment about the y-axis of the force $\mathbf{F} = 200\mathbf{i} + 400\mathbf{j}$ acting at (4, –6, 4).

a) 0 b) 200 c) 400 d) 600 e) 800

8.4 What total moment do the two forces $\mathbf{F}_1 = 50\mathbf{i} - 40\mathbf{k}$ and $\mathbf{F}_2 = 60\mathbf{j} + 80\mathbf{k}$ acting at (2, 0, –4) and (–4, 2, 0), respectively, produce about the x-axis?

a) 0 b) 80 c) 160 d) 240 e) 320

*8.5 The force system shown may be referred to as being

a) concurrent
b) coplanar
c) parallel
d) two-dimensional
e) non-concurrent, non-copla

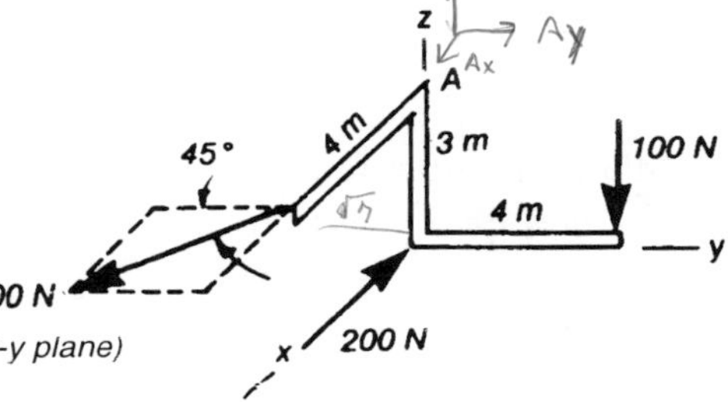

EQUILIBRIUM

8.6 If equilibrium exists due to a rigid support at A in the figure of Prob. 8.5, what reactive force must exist at A?

a) $-59\mathbf{i} - 141\mathbf{j} + 10\mathbf{k}$ b) $59\mathbf{i} + 141\mathbf{j} + 100\mathbf{k}$ c) $341\mathbf{i} - 141\mathbf{j} - 100\mathbf{k}$
d) $341\mathbf{i} + 141\mathbf{j} - 100\mathbf{k}$ e) $59\mathbf{i} - 141\mathbf{j} + 100\mathbf{k}$

*8.7 If equilibrium exists on the object in Prob. 8.5, what reactive moment must exist at the rigid support A?

a) $600\mathbf{i} + 400\mathbf{j} + 564\mathbf{k}$ b) $400\mathbf{i} + 564\mathbf{k}$ c) $400\mathbf{i} - 600\mathbf{j} + 564\mathbf{k}$
d) $400\mathbf{i} - 600\mathbf{j}$ e) $-600\mathbf{j} + 564\mathbf{j}$

*8.8 If three nonparallel forces hold a rigid body in equilibrium, they must

a) be equal in magnitude.
b) be concurrent.
c) be non-concurrent.
d) form an equilateral triangle.
e) be colinear.

*8.9 A truss member

a) is a two-force body.
b) is a three-force body.
c) resists forces in compression only.
d) may resist three concurrent forces.
e) resists lateral forces.

*8.10 Find the magnitude of the reaction at support B.

a) 400
b) 500
c) 600
d) 700
e) 800

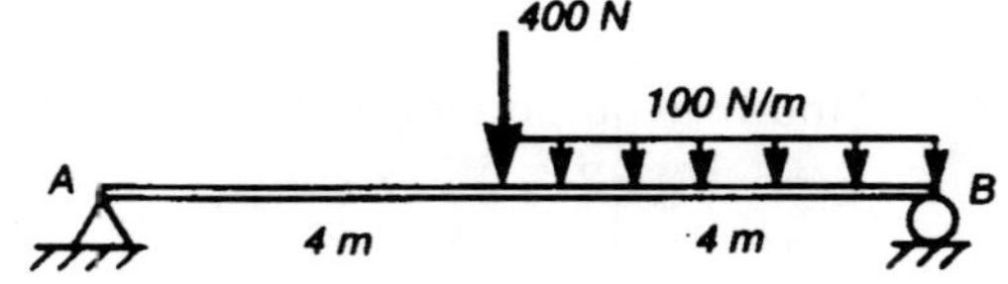

8.11 What moment M exists at support A?

a) 5600
b) 5000
c) 4400
d) 4000
e) 3600

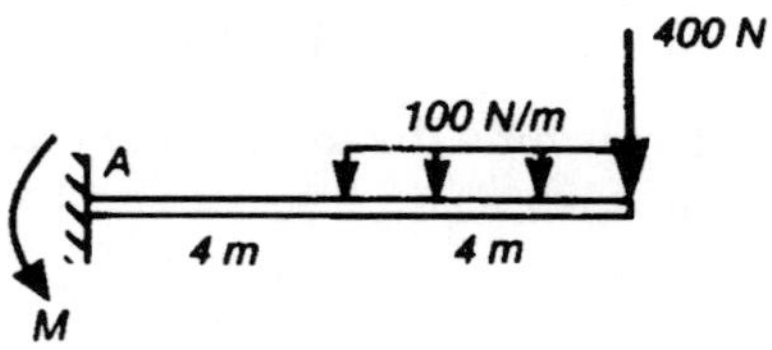

8.12 Calculate the reactive force at support A.

a) 250
b) 350
c) 450
d) 550
e) 650

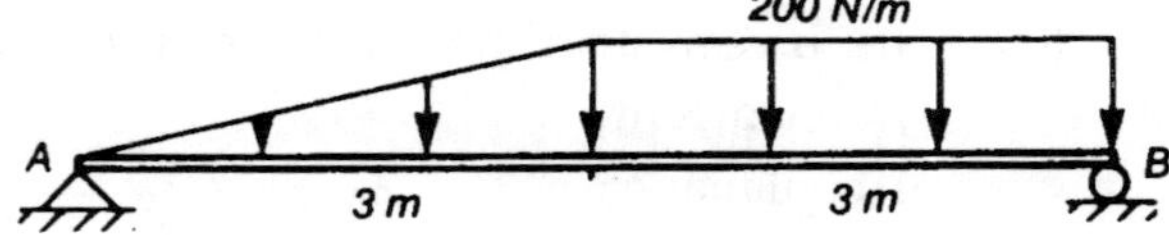

8.13 Find the support moment at A.

a) 66
b) 77
c) 88
d) 99
e) 111

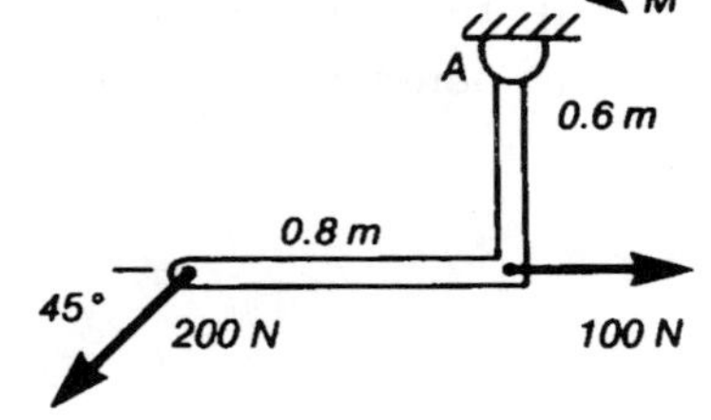

*8.14 To ensure equilibrium, what couple must be applied to this member?

a) 283 cw
b) 283 ccw
c) 400 cw
d) 400 ccw
e) 0

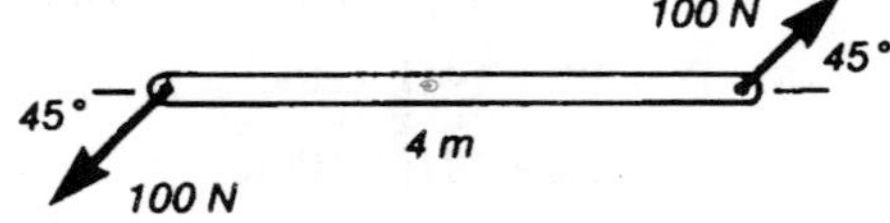

*8.15 Calculate the magnitude of the equilibrating force at A for the three-force body shown.

a) 217
b) 287
c) 343
d) 385
e) 492

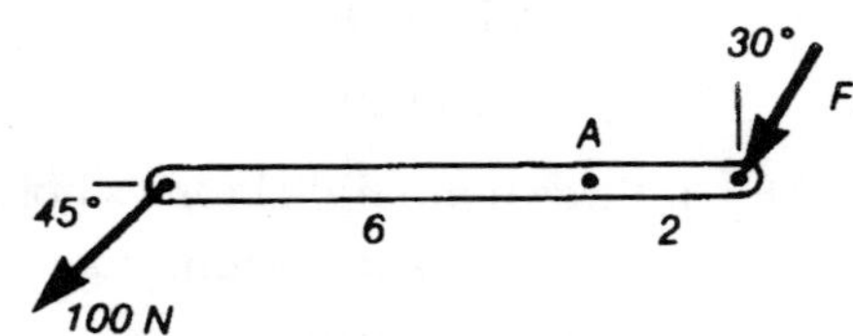

8.16 Find the magnitude of the equilibrating force at point A.

a) 187 d) 99
b) 142 e) 84
c) 114

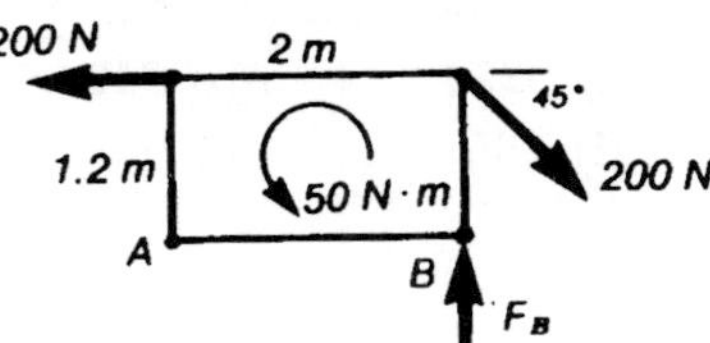

TRUSSES AND FRAMES

8.17 Find F_{DE} if all angles are equal.

a) 121 b) 142 c) 163
d) 176 e) 189

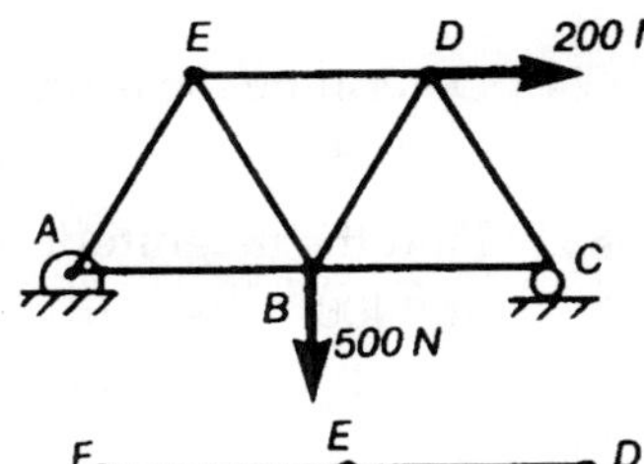

*8.18 Find F_{DE}.

a) 0 b) 1000 c) 2000
d) 2500 e) 5000

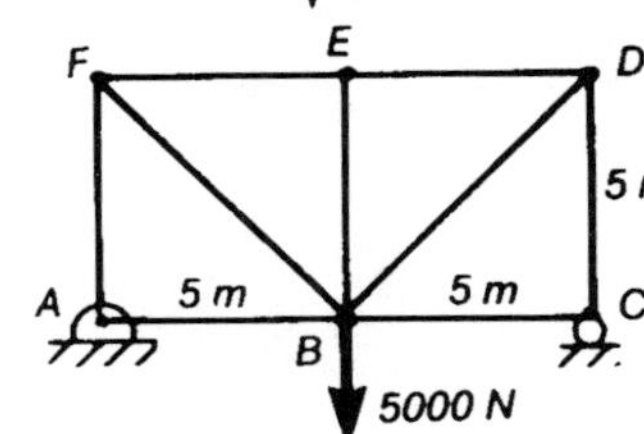

8.19 What is the force in member DE?

a) 1532 b) 1768 c) 18
d) 1946 e) 2231

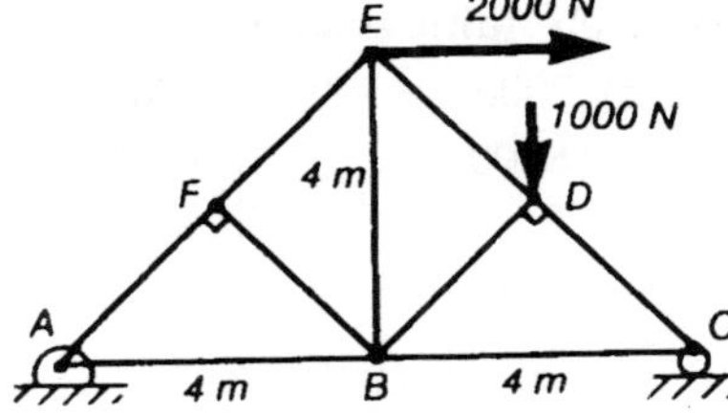

*8.20 Calculate F_{FB} in the truss of Problem 8.19.

a) 0 b) 932 c) 1561 d) 1732 e) 1887

8.21 Find the force in member IC.

a) 0
b) 1000
c) 1250
d) 1500
e) 2000

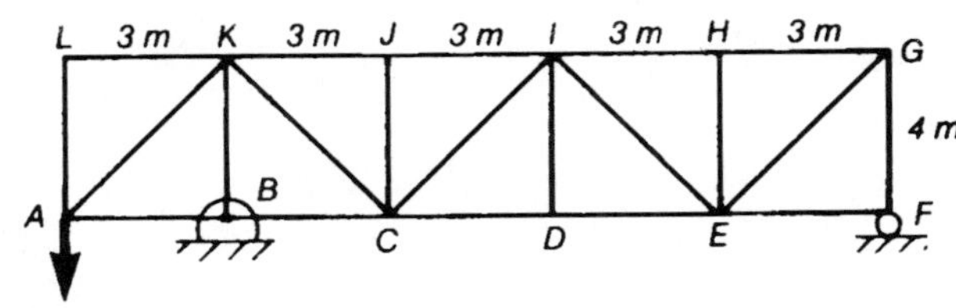

8.22 What force exists in member BC in the truss of Prob. 8.21?

a) 0 b) 1000 c) 2000 d) 3000 e) 4000

8.23 Find the force in member FC.

a) 5320 b) 3420 c) 2560
d) 936 e) 0

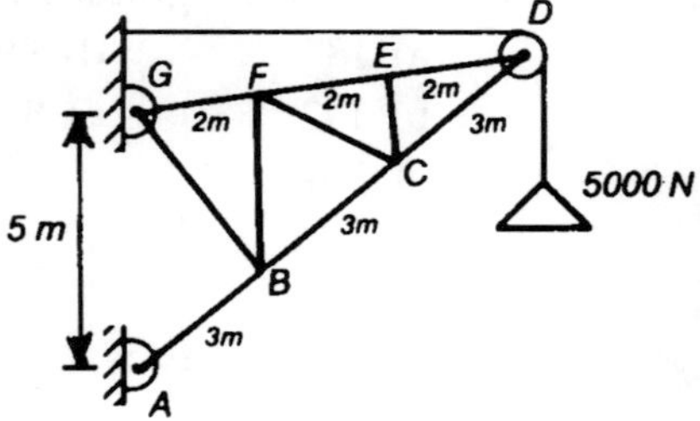

8.24 Determine the force in member BC in the truss of Prob. 8.23.

a) 3560 b) 4230 c) 4960 d) 5230 e) 5820

*8.25 Find the magnitude of the reactive force at support A.

a) 1400 b) 1300 c) 1200
d) 1100 e) 1000

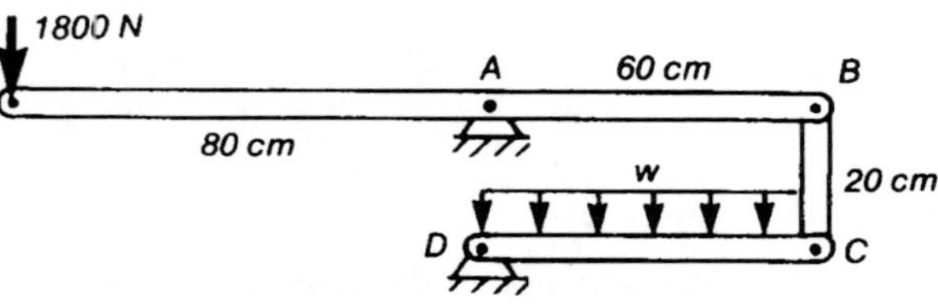

8.26 Determine the distributed force intensity w, in N/m, for equilibrium to exist.

a) 2000 b) 4000 c) 6000
d) 8000 e) 10 000

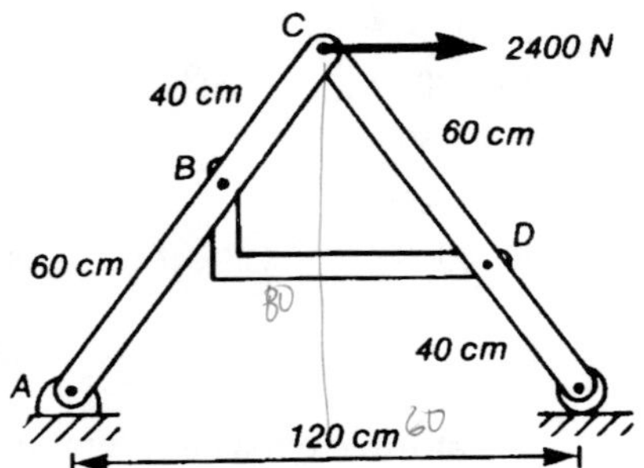

*8.27 Find the magnitude of the reactive force at support A.

a) 2580 b) 2670 c) 2790
d) 2880 e) 2960

8.28 Calculate the magnitude of the force in member BD of Prob. 8.27.

a) 2590 b) 2670 c) 2790 d) 2880 e) 2960

FRICTION

*8.29 What force, in Newtons, will cause impending motion up the plane?

a) 731 b) 821 c) 973
d) 1102 e) 1245

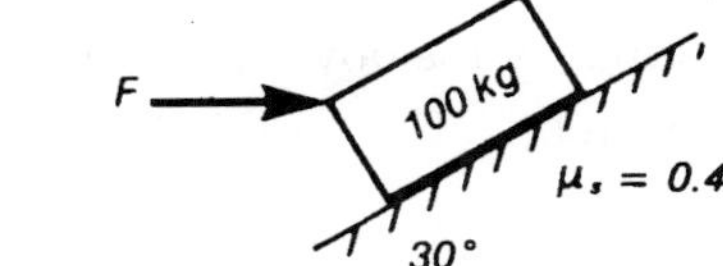

8.30 What is the maximum force F, in Newtons, that can be applied without causing motion to impend?

a) 184 b) 294 c) 316
d) 346 e) 392

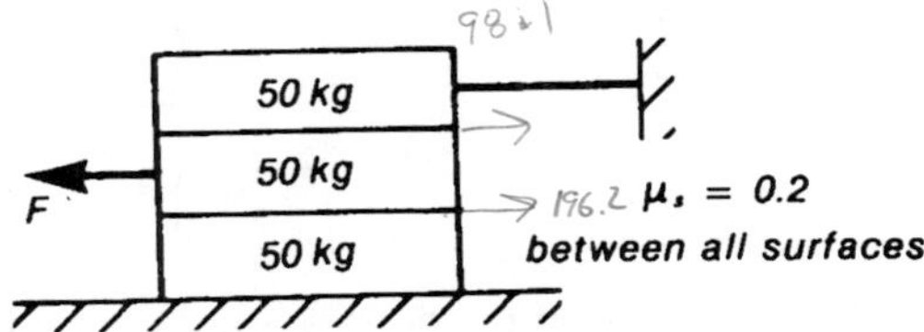

*8.31 Only the rear wheels provide braking. At what angle θ will the car slide if $\mu_s = 0.6$?

a) 10 b) 12 c) 16
d) 20 e) 24

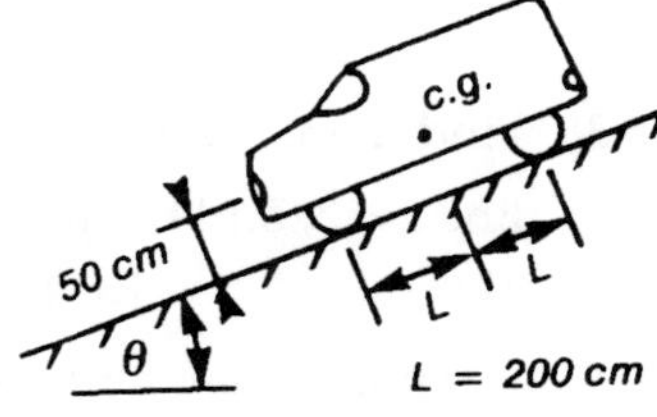

8.32 Find the minimum h value at which tipping will occur.

a) 8 b) 10 c) 12
d) 14 e) 16

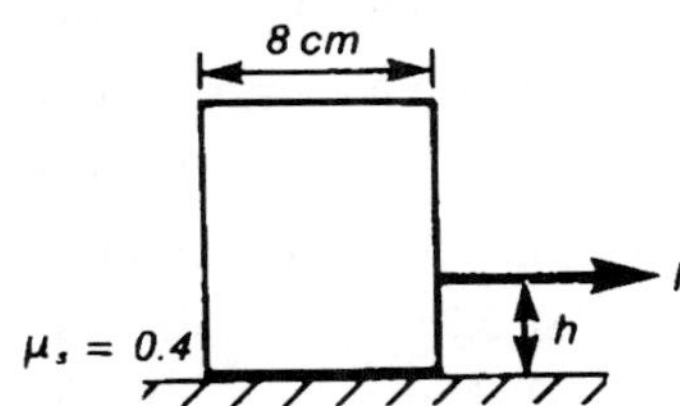

*8.33 What force F, in Newtons, will cause impending motion?

a) 240 b) 260 c) 280
d) 300 e) 320

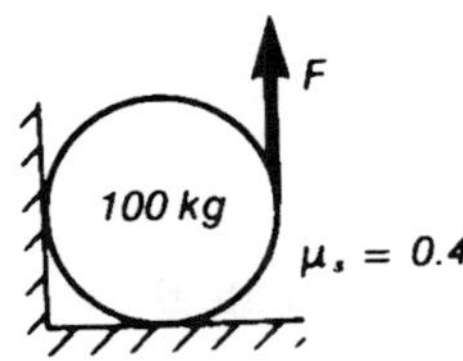

8.34 The angle θ at which the ladder is about to slip is

a) 50 b) 46 c) 42
d) 38 e) 34

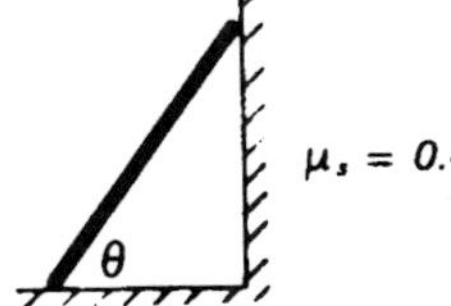

*8.35 A boy and his dad put a rope around a tree and stand side by side. What force by the boy can resist a force of 800 N by his dad? Use $\mu_s = 0.5$.

a) 166 b) 192 c) 231 d) 246 e) 297

8.36 What moment, in N impending motion?

a) 88
b) 99
c) 110
d) 121
e) 146

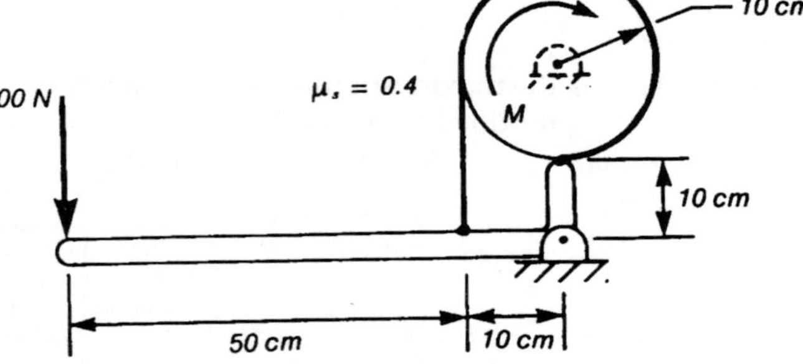

8.37 A 12-m-long rope is draped over a horizontal cylinder of 1.2-m-diameter so that both ends hang free. What is the length of the longer end at impending motion? Use $\mu_s = 0.5$.

a) 6.98 b) 7.65 c) 7.92 d) 8.37 e) 8.83

CENTROIDS AND MOMENTS OF INERTIA

*8.38 Find the x-coordinate of the centroid of the area bounded by the x-axis, the line $x = 3$, and the parabola $y = x^2$.

a) 2.0 b) 2.15 c) 2.20 d) 2.25 e) 2.30

8.39 What is the y-coordinate of the centroid of the area of Prob. 8.38?

a) 2.70 b) 2.65 c) 2.60 d) 2.55 e) 2.50

8.40 Calculate the x-coordinate of the centroid of the area enclosed by the parabolas $y = x^2$ and $x = y^2$.

a) 0.43 b) 0.44 c) 0.45 d) 0.46 e) 0.47

8.41 Find the y-component of the centroid of the area shown.

a) 3.35 b) 3.40 c) 3.45
d) 3.50 e) 3.55

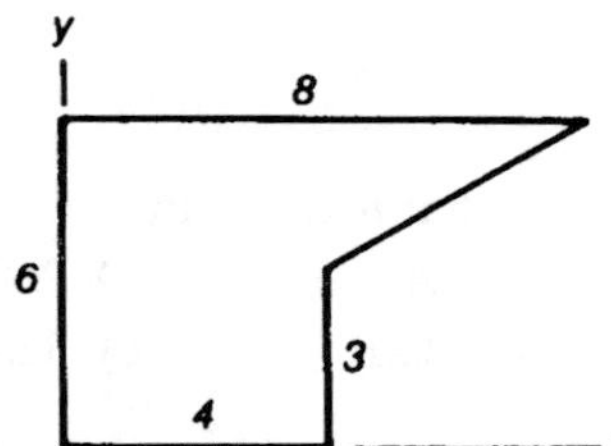

*8.42 Calculate the y-component of the centroid of the area shown.

a) 3.52 b) 3.56 c) 3.60
d) 3.64 e) 3.68

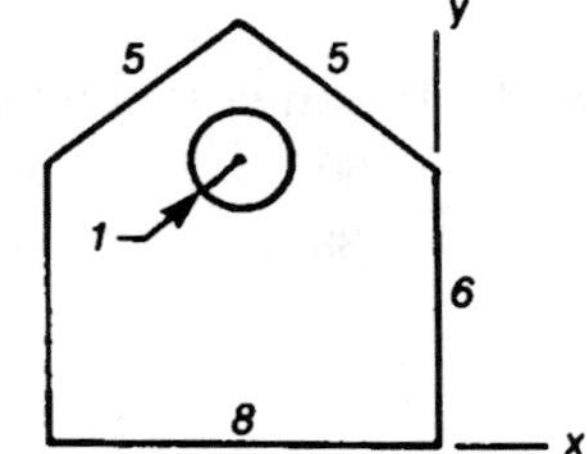

8.43 Find the x-component of the center of gravity of the three objects.

a) 2.33
b) 2.42
c) 2.84
d) 3.22
e) 3.64

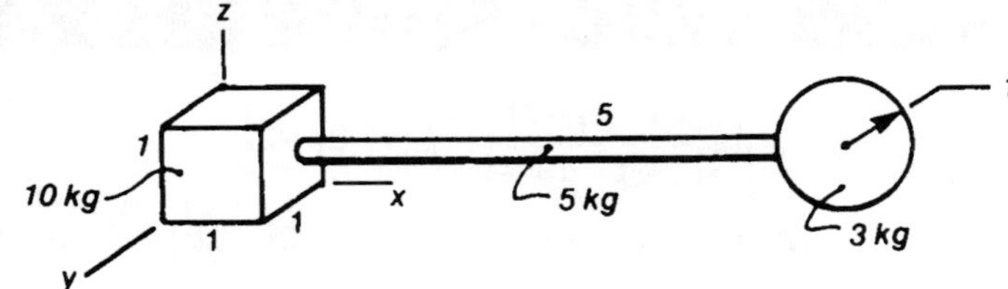

*8.44 Calculate the moment of inertia about the x-axis of the area of Prob. 8.38.

a) 94 b) 104 c) 112 d) 124 e) 132

8.45 What is I_x for the area of Prob. 8.42?

a) 736 b) 842 c) 936 d) 982 e) 1056

*8.46 Find I_y for the symmetrical area shown.

a) 4267 b) 4036 c) 3827
d) 3652 e) 3421

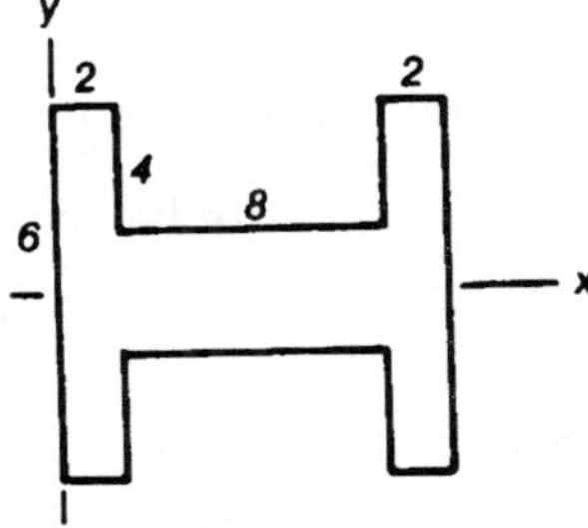

8.47 Determine the mass moment of inertia of a cube with edges of length b, about an axis passing through an edge.

a) $2mb^2/3$ b) $mb^2/6$ c) $3mb^2/2$ d) $mb^2/2$ e) $3mb^2/4$

*8.48 Find the mass moment of inertia about the x-axis if the mass of the rods per unit length is 1.0 kg/m.

a) 224 b) 268 c) 336
d) 386 e) 432

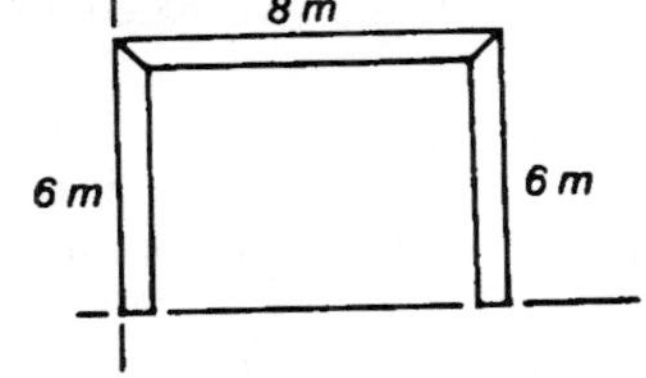

Solutions to Practice Problems

8.1 a) $\mathbf{i}_B = \dfrac{\mathbf{i} - 2\mathbf{j} - 2\mathbf{k}}{\sqrt{1+4+4}} = \dfrac{1}{3}(\mathbf{i} - 2\mathbf{j} - 2\mathbf{k})$

$\mathbf{A} \cdot \mathbf{i}_B = (15\mathbf{i} - 9\mathbf{j} + 15\mathbf{k}) \cdot \dfrac{1}{3}(\mathbf{i} - 2\mathbf{j} - 2\mathbf{k}) = 5 + 6 - 10 = 1$

8.2 c) $\mathbf{A} + \mathbf{B} + \mathbf{C} = (2\mathbf{i} + 5\mathbf{j}) + (6\mathbf{i} - 7\mathbf{k}) + (2\mathbf{i} - 6\mathbf{j} + 10\mathbf{k}) = 10\mathbf{i} - \mathbf{j} + 3\mathbf{k}$

magnitude $= \sqrt{10^2 + 1^2 + 3^2} = 10.49$

8.3 e) $\mathbf{M} = \mathbf{r} \times \mathbf{F} = (4\mathbf{i} - 6\mathbf{j} + 4\mathbf{k}) \times (200\mathbf{i} + 400\mathbf{j})$. $M_y = 4 \times 200 = 800$ since $\mathbf{k} \times \mathbf{i} = \mathbf{j}$

8.4 c) $\mathbf{M} = \mathbf{r}_1 \times \mathbf{F}_1 + \mathbf{r}_2 \times \mathbf{F}_2 = (2\mathbf{i} - 4\mathbf{k}) \times (50\mathbf{i} - 40\mathbf{k}) + (-4\mathbf{i} + 2\mathbf{j}) \times (60\mathbf{j} + 80\mathbf{k})$

$M_x = 2 \times 80 = 160$ since $\mathbf{j} \times \mathbf{k} = \mathbf{i}$

8.5 e) Concurrent $\Rightarrow$ all pass through a point.

Coplanar $\Rightarrow$ all in the same plane.

The forces are three - dimensional.

8.6 b) $\sum \mathbf{F} = 0. \quad \therefore \mathbf{R} + 141\mathbf{i} - 141\mathbf{j} - 200\mathbf{i} - 100\mathbf{k} = 0$

$\therefore \mathbf{R} = 59\mathbf{i} + 141\mathbf{j} + 100\mathbf{k}$

8.7 c) $\sum \mathbf{M} = 0. \quad \therefore \mathbf{M}_A + (4\mathbf{j} - 3\mathbf{k}) \times (-100\mathbf{k}) - 3\mathbf{k} \times (-200\mathbf{i}) + 4\mathbf{i} \times (141\mathbf{i} - 141\mathbf{j}) = 0$

$\therefore \mathbf{M}_A = 400\mathbf{i} - 600\mathbf{j} + 564\mathbf{k}$

8.8 b) They must be concurrent, otherwise a resultant moment would occur.

8.9 a) It is a two-force body.

8.10 b) $\sum M_A = 0. \quad F_B \times 8 = 400 \times 4 + 400 \times 6. \quad \therefore F_B = 500 \text{ N}$

8.11 a) $M_A = 400 \times 8 + 400 \times 6 = 5600 \text{ N} \cdot \text{m}$

8.12 b) $\sum M_B = 0. \quad 6F_A = 4 \times 300 + 600 \times 3/2. \quad \therefore F_A = 350 \text{ N}$

300 600

F_A 4 m F_B

8.13 c) $M_A = 0.6 \times 100 - 141 \times 0.6 + 141 \times 0.8 = 88.2$

8.14 a) $M = 100 \sin 45° \times 4 = 282.8$ cw

8.15 c) $\sum M_A = 0. \quad \therefore 6 \times 70.7 = 2 \times 0.866 F_1. \quad \therefore F_1 = 245$

$\sum F_x = 0. \quad \therefore -70.7 - 245 \times 0.5 + F_{Ax} = 0. \quad \therefore F_{Ax} = 193$

$\sum F_y = 0. \quad \therefore -70.7 - 245 \times 0.866 + F_{Ay} = 0. \quad \therefore F_{Ay} = 283$

$$\therefore F_A = \sqrt{F_{Ax}^2 + F_{Ay}^2} = \sqrt{193^2 + 283^2} = 343$$

8.16 e) $\sum M_A = 0. \quad \therefore 2F_B + 1.2 \times 200 - 141.4 \times 2 - 141.4 \times 1.2 + 50 = 0. \quad \therefore F_B = 81.2$

$\sum F_x = 0. \quad \therefore F_{Ax} - 200 + 141.4 = 0. \quad \therefore F_{Ax} = 58.6$

$\sum F_y = 0. \quad \therefore F_{Ay} + 81.2 - 141.4 = 0. \quad \therefore F_{Ay} = 60.2$

$$\therefore F_A = \sqrt{F_{Ax}^2 + F_{Ay}^2} = \sqrt{58.6^2 + 60.2^2} = 84.0$$

8.17 e) $\sum M_A = 0. \quad \therefore 500\ell + 200 \times 0.866\ell - F_C \times 2\ell = 0. \quad \therefore F_C = 337$

$0.866 F_{DC} = 337. \quad \therefore F_{DC} = 389$

$0.866 \times 389 = 0.866 F_{BD}. \quad \therefore F_{BD} = 389$

$-F_{DE} + 200 - 389 \times 0.5 - 389 \times 0.5 = 0. \quad \therefore F_{DE} = -189$

8.18 d) $\sum M_A = 0. \quad \therefore 5 \times 5000 = 10 \times F_C. \quad \therefore F_C = 2500 \quad \therefore F_{DC} = 2500$

$0.707 F_{BD} = 2500. \quad \therefore F_{BD} = 3536$

$0.707 \times 3536 = F_{DE}. \quad \therefore F_{DE} = 2500$

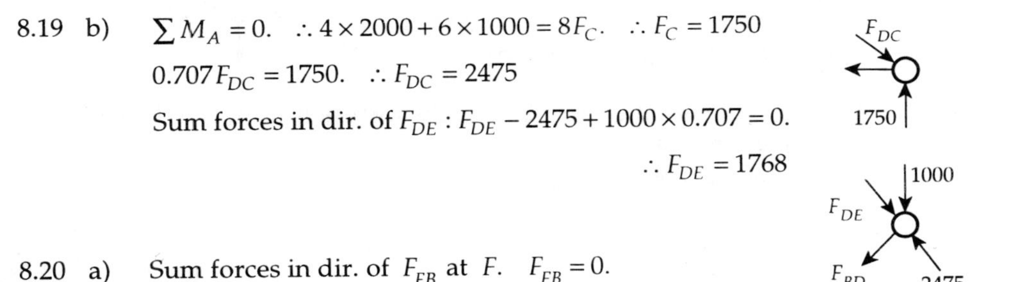

8.19 b) $\sum M_A = 0. \quad \therefore 4 \times 2000 + 6 \times 1000 = 8F_C. \quad \therefore F_C = 1750$

$0.707 F_{DC} = 1750. \quad \therefore F_{DC} = 2475$

Sum forces in dir. of F_{DE} : $F_{DE} - 2475 + 1000 \times 0.707 = 0.$

$\therefore F_{DE} = 1768$

8.20 a) Sum forces in dir. of F_{FB} at F. $F_{FB} = 0.$

8.21 c) $\sum M_B = 0. \quad \therefore 12F_F = 3 \times 4000. \quad \therefore F_F = 1000 \downarrow$

$\sum F_y = 0. \quad \therefore 0.8 F_{IC} = 1000. \quad \therefore F_{IC} = 1250$

8.22 d) Cut vertically through link KA. Then $F_{KA} = 5000.$
Obviously, $F_{AL} = 0. \quad \therefore F_{AB} = 3000. \quad \therefore F_{BC} = 3000$

8.23 e) At E we see that $F_{EC} = 0. \quad \therefore$ At C, $F_{FC} = 0$

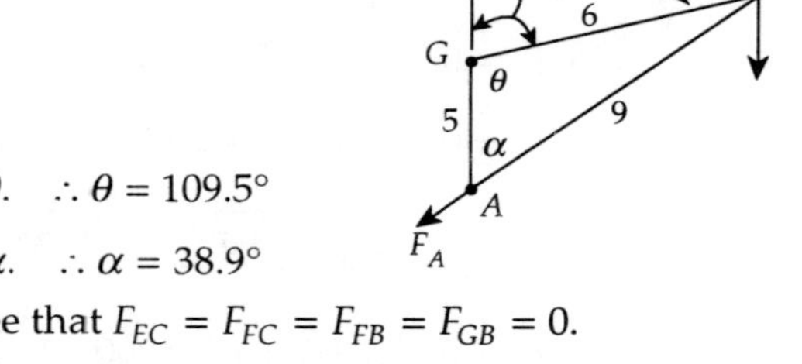

8.24 e) $9^2 = 6^2 + 5^2 - 2 \times 5 \times 6 \cos\theta. \quad \therefore \theta = 109.5°$

$6^2 = 9^2 + 5^2 - 2 \times 9 \times 5 \cos\alpha. \quad \therefore \alpha = 38.9°$

From pts E, C, F, B we see that $F_{EC} = F_{FC} = F_{FB} = F_{GB} = 0$.

Also, $F_A = F_{BC}$. $\sum M_G = 0$.

$\therefore 5 \times F_A \sin 38.9° + 5000 \times 6 \sin 70.5° = 5000 \times 6 \cos 70.5°.$

$\therefore F_A = -5817$

8.25 b) Recognize that link BC is a two-force member. $\sum M_A = 0$.

$\therefore 0.2 \times 1000 + 100 = 0.08 \times F_{BC} \times 0.8 + 0.2 \times F_{BC} \times 0.6$

$\therefore F_{BC} = 1630. \quad A_x = 1630 \times 0.8 = 1304. \quad A_y = 1630 \times 0.6 - 1000 = -22$

$\therefore F_A = \sqrt{1304^2 + 22^2} = 1304$ N

8.26 d) $1800 \times 0.8 = 0.6 F_{BC}. \quad \therefore F_{BC} = 2400$

$0.3 \times 0.6w = 0.6 \times 2400. \quad \therefore w = 8000$

8.27 d) $\sum M_A = 0. \quad \therefore 1.2 F_E = 0.8 \times 2400. \quad \therefore F_E = 1600$

$\therefore A_x = 2400. \quad A_y = 1600$

$\therefore F_A = \sqrt{2400^2 + 1600^2} = 2884$ N

8.28 a) Link BD is a two-force member. $\therefore$ the force acts from D to B. Hence, the angles are found.

$120^2 = 100^2 + 100^2 - 2 \times 100 \times 100 \cos\beta. \quad \therefore \beta = 73.7°$

$\overline{BD}^2 = 60^2 + 40^2 - 2 \times 60 \times 40 \cos 73.7°. \quad \therefore BD = 62.1$

$\frac{62.1}{\sin 73.7°} = \frac{40}{\sin\phi}. \quad \therefore \phi = 38.2°. \quad \alpha = (180 - 73.7)/2 = 53.2°$

$\sum M_C = 0. \quad 1600 \times 100 \cos 53.2° = 60 \times F_{BD} \sin 38.2°. \quad \therefore F_{BD} = 2587$

8.29 e) $\sum F_y = 0. \quad N \times 0.866 - 980 - 0.4N \times 0.5 = 0. \quad \therefore N = 1471$

$\sum F_x = 0. \quad F = 1471 \times 0.5 + 0.4 \times 1471 \times 0.866 = 1245$

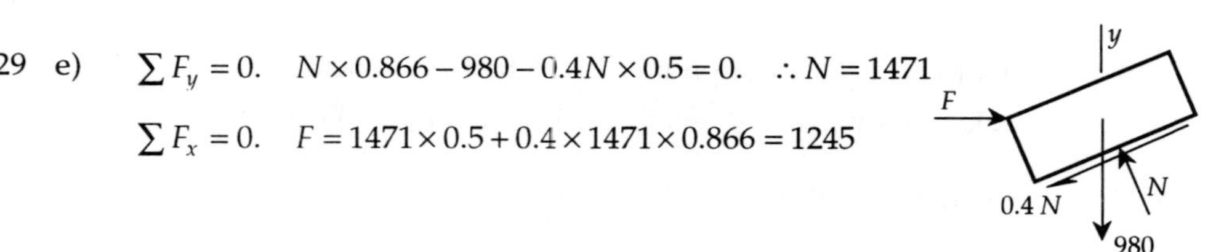

8.30 b) $N_1 = 490$. $N_2 = 980$. $\therefore F = 0.2(490 + 980) = 294$

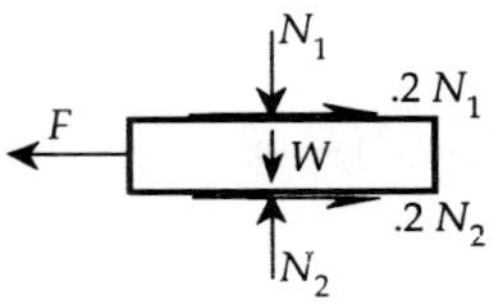

8.31 c) $\sum M_{\text{front wheel}} = 0$. $\therefore 400N_2 - W\cos\theta \times 200 + W\sin\theta \times 50 = 0$

$\sum F_x = 0$. $\therefore 0.6N_2 = W\sin\theta$.

$\therefore 400(W\sin\theta)/0.6 + 50W\sin\theta = 200W\cos\theta$

$\therefore \frac{\sin\theta}{\cos\theta} = \frac{200}{716.7} = \tan\theta$. $\therefore \theta = 15.6°$

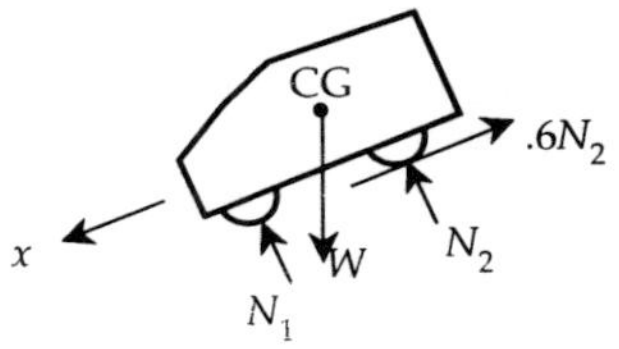

8.32 b) If $h < h_{\text{min}}$ then sliding occurs, and $F_f = 0.4N$.
If $h > h_{\text{min}}$ tipping occurs and $F_f < 0.4N$.
When $h = h_{\text{min}}$, $F_f = 0.4N = 0.4W = F$.
$\sum M_A = 0$. $\therefore 4W = hF = h \times 0.4W$. $\therefore h = 10$ cm

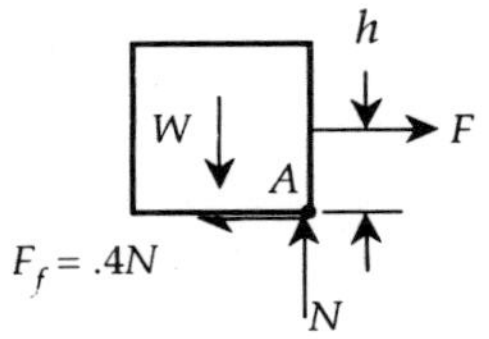

8.33 e) $\sum F_x = 0$. $\therefore N_2 = 0.4N_1$. Also, $W = 980$

$\sum M_A = 0$. $\therefore W \cdot r = (N_1 + 0.4N_1 + 2 \times 0.4N_2)r$.

$\therefore N_1 = 0.5814W = 570$

$\sum F_y = 0$. $\therefore F = 980 - 570 - 0.16 \times 570 = 319$

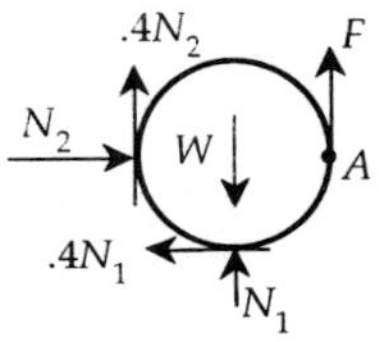

8.34 b) $\sum F_x = 0$. $\therefore N_2 = 0.4N_1$. $\sum F_y = 0$. $\therefore N_1 + 0.4N_2 = W$.

$\therefore N_2 = 0.345W$

$\sum M_A = 0$. $\therefore \frac{L}{2} \times W\cos\theta = N_2 \times L\sin\theta + 0.4N_2 \times L\cos\theta$.

This gives $\tan\theta = 1.049$. $\therefore \theta = 46.4°$

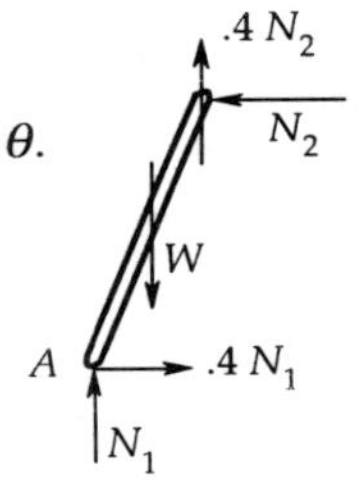

8.35 a) $F_B = F_D e^{-\mu\theta} = 800e^{-0.5\pi} = 166\text{N}$

8.36 a) $\sum M_A = 0$. $\therefore 200 \times 0.6 = 0.1 \times T_1 + 0.1 \times T_2$.

$T_1 = T_2 e^{0.4 \times 3\pi/2} = 6.59T_2$. Thus, $T_2 = 158$ and $T_1 = 1042$.

$\sum M_{\text{center}} = 0$. $\therefore M = 0.1 \times (1042 - 158) = 88.4\ \text{N} \cdot \text{m}$

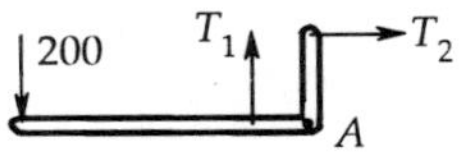

8.37 d) Let h = long end. m = mass/unit length. Then,
$(12 - 1.88 - h)mge^{0.5\pi} = hmg$. $\therefore h = 8.38$ m

8.38 d) $\bar{x} = \dfrac{\int_0^3 xy\,dx}{\int_0^3 y\,dx} = \dfrac{\int_0^3 x^3 dx}{\int_0^3 x^2 dx} = \dfrac{3^4/4}{3^3/3} = 2.25$

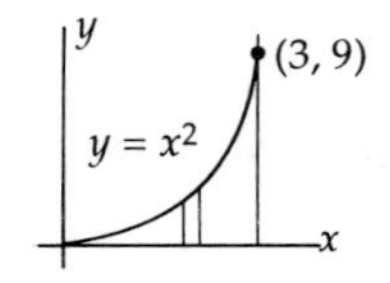

8.39 a) $\bar{y} = \dfrac{\int_0^3 \frac{y}{2} y\,dx}{\int_0^3 y\,dx} = \dfrac{\frac{1}{2}\int_0^3 x^4 dx}{\int_0^3 x^2 dx} = \dfrac{3^5/10}{3^3/3} = 2.7$

8.40 c) $\bar{x} = \dfrac{\int_0^1 \left(\sqrt{x} - x^2\right) x\,dx}{\int_0^1 \left(\sqrt{x} - x^2\right) dx} = \dfrac{\frac{1}{5/2} - \frac{1}{4}}{\frac{1}{3/2} - \frac{1}{3}} = 0.45$

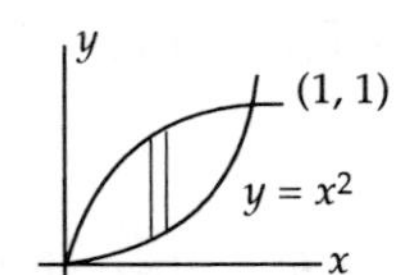

8.41 b) $\bar{y} = \dfrac{24 \times 3 + 6 \times 5}{6 \times 4 + 4 \times 3/2} = 3.4$

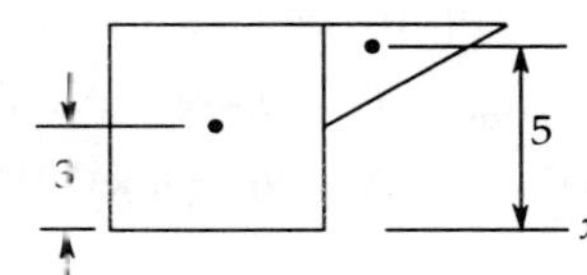

8.42 e) $\bar{y} = \dfrac{48 \times 3 + 12 \times 7 - \pi \times 6}{8 \times 6 + 3 \times 4 - \pi} = 3.68$

8.43 b) $\bar{x} = \dfrac{10 \times \frac{1}{2} + 5 \times 3.5 + 3 \times 7}{10 + 5 + 3} = 2.42$

8.44 b) $I_x = \int_0^3 y^3\,dx/3 = \int_0^3 x^6\,dx/3 = 3^7/21 = 104.1.$

With a horizontal strip: $I_x = \int_0^9 y^2(3 - x)dy = \int_0^9 y^2\left(3 - \sqrt{y}\right)dy = 9^3 - \dfrac{9^{7/2}}{7/2} = 104.1$

8.45 e) $I_x = 8 \times 6^3/3 + \left(8 \times 3^3/36 + 12 \times 7^2\right) - \left(\pi \times 1^4/4 + \pi \times 6^2\right) = 1056$

8.46 a) $I_y = 12 \times 12^3/3 - \left(8 \times 8^3/12 + 64 \times 6^2\right) = 4267.$ Or, alternatively:

$I_y = 8 \times 2^3/3 + 4 \times 12^3/3 + 8 \times 2^3/12 + 16 \times 11^2 = 4267$

8.47 a) $I_{\text{edge}} = I_{\text{c.g.}} + Md^2 = \dfrac{1}{12} M\left(b^2 + b^2\right) + M\dfrac{b^2}{2} = \dfrac{2}{3} Mb^2$

8.48 e) $I_x = \dfrac{1}{3}(6m) \times 6^2 \times 2 + 8m \times 6^2 = 432$ with $m = 1$

Dynamics

by Merle C. Potter

Overview

Dynamics is separated into two major divisions: *kinematics*, which is a study of motion without ref-erence to the forces causing the motion, and *kinetics*, which relates the forces on bodies to their resulting motions. Newton's laws of motion are necessary in relating forces to motions; they are:

1st law: A particle remains at rest or continues to move in a straight line with a constant velocity if no unbalanced force acts on it.

2nd law: The acceleration of a particle is proportional to the force acting on it and inversely proportional to the particle mass; the direction of acceleration is the same as the force direction.

3rd law: The forces of action and reaction between contacting bodies are equal in magnitude, opposite in direction, and colinear.

Law of gravitation: The force of attraction between two bodies is proportional to the product of their masses and inversely proportional to the square of the distance between their centers.

9.1 Kinematics

In kinematics we will consider three different kinds of particle motion: rectilinear motion, angular motion, and curvilinear motion. These will be followed by a review of motion of rigid bodies.

9.1.1 Rectilinear Motion

In rectilinear motion of a particle in which the particle moves in a straight line, the acceleration a, the velocity v, and the displacement s are related by

$$a = \frac{dv}{dt}, \qquad v = \frac{ds}{dt}, \qquad a = \frac{d^2s}{dt^2} = v\frac{dv}{ds}. \tag{9.1.1}$$

If the acceleration is a known function of time, the above can be integrated to give $v(t)$ and $s(t)$. For the important case of constant acceleration, integration yields

$$\begin{aligned} v &= v_o + at \\ s &= v_o t + at^2/2 \\ v^2 &= v_o{}^2 + 2as \end{aligned} \tag{9.1.2}$$

where at $t = 0$, $v = v_o$ and $s_o = 0$.

9.1.2 Angular Motion

Angular displacement is the angle θ that a line makes with a fixed axis, usually the positive x-axis. Counter clockwise motion is assumed to be positive, as shown in Fig. 9.1. The angular acceleration α, theangular velocity ω, and θ are related by

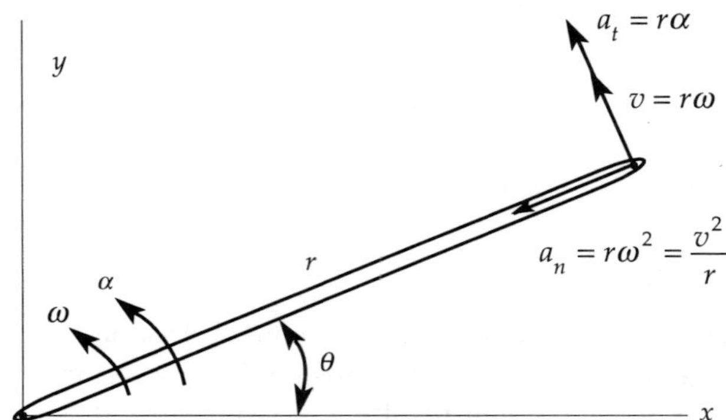

Figure 9.1 Angular motion.

$$\alpha = \frac{d\omega}{dt}, \qquad \omega = \frac{d\theta}{dt}, \qquad \alpha = \omega\frac{d\omega}{d\theta} = \frac{d^2\theta}{dt^2}. \tag{9.1.3}$$

If α is a constant, integration of these equations gives

$$\begin{aligned} \omega &= \omega_o + \alpha t \\ \theta &= \omega_o t + \alpha t^2/2 \\ \omega^2 &= \omega_o{}^2 + 2\alpha\theta \end{aligned} \tag{9.1.4}$$

where we have assumed that $\omega = \omega_o$ and $\theta_o = 0$ at $t = 0$.

9.1.3 Curvilinear Motion

When a particle moves on a plane curve as shown in Fig. 9.2, the motion may be described in terms of coordinates along the normal n and the tangent t to the curve at the instantaneous position of the particle.

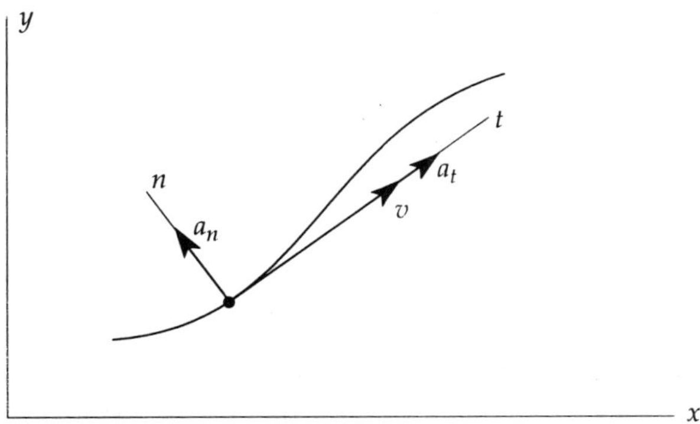

Figure 9.2 Motion on a plane curve.

The acceleration is the vector sum of the normal acceleration a_n and the tangential acceleration a_t. These components are given as

$$a_n = \frac{v^2}{r}, \qquad a_t = \frac{dv}{dt} \tag{9.1.5}$$

where r is the radius of curvature and v is the magnitude of the velocity. The velocity is always tangential to the curve so no subscript is necessary to identify the velocity.

It should be noted that rigid body traveling without rotation can be treated as particle motion.

Example 9.1

The velocity of a particle undergoing rectilinear motion is $v(t) = 3t^2 + 10t$ m/s. Find the acceleration and the displacement at $t = 10$ s, if $s_o = 0$ at $t = 0$.

Solution. The acceleration is found to be

$$a = \frac{dv}{dt}$$

$$= 6t + 10 = 6 \times 10 + 10 = 70 \text{ m/s}^2.$$

The displacement is found by integration

$$s = \int_0^{10} v\,dt$$

$$= \int_0^{10} (3t^2 + 10t)dt = t^3 + 5t^2 \Big|_0^{10} = 1500 \text{ m}.$$

Example 9.2

An automobile skids to a stop 60 m after its brakes are applied while traveling 25 m/s. What is its acceleration?

Solution. We use the relationship

$$v^2 = v_o^2 + 2as.$$

Letting $v = 0$, we find,

$$a = -\frac{v_o^2}{2s}$$

$$= -\frac{25^2}{2 \times 60} = -5.21 \text{ m/s}^2.$$

Example 9.3

A wheel, rotating at 100 rad/s ccw (counterclockwise), is subjected to an angular acceleration of 20 rad/s^2 cw. Find the total number of revolutions (cw plus ccw) through which the wheel rotates in 8 seconds.

Solution. The time at which the angular velocity is zero is found as follows:

$$\cancelto{0}{\omega} = \omega_o + \alpha t$$

$$\therefore t = -\frac{\omega_0}{\alpha} = -\frac{100}{-20} = 5 \text{ s}.$$

After three additional seconds the angular velocity is by

$$\omega = \cancelto{0}{\omega_o} + \alpha t$$

$$= -20 \times 3 = -60 \text{ rad / s}.$$

The angular displacement from 0 to 5 s is

$$\theta = \omega_o t + \alpha t^2/2$$

$$= 100 \times 5 - 20 \times 5^2/2 = 250 \text{ rad}.$$

During the next 3 s, the angular displacement is

$$\theta = \alpha t^2/2$$

$$= -20 \times 3^2/2 = -90 \text{ rad}.$$

The total number of revolutions rotated is

$$\theta = (250 + 90)/2\pi = 54.1 \text{ rev}.$$

Example 9.4

Consider idealized projectile motion (no air drag) in which $a_x = 0$ and $a_y = -g$. Find expressions for the range R and the maximum height H in terms of v_o and θ.

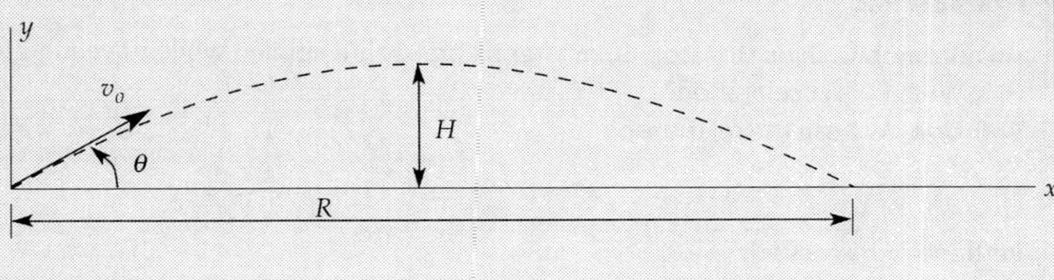

Solution. Using Eq. 9.1.2 for constant acceleration we have the point $(R, 0)$:

$$\overset{0}{\cancel{y}} = (v_o \sin\theta)t - gt^2/2.$$

$$\therefore t = 2v_o \sin\theta/g.$$

From the x-component equation:

$$x = (v_o \cos\theta)t + \overset{0}{\cancel{a_x}} t^2/2.$$

$$\therefore R = (v_o \cos\theta) 2v_o \sin\theta / g = v_o^2 \sin 2\theta / g.$$

Obviously, the maximum height occurs when the time is one-half that which yields the range R. Hence,

$$H = (v_o \sin\theta)\frac{v_o \sin\theta}{g} - \frac{g}{2}\left(\frac{v_o \sin\theta}{g}\right)^2$$

$$= \frac{v_o^2}{2g}\sin^2\theta.$$

Note: The maximum R for a given v_o occurs when $\sin 2\theta = 1$, which means $\theta = 45°$ for R_{max}.

Example 9.5

It is desired that the normal acceleration of a satellite be $9.6\ \text{m/s}^2$ at an elevation of 200 km. What should be the velocity be for a circular orbit? The radius of the earth is 6400 km.

Solution. The normal acceleration, which points toward the center of the earth, is

$$a_n = \frac{v^2}{r}.$$

$$\therefore v = \sqrt{a_n r}$$

$$= \sqrt{9.6 \times (6400 + 200) \times 1000} = 7960\ \text{m/s}.$$

The normal acceleration is essentially the value of gravity near the earth's surface. Gravity varies only slightly if the elevation is small with respect to the earth's radius.

9.1.4 Rigid Body Motion

The motion of a rigid body can be described using the relative velocity and relative acceleration equations

$$\mathbf{v}_A = \mathbf{v}_B + \mathbf{v}_{A/B}$$
$$\mathbf{a}_A = \mathbf{a}_B + \mathbf{a}_{A/B} \qquad (9.1.6)$$

where the velocity $\mathbf{v}_{A/B}$ is the velocity of point A with respect to point B and the acceleration $\mathbf{a}_{A/B}$ is the acceleration of point A with respect to point B. If points A and B are on the same rigid body then point A must move perpendicular to the line AB and

$$v_{A/B} = r\omega \tag{9.1.7}$$

$$\left(a_{A/B}\right)_n = r\omega^2$$

$$\left(a_{A/B}\right)_t = r\alpha$$

where ω is the angular velocity, α is the angular acceleration of the body, and r is the length of $\overline{AB}$. If point A is located on a body which moves with a constant velocity v relative to a coincident point B which is located on a second body rotating with an angular velocity ω (see Fig. 9.3), the acceleration of A

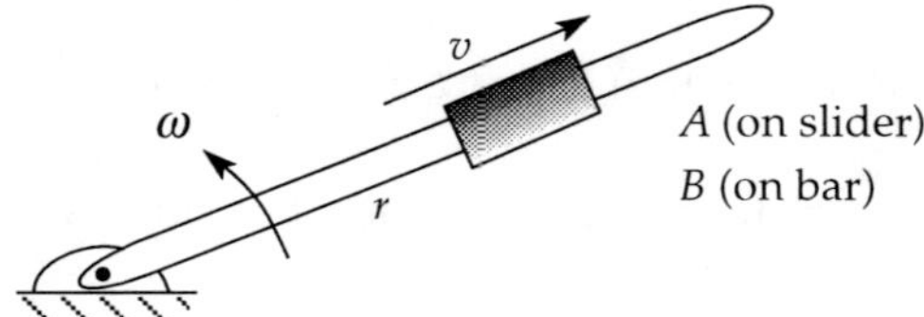

Figure 9.3 Coriolis acceleration.

with respect to B is called the *Coriolis acceleration*, given by

$$\mathbf{a}_{A/B} = 2\boldsymbol{\omega} \times \mathbf{v}. \tag{9.1.8}$$

We note that the Coriolis acceleration acts normal to both vectors $\boldsymbol{\omega}$ (use the right-hand rule) and $\mathbf{v}$. Thus it acts normal to the arm.

A final note regarding the instant center of zero velocity (a point which is often off the body, that has zero velocity): if such a point B can be located, then the magnitude of the velocity of point A is simply $r\omega$.

Example 9.6

Find the magnitude of the velocity and acceleration of point A.

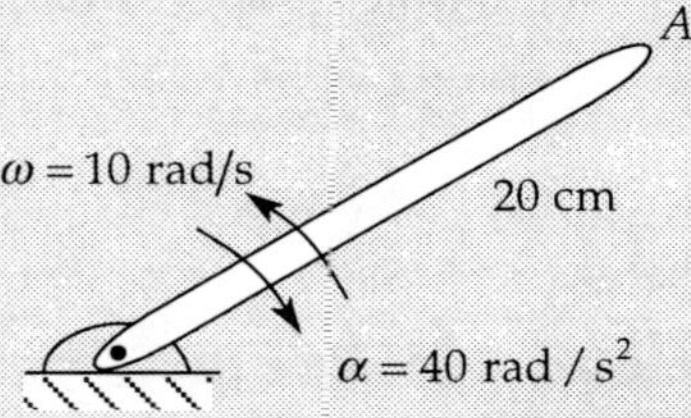

Solution. The velocity is found to be

$$v_A = r\omega$$
$$= 0.2 \times 10 = 2 \text{ m/s}.$$

The acceleration components are

$$a_n = r\omega^2 = 0.2 \times 10^2 = 20 \text{ m/s}^2$$

$$a_t = r\alpha = 0.2 \times 40 = 8 \text{ m/s}^2.$$

Thus,

$$a = \sqrt{a_n^2 + a_t^2} = \sqrt{20^2 + 8^2} = 21.5 \text{ m/s}^2.$$

Example 9.7

Find the velocity of C and ω_{BC}.

Solution. The velocity of B is normal to $\overline{AB}$ and is equal to

$$v_B = r\omega$$
$$= 0.4 \times 100 = 40 \text{ m/s}.$$

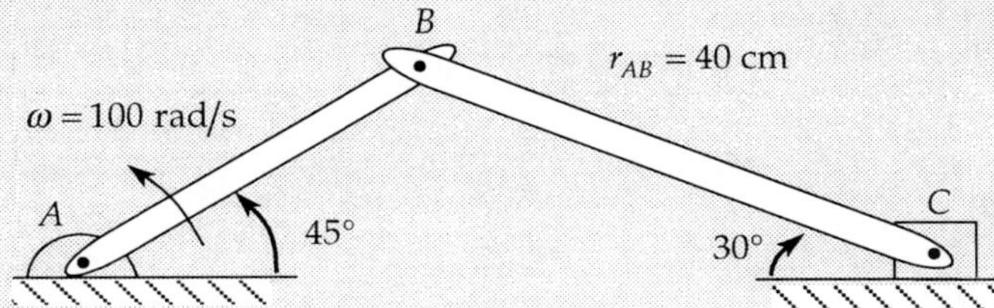

To find V_C we use the relative motion equation

$$\mathbf{v}_C = \mathbf{v}_B + \mathbf{v}_{C/B}.$$

The velocity of C must be horizontal and $\mathbf{v}_{C/B}$ must be normal to $\overline{BC}$. This can be displayed in a velocity polygon as follows:

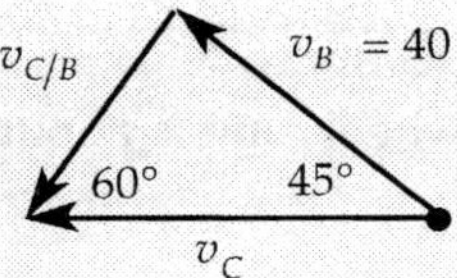

From the velocity polygon we use some simple trigonometry and find v_C to be to the left with magnitude

$$v_C = 44.6 \text{ m/s}.$$

The angular velocity of $\overline{BC}$ is found to be

$$\omega_{BC} = v_{C/B}/r_{BC}$$
$$= 32.7 / 0.5656 = 57.8 \text{ rad/s cw}.$$

Example 9.8

Find the acceleration of C in Example 9.7, assuming $\alpha_{AB} = 0$.

Solution. The acceleration of B is

$$(a_B)_n = r\omega^2$$
$$= 0.4 \times 100^2 = 4000 \text{ m/s}^2.$$

The relative acceleration equation

$$\mathbf{a}_C = \mathbf{a}_B + \mathbf{a}_{C/B}$$

can be displayed in an acceleration polygon, realizing that $\mathbf{a}_C$ must be horizontal, and $\mathbf{a}_{C/B}$ has both normal and tangential components. We find $(a_{C/B})_n$ to be

$$\left(a_{C/B}\right)_n = r_{BC}\omega_{BC}^{\ 2}$$

$$= 0.5656 \times 57.8^2 = 1890 \ \text{m/s}^2.$$

The acceleration polygon is shown. From the polygon we can find a_C to be to the left with magnitude

$$a_C = 3380 \ \text{m/s}^2.$$

Example 9.9

Find v_A and a_A.

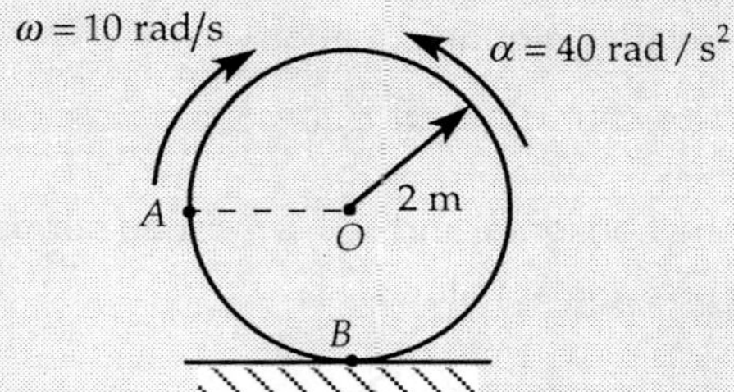

Solution. At this instant the wheel rotates about *B*. Hence, we can find v_A by using

$$v_A = r_{AB}\omega$$

$$= \frac{2}{\sin 45^\circ} \times 10 = 28.3 \ \text{m/s}.$$

To find the acceleration we relate *A* to the center *O* and use

$$\mathbf{a}_A = \mathbf{a}_O + \mathbf{a}_{A/O}.$$

The acceleration polygon follows using

$$a_O = r\alpha = 2 \times 40 = 80 \ \text{m/s}^2$$

$$(a_{A/O})_t = r\alpha = 2 \times 40 = 80 \ \text{m/s}^2$$

$$(a_{A/O})_n = r\omega^2 = 2 \times 10^2 = 200 \ \text{m/s}^2.$$

From the polygon we find a_A to be $a_A = 144 \ \text{m/s}^2$.

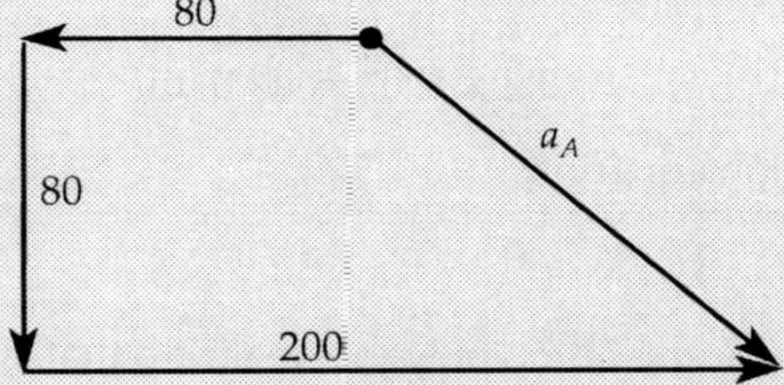

Example 9.10

Calculate a_A where A is on the slider.

Solution. The relative acceleration equation is

$$\mathbf{a}_A = \mathbf{a}_B + \mathbf{a}_{A/B}$$

$v = 5\,\text{m/s}$ $\omega = 10$ rad/s

40 cm

$a = 60\,\text{m/s}^2$ $\alpha = 40\ \text{rad/s}^2$

where point B is on the arm, coincident with point A. We know that

$$(a_B)_n = r\omega^2 = 0.4 \times 10^2 = 40\ \text{m/s}^2$$

$$(a_B)_t = r\alpha = 0.4 \times 40 = 16\ \text{m/s}^2$$

$$(a_{A/B})_t = 60\ \text{m/s}^2$$

$$(a_{A/B})_C = 2\omega v = 2 \times 10 \times 5 = 100\ \text{m/s}^2.$$

The acceleration polygon appears as follows:

The acceleration a_A is found to be $a_A = 86.4\ \text{m/s}^2$.

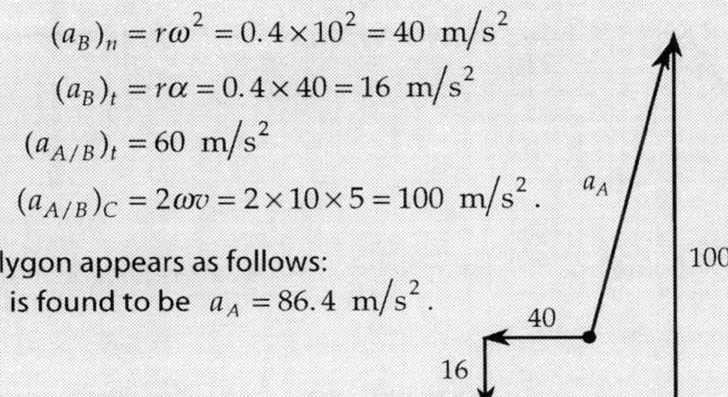

9.2 Kinetics

To relate the force acting on a body to the motion of that body we use Newton's laws of motion. Newton's 2nd law is used in the form

$$\sum \mathbf{F} = m\mathbf{a} \tag{9.2.1}$$

where the mass of the body is assumed to be constant and $\mathbf{a}$ is the acceleration of the center of mass (center of gravity) if the body is rotating. We also require

$$\sum \mathbf{M} = I\boldsymbol{\alpha} \tag{9.2.2}$$

where the moments must be summed about an axis passing through the center of mass. The mass moment of inertia I is often found by using the radius of gyration k and the relation $I = mk^2$.

The gravitational attractive force between one body and another is given by

$$F = K\frac{m_1 m_2}{r^2} \tag{9.2.3}$$

where $K = 6.67 \times 10^{-11}\ \text{N}\cdot\text{m}^2/\text{kg}^2$.

Note: Since metric units are used in the above relations, mass must be measured in kilograms. The weight is related to the mass by

$$W = mg \tag{9.2.4}$$

where we will use $g = 9.8\ \text{m/s}^2$, unless otherwise stated.

Example 9.11

Find the tension in the rope and the distance the 600 kg mass moves in 3 seconds. The mass starts from rest and the mass of the pulleys is negligible.

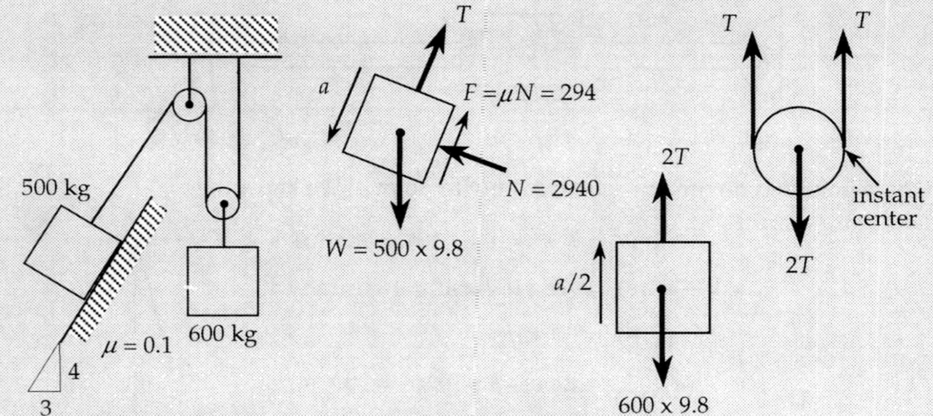

Solution. Applying Newton's 2nd law to the 500 kg mass gives

$$\sum F = ma$$

$$0.8 \times 500 \times 9.8 - 294 - T = 500\ a.$$

By studying the lower pulley we observe the 600 kg mass to be accelerating at $a/2$. Hence, we have

$$\sum F = ma$$

$$2T - 600 \times 9.8 = 600 \times a/2.$$

Solving the above equations simultaneously results in

$$a = 1.055\ \text{m/s}^2, \qquad T = 3100\ \text{N}.$$

The distance the 600 kg mass moves is

$$s = \frac{1}{2}\frac{a}{2}t^2$$

$$= \frac{1}{2} \times \frac{1.055}{2} \times 3^2 = 2.37\ \text{m}.$$

Example 9.12

Find the tension in the string if at the position shown, $v = 4$ m/s. Calculate the angular acceleration.

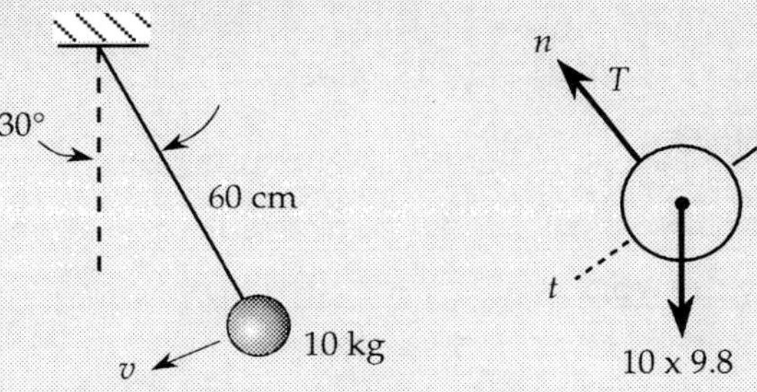

Solution. Sum forces in the normal direction and obtain

$$\sum F_n = ma_n = mv^2 / r$$

$$T - 10 \times 9.8 \cos 30^\circ = 10 \times 4^2 / 0.6.$$

$$\therefore\ T = 352 \text{ N}.$$

Sum forces in the tangential direction and find

$$\sum F_t = ma_t = mr\alpha$$

$$10 \times 9.8 \sin 30^\circ = 10 \times 0.6\alpha$$

$$\therefore\ \alpha = 8.17 \ \text{rad/s}^2.$$

Example 9.13

Find the angular acceleration of the 60 kg cylindrical pulley, and the tension in the rope.

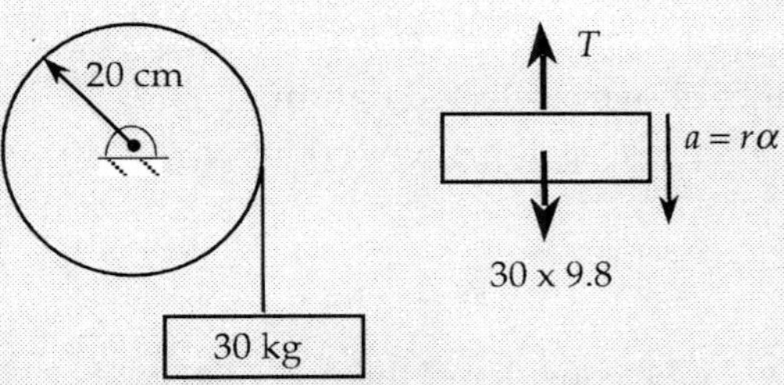

Solution. Summing forces on the 30 kg mass gives

$$\sum F = ma$$

$$-T + 30 \times 9.8 = 30 \times 0.2\alpha.$$

Summing moments about the center of the pulley yields

$$\sum M = I\alpha$$

$$T \times 0.2 = \frac{1}{2} \times 60 \times 0.2^2 \alpha$$

where $I = mr^2/2$ for a cylinder (see Table 8.2). A simultaneous solution results in

$$\alpha = 24.5 \ \text{rad/s}^2, \qquad T = 147 \text{ N}.$$

9.3 Work and Energy

Work is defined to be the dot product between a force and the distance it moves, that is,

$$W = \int \mathbf{F} \cdot d\mathbf{s}, \tag{9.3.1}$$

or, if the force is constant,

$$W = \mathbf{F} \cdot \Delta \mathbf{s}. \tag{9.3.2}$$

For a rotating body the work is

$$W = M\Delta\theta. \tag{9.3.3}$$

The *work-energy equation*, which results from integrating Newton's 2nd law, Eqs. 9.2.1 and 9.2.2, states that the net work done on a body (or several connected bodies) equals the change in energy of the body (or several bodies). This is expressed as

$$W_{net} = \Delta E \tag{9.3.4}$$

where E represents the kinetic energy, given by

$$E = \frac{1}{2} m v^2 \tag{9.3.5}$$

for a translating body, and

$$E = \frac{1}{2} I \omega^2 \tag{9.3.6}$$

for a rotating body. For a translating and rotating body, refer v and I to the mass center. Potential energy can be realized by allowing the body forces to do work; or they can be incorporated in the ΔE-term by using the potential energy as

$$E_p = mgh \tag{9.3.7}$$

where h is the distance above a selected datum.

By applying Eq. 9.3.1 to a spring, the work necessary to compress that spring a distance x is

$$W = \frac{1}{2} K x^2. \tag{9.3.8}$$

The quantity $Kx^2/2$ can be considered the potential energy stored in the spring.

Example 9.14

Neglecting friction, find v of the slider when it hits B if it starts from rest at A.

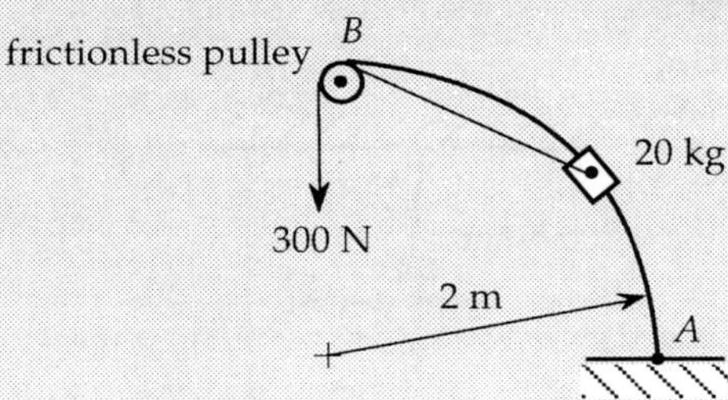

Solution. The distance the force F moves is

$$s = \frac{2}{\sin 45^\circ} = 2.828 \text{ m}.$$

The work-energy equation is written as

$$W_{net} = \frac{1}{2} m v^2$$

$$300 \times 2.828 - (20 \times 9.8) \times 2 = \frac{1}{2} \times 20 v^2.$$

$$\therefore v = 6.76 \text{ m/s}.$$

Note that the body force does negative work since the motion is up and the body force acts down.

Example 9.15

Neglect friction and estimate the angular velocity of the cylinder after the mass falls 2 m from rest.

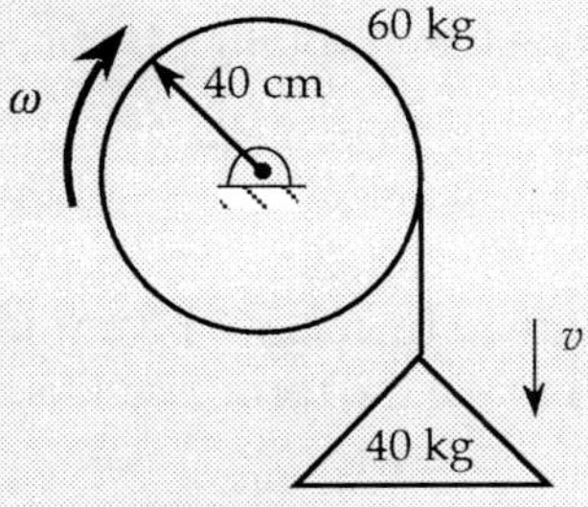

Solution. The work-energy equation provides

$$W_{net} = \frac{1}{2} m v^2 + \frac{1}{2} I \omega^2$$

$$(40 \times 9.8) \times 2 = \frac{1}{2} \times 40 \times (0.4\omega)^2 + \frac{1}{2}(\frac{1}{2} \times 60 \times 0.4^2)\omega^2.$$

$$\therefore \omega = 11.83 \text{ rad/s}.$$

Example 9.16

What is the velocity of the 40 kg mass after it falls 20 cm from rest? The spring is initially stretched 10 cm.

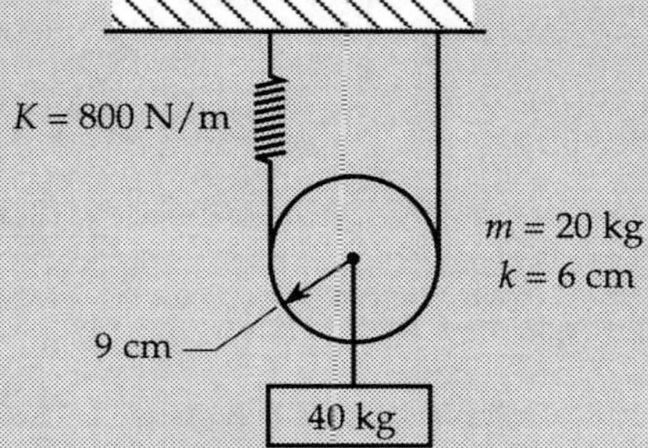

Solution. The spring will stretch an additional 40 cm. Thus, the work-energy equation results in

$$W_{net} = \frac{1}{2}mv^2 + \frac{1}{2}I\omega^2,$$

$$(20+40)\times 9.8\times 0.2 - \frac{1}{2}800\left(0.5^2 - 0.1^2\right) = \frac{1}{2}(20+40)v^2 + \frac{1}{2}\left(20\times 0.06^2\right)\left(\frac{v}{0.09}\right)^2.$$

$$\therefore v = 0.792 \text{ m/s}.$$

9.4 Impulse and Momentum

The impulse-momentum equations also result from integrating Newton's 2nd law. *Impulse* is defined for linear and rotating bodies, respectively, as

$$i_l = \int F dt$$
$$i_r = \int M dt. \qquad \textbf{(9.4.1)}$$

Momentum is velocity multiplied by mass. The impulse-momentum equations for a constant force and moment take the form (see Eqs. 9.2.1 and 9.2.2)

$$F\Delta t = m\Delta v$$
$$M\Delta t = I\Delta\omega. \qquad \textbf{(9.4.2)}$$

Objects impacting each other with no external forces acting experience a conservation of momentum. The *coefficient of restitution* e is used in such problems. It is defined as

$$e = \frac{\text{relative separation velocity}}{\text{relative approach velocity}}. \qquad \textbf{(9.4.3)}$$

If $e = 1$, the collision is *elastic* with no energy loss. If $e = 0$, the collision is *plastic* with maximum energy loss.

Example 9.17

Neglecting friction, estimate the angular velocity of the cylinder after 2 seconds if the motion starts from rest.

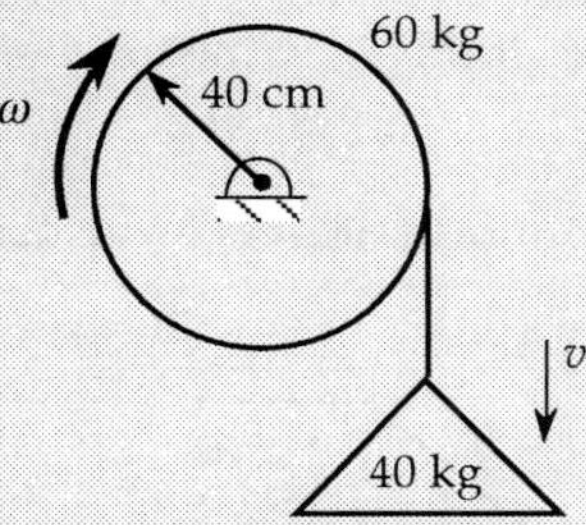

Solution. The impulse-momentum equation is used as follows:

$$M\Delta t = I\Delta\omega + m\Delta v \times r$$

$$0.4 \times (40 \times 9.8) \times 2 = \frac{1}{2} \times 60 \times 0.4^2 \omega + 40 \times 0.4\omega \times 0.4$$

$$\therefore \omega = 28.0 \text{ rad/s}.$$

Example 9.18

Find v' and θ if the coefficient of restitution is 0.8.

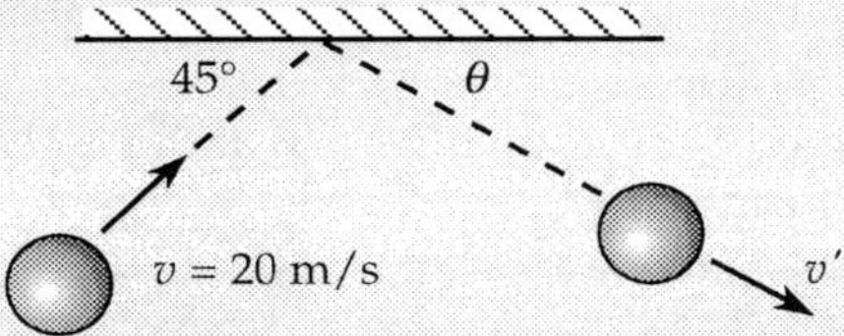

Solution. The coefficient of restitution is based on the normal components of velocity. Thus,

$$e = \frac{v' \sin\theta}{v \sin 45^\circ}.$$

$$\therefore v' \sin\theta = 0.8 \times 20 \times 0.707 = 11.31.$$

The tangential velocity component remains unchanged so that

$$v' \cos\theta = 20 \cos 45^\circ = 14.14.$$

Simultaneous solution of the above results in

$$v' = 18.11 \text{ m/s}, \qquad \theta = 38.65^\circ.$$

Practice Problems

(If you choose to work only a few problems, select those with a star.)

RECTILINEAR MOTION

*9.1 An object is moving with an initial velocity of 20 m/s. If it is decelerating at 5 m/s^2 how far does it travel, in meters, before it stops?

a) 10 b) 20 c) 30 d) 40 e) 50

9.2 If the particle starts from rest, what is its velocity, in m/s, at $t = 4$ s?

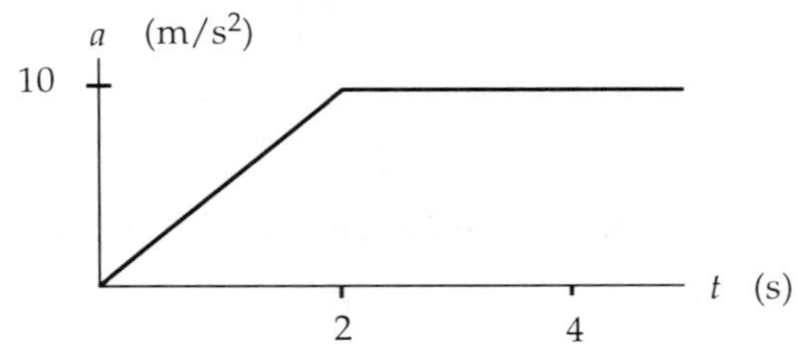

a) 10 b) 20 c) 30 d) 40 e) 50

*9.3 A projectile is shot straight up with $v_o = 40$ m/s. After how many seconds will it return if drag is neglected?

a) 4 b) 6 c) 8 d) 10 e) 12

*9.4 An automobile is traveling at 25 m/s. It takes 0.3 s to apply the brakes after which the deceleration is 6.0 m/s^2. How far does the automobile travel, in meters, before it stops?

a) 40 b) 45 c) 50 d) 55 e) 60

ANGULAR MOTION

9.5 A wheel accelerates from rest with $\alpha = 6$ rad/s^2. How many revolutions are experienced in 4 s?

a) 7.64 b) 9.82 c) 12.36 d) 25.6 e) 38.4

*9.6 A 2-m-long shaft rotates about one end at 20 rad/s. It begins to accelerates with $\alpha = 10$ rad/s^2. After how long, in seconds, will the velocity of the free end reach 100 m/s?

a) 7 b) 6 c) 5 d) 4 e) 3

CURVILINEAR MOTION

*9.7 A roller-coaster reaches a velocity of 20 m/s at a location where the radius of curvature is 40 m. Calculate the acceleration, in m/s^2.

a) 8 b) 9 c) 10 d) 11 e) 12

9.8 A bucket full of water is to be rotated in the vertical plane. What minimum angular velocity, in rad/s, is necessary to keep the water inside if the rotating arm is 120 cm?

a) 5.31 b) 4.26 c) 3.86 d) 3.15 e) 2.86

PROJECTILE MOTION

*9.9 Find the maximum height, in meters.

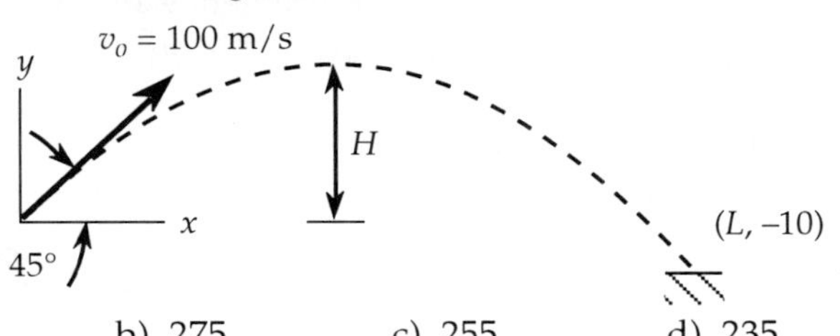

a) 295 b) 275 c) 255 d) 235 e) 215

9.10 Calculate the time, in seconds, it takes the projectile of Problem 9.9 to reach the low point.

a) 14.6 b) 12.2 c) 11.0 d) 10.2 e) 8.31

9.11 What is the distance L, in meters, in Problem 9.9?

a) 530 b) 730 c) 930 d) 1030 e) 1330

RIGID BODY MOTION

*9.12 The acceleration of the center O is given by

a) $r\omega^2$ b) v^2/r c) 0

d) ω^2/r e) rv^2

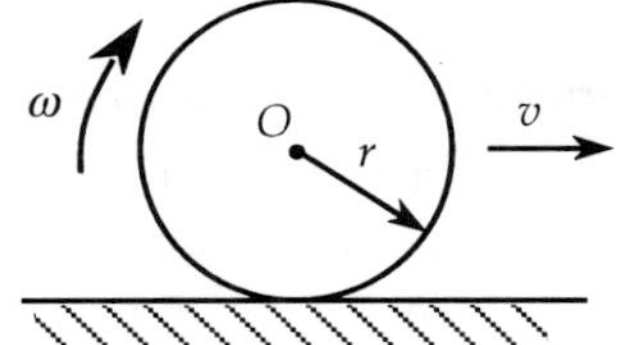

9.13 The spool rolls without slipping. Find v_o in m/s.

a) 30 b) 25 c) 20

d) 10 e) 5

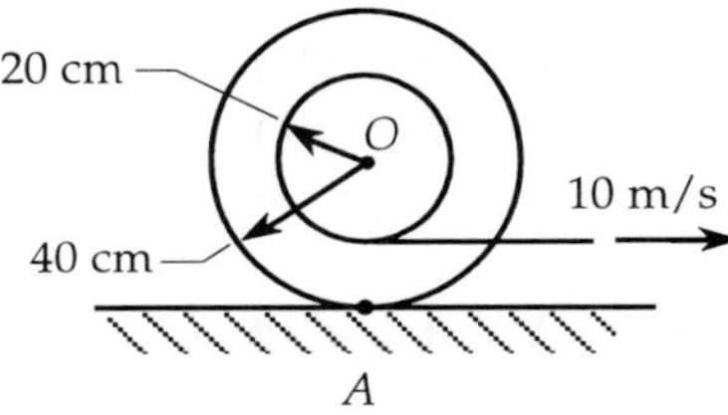

*9.14 The acceleration a_A, in m/s^2, in Problem 9.13 is

a) 1000 b) 800 c) 600 d) 400 e) 200

*9.15 If the acceleration of B is $60\mathbf{i} - 20\mathbf{j}$ m/s^2, find $\mathbf{a}_A$ in m/s^2.

a) $220\mathbf{i} + 60\mathbf{j}$

b) $100\mathbf{i} + 20\mathbf{j}$

c) $-100\mathbf{i} + 20\mathbf{j}$

d) $-100\mathbf{i} - 20\mathbf{j}$

e) $200\mathbf{i} + 20\mathbf{j}$

*9.16 Find ω_{AB} in rad/s.

a) 56.6 ccw

b) 56.6 cw

c) 34.1 ccw

d) 28.3 ccw

e) 28.3 cw

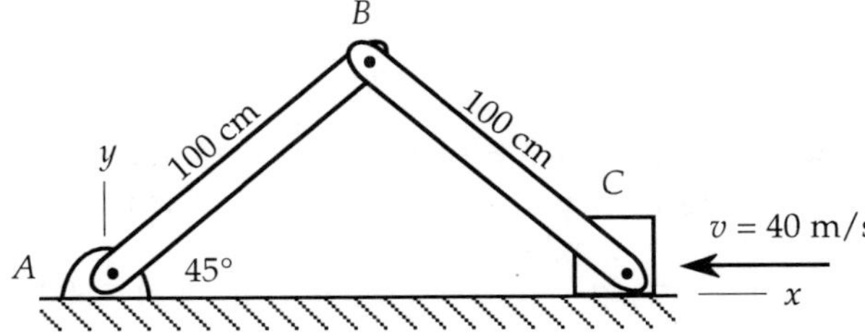

*9.17 What is ω_{BC}, in rad/s, for the linkage of Problem 9.16?

a) 56.6 ccw b) 56.6 cw c) 34.1 ccw d) 28.3 ccw e) 28.3 cw

9.18 Determine α_{AB}, in rad/s^2, if $a_C = 0$, in Problem 9.16.

a) 800 cw b) 800 ccw c) 1160 cw d) 3200 cw e) 3200 ccw

9.19 What is $\mathbf{a}_B$, in m/s^2, if $a_C = 0$, in Problem 9.16?

a) 1130 **i** b) –1130 **i** c) 1130 **j** d) –1130 **j** e) 1130 **k**

9.20 Find ω_{AB} in rad/s.

a) 20 ccw
b) 20 cw
c) 10 ccw
d) 10 cw
e) 0

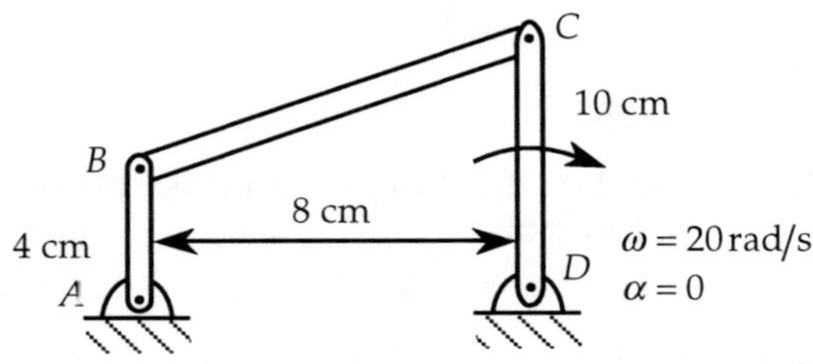

9.21 Find α_{AB}, in rad/s^2, if $a_C = 0$, in Problem 9.20.

a) 750 cw b) 750 ccw c) 400 cw d) 1000 cw e) 1000 ccw

*9.22 What is the acceleration, in m/s^2, of the bead if it is 10 cm from the center?

a) 20 **i** + 40 **j**
b) –40 **i** – 20 **j**
c) –40 **i** + 40 **j**
d) 20 **i** – 40 **j**
e) 40 **i** + 20 **j**

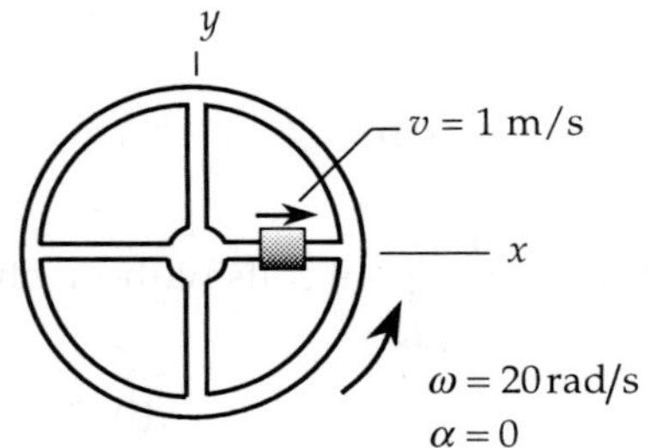

KINETICS

*9.23 What is a_A in m/s^2?

a) 2.09
b) 1.85
c) 1.63
d) 1.47
e) 1.22

A
500 kg
frictionless, massless pulleys
$\mu = 0.2$
B
400 kg

*9.24 How far, in meters, will the weight move in 10 s, if released from rest?

a) 350 b) 300 c) 250
d) 200 e) 150

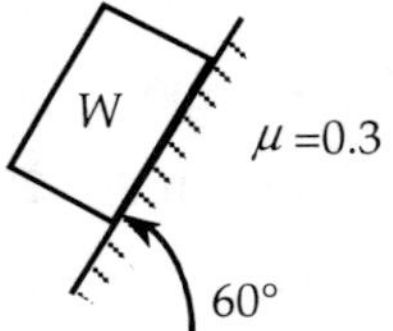

*9.25 At what angle, in degrees, should a road be slanted to prevent an automobile traveling at 25 m/s from tending to slip? The radius of curvature is 200 m.

a) 22 b) 20 c) 18 d) 16 e) 14

9.26 A satellite orbits the earth 200 km above the surface. What speed, in m/s, is necessary for a circular orbit? The radius of the earth is 6400 km and $g = 9.2\ \text{m/s}^2$.

a) 7800 b) 7200 c) 6600 d) 6000 e) 5400

9.27 Determine the mass of the earth, in kg, if the radius of the earth is 6400 km.

a) 6×10^{22} b) 6×10^{23} c) 6×10^{24} d) 6×10^{25} e) 6×10^{26}

9.28 The coefficient of sliding friction between rubber and asphalt is about 0.6. What minimum distance, in meters, can an automobile slide on a horizontal surface if it is traveling at 25 m/s?

a) 38 b) 43 c) 48 d) 53 e) 58

*9.29 The center of mass is 30 cm in front of the rear wheel of a motorcycle and 80 cm above the roadway. What maximum acceleration, in m/s^2, is possible?

a) 4.5 b) 4.3 c) 4.1 d) 3.9 e) 3.7

9.30 Find the ratio of the tension in the wire before and immediately after the string is cut.

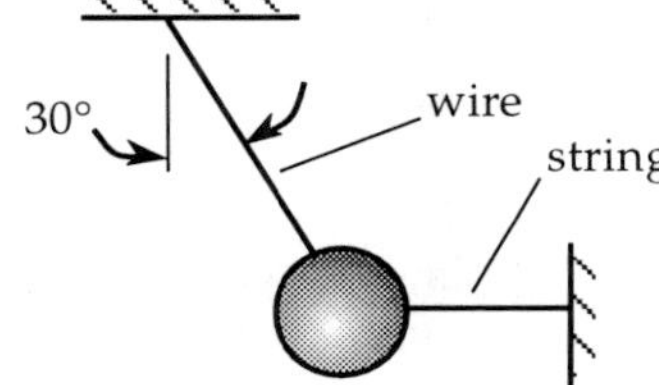

a) $2/\sqrt{3}$ b) $4/3$ c) $2/3$

d) $3/4$ e) $\sqrt{3}/2$

*9.31 Find the force, in kN, on the front wheels if $a = 2\ m/s^2$. The center of mass is at G.

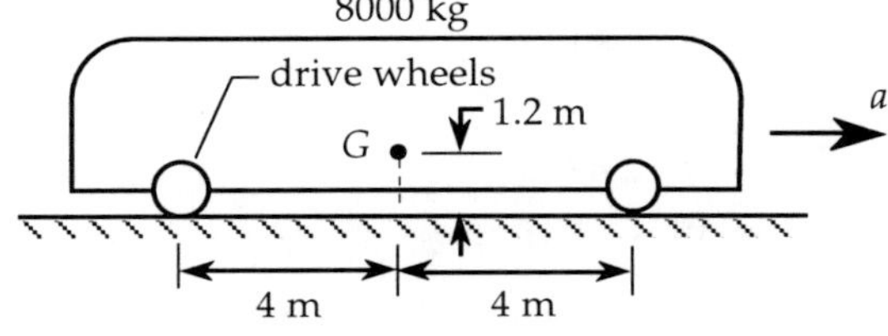

a) 58.2
b) 47.3
c) 41.6
d) 36.8
e) 22.8

9.32 The radius of gyration of the pulley is 10 cm. Calculate α in rad/s^2.

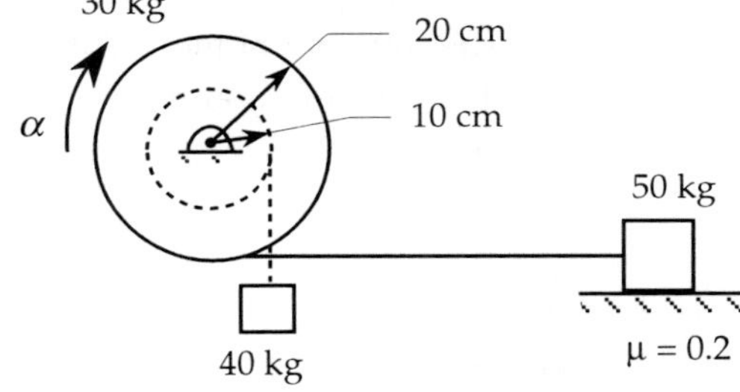

a) 8.52
b) 7.26
c) 6.58
d) 5.32
e) 4.69

9.33 What is the force at O immediately after the string is cut?

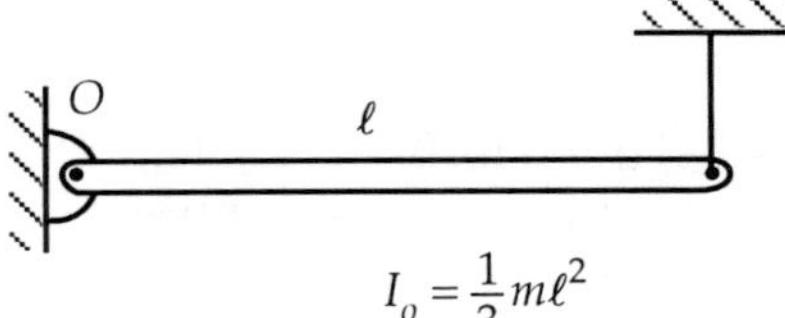

a) $\frac{mg}{2}$

b) $\frac{mg}{3}$

c) $\frac{mg}{4}$

d) $\frac{mg}{5}$

e) 0

WORK ENERGY

9.34 Find the velocity, in m/s, after the mass moves 10 m if it starts from rest.

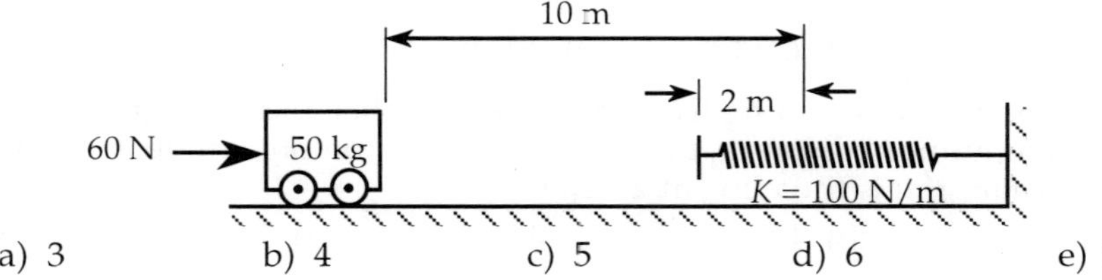

a) 3 b) 4 c) 5 d) 6 e) 7

*9.35 If the force acts through 4 m, what is the angular velocity, in rad/s, of the solid cylinder? Assume no slip and the cylinder starts from rest.

100 N, 20 cm, 100 kg, $I_o = \frac{1}{2}mr^2$

a) 18 b) 16 c) 14

d) 12 e) 10

*9.36 The spring is initially free. Calculate the velocity, in m/s, of the 2 kg mass after it falls 40 cm. It starts from rest.

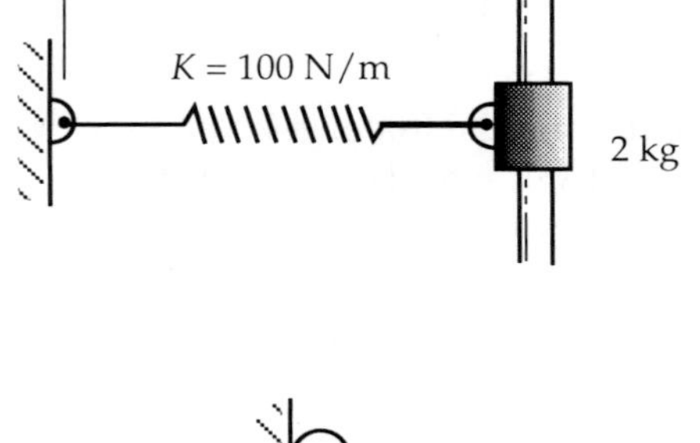

a) 4.62
b) 3.84
c) 2.96
d) 2.42
e) 1.95

9.37 Find the velocity, in m/s, of the end of the 10 kg bar as it passes *A*. The free spring length is 30 cm. The moment of inertia of a bar about its mass center is $m\ell^2/12$.

50 cm, K = 100 N/m, 40 cm, 30 cm, A

a) 5.2
b) 4.6
c) 3.5
d) 2.4
e) 1.2

IMPULSE-MOMENTUM

*9.38 If the force in Problem 9.35 acts for 4 seconds, find the angular velocity, in rad/s, assuming no slip. The cylinder starts from rest.

a) 2.4 b) 5.2 c) 8.6 d) 10.2 e) 26.7

*9.39 Find the velocity, in m/s, of a 100 kg mass at $t = 2$ s. The coefficient of friction is 0.2.

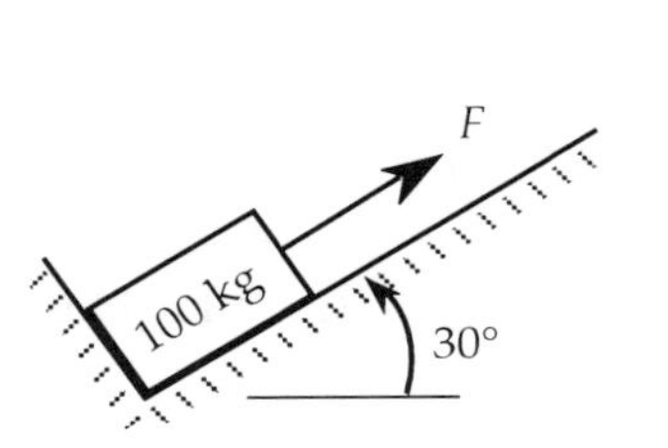

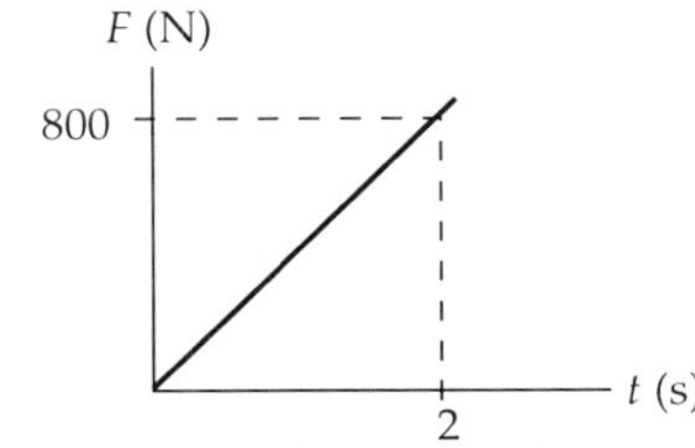

a) 0.245 b) 0.345 c) 0.456 d) 0.567 e) 0.678

*9.40 If the coefficient of restitution is 0.8, find v'_B in m/s.

a) 16
b) 13
c) 11
d) 9
e) 7

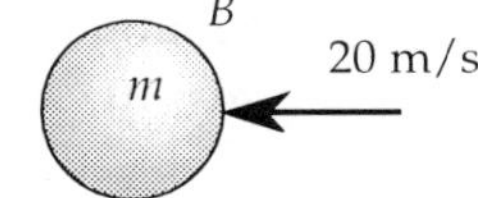

*9.41 Calculate the energy lost, in joules, in the collision of Problem 9.40 if $m = 2$ kg.

a) 200 b) 180 c) 160 d) 140 e) 120

Solutions to Practice Problems

9.1 d) $v^2 = v_o^2 + 2as.\quad 0 = 20^2 - 2\times 5s \quad \therefore\ s = 40$ m

9.2 c) $a = 5t \quad t \le 2 \quad v = \int a\,dt = \int_0^2 5t\,dt + \int_2^4 10\,dt = 5\frac{2^2}{2} + 10(4-2) = 30$ m/s

9.3 c) $v = v_o + at.\quad 0 = 40 - 9.8t \quad \therefore\ t = 4.08$ s. $\therefore\ t_{total} = 2t = 8.16$ s

9.4 e) $\Delta s = v\Delta t = 25\times 0.3 = 7.5$ m

$v^2 = v_o^2 + 2as.\quad 0 = 25^2 - 2\times 6s \quad \therefore\ s = 52.1$ m

$\therefore\ s_{total} = 7.5 + 52.1 = 59.6$ m

9.5 a) $\theta = \alpha t^2/2 = 6\times 4^2/2 = 48$ rad. $48/2\pi = 7.64$ rev.

9.6 e) $\omega = \omega_o + \alpha t = 20 + 10t.\quad \omega = v/L = 100/2 = 50.$

$\therefore\ 20 + 10t = 50.\quad \therefore\ t = 3$ s.

9.7 c) $a = \frac{v^2}{r} = \frac{20^2}{40} = 10$ m/s^2

9.8 e) $\frac{v^2}{r} = g$. $\frac{v^2}{1.20} = 9.8$. $\therefore\ v = 3.43\ \text{m/s}$.

$v = r\omega$ $\therefore\ \omega = 3.43/1.2 = 2.86\ \text{rad/s}$

9.9 c) $H = v_0^2 \sin^2\theta / 2g = 100^2 \times .707^2 2 \times 9.8 = 255\ \text{m}$

9.10 a) $y = v_o t \sin\theta - gt^2/2$. $-10 = 100t \times .707 - 9.8t^2/2$. $\therefore t = 14.6\ \text{s}$

9.11 d) $x = v_o t \cos\theta = 100 \times 14.6 \times 0.707 = 1032\ \text{m}$

9.12 c) The acceleration of the point of contact is v^2/r or $r\omega^2$. The acceleration of the center is 0.

9.13 c) Motion is about the point of contact. Thus,
$\omega = 10/.02 = 50\ \text{rad/s}$. $v = r\omega = 0.4 \times 50 = 20\ \text{m/s}$.

9.14 a) $a = v^2/r = 20^2/0.4 = 1000\ \text{m/s}^2$

9.15 c) $\mathbf{a}_A = \mathbf{a}_B + \mathbf{a}_{A/B}$

$= 60\mathbf{i} - 20\mathbf{j} - .4 \times 20^2\mathbf{i} + .4 \times 100\mathbf{j} = -100\mathbf{i} + 20\mathbf{j}$

9.16 d) $v_B = 40\cos 45° = 28.3 = r\omega_{AB} = 1.0\omega_{AB}$

$\therefore\ \omega_{AB} = 28.3$ ccw.

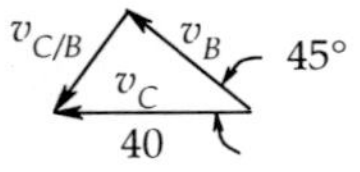

9.17 e) $v_{C/B} = 40\sin 45° = 28.3 = r_{BC}\omega_{BC} = 1.0\omega_{BC}$. $\therefore\ \omega_{BC} = 28.3$ cw

9.18 a) $\mathbf{a}_C = \mathbf{a}_B + \mathbf{a}_{C/B}$

$r_{AB}\omega_{AB}^2 = 1.0 \times 28.3^2 = 800$

$r_{BC}\omega_{BC}^2 = 1.0 \times 28.3^2 = 800$

Then, $r_{AB}\alpha_{AB} = 800$. $\therefore\ \alpha_{AB} = 800$ cw

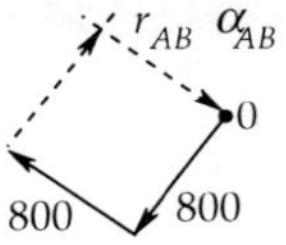

9.19 d) $\mathbf{a}_B = \cancelto{0}{\mathbf{a}_A} + \mathbf{a}_{B/A}$

$r\omega^2 = 1.0 \times 28.3^2 = 800$

$r\alpha = 1.0 \times 800 = 800$

$\therefore\ \mathbf{a}_B = -1130j$

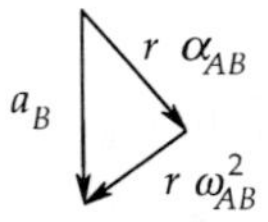

9.20 e) $\mathrm{v}_A = \mathrm{v}_B + \mathrm{v}_{A/B}$. $v_{A/B} = r\omega_{AB} \perp \overline{AB}$. But, v_A and v_B are both horizontal. Thus, $\mathrm{v}_{A/B} = 0$ and $\omega_{AB} = 0$.

9.21 b) $v_A = v_B = 20 \times 0.1 = 0.04\,\omega_A$.

$\therefore\ \omega_A = 50.\quad r_B\omega_B^2 = .1 \times 20^2 = 40$

$r_A\omega_A^2 = .04 \times 50^2 = 100.$

$\therefore\ r_{AB}\alpha_{AB} = \dfrac{60}{0.8} = 75.$

But, $r_{AB} = 0.1.\quad \therefore\ \alpha_{AB} = 750$ ccw.

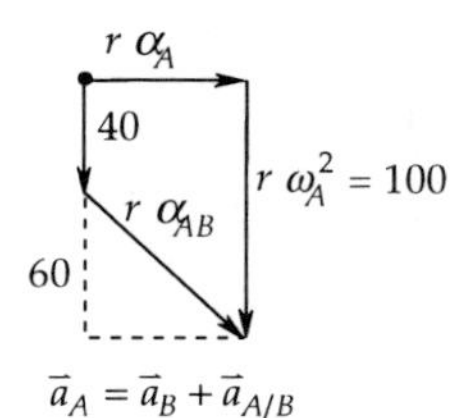

9.22 c) $\mathbf{a} = -r\omega^2\mathbf{i} + 2\omega v\mathbf{j} = -0.1 \times 20^2\mathbf{i} + 2 \times 20 \times 1.0\mathbf{j} = -40\mathbf{i} + 40\mathbf{j}$

9.23 c) $a_A = 2a_B$(from small pulley above B)

$400 \times 9.8 - 2T = 400\dfrac{a_A}{2}$ (body B)

$T - 0.2 \times 500 \times 9.8 = 500\ a_A$ (body A), $\quad\therefore\ a_A = 1.63\ \text{m/s}^2$

9.24 a) $W\cos 60° - W\mu \sin 60° = \dfrac{W}{g}a.$

$\therefore\ a = 9.8(0.866 - 0.3 \times 0.5) = 7.02\ \text{m/s}^2.\quad \therefore\ s = at^2/2 = 7.02 \times 10^2/2 = 351\ \text{m}$

9.25 c) $W\sin\theta = \dfrac{W}{g}\dfrac{v^2}{r}\cos\theta \qquad \tan\theta = \dfrac{25^2}{200 \times 9.8}.\quad \therefore\ \theta = 17.7°$

9.26 a) $W = \dfrac{W}{g}\dfrac{v^2}{r}.\quad \therefore\ v = \sqrt{9.2 \times 6\ 600\ 000} = 7790\ \text{m/s}$

9.27 c) $F = k\dfrac{m_1 m_2}{r^2}.\quad W = k\dfrac{m_e W/g}{r^2}.$

$\therefore\ m_e = r^2 g/k = 6400000^2 \times 9.8/6.67 \times 10^{-11} = 6 \times 10^{24}\ \text{kg}$

9.28 d) $v^2 = v_o^2 + 2as.\quad -0.6W = \dfrac{W}{g}a.\quad \therefore\ a = -.06g.\quad 0 = 25^2 - 2 \times .6 \times 9.8s.$

$\therefore\ s = 53.2\ \text{m}$

9.29 e) $\Sigma M = 0$

For maximum accel. the force on the front wheel = 0.

Thus, $30W = 80\ ma.\quad \therefore\ a = \dfrac{30 \times 9.8}{80} = 3.68\ \text{m/s}^2$

9.30 b) Before : $\Sigma F_y = 0$. $T\cos 30° = W$. $\therefore T = \dfrac{W}{0.866}$

After : $F = 0$, $\Sigma F_n = ma_n$. $n \perp t$. $\therefore$ $T = W \cos$ $= 0.866$ W.

$$\text{ratio} = \frac{\text{before}}{\text{after}} = \frac{1}{0.866 \times 0.866} = 1.33$$

9.31 d) Take moments about the back wheels.

$$4W - 8N_2 = 1.2\frac{W}{g} \times 2. \quad \therefore N_2 = 0.469W = .469 \times 8000 \times 9.8 = 36\ 800 \text{ N}$$

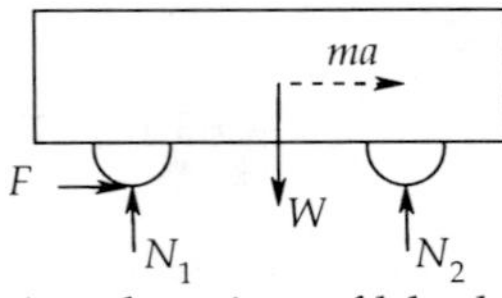

9.32 b) Acceleration of block = 0.2α. Thus, $T = 50 \times 0.2\alpha + 50 \times 9.8 \times 0.2$.

$\Sigma M = I\alpha$. Use $I = mk^2$. $(40 \times 9.8 - 40\alpha) \times 0.1 - (10\alpha + 9.8) \times 0.2 = 30 \times .1^2 \alpha$.

$\therefore \alpha = 7.26 \text{ rad/s}^2$

9.33 c) $\Sigma M_o = I_o \alpha$. $mg \times \dfrac{\ell}{2} = \dfrac{1}{3} m\ell^2 \alpha$. $\therefore \alpha = \dfrac{3g}{2\ell}$.

$\Sigma F_y = m\bar{a}$ ($\bar{a}$ is acc. of mass center).

$$mg - F_o = m\frac{\ell}{2}\alpha = m\frac{3}{4}g. \quad \therefore F_o = \frac{mg}{4}.$$

9.34 b) $W_{\text{net}} = \Delta E = \dfrac{1}{2}mv^2$. $60 \times 10 - \dfrac{1}{2} \times 100 \times 2^2 = \dfrac{1}{2}50v^2$. $\therefore v = 4$ m/s

9.35 d) $W_{\text{net}} = \Delta KE = \dfrac{1}{2}mv^2 + \dfrac{1}{2}I\omega^2$. $v = r\omega$.

$$\therefore\ 100 \times 4 = \frac{1}{2} \times 100(0.2\omega)^2 + \frac{1}{2}\left(\frac{1}{2} \times 100 \times .2^2\right)\omega^2. \quad \therefore \omega = 11.55 \text{ rad/s}$$

9.36 d) $W_{\text{net}} = \Delta KE = \dfrac{1}{2}mv^2$. $\Delta x = 50 - 30 = 20$.

$$\therefore\ 2 \times 9.8 \times 0.4 - \frac{1}{2} \times 100 \times .2^2 = \frac{1}{2} \times 2v^2. \quad \therefore v = 2.42 \text{ m/s}$$

9.37 e) $W_{\text{net}} = \Delta KE = \dfrac{1}{2}mv^2 + \dfrac{1}{2}I\omega^2$.

$$\frac{1}{2}100\left(.4^2 - .1^2\right) - 10 \times 9.8(.25 - .20) = \frac{1}{2}10\left(\frac{v}{2}\right)^2 + \frac{1}{2}\left(10 \times .5^2 / 12\right)\left(\frac{v}{.5}\right)^2$$

where $\bar{v} = v/2$ and $\omega = v/.5$. The above gives $v = 1.25$ m/s.

9.38 b) $\Sigma M_C \Delta t = I_C \omega$ where C is the point of contact. $I_C = I_o + mr^2 = \dfrac{3}{2}mr^2$.

$$100 \times 0.4 \times 4 = \frac{3}{2} \times 100 \times .2^2 \omega. \quad \therefore \omega = 26.7 \text{ rad/s}.$$

9.39 a) Integrate Eq. 9.2.1 as follows: $\int F dt = m\Delta v$. $F = 400t$.

$F_f = \mu N = 0.2 \times 100 \times 0.866 \times 9.8 = 170$

$400t = 100 \times 0.5 \times 9.8 + 170$. $\therefore t = 1.65$ s when motion initiates.

$$\int_{1.65}^{2.0} 400t dt - (170 + 50 \times 9.8) \times 0.35 = 100v. \quad \therefore v = 0.245 \text{ m/s}$$

9.40 e) $10\,m - 20\,m = v'_A m + v'_B m$. $\therefore$ $v'_A + v'_B = -10$

$0.8 = \dfrac{v'_B - v'_A}{20 + 10}$. $\therefore v'_B - v'_A = 24$. Simultaneous solutions yields $v'_B = 7$ m/s,

$v'_A = -17$ m/s.

9.41 c) $\Delta KE = \frac{1}{2} m v_A^2 + \frac{1}{2} m v_B^2 - \frac{1}{2} m v'^2_A - \frac{1}{2} m v'^2_B = \frac{2}{2}(10^2 + 20^2) - \frac{2}{2}(7^2 + 17^2) = 162$ J.

Mechanics of Materials

by George E. Mase

The mechanics of materials is one of a number of names given to the study of deformable solids sub-jected to applied forces and moments. The foundations of this subject reside in three basic topics:

1. internal equilibrium (stress concepts)
2. geometry of deformation (strain concepts)
3. mechanical and thermal properties (by which stress and strain are related)

Additionally, we assume homogeneity (properties are independent of position) and isotropy (absence of directional properties) in the materials considered.

10.1 Stress and Strain, Elastic Behavior

Consider a prismatic bar of length L and a cross sectional area A situated along the x-axis as shown in Fig. 10.1a. Let the bar be subjected to a constant axial force P applied at the centroids of the end faces so as to stretch the bar by an amount δ, Fig. 10.1b.

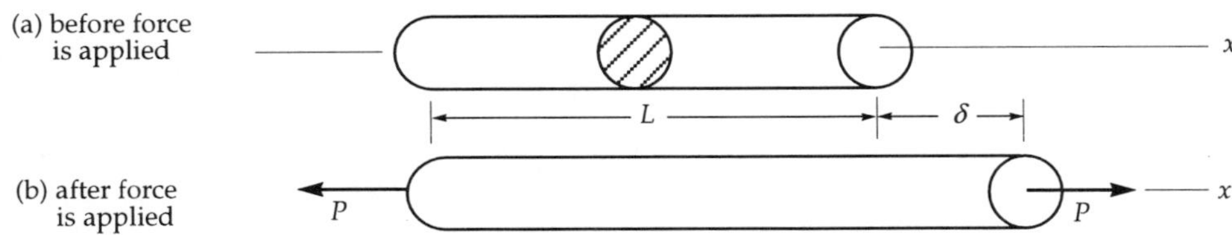

Figure 10.1 Normal stress and strain.

We define the *longitudinal,* or *normal strain* ε_x by the ratio

$$\varepsilon_x = \delta/L. \tag{10.1.1}$$

Strain is dimensionless, having the units of m/m, or in/in, etc. Normal strains are positive if due to elongation, negative if the result is shortening. The stress

$$\sigma_x = P/A \tag{10.1.2}$$

is the *normal,* or *axial stress* in the bar. Stress has units of N/m^2, lbs/in^2 (psi) or kips/in^2 (ksi). One Newton per square meter is called a *Pascal,* abbreviated Pa. Note that the stress and strain defined here are averages, constant over the length of the bar, and uniform over its cross section. By contrast, for a tapered bar hanging from the ceiling under its own weight, the stress and strain would vary along the bar. Also, it is natural that a positive longitudinal strain will be accompanied by a negative lateral strain. Indeed, this ratio

$$\nu = -\frac{\text{lateral strain}}{\text{longitudinal strain}} \tag{10.1.3a}$$

is called *Poisson's ratio*, an important property of a given material. Thus, for the bar in Fig. 10.1,

$$\varepsilon_y = \varepsilon_z = -\nu\varepsilon_x. \tag{10.1.3b}$$

If the bar in Fig. 10.1 is made of a *linear elastic* material, its axial stress and strain are related by the formula, often called *Hooke's law,*

$$\sigma_x = E\varepsilon_x \tag{10.1.4}$$

where E is a material constant called *Young's modulus,* or the *modulus of elasticity.* The units of E are the same as those of stress. By inserting Eqs. 10.1.1 and 10.1.2 into Eq. 10.1.4 and solving for δ we obtain the useful formula

$$\delta = \frac{PL}{AE}. \tag{10.1.5}$$

Stress is always accompanied by strain, but strain may occur without stress. In particular, a temperature change in an unconstrained bar will cause it to expand (or shrink) inducing a thermal deformation

$$\delta_t = \alpha L(T - T_o) \tag{10.1.6}$$

where α is the *coefficient of thermal expansion,* and $(T - T_o)$ the temperature change. Typical units of α are meters per meter per degree Celsius ($°\text{C}^{-1}$). Important properties of several materials are listed in Table 10.1.

Table 10.1 Average Material Properties.

	Modulus of Elasticity E		Shear Modulus G		Poisson's Ratio ν	Density ρ		Coefficient of Thermal Expansion α	
	$\times 10^6$ kPa	$\times 10^6$ psi	$\times 10^6$ kPa	$\times 10^6$ psi		kg/m³	lb/ft³	$\times 10^{-6}$ °C⁻¹	$\times 10^{-6}$ °F⁻¹
steel	210	30	83	12	.28	7850	490	11.7	6.5
aluminum	70	10	27	3.9	.33	2770	173	23.0	12.8
magnesium	45	6.5	17	2.4	.35	1790	112	26.1	14.5
cast iron	140	20	55	8	.27	7080	442	10.1	5.6
titanium	106	15.4	40	6	.34	4520	282	8.8	4.9
brass	100	15	40	6	.33	8410	525	21.2	11.8
concrete	20	3	—	—	—	2400	150	11.2	6.2

Next consider a material cube subjected to a pair of equilibrating couples acting in the plane of the faces of the cube as shown pictorially in Fig. 10.2a, and schematically in Fig. 10.2b. For a cube whose faces have an area A we define the *shear stress* in a plane parallel to those on which the forces act as

$$\tau_{xy} = F/A. \quad \textbf{(10.1.7)}$$

If the material of the cube is linearly elastic, the top will be displaced relative to the bottom as shown in Fig. 10.2c. The angle γ_{xy} measures the *shear strain* of the cube, and since for elastic behavior this angle is very small, we define the shear strain as

$$\gamma_{xy} \approx \tan \gamma_{xy} = \Delta x/h. \quad \textbf{(10.1.8)}$$

Also, for elastic behavior,

$$\tau_{xy} = G\gamma_{xy} \quad \textbf{(10.1.9)}$$

where G is the *shear modulus*, or *modulus of rigidity*, having the units of Pa or psi. It is related to E and ν by

$$G = \frac{E}{2(1+\nu)}. \quad \textbf{(10.1.10)}$$

There are only two independent material properties in an isotropic, elastic solid. These properties may depend on position as in a heat-treated steel.

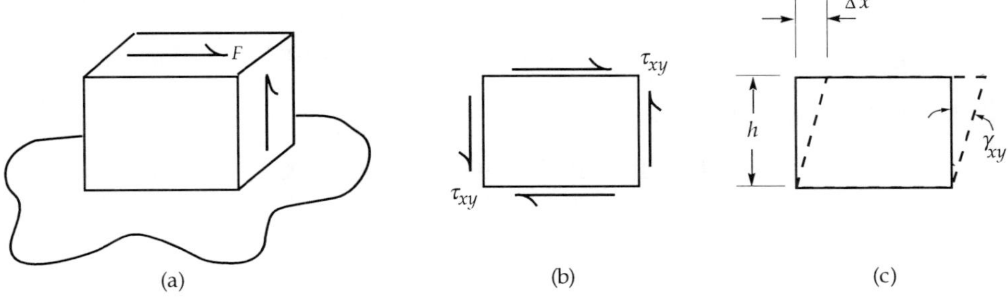

Figure 10.2 Shear stress and strain.

Example 10.1

A 2 cm × 2 cm square aluminum bar *AB* supported by 1.25 cm diameter steel cable *BC* carries a 7000 N load in the arrangement shown. Determine the stresses in the steel and in the aluminum. Also, calculate the elongation of the cable and the shortening of the bar.

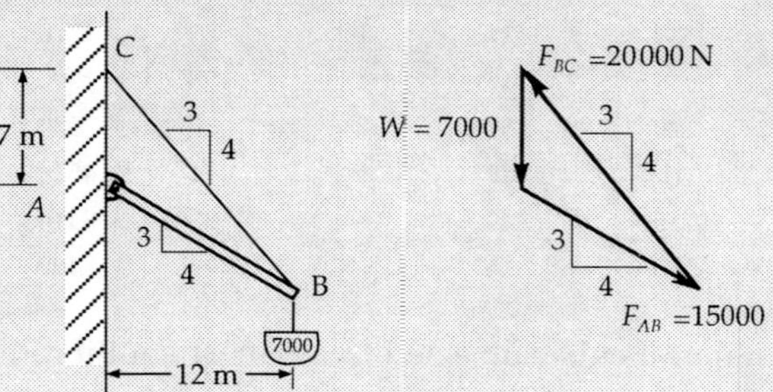

Solution. As shown by the force polygon for the equilibrium, $F_{AB} = -15\,000$ N and $F_{BC} = 20\,000$ N. Thus there results

$$\sigma_{al} = \frac{P}{A} = -\frac{15\,000}{0.02\times 0.02} = -37.5\times 10^6 \text{ Pa} \quad \text{or} \quad -37.5 \text{ MPa.}$$

$$\sigma_{st} = \frac{P}{A} = \frac{20\,000}{\pi(0.0125)^2/4} = 163\times 10^6 \text{ Pa} \quad \text{or} \quad 163 \text{ MPa.}$$

From Table 10.1 and Eq. 10.1.5,

$$\delta_{al} = \frac{PL}{AE} = \frac{-15\,000(15)}{0.02\times 0.02\times 70\times 10^9} = -0.00804 \text{ m} \quad \text{or} \quad -8.04 \text{ mm.}$$

$$\delta_{st} = \frac{PL}{AE} = -\frac{20\,000(20)}{(\pi\times 0.0125^2/4)\times 210\times 10^9} = 0.0155 \text{ m} \quad \text{or} \quad 15.5 \text{ mm.}$$

Example 10.2

A 5-cm-dia, 80-cm-long steel bar is restrained from moving. If its temperature is increased 100°C, what compressive stress is induced?

Solution. The strain can be calculated using the deformation of Eq. 10.1.6:

$$\varepsilon = \frac{\delta_t}{L} = \alpha(T - T_o)$$
$$= 11.7\times 10^{-6}\times 100 = 11.7\times 10^{-4}.$$

Hence, the induced stress is

$$\sigma = \varepsilon E = 11.7\times 10^{-4}\times 210\times 10^6 = 246\ 000 \text{ kPa.}$$

Example 10.3

The steel block ($G = 83\times10^6$ kPa) is welded securely to a horizontal platen and subjected to 1000 kN horizontal force as shown. Determine the shear stress in a typical horizontal plane of the block, and the horizontal displacement of the top edge *AB*.

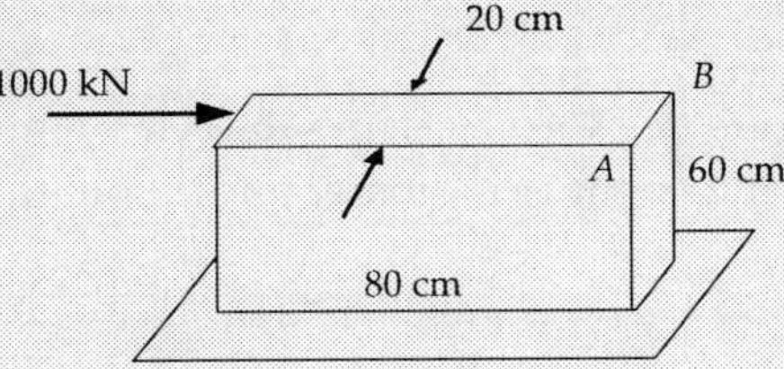

Solution. From Eq. 10.1.7

$$\tau = \frac{F}{A} = \frac{1000}{(0.8)(0.2)} = 6250 \text{ kPa.}$$

From Eq. 10.1.9

$$\gamma_{xy} = \frac{\tau_{xy}}{G} = \frac{6250}{83\times10^6} = 7.53\times10^{-5} \text{ rad.}$$

From Eq. 10.1.8 the horizontal displacement is

$$\Delta x = h\gamma_{xy} = 600(7.53\times10^{-5}) = 0.0452 \text{ mm.}$$

10.2 Torsion

A straight member of constant circular cross section subjected to a twisting couple at each end is said to be in *torsion*, and such a member is called a *shaft*. For an elastic shaft of length L and radius c subjected to a *torque* T (pair of equilibrium couples), as shown in Fig. 10.3a, the angular displacement of one end relative to the other is given by the angle ϕ (in radians) as

$$\phi = \frac{TL}{JG} \qquad \textbf{(10.2.1)}$$

where $J = \pi a^4/2$ is the *polar moment of inertia* of the circular cross section. Also, the *torsional shear stress* at the radial distance r from the axis of the shaft in a given cross section will be

$$\tau = \frac{Tr}{J} \qquad \textbf{(10.2.2)}$$

which increases linearly as shown in Fig. 10.3b. Thus, the maximum shear stress occurs at $r = a$,

$$\tau_{\max} = \frac{Ta}{J}. \qquad \textbf{(10.2.3)}$$

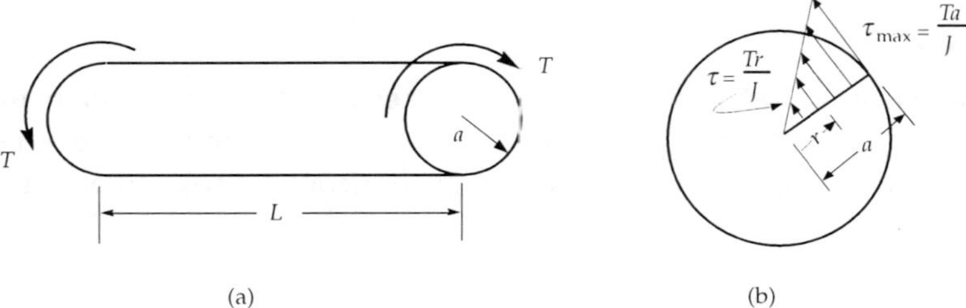

(a) (b)

Figure 10.3 Circular shaft subject to a torque.

For a hollow shaft having an inner radius a_i and an outer a_o the above formulas are all valid, but with

$$J = \pi(a_o^4 - a_i^4)/2 \tag{10.2.4}$$

Example 10.4

A 6 cm diameter, 2 m long magnesium ($G = 17 \times 10^9$ Pa) shaft is welded to a hollow ($c_o = 3$ cm and $c_i = 1.5$ cm) aluminum ($G = 27 \times 10^9$ Pa) shaft 1.2 m long. A moment of 2000 $\text{m} \cdot \text{N}$ is applied at end *A*. Determine the maximum torsional stress in each material and the angle of twist of end *A* relative to fixed end *B*.

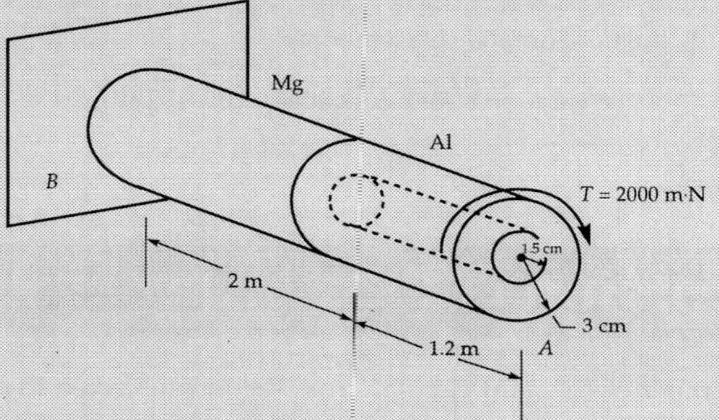

Solution. The polar moments of inertia are

$$J_{mg} = \pi a^4/2 = \pi(0.03)^4/2 = 1.272 \times 10^{-6} \text{ m}^4.$$

$$J_{al} = \pi(a_o^4 - a_i^4)/2 = \pi(0.03^4 - 0.015^4)/2 = 1.193 \times 10^{-6} \text{ m}^4.$$

From Eq. 10.2.3

$$\tau_{mg} = \frac{Ta}{J} = \frac{2000 \times 0.03}{1.272 \times 10^{-6}} = 47.17 \times 10^6 \text{ Pa.}$$

$$\tau_{al} = \frac{2000 \times 0.03}{1.193 \times 10^{-6}} = 50.29 \times 10^6 \text{ Pa.}$$

From Eq. 10.2.1 the angle of twist is

$$\phi = \phi_{mg} + \phi_{al} = \left(\frac{TL}{JG}\right)_{mg} + \left(\frac{TL}{JG}\right)_{al}$$

$$= \frac{2000 \times 2}{1.272 \times 10^{-6} \times 17 \times 10^9} + \frac{2000 \times 1.2}{1.193 \times 10^{-6} \times 27 \times 10^9}$$

$$= 0.1850 + 0.0745 = 0.2595 \text{ rad.}$$

10.3 Beam Theory

The usual geometry of a beam is that of a member having the length much larger than the depth with the forces applied perpendicular to this long dimension. The beams considered here have a longitudinal plane of symmetry in which the forces act and in which beam deflections occur. To illustrate we consider a *T* shaped beam having the cross section shown in Fig. 10.4b, supported either as a cantilever beam, Fig. 10.4a, or as a simply-supported beam, Fig. 10.4c. The longitudinal axis of the beam (x-axis here) passes through the centroidal points of all cross sections. The xy-plane is the plane of symmetry. Any combination of concentrated and distributed loads may act on the beam.

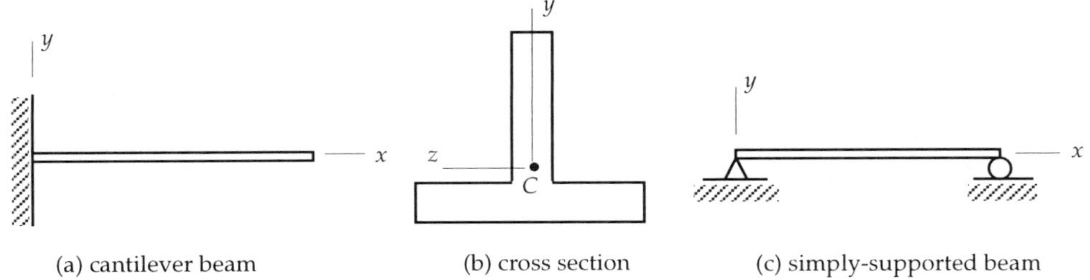

Figure 10.4 Beam geometry.

At the typical cross section of the loaded beam there is an internal force V called the *shear force*, and an internal moment M called the *bending moment*. Both V and M may be determined by a free-body diagram of the left hand portion of the beam, and are, in general, functions of x as shown in Fig. 10.5, where positive values of V and M are displayed. A plot of $V(x)$ is called a *shear diagram* and a plot of $M(x)$ is a *moment diagram*. At a given cross section where the moment has the value M, the (longitudinal) bending stress acting normal to the cross section is

$$\sigma_x = -\frac{My}{I} \tag{10.3.1}$$

where I is the plane moment of inertia of the cross sectional area relative to the centroidal axis. The minus sign is needed to assure a compressive stress for positive y values when the moment M is positive. The stress is a linear function of y as is shown in Fig. 10.6b, with the maximum compression occurring at the top of the beam, and the maximum tension at the bottom for the positive M. The bending stress is zero at $y = 0$, the so-called *neutral axis.*

The stress due to the shear force V is a vertical shear stress

$$\tau_{xy} = \frac{VQ}{Ib} \tag{10.3.2}$$

where b is breadth, or thickness of the beam at the position (y coordinate) at which the shear stress is calculated. The symbol Q stands for the *statical moment* about the neutral axis of the area between the position and the top of the beam. For a rectangular beam τ_{xy} is parabolic. For the *T* beam it has the shape shown in Fig. 10.6c. In both cases the maximum shear stress occurs at the neutral axis (centroidal position).

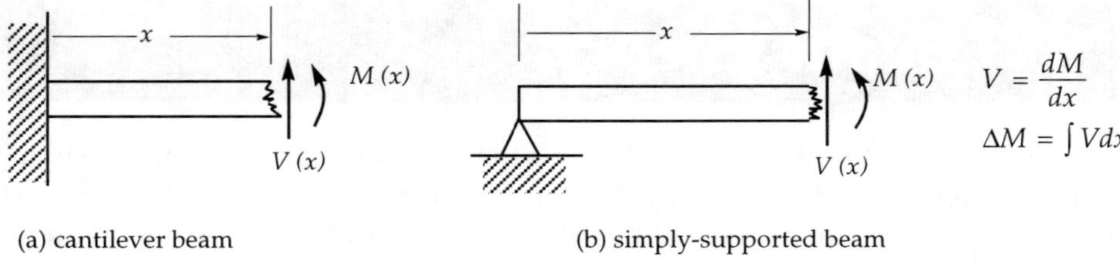

(a) cantilever beam (b) simply-supported beam

Figure 10.5 Internal shear force and bending moment.

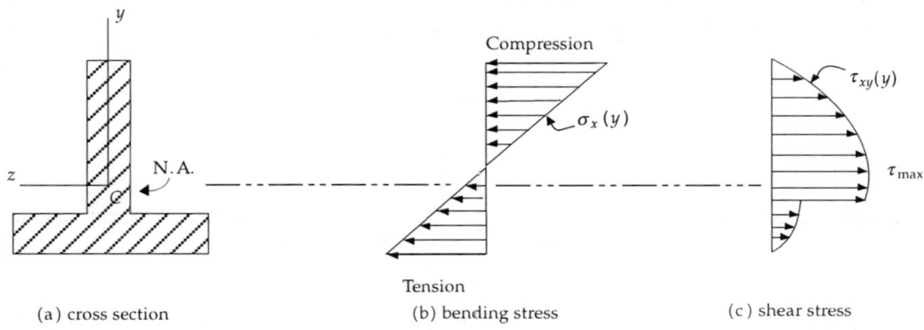

Figure 10.6 Beam stresses for positive *M*.

The vertical displacement of the x-axis of a loaded beam measures the beam deflection. The curve of this deflection $v = v(x)$ is called the *equation of the elastic line*, shown in Figure 10.7. Also, the slope $\theta = \theta(x) = dv / dx$ of the deflection curve is an important quantity in beam theory. Table 10.2 lists some useful formulas for basic beams.

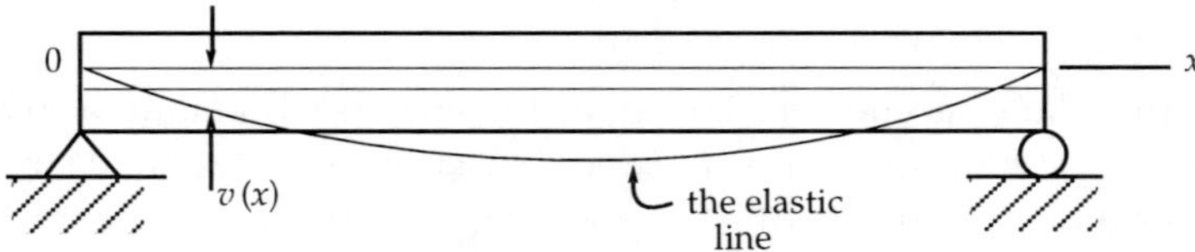

Figure 10.7 Beam deflection.

TABLE 10.2 Beam Formulas.

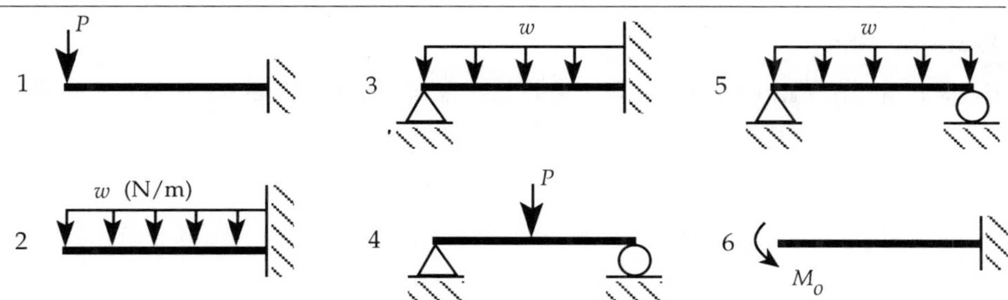

Number	Max Shear	Max Moment	Max Deflection	Max Slope
1	P	PL	$PL^3/3EI$	$PL^2/2EI$
2	wL	$wL^2/2$	$wL^4/8EI$	$wL^3/6EI$
3	$5wL/8$	$wL^2/8$	$wL^4/185EI$	
4	$P/2$	$PL/4$	$PL^3/48EI$	$PL^2/16EI$
5	$wL/2$	$wL^2/8$	$5wL^4/384EI$	$wL^3/24EI$
6		M_o	$M_oL^2/2EI$	M_oL/EI

Example 10.5

Sketch the shear and moment diagrams for the beam shown.

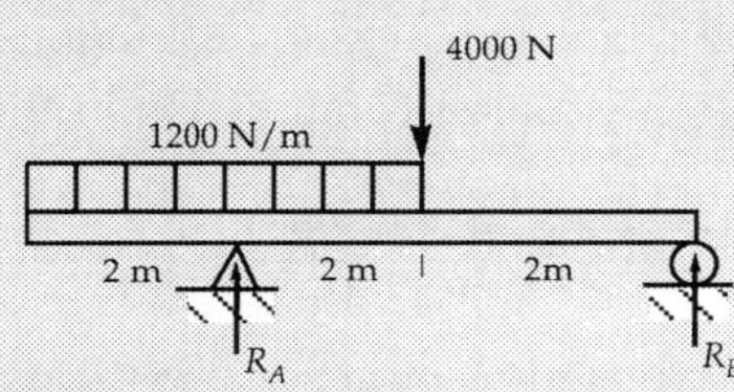

Solution.

First, determine the support reactions R_A and R_B:

$$\sum M_B = 4R_A - 4000(2) - 4(1200)4 = 0$$

$$\therefore\ R_A = 6800 \text{ N}$$

$$\sum M_A = 4R_B - 4000(2) = 0$$

$$\therefore\ R_B = 2000 \text{ N}$$

V: 4400, 2000, -2400, -2000, -2000

M: 4000, -2400

The values of *V* and *M* as functions of *x* are shown in the sketches; *V* is the resultant of all forces acting on the portion of the beam to the left of the *x*-location. The change in moment is the area under the shear diagram, $\Delta M = \int V dx$. Note from the diagrams that the maximum positive and negative moments occur at locations where the shear plot crosses the *x*-axis; the values of −2400 and 4000 are simply the appropriate areas under the shear diagram. Note that $M = 0$ at both ends and $V = 0$ at the left end.

Example 10.6

If the beam of Example 10.5 has the cross section shown below, determine the maximum tensile and compressive stresses, and the maximum shear stress.

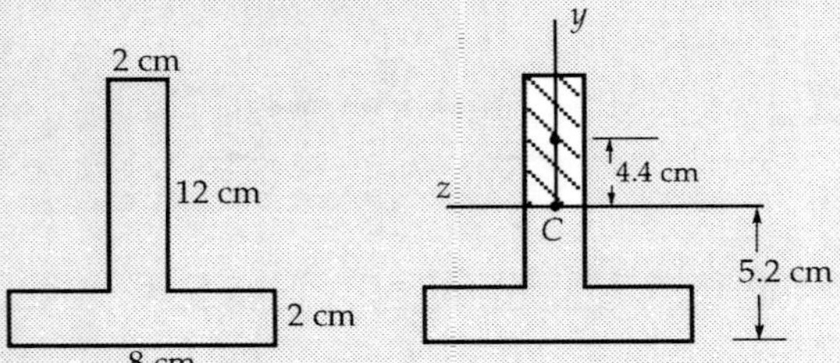

Solution. First, we locate the neutral axis by determining C relative to the bottom of the beam, as shown above:

$$\bar{y} = \frac{\sum y_i A_i}{\sum A_i} = \frac{16(1)+24(8)}{16+24} = 5.2 \text{ cm}.$$

The moment of inertia is (use $I = \bar{I} + Ad^2$)

$$I = \frac{8(2^3)}{12} + 16(4.2)^2 + \frac{2(12)^3}{12} + 24(2.8)^2$$

$$= 763.7 \text{ cm}^4.$$

The maximum positive M is $4000 \text{ N}\cdot\text{m}$ at $x = 4$, so the maximum compressive stress is (intuitively, we can visualize compression in the top fibers under the 4000 N force)

$$\left(\sigma_c\right)_{\max} = \frac{My}{I} = \frac{4000(0.088)}{764\times10^{-8}} = 46.1\times10^6 \text{ Pa}.$$

The maximum negative M is $2400 \text{ N}\cdot\text{m}$ at $x = 2$, so the maximum tensile stress is (intuitively, we know that tension occurs in the top fibers to the left of R_A)

$$\left(\sigma_t\right)_{\max} = \frac{My}{I} = \frac{2400(0.088)}{764\times10^{-8}} = 27.6\times10^6 \text{ Pa}.$$

The maximum V is 4400 N. The moment of the shaded area with respect to the neutral axis is Q. Therefore,

$$Q = \bar{y}A = 0.044\times(0.02\times0.088) = 7.744\times10^{-5} \text{ m}^3.$$

Since $b = 0.02$ at $y = 0$, the maximum shear stress is

$$\tau_{\max} = \frac{VQ}{Ib} = \frac{4400\times7.744\times10^{-5}}{764\times10^{-8}\times0.02} = 2.23\times10^6 \text{ Pa}.$$

Example 10.7

Determine the maximum deflection of a 3 cm × 24 cm rectangular aluminum beam, 5 m long, if a concentrated load of 800 N acts downward at its mid-point.

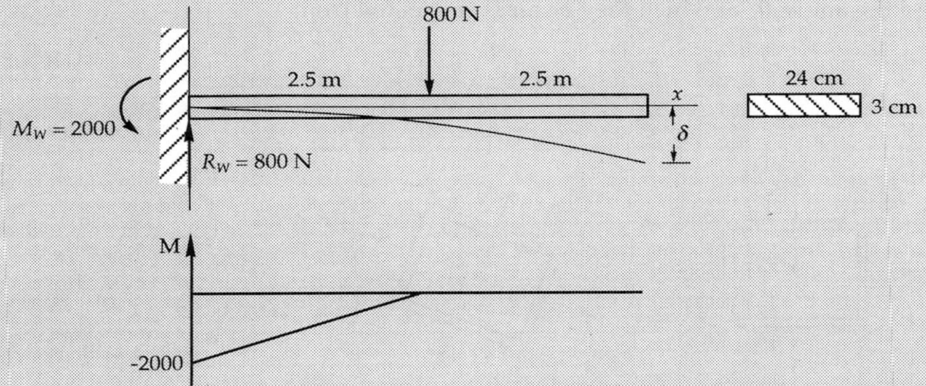

Solution. As the moment diagram shows, *M* is zero for the right-hand half of the beam. The right-hand half remains straight, but is inclined at the slope of the beam at mid-point. The left hand half is a simple end-loaded cantilever. From Table 10.2 with $L = 2.5$ m,

$$\delta = \delta_{\text{middle}} + \theta L = \frac{PL^3}{3EI} + \frac{PL^2}{2EI}(L) = \frac{5PL^3}{6EI}$$

$$= \frac{5(800)(2.5)^3}{6\times(70\times10^9)(0.24)0.03^3/12} = 0.276 \text{ m}$$

10.4 Combined Stress

It often happens that structural members are simultaneously subjected to some combination of axial, torsional and bending loads. In such cases the state of stress at points on the surface of the member consists of both normal and shear components, and is called *combined stress*. At any given point of interest on the surface we introduce a local set of coordinate axes and focus attention on the stresses acting on a very small rectangular element of material at the same point *P* as shown by Fig. 10.8a. For an element aligned with a rotated set of $x'y'$-axes at the same point, located by the c.c.w. angle θ, shown in Fig. 10.8b, the primed stresses will differ from the original unprimed stresses, the relationship being a function of θ. At a certain angle θ_p, with which we associate the axes x^* and y^*, Fig. 10.8c, the normal stresses will reach their maximum (x^*-direction) and minimum (y^*-direction) values, while the shear stresses vanish. These axes are called *principal axes of stress*, and the values σ_1 and σ_2 are called the *principal stresses*. It turns out that

$$\sigma_1 = \sigma_{\max} = \frac{\sigma_x + \sigma_y}{2} + \sqrt{\left(\frac{\sigma_x - \sigma_y}{2}\right)^2 + \tau_{xy}^2} \qquad \textbf{(10.4.1a)}$$

$$\sigma_2 = \sigma_{\min} = \frac{\sigma_x + \sigma_y}{2} - \sqrt{\left(\frac{\sigma_x - \sigma_y}{2}\right)^2 + \tau_{xy}^2} \qquad \textbf{(10.4.1b)}$$

and the angle θ_p at which they occur is calculated from

$$\tan 2\theta_p = \frac{2\tau_{xy}}{\sigma_x - \sigma_y}. \qquad \textbf{(10.4.2)}$$

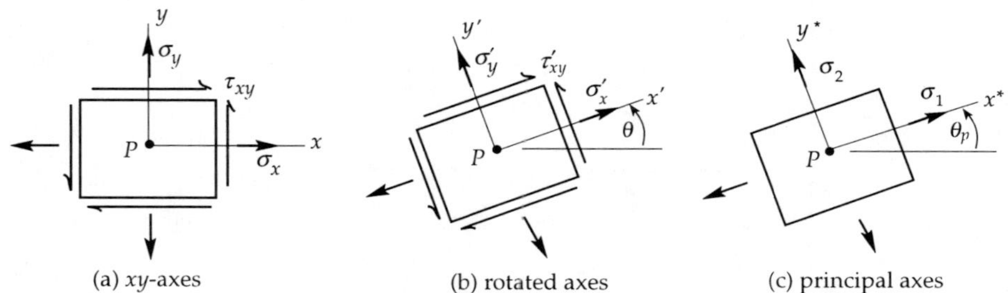

Figure 10.8 State of stress.

As stated, relative to the x^*y^*-axes the shear stresses are zero. The maximum shear stress occurs with respect to axes rotated 45° relative to the principal axes; its value is

$$\tau_{\max} = \frac{\sigma_1 - \sigma_2}{2} = \sqrt{\left(\frac{\sigma_x - \sigma_y}{2}\right)^2 + \tau_{xy}^2} \qquad \textbf{(10.4.3)}$$

A graphical method is often used when obtaining stresses on a particular plane; it utilizes *Mohr's circle*. Mohr's circle is sketched by locating both ends of a diameter, whose center is always on the horizontal axis. For the stress state of Fig. 10.8a, which shows positive stresses, we plot the diameter ends as $(\sigma_x, -\tau_{xy})$ and (σ_y, τ_{xy}). The stresses on any plane, oriented at an angle θ with respect to the stresses on any known plane, are then the coordinates of a point on Mohr's circle located an angle 2θ from the known point. Once Mohr's circle is sketched, it is relatively obvious that the maximum shear stress is the circle's

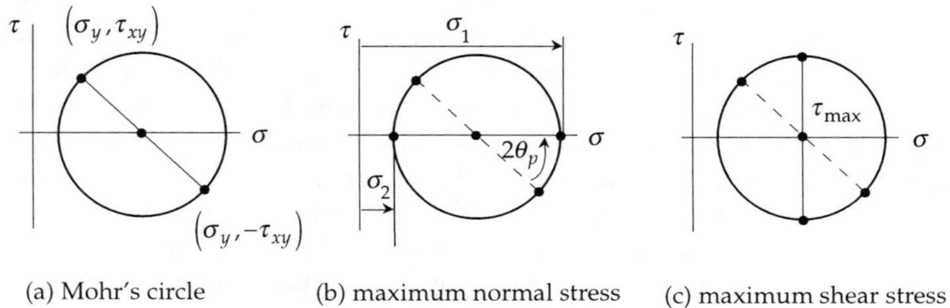

Figure. 10.9 Mohr's circle and maximum stresses.

radius, and the maximum normal stress is the circle's radius plus $(\sigma_x + \sigma_y)/2$, as observed in Fig. 10.9a. These are, in fact, equivalent to the formulas in the above equations. Rather than refer to the formulas, we can simply sketch Mohr's circle and easily find $\tau_{\max}$ and σ_1, the quantities often of interest, since they may lead to failure.

Example 10.8

A solid circular shaft of radius 5 cm and length 3 m has a 2 m rigid bar welded to end *A*, and is "built in" to the vertical wall at *B*. A load of 8 kN acts at end *C*, and an axial force of 80 kN compresses the shaft as shown. Determine the maximum normal, and maximum shear stress at point *P* on the top of the shaft, midway between *A* and *B*.

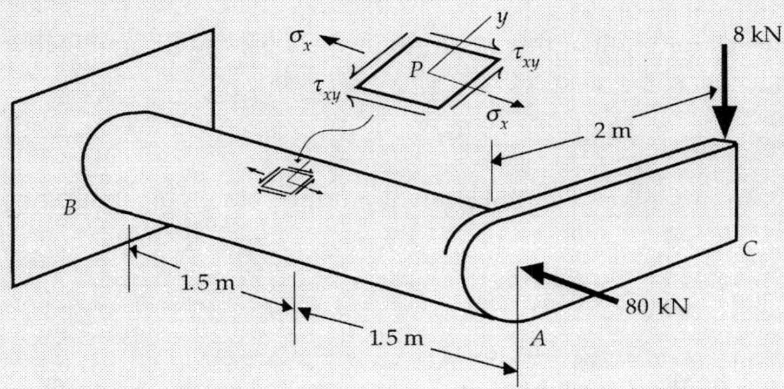

Solution. The torque on the shaft is $T = (8000)(2) = 16\ 000$ N.m. The torsional shear stress on the element at *P* is

$$\tau_{xy} = \frac{Tr}{J} = \frac{16\ 000(0.05)}{\pi(0.05)^4/2} = 81.5\times10^6 \text{ Pa or } 81.5 \text{ MPa.}$$

The axial compressive stress is

$$\sigma_x = \frac{F}{A} = \frac{-80\ 000}{\pi(0.05)^2} = -10.2\times10^6 \text{ Pa or } -10.2 \text{ MPa.}$$

The tensile bending stress on the element at *P* (also a σ_x stress) is

$$\sigma_x = \frac{My}{I} = \frac{8000(1.5)(0.05)}{\pi(0.05)^4/4} = 122.2\times10^6 \text{ Pa or } 122.2 \text{ MPa.}$$

Mohr's circle

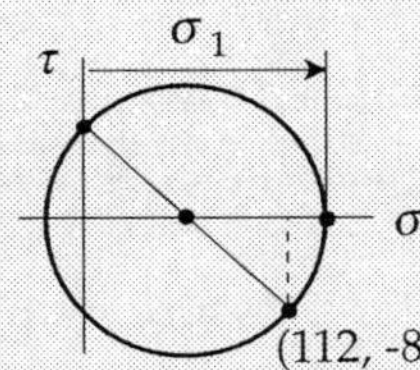

$$\text{radius} = \sqrt{(112/2)^2 + 81.5^2} = 98.9 = \tau_{max}$$

$$\sigma_1 = \text{radius} + 112/2 = 154.9$$

Thus, the stress components on the element at *P* (in MPa) are

$$\sigma_x = 122.2 - 10.2 = 112, \quad \sigma_y = 0, \quad \tau_{xy} = 81.5$$

so that from Eq. 10.4.1a the maximum normal stress is

$$\sigma_{max} = \frac{112+0}{2} + \sqrt{\left(\frac{112-0}{2}\right)^2 + (81.5)^2} = 155 \text{ MPa.}$$

From Eq. 10.4.3 the maximum shear stress is

$$\tau_{max} = \sqrt{\left(\frac{112-0}{2}\right)^2 + (81.5)^2} = 98.9 \text{ MPa.}$$

As another case illustrating the ideas of combined stress, let us consider a cylindrical vessel of inside diameter D and wall thickness t (with $t/D << 0.1$) containing a fluid under a pressure p, and subjected to a torque T as shown in Fig. 10.9. We consider the stresses acting upon a small element of the wall having sides parallel and perpendicular, respectively, to the axis of the cylinder. By sectioning the cylinder perpendicular to its axis at the element, we find from axial equilibrium that the *longitudinal stress*, also called *axial stress*, in the wall is

$$\sigma_a = \frac{pD}{4t}. \tag{10.4.4}$$

Similarly by sectioning lengthwise through the axis, radial equilibrium requires the *circumferential stress*, also called *hoop stress*, to be

$$\sigma_t = \frac{pD}{2t}. \tag{10.4.5}$$

And finally, from torsional equilibrium about the vessel's axis, the shear stress in the wall is

$$\tau = \frac{Tr}{J} \tag{10.4.6}$$

where $J = 2\pi r^3 t$, the approximate polar moment of inertia. From these formulas we may calculate the maximum normal and shear stresses as a problem in combined stress.

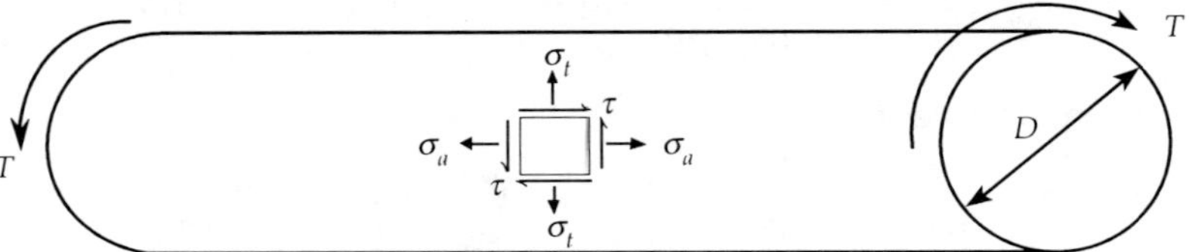

Figure 10.10 Pressurized cylinder under torque.

Finally, we note that in the absence of the torque T in Fig. 10.10, the maximum tensile stress is σ_t, the minimum tensile stress is σ_a, and the maximum shear stress is

$$\tau_{\max} = \frac{\sigma_t - \sigma_a}{2}. \tag{10.4.7}$$

Also, for a thin-walled spherical container under pressure p the normal stress in the wall is

$$\sigma = \frac{pD}{4t} \tag{10.4.8}$$

in every direction, and the shear stress in the wall is zero everywhere.

Example 10.9

A cylindrical tank of radius 40 cm and wall thickness 3 mm is subjected to an internal pressure of 2 MPa, and a torque of 0.5 $\text{MN} \cdot \text{m}$. Determine the maximum normal and shear stresses in the cylinder wall.

Solution. From Eq. 10.4.4 the longitudinal stress is

$$\sigma_\ell = \frac{pr}{2t} = \frac{2(0.4)}{2(0.003)} = 133 \text{ MPa}.$$

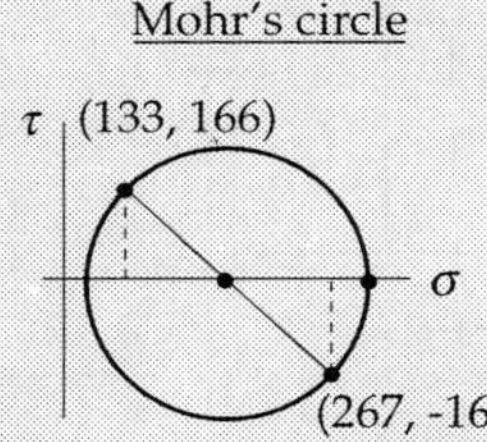

$$\text{radius} = \left[\left(\frac{267 - 133}{2}\right)^2 + 166^2\right]^{1/2}$$
$$= 179 \text{ MPa} = \tau_{\max}$$

$$\sigma_1 = \text{radius} + (267 + 133)/2$$
$$= 379 \text{ MPa}$$

From Eq. 10.4.5 the circumferential stress is

$$\sigma_c = \frac{pr}{t} = \frac{(2)(0.4)}{(0.003)} = 267 \text{ MPa}.$$

From Eq. 10.4.6 the shear stress is

$$\tau_{xy} = \frac{Tr}{J} = \frac{(0.5)(0.4)}{2\pi(0.4)^3(0.003)} = 166 \text{ MPa}.$$

Thus, from Eq. 10.4.1a

$$\sigma_{\max} = \frac{133 + 267}{2} + \sqrt{\left(\frac{133 - 267}{2}\right)^2 + (166)^2} = 379 \text{ MPa}$$

and from Eq. 10.4.3

$$\tau_{\max} = \sqrt{\left(\frac{133 - 267}{2}\right)^2 + (166)^2} = 179 \text{ MPa}.$$

10.5 Composite Bars and Beams

(May not be tested.)

Consider a member composed of several parallel portions, each of a particular material, securely bonded together and loaded axially. As an example, we show in Fig. 10.11 a composite bar of three materials subjected through rigid and parallel end plates to an axial force P. Let the portion of the bar have cross-sectional areas A_1, A_2 and A_3, as well as moduli of elasticity E_1, E_2, and E_3, respectively. Furthermore, let $E_1 \le E_2 \le E_3$ and form the ratios

$$m = E_2/E_1 \quad \text{and} \quad n = E_3/E_1. \tag{10.5.1}$$

Since the axial deformation is the same for each material, Hooke's law requires

$$\sigma_2 = m\sigma_1 \quad \text{and} \quad \sigma_3 = n\sigma_1 \tag{10.5.2}$$

and also that $\sigma_1 = P/A_T$ where A_T is the "transformed area" such that

$$\sigma_1 = \frac{P}{A_T} = \frac{P}{A_1 + mA_2 + nA_3}. \tag{10.5.3}$$

The generalization to a bar of any number of materials is obvious.

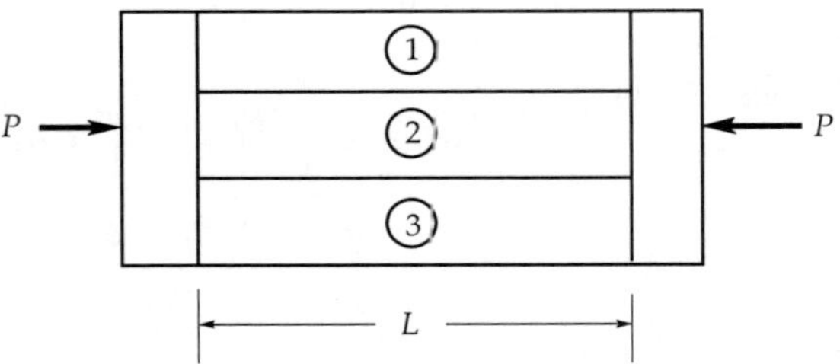

Figure 10.11 Composite bar.

Example 10.10

Let a composite bar made of aluminum, steel, and brass be subjected to an axial load of 500 kN. Determine the stress in each material if $E_{al} = 70$ GPa, $E_{st} = 210$ GPa, and $E_{br} = 105$ GPa, together with $A_{al} = 0.04\ \text{m}^2$, $A_{st} = 0.006\ \text{m}^2$, and $A_{br} = 0.08\ \text{m}^2$.

Solution. From Eq. 10.5.1, with reference to Fig. 10.10,

$$m = \frac{E_{br}}{E_{al}} = \frac{105}{70} = 1.5.$$

$$n = \frac{E_{st}}{E_{al}} = \frac{210}{70} = 3.0.$$

From Eq. 10.5.3 we have

$$\sigma_{al} = \frac{P}{A_1 + mA_2 + nA_3} = \frac{500 \times 10^3}{0.04 + 1.5(0.08) + 3(0.006)} = 2.8\ \text{MPa}.$$

From Eq. 10.5.2 there results

$$\sigma_{br} = m\sigma_1 = 1.5(2.8) = 4.2\ \text{MPa}.$$

$$\sigma_{st} = n\sigma_1 = 3.0(2.8) = 8.4\ \text{MPa}.$$

For a beam having a composite section and subjected to a bending moment M we again determine the ratios $E_2/E_1 = m$, $E_3/E_1 = n$, etc., and from them construct a "transformed cross section" by multiplying the width of each material by the corresponding ratio. We then determine the centroid of the transformed cross section, and calculate the moment of inertia I_T about the neutral axis of the transformed section. Thus,

$$\sigma_1 = -\frac{My}{I_T} \tag{10.5.4}$$

with $\sigma_2 = m\sigma_1$ and $\sigma_3 = n\sigma_1$, etc. An example illustrates the method.

Example 10.11

Let a composite steel and aluminum beam having the cross section shown be subjected to a positive bending moment of $90\ \text{N}\cdot\text{m}$. Determine the maximum bending stress in each material.

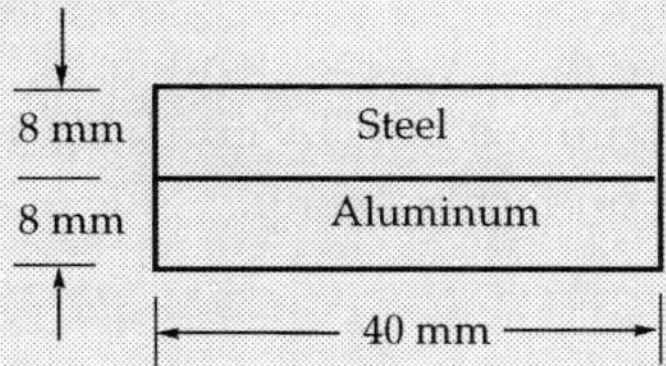

Solution. Here $m = E_s/E_a = 210/70 = 3$ so that the area has the geometry shown below with the neutral axis (N.A.) calculated to be 10 mm from the bottom.

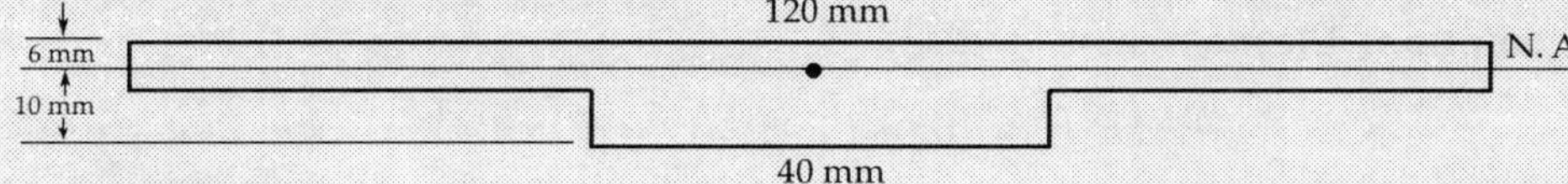

The moment of inertia about the N.A. is

$$I_T = \frac{(0.12)(0.006)^3}{3} + \frac{(0.04)(0.01)^3}{3} + \frac{(0.08)(0.002)^3}{3}$$

$$= 22 \times 10^{-9}\ \text{m}^4.$$

Thus, the maximum bending stress in the aluminum is

$$(\sigma_{al})_{\max} = -\frac{My}{I_T} = -\frac{90(-0.01)}{22 \times 10^{-9}} = 41\ \text{MPa}$$

and the maximum bending stress in the steel is

$$(\sigma_{st})_{\max} = -\frac{My}{I_t} = -3\left(\frac{90(0.006)}{22 \times 10^{-9}}\right) = -74\ \text{MPa}$$

10.6 Columns

Long slender members loaded axially in compression are referred to as *columns*. Such members frequently fail by *buckling* (excessive lateral deflection) rather than by crushing. Buckling onset depends not only on the material properties but also the geometry and type of end supports of the column. The axial load at the onset of buckling is called the *critical load*.

If the *slenderness ratio* of the column, defined as L/r (length divided by least radius of gyration r where $r = \sqrt{I/A}$), is greater than 120, the critical load for a column is the Euler load

$$P_{cr} = \pi^2 EI/k^2 L^2 \tag{10.6.1}$$

Values of k, with end supports shown in parentheses, are given as:

$$\begin{aligned} k &= 1 && \text{(pinned - pinned)} \\ k &= 0.5 && \text{(fixed - fixed)} \\ k &= 0.7 && \text{(pinned - fixed)} \\ k &= 2 && \text{(free - fixed)} \end{aligned} \qquad \textbf{(10.6.2)}$$

Intermediate columns are those whose slenderness ratios are less than 120 but greater than that at which failure occurs by crushing. For these, empirical formulas have been developed to predict buckling.

Example 10.12

Determine the critical load for a square steel (E = 210 GPa) strut 8 cm × 8 cm if its length is 6 m under (a) pinned ends, (b) fixed ends.

Solution. The moment of inertia is

$$I = bh^3/12 = (0.08)(0.08)^3/12 = 3.4 \times 10^{-6} \text{ m}^4.$$

a) The critical load for pinned ends is

$$P_{cr} = \frac{\pi^2 EI}{L^2} = \frac{\pi^2(210 \times 10^9)(3.4 \times 10^{-6})}{6^2} = 195\ 000 \text{ N}.$$

The normal stress, which must not exceed the yield stress, is

$$\sigma = \frac{F}{A} = \frac{195 \times 10^3}{0.0064} = 30.5 \times 10^6 \text{ Pa}.$$

This is substantially less than the yield stress for all steels.

b) The critical load for fixed ends is

$$P_{cr} = \frac{\pi^2(210 \times 10^9)(3.4 \times 10^{-6})}{0.5^2 \times 6^2} = 780\ 000 \text{ N}.$$

The normal stress for this case is

$$\sigma = \frac{780 \times 10^3}{0.0064} = 122 \times 10^6 \text{ Pa}.$$

Practice Problems (Metric Units)

(If you choose to work only a few problems, select those with a star.)

STRESS AND STRAIN

*10.1 A structural member with the same material properties in all directions at any particular point is

a) homogeneous
b) isotropic
c) isentropic
d) holomorphic
e) orthotropic

*10.2 The amount of lateral strain in a tension member can be calculated using

a) the bulk modulus.
b) the moment of inertia.
c) the yield stress.
d) Hooke's law.
e) Poisson's ratio.

*10.3 Find the allowable load, in kN, on a 2-cm-dia, 1-m-long, steel rod if its maximum elongation cannot exceed 0.1 cm.

a) 35 b) 45 c) 55 d) 66 e) 76

10.4 An elevator is suspended by a 2-cm-dia, 30-m-long steel cable. Twenty people, with a total weight of 14 000 N, enter. How far, in millimeters, does the elevator drop?

a) 3.5 b) 4.5 c) 5.5 d) 6.4 e) 7.4

10.5 A hole, one meter from the end of a structural steel member fixed at one end, is 0.8 mm shy of matching another hole for possible connection. What force, in kN, is necessary to stretch it for connection? The cross section is 25 mm × 3 mm.

a) 12.6 b) 13.6 c) 14.7 d) 15.8 e) 17.2

10.6 As the load is applied, edge *AB* moves 0.03 mm to the right. Determine the shear modulus, in MPa.

a) 50 300
b) 41 700
c) 38 600
d) 32 500
e) 26 300

10.7 A 5-cm-dia steel shaft is subjected to an axial tensile force of 600 kN. What is the diameter, in cm, after the force is applied? Use $\nu = 0.28$.

a) 4.998 b) 4.996 c) 4.994 d) 4.992 e) 4.990

*10.8 An aluminum cylinder carries an axial compressive load of 1500 kN. Its diameter measures exactly 12.015 cm and its height 19.311 cm. What was its original diameter, in cm?

a) 12.010 b) 12.008 c) 12.006 d) 12.004 e) 12.002

THERMAL STRESS

10.9 A tensile stress of 100 MPa exists in a 2-cm-dia steel rod that is fastened securely between two rigid walls. If the temperature increases by 30° C, determine the final stress, in MPa in the rod.

a) 46.7 b) 41.2 c) 36.9 d) 31.2 e) 26.2

*10.10 A steel bridge span is normally 300 m long. What is the difference in length, in cm, between January (–35° C) and August (40° C)?

a) 26 b) 28 c) 30 d) 32 e) 34

10.11 An aluminum bar at 30° C is inserted between two rigid stationary walls by inducing a compressive stress of 70MPa. At what temperature, in ° C, will the bar drop out?

a) 10 b) 0 c) –8 d) –14 e) –20

*10.12 Brass could not be used to reinforce concrete because

a) its density is too large.
b) its density is too low.
c) it is too expensive.
d) its coefficient of thermal expansion is not right.
e) it does not adhere well to concrete.

TORSION

*10.13 The maximum shearing stress, in MPa, that exists in a 6-cm-dia shaft subjected to a 200 N·m torque is

a) 4.72 b) 5.83 c) 7.29 d) 8.91 e) 9.97

10.14 The shaft of Prob. 10.13 is replaced with a 6-cm-outside diameter, 5-cm inside diameter hollow shaft. What is the maximum shearing stress, in MPa?

a) 5.5 b) 6.4 c) 7.3 d) 8.2 e) 9.1

10.15 The maximum allowable shear stress in a 10-cm-dia shaft is 140 MPa. What maximum torque, in N·m, can be applied?

a) 27 500 b) 21 400 c) 19 300 d) 17 100 e) 15 300

10.16 A builder uses a 50-cm-long, 1-cm-dia steel drill. If two opposite forces of 200 N are applied normal to the shaft, each with a moment arm of 15 cm, what angle of twist, in degrees, occurs in the drill?

a) 29.3 b) 24.6 c) 22.8 d) 21.1 e) 19.2

BENDING MOMENTS IN BEAMS

10.17 The maximum bending stress at a given cross section of an I-beam occurs

a) where the shearing stress is maximum.
b) at the outermost fiber.
c) at the joint of the web and the flange.
d) at the neutral axis.
e) just below the joint of the web and the flange

10.18 The moment diagram for a simply-supported beam with a load at the midpoint is

a) a triangle. b) a parabola. c) a trapezoid. d) a rectangle. e) a semicircle.

10.19 Find the bending moment, in N·m, at A.

a) 12 000 b) 14 000 c) 16 000
d) 18 000 e) 20 000

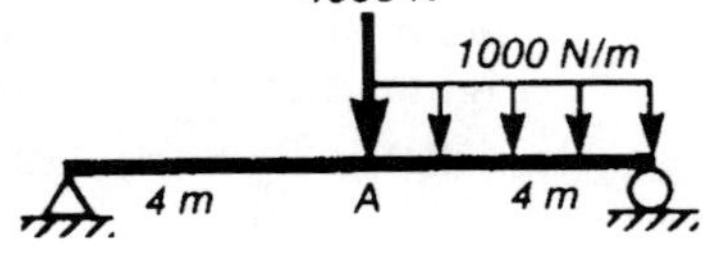

*10.20 What is the bending moment, in N·m, at A?

a) 26 000 b) 24 000 c) 22 000
d) 20 000 e) 18 000

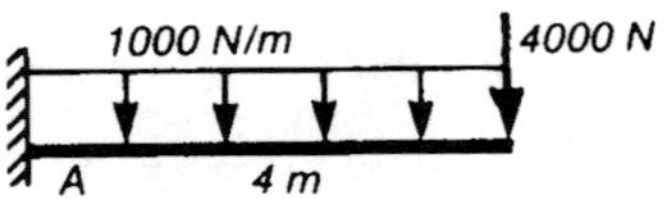

STRESSES IN BEAMS

*10.21 Find the maximum tensile stress, in MPa.

a) 94 b) 86 c) 82
d) 76 e) 72

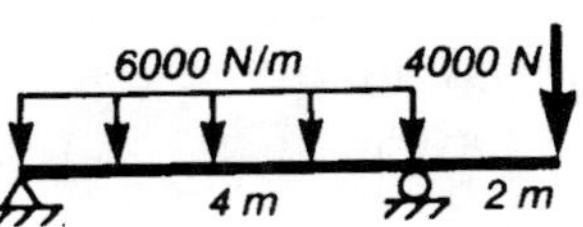

*10.22 What is the maximum compressive stress, in MPa, in the beam of Prob. 10.21?

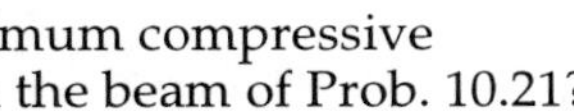

a) 96 b) 90 c) 82
d) 76 e) 72

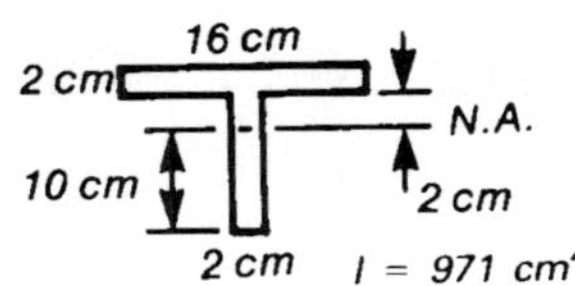

*10.23 What is the maximum shearing stress, in MPa, in the beam of Prob. 10.21?

a) 13.8 b) 11.3 c) 9.6 d) 8.2 e) 7.2

*10.24 The shearing stress distribution $\tau = VQ/Ib$ on the cross section of the *T*-beam in Prob. 10.21 most resembles which sketch?

a) 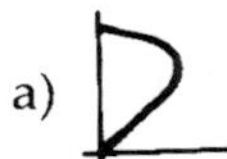 b) c) d) e)

10.25 If the allowable bending stress is 140 MPa in the beam of Prob. 10.20, calculate the *section modulus* defined by I/y, in cm^3.

a) 196 b) 184 c) 171 d) 162 e) 153

10.26 Find the maximum bending stress, in MPa, if the 10-cm-wide beam is 5 cm deep.

a) 200 b) 180 c) 160
d) 140 e) 120

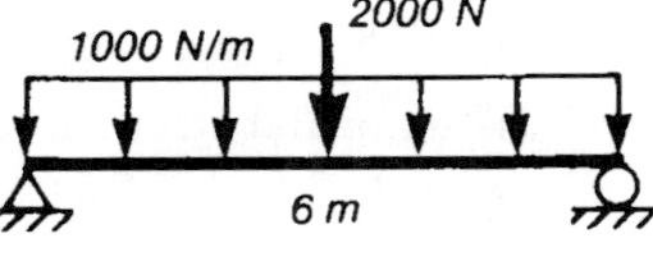

10.27 If the beam of Prob. 10.26 were 5 cm wide and 10 cm deep, find the maximum bending stress, in MPa.

a) 130 b) 120 c) 110 d) 100 e) 90

10.28 Find the maximum shearing stress, in MPa, of a simply supported, 6-m-long beam with a 5 cm × 5 cm cross section if it has a 2000 N load at the mid point.

a) 0.6 b) 0.9 c) 1.2 d) 1.6 e) 2.4

DEFLECTION OF BEAMS

*10.29 What is the maximum deflection, in cm, of a simply supported, 6-m-long steel beam with a 5 cm x 5 cm cross-section if it has a 2000 N load at the midpoint?

a) 6.35 b) 7.02 c) 7.63 d) 8.23 e) 8.92

10.30 Find the maximum deflection, in cm, for the steel beam of Prob. 10.26.

a) 39.7 b) 32.4 c) 28.3 d) 19.2 e) 11.8

10.31 If the deflection of the right end of the 5-cm-dia steel beam is 10 cm, what is the load *P*, in N?

a) 403 b) 523 c) 768
d) 872 e) 935

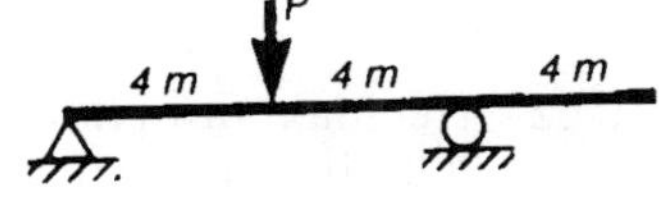

COMBINED STRESSES

*10.32 Find the maximum shearing stress, in MPa.

a) 80 b) 70 c) 60
d) 50 e) 40

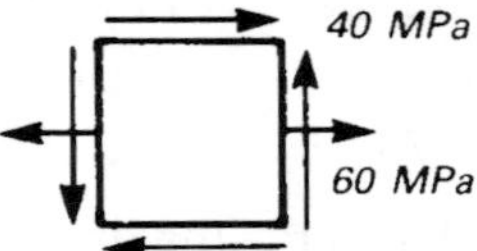

10.33 What is the maximum tensile stress, in MPa?

a) 40 b) 30 c) 20
d) 10 e) 0

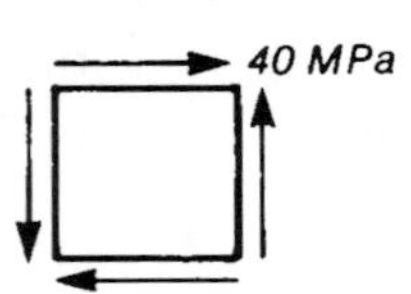

10.34 Determine the maximum shearing stress, in MPa.

a) 80 b) 60 c) 50
d) 40 e) 30

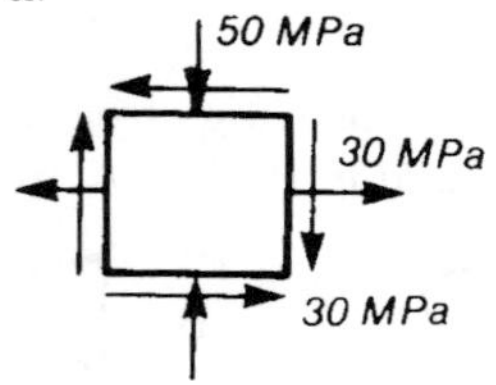

*10.35 Find the maximum shearing stress, in MPa, in the shaft.

a) 29.5 b) 28.5 c) 27.5
d) 26.5 e) 25.5

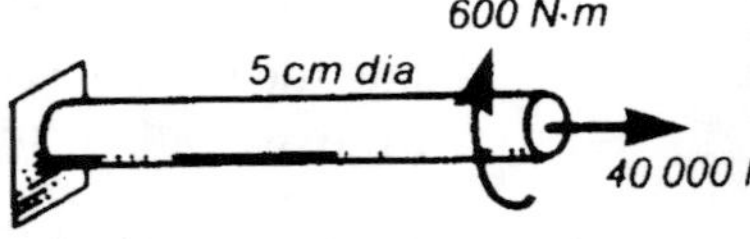

10.36 The maximum normal stress, in MPa, in the shaft of Prob. 10.35 is

a) 52.8 b) 41.7 c) 36.7 d) 30.1 e) 25.3

10.37 The normal stress, in MPa, at pt. A is

a) 263 b) 241 c) 228
d) 213 e) 201

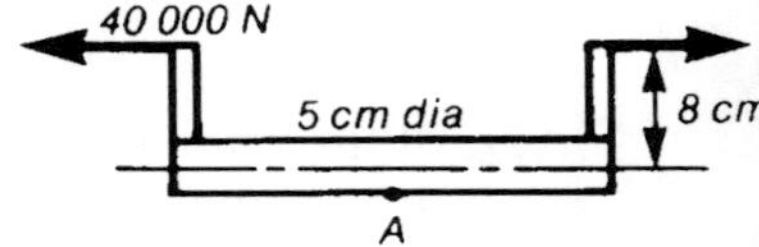

10.38 The maximum shearing stress, in MPa, at pt. A in Prob. 10.37 is

a) 140 b) 130 c) 120 d) 110 e) 100

10.39 The maximum shearing stress, in MPa, in the circular shaft is

a) 171 b) 167 c) 154
d) 142 e) 133

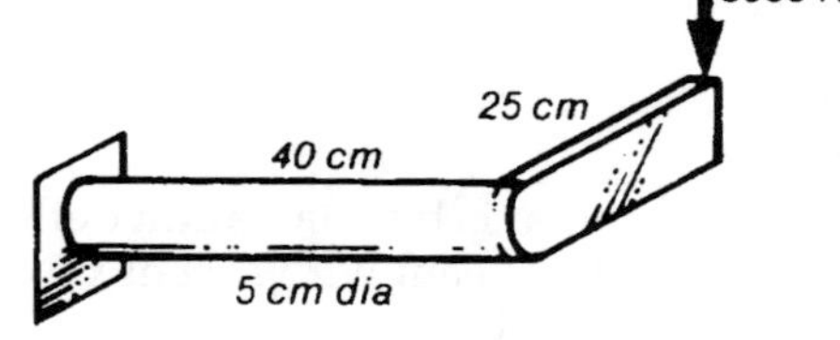

10.40 The maximum tensile stress, in MPa, in the circular shaft of Prob. 10.39 is

a) 284 b) 248 c) 223 d) 212 e) 197

THIN-WALLED PRESSURE VESSELS

*10.41 The allowable tensile stress for a pressurized cylinder is 180 MPa. What maximum pressure, in kPa, is allowed if the 80-cm-dia cylinder is made of 0.5 cm thick material?

a) 2400 b) 2250 c) 2150 d) 2050 e) 1950

10.42 The maximum normal stress that can occur in a 120-cm-dia steel sphere is 200 MPa. If it is to contain a pressure of 8000 kPa, what must be the minimum thickness, in cm?

a) 1.6 b) 1.4 c) 1.2 d) 1.0 e) 0.8

10.43 What is the maximum shearing stress, in MPa, in the sphere of Prob. 10.42?

a) 200 b) 150 c) 100 d) 50 e) 0

COMPOSITE SECTIONS

10.44 A compression member, composed of 1.2-cm-thick steel pipe with 25-cm-inside diameter, is filled with concrete. Find the stress, in MPa, in the steel if the load is 2000 kN.

a) 137 b) 145 c) 155 d) 165 e) 175

10.45 If the flanges are aluminum and the rib is steel, find the maximum tensile stress, in MPa, in the beam.

a) 7.89 b) 6.31 c) 5.72
d) 4.91 e) 3.88

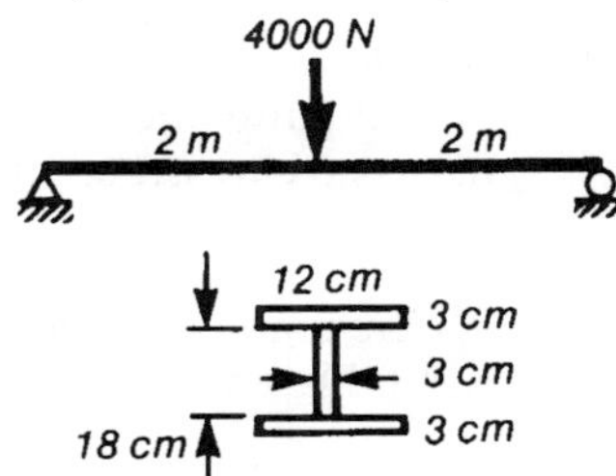

10.46 If the flanges of the I-beam of Prob. 10.45 are steel and the rib is aluminum, what is the maximum tensile stress, in MPa, in the beam?

a) 7.89 b) 6.31 c) 5.67 d) 4.91 e) 3.88

COLUMNS

10.47 What is the minimum length, in meters, for which a 10 cm × 10 cm wooden post can be considered a long column? Assume a maximum slenderness ratio of 60.

a) 4.03 b) 3.12 c) 2.24 d) 1.73 e) 1.12

*10.48 A free-standing platform, holding 2000 N, is to be supported by a 10-cm-dia vertical aluminum strut. How long, in meters, can it be if a safety factor of 2 is used?

a) 18.3 b) 16.6 c) 14.6 d) 12.2 e) 9.32

10.49 What increase in temperature, in °C, is necessary to cause a 2-cm-dia, 4-m-long, steel rod with fixed ends to buckle? There is no initial stress.

a) 9.38 b) 8.03 c) 7.12 d) 6.34 e) 5.27

10.50 A column with both ends fixed buckles when subjected to a force of 30 000 N. One end is then allowed to be free. At what force, in Newtons, will it buckle?

a) 2025 b) 1875 c) 1725 d) 1650 e) 1575

Practice Problems (English Units)

(If you choose to work only a few problems, select those with a star.)

STRESS AND STRAIN

*10.1 A structural member with the same material properties in all directions at any particular point is

a) homogeneous b) isotropic c) isentropic d) holomorphic e) orthotropic

*10.2 The amount of lateral strain in a tension member can be calculated using

a) the bulk modulus.
b) the moment of inertia.
c) the yield stress.
d) Hooke's law.
e) Poisson's ratio.

*10.3 Find the allowable load, in pounds, on a 1/2" dia, 4-ft-long, steel rod if its maximum elongation cannot exceed 0.04 inches.

a) 9290 b) 6990 c) 5630 d) 4910 e) 3220

10.4 An elevator is suspended by a 1/2" dia, 100-ft-long steel cable. Twenty people, with a total weight of 3500 lbs, enter. How far, in inches, does the elevator drop?

a) 1.3 b) 1.1 c) 0.9 d) 0.7 e) 0.5

10.5 A hole in a 3-ft-long structural steel member fixed at one end is 1/32" shy of matching another hole for possible connection. What force, in pounds, is necessary to stretch it for connection? The cross section is 1/8" × 1".

a) 3300 b) 3000 c) 2700 d) 2400 e) 2100

10.6 As the load is applied, edge *AB* moves 0.0012" to the right. Determine the shear modulus, in psi.

5000 lb
½"
A
B
6"
8"

a) 7.5×10^6

b) 6.2×10^6

c) 5.7×10^6

d) 5.2×10^6

e) 4.5×10^6

10.7 A 2" dia steel shaft is subjected to an axial tensile force of 150,000 lbs. What is the diameter, in inches, after the force is applied? Use $\nu = 0.28$.

a) 1.999 b) 1.998 c) 1.997 d) 1.996 e) 1.995

*10.8 An aluminum cylinder carries an axial compressive load of 400, 000 lbs. Its diameter measures exactly 5.923" and its height 8.314". What was its original diameter, in inches?

a) 5.922 b) 5.920 c) 5.918 d) 5.916 e) 5.914

THERMAL STRESS

10.9 A tensile stress of 16,000 psi exists in a 1" dia steel rod that is fastened securely between two rigid walls. If the temperature increases by 50° F, determine the final stress, in psi, in the rod.

a) 12,200 b) 9400 c) 8600 d) 7400 e) 6200

*10.10 A steel bridge span is normally 1000 ft long. What is the difference in length, in inches, between January (–30° F) and August (100° F)?

a) 10 b) 9 c) 8 d) 7 e) 6

10.11 An aluminum bar at 80° F is inserted between two rigid stationary walls by inducing a compressive stress of 10,000 psi. At what temperature, in ° F, will the bar drop out?

a) 36 b) 24 c) 10 d) 2 e) –10

*10.12 Brass could not be used to reinforce concrete because

a) it is not sufficiently strong.

b) its density is too large.

c) it is too expensive.

d) its coefficient of thermal expansion is not right.

e) it does not adhere well to concrete.

TORSION

*10.13 The maximum shearing stress, in psi, that exists in a 2″ dia shaft subjected to a 2000 in-lb torque is

a) 1270 b) 1630 c) 1950 d) 2610 e) 3080

10.14 The shaft of Prob. 10.13 is replaced with a 2″ outside diameter, 1.75″ inside diameter hollow shaft. What is the maximum shearing stress, in psi?

a) 1270 b) 1630 c) 1950 d) 2610 e) 3080

10.15 The maximum allowable shear stress in a 4″ dia shaft is 20,000 psi. What maximum torque, in ft-lb, can be applied?

a) 20,900 b) 15,600 c) 11,200 d) 8,600 e) 4,210

10.16 A builder uses an 18″ long, 7/8″ dia steel drill. If two opposite forces of 160 lbs are applied normal to the shaft, each with a moment arm of 12", what angle of twist, in degrees, occurs in the drill?

a) 10.3 b) 8.29 c) 6.95 d) 5.73 e) 4.68

BENDING MOMENTS IN BEAMS

10.17 The maximum bending stress at a given cross section of an I-beam occurs

a) where the shearing stress is maximum.
b) at the outermost fiber.
c) just below the joint of the web and the flange.
d) at the neutral axis.
e) at the joint of the web and the flange.

10.18 The moment diagram for a simply supported beam with a load at the mid-point is a

a) triangle. b) parabola. c) trapezoid. d) rectangle. e) semicircle.

10.19 Find the bending moment, in ft-lb, at A.

a) 7500 b) 7000 c) 6500
d) 6000 e) 5000

*10.20 What is the bending moment, in ft-lb, at A?

a) 18,000 b) 15,000 c) 12,000
d) 10,000 e) 8000

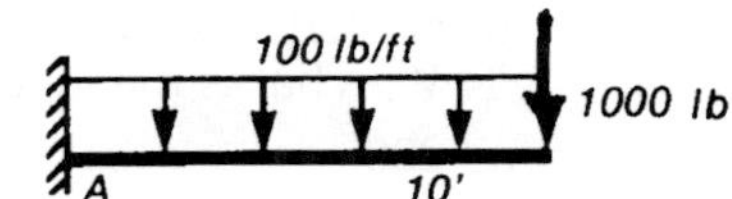

STRESSES IN BEAMS

*10.21 Find the maximum tensile stress, in psi.

a) 4360 b) 3960 c) 3240
d) 2860 e) 2110

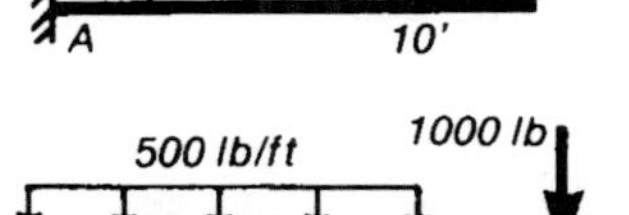

*10.22 What is the maximum compressive stress, in psi, in the beam of Prob. 10.21?

a) 4360 b) 3960 c) 3240 d) 2860 e) 2110

*10.23 What is the maximum shearing stress, in psi, in the beam of Prob. 10.21?

a) 1000 b) 900 c) 800 d) 700 e) 600

10.24 The shearing stress distribution $\tau = VQ/Ib$ on the cross section of the *T*-beam in Prob. 10.21 most resembles which sketch?

a) b) c) d) e)

10.25 If the allowable bending stress is 20,000 psi in the beam of Prob. 10.20, calculate the *section modulus* defined by I/y, in in^3.

a) 11 b) 10 c) 9 d) 8 e) 7

10.26 Find the maximum bending stress, in psi, if the 4″ wide beam is 2″ deep.

a) 29,200 b) 23,400 c) 18,600
d) 15,600 e) 11,700

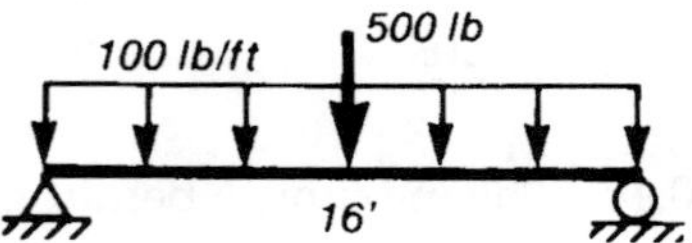

10.27 If the beam of Prob. 10.26 were 2″ wide and 4″ deep, find the maximum bending stress, in psi.

a) 29,200 b) 23,400 c) 18,600 d) 15,600 e) 11,700

10.28 Find the maximum shearing stress, in psi, of a simply supported, 20-ft-long beam with a 2″ × 2″ cross section if it has a 500-lb load at the mid point.

a) 188 b) 152 c) 131 d) 109 e) 94

DEFLECTION OF BEAMS

*10.29 What is the maximum deflection, in inches, of a simply supported, 20-ft-long steel beam with a 2″ × 2″ cross section if it has a 500-lb load at the midpoint?

a) 7.2 b) 6.0 c) 4.8 d) 3.6 e) 2.4

10.30 Find the maximum deflection, in inches, for the steel beam of Prob. 10.26.

a) 4.22 b) 4.86 c) 3.52 d) 2.98 e) 2.76

10.31 If the deflection of the right end of the 2″ dia steel beam is 4″, what is the load *P*, in pounds?

a) 220 b) 330 c) 440
d) 550 e) 660

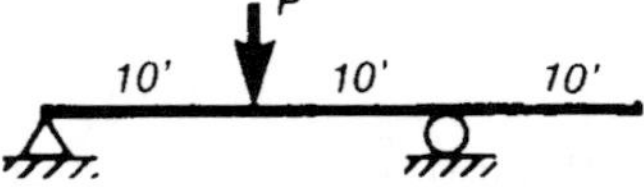

COMBINED STRESSES

*10.32 Find the maximum shearing stress, in psi.

a) 8000 b) 7000 c) 6000
d) 5000 e) 4000

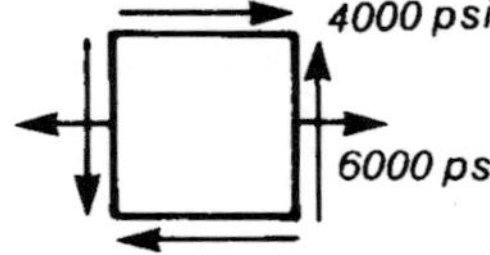

10.33 What is the maximum tensile stress, in psi?

a) 4000 b) 3000 c) 2000
d) 1000 e) 0

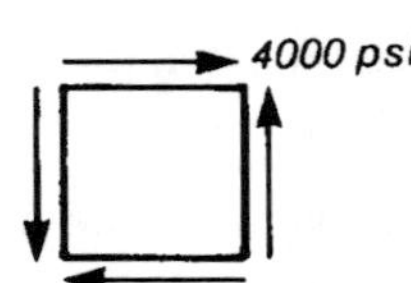

10.34 Determine the maximum shearing stress, in psi.

a) 8000 b) 6000 c) 5000
d) 4000 e) 3000

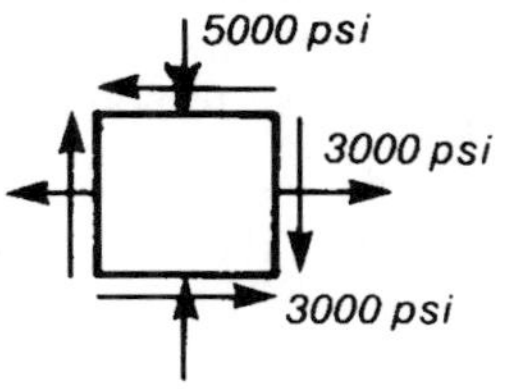

5000 psi
3000 psi
3000 psi

*10.35 Find the maximum shearing stress, in psi, in the shaft.

a) 7340 b) 6520 c) 5730
d) 4140 e) 3160

500 ft-lb
2" dia
10,000 lb

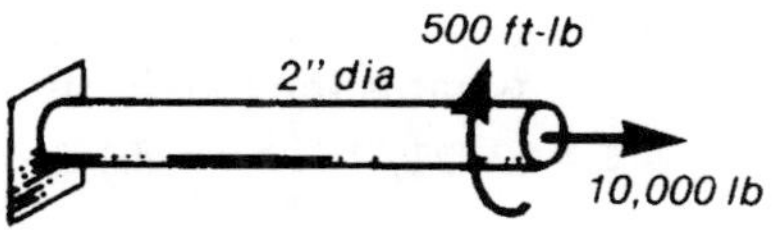

10.36 The maximum normal stress, in psi, in the shaft of Prob. 10.35 is

a) 7340 b) 6520 c) 5730 d) 4140 e) 3160

10.37 The normal stress, in psi, at pt. *A* is

a) 25,000 b) 35,000 c) 41,000
d) 46,000 e) 55,000

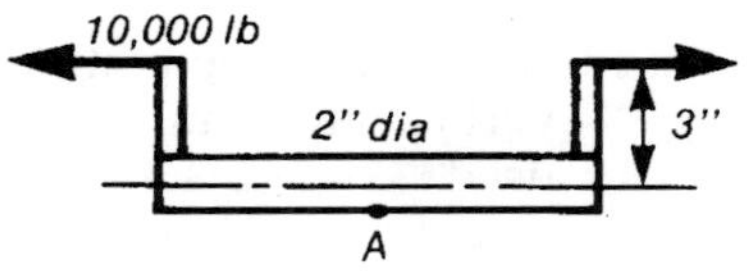

10,000 lb
2" dia
3"
A

10.38 The maximum shearing stress, in psi, at pt. *A* in Prob. 10.37 is

a) 12,500 b) 15,000 c) 17,500 d) 20,500 e) 23,000

10.39 The maximum shearing stress, in psi, in the circular shaft is

a) 12,000 b) 18,000 c) 24,000
d) 30,000 e) 36,000

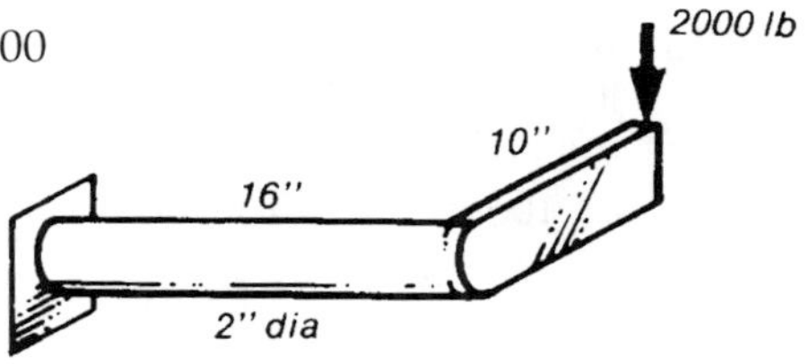

2000 lb
10"
16"
2" dia

10.40 The maximum tensile stress, in psi, in the circular shaft of Prob. 10.39 is

a) 11,100 b) 22,200 c) 33,300 d) 44,400 e) 55,500

THIN-WALLED PRESSURE VESSELS

*10.41 The allowable tensile stress for a pressurized cylinder is 24,000 psi. What maximum pressure, in psi, is allowed if the 2-ft-dia cylinder is made of 1/4" thick material?

a) 250 b) 500 c) 1000 d) 1500 e) 2000

10.42 The maximum normal stress that can occur in a 4-ft-dia steel sphere is 30,000 psi. If it is to contain a pressure of 2000 psi, what must be the minimum thickness, in inches?

a) 1.0 b) 0.9 c) 0.8 d) 0.7 e) 0.6

10.43 What is the maximum shearing stress, in psi, in the sphere of Prob. 10.42?

a) 30,000 b) 20,000 c) 15,000 d) 10,000 e) 0

COMPOSITE SECTIONS

*10.44 A compression member, composed of 1/2" thick steel pipe with 10" inside diameter is filled with concrete. Find the stress, in psi, in the steel if the load is 400,000 lbs.

a) 16,400 b) 14,300 c) 12,600 d) 10,100 e) 8,200

10.45 If the flanges are aluminum and the rib is steel, find the maximum tensile stress, in psi, in the beam.

a) 2170 b) 1650 c) 1320
d) 1150 e) 1110

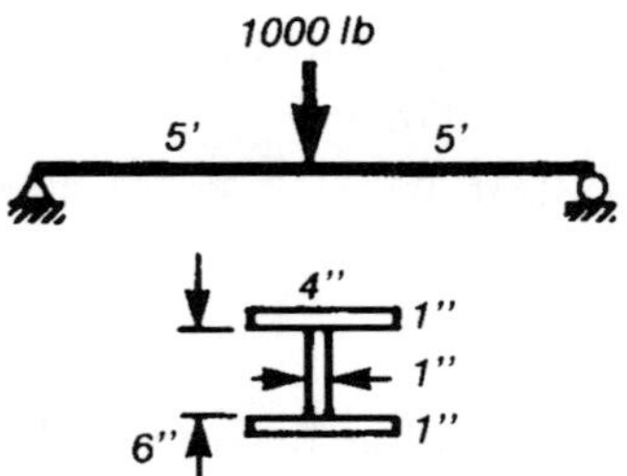

10.46 If the flanges of the I-beam of Prob. 10.45 are steel and the rib is aluminum, what is the maximum tensile stress, in psi?

a) 2170 b) 1650 c) 1320 d) 1150 e) 1110

COLUMNS

10.47 What is the minimum length, in ft, for which a 4" ×4" wooden post can be considered a long column? Assume a maximum slenderness ratio of 60.

a) 9.2 b) 7.6 c) 6.8 d) 5.8 e) 4.2

*10.48 A free-standing platform, holding 500 lb, is to be supported by a 4" dia vertical aluminum strut. How long, in ft, can it be if a safety factor of 2 is used?

a) 66 b) 56 c) 46 d) 36 e) 26

10.49 What increase in temperature, in °F, is necessary to cause a 1" dia, 10 ft long, steel rod with fixed ends to buckle? There is no initial stress.

a) 66 b) 56 c) 46 d) 36 e) 26

10.50 A column with both ends fixed buckles when subjected to a force of 8000 lbs. One end is then allowed to be free. At what force, in pounds, will it buckle?

a) 100 b) 500 c) 2000 d) 8000 e) 32,000

Solutions to Practice Problems

(Metric Units)

10.1 b) Isotropic

10.2 e) Poisson's Ratio

10.3 d) $\sigma = E\delta/L = P/\pi r^2$

$\therefore P = E\delta\pi r^2/L = 210\times10^9\times0.001\times\pi\times0.01^2/1 = 66\,000$ N

10.4 d) $\dfrac{P}{\pi r^2} = E\dfrac{\delta}{L}. \quad \therefore \delta = \dfrac{14\,000\times30}{\pi\times0.01^2\times210\times10^9} = 0.0064$ m

10.5 a) $\dfrac{P}{A} = E\dfrac{\delta}{L}. \quad \therefore P = \dfrac{AE\delta}{L} = 0.025\times0.003\times\left(210\times10^9\right)\times\dfrac{0.0008}{1} = 12\,600$ N

10.6 b) $\dfrac{P}{A} = G\dfrac{\delta}{L}. \quad \therefore G = \dfrac{PL}{A\delta} = \dfrac{20\,000\times0.15}{0.012\times0.2\times0.00003} = 41.7\times10^9$ Pa

10.7 a) $P/A = E\varepsilon. \quad \varepsilon = \dfrac{600\,000}{\pi \times 0.025^2 \times 210 \times 10^9} = 0.001455$

$\Delta d = \nu \varepsilon d = 0.28 \times 0.001455 \times 5 = 0.00204$ cm

$d = d - \Delta d = 5 - 0.00204 = 4.9980$ cm

10.8 b) $P/A = E\varepsilon. \quad \varepsilon = \dfrac{1500\,000}{\pi \times 0.06^2 \times 70 \times 10^9} = 0.00190$

$\Delta d = \nu \varepsilon d = 0.33 \times 0.0019 \times 12.015 = 0.0075$ cm

$d = d - \Delta d = 12.015 - 0.0075 = 12.008$ cm

10.9 e) $\sigma = E\delta/L. \quad 100 \times 10^6 = 210 \times 10^9 (\delta/L). \quad \therefore \; \delta = 4.76 \times 10^{-4} L$

$\delta_T = \alpha L \Delta T = 11.7 \times 10^{-6} \times 30L = 3.51 \times 10^{-4} L. \quad \delta_{final} = \delta - \delta_T = 1.25 \times 10^{-4} L$

$\therefore \sigma = \left(210 \times 10^9\right) \times \left(1.25 \times 10^{-4}\right) = 26.2 \times 10^6$ Pa

10.10 a) $\delta = \alpha L \Delta T = \left(11.7 \times 10^{-6}\right) \times 300 \times 75 = 0.263$ m

10.11 d) $\dfrac{\delta}{L} = \dfrac{\sigma}{E} = \alpha \Delta T. \quad \dfrac{70 \times 10^6}{70 \times 10^9} = 23 \times 10^{-6}(30 - T). \quad \therefore T = -13.5°\text{C}$

10.12 d) It expands at a different rate.

10.13 a) $\tau = \dfrac{Tr}{J} = \dfrac{200 \times 0.03}{\pi \times 0.06^4/32} = 4.72 \times 10^6$ Pa

10.14 e) $J = \pi\left(a_1^4 - a_2^4\right)/2 = \pi\left(0.03^4 - .025^4\right)/2 = 65.9 \times 10^{-8}$

$\tau = \dfrac{Tr}{J} = \dfrac{200 \times 0.03}{65.9 \times 10^{-8}} = 9.10 \times 10^6$ Pa

10.15 a) $T = \dfrac{\tau J}{r} = \dfrac{\left(140 \times 10^6\right) \times \pi \times .05^4/2}{0.05} = 27\,500 \text{ N} \cdot \text{m}$

10.16 d) $\theta = \dfrac{TL}{JG} = \dfrac{200 \times 0.3 \times 0.5}{\left(83 \times 10^9\right) \times \pi \times .05^4/2} = 0.368$ rad or 21.1°

10.17 b) $\sigma = My/I. \quad \sigma_{\max}$ occurs at $y = y_{\max}$

10.18 a) A triangle.

V M

10.19 a) $\sum M_{right} = 0. \quad \therefore 8F_{left} = 4\,000 \times 4 + 4\,000 \times 2. \quad \therefore F = 3000.$

$M_A = 3000 \times 4 = 12\,000 \text{ N} \cdot \text{m}$

10.20 b) $M_A = 4\,000 \times 4 + 4\,000 \times 2 = 24\,000 \text{ N} \cdot \text{m}$

10.21 b) $4F_{right} = 24\,000 \times 2 + 4\,000 \times 6$. $\therefore F_{right} = 18\,000$ N. $4F_{left} = 24\,000 \times 2 - 4\,000 \times 2$.

$\therefore F_{left} = 10\,000$ N. M_{max} = area under diagram $= 10\,000 \times 1.667/2 = 8\,330$ N·m.

$$\sigma = \frac{My}{I} = \frac{8\,330 \times 0.1}{971 \times 10^{-8}} = 85.8 \times 10^6 \text{ Pa}$$

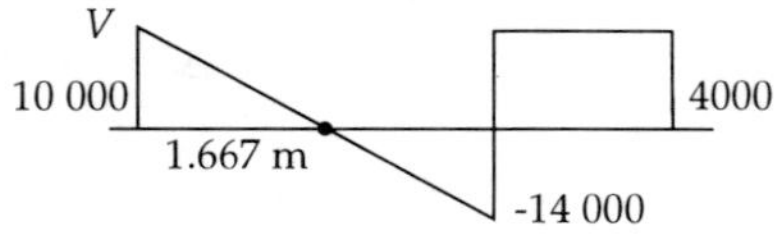

V
10 000
1.667 m
-14 000
4000

10.22 c) Compression occurs in bottom fibers over right support.

There, $M = 4\,000 \times 2 = 8\,000$ N·m. $\sigma = \dfrac{My}{I} = \dfrac{8\,000 \times 0.1}{971 \times 10^{-8}} = 82.4 \times 10^6$ Pa

10.23 e) $\tau_{max} = \dfrac{VQ}{Ib} = \dfrac{14\,000(0.002 \times 0.05)}{\left(971 \times 10^{-8}\right) \times 0.02} = 7.21 \times 10^6$ Pa

10.24 c) τ_{max} occurs on the N.A. with a sudden decrease when b goes from 2 to 16 cm.

Also, it is a parabolic distribution.

10.25 c) $\sigma = \dfrac{My}{I}$. $\dfrac{I}{y} = \dfrac{M}{\sigma} = \dfrac{24\,000}{140 \times 10^6} = 171 \times 10^{-6}$ m^3

10.26 b) Using the area under the curve: $M_{max} = 1000 \times 3 + 3000 \times 3/2 = 7500$ N·m

$$\sigma_{max} = \frac{My}{I} = \frac{7500 \times 0.025}{0.1 \times 0.05^3/12} = 180 \times 10^6 \text{ Pa}$$

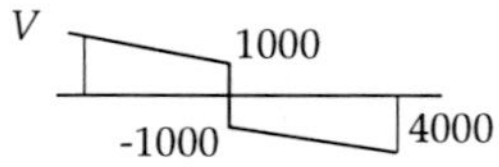

V
1000
-1000
4000

10.27 e) $\sigma_{max} = \dfrac{My}{I} = \dfrac{7500 \times 0.05}{0.05 \times 0.1^3/12} = 90 \times 10^6$ Pa

10.28 a) $V_{max} = 1000$. $\tau_{max} = \dfrac{VQ}{Ib} = \dfrac{1000(0.025 \times 0.05 \times 0.0125)}{\left(0.05 \times 0.05^3/12\right) \times 0.05} = 600 \times 10^3$ Pa

10.29 d) $\delta = \dfrac{PL^3}{48EI} = \dfrac{2000 \times 6^3}{48 \times \left(210 \times 10^9\right) \times 0.05^4/12} = 0.0823$ m

10.30 e) $\delta = \dfrac{PL^3}{48EI} + \dfrac{5wL^4}{384EI}$. $I = \dfrac{bh^3}{12} = \dfrac{0.1 \times 0.05^3}{12} = 1.04 \times 10^{-6}$

$$= \frac{2000 \times 6^3}{48 \times \left(210 \times 10^9\right) \times 1.04 \times 10^{-6}} + \frac{1000 \times 5 \times 6^4}{384 \times \left(210 \times 10^9\right) \times 1.04 \times 10^{-6}} = 0.118 \text{ m}$$

10.31 a) $\delta = \theta L_2 = \dfrac{PL^2}{16EI} \times 4$. $0.1 = \dfrac{P \times 8^2 \times 4}{16 \times \left(210 \times 10^9\right)\pi \times .025^4/4}$. $\therefore P = 403$ N

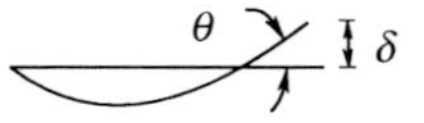

θ δ

10.32 d) $\tau_{max} = \dfrac{1}{2}\sqrt{\left(\sigma_x - \sigma_y\right)^2 + 4\tau^2} = \dfrac{1}{2}\sqrt{60^2 + 4 \times 40^2} = 50$ MPa

10.33 a) $\tau_{max} = \frac{1}{2}\left(\sigma_x + \sigma_y\right) + \tau_{max} = 0 + 40 = 40 \text{ MPa}$

10.34 c) $\tau_{max} = \frac{1}{2}\sqrt{(30+50)^2 + 4\times 30^2} = 50 \text{ MPa}$

10.35 d) $\tau = Tr/J = 600 \times 0.025 \Big/ \frac{\pi \times .025^4}{2} = 24.45 \times 10^6 \text{ Pa}$

$\sigma = P/A = 40\,000/\pi \times 0.025^2 = 20.37 \times 10^6 \text{ Pa}$

$\therefore \tau_{max} = \frac{1}{2}\sqrt{20.37^2 + 4 \times 24.45^2} = 26.5 \times 10^6 \text{ Pa}$

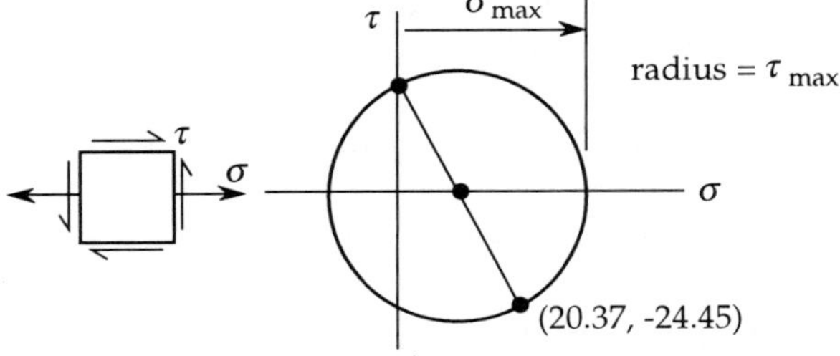

10.36 c) $\sigma_{max} = \left(\frac{1}{2} \times 20.37 + 26.5\right) \times 10^6 = 36.7 \times 10^6 \text{ Pa}$

10.37 b) $My/I = 3\,200 \times 0.025 \Big/ \frac{\pi \times .025^4}{4} = 261 \times 10^6$ comp.

$P/A = 40\,000/\pi \times 0.025^2 = 20.4 \times 10^6$ Pa tension.

$\sigma_A = (261 - 20.4) \times 10^6 = 241 \times 10^6 \text{ Pa}$

10.38 c) $\tau_{max} = \sigma/2 = 120 \times 10^6$ Pa. $VQ/Ib = 0$ on outer fibers.

10.39 c) $\tau = \frac{Tr}{J} = \frac{8\,000 \times 0.25 \times 0.025}{\pi \times .025^4/2} = 81.5 \times 10^6 \text{ Pa}$

$\sigma = \frac{My}{I} = \frac{8\,000 \times 0.4 \times 0.025}{\pi \times .025^4/4} = 261 \times 10^6 \text{ Pa}$

$\tau_{max} = \frac{1}{2}\sqrt{261^2 + 4 \times 81.5^2} \times 10^6 = 154 \times 10^6 \text{ Pa}$

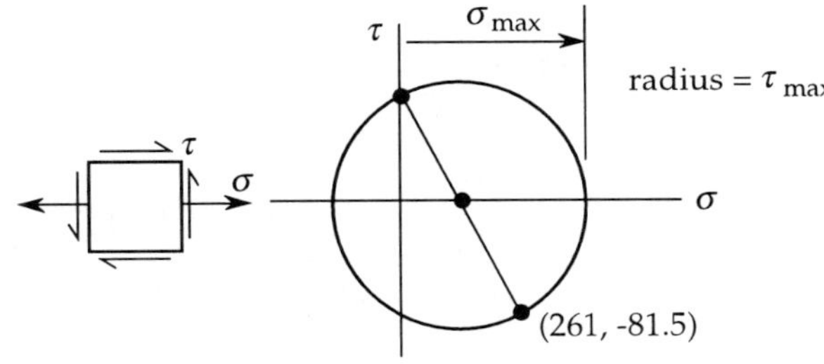

10.40 a) $\sigma_{max} = \left(\frac{1}{2} \times 261 + 154\right) \times 10^6 = 284 \times 10^6 \text{ Pa}$

10.41 b) $\sigma_t = pD/2t. \quad \therefore p = 180 \times 10^6 \times 2 \times 0.005/0.8 = 2250 \times 10^3 \text{ Pa}$

10.42 c) $\sigma_a = pD/4t. \quad \therefore t = \dfrac{8000\times 10^3 \times 1.2}{4\times 200\times 10^6} = 0.012 \text{ m}$

10.43 e) $\tau_{\max} = \dfrac{1}{2}\sqrt{(200-200)+0\times 4} = 0$

10.44 a) $\left(\dfrac{\Delta L}{L}\right)_s = \left(\dfrac{\Delta L}{L}\right)_c. \quad \therefore \varepsilon_s = \varepsilon_c. \quad \therefore \sigma_s = \dfrac{E_s}{E_c}\sigma_c = 10.5\,\sigma_c.$

$F_s + F_c = 2000000$ or $A_s\sigma_s + A_c\sigma_c = 2000000.$

$\sigma_s\left[\pi\left(0.137^2 - 0.125^2\right) + \pi\times\dfrac{0.125^2}{10.5}\right] = 2\times 10^6. \quad \therefore \sigma_s = 137\times 10^6 \text{ Pa.}$

10.45 e) $n = E/E_{\min} = 3.$ The area is transformed:

$I_t = \dfrac{0.12\times 0.24^3}{12} - \dfrac{0.03\times 0.18^3}{12} = 1.237\times 10^{-4}\ \text{m}^4.$

3, 18, 9, 3, 12

$\therefore \sigma_{al} = \dfrac{My}{I} = \dfrac{2000\times 2\times 0.12}{1.237\times 10^{-4}} = 3.88\times 10^6 \text{ Pa}$

10.46 c) $n = E/E_{\min} = 3.$ The area is transformed:

$I_t = \dfrac{0.36\times 0.24^3}{12} - \dfrac{0.33\times 0.18^3}{12} = 2.54\times 10^{-4}\ \text{m}^4.$

$\therefore \sigma_s = \dfrac{nMy}{I} = \dfrac{3\times 4000\times 0.12}{2.54\times 10^{-4}} = 5.67\times 10^6 \text{ Pa}$

3, 18, 3, 36

10.47 d) $60 = \dfrac{L}{r} = \dfrac{L}{\sqrt{I/A}} = \dfrac{L}{\sqrt{0.1\times\left(0.1^3/12\right)/0.01}}. \quad \therefore L = 1.73 \text{ m}$

10.48 c) Assume $P = 4000$ N using a factor of safety of 2.

$4000 = \dfrac{\pi^2\times 70\times 10^9\times\pi\times\left(0.1^4/64\right)}{4L^2}. \quad \therefore L = 14.55 \text{ m}$

10.49 e) $P_{cr} = 4\pi^2 EI/L^2 = \alpha\Delta TEA. \quad \Delta T = \dfrac{4\pi^2 I}{\alpha AL^2} = \dfrac{4\pi^2\times\pi\times\left(0.02^4/64\right)}{11.7\times 10^{-6}\times\pi\times 0.01^2\times 4^2} = 5.27\ °\text{C}$

10.50 b) $P_{cr} = 4\pi^2 EI/L^2 = 30000. \quad \therefore \pi^2 EI/L^2 = 7500. \quad \therefore P_{cr} = \pi^2 EI/4L^2 = 7500/4 = 1875 \text{ N}$

Solutions to Practice Problems

(English Units)

10.1 b) Isotropic

10.2 e) Poisson's Ratio

10.3 d) $\sigma = E\delta/L = P/\pi r^2$. $\therefore P = E\delta\pi r^2/L = 30\times10^6\times0.04\times\pi(1/4)^2/48 = 4909$ lb.

10.4 d) $\frac{P}{\pi r^2} = E\frac{\delta}{L}$. $\therefore \delta = \frac{PL}{\pi r^2 E} = \frac{3500\times100\times12}{\pi\times(1/4^2)\times30\times10^6} = 0.71$ in

10.5 a) $\frac{P}{A} = E\frac{\delta}{L}$. $\therefore P = \frac{AE\delta}{L} = \frac{1}{8}\times1\times30\times10^6\times\frac{1/32}{36} = 3255$ lb.

10.6 b) $\frac{P}{A} = G\frac{\delta}{L}$. $\therefore G = \frac{PL}{A\delta} = \frac{5000\times6}{8\times(1/2)\times0.0012} = 6.25\times10^6$

10.7 a) $\frac{P}{A} = E\varepsilon$. $\therefore \varepsilon = \frac{150,000}{\pi\times1^2\times30\times10^6} = 0.00159$

$\therefore \Delta d = v\varepsilon d = 0.28\times0.00159\times2 = 0.00089$. $\therefore d_{after} = 2-0.00095 = 1.9991$ in

10.8 b) $\frac{P}{A} = E\varepsilon$. $\therefore \varepsilon = \frac{400,000}{\pi\times(5.923/2)^2\times10\times10^6} = 0.00145$

$\Delta d = v\varepsilon d = 0.33\times0.00145\times5.923 = 0.00283$.

$\therefore d_f = d - \Delta d = 5.923 - 0.00283 = 5.920$ in

10.9 e) $\sigma = E\delta/L$. $16,000 = 30\times10^6(\delta/L)$. $\therefore \delta = 5.33\times10^{-4}L$.

$\delta_T = \alpha L\Delta T = 6.5\times10^{-6}\times50L = 3.25\times10^{-4}L$.

$\therefore \delta_{final} = 2.08\times10^{-4}L$. $\therefore \sigma = 30\times10^6\times2.08\times10^{-4} = 6,240$ psi

10.10 a) $\delta = \alpha L\Delta T = 6.5\times10^{-6}\times1000\times12\times130 = 10.1''$

10.11 d) $\frac{\delta}{L} = \frac{\sigma}{E} = \alpha\Delta T$. $\therefore \frac{10,000}{10\times10^6} = 12.8\times10^{-6}(80-T)$. $\therefore T = 1.88°F$

10.12 d) It expands at a different rate.

10.13 a) $\tau = \frac{Tr}{J} = \frac{2000\times1}{\pi\times1^4/2} = 1,273$ psi

10.14 e) $J = \pi(a_1^4 - a_2^4)/2 = \pi(1^4 - .875^4)/2 = 0.650\text{ in}^4$

$\tau = \frac{Tr}{J} = 2,000\times1/0.650 = 3,077$ psi

10.15 a) $T = \frac{\tau J}{r} = 20,000\times(\pi2^4/2)/2 = 251,300\text{ in}\cdot\text{lb or }20,940\text{ ft}\cdot\text{lb}$

10.16 d) $\theta = \frac{TL}{JG} = \frac{160 \times 24 \times 18}{\left[\pi (7/16)^4 /2\right] \times 12 \times 10^6} = 0.1001 \text{ rad } (5.73°)$

10.17 b) $\sigma = My/I. \quad \therefore \sigma = \sigma_{\max}$ at $y = y_{\max}$

10.18 a) A triangle.

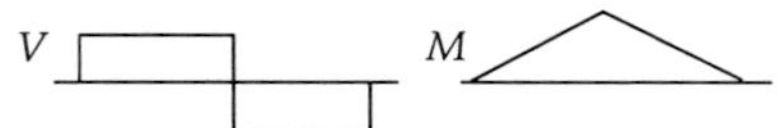

10.19 a) $\sum M_{right\ end} = 0. \quad \therefore 20F_{left} = 1,000 \times 10 + 1,000 \times 5. \quad \therefore F_{left} = 750.$

$M_A = 750 \times 10 = 7,500 \text{ lb} \cdot \text{ft}$

10.20 b) $M_A = 1,000 \times 10 + 1,000 \times 5 = 15,000 \text{ ft} \cdot \text{lb}$

10.21 a) $10F_{right} = 5,000 \times 5 + 1,000 \times 14. \quad \therefore F_{right} = 3,900 \text{ lb}.$

$10F_{left} = 5,000 \times 5 - 1,000 \times 4. \quad \therefore F_{left} = 2,100 \text{ lb}.$

$M_{\max} = 2,100 \times (4.2/2) = 4,410 \text{ ft} \cdot \text{lb} = \text{area under V - diagram.}$

$\sigma = \frac{My}{I} = \frac{4,410 \times 12 \times 5}{60.67} = 4,361 \text{ psi}$

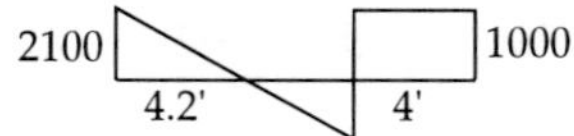

10.22 b) Compression occurs in bottom fibers over right support.

$\sigma = \frac{My}{I} = \frac{4,000 \times 12 \times 5}{60.67} = 3,956 \text{ psi}$

10.23 e) $\tau_{\max} = \frac{VQ}{Ib} = \frac{2,900 \times (5 \times 2.5)}{60.67 \times 1} = 597 \text{ psi}$

10.24 c) $\tau_{\max}$ occurs on the N.A. with a sudden decrease when b goes from 1" to 8". Also, it is parabolic.

10.25 c) $\sigma = \frac{My}{I}. \quad \therefore \frac{I}{y} = \frac{M}{\sigma} = \frac{15,000 \times 12}{2,000} = 9 \text{ in}^3$

10.26 b) Area under curve: $M_{\max} = 250 \times 8 + 800 \times (8/2) = 5,200$

$\sigma = \frac{My}{I} = \frac{5,200 \times 12 \times 1}{4 \times 2^3/12} = 23,4000 \text{ psi}$

V 250 -250 1050

10.27 e) $\sigma_{\max} = \frac{5,200 \times 12 \times 2}{2 \times 4^3/12} = 11,700 \text{ psi}$

10.28 e) $V_{\max} = 250. \quad \tau_{\max} = \frac{VQ}{Ib} = \frac{250 \times 2 \times (1/2)}{2 \times (2^3/12) \times 2} = 93.8$

10.29 d) $\delta = PL^3/48EI = 500 \times 240^3 / \left(48 \times 30 \times 10^6 \times 2^4/12\right) = 3.6"$

10.30 e) $\delta = \dfrac{PL^3}{48EI} + \dfrac{5wL^4}{384EI}$ $\qquad I = \dfrac{4 \times 2^3}{12} = 2.67 \text{ in}^4$

$$= \frac{500 \times (16 \times 12)^3}{48 \times 30 \times 10^6 \times 2.67} + \frac{5 \times (100/12) \times (16 \times 12)^4}{384 \times 30 \times 10^6 \times 2.67} = 2.76"$$

10.31 a) $\delta = \theta L_2 = \dfrac{PL^2}{16EI} \times L_2. \quad \therefore 4 = \dfrac{P \times 240^2 \times 120}{16 \times 30 \times 10^6 \times \pi \times 1^4/4}. \quad \therefore P = 218 \text{ lb}$

10.32 d) $\tau_{\max} = \dfrac{1}{2}\sqrt{\left(\sigma_x - \sigma_y\right)^2 + 4\tau^2} = \dfrac{1}{2}\sqrt{6{,}000^2 + 4 \times 4{,}000^2} = 5{,}000 \text{ psi}$

10.33 a) $\sigma_{\max} = \dfrac{1}{2}\left(\sigma_x + \sigma_y\right) + \tau_{\max} = 0 + 4{,}000 = 4{,}000 \text{ psi}$

10.34 c) $\tau_{\max} = \dfrac{1}{2}\sqrt{(3{,}000 + 5{,}000)^2 + 4 \times 3{,}000^2} = 5{,}000 \text{ psi}$

10.35 d) $\tau = Tr/J = 500 \times 12 \times 1 \big/ \left(\pi \times 1^4/2\right) = 3{,}820 \text{ psi.}$

$$\sigma = P/A = 10{,}000 \big/ \pi \times 1^2 = 3{,}180 \text{psi.}$$

$$\tau_{\max} = \frac{1}{2}\sqrt{3{,}180^2 + 4 \times 3{,}820^2} = 4{,}137 \text{ psi}$$

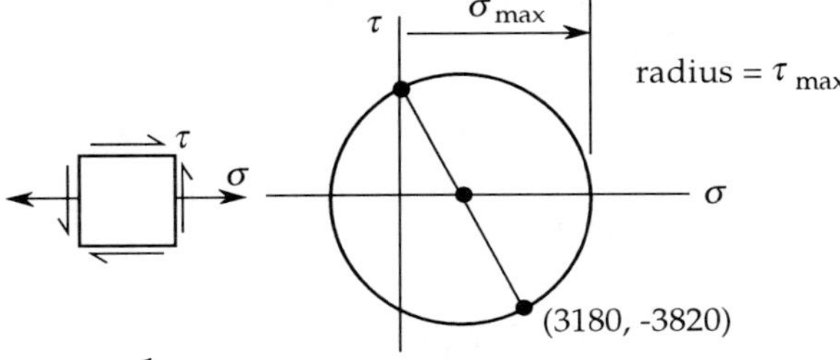

10.36 c) $\sigma_{\max} = \dfrac{1}{2} \times 3{,}180 + 4{,}137 = 5{,}727 \text{ psi}$

10.37 b) $M = 30{,}000 \text{ in} \cdot \text{lb.} \quad \dfrac{My}{I} = \dfrac{30{,}000 \times 1}{\pi \times \left(1^4/4\right)} = 38{,}200 \text{ comp.}$

$$\frac{P}{A} = \frac{10{,}000}{\pi \times 1^2} = 3{,}183 \text{ tens.}$$

$$\sigma_A = 38{,}200 - 3{,}183 = 35{,}000 \text{ comp.}$$

10.38 c) $\tau_{\max} = \sigma/2 = 17{,}500 \text{ psi.} \quad VQ/Ib = 0$ on outer fibers

10.39 c) $\tau = \dfrac{Tr}{J} = \dfrac{2{,}000 \times 10 \times 1}{\pi \times \left(1^4/2\right)} = 12{,}732 \text{ psi}$

$$\sigma = \frac{My}{I} = \frac{2{,}000 \times 16 \times 1}{\pi \times \left(1^4/2\right)} = 40{,}744 \text{ psi}$$

$$\tau_{\max} = \frac{1}{2}\sqrt{40{,}744^2 + 4 \times 12{,}732^2} = 24{,}023 \text{ psi}$$

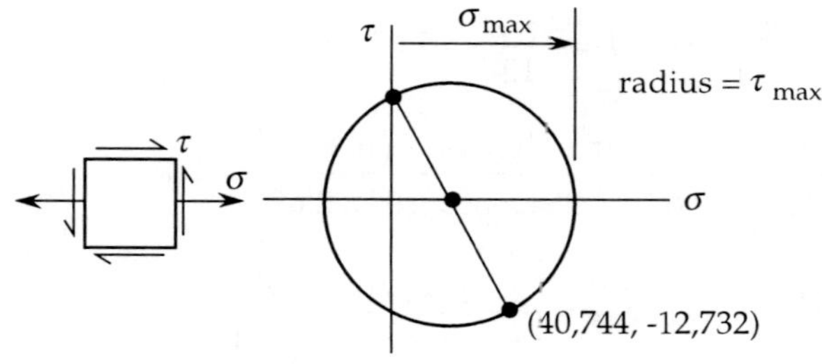

10.40 d) $\sigma_{\max} = \frac{1}{2} \times 40,744 + 24,023 = 44,395 \text{ psi}$

10.41 b) $\sigma_t = pD/2t. \quad \therefore p = 24,000 \times 2 \times \frac{1}{4}\Big/24 = 500 \text{ psi}$

10.42 c) $\sigma_a = pD/4t. \quad \therefore t = \frac{2000 \times 48}{4 \times 30,000} = 0.800 \text{ in}$

10.43 e) $\tau_{\max} = \frac{1}{2}\sqrt{(30,000 - 30,000)^2 + 4 \times 0^2} = 0$

10.44 a) $\left(\frac{\Delta L}{L}\right)_s = \left(\frac{\Delta L}{L}\right)_c. \quad \therefore \varepsilon_s = \varepsilon_c. \quad \therefore \sigma_s = \frac{E_s}{E_c}\sigma_c = 10\,\sigma_c.$

$F_s + F_c = 400,000 \text{ or } A_s\sigma_s + A_c\sigma_c = 400,000.$

$\sigma_s\left[\pi\left(5.5^2 - 5^2\right) + \pi \times \frac{5^2}{10}\right] = 400,000. \quad \therefore \sigma_s = 16,430 \text{ psi}$

10.45 e) $n = E/E_{\min} = 3.$ The area is transformed:

$I_t = \frac{4 \times 8^3}{12} - \frac{1 \times 6^3}{12} = 152.7 \text{ in}^4.$

$\therefore \sigma_{al} = \frac{My}{I} = \frac{2,500 \times 12 \times 4}{152.7} = 786 \text{ psi}$

10.46 d) $n = 3.$ The area is transformed:

$I_t = \frac{12 \times 8^3}{12} - \frac{11 \times 6^3}{12} = 314 \text{ in}^4.$

$\therefore \sigma_s = \frac{nMy}{I} = \frac{3 \times 2,500 \times 12 \times 4}{314} = 1,146 \text{ psi.}$

10.47 d) $60 = \frac{L}{r} = \frac{L}{\sqrt{I/A}} = \frac{L}{\sqrt{4 \times \left(4^3/12\right)/16}}. \quad \therefore L = 69.3\text{" or } 5.77\text{'}$

Assume $P = 1,000$ lb using a factor of safety of 2.

10.48 c) $1,000 = \frac{\pi^2 \times 10 \times 10^6 \times \pi \times 4^4/64}{4L^2}. \quad \therefore L = 557\text{" or } 46.4\text{'}$

10.49 e) $P_{cr} = 4\pi^2 EI/L^2 = \alpha\Delta TEA. \quad \Delta T = \frac{4\pi^2 I}{\alpha AL^2} = \frac{4\pi^2 \times \pi \times 1^4/64}{6.5 \times 10^{-6} \times \pi \times 0.5^2 \times 120^2} = 26.4\ °\text{F}$

10.50 b) $P_{cr} = 4\pi^2 EI/L^2 = 8,000.\quad \therefore \pi^2 EI/L^2 = 2,000.$

$\therefore P_{cr} = \pi^2 EI/4L^2 = 2,000/4 = 500$ lb.

Fluid Mechanics

by Merle C. Potter

Fluid Mechanics deals with the statics, kinematics and dynamics of fluids, including both gases and liquids. Most fluid flows can be assumed to be incompressible (constant density); such flows include liquid flows as well as low speed gas flows (with velocities less than about 100 m/s). In addition, particular flows are either viscous or inviscid. Viscous effects dominate internal flows—such as flow in a pipe—and must be included near the boundaries of external flows (flow near the surface of an airfoil). Viscous flows are laminar if well-behaved, or turbulent if chaotic and highly fluctuating. Inviscid flows occur primarily as external flows outside the boundary layers that contain viscous effects. This review will focus on *Newtonian fluids,* that is, fluids which exhibit linear stress-strain-rate relationships; Newtonian fluids include air, water, oil, gasoline and tar. Please note that we have used upper case P for pressure in thermo, whereas in fluids we use lower case p; that's simply conventional.

11.1 Fluid Properties

Some of the more common fluid properties are defined below and listed in Tables 11.1 and 11.2 for water and air at standard conditions.

density $$\rho = \frac{M}{\forall} \qquad \textbf{(11.1.1)}$$

specific weight $$\gamma = \rho g = \frac{W}{\forall} \qquad \textbf{(11.1.2)}$$

viscosity $$\mu = \frac{\tau}{du/dy} \qquad \textbf{(11.1.3)}$$

kinematic viscosity $\nu = \frac{\mu}{\rho}$ **(11.1.4)**

specific gravity $SG = \frac{\rho_x}{\rho_{H_2O}}$ **(11.1.5)**

bulk modulus $K = -\forall \frac{\Delta p}{\Delta \forall}$ **(11.1.6)**

speed of sound $c_{liquid} = \sqrt{K/\rho}$ $c_{gas} = \sqrt{kRT}$ $(k_{air} = 1.4)$ **(11.1.7)**

Example 11.1

A velocity difference of 2.4 m/s is measured between radial points 2 mm apart in a pipe in which 20°C water is flowing. What is the shear stress?

Solution. Using Eq. 11.1.3 we find, with $\mu = 10^{-3}\ \text{N}\cdot\text{s/m}^2$ from Table 5.1,

$$\tau = \mu \frac{du}{dy} \cong \mu \frac{\Delta u}{\Delta y} = 10^{-3} \frac{2.4}{0.002} = 1.2 \text{ Pa}.$$

TABLE 11.1 Properties

Property	Symbol	Definition	Water (20°C, 68°F)	Air (STP)
density	ρ	$\frac{\text{mass}}{\text{volume}}$	$1000\ \text{kg/m}^3$ $1.94\ \text{slug/ft}^3$	$1.23\ \text{kg/m}^3$ $0.0023\ \text{slug/ft}^3$
viscosity	μ	$\frac{\text{shear stress}}{\text{velocity gradient}}$	$10^{-3}\ \text{N}\cdot\text{s/m}^2$ $2\times10^{-5}\ \text{lb-sec/ft}^2$	$2.0\times10^{-5}\ \text{N}\cdot\text{s/m}^2$ $3.7\times10^{-7}\ \text{lb-sec/ft}^2$
kinematic viscosity	ν	$\frac{\text{viscosity}}{\text{density}}$	$10^{-6}\ \text{m}^2/\text{s}$ $10^{-5}\ \text{ft}^2/\text{sec}$	$1.6\times10^{-5}\ \text{m}^2/\text{s}$ $1.6\times10^{-4}\ \text{ft}^2/\text{sec}$
speed of sound	c	velocity of propagation of a small wave	1480 m/s 4900 ft/sec	343 m/s 1130 ft/sec
specific weight	γ	$\frac{\text{weight}}{\text{volume}}$	$9800\ \text{N/m}^3$ $62.4\ \text{lb/ft}^3$	$12\ \text{N/m}^3$ $0.077\ \text{lb/ft}^3$
surface tension	σ	stored energy per unit area	$0.073\ \text{J/m}^2$ 0.005 lb/ft	
bulk modulus	K	$-\text{volume}\frac{\Delta\ \text{pressure}}{\Delta\ \text{volume}}$	220×10^4 kPa 323,000 psi	
vapor pressure	p_v	pressure at which liquid & vapor are in equilibrium	2.45 kPa 0.34 psia	

Notes:
- Kinematic viscosity is used because the ratio μ/ρ occurs frequently.
- Surface tension is used primarily for calculating capillary rise.
- Vapor pressure is used to predict *cavitation* which exists whenever the local pressure falls below the vapor pressure (for water at standard temperatures, the vapor pressure is close to absolute zero).

TABLE 11.2 Properties of Water and Air

Properties of Water (Metric)

Temperature °C	Density kg/m³	Viscosity N·s/m²	Kinematic Viscosity m²/s	Bulk Modulus kPa	Surface Tension N/m	Vapor Pressure kPa
0	999.9	1.792×10^{-3}	1.792×10^{-6}	204×10^{4}	7.62×10^{-2}	0.588
5	1000.0	1.519	1.519	206	7.54	0.882
10	999.7	1.308	1.308	211	7.48	1.176
15	999.1	1.140	1.141	214	7.41	1.666
20	998.2	1.005	1.007	220	7.36	2.447
30	995.7	0.801	0.804	223	7.18	4.297
40	992.2	0.656	0.661	227	7.01	7.400
50	988.1	0.549	0.556	230	6.82	12.220
60	983.2	0.469	0.477	228	6.68	19.600
70	977.8	0.406	0.415	225	6.50	30.700
80	971.8	0.357	0.367	221	6.30	46.400
90	965.3	0.317	0.328	216	6.12	68.200
100	958.4	0.284×10^{-3}	0.296×10^{-6}	207×10^{4}	5.94×10^{-2}	97.500

Properties of Air at Standard Pressure (Metric)

Temperature	Density kg/m³	Specific Weight N/m³	Viscosity N·s/m²	Kinematic Viscosity m²/s
–20°C	1.39	13.6	1.56×10^{-5}	1.13×10^{-5}
–10°C	1.34	13.1	1.62×10^{-5}	1.21×10^{-5}
0°C	1.29	12.6	1.68×10^{-5}	1.30×10^{-5}
10°C	1.25	12.2	1.73×10^{-5}	1.39×10^{-5}
20°C	1.20	11.8	1.80×10^{-5}	1.49×10^{-5}
40°C	1.12	11.0	1.91×10^{-5}	1.70×10^{-5}
60°C	1.06	10.4	2.03×10^{-5}	1.92×10^{-5}
80°C	0.99	9.71	2.15×10^{-5}	2.17×10^{-5}
100°C	0.94	9.24	2.28×10^{-5}	2.45×10^{-5}

Properties of the Atmosphere (Metric)

Altitude m	Temperature K	p/p_0 (p_0=101 kPa)	ρ/ρ_0 (ρ_0=1.23 kg/m³)
0	288.2	1.000	1.000
1 000	281.7	0.8870	0.9075
2 000	275.2	0.7846	0.8217
4 000	262.2	0.6085	0.6689
6 000	249.2	0.4660	0.5389
8 000	236.2	0.3519	0.4292
10 000	223.3	0.2615	0.3376
12 000	216.7	0.1915	0.2546
14 000	216.7	0.1399	0.1860
16 000	216.7	0.1022	0.1359
18 000	216.7	0.07466	0.09930
20 000	216.7	0.05457	0.07258
22 000	218.6	0.03995	0.05266
26 000	222.5	0.02160	0.02797
30 000	226.5	0.01181	0.01503
40 000	250.4	0.2834×10^{-2}	0.3262×10^{-2}
50 000	270.7	0.7874×10^{-3}	0.8383×10^{-3}
60 000	255.8	0.2217×10^{-3}	0.2497×10^{-3}
70 000	219.7	0.5448×10^{-4}	0.7146×10^{-4}
80 000	180.7	0.1023×10^{-4}	0.1632×10^{-4}
90 000	180.7	0.1622×10^{-5}	0.2588×10^{-5}

TABLE 11.2(E) Properties of Water and Air
Properties of Water (English)

Temperature °F	Density slugs/ft³	Viscosity lb-sec/ft²	Surface Tension lb/ft	Vapor Pressure lb/in²	Bulk Modulus lb/in²
32	1.94	3.75×10^{-5}	0.518×10^{-2}	0.089	293,000
40	1.94	3.23×10^{-5}	0.514×10^{-2}	0.122	294,000
50	1.94	2.74×10^{-5}	0.509×10^{-2}	0.178	305,000
60	1.94	2.36×10^{-5}	0.504×10^{-2}	0.256	311,000
70	1.94	2.05×10^{-5}	0.500×10^{-2}	0.340	320,000
80	1.93	1.80×10^{-5}	0.492×10^{-2}	0.507	322,000
90	1.93	1.60×10^{-5}	0.486×10^{-2}	0.698	323,000
100	1.93	1.42×10^{-5}	0.480×10^{-2}	0.949	327,000
120	1.92	1.17×10^{-5}	0.465×10^{-2}	1.69	333,000
140	1.91	0.98×10^{-5}	0.454×10^{-2}	2.89	330,000
160	1.90	0.84×10^{-5}	0.441×10^{-2}	4.74	326,000
180	1.88	0.73×10^{-5}	0.426×10^{-2}	7.51	318,000
200	1.87	0.64×10^{-5}	0.412×10^{-2}	11.53	308,000
212	1.86	0.59×10^{-5}	0.404×10^{-2}	14.7	300,000

Properties of Air at Standard Pressure (English)

Temperature °F	Density slugs/ft³	Viscosity lb-sec/ft²	Kinematic Viscosity ft²/sec
0	0.00268	3.28×10^{-7}	12.6×10^{-5}
20	0.00257	3.50×10^{-7}	13.6×10^{-5}
40	0.00247	3.62×10^{-7}	14.6×10^{-5}
60	0.00237	3.74×10^{-7}	15.8×10^{-5}
68	0.00233	3.81×10^{-7}	16.0×10^{-5}
80	0.00228	3.85×10^{-7}	16.9×10^{-5}
100	0.00220	3.96×10^{-7}	18.0×10^{-5}
120	0.00215	4.07×10^{-7}	18.9×10^{-5}

Properties of the Atmosphere (English)

Altitude ft	Temperature °F	Pressure lb/ft²	Density slugs/ft³	Kinematic Viscosity ft²/sec	Velocity of Sound ft/sec
0	59.0	2116	0.00237	1.56×10^{-4}	1117
1,000	55.4	2041	0.00231	1.60×10^{-4}	1113
2,000	51.9	1968	0.00224	1.64×10^{-4}	1109
5,000	41.2	1760	0.00205	1.77×10^{-4}	1098
10,000	23.4	1455	0.00176	2.00×10^{-4}	1078
15,000	5.54	1194	0.00150	2.28×10^{-4}	1058
20,000	−12.3	973	0.00127	2.61×10^{-4}	1037
25,000	−30.1	785	0.00107	3.00×10^{-4}	1016
30,000	−48.0	628	0.000890	3.47×10^{-4}	995
35,000	−65.8	498	0.000737	4.04×10^{-4}	973
36,000	−67.6	475	0.000709	4.18×10^{-4}	971
40,000	−67.6	392	0.000586	5.06×10^{-4}	971
50,000	−67.6	242	0.000362	8.18×10^{-4}	971
100,000	−67.6	22.4	3.31×10^{-5}	89.5×10^{-4}	971
110,000	−47.4	13.9	1.97×10^{-5}	1.57×10^{-6}	996
150,000	113.5	3.00	3.05×10^{-6}	13.2×10^{-6}	1174
200,000	160.0	0.665	6.20×10^{-7}	68.4×10^{-6}	1220
260,000	−28	0.0742	1.0×10^{-7}	321×10^{-6}	1019

Example 11.2

Find the speed of sound in air at an elevation of 1000 m.

Solution. From Table 11.2 we find T = 281.7 K. Using Eq. 11.1.7, with R = 287 J/kg·K, there results

$$c = \sqrt{kRT} = \sqrt{1.4 \times 287 \times 281.7} = 336.4\,\text{m/s}.$$

Note: Temperature must be absolute.

11.2 Fluid Statics

Typical problems in fluid statics involve manometers, forces on plane and curved surfaces, and buoyancy. All of these problems are solved by using the pressure distribution derived from summing forces on an infinitesimal element of fluid; in differential form with h positive downward, it is

$$dp = \gamma dh. \tag{11.2.1}$$

For constant specific weight, assuming $p = 0$ at $h = 0$, we have

$$p = \gamma h. \tag{11.2.2}$$

Equation 11.2.2 can be used to interpret manometer readings directly. By summing forces on elements of a plane surface, we would find the magnitude and location of a force acting on one side (refer to Fig. 11.1) to be

$$F = \gamma h_c A \tag{11.2.3}$$

NCEES form

$$y_p = y_c + \frac{I_c}{y_c A} \qquad z^* = \frac{I}{AZ_c} \tag{11.2.4}$$

where y_c (Z_c in NCEES handbook) locates the centroid and I_c is the second moment* of the area about the centroidal axis.

To solve problems involving curved surfaces, we simply draw a free-body diagram of the liquid contained above the curved surface and, using the above formulas, solve the problem.

To solve buoyancy-related problems we use Archimedes' principle which states: the buoyant force on a submerged object is equal to the weight of displaced liquid; that is,

$$F_b = \gamma \forall_{displaced}. \tag{11.2.5}$$

*The second moment I_c (often symbolized by $\bar{I}$) of three common areas:

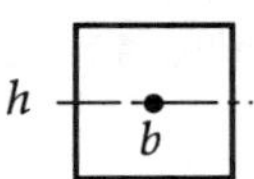

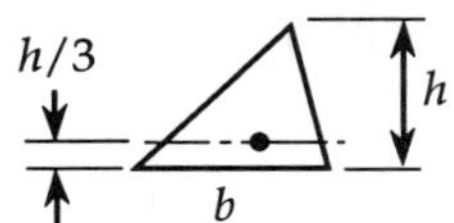

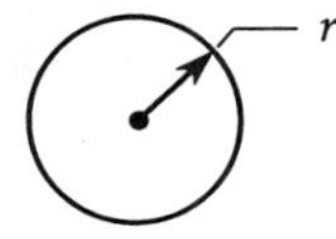

$$I_c = \frac{bh^3}{12} \qquad I_c = \frac{bh^3}{36} \qquad I_c = \frac{br^4}{4}$$

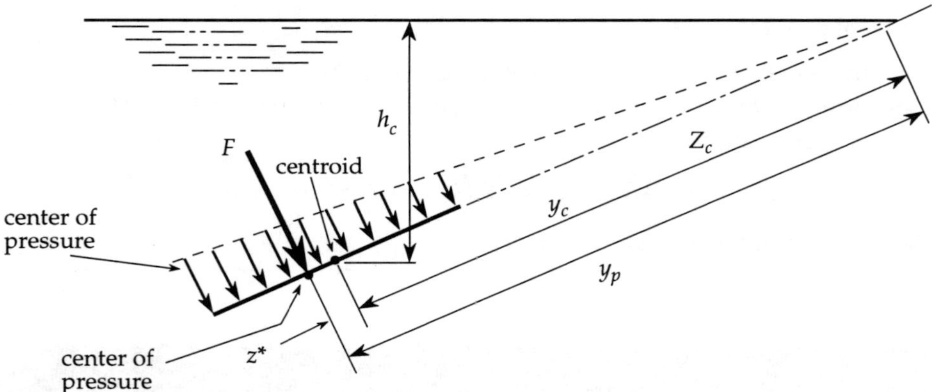

Figure 11.1 Force on a plane surface.

Example 11.3

Find the pressure difference between the air pipe and the water pipe.

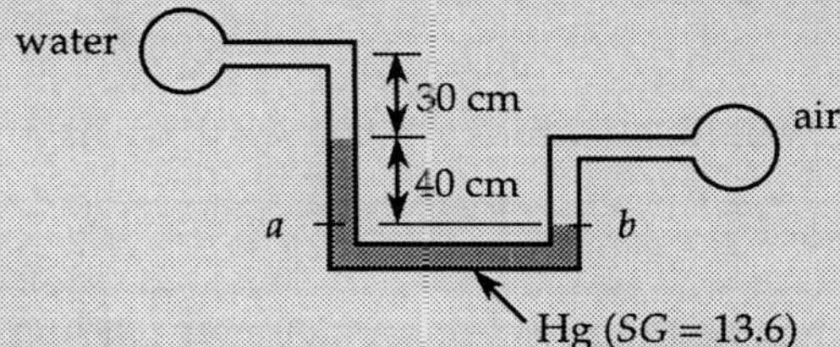

Solution. We first locate points "*a*" and "*b*" in the same fluid where $p_a = p_b$; then using Eq. 11.2.2

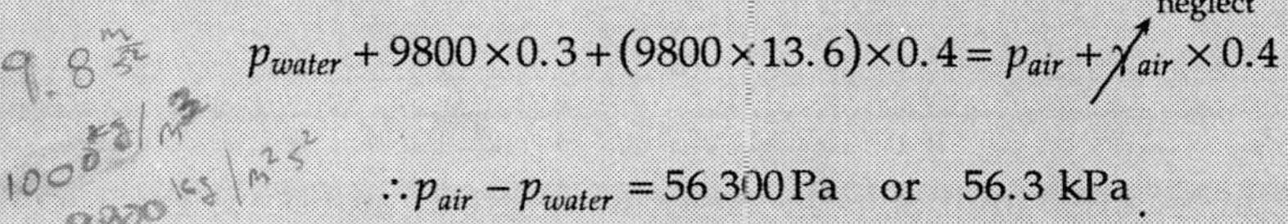

$$p_{water} + 9800 \times 0.3 + (9800 \times 13.6) \times 0.4 = p_{air} + \gamma_{air} \times 0.4$$

$$\therefore p_{air} - p_{water} = 56\,300\,\text{Pa} \quad \text{or} \quad 56.3\ \text{kPa}.$$

Example 11.4

What is the pressure in pipe *A*?

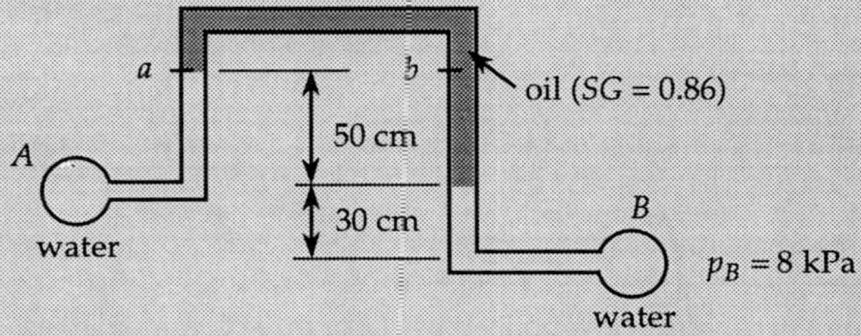

Solution. Locate points "*a*" and "*b*" so that $p_a = p_b$. Then, using Eq. 11.2.2 there results

$$p_A - 9800 \times 0.5 = 8000 - 9800 \times 0.3 - (9800 \times 0.86) \times 0.5$$

$$\therefore p_A = 5750\ \text{Pa} \quad \text{or} \quad 5.75\ \text{kPa}.$$

Example 11.5

Find the force *P* needed to hold the 5-m-wide gate closed.

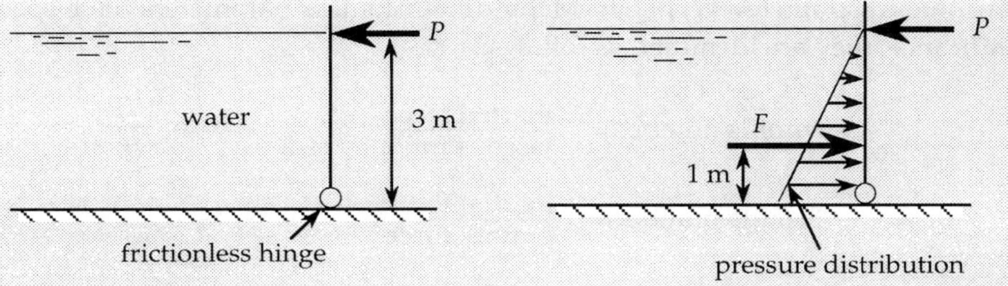

Solution. First, we note that the pressure distribution is triangular, as shown. Hence, the resultant force *F* acts 1/3 up from the hinge through the centroid of the triangular distribution. Summing moments about the hinge gives

$$F \times 1 = P \times 3$$

$$\therefore P = F/3 = \gamma h_c A/3$$

$$= 9800 \times \frac{3}{2} \times (5 \times 3)/3 = 73\,500 \text{ N}.$$

Note: If the top of the gate were not at the free surface, we would find y_p using Eq. 11.2.4.

Example 11.6

A rectangular 4 m × 20 m vessel has a mass of 40 000 kg. How far will it sink in water when carrying a load of 100 000 kg?

Solution. The total weight of the loaded vessel must equal the weight of the displaced water. This is expressed as

$$W = \gamma V\!\!\!\!-$$

$$(40\,000 + 100\,000) \times 9.8 = 9800 \times 4 \times 20 \times h$$

$$\therefore h = 1.75 \text{m}.$$

11.3 Dimensionless Parameters and Similitude

Information involving phenomena encountered in fluid mechanics is often presented in terms of dimensionless parameters. For example, the lift force F_L on a streamlined body can be represented by a lift coefficient C_L, a dimensionless parameter. Rather than plotting the lift force as a function of velocity, the lift coefficient could be plotted as a function of the Reynolds number, or the Mach number—two other dimensionless parameters.

To form dimensionless parameters, we first list various quantities encountered in fluid mechanics in Table 11.3. A dimensionless parameter involving several quantities is then formed by combining the quantities so that the combination of quantities is dimensionless. If all units are present in the quantities to be combined, this usually requires four quantities. For example, the four quantities power $\dot{W}$, flow rate Q, specific weight γ, and head H can be arranged as the dimensionless parameter $\dot{W}/\gamma QH$. Many dimensionless parameters have special significance; they are identified as follows:

$$\text{Reynolds number} = \frac{\text{inertial force}}{\text{viscous force}} \qquad \text{Re} = \frac{V\ell\rho}{\mu}$$

$$\text{Froude number} = \frac{\text{inertial force}}{\text{gravity force}} \qquad \text{Fr} = \frac{V^2}{\ell g}$$

$$\text{Mach number} = \frac{\text{inertial force}}{\text{compressibility force}} \qquad \text{M} = \frac{V}{c}$$

$$\text{Weber number} = \frac{\text{inertial force}}{\text{surface tension force}} \qquad \text{We} = \frac{V^2\ell\rho}{\sigma}$$

$$\text{Pressure coefficient} = \frac{\text{pressure force}}{\text{inertial force}} \qquad C_p = \frac{\Delta p}{\frac{1}{2}\rho V^2}$$

$$\text{Drag coefficient} = \frac{\text{drag force}}{\text{inertial force}} \qquad C_D = \frac{\text{drag}}{\frac{1}{2}\rho V^2 A}$$

So, rather than writing the drag force on a cylinder as a function of length ℓ, diameter D, velocity V, viscosity μ, and density ρ, i.e.,

$$F_D = f(\ell, D, V, \mu, \rho) \tag{11.3.2}$$

we express the relationship using dimensionless parameters as

$$C_D = f\left(\frac{V\rho D}{\mu}, \frac{\ell}{D}\right). \tag{11.3.3}$$

The subject of similarity is encountered when attempting to use the results of a model study in predicting the performance of a prototype. We always assume *geometric similarity*, that is, the model is constructed to scale with the prototype; the length scale $\ell_p/\ell_m = \lambda$ is usually designated. The primary notion is simply stated: *Dimensionless quantities associated with the model are equal to corresponding dimensionless quantities associated with the prototype*. For example, if viscous effects dominate we would require

$$\text{Re}_m = \text{Re}_p. \tag{11.3.4}$$

Then if we are interested in, for example, the drag force, we would demand the dimensionless forces to be equal:

$$(F_D)_m^* = (F_D)_p^* \tag{11.3.5}$$

where the asterisk * denotes a dimensionless quantity. Since force is pressure (ρV^2) times area (ℓ^2), the above equation can be expressed in terms of dimensional quantities:

$$\frac{(F_D)_m}{\rho_m V_m^2 \ell_m^2} = \frac{(F_D)_p}{\rho_p V_p^2 \ell_p^2}. \tag{11.3.6}$$

TABLE 11.3 Symbols and Dimensions of Quantities Used in Fluid Mechanics

Quantity	Symbol	Dimensions	Quantity	Symbol	Dimensions
Length	ℓ	L	Pressure	p	M/LT^2
Time	t	T	Stress	τ	M/LT^2
Mass	m	M	Density	ρ	M/L^3
Force	F	ML/T^2	Specific Weight	γ	M/L^2T^2
Velocity	V	L/T	Viscosity	μ	M/LT
Acceleration	a	L/T^2	Kinematic Viscosity	ν	L^2/T
Frequency	ω	T^{-1}	Work	W	ML^2/T^2
Gravity	g	L/T^2	Power	$\dot{W}$	ML^2/T^3
Area	A	L^2	Heat Flux	$\dot{Q}$	ML^2/T^3
Flow Rate	Q	L^3/T	Surface Tension	σ	M/T^2
Mass Flux	$\dot{m}$	M/T	Bulk Modulus	K	M/LT^2

This would allow us to predict the drag force expected on the prototype as

$$(F_D)_p = (F_D)_m \frac{\rho_p \ell_p{}^2 V_p{}^2}{\rho_m \ell_m{}^2 V_m{}^2}. \qquad (11.3.7)$$

The same strategy is used for other quantities of interest.

Example 11.7

Combine $\dot{W}$, Q, γ, and H as a dimensionless parameter.

Solution. First, let us note the dimensions on each variable:

$$[\dot{W}] = \frac{ML^2}{T^3} \qquad [Q] = \frac{L^3}{T} \quad [\gamma] = \frac{M}{L^2T^2} \qquad [H] = L$$

Now, by inspection we simply form the dimensionless parameter. Note that to eliminate the mass unit, $\dot{W}$ and γ must appear as the ratio, $\dot{W}/\gamma$. This puts an extra time unit in the denominator; hence, Q must appear with γ as $\dot{W}/\gamma Q$. Now, we inspect the length unit and find one length unit still in the numerator. This requires H in the denominator giving the dimensionless parameter as

$$\frac{\dot{W}}{\gamma QH}$$

Example 11.8

If a flow rate of 0.2 m^3/s is measured over a 9-to-1 scale model of a weir, what flow rate can be expected on the prototype?

Solution. First, we recognize that gravity forces dominate (as they do in all problems involving weirs, dams, ships, and open channels), and demand that

$$\text{Fr}_p = \text{Fr}_m \quad \text{or} \quad \frac{V_p^{\,2}}{\ell_p g_p} = \frac{V_m^{\,2}}{\ell_m g_m}.$$

$$\therefore \frac{V_p}{V_m} = \sqrt{\frac{\ell_p}{\ell_m}} = 3.$$

The dimensionless flow rates are now equated:

$$Q_p{}^* = Q_m{}^*$$

$$\frac{Q_p}{V_p \ell_p^{\,2}} = \frac{Q_m}{V_m \ell_m^{\,2}}$$

recognizing that velocity times area ($V \times \ell^2$) give the flow rate. We have

$$Q_p = Q_m \frac{V_p \ell_p^{\,2}}{V_m \ell_m^{\,2}}$$

$$= 0.2 \times 3 \times 9^2 = 48.6 \ \text{m}^3/\text{s}.$$

11.4 Control Volume Equations

When solving problems in fluid dynamics, we are most often interested in volumes into which and from which fluid flows; such volumes are called *control volumes*. The control volume equations include the conservation of mass (the continuity equation), Newton's second law (the momentum equation), and the first law of thermodynamics (the energy equation). We will not derive the equations but simply state them and then apply them to some situations of interest. We will assume *steady, incompressible flow* with *uniform velocity profiles*. The equations take the following forms:

continuity: $$A_1 V_1 = A_2 V_2 \qquad \textbf{(11.4.1)}$$

momentum: $$\Sigma \mathbf{F} = \rho Q(\mathbf{V}_2 - \mathbf{V}_1) \qquad \textbf{(11.4.2)}$$

energy $$-\frac{\dot{W}_S}{\gamma Q} = \frac{V_2^{\,2} - V_1^{\,2}}{2g} + \frac{p_2 - p_1}{\gamma} + z_2 - z_1 + h_f \qquad \textbf{(11.4.3)}$$

where

$$Q = AV = \text{flow rate} \qquad \textbf{(11.4.4)}$$

$\dot{W}_S$ = shaft work (positive for a turbine)

h_f = head loss.

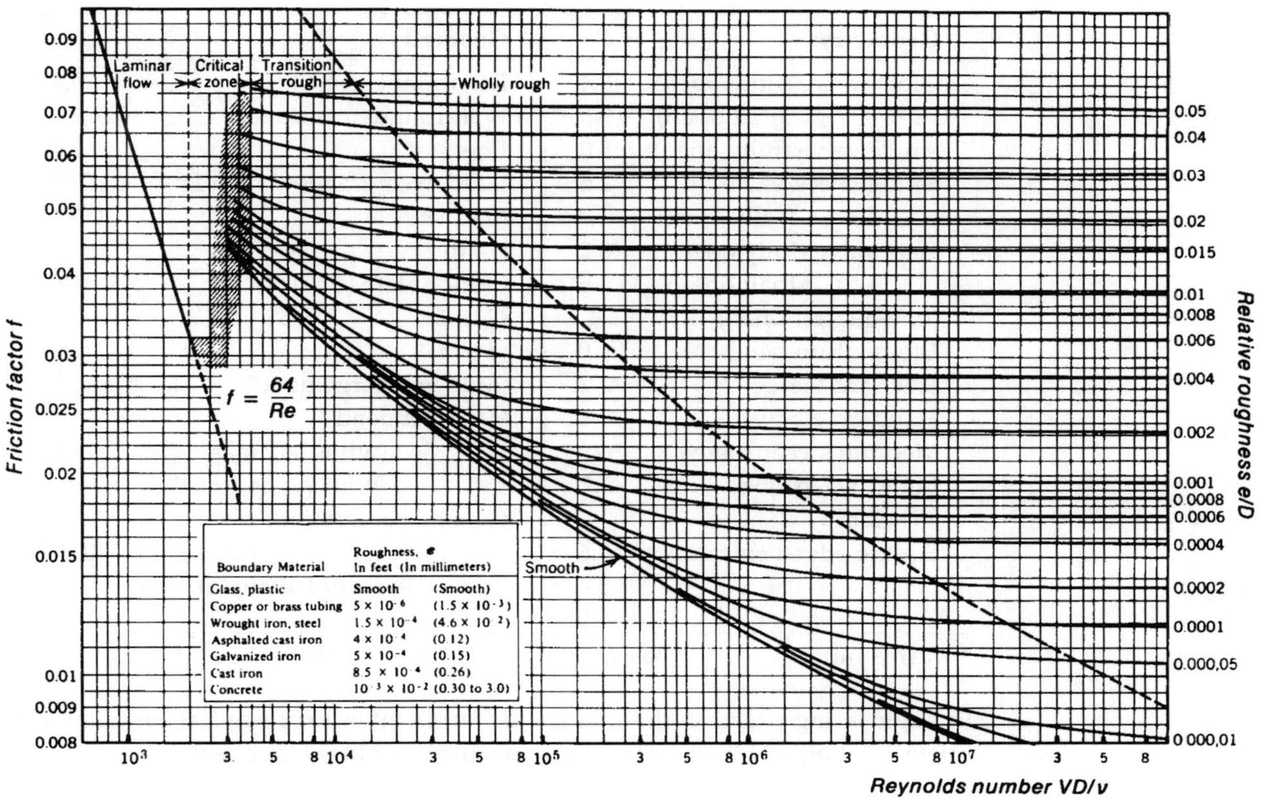

Boundary Material	Roughness, e In feet	(In millimeters)
Glass, plastic	Smooth	(Smooth)
Copper or brass tubing	5×10^{-6}	(1.5×10^{-3})
Wrought iron, steel	1.5×10^{-4}	(4.6×10^{-2})
Asphalted cast iron	4×10^{-4}	(0.12)
Galvanized iron	5×10^{-4}	(0.15)
Cast iron	8.5×10^{-4}	(0.26)
Concrete	$10^{-3} \times 10^{-2}$	(0.30 to 3.0)

Figure 11.2 The Moody Diagram.

If there is no shaft work term $\dot{W}_S$ (due to a pump or turbine) between the two sections and the losses are zero, then the energy equation reduces to the Bernoulli equation, namely,

$$\frac{V_2^2}{2g} + \frac{p_2}{\gamma} + z_2 = \frac{V_1^2}{2g} + \frac{p_1}{\gamma} + z_1. \tag{11.4.5}$$

For flow in a pipe, the head loss can be related to the friction factor by the Darcy-Weisbach equation,

$$h_f = f \frac{L}{D} \frac{V^2}{2g} \tag{11.4.6}$$

where the friction factor is related to the Reynolds number, $\mathrm{Re} = VD/\nu$, and the relative roughness e/D by the Moody diagram, Fig. 11.2; the roughness e is given for various materials. Note that for completely turbulent flows, the friction factor is constant so that the head loss varies with the square of the velocity. For laminar flow the head loss is directly proportional to the velocity.

For sudden geometry changes, such as valves, elbows, and enlargements, the head loss (often called a minor loss) is determined by using a loss coefficient C; that is,

$$h_f = C \frac{V^2}{2g} \tag{11.4.7}$$

where V is the characteristic velocity associated with the device. Typical values are given in Table 11.4.

TABLE 11.4 Loss Coefficients

Geometry	C	Geometry	C
Globe valve (fully open)	6.4	Reentrant entrance	0.8
(half open)	9.5	Well-rounded entrance	0.03++
Angle valve (fully open)	5.0	Pipe exit	1.0
Swing check valve	2.5	Sudden contraction (2 to 1)*	0.25++
(fully open)		(5 to 1)*	0.41++
Gate valve (fully open)	0.2	(10 to 1)*	0.46++
(half open)	5.6	Orifice plate (1.5 to 1)*	0.85
(one-quarter open)	24.	(2 to 1)*	3.4
Close return bend	0	(4 to 1)*	29
Standard tee	2.2	Sudden enlargement+	$(1-A_1/A_2)^2$
Standard elbow	1.8	90° miter bend	
Medium sweep elbow	0.9	(without vanes)	1.1
Long sweep elbow	0.7	(with vanes)	0.2
45° elbow	5	General contraction	
Square-edged entrance	0.6	(30° included angle)	0.02
	0.4	(70° included angle)	0.07
	0.5		

*Area ratio   +Based on V_1 ++Based on V_2

In engineering practice, the loss coefficient is often expressed as an *equivalent length* L_e of pipe; if that is done, the equivalent length is expressed as

$$L_e = C\frac{D}{f}. \tag{11.4.8}$$

The above analysis, using the Moody diagram and the loss coefficients, can be applied directly to only circular cross-section conduits; if the cross section is non-circular but fairly "open" (rectangular with aspect ratio less than four, oval, or triangular), a good approximation can be obtained by using the *hydraulic radius* defined by

$$R_H = A/P \tag{11.4.9}$$

where A is the cross sectional area and P is the *wetted perimeter* (that perimeter where the fluid is in contact with the solid boundary). Using this formula the diameter of a pipe is $D = 4R_H$. The Reynolds number then takes the form

$$\mathrm{Re} = \frac{4VR_H}{\nu}. \tag{11.4.10}$$

If the shape is not "open," such as flow in an anulus, the error in using the above relationships will be quite significant.

A final note in this article defines the energy grade line (*EGL*) and the hydraulic grade line (*HGL*). The distance $(z+p/\gamma)$ above the datum (the zero elevation line) locates the *HGL*, and the distance $\left(z+p/\gamma+V^2/2g\right)$ above the datum locates the *EGL*. These are shown in Fig. 11.3. Note that the pump head H_P is given by

$$H_P = -\frac{\dot{W}_P}{\gamma Q}. \tag{11.4.11}$$

The negative sign is necessary since the pump power $\dot{W}_P$ is negative.

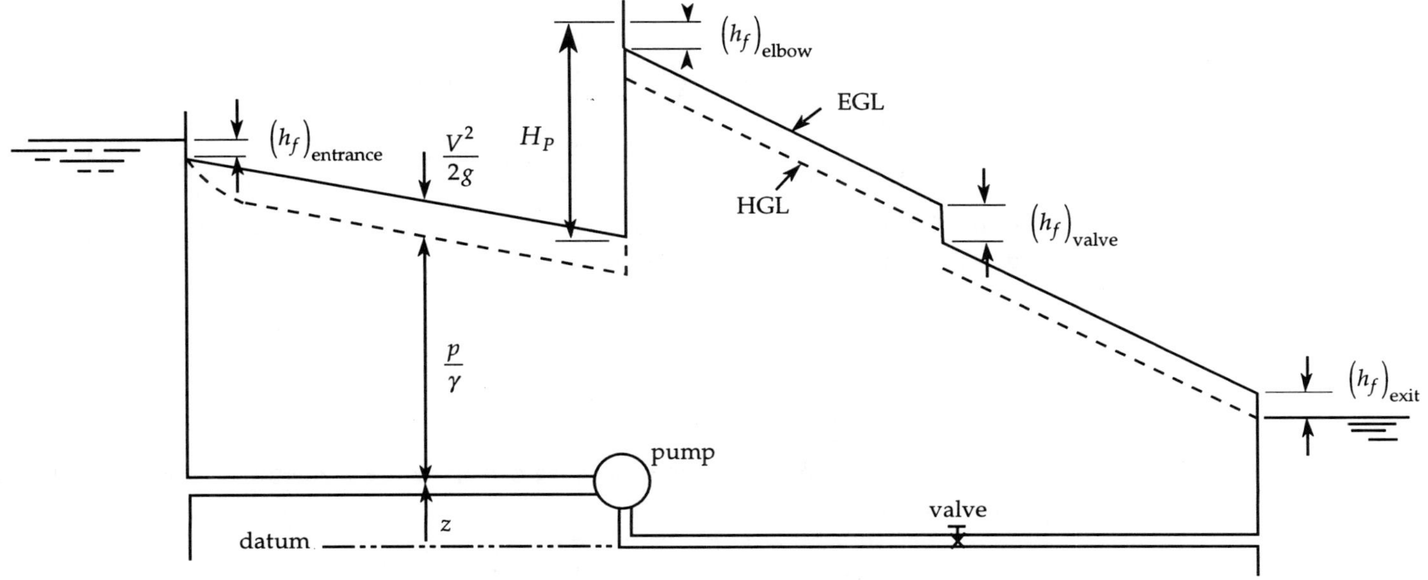

Figure 11.3 The energy grade line *(EGL)* and the hydraulic grade line *(HGL)*.

Example 11.9

The velocity in a 2-cm-dia pipe is 10 m/s. If the pipe enlarges to 4-cm-dia, find the velocity and the flow rate.

Solution. The continuity equation is used as follows:

$$A_1V_1 = A_2V_2$$

$$\frac{\pi D_1^{\ 2}}{4}V_1 = \frac{\pi D_2^{\ 2}}{4}V_2.$$

$$\therefore V_2 = V_1\frac{D_1^{\ 2}}{D_2^{\ 2}} = 10 \times \frac{2^2}{4^2} = 2.5 \text{ m / s.}$$

The flow rate is

$$Q = A_1V_1$$

$$= \pi \times 0.01^2 \times 10 = 0.00314 \ \text{m}^3/\text{s}.$$

Example 11.10

What force is exerted on the joint if the flow rate of water is 0.01 m^3/s?

Solution. The velocities are found to be

$$V_1 = \frac{Q}{A_1} = \frac{0.01}{\pi \times 0.02^2} = 7.96 \text{ m/s}$$

$$V_2 = \frac{Q}{A_2} = \frac{0.01}{\pi \times 0.01^2} = 31.8 \text{ m/s}.$$

Bernoulli's equation is used to find the pressure at section 1. There results, using $p_2 = 0$ (atmospheric pressure is zero gage),

$$\frac{V_1^2}{2g} + \frac{p_1}{\gamma} = \frac{V_2^2}{2g} + \frac{p_2}{\gamma}$$

$$\frac{7.96^2}{2 \times 9.8} + \frac{p_1}{9800} = \frac{31.8^2}{2 \times 9.8}. \qquad \therefore p_1 = 474\,000 \text{ Pa}.$$

Now, using the control volume shown, we can apply the momentum equation (11.4.2) in the x-direction:

p_1A_1 → F_j ← x

F_j = the force of the contraction on the water

$$p_1A_1 - F_j = \rho Q(V_2 - V_1)$$

$$474\,000 \times \pi \times 0.02^2 - F_j = 1000 \times 0.01(31.8 - 7.96)$$

$$\therefore F_j = 357 \text{ N}.$$

Note: Remember, all forces on the control volume must be included. Never forget the pressure force.

Example 11.11

What is the force F needed to hold the plate as shown?

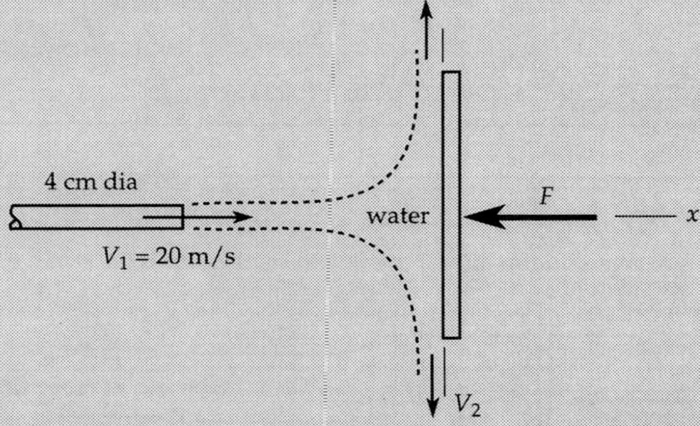

Solution. The momentum equation (11.4.2) is a vector equation; applying it in the x-direction results in

$$-F = \rho Q(\overset{0}{\cancel{V_{2x}}} - V_{1x}).$$

$$\therefore F = \rho A_1 V_1^2$$

$$= 1000 \times \pi \times 0.02^2 \times 20^2 = 503\text{ N}.$$

Note: Since the water is open to the atmosphere, $p_2 = p_1$, and if we neglect elevation changes, Bernoulli's equation requires $V_2 = V_1$. However, here $V_{2x} = 0$ so V_2 was not necessary.

Example 11.12

What is the pump power needed to increase the pressure by 600 kPa in a 8-cm-dia pipe transporting 0.04 m^3/s of water?

Solution. The energy equation (11.4.3) is used:

$$-\frac{\dot{W}_S}{\gamma Q} = \overset{0}{\cancel{\frac{V_2^2 - V_1^2}{2g}}} + \frac{p_2 - p_1}{\gamma} + \overset{0}{\cancel{z_2 - z_1}}$$

$$-\frac{-\dot{W}_P}{9800 \times 0.04} = \frac{600\ 000}{9800}. \qquad \therefore \dot{W}_P = 24\ 000\text{ W} \quad \text{or} \quad 24\text{ kW}.$$

Example 11.13

A pitot tube is used to measure the velocity in the pipe. If $V = 15$ m/s, what is H?

Solution. Bernoulli's equation can be used to relate the pressure at pt. 2, which is just inside the tube, to the velocity V. It gives

$$\overset{0}{\cancel{\frac{V_2^2}{2g}}} + \frac{p_2}{\gamma} + z_2 = \frac{V_1^2}{2g} + \frac{p_1}{\gamma} + z_1$$

$$\therefore p_2 = p_1 + \gamma \frac{V_1^2}{2g}.$$

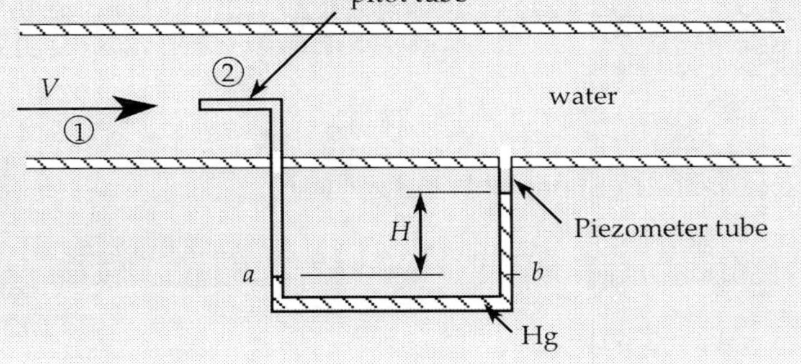

pitot tube
V
②
①
water
H
a
b
Piezometer tube
Hg

The manometer allows us to write

$$p_a = p_b$$

$$\gamma H + p_2 = \gamma_{Hg} H + p_1$$

where the distance from pt. 2 to the top of *H* balances on each side. Substituting for p_2 we have

$$\gamma H + p_1 + \gamma \frac{V_1^2}{2g} = \gamma_{Hg} H + p_1$$

$$\therefore H = \frac{\gamma}{\gamma_{Hg} - \gamma} \cdot \frac{V_1^2}{2g}$$

$$= \frac{9800}{13.6 \times 9800 - 9800} \cdot \frac{15^2}{2 \times 9.8} = 0.91 \text{ m}.$$

Note: The piezometer tube on the right leg measures the pressure p_1 in the pipe.

Example 11.14

For a flow rate of 0.02 m^3/s, find the turbine output if it is 80% efficient.

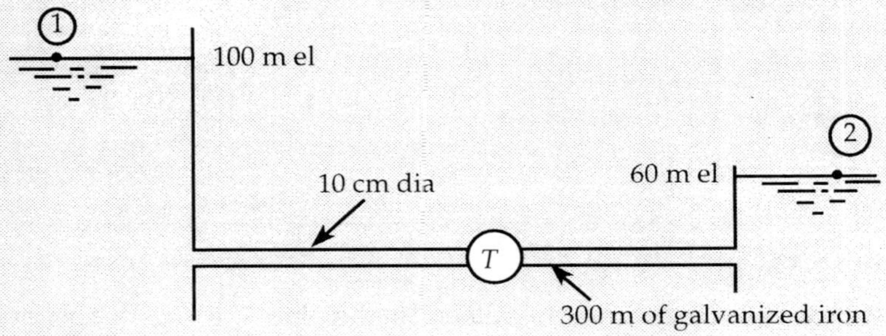

Solution. The energy equation (11.4.3) takes the form:

$$-\frac{\dot{W}_T}{\gamma Q} = \frac{\cancelto{0}{V_2^2 - V_1^2}}{2g} + \frac{\cancelto{0}{p_2 - p_1}}{\gamma} + z_2 - z_1 + \left(C_{inlet} + C_{exit} + f\frac{L}{D} \right) \frac{V^2}{2g}$$

where Eqs. 11.4.6 and 11.4.7 have been used for the head loss. To find *f*, using the Moody diagram, we need

$$V = \frac{0.02}{\pi \times 0.05^2} = 2.55 \text{ m/s} \quad \text{Re} = \frac{2.55 \times 0.1}{10^{-6}} = 2.55 \times 10^5 \qquad \frac{e}{D} = \frac{0.15}{100} = 0.0015$$

$$\therefore f = 0.022.$$

Using the loss coefficients from Table 11.4 we have

$$-\frac{\dot{W}_T / 0.8}{9800 \times 0.02} = 60 - 100 + \left(0.5 + 1.0 + 0.022 \frac{300}{0.1} \right) \frac{2.55^2}{2 \times 9.8}$$

$$\therefore \dot{W}_T = 2760 \text{ W} \quad \text{or} \quad 2.76 \text{ kW}.$$

Example 11.15

The pressure drop over a 4-cm-dia, 300-m-long section of pipe is measured to be 120 kPa. If the elevation drops 25 m over that length of pipe and the flow rate is 0.003 m³/s, calculate the friction factor and the power loss.

Solution. The velocity is found to be

$$V = \frac{Q}{A} = \frac{0.003}{\pi \times 0.02^2} = 2.39 \text{ m/s}.$$

The energy equation (11.4.3) with Eq. 11.4.6 then gives

$$-\frac{\dot{W}_S}{\gamma Q}^{\nearrow 0} = \frac{V_2^2 - V_1^2}{2g}^{\nearrow 0} + \frac{p_2 - p_1}{\gamma} + z_2 - z_1 + f\frac{L}{D}\frac{V^2}{2g}$$

$$0 = -\frac{120\,000}{9800} - 25 + f\frac{300}{0.04}\frac{2.39^2}{2 \times 9.8}$$

$$\therefore f = 0.0170, \qquad h_f = 37.2 \text{ m}$$

The power loss is

$$\begin{aligned} \dot{W}_{friction} &= h_f \gamma Q \\ &= 37.2 \times 9800 \times 0.003 = 1095 \text{ W} \end{aligned}$$

where the head loss is analogous to the pump head in Eq. 11.4.11.

Example 11.16

Estimate the loss coefficient for an orifice plate if $A_1/A_0 = 2$.

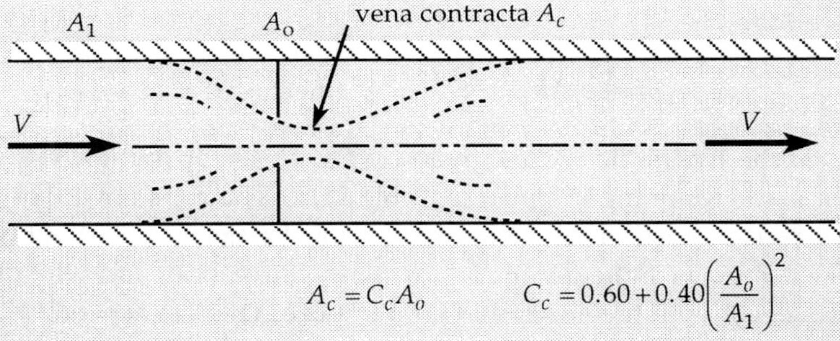

$$A_c = C_c A_o \qquad C_c = 0.60 + 0.40\left(\frac{A_o}{A_1}\right)^2$$

Solution. We approximate the flow situation shown as a gradual contraction up to A_c and a sudden enlargement from A_c back to A_1. The loss coefficient for the contraction is very small so it will be neglected. For the enlargement, we need to know A_c; it is

$$A_c = C_c A_0 = \left[0.6 + 0.4\left(\frac{A_0}{A_1}\right)^2\right] A_0 = \left[0.6 + 0.4\left(\frac{1}{2}\right)^2\right]\frac{A_1}{2} = 0.35 A_1.$$

Using the loss coefficient for an enlargement from Table 11.4, there results

$$h_f = C\frac{V_c^2}{2g}$$

$$= \left(1-\frac{A_c}{A_1}\right)^2 \frac{V_c^2}{2g} = (1-0.35)^2 \frac{1}{0.35^2}\frac{V_1^2}{2g} = 3.45\frac{V_1^2}{2g}$$

where the continuity equation $A_cV_c = A_1V_1$ has been used. The loss coefficient for the orifice plate is thus

$$C = 3.4$$

Note: Two-place accuracy is assumed since C_c is known to only two significant figures.

11.5 Open Channel Flow

If liquid flows down a slope in an open channel at a constant depth, the energy equation (11.4.3) takes the form

$$-\cancelto{0}{\frac{\dot{W}_S}{\gamma Q}} = \cancelto{0}{\frac{V_2^2-V_1^2}{2g}} + \cancelto{0}{\frac{p_2-p_1}{\gamma}} + z_2 - z_1 + h_f \qquad \textbf{(11.5.1)}$$

which shows that the head loss is given by

$$h_f = z_1 - z_2 = LS \qquad \textbf{(11.5.2)}$$

where L is the length of the channel between the two sections and S is the slope. Since we normally have small angles, we can use $S = \tan\theta = \sin\theta = \theta$ where θ is the angle that the channel makes with the horizontal.

The Chezy-Manning equation is used to relate the flow rate to the slope and the cross section; it is

$$Q = \frac{1.0}{n} A R_H{}^{2/3} S^{1/2} \qquad \text{(metric-SI)} \qquad \textbf{(11.5.3)}$$

where R_H is the hydraulic radius given by Eq. 11.4.9, A is the cross sectional area, and n is the Manning n, given in Table 11.5. The constant 1.0 must be replaced by 1.49 if English units are used. The most efficient cross section occurs when the width is twice the depth for a rectangular section, and when the sides make angles of 60° with the horizontal for a trapezoidal cross section.

TABLE 11.5 Average Values* of the Manning *n*

Wall Material	Manning *n*	Wall Material	Manning *n*
Planed wood	.012	Concrete pipe	.015
Unplaned wood	.013	Riveted steel	.017
Finished concrete	.012	Earth, straight	.022
Unfinished concrete	.014	Corrugated metal flumes	.025
Sewer Pipe	.013	Rubble	.03
Brick	.016	Earth with stones and weeds	.035
Cast iron, wrought iron	.015	Mountain streams	.05

*If $R_H > 3$ m, increase n by 15%.

Example 11.17

A 2-m-dia concrete pipe transports water at a depth of 0.8 m. What is the flow rate if the slope is 0.001?

Solution. Calculate the geometric properties:

$$\alpha = \sin^{-1}\frac{0.2}{1.0} = 11.54°$$

$$\therefore \theta = 156.9°.$$

$$A = \pi \times 1^2 \times \frac{156.9}{360} - 0.2 \times \cos 11.54° \times \frac{1}{2} \times 2 = 1.174 \text{ m}^2$$

$$P = \pi \times 2 \times \frac{156.9}{360} = 2.738 \text{ m}.$$

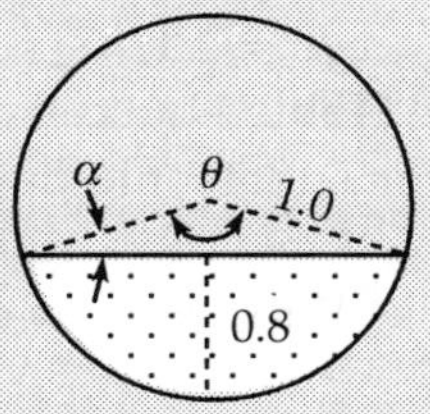

For concrete pipe, $n = 0.015$, so

$$Q = \frac{1.0}{n} A R_H{}^{2/3} S^{1/2} = \frac{1.0}{0.015} \times 1.174 \times \left(\frac{1.174}{2.738}\right)^{2/3} \times 0.001^{1/2} = 1.41 \text{ m}^3/\text{s}.$$

11.6 Compressible Flow

(May not be tested)

A gas flow with a Mach number below 0.3 (at standard conditions this means velocities less than about 100 m/s) can be treated as an incompressible flow, as in previous articles. If the Mach number is greater than 0.3, the density variation must be accounted for. For such problems, we use the control volume equation as

continuity: $$\rho_1 A_1 V_1 = \rho_2 A_2 V_2 \qquad \textbf{(11.6.1)}$$

momentum (x-direction): $$\sum F = \dot{m}(V_2 - V_1) \qquad \textbf{(11.6.2)}$$

energy: $$\frac{\dot{Q} - \dot{W}_s}{\dot{m}} = \frac{V_2{}^2 - V_1{}^2}{2} + C_p(T_2 - T_1) \qquad \textbf{(11.6.3)}$$

where

$$\dot{m} = \rho A V = \text{mass flux} \qquad \textbf{(11.6.4)}$$

$$C_p = \text{constant pressure specific heat}$$

We often use the ideal gas relations (they become inaccurate at high pressure or low temperature)

$$p = \rho R T \qquad C_p = C_v + R \qquad k = C_p/C_v \qquad \textbf{(11.6.5)}$$

The energy equation for an ideal gas takes the form

$$\frac{\dot{Q} - \dot{W}_S}{\dot{m}} = \frac{V_2{}^2 - V_1{}^2}{2} + \frac{k}{k-1}\left(\frac{p_2}{\rho_2} - \frac{p_1}{\rho_1}\right) \qquad \textbf{(11.6.6)}$$

We recall that the speed of sound and Mach number are given by

$$c = \sqrt{kRT} \qquad \mathrm{M} = V/c \tag{11.6.7}$$

Subsonic flow occurs whenever M < 1 and supersonic flow whenever M > 1. In subsonic flows, losses are quite small and isentropic flows can usually be assumed; thus, we can relate the properties by the isentropic relations

$$\frac{T_2}{T_1} = \left(\frac{p_2}{p_1}\right)^{\frac{k-1}{k}} \qquad \frac{p_2}{p_1} = \left(\frac{\rho_2}{\rho_1}\right)^{k} \tag{11.6.8}$$

For air $k = 1.4$, $C_p = 1.00\ \text{kJ/kg}\cdot\text{K}$, $R = 0.287\ \text{kJ/kg}\cdot\text{K}$. The isentropic flow table (Table 11.6) can also be used.

For supersonic flows, shock waves are encountered; across a normal shock wave, Table 11.7 can be used with isentropic flow assumed before and after the shock. If the entropy change is desired, it is given by

$$\Delta s = C_p \ln\frac{T_2}{T_1} - R\ln\frac{p_2}{p_1} \tag{11.6.9}$$

Supersonic flow behaves quite differently from subsonic flow; its velocity increases with increasing area and decreases with decreasing area. Hence, to obtain a supersonic flow from a reservoir, the flow must first accelerate through a converging section to a throat where $M = 1$; then an enlarging section will allow it to reach supersonic speed with $M > 1$. Supersonic flow cannot occur in a converging section only.

Example 11.18

Air is flowing from a 20°C reservoir to the atmosphere through the converging-diverging nozzle shown. What reservoir pressure will locate a normal shock wave at the exit? Also, find V_C and the mass flux.

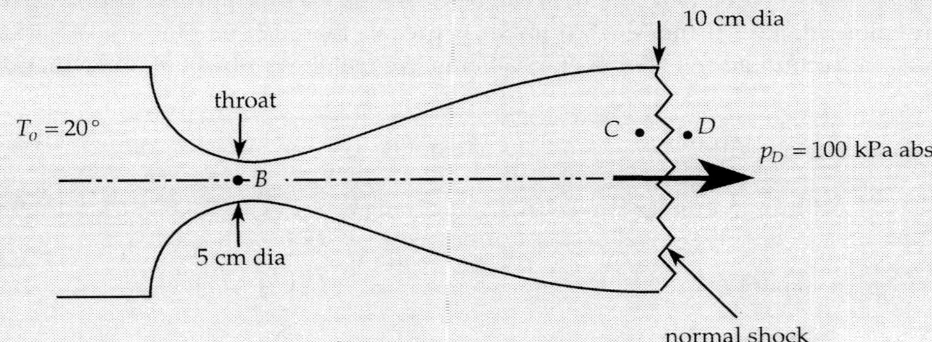

Solution. We know that at the throat the Mach number is unity. Such an area is the *critical area* and is designated A^* in the isentropic flow table. Between points B and C we have

$$\frac{A_C}{A_B} = \frac{A}{A^*} = \frac{\pi \times 5^2}{\pi \times 2.5^2} = 4.$$

Using Table 11.6 we find

$$\mathrm{M}_C = 2.94.$$

Then from Table 11.7 we find, across the shock,

$$\frac{p_C}{p_D} = 9.92 \qquad \therefore p_C = 9.92 \times 100 = 992 \text{ kPa}$$

The isentropic flow Table 11.6 then gives, at M = 2.94,

$$\frac{p_C}{p_o} = 0.0298 \qquad \therefore p_o = \frac{992}{0.0298} = 33\,300 \text{ kPa}$$

To find V_C we must determine T_C. Using Table 11.6 at M = 2.94 we find

$$\frac{T_C}{T_o} = 0.3665 \qquad \therefore T_C = (273 + 20) \times 0.3665 = 107.4 \text{ K.}$$

The velocity is then found to be

$$\begin{aligned} V_C &= \mathrm{M}_C \sqrt{kRT_C} \\ &= 2.94\sqrt{1.4 \times 287 \times 107.4} = 611 \text{ m/s.} \end{aligned}$$

The mass flux is

$$\begin{aligned} \dot{m} &= \rho_C A_C V_C \\ &= \frac{p_C}{RT_C} A_C V_C = \frac{992\,000}{287 \times 107.4} \times \pi \times 0.05^2 \times 611 = 154 \text{ kg/s.} \end{aligned}$$

TABLE 11.6 Isentropic Flow (air, k = 1.4)

M_1	A/A^*	p/p_0	ρ/ρ_0	T/T_0	V/V^*
0.00	∞	1.000	1.000	1.000	0.000
0.05	11.591	0.998	0.999	1.000	0.055
0.10	5.822	0.993	0.995	0.998	0.109
0.15	3.910	0.984	0.989	0.996	0.164
0.20	2.964	0.972	0.980	0.992	0.218
0.25	2.403	0.957	0.969	0.988	0.272
0.30	2.035	0.939	0.956	0.982	0.326
0.35	1.778	0.919	0.941	0.976	0.379
0.40	1.590	0.896	0.924	0.969	0.431
0.45	1.449	0.870	0.906	0.961	0.483
0.50	1.340	0.843	0.885	0.952	0.535
0.55	1.255	0.814	0.863	0.943	0.585
0.60	1.188	0.784	0.840	0.933	0.635
0.65	1.136	0.753	0.816	0.922	0.684
0.70	1.094	0.721	0.792	0.911	0.732
0.75	1.062	0.689	0.766	0.899	0.779
0.80	1.038	0.656	0.740	0.887	0.825
0.85	1.021	0.624	0.714	0.874	0.870
0.90	1.009	0.591	0.687	0.861	0.915
0.95	1.002	0.559	0.660	0.847	0.958
1.00	1.000	0.528	0.634	0.833	1.000
1.10	1.008	0.468	0.582	0.805	1.081
1.20	1.030	0.412	0.531	0.776	1.158
1.30	1.066	0.361	0.483	0.747	1.231
1.40	1.115	0.314	0.437	0.718	1.300
1.50	1.176	0.272	0.395	0.690	1.365
1.60	1.250	0.235	0.356	0.661	1.425
1.70	1.338	0.203	0.320	0.634	1.482
1.80	1.439	0.174	0.287	0.607	1.536
1.90	1.555	0.149	0.257	0.581	1.586
2.00	1.687	0.128	0.230	0.556	1.633
2.10	1.837	0.109	0.206	0.531	1.677
2.20	2.005	0.094	0.184	0.508	1.718
2.30	2.193	0.080	0.165	0.486	1.756
2.40	2.403	0.068	0.147	0.465	1.792
2.50	2.637	0.059	0.132	0.444	1.826
2.60	2.896	0.050	0.118	0.425	1.857
2.70	3.183	0.043	0 106	0.407	1.887
2.80	3.500	0.037	0.0g5	0.389	1.914
2.90	3.850	0.032	0.085	0.373	1.940
3.00	4.235	0.027	0.076	0.357	1.964
3.50	6.790	0.013	0.045	0.290	2.064
4.00	10.719	0.007	0.028	0.238	2.138
4.50	16.562	0.003	0.017	0.198	2.194
5.00	25.000	0.002	0.011	0.167	2.236
5.50	36.869	0.001	0.008	0.142	2.269
6.00	53.180	0.001	0.005	0.122	2.295
6.50	75.134	0.000	0.004	0.106	2.316
7.00	104.143	0.000	0.003	0.093	2.333
7.50	141.842	0.000	0.002	0.082	2.347
8.00	190.110	0.000	0.001	0.072	2.359
8.50	251.086	0.000	0.001	0.065	2.369
8.00	327.189	0.000	0.001	0.058	2.377
9.50	421.130	0.000	0.001	0.052	2.384
10.00	535.936	0.000	0.000	0.048	2.390
∞	∞	0.000	0.000	0.000	∞

TABLE 11.7 Normal Shock Wave (air, $k = 1.4$)

M_1	M_2	p_2/p_1	T_2/T_1	ρ_2/ρ_1	p_{02}/p_{01}
1.00	1.000	1.000	1.000	1.000	1.000
1.05	0.953	1.120	1.033	1.084	1.000
1.10	0.912	1.245	1.065	1.169	0.999
1.15	0.875	1.376	1.097	1.255	0.997
1.20	0.842	1.513	1.128	1.342	0.993
1.25	0.813	1.656	1.159	1.429	0.987
1.30	0.786	1.805	1.191	1.516	0.979
1.35	0.762	1.960	1.223	1.603	0.970
1.40	0.740	2.120	1.255	1.690	0.958
1.45	0.720	2.286	1.287	1.776	0.945
1.50	0.701	2.458	1.320	1.862	0.930
1.55	0.684	2.636	1.354	1.947	0.913
1.60	0.668	2.820	1.388	2.032	0.895
1.65	0.654	3.010	1.423	2.115	0.876
1.70	0.641	3.205	1.458	2.198	0.856
1.75	0.628	3.406	1.495	2.279	0.835
1.80	0.617	3.613	1.532	2.359	0.813
1.85	0.606	3.826	1.569	2.438	0.790
1.90	0.596	4.045	1.608	2.516	0.767
1.95	0.586	4.270	1.647	2.592	0.744
2.00	0.577	4.500	1.687	2.667	0.721
2.05	0.569	4.736	1.729	2.740	0.698
2.10	0.561	4.978	1.770	2.812	0.674
2.15	0.554	5.226	1.813	2.882	0.651
2.20	0.547	5.480	1.857	2.951	0.628
2.25	0.541	5.740	1.901	3.019	0.606
2.30	0.534	6.005	1.947	3.085	0.583
2.35	0.529	6.276	1.993	3.149	0.561
2.40	0.523	6.553	2.040	3.212	0.540
2.45	0.518	6.836	2.088	3.273	0.519
2.50	0.513	7.125	2.137	3.333	0.499
2.55	0.508	7.420	2.187	3.392	0.479
2.60	0.504	7.720	2.238	3.449	0.460
2.65	0.500	8.026	2.290	3.505	0.442
2.70	0.496	8.338	2.343	3.559	0.424
2.75	0.492	8.656	2.397	3.612	0.406
2.80	0.488	8.980	2.451	3.664	0.389
2.85	0.485	9.310	2.507	3.714	0.373
2.90	0.481	9.645	2.563	3.763	0.358
2.95	0.478	9.986	2.621	3.811	0.343
3.00	0.475	10.333	2.679	3.857	0.328
3.50	0.451	14.125	3.315	4.261	0.213
4.00	0.435	18.500	4.047	4.571	0.139
4.50	0.424	23.458	4.875	4.812	0.092
5.00	0.415	29.000	5.800	5.000	0.062
5.50	0.409	35.125	6.822	5.149	0.042
6.00	0.404	41.833	7.941	5.268	0.030
6.50	0.400	49.125	9.156	5.365	0.021
7.00	0.397	57.000	10.469	5.444	0.015
7.50	0.395	65.458	11.879	5.510	0.011
8.00	0.393	74.500	13.387	5.565	0.008
8.50	0.391	84.125	14.991	5.612	0.006
9.00	0.390	94.333	16.693	5.651	0.005
9.50	0.389	105.125	18.492	5.685	0.004
10.00	0.388	116.500	20.387	5.714	0.003
∞	0.378	∞	∞	6.000	0.000

Practice Problems (Metric Units)

(If you choose to work only a few problems, select those with a star.)

GENERAL

*11.1 A fluid is a substance that
a) is essentially incompressible.
b) always moves when subjected to a shearing stress.
c) has a viscosity that always increases with temperature.
d) has a viscosity that always decreases with temperature.
e) expands until it fills its space.

*11.2 Viscosity has dimensions of
a) FT^2/L b) F/TL^2 c) M/LT^2 d) M/LT e) ML/T

*11.3 The viscosity of a fluid varies with
a) temperature.
b) pressure.
c) density.
d) temperature and pressure.
e) temperature, pressure, and density.

*11.4 In an isothermal atmosphere the pressure
a) is constant with elevation.
b) decreases linearly with elevation.
c) cannot be related to elevation.
d) decreases near the surface but approaches a constant value.
e) decreases exponentially with elevation.

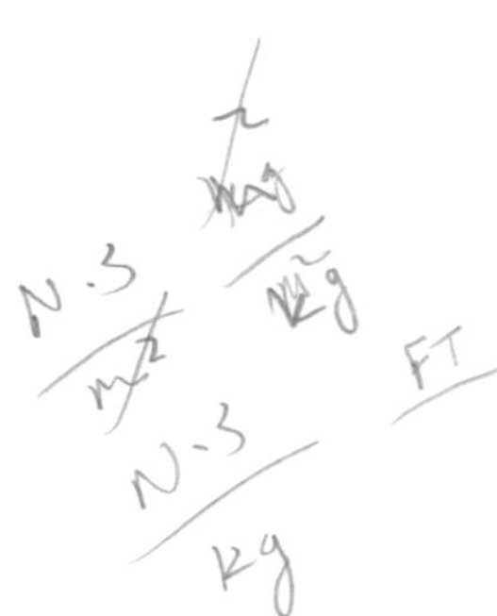

*11.5 A torque of 1.6 N·m is needed to rotate the cylinder at 1000 rad/s. Estimate the viscosity ($N \cdot s/m^2$).
a) 0.1 b) 0.2 c) 0.3
d) 0.4 e) 0.5

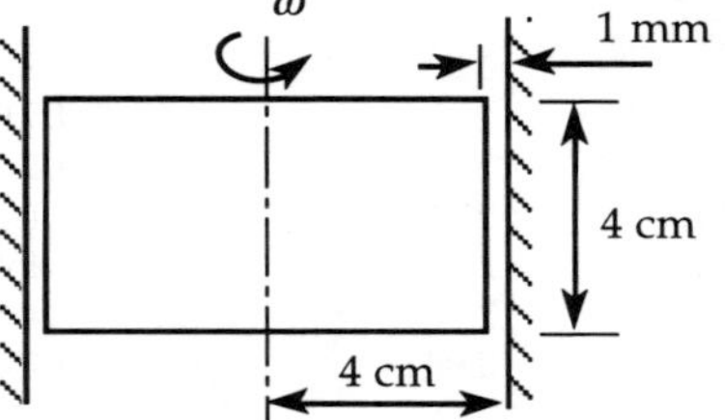

11.6 A pressure of 500 kPa applied to 2 m^3 of liquid results in a volume change of 0.004 m^3. The bulk modulus, in MPa, is
a) 2.5 b) 25 c) 250 d) 2500 e) 2.5×10^6

*11.7 Water at 20°C will rise, in a clean 1-mm-dia glass tube, a distance, in cm, of
a) 1 b) 2 c) 3 d) 4 e) 5

11.8 Water at 20°C flows in a piping system at a low velocity. At what pressure, in kPa abs, will cavitation result?
a) 35.6 b) 20.1 c) 10.6 d) 5.67 e) 2.45

11.9 A man is observed to strike an object and 1.2 s later the sound is heard. How far away, in meters, is the man?
a) 220 b) 370 c) 410 d) 520 e) 640

11.10 The viscosity of a fluid with specific gravity 1.3 is measured to be 0.0034 $\text{N}\cdot\text{s/m}^2$. Its kinematic viscosity, in m^2/s, is

a) 2.6×10^{-6} b) 4.4×10^{-6} c) 5.8×10^{-6} d) 7.2×10^{-6} e) 9.6×10^{-6}

FLUID STATICS

11.11 Fresh water 2 m deep flows over the top of 4 m of salt water $(SG = 1.04)$. The pressure at the bottom, in kPa, is

a) 60.4 b) 58.8 c) 55.2 d) 51.3 e) 47.9

*11.12 What pressure, in kPa, is equivalent to 600 mm of Hg?

a) 100 b) 95.2 c) 80.0 d) 55.2 e) 51.3

11.13 What pressure, in MPa, must be maintained in a diving bell, at a depth of 1200 m, to keep out the ocean water $(SG = 1.03)$?

a) 1.24 b) 5.16 c) 9.32 d) 12.1 e) 14.3

11.14 Predict the pressure, in kPa, at an elevation of 2000 m in an isothermal atmosphere assuming $T = 20°\text{C}$. Assume $p_{atm} = 100$ kPa.

a) 87 b) 82 c) 79 d) 71 e) 63

*11.15 The force F, in Newtons, is

a) 25
b) 8.9
c) 2.5
d) 1.5
e) 0.36

1000 N
F
10 cm dia
oil
SG = 0.86
5 mm dia
20 cm
40 cm

11.16 A U-tube manometer, attached to an air pipe, measures 20 cm of mercury. The pressure, in kPa, in the air pipe is

a) 26.7 b) 32.4 c) 38.6 d) 42.5 e) 51.3

*11.17 The pressure p, in kPa, is

a) 51.3
b) 48.0
c) 45.2
d) 40.0
e) 37.0

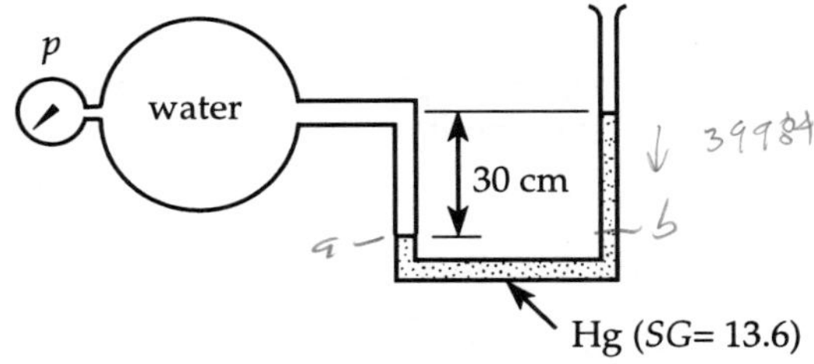

11.18 A 2-m-dia, 3-m-high, cylindrical water tank is pressurized such that the pressure at the top is 20 kPa. The force, in kN, acting on the bottom is

a) 195 b) 176 c) 155 d) 132 e) 106

11.19 The force, in kN, acting on one of the 1.5-m sides of an open cubical water tank (which is full) is

a) 18.2 b) 16.5 c) 15.3 d) 12.1 e) 10.2

*11.20 The force P, in kN, to just open the 3-m-wide gate is

a) 55
b) 60
c) 65
d) 70
e) 75

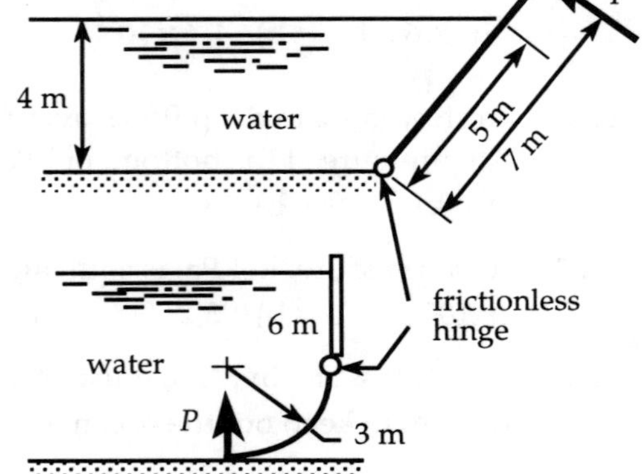

11.21 The force P, in kN, to just open the 4-m-wide gate is

a) 710
b) 762
c) 831 d) 983
e) 1220

*11.22 The force P, in kN, on the 5-m-wide gate is

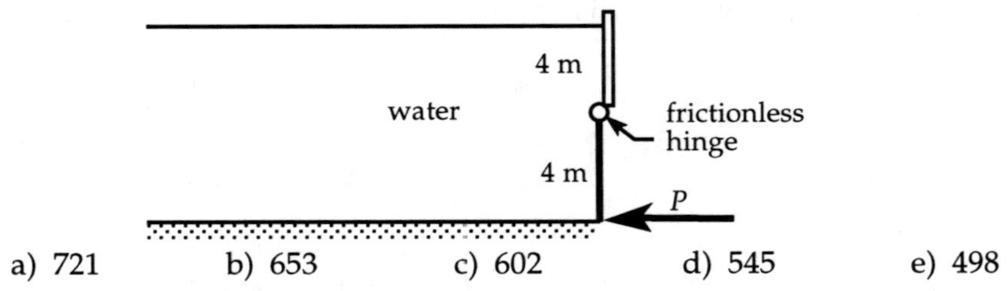

a) 721 b) 653 c) 602 d) 545 e) 498

11.23 Four cars, with a mass of 1500 kg each, are loaded on a 6-m-wide, 12-m-long small-car ferry. How far, in cm, will it sink in the water?
a) 15.2 b) 11.5 c) 10.2 d) 9.6 e) 8.3

*11.24 An object weighs 100 N in air and 25 N when submerged in water. Its specific gravity is
a) 1.11 b) 1.22 c) 1.33 d) 1.44 e) 1.55

11.25 What pressure differential, in Pascals, exists at the bottom of a 3 m vertical wall if the temperature inside is 20°C and outside it is –20°C? Assume equal pressures at the top.
a) 15 b) 12 c) 9 d) 6 e) 3

DIMENSIONLESS PARAMETERS

11.26 Arrange pressure p, flow rate Q, diameter D, and density ρ into a dimensionless group.

a) $pQ^2/\rho D^4$ b) $p/\rho Q^2 D^4$ c) $pD^4\rho/Q^2$ d) $pD^4/\rho Q^2$ e) $p/\rho Q^2$

*11.27 Combine surface tension σ, density ρ, diameter D, and velocity V into a dimensionless para-meter.

a) $\sigma/\rho V^2 D$ b) $\sigma D/\rho V$ c) $\sigma\rho/VD$ d) $\sigma V/\rho D$ e) $\sigma D^2/\rho V$

*11.28 The Reynolds number is a ratio of

a) velocity effects to viscous effects.
b) inertial forces to viscous forces.
c) mass flux to viscosity.
d) flow rate to kinematic viscosity.
e) mass flux to kinematic viscosity.

SIMILITUDE

*11.29 What flow rate, in m^3/s, is needed using a 20:1 scale model of a dam over which 4 m^3/s of water flows?

a) 0.010 b) 0.0068 c) 0.0047 d) 0.0022 e) 0.0015

11.30 It is proposed to model a submarine moving at 10 m/s by testing a 10:1 scale model. What velocity, in m/s, would be needed in the model study?

a) 1 b) 10 c) 40 d) 80 e) 100

11.31 The drag force on a 40:1 scale model of a ship is measured to be 10 N. What force, in kN, is expected on the ship?

a) 640 b) 520 c) 320 d) 160 e) 80

11.32 The power output of a 10:1 scale model of a water wheel is measured to be 20 W. The power output in kW expected from the prototype is

a) 200 b) 150 c) 100 d) 63 e) 2

CONTINUITY

11.33 The velocity in a 2-cm-dia pipe is 20 m/s. If the pipe enlarges to 5-cm-dia, the velocity, in m/s, will be

a) 8.0 b) 6.4 c) 5.2 d) 4.8 e) 3.2

*11.34 A 2-cm-dia pipe transports water at 20 m/s. If it exits out 100 small 2-mm-dia holes, the exiting velocity, in m/s, will be

a) 120 b) 80 c) 40 d) 20 e) 10

11.35 Water flows through a 2-cm-dia pipe at 20 m/s. It then flows radially outward between two discs, 2 mm apart. When it reaches a radius of 40 cm, its velocity, in m/s, will be

a) 5.0 b) 2.5 c) 2.25 d) 1.85 e) 1.25

BERNOULLI'S EQUATION

11.36 The pressure force, in Newtons, on the 15-cm-dia headlight of an automobile traveling at 25 m/s is

a) 10.4 b) 6.8 c) 5.6 d) 4.8 e) 3.2

*11.37 The pressure inside a 4-cm-dia hose is 700 kPa. If the water exits through a 2-cm-dia nozzle, what velocity, in m/s, can be expected inside the hose?

a) 20.4 b) 16.3 c) 12.4 d) 10.5 e) 9.7

11.38 Calculate V, in m/s.

a) 8
b) 7
c) 6
d) 5
e) 4

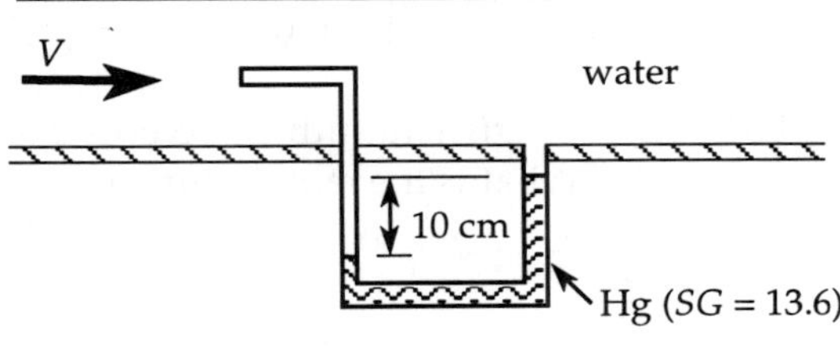

11.39 Water enters a turbine at 900 kPa with negligible velocity. What maximum speed, in m/s, can it reach before it enters the turbine rotor?

a) 52 b) 47 c) 45 d) 42 e) 38

MOMENTUM

*11.40 If the density of the air is 1.2 kg/m^3, find F, in Newtons.

a) 2.4
b) 3.6
c) 4.8
d) 7.6
e) 9.6

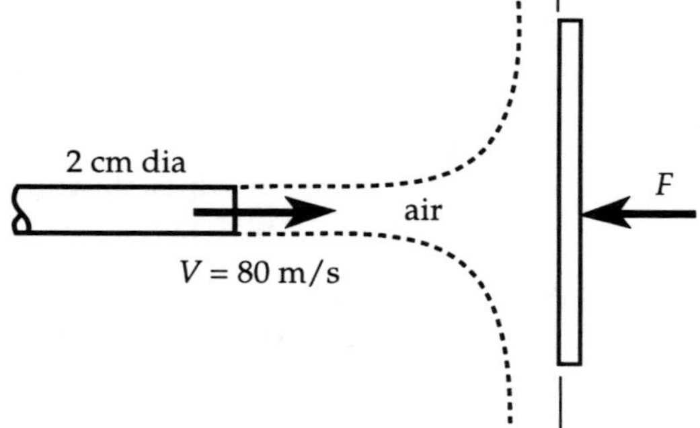

11.41 A rocket exits exhaust gases with $\rho = 0.5$ kg/m^3 out a 50-cm-dia nozzle at a velocity of 1200 m/s. Estimate the thrust, in kN.

a) 420 b) 280 c) 140 d) 90 e) 40

11.42 A high-speed vehicle, traveling at 50 m/s, dips an 80-cm-wide scoop into water and deflects the water 180°. If it dips 5 cm deep, what force, in kN, is exerted on the scoop?

a) 200 b) 150 c) 100 d) 50 e) 25

11.43 What force, in Newtons, acts on the nozzle?

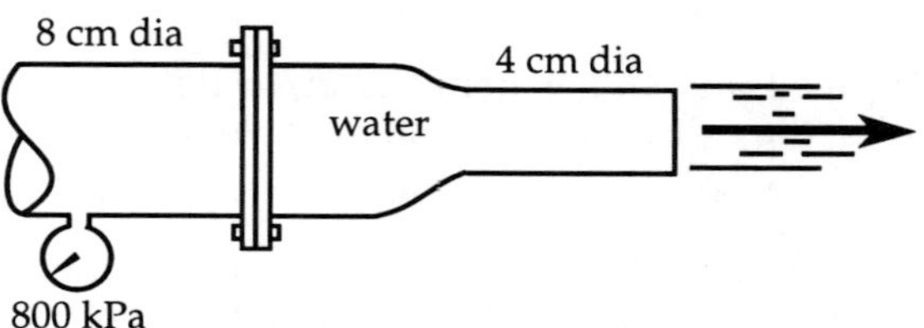

a) 4020 b) 3230 c) 2420 d) 1830 e) 1610

ENERGY

*11.44 The locus of elevations that water will rise in a series of pitot tubes is called

a) the hydraulic grade line.
b) the energy grade line.
c) the velocity head.
d) the pressure head.
e) the head loss.

*11.45 A pressure rise of 500 kPa is needed across a pump in a pipe transporting 0.2 m^3/s of water. If the pump is 85% efficient , the power needed, in kW, would be

a) 118 b) 100 c) 85 d) 65 e) 60

11.46 An 85% efficient turbine accepts 0.8 m^3/s of water at a pressure of 600 kPa. What is the max-imum power output, in kW?

a) 820 b) 640 c) 560 d) 480 e) 410

11.47 If the turbine is 88% efficient, the power output, in kW, is

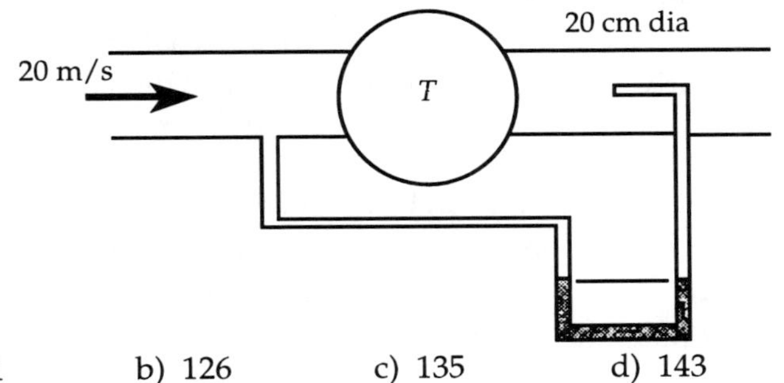

a) 111 b) 126 c) 135 d) 143 e) 176

LOSSES

*11.48 In a completely turbulent flow the head loss
- a) increases with the velocity.
- b) increases with the velocity squared.
- c) decreases with wall roughness.
- d) increases with diameter.
- e) increases with flow rate.

*11.49 The shear stress in a turbulent pipe flow
- a) varies parabolically with the radius.
- b) is constant over the pipe radius.
- c) varies according to the 1/7th power law.
- d) is zero at the center and increases linearly to the wall.
- e) is zero at the wall and increases linearly to the center.

*11.50 The velocity distribution in a turbulent flow in a pipe is often assumed to
- a) vary parabolically.
- b) be zero at the wall and increase linearly to the center.
- c) vary according to the 1/7th power law.
- d) be unpredictable and is thus not used.
- e) be maximum at the wall and decrease linearly to the center.

*11.51 The Moody diagram is sketched. The friction factor for turbulent flow in a smooth pipe is given by curve
- a) *A*
- b) *B*
- c) *C*
- d) *D*
- e) *E*

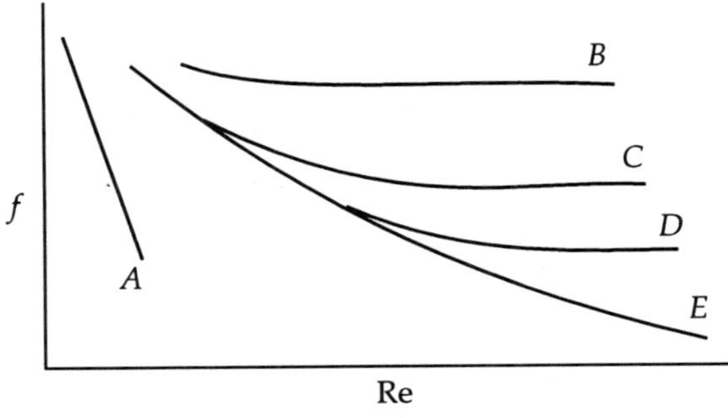

*11.52 For the Moody diagram given in Problem 11.51, the completely turbulent flow is best represented by curve

a) *A* b) *B* c) *C* d) *D* e) *E*

*11.53 The pressure gradient $(\Delta p/\Delta x)$ in a developed turbulent flow in a horizontal constant diameter pipe
a) is constant.
b) varies linearly with axial distance.
c) is zero.
d) decrease exponentially.
e) varies directly with the average velocity.

*11.54 The head loss in a pipe flow can be calculated using
a) the Bernoulli equation.
b) Darcy's law.
c) the Chezy-Manning equation.
d) the Momentum equation.
e) the Darcy-Weisbach equation.

*11.55 Minor losses in a piping system are
a) less than the friction factor losses, $f\frac{L}{D}\frac{V^2}{2g}$.
b) due to the viscous stresses.
c) assumed to vary linearly with the velocity.
d) found by using loss coefficients.
e) independent of the flow rate.

*11.56 In a turbulent flow in a pipe we know the
a) Reynolds number is greater than 10 000.
b) fluid particles move in straight lines.
c) head loss varies linearly with the flow rate.
d) shear stress varies linearly with radius.
e) viscous stresses dominate.

*11.57 Water flows through a 10-cm-dia, 100-m-long pipe connecting two reservoirs with an elevation difference of 40 m. The average velocity is 6 m/s. Neglecting minor losses, the friction factor is
a) 0.020 b) 0.022 c) 0.024 d) 0.026 e) 0.028

*11.58 Find the energy required, in kW, by the 85% efficient pump if $Q = 0.02\,\text{m}^3/\text{s}$.

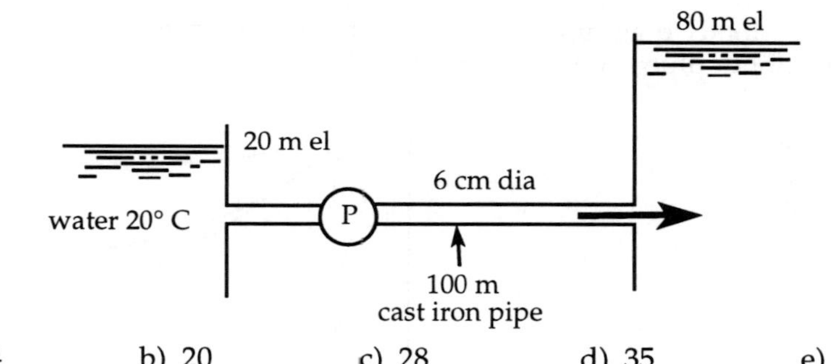

a) 14 b) 20 c) 28 d) 35 e) 44

11.59 The pressure at section A in a 4-cm-dia, wrought-iron, horizontal pipe is 510 kPa. A fully open globe valve, two elbows, and 50 meters of pipe connect section B. If $Q = 0.006\ \text{m}^3/\text{s}$ of water, the pressure p_B, in kPa, is
a) 300 b) 250 c) 200 d) 150 e) 100

11.60 Air at 20°C and 100 kPa abs is transported through 500 m of smooth, horizontal, 15 cm × 40 cm rectangular duct with a flow rate of 0.3 m^3/s. The pressure drop, in Pascals, is
a) 300 b) 400 c) 500 d) 600 e) 700

11.61 Estimate the loss coefficient C in a sudden contraction $A_1/A_2 = 2$ by neglecting the losses up to the vena contracta A_c. Assume that $A_c/A_2 = 0.62 + 0.38(A_2/A_1)^3$ and $h_f = CV_2^2/2g$.
a) 0.40 b) 0.35 c) 0.30 d) 0.25 e) 0.20

11.62 An elbow exists in a 6-cm-dia galvanized iron pipe transporting 0.02 m^3/s of water. Find the equivalent length of the elbow, in meters.
a) 6.3 b) 4.5 c) 3.6 d) 2.8 e) 2.2

OPEN CHANNEL FLOW

*11.63 The depth of water in a 3-m-wide, rectangular, finished concrete channel is 2 m. If the slope is 0.001, estimate the flow rate, in m^3/s.
a) 14 b) 13 c) 12 d) 11 e) 10

11.64 At what depth, in meters, will 10 m^3/s of water flow in a 4-m-wide, rectangular, brick channel if the slope is 0.001?
a) 1.3 b) 1.4 c) 1.5 d) 1.6 e) 1.7

11.65 A 2-m-dia, brick storm sewer transports 10 m^3/s when it's nearly full. Estimate the slope of the sewer.
a) 0.0070 b) 0.0065 c) 0.0060 d) 0.0055 e) 0.0050

COMPRESSIBLE FLOW

*11.66 The pressure and temperature in a 10-cm-dia pipe are 500 kPa abs and 40°C, respectively. What is the mass flux, in kg/s, if the velocity is 100 m/s?
a) 6.63 b) 5.81 c) 5.36 d) 4.92 e) 4.37

*11.67 Air in a reservoir at 20°C and 500 kPa abs exits a hole with a velocity, in m/s, of
a) 353 b) 333 c) 313 d) 293 e) 273

11.68 A farmer uses 20°C nitrogen pressurized to 800 kPa abs . Estimate the temperature, in °C, in the nitrogen as it exits a short hose fitted to the tank.
a) –110 b) –90 c) –70 d) –50 e) –30

11.69 Estimate p_1 in kPa.

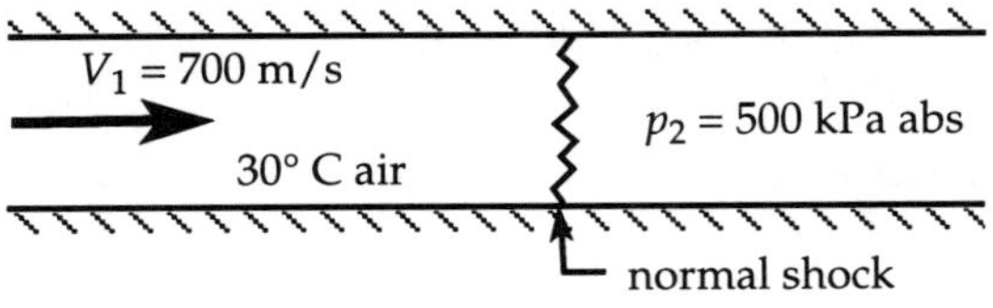

$V_1 = 700$ m/s

30° C air

$p_2 = 500$ kPa abs

normal shock

a) 110 b) 120 c) 130 d) 140 e) 150

11.70 Air leaves a reservoir and accelerates until a shock wave is encountered at a diameter of 10 cm. If the throat diameter is 6 cm, what is the Mach number before the shock wave?
a) 2.03 b) 2.19 c) 2.25 d) 2.40 e) 2.56

11.71 A supersonic aircraft flies at M=2 at an elevation of 1000 m. How long, in seconds, after it passes overhead is the aircraft heard?
a) 2.3 b) 2.5 c) 2.7 d) 2.9 e) 3.1

11.72 Air at 20°C is to exit a nozzle from a reservoir. What maximum pressure, in kPa abs, can the reservoir have if compressibility effects can be neglected? $p_{atm} = 100$ kPa.

a) 115 b) 111 c) 109 d) 106 e) 104

Practice Problems (English Units)

(If you choose to work only a few problems, select those with a star.)

GENERAL

*11.1 A fluid is a substance that

a) is essentially incompressible.
b) always moves when subjected to a shearing stress.
c) has a viscosity that always increases with temperature.
d) has a viscosity that always decreases with temperature.
e) expands until it fills its space.

*11.2 Viscosity has dimensions of

a) FT^2/L b) F/TL^2 c) M/LT^2 d) M/LT e) ML/T

*11.3 The viscosity of a fluid varies with

a) temperature.
b) pressure.
c) density.
d) temperature and pressure.
e) temperature, pressure, an density.

*11.4 In an isothermal atmosphere the pressure

a) is constant with elevation.
b) decreases linearly with elevation.
c) cannot be related to elevation.
d) decreases near the surface but approaches a constant value.
e) decreases exponentially with elevation.

*11.5 A torque of 1.2 ft-lb is needed to rotate the cylinder at 1000 rad/sec. Estimate the viscosity (lb-sec/ft^2).

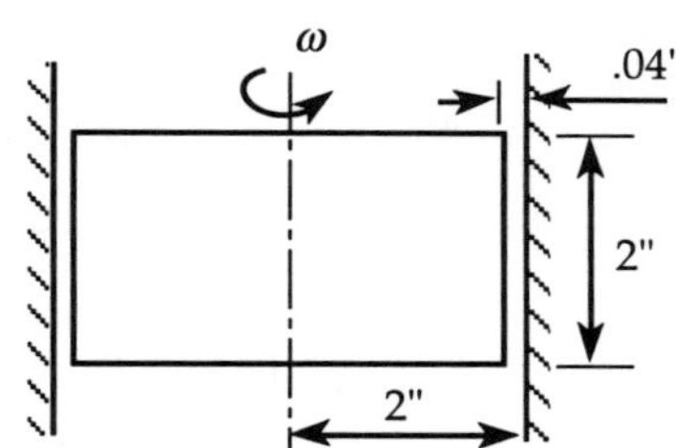

a) 8.25×10^{-4}
b) 7.16×10^{-4}
c) 6.21×10^{-4}
d) 5.27×10^{-4}
e) 4.93×10^{-4}

11.6 A pressure of 80 psi applied to 60 ft^3 of liquid results in a volume change of 0.12 ft^3. The bulk modulus, in psi, is
a) 20,000 b) 30,000 c) 40,000 d) 50,000 e) 60,000

*11.7 Water at 70°F will rise, in a clean 0.04" radius glass tube, a distance, in inches, of
a) 1.21 b) 0.813 c) 0.577 d) 0.401 e) 0.311

11.8 Water at 70°F flows in a piping system at a low velocity. At what pressure, in psia, will cavitation result?
a) 0.79 b) 0.68 c) 0.51 d) 0.42 e) 0.34

11.9 A man is observed to strike an object and 1.2 sec later the sound is heard. How far away, in feet, is the man?
a) 1750 b) 1550 c) 1350 d) 1150 e) 950

11.10 The viscosity of a fluid with specific gravity 1.3 is measured to be 7.2×10^{-5} lb-sec/ft^2. Its kinematic viscosity, in ft^2/sec, is
a) 2.85×10^{-5} b) 1.67×10^{-5} c) 1.02×10^{-5} d) 9.21×10^{-4} e) 8.32×10^{-4}

FLUID STATICS

11.11 Fresh water 6 ft deep flows over the top of 12 ft of salt water $(SG = 1.04)$. The pressure at the bottom, in psi, is
a) 8 b) 9 c) 10 d) 11 e) 12

*11.12 What pressure, in psi, is equivalent to 28" of Hg?
a) 14.6 b) 14.4 c) 14.2 d) 14.0 e) 13.8

11.13 What pressure, in psi, must be maintained in a diving bell, at a depth of 4000 ft, to keep out ocean water $(SG = 1.03)$?
a) 1480 b) 1540 c) 1660 d) 1780 e) 1820

11.14 Predict the pressure, in psi, at an elevation of 6000 ft in an isothermal atmosphere assuming $T = 70°\text{F}$. Assume $p_{atm} = 14.7$ psi.
a) 10.7 b) 11.2 c) 11.9 d) 12.3 e) 12.8

*11.15 The force F, in lbs, is

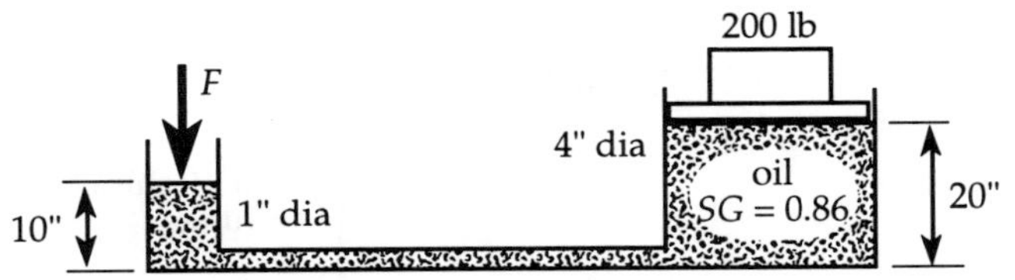

200 lb
F
4" dia
oil
SG = 0.86
20"
10"
1" dia

a) 15.7 b) 13.5 c) 12.7 d) 11.7 e) 10.2

11.16 A U-tube manometer, attached to an air pipe, measures 10" of Hg. The pressure, in psi, in the air pipe is
a) 4.91 b) 4.42 c) 4.01 d) 3.81 e) 3.12

*11.17 The pressure p, in psi, is

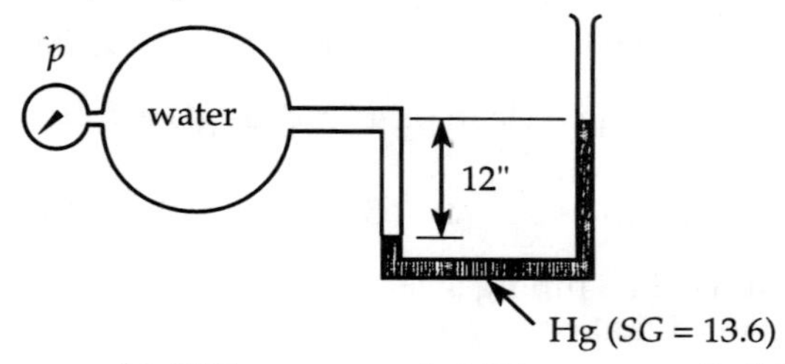

a) 8.32 b) 7.51 c) 6.87 d) 6.21 e) 5.46

11.18 A 2-ft dia, 10-ft high, cylindrical water tank is pressurized such that the pressure at the top is 3 psi. The force, in lb, acting on the bottom is

a) 4250 b) 3960 c) 3320 d) 2780 e) 2210

11.19 The force, in lb, acting on one of the 5-ft sides of an open cubical water tank which is full is

a) 4200 b) 3900 c) 3600 d) 3300 e) 3000

*11.20 The force P, in lb, for the 10-ft wide gate is

a) 17,000
b) 16,000
c) 15,000
d) 14,000
e) 13,000

11.21 The force P, in lb, to just open the 12-ft wide gate is

a) 88,600
b) 82,500
c) 79,100
d) 73,600
e) 57,100

*11.22 The force P, in lb, on the 15-ft wide gate is

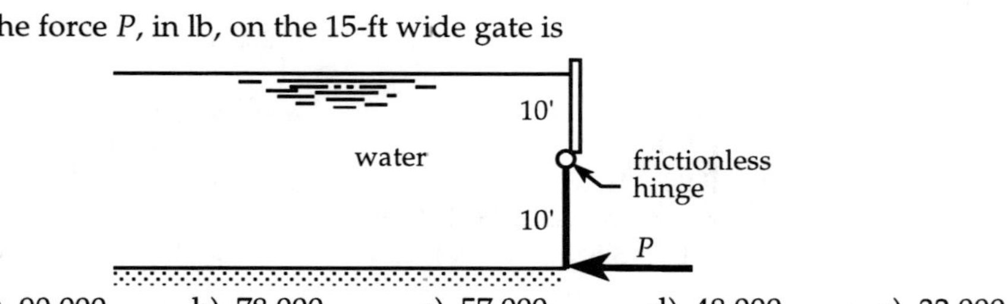

a) 90,000 b) 78,000 c) 57,000 d) 48,000 e) 32,000

11.23 Four cars, with a mass of 3200 lb each, are loaded on a 20-ft-wide, 40-ft-long small-car ferry. How far, in inches, will it sink in the water?

a) 7 b) 6 c) 5 d) 4 e) 3

*11.24 An object weighs 25 lb in air and 6 lb when submerged in water. Its specific gravity is

a) 1.5 b) 1.4 c) 1.3 d) 1.2 e) 1.1

11.25 What pressure differential, in psf, exists at the bottom of a 10-ft-vertical wall if the temperature inside is 70°F and outside is –10°F? Assume equal pressures at the top.
a) 0.478 b) 0.329 c) 0.211 d) 0.133 e) 0.101

DIMENSIONLESS PARAMETERS

11.26 Arrange pressure p, flow rate Q, diameter D, and density ρ into a dimensionless group.
a) $pQ^2/\rho D^4$ b) $p/\rho Q^2 D^4$ c) $pD^4\rho/Q^2$ d) $pD^4/\rho Q^2$ e) $p/\rho Q^2$

*11.27 Combine surface tension σ, density ρ, diameter D, and velocity V into a dimensionless para-meter.
a) $\sigma/\rho V^2 D$ b) $\sigma D/\rho V$ c) $\sigma\rho/VD$ d) $\sigma V/\rho D$ e) $\sigma D^2/\rho V$

*11.28 The Reynolds number is a ratio of
a) velocity effects to viscous effects.
b) inertial forces to viscous forces.
c) mass flux to viscosity.
d) flow rate to kinematic viscosity.
e) mass flux to kinematic viscosity.

SIMILITUDE

*11.29 What flow rate, in ft^3/sec, is needed using a 20:1 scale model of a dam over which 120 ft^3/sec of water flows?
a) 0.20 b) 0.18 c) 0.092 d) 0.067 e) 0.052

11.30 It is proposed to model a submarine moving at 30 ft/sec by testing a 10:1 scale model. what velocity, in ft/sec, would be needed in the model study?
a) 300 b) 400 c) 500 d) 700 e) 900

11.31 The drag force on a 40:1 scale model of a ship is measured to be 2 lb. What force, in lb, is expected on the ship?
a) 128,000 b) 106,000 c) 92,000 d) 80,000 e) 60,000

11.32 The power output of a 10:1 scale model of a water wheel is measured to be 0.06 Hp. The power output, in Hp, expected from the prototype is
a) 130 b) 150 c) 170 d) 190 e) 210

CONTINUITY

11.33 The velocity in a 1″-dia pipe is 60 ft/sec. If the pipe enlarges to 2.5″-dia the velocity, in ft/sec, will be
a) 24.0 b) 20.0 c) 16.2 d) 12.4 e) 9.6

11.34 A 1″-dia pipe transports water at 60 ft/sec. If it exits out 100 small 0.1″-dia holes, the exiting velocity, in ft/sec, will be
a) 180 b) 120 c) 90 d) 60 e) 30

11.35 Water flows through a 1″-dia pipe at 60 ft/sec. It then flows radially outward between two discs, 0.1″ apart. When it reaches a radius of 20″, its velocity, in ft/sec, will be
a) 30 b) 22.5 c) 15.0 d) 7.50 e) 3.75

BERNOULLI'S EQUATION

11.36 The pressure force, in lb, on the 6″-dia headlight of an automobile traveling at 90 ft/sec is
a) 1.02 b) 1.83 c) 2.56 d) 3.75 e) 4.16

*11.37 The pressure inside a 2″-dia hose is 100 psi. If the water exits through a 1″-dia nozzle, what velocity, in ft/sec, can be expected inside the hose?
a) 39.8 b) 37.6 c) 35.1 d) 33.7 e) 31.5

11.38 Calculate V, in ft/sec.

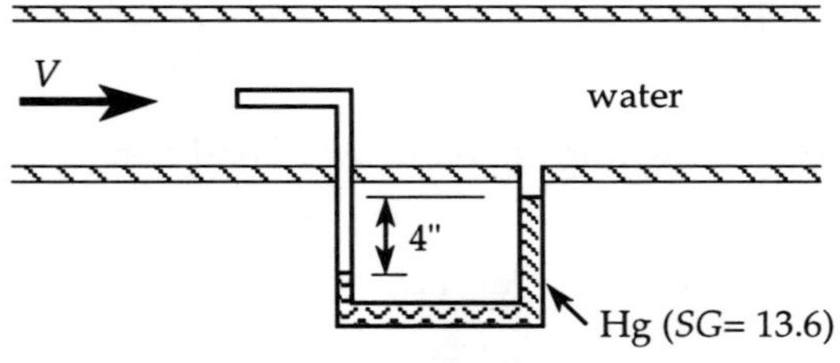

a) 19.7 b) 18.9 c) 17.4 d) 16.4 e) 15.2

11.39 Water enters a turbine at 150 psi with negligible velocity. What maximum speed, in ft/sec, can it reach before it enters the turbine rotor?

a) 119 b) 131 c) 156 d) 168 e) 183

MOMENTUM

*11.40 If the density of the air is 0.0024 slug/ft^3, find F, in lb.

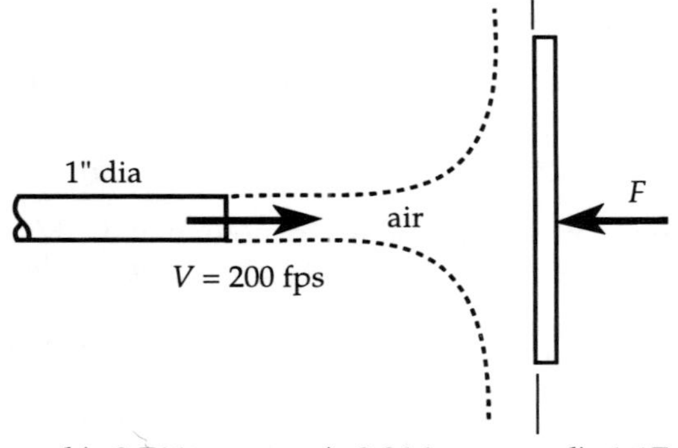

a) 0.524 b) 0.711 c) 0.916 d) 1.17 e) 2.25

11.41 A rocket exits exhaust gases with $\rho = 0.001$ slug/ft^3 out a 20"-dia nozzle at a velocity of 4000 ft/sec. Estimate the thrust, in lb.

a) 24,600 b) 30,100 c) 34,900 d) 36,200 e) 41,600

11.42 A high-speed vehicle, traveling at 150 ft/sec, dips a 30"-wide scoop into water and deflects the water 180°. If it dips 2" deep, what force, in lb, is exerted on the scoop?

a) 36,400 b) 32,100 c) 26,200 d) 22,100 e) 19,900

11.43 What force, in lb, acts on the nozzle?

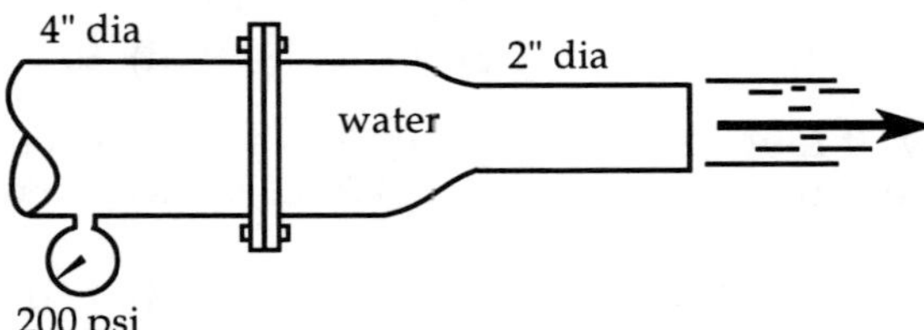

a) 1700 b) 1600 c) 1500 d) 1400 e) 1300

ENERGY

*11.44 The locus of elevations that water will rise in a series of pitot tubes is called

a) the hydraulic grade line.
b) the energy grade line.
c) the velocity head.
d) the pressure head.
e) the head loss.

*11.45 A pressure rise of 75 psi is needed across a pump in a pipe transporting 6 cfs of water. If the pump is 85% efficient, the power needed, in Hp, would be

a) 140 b) 130 c) 120 d) 110 e) 100

11.46 An 85% efficient turbine accepts 3 cfs of water at a pressure of 90 psi. What is the maximum power output, in Hp?

a) 100 b) 90 c) 80 d) 70 e) 60

11.47 If the turbine is 88% efficient, the power output, in Hp, is

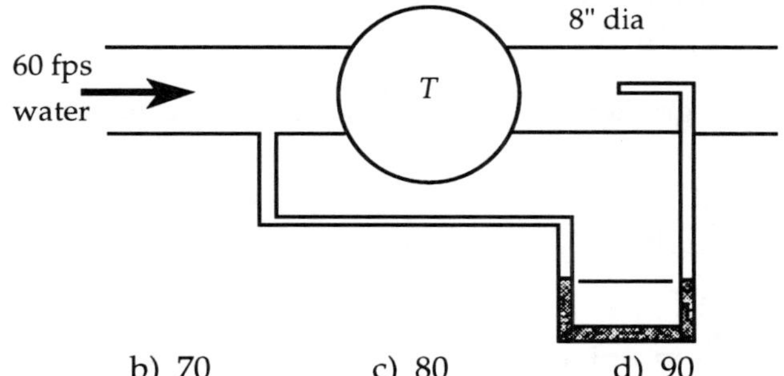

a) 60 b) 70 c) 80 d) 90 e) 120

LOSSES

*11.48 In a completely turbulent flow the head loss

a) increases with the velocity.
b) increases with the velocity squared.
c) decreases with wall roughness.
d) increases with diameter.
e) increases with flow rate.

*11.49 The shear stress in a turbulent pipe flow

a) varies parabolically with the radius.
b) is constant over the pipe radius.
c) varies according to the 1/7th power law.
d) is zero at the center and increases linearly to the wall.
e) is zero at the wall and increases linearly to the center.

*11.50 The velocity distribution in a turbulent flow in a pipe is often assumed to

a) vary parabolically.
b) be zero at the wall and increase linearly to the center.
c) vary according to the 1/7th power law.
d) be unpredictable and is thus not used.
e) be maximum at the wall and decrease linearly to the center.

*11.51 The Moody diagram is sketched. The friction factor for turbulent flow in a smooth pipe is given by curve?

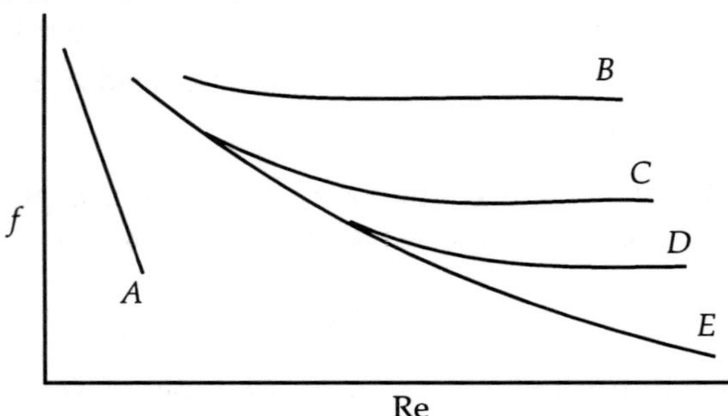

a) A b) B c) C d) D e) E

*11.52 For the Moody diagram given in Problem 11.51, the completely turbulent flow is best represented by curve

a) A b) B c) C d) D e) E

*11.53 The pressure gradient $(\Delta p/\Delta x)$ in a developed turbulent flow in a horizontal pipe

a) is constant.
b) varies linearly with axial distance.
c) is zero.
d) decreases exponentially.
e) varies directly with the average velocity.

*11.54 The head loss in a pipe flow can be calculated using

a) the Bernoulli equation.
b) Darcy's law.
c) the Chezy-Manning equation.
d) the momentum equation.
e) the Darcy-Weisbach equation.

*11.55 Minor losses in a piping system are

a) less than the friction factor losses, $f\frac{L}{D}\frac{V^2}{2g}$.
b) due to the viscous stresses.
c) assumed to vary linearly with the velocity.
d) found by using loss coefficients.
e) independent of the flow rate.

11.56 In a turbulent flow in a pipe we know the

a) Reynolds number is greater than 10,000.
b) fluid particles move in straight lines.
c) head loss varies linearly with flow rate.
d) shear stress varies linearly with radius.
e) viscous stresses dominate.

*11.57 Water flows through a 4"-dia, 300-ft-long pipe connecting two reservoirs with an elevation difference of 120 ft. The average velocity is 20 fps. Neglecting minor losses, the friction factor is

a) 0.0257 b) 0.0215 c) 0.0197 d) 0.0193 e) 0.0182

*11.58 Find the power required, in Hp, by the 85% efficient pump if Q=0.6 cfs.

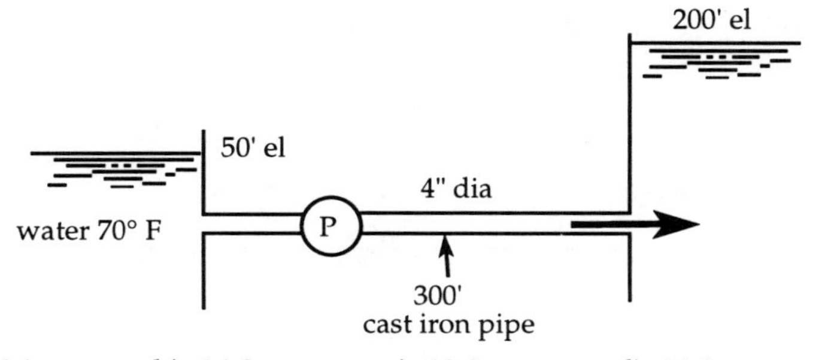

a) 16.1 b) 14.9 c) 13.3 d) 11.2 e) 10.1

11.59 The pressure at section A in a 2"-dia, wrought-iron, horizontal pipe is 70 psi. A fully open globe valve, two elbows, and 150 ft of pipe connect section B. If Q=0.2 cfs of water, the pressure p_B, in psi, is

a) 35 b) 40 c) 45 d) 50 e) 55

11.60 Air at 70°F and 14.7 psia is transported through 1500 ft of smooth, horizontal, 40"x15" rectangular duct with a flow rate of 40 cfs. The pressure drop, in psf, is

a) 2.67 b) 1.55 c) 0.96 d) 0.52 e) 0.17

11.61 Estimate the loss coefficient C in a sudden contraction $A_1/A_2 = 2$ by neglecting the losses up to the vena contracta A_c. Assume that $A_c/A_2 = 0.62 + 0.38\,(A_2/A_1)^3$ and $h_f = CV_2^{\,2}/2g$.

a) 0.28 b) 0.27 c) 0.26 d) 0.25 e) 0.24

11.62 An elbow exists in a 4"-dia galvanized iron pipe transporting 0.6 cfs of water. Find the equivalent length of the elbow in ft.

a) 9.7 b) 10.3 c) 11.0 d) 12.1 e) 13.6

OPEN CHANNEL FLOW

*11.63 The depth of water in a 10-ft-wide, rectangular, finished concrete channel is 6 ft. If the slope is 0.001, estimate the flow rate, in cfs.

a) 460 b) 350 c) 290 d) 280 e) 270

11.64 At what depth, in ft, will 300 cfs of water flow in a 12-ft-wide, rectangular, brick channel if the slope is 0.001?

a) 7.5 b) 6.5 c) 5.5 d) 4.5 e) 3.5

11.65 A 6-ft-dia, brick storm sewer transports 100 cfs when it's nearly full. Estimate the slope of the sewer.

a) 0.0008 b) 0.002 c) 0.003 d) 0.004 e) 0.005

COMPRESSIBLE FLOW

*11.66 The pressure and temperature in a 4"-dia pipe are 70 psia and 100°F, respectively. What is the mass flux, in slug/sec, if the velocity is 300 fps?

a) 2.43 b) 1.02 c) 0.763 d) 0.487 e) 0.275

*11.67 Air in a tank at 70°F and 75 psia exits a hole with a velocity, in fps, of

a) 1050 b) 1000 c) 960 d) 920 e) 900

11.68 A farmer uses 70°F nitrogen pressurized to 120 psia. Estimate the temperature, in °F, in the nitrogen as it exits a short hose fitted to the tank.

a) –170 b) –150 c) –130 d) –110 e) –90

11.69 Estimate p_1, in psia.

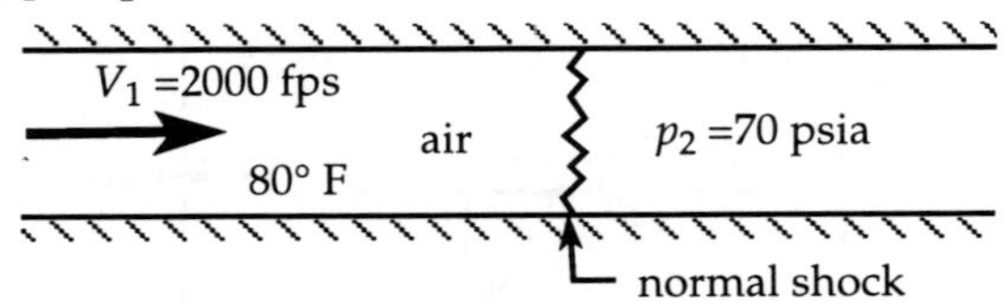

a) 20 b) 25 c) 30 d) 35 e) 40

11.70 Air leaves a reservoir and accelerates until a shock wave is encountered at a diameter of 4″. If the throat diameter is 3″, what is the Mach number before the shock wave?

a) 2.28 b) 2.22 c) 2.18 d) 2.12 e) 2.07

11.71 A supersonic aircraft flies at M=2 at an elevation of 3000 ft. How long, in seconds, after it passes overhead is the aircraft heard? Assume T = 70°F.

a) 2.4 b) 2.3 c) 2.2 d) 2.1 e) 2.0

11.72 Air at 70°F is to exit a nozzle from a reservoir. What maximum pressure, in psia, can the reservoir have if compressibility effects can be neglected? $p_{atm} = 14.7$ psi.

a) 15.0 b) 15.2 c) 15.4 d) 15.6 e) 15.8

Solutions to Practice Problems

(Metric Units)

11.1 b) a) is true for a liquid and low speed gas flows. c) and e) are true of gases. d) is true for a liquid.

11.2 d) $\tau = \mu du/dy. \quad \therefore \mu = \tau/du/dy. \quad [\mu] = \dfrac{F/L^2}{\frac{L}{T}/L} = \dfrac{FT}{L^2} = \dfrac{\left(ML/T^2\right)T}{L^2} = \dfrac{M}{LT}$

11.3 a) Viscosity μ varies with temperature only.

11.4 e) $dp = -\gamma dz = -\rho g dz. \quad p = \rho RT$ (ideal gas)

$\therefore \ dp = -\dfrac{p}{RT} g dz$ or $\dfrac{dp}{p} = -\dfrac{g}{RT} dz.$

$\int \dfrac{dp}{p} = -\dfrac{g}{RT} \int dz. \quad \therefore \ \ln p = -Cz. \quad \therefore \ p = e^{-Cz}.$

11.5 a) $T = \tau A r = \mu \dfrac{du}{dy} A r = \mu \dfrac{r\omega}{t} 2\pi r L r.$

$1.6 = \mu \dfrac{0.04 \times 1000}{0.001} 2\pi \times 0.04 \times 0.04 \times 0.04. \quad \therefore \ \mu = 0.1 \text{ N}\cdot\text{s/m}^2$

11.6 c) $K = -\forall \dfrac{\Delta p}{\Delta \forall} = -2 \dfrac{500}{-0.004} = 250\ 000$ or 250 MPa.

11.7 c) $\sigma 2\pi D = \gamma \pi r^2 L$

$0.0736 \times \pi \times 0.001 = 9800\pi \times 0.0005^2 L$

$\therefore \ L = 0.03$ m or 3 cm.

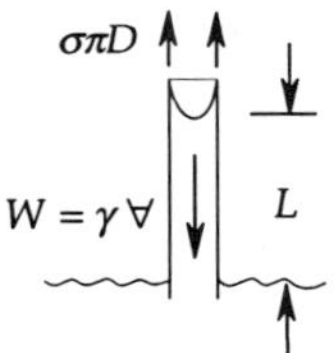

11.8 e) Cavitation occurs when the pressure reaches the vapor pressure = 2.45 kPa. (See Table 11.2.)

11.9 c) $L = V\Delta t = \sqrt{kRT}\Delta t = \sqrt{1.4 \times 287 \times 288} \times 1.2 = 408$ m.

Assume $T = 15°$C. T must be in absolute. $\therefore \ T = 288$.

11.10 a) $\nu = \dfrac{\mu}{\rho} = \dfrac{0.0034}{1.3 \times 1000} = 2.6 \times 10^{-6}.$

11.11 a) $p = \gamma_1 \Delta h_1 + \gamma_2 \Delta h_2$ where $\gamma_2 = SG \times \gamma_{H_2O}$

$= 9800 \times 2 + 1.04 \times 9800 \times 4 = 60\ 400$ Pa or 60.4 kPa.

11.12 c) $p = \gamma h$ where $\gamma_{Hg} = 13.6\ \gamma_{H_2O}$

$= 13.6 \times 9800 \times 0.6 = 80\ 000$ Pa.

11.13 d) $p = \gamma h = 1.03 \times 9800 \times 1200 = 12.1 \times 10^6$ Pa or 12.1 MPa.

11.14 c) $dp=-\gamma dz=-\rho g dz=-p\frac{g}{RT}dz$ using $p=\rho RT$

$\therefore \frac{dp}{p}=-\frac{g}{RT}dz.\quad \int_{100}^{p} dp/p=-\frac{g}{RT}\int_{0}^{2000} dz.$

$\therefore \ln\frac{p}{100}=-\frac{9.8}{287\times 293}\times 2000.\quad \therefore\ p=79.2\text{ kPa}.$

11.15 c) $F=pA.\ \ p=\frac{1000}{\pi\times 0.05^2}+0.86\times 9800\times 0.2=129\ 000\text{ Pa}.$

$\therefore\ F=129\ 000\times\pi\times 0.0025^2=2.53\text{ N}.$

11.16 a) $p=\gamma h=(13.6\times 9800)\times 0.2=26\ 700\text{ Pa}$ or 26.7 kPa.

11.17 e) $p+9800\times 0.3=13.6\times 9800\times 0.3.\quad \therefore p=37\ 000\text{ Pa}.$

11.18 c) $F=pA=(20\ 000\times 9800\times 3)\pi\times 1^2=155\ 000\text{ N}.$

11.19 b) $F=p_cA=9800\times\frac{1.5}{2}\times 1.5^2=16\ 500\text{ N}.$

11.20 d) $7P=\frac{5}{3}F=\frac{5}{3}\gamma h_cA.\quad \therefore\ P=\frac{5}{21}\times 9800\times 2\times 15=70\ 000\text{ N}.$

11.21 d) All pressures on the curved section pass through the center. Moments about the hinge give
$P=F_v=\gamma\times\text{Volume}=9800\times(9\pi/4)\times 4+9800\times 6\times 3\times 4=983\ 000\text{ N}.$

11.22 b) $y_p=y_c+\frac{I_c}{y_c}=6+\frac{5\times 4^3/12}{6\times 20}=6.22\text{ m}.$

$4P=F\times 2.22=\gamma h_cA\times 2.22.\quad \therefore\ P=9800\times 6\times 20\times 2.22/4=653\ 000\text{ N}.$

11.23 e) $W=\gamma V\!\!\!\!/.\quad 4\times 1500\times 9.8=9800\times 6\times 12\times h.\quad \therefore\ h=0.0833\text{ m}.$

11.24 c) $25=100-9800V\!\!\!\!/.\quad \therefore\ V\!\!\!\!/=0.00765\text{ m}^3.$

$100=9800(SG)\times 0.00765.\quad \therefore\ SG=1.33.$

11.25 d) $\Delta p=\Delta\gamma\times h=\left(\frac{1}{253}-\frac{1}{293}\right)\times\frac{100}{0.287}\times 3\times 9.8=5.53\text{ Pa}.$

11.26 d) $[p]=\frac{M}{LT^2}\quad [Q]=\frac{L^3}{T}\quad [D]=L\quad [\rho]=\frac{M}{L^3}.$

First, eliminate M, then T, then L:

$$\frac{M}{LT^2}\cdot\frac{L^3}{M}\cdot\frac{T^2}{L^6}\cdot L^4=p\cdot\frac{1}{\rho}\cdot\frac{1}{Q^2}\cdot D^4=\frac{pD^4}{\rho Q^2}.$$

11.27 a) $[\sigma]=\frac{M}{T^2}\quad [\rho]=\frac{M}{L^3}\quad [D]=L\quad [V]=\frac{L}{T}.$ Combine: $\frac{\sigma}{\rho}\frac{1}{V^2}\frac{1}{D}$

11.28 b) Inertial force to viscous forces.

11.29 d) $(\text{Fr})_m = (\text{Fr})_p$. $\therefore \frac{V_m^2}{l_m g} = \frac{V_p^2}{l_p g}$. $\therefore \frac{V_m^2}{V_p^2} = \frac{1}{20}$.

$Q_m^* = Q_p^*$ or $\frac{Q_m}{V_m l_m^2} = \frac{Q_p}{V_p l_p^2}$. $\therefore Q_m = 4 \times \frac{1}{20^2} \times \frac{1}{\sqrt{20}} = 0.0022$.

11.30 a) $\text{Re}_m = \text{Re}_p$. $\left(\frac{Vl}{\nu}\right)_m = \left(\frac{Vl}{\nu}\right)_p$. $\therefore \frac{V_m}{V_p} = \frac{l_p}{l_m} = 10$.

$\therefore V_m = 10\, V_p = 10 \times 10 = 100$ m/s.

11.31 a) $\text{Fr}_m = \text{Fr}_p$. $\therefore \left(\frac{V^2}{lg}\right)_m = \left(\frac{V^2}{lg}\right)_p$. $\therefore \frac{V_p^2}{V_m^2} = \frac{\ell_p}{\ell_m}$.

$(F_D)_m^* = (F_D)_p^*$ or $\frac{(F_D)_m}{\rho_m V_m^2 l_m^2} = \frac{(F_D)_p}{\rho_p V_p^2 l_p^2}$. $\therefore (F_D)_p = 10 \frac{\rho_p}{\rho_m} \frac{V_p^2}{V_m^2} \frac{l_p^2}{l_m^2} = 10 \frac{l_p^3}{l_m^3}$.

$\therefore (F_D)_p = 10 \times 40^3 = 640\,000$ N.

11.32 d) $\text{Fr}_m = \text{Fr}_p$. $\left(\frac{V^2}{lg}\right)_m = \left(\frac{V^2}{lg}\right)_p$. $\therefore \frac{V_p^2}{V_m^2} = 10$.

$\dot{W}_p^* = \dot{W}_m^*$. $\frac{\dot{W}_m}{\rho_m V_m^3 l_m^2} = \frac{\dot{W}_p}{\rho_p V_p^3 l_p^2}$.

$\therefore \dot{W}_p = \frac{V_p^3}{V_m^3} \frac{l_p^2}{l_m^2} \dot{W}_m = 10^3 \sqrt{10} \times 20 = 63\,250$ W.

11.33 e) $V_2 = 20\pi \times 2^2 / \pi \times 5^2 = 3.2$ m/s.

11.34 d) $V_2 = 20\pi \times 1^2 / 100 \times \pi \times 0.1^2 = 20$ m/s.

11.35 e) $V_2 \times 2\pi \times 40 \times 0.2 = 20 \times \pi \times 1^2$. $\therefore V_2 = 1.25$ m/s.

11.36 b) $p = \rho V^2/2 = 1.23 \times 25^2/2 = 384$ Pa. $F = pA = 384 \times \pi \times 0.075^2 = 6.79$ N.

11.37 e) $V_2 A_2 = V_1 A_1$. $\therefore V_2 = V_1 \times 4^2/2^2 = 4V_1$.

$\frac{p_1}{\rho} + \frac{V_1^2}{2} = \cancelto{0}{\frac{p_2}{\rho}} + \frac{V_2^2}{2}$. $\frac{700\,000}{1000} + \frac{V_1^2}{2} = \frac{16V_1^2}{2}$. $\therefore V_1 = 9.66$ m/s

11.38 d) $p + \rho \frac{V^2}{2} + 9800 \times 0.1 = p + 13.6 \times 9800 \times 0.1$.

$\therefore V^2 = 12.6 \times 9800 \times 0.1 \times 2 / 1000$. $\therefore V = 4.97$ m/s.

11.39 c) Cavitation results if $p_2 = -100$ kPa.

$\frac{p_1}{\rho} + \cancelto{0}{\frac{V_1^2}{2}} = \frac{p_2}{\rho} + \frac{V_2^2}{2}$. $900\,000/1000 = -100\,000/1000 + V_2^2/2$. $\therefore V_2 = 44.7$ m/s.

11.40 a) $F = \rho AV^2 = 1.2 \times \pi \times 0.01^2 \times 80^2 = 2.41.$

11.41 c) $F = \rho AV^2 = .5 \times \pi \times 0.25^2 \times 1200^2 = 141\ 000$ N.

11.42 a) $-F = \rho AV(-V - V)$. $\therefore F = 2\rho AV^2$.

$\therefore F = 2 \times 1000 \times 0.05 \times 0.8 \times 50^2 = 200\ 000$ N.

11.43 c) $p_1 A_1 - F = \rho A_1 V_1 (V_2 - V_1)$. $V_2 = 4V_1$. $\frac{p_1}{\rho} + \frac{V_1^2}{2} = \frac{\overset{0}{\cancel{p_2}}}{\rho} + \frac{V_2^2}{2} = \frac{16V_1^2}{2}$.

$\therefore V_1 = \sqrt{\frac{800\ 000}{1000} \times \frac{2}{15}} = 10.3.$ $V_2 = 41.2$.

$\therefore F = 800\ 000 \times \pi \times 0.04^2 - 1000 \times \pi \times 0.04^2 \times 10.3 \times (41.2 - 10.3) = 2420$ N.

11.44 b) The energy grade line.

11.45 a) $\dot{W}_P = \gamma Q \frac{\Delta p}{\gamma} \Big/ 0.85 = 0.2 \times 500 / 0.85 = 117.6$ kW.

11.46 e) $\dot{W}_T = \gamma Q \frac{\Delta p}{\gamma} \times 0.85 = 0.8 \times 600 \times 0.85 = 408$ kW.

11.47 a) manometer: $p_1 = p_2 + \rho V_2^2 / 2$. $-\frac{\dot{W}_T}{\gamma Q} = \frac{V_2^2}{2g} + \frac{p_2}{\gamma} - \frac{V_1^2}{2g} - \frac{p_1}{\gamma}$ (100% efficient)

$\therefore \dot{W}_T = Q \frac{V_2^2}{2} \rho \eta = \left(20 \times \pi \times 0.1^2\right) \frac{20^2}{2} \times 1000 \times .88 = 111\ 000$ W.

11.48 b) Increases with the velocity squared.

11.49 d) Increases linearly to the wall.

11.50 c) Vary as the 1/7th power law.

11.51 e) By curve E.

11.52 b) By curve B.

11.53 a) Pressure varies linearly. $\therefore \frac{\Delta p}{\Delta x} = \text{Const.}$

11.54 e) The Darcy-Weisbach equation.

11.55 d) Found by using loss coefficients.

11.56 d) Shear stress varies linearly with radius.

11.57 b) $h_f = f \frac{L}{D} \frac{V^2}{2g}$. $\therefore f = 40 \frac{0.1}{100} \frac{2 \times 9.8}{6^2} = 0.0218.$

11.58 e) $V = Q/A = \frac{0.02}{\pi \times 0.03^2} = 7.07$ m/s. $\text{Re} = \frac{VD}{\nu} = 7.07 \times 0.06 / 10^{-6} = 4.2 \times 10^5$.

$\frac{e}{D} = \frac{.26}{60} = .0043$. From Fig. 11.2 $f = 0.03$.

$$\dot{W}_p = \frac{\gamma Q}{\eta}\left(\frac{V_2^2}{2g}^{0} + \frac{p_2}{\gamma}^{0} + z_2 - \frac{V_1^2}{2g}^{0} - \frac{p_1}{\gamma}^{0} - z_1 + f\frac{L}{D}\frac{V^2}{2g} + C\frac{V^2}{2g}\right)$$

$$= \frac{9800\times 0.02}{0.85}\left[80-20+\left(0.03\frac{100}{.06}+1+.5\right)\frac{7.07^2}{2\times 9.8}\right] = 44\ 000\ \text{W}.$$

11.59 e) $V = Q/A = \dfrac{0.006}{\pi\times 0.03^2} = 4.77\ \text{m/s}$. $\text{Re} = \dfrac{VD}{\nu} = 4.77\times .04/10^{-6} = 1.9\times 10^5$.

$\dfrac{e}{D} = \dfrac{0.046}{40} = 0.0011$. From Fig. 4.2 $f = 0.022$. $0 = \dfrac{p_B - p_A}{\gamma} + f\dfrac{L}{D}\dfrac{V^2}{2g} + C\dfrac{V^2}{2g}$.

$\therefore\ p_B = 510\ 000 - \left(.022\dfrac{50}{.04} + 6.4 + 2\times .9\right)\dfrac{4.77^2}{2\times 9.8}\times 9800 = 104\ 000\ \text{kPa}$.

11.60 e) $V = Q/A = 0.3/(.15\times .4) = 5\ \text{m/s}$. $R_H = \dfrac{40\times 15}{110} = 5.45\ \text{cm}$.

$\text{Re} = \dfrac{5\times 4\times .0545}{1.6\times 10^{-5}} = 6.8\times 10^4$. With $\dfrac{e}{D} = 0,\ f = 0.02$.

$\Delta p = f\dfrac{L}{4R_H}\dfrac{V^2}{2g}\gamma = .02\dfrac{500}{4\times .0545}\dfrac{5^2}{2}\times 1.23 = 705\ \text{Pa}$.

11.61 d) $A_c/A_2 = .62 + .38(.5)^3 = 0.668$. $C_1 = (1-.668)^2 = 0.11$.

$0.11\ V_c^2/2g = C\ V_2^2/2g$. $\therefore\ C = 0.11\left(\dfrac{A_2}{A_c}\right)^2 = .11\times\dfrac{1}{.668^2} = 0.25$.

11.62 e) $V = Q/A = 0.02/\pi\times .03^2 = 7.07\ \text{m/s}$. $e/D = .15/60 = .0025$

$\text{Re} = \dfrac{7.07\times .06}{10^{-6}} = 4.2\times 10^5$. $\therefore\ f = 0.024$.

$L_e = CD/f = 0.9\times 0.06/0.025 = 2.16\ \text{m}$.

11.63 a) $Q = \dfrac{1}{n}AR_H{}^{2/3}S^{1/2} = \dfrac{1}{.012}\times 6\times .86^{2/3}\times .001^{1/2} = 14.3$

where $R_H = 6/(3+4) = 0.86\ \text{m}$.

11.64 b) $Q = \dfrac{1}{n}AR_H{}^{2/3}S^{1/2} = \dfrac{1}{.016}4h\left(\dfrac{4h}{4+2h}\right)^{2/3}\times .001^{1/2} = 10$.

Trial-and-error: $h = 1.4\ \text{m}$.

11.65 b) $Q = \dfrac{1}{n}AR_H{}^{2/3}S^{1/2} = \dfrac{1}{.016}\pi\times 1^2\times .5^{2/3}S^{1/2} = 10$, where $R_H = \dfrac{A}{P} = \dfrac{\pi\times 1^2}{2\pi} = .5$.

$\therefore\ S = 0.00654$.

11.66 e) $\rho = p/RT = 500/.287\times 313 = 5.57\ \text{kg/m}^3$

$\dot{m} = \rho AV = 5.57\times\pi\times .05^2\times 100 = 4.37\ \text{kg/s}$

11.67 e) $T_e = T_o(p_e/p_o)^{k-1/k} = 293\left(\frac{100}{500}\right)^{.286} = 185\,\text{K}.$

$V = \text{M}c = 1\sqrt{1.4 \times 287 \times 185} = 273\ \text{m/s}.$

11.68 a) $T_e = T_o(p_e/p_o)^{k-1/k} = 293\left(\frac{100}{800}\right)^{.286} = 162\,\text{K}. \quad \therefore T_e = 162 - 273 = -111°\text{C}.$

11.69 a) $\text{M}_1 = \frac{V_1}{c_1} = \frac{700}{\sqrt{1.4 \times 287 \times 303}} = 2.01 \quad \therefore \frac{p_2}{p_1} = 4.54$ (from Normal Shock Table)

$\therefore p_1 = p_2/4.54 = 500/4.54 = 110$ kPa

11.70 e) $\frac{A}{A^*} = \frac{10^2}{6^2} = 2.78. \quad \therefore 2.5 < \text{M}_1 < 2.6$ (Isentropic flow Table)

11.71 b) $\sin\phi = \frac{1}{\text{M}} = \frac{1}{2}. \quad \therefore \phi = 30°$

$\tan 30° = \frac{1000}{L}. \quad \therefore L = 1732$ m

$\Delta t = \frac{L}{V} = \frac{1732}{2\sqrt{1.4 \times 287 \times 293}} = 2.52$ sec.

V
1000 m
ϕ
L

11.72 d) Assume $\text{M}_e = 0.3$, the maximum if the density is assumed constant (i.e., $\rho_e = 0.97\rho_o$). $V_e = \text{M}_e c_e = 0.3\sqrt{1.4 \times 287\, T_e}$.

$\therefore V_e^2 = 36.2\, T_e$. energy: $0 = \frac{V_e^2 - V_o^2}{2} + C_p(T_e - T_o).$

$36.2 T_e = 2 \times 1000(293 - T_e). \quad \therefore T_e = 287.8$

$p_o = p_e(T_o/T_e)^{\frac{k}{k-1}} = 100\left(\frac{293}{287.8}\right)^{\frac{1.4}{.4}} = 106.$

Solutions to Practice Problems

(English Units)

11.1 b) a) is true for a liquid and low speed gas flows. c) and e) are true of gases. d) is true for a liquid.

11.2 d) $\tau = \mu\, du/dy. \quad \therefore \mu = \frac{\tau}{du/dy}. \quad [\mu] = \frac{F/L^2}{\frac{L}{T}/L} = \frac{FT}{L^2} = \frac{(ML/T^2)T}{L^2} = \frac{M}{LT}.$

11.3 a) Viscosity μ varies with temperature only.

11.4 e) $dp = -\gamma dz = -\rho g dz.\ \ p = \rho RT$ (ideal gas)

$\therefore\ dp = -\frac{p}{RT} g dz$ or $\frac{dp}{p} = -\frac{g}{RT} dz.$

$\int \frac{dp}{p} = -\frac{g}{RT} \int dz.\ \ \therefore\ \ln p = -Cz.\ \ \therefore\ p = e^{-Cz}$

11.5 a) $T = \tau A r = \mu \frac{du}{dy} A r = \mu \frac{r\omega}{t} 2\pi r L r.$

$1.2 = \frac{2/12 \times 1000}{.04/12} \times 2\pi \times \frac{2}{12} \times \frac{2}{12} \times \frac{2}{12} \mu.\ \ \therefore\ \mu = 8.25 \times 10^{-4}$

11.6 c) $K = -\forall \frac{\Delta p}{\Delta \forall} = -60 \frac{80}{0.12} = 40{,}000$ psi

11.7 c) $\sigma 2\pi r = \gamma \pi r^2 L$

$0.005 \times 2\pi \times \frac{.04}{12} = 62.4\pi \frac{.04^2}{144} L$

$\therefore\ L = 0.0481'$ or $0.577''$

$\sigma\pi D$

$W = \gamma\forall$

L

11.8 e) Cavitation occurs when the pressure reaches the vapor pressure = 0.34 psi abs. (See Table 11.1.)

11.9 c) $L = V\Delta t = \sqrt{kRT}\Delta t = \sqrt{1.4 \times 53.3 \times 32.2 \times 530} \times 1.2 = 1354'$

Assume $T = 70°F = 530°R$

11.10 a) $\nu = \frac{\mu}{\rho} = \frac{7.2 \times 10^{-5}}{1.3 \times 1.94} = 2.85 \times 10^{-5}$. ρ must be in slug/ft^3.

11.11 a) $p = \gamma_1 \Delta h_1 + \gamma_2 \Delta h_2 = 62.4 \times 6 + (62.4 \times 1.04) \times 12 = 1153$ psf or 8.01 psi.

11.12 c) $p = \gamma h = (13.6 \times 62.4) \times 28/12 = 1980$ psf or 13.75 psi.

11.13 d) $p = \gamma h = (1.03 \times 62.4) \times 4000 = 257{,}000$ psf or 1785 psi.

11.14 c) $dp = -\gamma dz = -\rho g dz = -p \frac{g}{RT} dz$ using $p = \rho RT$

$\therefore\ \frac{dp}{p} = -\frac{g}{RT} dz.\ \ \int_{14.7}^{p} dp/p = -\frac{g}{RT} \int_0^{6000} dz.$

$\therefore\ \ln \frac{p}{14.7} = -\frac{32.2}{53.3 \times 32.2 \times 530} \times 6000.\ \ \therefore\ p = 11.89$ psia.

11.15 c) $F = pA.\ \ p = \frac{200}{\pi \times 2^2} + 0.86 \times 62.4 \times \frac{10}{12} \Big/ 144 = 16.2$ psi.

$\therefore\ F = 16.2 \times \pi \times (1/2)^2 = 12.74$ lb.

11.16 a) $p = \gamma h = (13.6 \times 62.4) \times \frac{10}{12} \Big/ 144 = 4.91$ psi.

11.17 e) $p + 62.4 \times 1 = 13.6 \times 62.4 \times 1.\ \ \therefore p = 786$ psf or $p = 5.46$ psi.

11.18 c) $F = pA = (3\times 144 + 62.4\times 10)\pi \times 1^2 = 3318$ lb.

11.19 b) $F = p_c A = 62.4\times \frac{5}{2}\times 5^2 = 3900$ lb.

11.20 d) $20P = \frac{15}{3}(62.4\times 6\times 150)$. $\therefore\ P = 14{,}040$ lb.

11.21 d) All pressures on the curved section pass through the center. Moments about the hinge give
$P = F_v = \gamma\,\text{Volume} = 62.4\times 12\times \pi 6^2/4 + 62.4\times 6\times 8\times 12 = 57{,}000$ lb.

11.22 b) $y_p = y_c + \frac{I_c}{y_c A} = 15 + \frac{15\times 10^3/12}{150\times 15} = 15.56'$. $10P = 62.4\times 15\times 150\times 5.56$.

$\therefore\ P = 78{,}000$ lb.

11.23 e) $W = \gamma V\!\!\!/$. $4\times 3200 = 62.4\times 20\times 40h$ $\therefore\ h = .256'$ or $3.08''$.

11.24 c) $6 = 25 - 62.4 V\!\!\!/$. $\therefore\ V\!\!\!/ = 0.3045\ \text{ft}^3$. $25 = 62.4(SG)\times 0.3045$.

$\therefore\ SG = 1.316$.

11.25 d) $\Delta p = \Delta\gamma\times h = \left(\frac{1}{450} - \frac{1}{530}\right)\frac{14.7\times 144}{53.3}\times 10 = 0.1332$ psf.

11.26 d) $[p] = \frac{M}{LT^2}$ $[Q] = \frac{L^3}{T}$ $[D] = L$ $[\rho] = \frac{M}{L^3}$. First, eliminate M, then T,

then L: $\frac{M}{LT^2}\cdot\frac{L^3}{M}\cdot\frac{T^2}{L^6}\cdot L^4 = p\cdot\frac{1}{\rho}\cdot\frac{1}{Q^2}\cdot D^4 = \frac{pD^4}{\rho Q^2}$.

11.27 a) $[\sigma] = \frac{M}{T^2}$ $[\rho] = \frac{M}{L^3}$ $[D] = L$ $[V] = \frac{L}{T}$.

$$\frac{M}{T^2}\cdot\frac{T^2}{L^2}\cdot\frac{L^3}{M}\cdot\frac{1}{L} = \sigma\cdot\frac{1}{V^2}\cdot\frac{1}{\rho}\cdot\frac{1}{D} = \frac{\sigma}{\rho D V^2}.$$

11.28 b) Inertial force to viscous forces.

11.29 d) $(\text{Fr})_m = (\text{Fr})_p$. $\therefore\ \frac{V_m^2}{l_m g} = \frac{V_p^2}{l_p g}$. $\therefore\ \frac{V_m^2}{V_p^2} = \frac{1}{20}$.

$Q_m^* = Q_p^*$ or $\frac{Q_m}{V_m l_m^2} = \frac{Q_p}{V_p l_p^2}$. $\therefore\ Q_m = 120\times\frac{1}{20^2}\times\frac{1}{\sqrt{20}} = .0671$ cfs.

11.30 a) $\text{Re}_m = \text{Re}_p$. $\left(\frac{Vl}{\nu}\right)_m = \left(\frac{Vl}{\nu}\right)_p$. $\therefore\ \frac{V_m}{V_p} = \frac{l_p}{l_m} = 10$.

$\therefore\ V_m = 10\,V_p = 10\times 30 = 300$ fps.

11.31 a) $\text{Fr}_m = \text{Fr}_p$. $\therefore \left(\frac{V^2}{\ell g}\right)_m = \left(\frac{V^2}{\ell g}\right)_p$. $\therefore \frac{V_p^2}{V_m^2} = \frac{\ell_p}{\ell_m}$.

$(F_D)^*_m = (F_D)^*_p$ or $\frac{(F_D)_m}{\rho_m V_m^2 \ell_m^2} = \frac{(F_D)_p}{\rho_p V_p^2 \ell_p^2}$. $\therefore (F_D)_p = 2\frac{\rho_p}{\rho_m}\frac{V_p^2}{V_m^2}\frac{\ell_p^2}{\ell_m^2} = 2\frac{\ell_p^3}{\ell_m^3}$.

$\therefore (F_D)_p = 2 \times 40^3 = 128{,}000$ lb.

11.32 d) $\text{Fr}_m = \text{Fr}_p$. $\left(\frac{V^2}{lg}\right)_m = \left(\frac{V^2}{lg}\right)_p$. $\therefore \frac{V_p^2}{V_m^2} = 10$.

$\dot{W}_p^* = \dot{W}_m^*$. or $\frac{\dot{W}_m}{\rho_m V_m^3 l_m^2} = \frac{\dot{W}_p}{\rho_p V_p^3 l_p^2}$.

$\therefore \dot{W}_p = .06\frac{V_p^3}{V_m^3}\frac{l_p^2}{l_m^2}\dot{W}_m = .06 \times 10\sqrt{10} \times 10^2 = 189.7$ Hp.

11.33 e) $V_2 = 60\pi \times \left(\frac{1}{2}\right)^2 \Big/ \pi \times 1.25^2 = 9.6$ fps.

11.34 d) $V_2 = 60\pi \times \left(\frac{1}{2}\right)^2 \Big/ 100\pi \times .05^2 = 60$ fps.

11.35 e) $V_2 \times 0.1 \times 2\pi \times 20 = 60 \times \pi \times \left(\frac{1}{2}\right)^2$. $\therefore V_2 = 3.75$ fps.

11.36 b) $p = \rho V^2/2 = .0023 \times 90^2/2 = 9.32$ psf. $F = pA = 9.32\pi \times 3^2/144 = 1.83$ lb.

11.37 e) $V_2 A_2 = V_1 A_1$. $\therefore V_2 = V_1 \times 2^2/1^2 = 4V_1$.

$\frac{p_1}{\rho_1} + \frac{V_1^2}{2} = \frac{p_2}{\rho} + \frac{V_2^2}{2}$ (with $p_2 = 0$). $\frac{100 \times 144}{1.94} + \frac{V_1^2}{2} = \frac{16V_1^2}{2}$. $\therefore V_1 = 31.5$ fps.

11.38 d) $p + \rho\frac{V^2}{2} + 62.4 \times \frac{4}{12} = p + 13.6 \times 62.4 \times \frac{4}{12}$. Using $\rho = 1.94$ slug/ft^3,

$V = 16.44$ fps.

11.39 c) Cavitation results if $p_2 = -14.7$ psi.

$p_1/\rho + V_1^2/2 = p_2/\rho + V_2^2/2$ (with $V_1 = 0$).

$150 \times 144/1.94 = -14.7 \times 144/1.94 + V_2^2/2$. $\therefore V_2 = 156$ fps.

11.40 a) $F = \rho A V^2 = .0024 \times \pi \times \frac{.5^2}{144} \times 200^2 = 0.524$ lb.

11.41 c) $F = \rho A V^2 = .001 \times \pi \times \frac{10^2}{144} \times 4000^2 = 34{,}900$ lb.

11.42 a) $-F = \rho AV(-V - V). \quad \therefore F = 2\rho AV^2.$

$\therefore F = 2 \times 1.94 \times \frac{2 \times 30}{144} \times 150^2 = 36{,}400$ lb.

11.43 c) $V_2 = 4V_1. \quad \frac{p_1}{\rho} + \frac{V_1^2}{2} = \frac{p_2}{\rho}^{0} + \frac{V_2^2}{2} = \frac{16V_1^2}{2}. \quad \therefore V_1^2 = \frac{2p_1}{15\rho}.$

$\therefore V_1 = \sqrt{\frac{2 \times 200 \times 144}{15 \times 1.94}} = 44.5. \quad \therefore V_2 = 178.$

$p_1 A_1 - F = \rho A_1 V_1 (V_2 - V_1).$

$\therefore F = 200\pi \times 2^2 - 1.94 \times \pi \times \frac{2^2}{144} \times 44.5 \times 133.5 = 1508$ lb.

11.44 b) The energy grade line.

11.45 a) $\dot{W}_P = \gamma Q \frac{\Delta p}{\gamma} \Big/ 0.85 = 6 \times (75 \times 144)/.85 = 76{,}200$ ft-lb/sec or 139 Hp.

11.46 e) $\dot{W}_T = \gamma Q \frac{\Delta p}{\gamma} \times 0.85 = 3 \times (90 \times 144) \times 0.85 = 33{,}050$ ft-lb/sec or 60.1 Hp.

11.47 a) manometer: $p_1 = p_2 + \rho V_2^2/2.$

$-\frac{\dot{W}_T}{\gamma Q} = \frac{V_2^2}{2g} + \frac{p_2}{\gamma} - \frac{V_1^2}{2g} - \frac{p_1}{\gamma}$ (100% efficient)

$\therefore \dot{W}_T = Q \frac{V_2^2}{2} \rho\eta = \left(60 \times \pi \times \frac{4^2}{144}\right) \frac{60^2}{2} \times 1.94 \times .88 = 64{,}400$ ft-lb/sec or 117 Hp.

11.48 b) Increases with the velocity squared.

11.49 d) Increases linearly to the wall.

11.50 c) Vary as the 1/7th power law.

11.51 e) By curve E.

11.52 b) By curve B.

11.53 a) Pressure varies linearly $\therefore \frac{\Delta p}{\Delta x} = \text{Const.}$

11.54 e) The Darcy-Weisbach equation.

11.55 d) Found by using loss coefficients.

11.56 d) Shear stress varies linearly with radius.

11.57 b) $h_f = f \frac{L}{D} \frac{V^2}{2g}. \quad \therefore f = 120 \frac{4/12}{300} \frac{2 \times 32.2}{20^2} = .0215$

11.58 c) $V = Q/A = \dfrac{0.6}{\pi \times 2^2/144} = 6.875 \text{ fps}.$

$\text{Re} = \dfrac{VD}{\nu} = 6.875 \times \frac{4}{10}\Big/10^{-5} = 2.3 \times 10^5 \quad \dfrac{e}{D} = \dfrac{.00085}{4/12} = .0025$

from Fig. 11.2 $f = .02$

$\dot{W}_p = \dfrac{\gamma Q}{\eta}\left(\dfrac{V_2^2}{2g} + \dfrac{p_2}{\gamma} + z_2 - \dfrac{V_1^2}{2g} - \dfrac{p_1}{\gamma} - z_1 + f\dfrac{L}{D}\dfrac{V^2}{2g} + C\dfrac{V^2}{2g}\right)$ (with $\frac{V_2^2}{2g}$, $\frac{p_2}{\gamma}$, $\frac{V_1^2}{2g}$, $\frac{p_1}{\gamma}$ struck out as 0)

$= \dfrac{62.4 \times .6}{.85}\left[200 - 50 + \left(.02\dfrac{300}{4\sqrt{12}} + 1 + .5\right)\dfrac{6.875^2}{64.4}\right] = 73.0 \text{ ft-lb/sec} \quad \text{or } 13 \text{ Hp}.$

11.59 e) $V = Q/A = \dfrac{.02}{\pi \times 1^2/144} = 9.167 \text{ fps}$

$\text{Re} = \dfrac{VD}{\nu} = 9.167 \times \frac{2}{12}\Big/10^{-5} = 1.53 \times 10^5. \quad \dfrac{e}{D} = \dfrac{.00015}{2/12} = .0009$

From Fig. 4.2 $f = .021. \quad 0 = \dfrac{p_B - p_A}{\gamma} + f\dfrac{L}{D}\dfrac{V^2}{2g} + C\dfrac{V^2}{2g}.$

$\therefore p_B = 70 - \left(.021\dfrac{150}{2/12} + 6.4 + 2 \times .9\right)\dfrac{9.167^2}{2 \times 32.2} \times 62.4/144 = 54.7 \text{ psi}.$

11.60 b) $V = Q/A = 40\Big/\left(\frac{40}{12} \times \frac{15}{12}\right) = 9.6 \text{ fps}. \quad R_H = \dfrac{40 \times 15}{110} = 5.455''.$

$\text{Re} = \dfrac{4 \times 9.6 \times 5.45/12}{1.6 \times 10^{-4}} = 1.10 \times 10^5. \quad \text{With } \dfrac{e}{D} = 0, \; f = 0.0175.$

$\gamma = p/RT = \dfrac{14.7 \times 144}{53.3 \times 530} = .075$

$\Delta p = f\dfrac{L}{4R_H}\dfrac{V^2}{2g}\gamma = .0175\dfrac{1500}{4 \times 5.45/12}\dfrac{9.6^2}{64.4} \times .075 = 1.55 \text{ psf}.$

11.61 d) $A_c/A_2 = .62 + .38(.5)^3 = .6675. \quad C_1 = (1 - .6675)^2 = .111.$

$.111\, V_c^2/2g = C\, V_2^2/2g. \quad \therefore C = .111(A_2/A_c)^2 = .111/.6675^2 = .249.$

11.62 e) $V = Q/A = .6/\left(\pi \times 2^2/144\right) = 6.875. \quad e/D = .0005\Big/\frac{1}{3} = .0015.$

$\text{Re} = \dfrac{6.875 \times 4/12}{10^{-5}} = 2.3 \times 10^5. \quad \therefore f = .022. \quad L_e = CD/f = .9 \times \frac{4}{12}\Big/.022 = 13.6'.$

11.63 a) $Q = \dfrac{1.49}{n}AR_H^{2/3}S^{1/2} = \dfrac{1.49}{.012}60 \times 2.73^{2/3} \times .001^{1/2} = 460 \text{ cfs}$

where $R_H = 60/22 = 2.73'$.

11.64 b) $Q = \dfrac{1.49}{n}AR_H^{2/3}S^{1/2} = \dfrac{1.49}{.016}12h\left(\dfrac{12h}{12 + 2h}\right)^{2/3} \times .001^{1/2} = 300 \text{ cfs}.$

Trial-and-error: $h = 4.52'$.

11.65 b) $Q = \dfrac{1.49}{n}AR_H^{2/3}S^{1/2} = \dfrac{1.49}{.016}\pi \times 3^2 \times 1.5^{2/3}S^{1/2} = 100,$

where $R_H = \dfrac{A}{P} = \dfrac{\pi r^2}{2\pi r} = \dfrac{r}{2} = 1.5'. \quad \therefore S = .00084.$

11.66 e) $\rho = p/RT = 70\times 144/53.3\times 32.2\times 560 = .0105\ \text{slug/ft}^3$

$$\dot{m} = \rho AV = .0105\times\frac{\pi\times 2^2}{144}\times 300 = 0.275\ \text{slug/sec}.$$

11.67 e) $T_e = T_o(p_e/p_o)^{k-1/k} = 530\left(\frac{14.7}{75}\right)^{1.4-1/1.4} = 339°\text{R}.$

$$V = \text{M}c = 1\sqrt{1.4\times 53.3\times 32.2\times 540} = 903\ \text{fps}.$$

11.68 a) $T_e = T_o(p_e/p_o)^{k-1/k} = 530\left(\frac{14.7}{120}\right)^{.2857} = 291°\text{R}\quad \text{or}\ 291-460 = -169°\text{F}.$

11.69 a) $\text{M}_1 = \frac{V_1}{c_1} = \frac{2000}{\sqrt{1.4\times 53.3\times 32.2\times 540}} = 1.76.$

$$\therefore \frac{p_2}{p_1} = 3.44\ \text{(from Normal Shock Table)}.$$

$$\therefore p_1 = p_2/3.44 = 70/3.44 = 20.3\ \text{psia}$$

11.70 e) $\frac{A}{A^*} = \frac{4^2}{3^2} = 1.78.\quad \therefore 2.0 < \text{M}_1 < 2.1\ \text{(Isentropic flow Table)}.$

11.71 b) $\sin\phi = \frac{1}{\text{M}} = \frac{1}{2}.\quad \therefore \phi = 30°$

$$\tan 30° = \frac{3000}{L}.\quad \therefore L = 5196'$$

$$\Delta t = \frac{L}{V} = 2\frac{5196}{\sqrt{1.4\times 53.3\times 32.2\times 530}} = 2.3\ \text{sec}.$$

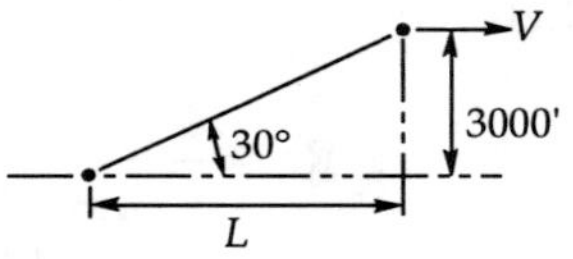

11.72 d)

Assume $\text{M}_e = 0.3$, the maximum is the density if assumed constant (i.e., $\rho_e = 0.97\rho_o$).
$V_e = \text{M}_e c_e = 0.3\sqrt{1.4\times 53.3\times 32.2 T_e}\quad \therefore V_e^2 = 216 T_e.$

energy: $0 = \frac{V_e^2 - V_o^2 \nearrow^0}{2} + C_p(T_e - T_o).$

unit conversions $\frac{\text{BTU}}{\text{lb}} \rightarrow \frac{\text{ft - lb}}{\text{slug}}$

$$\therefore \frac{216 T_e}{2} = 0.24\times(778\times 32.2)(530 - T_e).\quad \therefore T_e = 520.6$$

$$p_o = p_e(T_o/T_e)^{\frac{k}{k-1}} = 14.7\left(\frac{530}{520.6}\right)^{\frac{1.4}{.4}} = 15.6\ \text{psia}.$$

FE/EIT Final Practice Exam

Do not look or even glance at the material in this practice examination until you're ready to sit for it. Spend at most four hours on each part.

You may use only the Reference Handbook supplied by NCEES. One is bound with this book; another will be sent to you after you register for the exam. (You may also order additional copies by calling Great Lakes Press at 314-273-6016.) If you are studying without the NCEES Handbook, use the equation summaries provided in this manual to familiarize yourself with the most important equations.

Battery operated silent calculators are also allowed. Indicate your answers on the separate answer sheet. Be sure and answer all questions. Guess at those you do not have time to work through!

You may write on scratch paper now, but remember, scratch paper is not allowed during the actual examination; large margins will provide sufficient work space.

To score yourself, add your correct responses from the afternoon part to one half of your correct responses from the morning part (a maximum score of 140). If you score less than 70 (50%) you may have difficulty with the actual examination. If your score is substantially greater than 70, you should be ready to pass the exam! If your score is marginal, study those areas that have a large numbers of problems in which you scored the lowest.

tear out answer sheet

Morning Session Answer Sheet

Math/Computers

1. A B C D E
2. A B C D E
3. A B C D E
4. A B C D E
5. A B C D E
6. A B C D E
7. A B C D E
8. A B C D E
9. A B C D E
10. A B C D E
11. A B C D E
12. A B C D E
13. A B C D E
14. A B C D E
15. A B C D E
16. A B C D E
17. A B C D E
18. A B C D E
19. A B C D E
20. A B C D E

score ____

Materials

21. A B C D E
22. A B C D E
23. A B C D E
24. A B C D E
25. A B C D E
26. A B C D E
27. A B C D E
28. A B C D E
29. A B C D E
30. A B C D E
31. A B C D E
32. A B C D E
33. A B C D E
34. A B C D E

score______

Chemistry

35. A B C D E
36. A B C D E
37. A B C D E
38. A B C D E
39. A B C D E
40. A B C D E
41. A B C D E
42. A B C D E
43. A B C D E
44. A B C D E
45. A B C D E
46. A B C D E
47. A B C D E
48. A B C D E

score_______

Economics

49. A B C D E.
50. A B C D E
51. A B C D E
52. A B C D E
53. A B C D E
54. A B C D E
55. A B C D E
56. A B C D E
57. A B C D E
58. A B C D E
59. A B C D E

score_______

Electrical Theory

60. A B C D E
61. A B C D E
62. A B C D E
63. A B C D E
64. A B C D E
65. A B C D E
66. A B C D E
67. A B C D E
68. A B C D E
69. A B C D E
70. A B C D E
71. A B C D E
72. A B C D E
73. A B C D E

score______

Thermodynamics

74. A B C D E
75. A B C D E
76. A B C D E
77. A B C D E
78. A B C D E
79. A B C D E
80. A B C D E
81. A B C D E
82. A B C D E
83. A B C D E
84. A B C D E
85. A B C D E
86. A B C D E
87. A B C D E

score_______

Statics

88. A B C D E
89. A B C D E
90. A B C D E
91. A B C D E
92. A B C D E
93. A B C D E
94. A B C D E
95. A B C D E
96. A B C D E
97. A B C D E
98. A B C D E
99. A B C D E
100. A B C D E
101. A B C D E

score______

Dynamics

102. A B C D E
103. A B C D E
104. A B C D E
105. A B C D E
106. A B C D E
107. A B C D E
108. A B C D E
109. A B C D E
110. A B C D E
111. A B C D E
112. A B C D E
113. A B C D E
114. A B C D E
115. A B C D E

score_______

Mechanics of Materials

116. A B C D E
117. A B C D E
118. A B C D E
119. A B C D E
120. A B C D E
121. A B C D E
122. A B C D E
123. A B C D E
124. A B C D E
125. A B C D E
126. A B C D E

score______

Fluid Mechanics

127. A B C D E
128. A B C D E
129. A B C D E
130. A B C D E
131. A B C D E
132. A B C D E
133. A B C D E
134. A B C D E
135. A B C D E
136. A B C D E
137. A B C D E
138. A B C D E
139. A B C D E
140. A B C D E

score______

Morning Session—Practice Exam

4 hours maximum

Each of the 140 problems is followed by five answers. Select the response that is best and circle the corresponding letter on the answer sheet.

---

1. The equation of a straight line passing through (1,4) and (4,1) is
(A) $x - y = 3$ (B) $x + y = 5$ (C) $x + y = 3$ (D) $x - y = 5$ (E) $x + y = 4$

2. The radius of a circle whose equation is $x^2 - 6(x + y) + y^2 = -2$ is
(A) 1 (B) 2 (C) 3 (D) 4 (E) 5

3. Evaluate the determinant of this matrix: $\begin{bmatrix} 0 & 1 & 0 \\ 2 & 3 & 3 \\ 3 & 4 & 6 \end{bmatrix}$
(A) 16 (B) 8 (C) 3 (D) 0 (E) –3

4. Find the value of this limit: $\lim_{x \to 0} \frac{\sin x}{x}$
(A) 0 (B) 1/2 (C) 1 (D) 2 (E) ∞

5. Find the area bounded by $x = y^2$ and $y = x^2$.
(A) 1/6 (B) 1/3 (C) 1/2 (D) 2 (E) 5/6

6. One of the roots of $i^{1/3}$ is given by
(A) $-\frac{\sqrt{3}}{2} + \frac{1}{2}i$ (B) $-\frac{1}{2} + \frac{\sqrt{3}}{2}i$ (C) $\frac{\sqrt{3}}{2} + \frac{1}{2}i$ (D) $\frac{1}{2} + \frac{\sqrt{3}}{2}i$ (E) –1

7. Find the value of x that results in a maximum value for y if $x^2 - 4x + y^2 + 4y = -2$.
(A) 1 (B) 2 (C) 3 (D) 4 (E) 5

8. Given: $\dot{y} + 4y = 0,\ y(0) = 2$. Find $y(t)$.
(A) $2e^{-4t}$ (B) $2e^{4t}$ (C) $2 \sin 4t$ (D) $2 \cos 4t$ (E) $2e^{4t} - 2e^{-4t}$

9. The solution, as $t \to \infty$, of $\ddot{y} + 4\dot{y} + 4y = 20$ is
(A) $y = 0$ (B) $y = e^{-2t}$ (C) $y = te^{-2t}$ (D) $y = 5$ (E) $y = \infty$

10. Find the solution (x_1, x_2, x_3) of the following set of equations.

$$\begin{aligned} x_1 + x_2 \quad &= 3 \\ x_2 + x_3 &= 1 \\ x_1 + x_2 + x_3 &= 3 \end{aligned}$$

(A) (1, 2, –1) (B) (1, 2, 1) (C) (2, 1, 0) (D) (0, 3, –2) (E) (3, 0, 1)

11. A matrix **A** has eigenvalues (characteristic values) λ found from which of the following equations?

(A) $|\mathbf{A} - \lambda\mathbf{I}| = 0$ (B) $\mathbf{AI} = \lambda\mathbf{x}$ (C) $|\mathbf{A}| = \lambda$ (D) $\mathbf{Ax} - \lambda\mathbf{I} = 0$ (E) $\mathbf{Ax} = \lambda\mathbf{x}$

12. Find the projection of $\mathbf{A} = 10\mathbf{i} - 2\mathbf{j} + 8\mathbf{k}$ in the direction of $\mathbf{B} = 2\mathbf{i} - 6\mathbf{j} + 3\mathbf{k}$.

(A) –6 (B) 0 (C) 4 (D) 8 (E) 10

13. The divergence of the vector function $\mathbf{u} = xy\mathbf{i} + 2y^2\mathbf{j} - yz\mathbf{k}$ at the point (0, 1, 1) is

(A) –4 (B) –1 (C) 0 (D) 1 (E) 4

14. The centroid $\bar{x}$ of the area formed by the x-axis, $x^2 = y$, and the line $x = 2$ is

(A) 1.4 (B) 1.2 (C) 1.1 (D) 1.0 (E) 0.8

15. A paraboloid 4 units high is formed by rotating $x^2 = y$ about the y-axis. Its volume is

(A) 4π (B) 6π (C) 8π (D) 10π (E) 12π

16. How many times out of 100 attempts are you likely to select a red ball first and a black ball second from a bucket of balls containing 40 red balls, 25 black balls, and 35 white balls? (The first ball is to be reinserted.)

(A) 14 (B) 12 (C) 10 (D) 8 (E) 4

17. Find the sample standard deviation of 5, 2, 6, 3.

(A) 0.80 (B) 1.83 (C) 2.54 (D) 3.26 (E) 4.00

18. What is the base 16 equivalent of 739?

(A) 739 (B) 397 (C) 2143 (D) 1343 (E) 2E3

19. Suppose that a state variable equation for a certain application is

$$W_{next} = (A + B)W + \overline{A}\,\overline{B}\,\overline{W}$$

If the state variable were to be implemented with a set-reset (S-R) flip-flop, the simplest inputs would be

(A) $S = \overline{A}\,\overline{B}\,\overline{W}$, $R = \overline{A}\,\overline{B}W$

(B) $S = \overline{A}\,\overline{B}$, $R = A + B$

(C) $S = R = A + B$

(D) $S = A + B$, $R = \overline{A}\,\overline{B}$

(E) $S = R = (A + B)W + \overline{A}\,\overline{B}\,\overline{W}$

20. The Boolean equation and DON'T CARE condition

$$Y = AC + \overline{A}\overline{B}\overline{C}D + ABC; \quad \overline{B}\overline{C}D + \overline{B}\overline{C}\overline{D} = 0$$

simplify to

(A) $Y = AC + \overline{A}\overline{C}D$
(B) $Y = AC + \overline{A}B\overline{C}D + \overline{A}\overline{B}\overline{C}$
(C) $Y = A\overline{B}\overline{C}$
(D) $Y = \overline{A}D$
(E) $Y = \overline{A}B\overline{C}D$

21. The development of the mass spectrometer provided the first evidence of
(A) isotopes (B) neutrons (C) deuterons (D) isobars (E) isotones

22. The work function of a photo sensitive material is 1.9 eV. If the incident radiation on this surface has a wavelength of 4000 Å, what is the threshold wavelength?
(A) 7321 Å (B) 6526 Å (C) 5345 Å (D) 3250 Å (E) 1986 Å

23. The key postulate of the Bohr model of the atom is the
(A) quantization of charge.
(B) quantization of angular momentum.
(C) quantization of particles.
(D) quantization of linear momentum.
(E) quantization of mass.

24. Schrödinger quantum mechanics for the hydrogen atom provides evidence for the existence of how many quantum numbers?
(A) 1 (B) 2 (C) 3 (D) 4 (E) 5

25. The non-linear portion of an engineering stress-strain diagram represents
(A) ultimate tensile strength.
(B) reversible deformation.
(C) plastic deformation.
(D) engineering modulus.
(E) elastic deformation.

26. The 0.2% offset yield stress is the
(A) stress at 0.2% elastic strain.
(B) yield point of a carbon steel.
(C) stress recovery after plastic deformation.
(D) ultimate tensile strength
(E) stress at the intersection point of a line drawn parallel to the elastic portion of the stress-strain diagram at a total strain of 0.2%.

27. Upper and lower yield points are observed in
(A) all pure metals.
(B) carbon steels.
(C) brittle metals.
(D) an alpha-beta brass.
(E) aluminum after cold working.

28. Creep failure occurs due to
    (A) gradual application of a load at ordinary temperatures.
    (B) a very rapid rate of loading at subzero temperatures.
    (C) formation of voids under a steady load at elevated temperatures.
    (D) excessive work hardening.
    (E) a very rapid rate of loading at ordinary temperatures.

29. Which of the following statements regarding ductile-brittle transition is not true?
    (A) There is a sudden loss of ductility below a critical temperature.
    (B) There is a loss of toughness at lower temperatures.
    (C) Carbon steels are less susceptible to ductile-brittle transition.
    (D) Some pure metals, such as aluminum, copper, or nickel, do not exhibit a ductile-brittle transition.
    (E) The ductile-brittle transition is independent of any imposed pressure.

30. Metal fatigue can be defined as
    (A) failure due to repeated application of stress higher than the ultimate tensile strength.
    (B) failure due to repeated cyclic stress below the endurance limit.
    (C) failure due to repeated cyclic stress after necking has just been observed.
    (D) failure after repeated necking.
    (E) failure due to repeated cyclic stress, even below the yield stress.

31. For a plain carbon steel, all of the following are true except:
    (A) a steel can be hardened without carburizing.
    (B) yield strength of a steel can be increased by cold rolling.
    (C) high carbon steel is less ductile than medium carbon steel.
    (D) hardness of steel can be increased by rapid quenching from high temperature.
    (E) larger grain size steel is stronger than small grain size steel.

32. All of the following conditions strengthen metals and alloys except:
    (A) presence of second phase precipitates.
    (B) presence of dispersed fibers or particles.
    (C) presence of martensite phase in steel.
    (D) annealing of cold worked metal above its recrystallization temperature.
    (E) a homogeneous solid solution of two chemical elements.

33. A plain carbon steel containing 0.18 wt.% carbon (SAE 1018 steel) is heated to 1100° C and then slow cooled to room temperature. Metallographic examination of this steel will show:
    (A) Ferrite + Pearlite.
    (B) Bainite + Martensite
    (C) Cementite + Ferrite.
    (D) Martensite + Cementite.
    (E) Pearlite + Carbide.

34. A peritectoid reaction is one in which:
(A) a solid phase reacts with a liquid phase to form a different solid phase.
(B) a solid phase reacts with a second solid phase to form a third solid phase.
(C) a liquid phase reacts with a second liquid phase to form a new solid phase.
(D) a solid phase decomposes to two different solid phases.
(E) a liquid phase reacts with a second liquid phase to form a third liquid phase.

35. Astatine (At, element no. 85) is in Group VIIA of the periodic table, as are chlorine and iodine. Which of the following is not characteristic of astatine?
(A) It is a deeply colored, volatile substance.
(B) It reacts with sodium vigorously to give $NaAt$.
(C) It reacts with hydrogen to give H_2At.
(D) It is less electronegative than chlorine.
(E) It is a weaker oxidizing agent than fluorine.

36. Which of the following elements is least likely to form a negative ion?
(A) chlorine (Cl, no. 17)
(B) strontium (Sr, no. 38)
(C) sulfur (S, no. 16)
(D) phosphorous (P, no. 15)
(E) oxygen (O, no. 8)

37. In the electrolysis of an aqueous solution of potassium bromide, what species migrates towards the positive electrode (the anode)?
(A) K^+ (B) H_2O (C) H_2O^- (D) Br^- (E) KBr

38. In order to determine the number of moles of N_2 in 3.7 liters of the pure gas at STP, you would use
(A) the law of definite proportions.
(B) Avogadro's Law
(C) the equilibrium constant
(D) the periodic table.
(E) Gay-Lussac's law.

39. The oxidation number (oxidation state of valence) of chlorine in potassium perchlorate, $KClO_4$, is
(A) –4 (B) –1 (C) 0 (D) +3 (E) +7

40. Fewer collisions between molecules of gases at lower temperature generally results in
(A) a smaller equilibrium constant.
(B) a smaller rate constant.
(C) Boyle's law.
(D) a larger equilibrium constant.
(E) a larger rate constant.

41. Which of the following would not increase the amount of the product in the given reaction?

$$CH_3OH + CH_3CO_2H \overset{HCl}{\leftrightarrows} CH_3CO_2CH_3 + H_2O$$

(A) Addition of more CH_3OH
(B) Addition of more CH_3CO_2H
(C) Addition of more HCl
(D) Removal of H_2O
(E) Removal of $CH_3CO_2CH_3$

42. If the endothermic reaction given were at equilibrium, which of the following would shift the equilibrium to the right?

$$\text{Heat} + Br_2(g) + 2Cl^- \leftrightarrows Cl_2(g) + 2Br^-$$

(A) Increasing P at constant T
(B) Decreasing P at constant T
(C) Increasing T at constant V
(D) Decreasing T at constant V
(E) Letting the reaction run longer

43. Oxidation of propene with potassium permanganate proceeds as follows (not balanced):

$$CH_3CHCH_2 + KMnO_4 + H_2SO_4 \rightarrow CH_3CO_2H + CO_2(g) + K_2SO_4 + MnO_2 + H_2O$$

In the balanced equation, how many liters of CO_2 would be formed from 1 mol of propene?

(A) 1 (B) 22.4 (C) 44.0 (D) 44.8 (E) 89.6

44. Iron reacts with copper sulfate according to the following unbalanced equation:

$$Fe + CuSO_4 \rightarrow Fe_2\,(SO_4)_3 + Cu$$

Atomic weights are $Fe = 56$, $Cu = 63.5$, $S = 32$, $O = 16$. How much copper will be formed from reaction of one gram-atom of iron with excess copper sulfate?

(A) 0.5 g-atom (B) 1.0 g-atom (C) 127 g (D) 95 g (E) 56 g

45. The valence electron configuration of Sulfur (atomic number 16) is

(A) $1s(2), 2s(2), 2p_x(2), 2p_y(2), 2p_z(2), 3s(2), 3p_x(2), 3p_y(1), 3p_z(1)$.
(B) $2s(2), 2p_x(2), 2p_y(1), 2p_z(1)$.
(C) $3s(2), 3p_x(2), 3p_y(1), 3p_z(1)$.
(D) $3s(2), 3p_x(1), 3p_y(1)$.
(E) $3s(2), 3p_x(2), 3p_y(2), 3p_z(2)$.

46. The product of the reaction of Mg (atomic number 12) with chlorine (atomic number 17) is

(A) $MgCl$ (B) $MgCl_2$ (C) $MgCl_3$ (D) Mg_2Cl (E) Mg_2Cl_3

47. The type of bond present in ethane, C_2H_6, is

(A) metallic (B) ionic (C) Van der Waals
(D) covalent (E) polar

48. What is the equilibrium constant for the following reaction?

$$C(s) + 2Cl_2(g) = CCl_4(g)$$

(A) $[CCl_4]/[Cl_2]$ (B) $[CCl_4]/[Cl_2]^2$ (C) $[Cl_2]/[CCl_4]$
(D) $[Cl_2]^2/[CCl_4]$ (E) $[CCl_4]/[Cl_2]^2[C]$

49. At the end of each of the next ten years, a payment of $200 is due. At an interest rate of 6%, what is the present worth of the payments?
(A) $27 (B) $200 (C) $1472 (D) $2000 (E) $2636

50. An amount P is invested at interest rate i per compounding period. F is the account balance after n compounding periods. Select the formula that relates F to P.

(A) $F = P(1+i)^{n-1}$
(B) $F = P(1+i)^n$
(C) $F = P(1+i)^i$
(D) $F = P(1+i)^{-n}$
(E) $F = P(1+ni)$

51. $12,000 is borrowed now at 12% interest. The first payment is $4000 and is made 3 years from now. The balance of the debt immediately after the payment is
(A) $4000 (B) $8000 (C) $12,000 (D) $12,860 (E) $16,860

52. A truck is purchased for $20,000. At the end of its 5 year life its salvage value will be $2000. Using general straight line depreciation, compute the book value of the truck after 3 years.
(A) $0 (B) $3600 (C) $7200 (D) $8000 (E) $9200

53. A machine costs $10,000, has an estimated life of 10 years and a scrap value of $1500. Assuming no inflation and an interest rate of 4%, what uniform annual amount must be invested at the end of each of the 10 years in order to replace the machine?
(A) $708 (B) $850 (C) $1000 (D) $1152 (E) $1278

54. An investment has infinite life and annual costs of $3000 for the first 5 years and $1600 per year thereafter. Using 6% interest per annum, compute the present worth of the annual disbursements.
(A) $15,000 (B) $25,000 (C) $32,600 (D) $50,200 (E) $86,100

55. A $1000 debt is to be repaid in four weekly payments of $300 each, beginning one week after the debt is incurred. Compute the nominal annual interest rate.
(A) 8% (B) 20% (C) 240% (D) 400% (E) 5000%

56. The purchase price of an instrument is $12,000 and its estimated maintenance costs are $500 for the first year, $1500 for the second and $2500 for the third year. After three years of use the instrument is replaced; it has no salvage value. Compute the present equivalent cost of the instrument using 10% interest.
(A) $14,070 (B) $15,570 (C) $15,730 (D) $16,500 (E) $20,720

57. If an amount invested five years ago has doubled, what is the annual interest rate?

(A) 5% (B) 6% (C) 10% (D) 12% (E) 15%

58. Interest on a debt is 12% per year compounded monthly. Compute the effective annual interest rate.

(A) 1% (B) 12% (C) 12.7% (D) 13.2% (E) 144%

59. An alumnus establishes a perpetual endowment fund to help Saint Louis University. What amount must be invested now to produce income of $100,000 one year from now and at one-year intervals forever? Interest rate is 8%.

(A) $8000 (B) $100,000 (C) $1,250,000
(D) $10,000,000 (E) Cannot be done.

60. The current through the battery in the circuit shown is

(A) 1.0 A
(B) 1.5 A
(C) 2.0 A
(D) 2.5 A
(E) 3.0 A

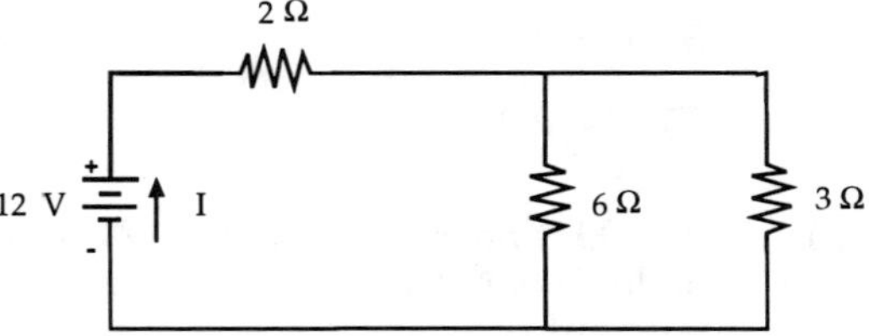

61. The voltage across the 5 ohm resistor in the circuit shown is

(A) 1.0 V
(B) 2.5 V
(C) 3.0 V
(D) 5.83 V
(E) 8.75 V

3 Ω
5 Ω
10 Ω
15 V
6 V

62. The equivalent resistance between points *a* and *b* is

(A) 4 Ω
(B) 6 Ω
(C) 8 Ω
(D) 10 Ω
(E) 12 Ω

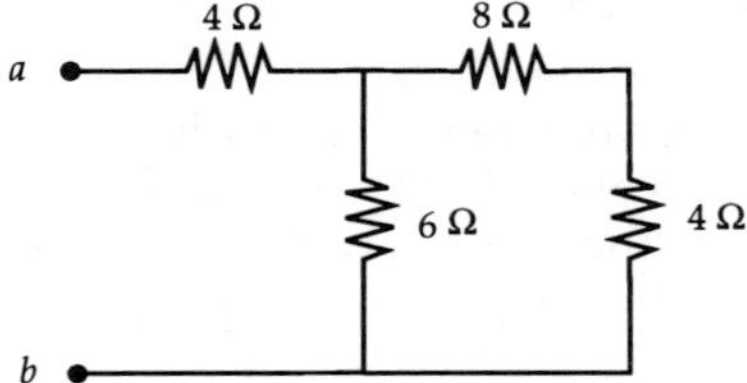

63. The power delivered to the 5 ohm resistor is

(A) 1.5 W
(B) 2.15 W
(C) 2.85 W
(D) 3.2 W
(E) 5.0 W

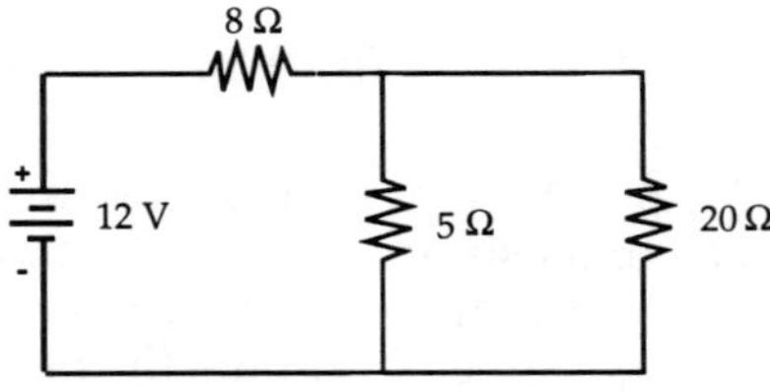

64. If $v(t) = 100 \sin 377t$, the magnitude of the rms voltage across the 20 ohm resistor is

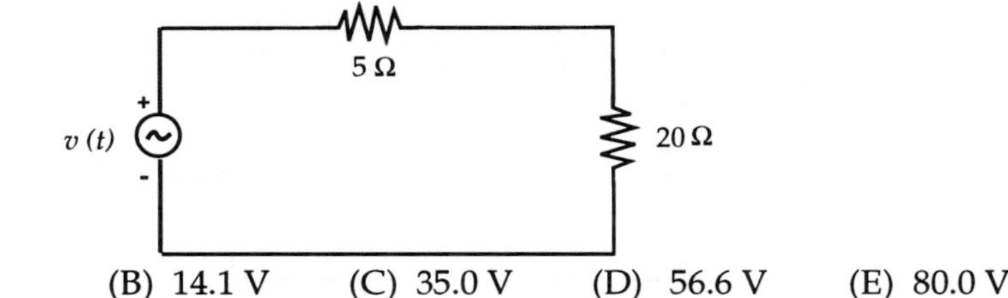

(A) 4.0 V (B) 14.1 V (C) 35.0 V (D) 56.6 V (E) 80.0 V

65. With a sinusoidal voltage of 100 volts rms, find the capacitive reactance X_C so that the source current is in phase with the source voltage.

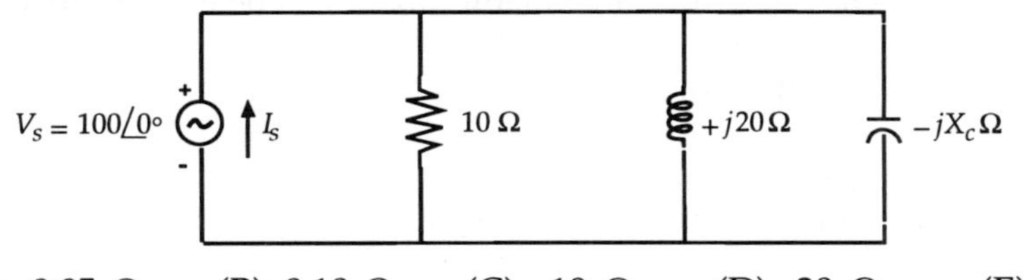

(A) 0.05 Ω (B) 0.10 Ω (C) 10 Ω (D) 20 Ω (E) 40 Ω

66. If $X_C = 25$ ohms in Prob. 65, what is the power delivered to the 10 ohm resistor?

(A) 10 W (B) 100 W (C) 200 W (D) 400 W (E) 1000 W

67. The square-wave voltage is measured with a true rms reading voltmeter. The voltmeter would read

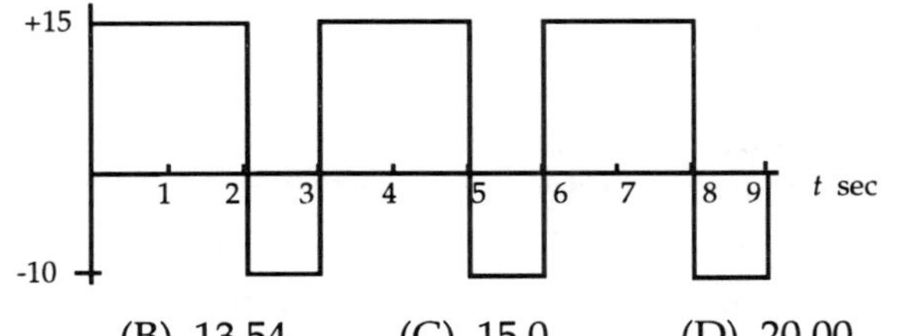

(A) 12.25 (B) 13.54 (C) 15.0 (D) 20.00 (E) 183.33

68. If $v(t) = 141.1 \sin 377t$, find R so that the capacitor's maximum voltage rating of 50 volts rms is not exceeded.

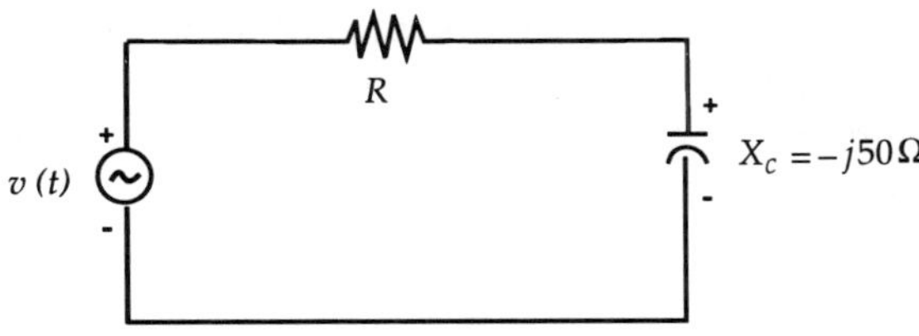

(A) 50.0 Ω (B) 86.6 Ω (C) 92.1 Ω (D) 100 Ω (E) 141.1 Ω

69. In a series RLC network the applied voltage is $v(t) = 70.7 \sin 377t$ and the circuit current is $i(t) = 5.1 \sin(377t + 0.16)$. The power delivered to the network at $t = 10^{-3}$ seconds is

(A) 67.9 W (B) 83.3 W (C) 95.6 W (D) 180.3 W (E) 360.5 W

70. If the initial current is zero and the switch is closed at $t = 0$, find the current at $t = 10^{-6}$ sec.

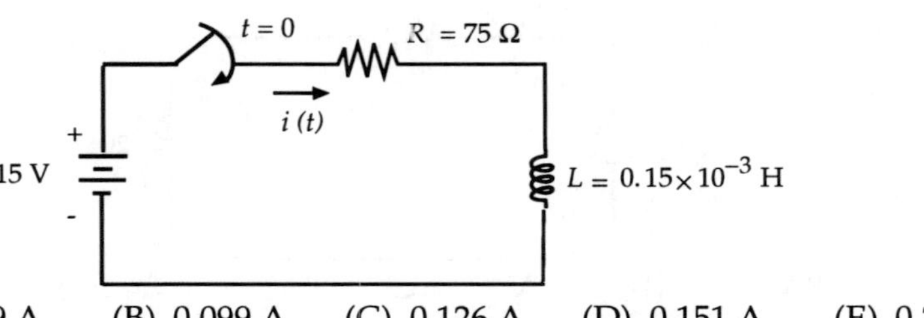

(A) 0.079 A (B) 0.099 A (C) 0.126 A (D) 0.151 A (E) 0.203 A

71. A 2-cm-dia., 10-cm-long air-core solenoid is used to create a DC magnetic flux density of 0.05 tesla. How many turns must be on the solenoid if the power supply can deliver only 2.5 amperes?
(A) 682 (B) 975 (C) 1590 (D) 2170 (E) 2650

72. Calculate the magnitude and direction of the force, in newtons, on an electron placed between the parallel plates shown.

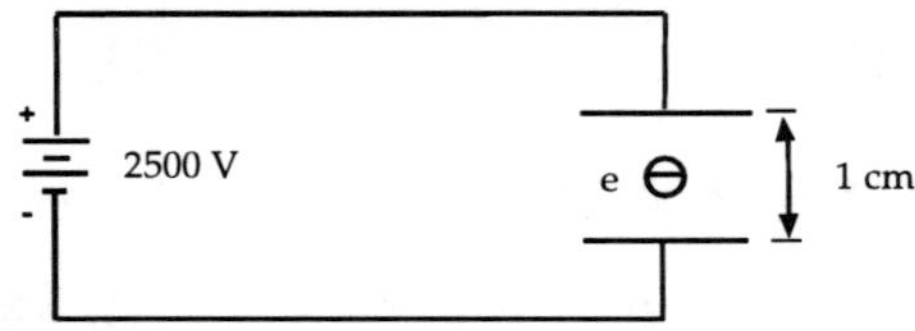

(A) 4×10^{-14} down
(B) 4×10^{-14} up
(C) 4×10^{-17} down
(D) 4×10^{-17} up
(E) 3.5×10^{-25} up

73. Two parallel conductors (a transmission line), 3 m apart in air, are oppositely charged to a charge density of $5.6 \text{x} 10^{-6}$ coulombs/m. Find the electrical force of attraction between the conductors.

(A) $1.88 \text{x} 10^{-5}$ N/m
(B) 0.19 N/m
(C) 0.336 N/m
(D) 7.95 N/m
(E) 4.56×10^4 N/m

74. A thermodynamic system, contained in a rigid vessel with a paddle wheel, undergoes an adiabatic process. Select the correct equation if kinetic and potential energy changes are negligible. (The energy equation is $Q - W = \Delta U$.)

(A) $Q = \Delta U$
(B) $W = (P_2 - P_1)V$
(C) $-W = \Delta U$
(D) $U = mc_p \Delta T$
(E) $\Delta H = 0$

75. A container contains half liquid water and half vapor by volume. Select the best estimate of the quality if the pressure is atmospheric.
(A) $x = 0$ (B) $x = 0.0006$ (C) $x = 0.1$ (D) $x = 0.5$ (E) $x = 1.0$

76. The reversible work done during an adiabatic process in a rigid container is equal to

(A) 0 (B) $(P_2 - P_1)V$ (C) $c_p \Delta T$ (D) $T \Delta S$ (E) $H - T \Delta S$

77. Water with quality $x = 50\%$ is contained in a rigid vessel. If more heat is added

(A) x increases.
(B) x decreases.
(C) x remains relatively constant.
(D) x may increase or decrease.
(E) superheating occurs.

78. During an ideal isentropic process

(A) the work is zero.
(B) the heat transfer is zero.
(C) the entropy increases.
(D) the entropy decreases.
(E) the enthalpy remains constant.

79. If both the volume and the pressure of an ideal gas contained in a cylinder are doubled, the temperature is

(A) constant.
(B) doubled.
(C) quadrupled.
(D) halved.
(E) quartered.

80. Select the correct statement about a Carnot cycle.

(A) It has four isentropic processes.
(B) It has two isentropic processes and two isothermal processes.
(C) It has two reversible processes and two irreversible processes.
(D) It has all irreversible processes.
(E) It has two constant volume processes.

<u>**Select either of the following two problems:**</u>

81. (metric) An inventor proposes to produce energy from 90 kg/s of geothermal water at 100° C. Calculate the maximum possible energy output if $T_{atm} = 20°$ C. $(c_p)_{water} = 4.18$ kJ/kg·°C.

(A) 1.2 MW (B) 3.6 MW (C) 5.8 MW (D) 6.4 MW (E) 9.2 MW

81. (English) An inventor proposes to produce energy from 200 lb/sec of geothermal water at 210° F. Calculate the maximum possible energy output if $T_{atm} = 70°$ F. $(c_p)_{water} = 1.00$ BTU/lbm - °F.

(A) 1200 BTU/sec (B) 2600 BTU/sec (C) 4100 BTU/sec
(D) 5900 BTU/sec (E) 6400 BTU/sec

<u>**Select either of the following two problems:**</u>

82. (metric) 40 J of work is done on a system in a rigid container while the internal energy increases by 20 J. The heat transfer is

(A) 20 J (B) –20 J (C) 60 J (D) –60 J (E) 0

82. (English) 4000 ft-lb of work is done on a system in a rigid container while the internal energy increases by 2 BTU. The heat transfer is

(A) 3.14 BTU (B) –3.14 BTU (C) 7.14 BTU (D) –7.14 BTU (E) 0 BTU

83. An ideal gas undergoes an isothermal expansion in a cylinder. Select the true statement. ($Q - W = \Delta U$)

(A) $W = (P_2 - P_1)V$ (B) $W = 0$ (C) $W = -\Delta U$ (D) $Q = 0$ (E) $W = Q$

84. The net entropy change in the universe during an adiabatic, irreversible process is

(A) positive (B) negative (C) zero (D) $\int \frac{dQ}{T}$ (E) indeterminate

85. Select the incorrect statement. ($Q - W = \Delta U$)

(A) Work and heat transfer represent energy crossing a boundary.
(B) The differentials of work and heat transfer are exact.
(C) Work and heat transfer are path integrals.
(D) Net work and net heat transfer are equal for a cycle.
(E) On a molecular level, heat transfer results from work being done.

86. The change in thermodynamic availability is equal to

(A) the quantity $T\Delta S$.
(B) the change in enthalpy.
(C) the actual work output.
(D) the maximum possible work output.
(E) the reversible work output minus the actual work output.

Select either of the following two problems:

87. (metric) Refrigerant-134a is heated from -40° C until it is completely vaporized while the pressure is held constant at 200 kPa absolute. The heat transfer is

(A) 400 kJ/kg (B) 360 kJ/kg (C) 320 kJ/kg (D) 280 kJ/kg (E) 240 kJ/kg

87. (English) Water is heated from -40° F until it is completely vaporized while the pressure is held constant at 20 psia absolute. The heat transfer is

(A) 141 BTU/lb (B) 132 BTU/lb (C) 124 BTU/lb
(D) 112 BTU/lb (E) 103 BTU/lb

---

88. The moment about the x-axis produced by the force $\mathbf{F} = 10\mathbf{i} - 20\mathbf{j} + 40\mathbf{k}$ acting at the point (2, 1, 1) is

(A) 20 (B) 30 (C) 40 (D) 50 (E) 60

89. The reaction at A is

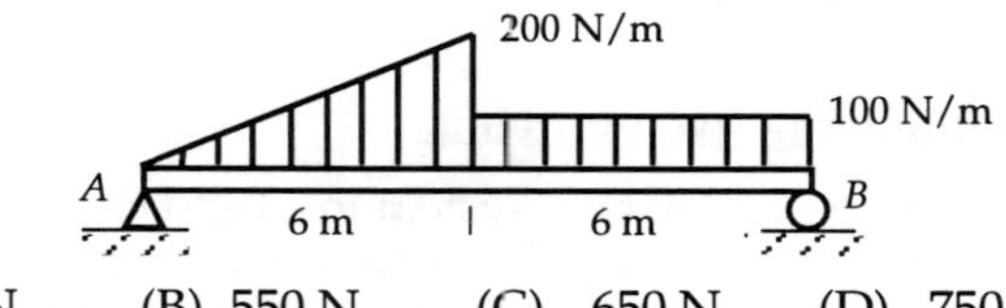

(A) 450 N (B) 550 N (C) 650 N (D) 750 N (E) 850 N

90. The force in link AB is
(A) 800 N
(B) 900 N
(C) 1000 N
(D) 1100 N
(E) 1200 N

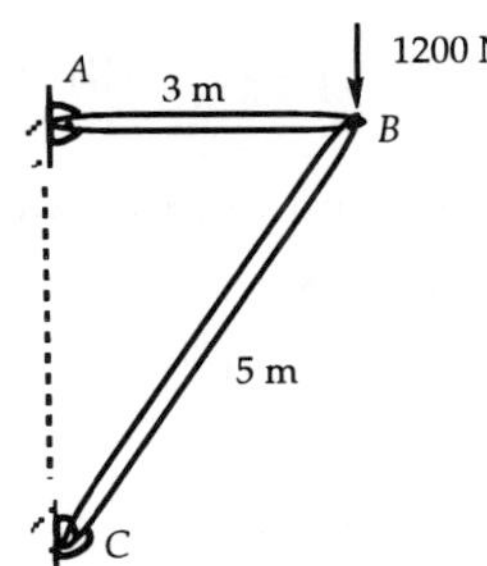

91. Determine the tension in cable AB.

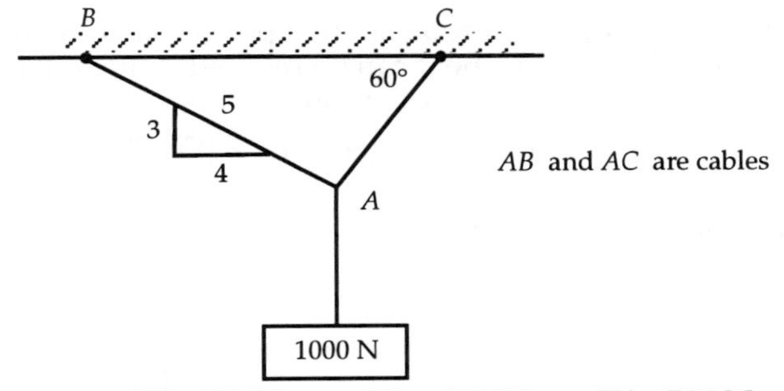

(A) 392 N (B) 451 N (C) 482 N (D) 504 N (E) 522 N

92. Find the total reactive force at A if all angles are equal.

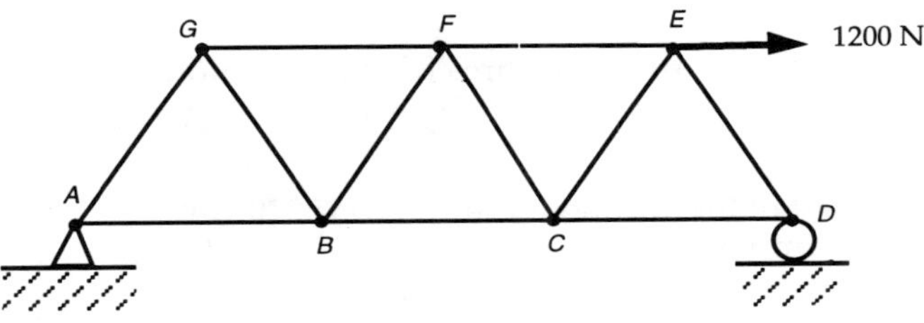

(A) 350 N (B) 550 N (C) 850 N (D) 1150 N (E) 1250 N

93. The force in link BF of Prob. 92 is
(A) 800 N (B) 700 N (C) 600 N (D) 500 N (E) 400 N

94. A block will slide down a 30° incline if the coefficient of friction is less than
(A) 0.577 (B) 0.591 (C) 0.625 (D) 0.674 (E) 0.711

95. What force F is needed for equilibrium? Neglect friction.
(A) $W/4$
(B) $W/5$
(C) $W/6$
(D) $W/7$
(E) $W/8$

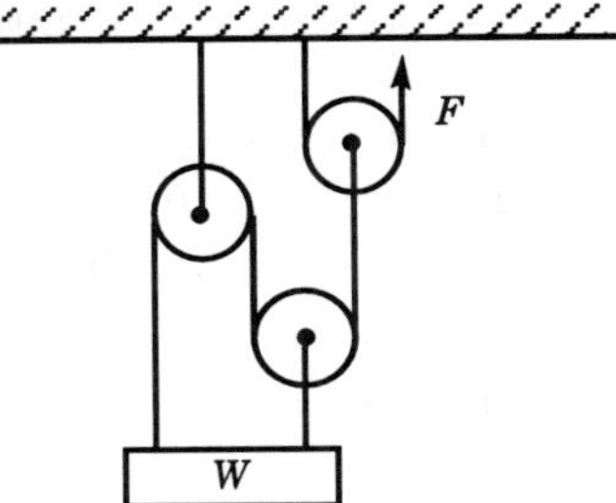

96. The scalar product of $\mathbf{F} = 2\mathbf{i} - 3\mathbf{j} + \mathbf{k}$ and $\mathbf{d} = \mathbf{i} + 2\mathbf{j} - \mathbf{k}$ is

(A) $2\mathbf{i} - 6\mathbf{j} - \mathbf{k}$ (B) $\mathbf{i} + 3\mathbf{j} + 7\mathbf{k}$ (C) 5 (D) –5 (E) 9

97. The maximum moment in the beam is

(A) 4200 N·m
(B) 5600 N·m
(C) 6400 N·m
(D) 6800 N·m
(E) 7200 N·m

1000 N
200 N/m
4 m

98. One complete revolution of a rope about a post is used to hold a boat. What maximum force by the boat can be resisted if the holder provides a force of 100 N? The coefficient of friction is 0.5.

(A) 200 N (B) 900 N (C) 1800 N (D) 2300 N (E) 3600 N

99. Find the y-coordinate of the centroid.

(A) 2
(B) 2.25
(C) 2.5
(D) 2.75
(E) 3.0

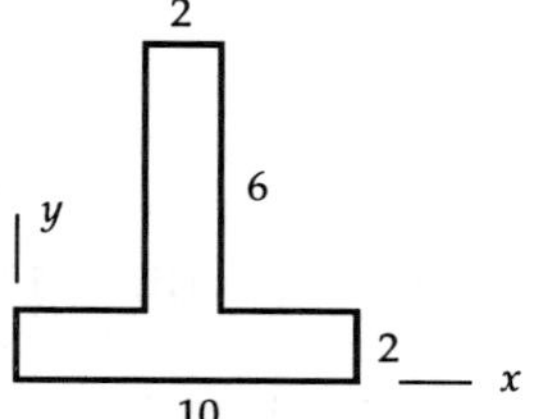

100. The moment of inertia of the area of Prob. 99 about the x-axis is

(A) 81 (B) 187 (C) 289 (D) 341 (E) 363

101. The moment of inertia about the x-axis is 320 cm^4. What is the moment of inertia about the x'-axis?

(A) 1220 cm^4 (D) 2160 cm^4
(B) 1760 cm^4 (E) 2240 cm^4
(C) 1920 cm^4

b
4 cm
x
4 cm
x'

102. A ball is given an initial velocity of 40 m/s straight up. How high will it go? Neglect friction. $g = 9.8$ m/s^2.

(A) 20 m (B) 40 m (C) 60 m (D) 80 m (E) 100 m

103. Neglecting the change of gravity with elevation, estimate the speed a satellite must have to orbit the earth at an elevation of 100 km. Earth's radius = 6400 km.

(A) 4000 m/s (B) 6000 m/s (C) 8000 m/s (D) 10 000 m/s (E) 12 000 m/s

104. An object experiences the velocity shown. How far will it move in 6 seconds?

(A) 40 m (D) 100 m
(B) 60 m (E) 120 m
(C) 80 m

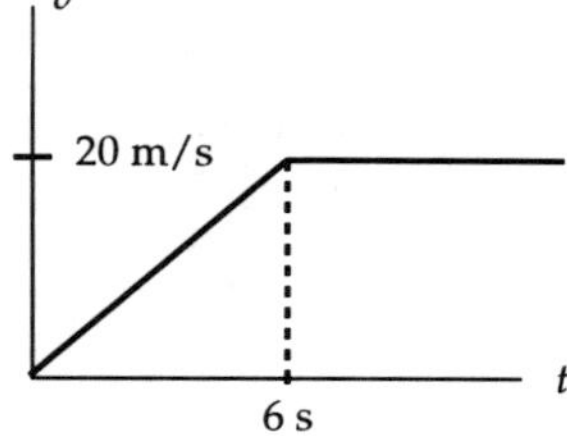

105. A wheel rotates at 20 rad/s. How many total revolutions will it rotate in 4 s, after it begins to decelerate at 10 rad/s^2?
(A) 3.18 (B) 4.25 (C) 5.91 (D) 6.37 (E) 7.4

106. An object is accelerating to the right along a straight path at 2 m/s^2. The object begins with a velocity of 10 m/s to the left. How far does it travel (total distance) in 15 seconds?
(A) 125 (B) 115 (C) 105 (D) 90 (E) 75

107. A cannon can make a maximum angle of 30° with the horizon. What is the minimum speed of a cannon ball if it must clear a 10-m-high obstacle?
(A) 28.0 m/s (B) 30.5 m/s (C) 32.8 m/s (D) 34.1 m/s (E) 36.8 m/s

108. Find ω_{AB} if V_C is 20 m/s to the right.
(A) 20 rad/s
(B) 30 rad/s
(C) 40 rad/s
(D) 50 rad/s
(E) 60 rad/s

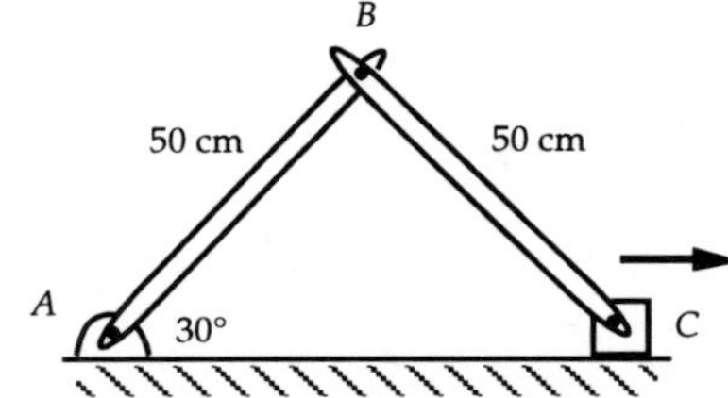

109. Find the angular acceleration of link AB in Prob. 108 if $a_C = 0$ and $\omega_{AB} = 40$ rad/s.
(A) 120 rad/s^2 (B) 620 rad/s^2 (C) 980 rad/s^2 (D) 1720 rad/s^2 (E) 2770 rad/s^2

110. A boy moves toward the center of a merry-go-round at 4 m/s. If the merry-go-round rotates at 5 rev/min., calculate the acceleration component of the boy normal to the radius.
(A) 6.7 m/s^2 (B) 6.1 m/s^2 (C) 5.8 m/s^2 (D) 5.2 m/s^2 (E) 4.2 m/s^2

111. Find the tension in the string which is wrapped around the 50 kg cylinder.

$I_{cylinder} = \frac{1}{2}mr^2.$

(A) 80 N (D) 140 N
(B) 100 N (E) 160 N
(C) 110 N

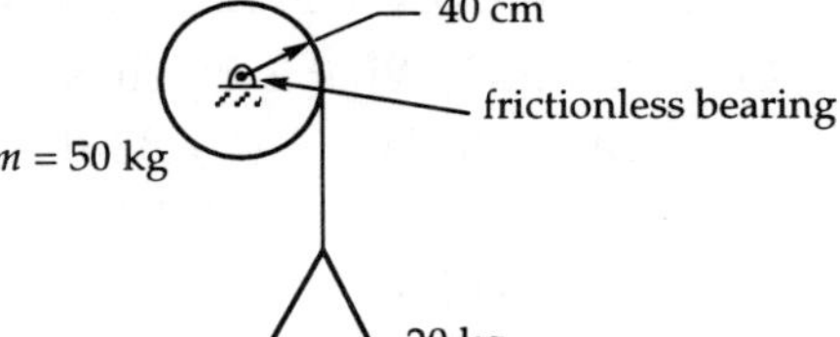

112. Find the maximum acceleration of a mass at the end of a 2-m-long string. It swings like a pendulum with a maximum angle from the vertical of 30°.
(A) 4.9 m/s^2 (B) 5.6 m/s^2 (C) 6.2 m/s^2 (D) 7.5 m/s^2 (E) 8.7 m/s^2

113. The mass in Prob. 111 drops from rest. If it drops 2 m, what is the angular velocity of the cylinder?
(A) 6.2 rad/s (B) 8.4 rad/s (C) 9.3 rad/s (D) 10.5 rad/s (E) 12.2 rad/s

114. The 50 kg object strikes the unstretched spring. The maximum deflection is

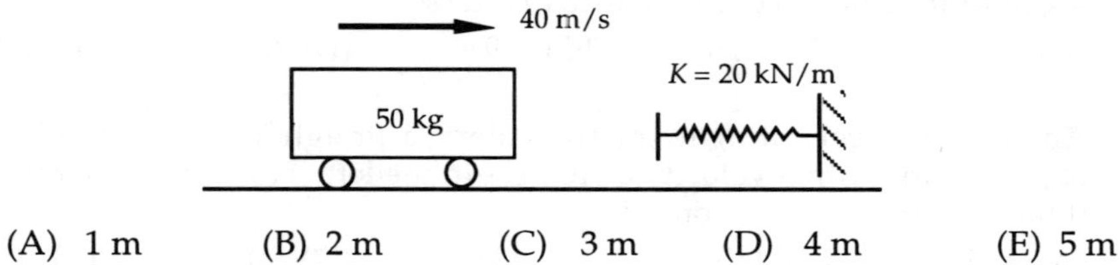

(A) 1 m (B) 2 m (C) 3 m (D) 4 m (E) 5 m

115. A ball strikes a wall as shown. If the coefficient of restitution is 0.8, find θ.

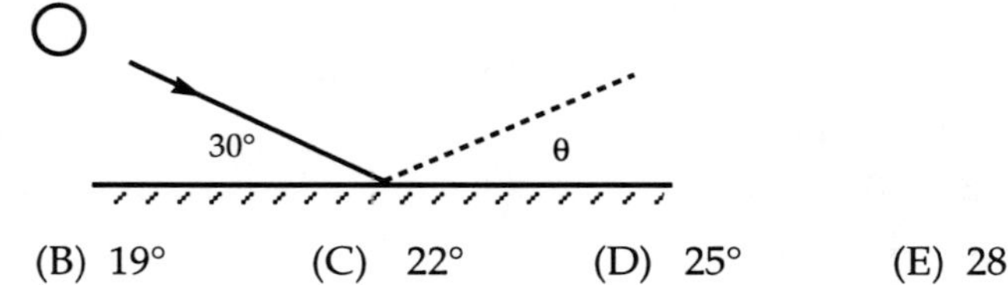

(A) 16° (B) 19° (C) 22° (D) 25° (E) 28°

Select either of the following two problems:

116. (metric) A weight of 40 kN is suspended from a 20-m-long, 2-cm-dia steel cable. How far does the cable stretch? $E = 210 \times 10^6$ kPa.
(A) 1.2 cm (B) 2.5 cm (C) 3.5 cm (D) 5 cm (E) 6 cm

116. (English) A weight of 10,000 lb is suspended from a 60-ft-long, 0.8-in-dia steel cable. How far does the cable stretch? $E = 30 \times 10^6$ psi.
(A) 0.5" (B) 1" (C) 1.5" (D) 2" (E) 2.5"

Select either of the following two problems:

117. (metric) What is the maximum normal stress?
(A) 3 MPa (D) 8 MPa
(B) 4 MPa (E) 10 MPa
(C) 6 MPa

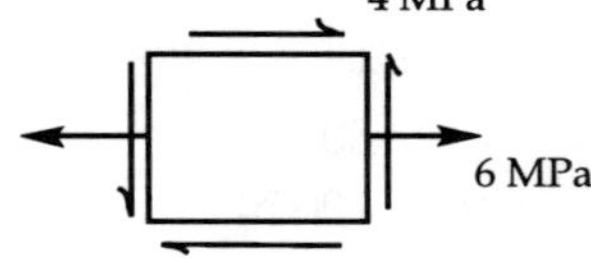

117. (English) What is the maximum normal stress?
(A) 3000 psi (D) 8000 psi
(B) 4000 psi (E) 10,000 psi
(C) 6000 psi

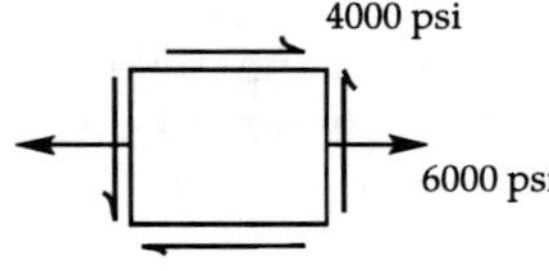

Select either of the following two problems:

118. (metric) A steel rod just fits into position. What stress is induced if $\Delta T_{rod} = +40° C$? $E = 210 \times 10^6$ kPa, $\alpha = 11.7 \times 10^{-6}° C^{-1}$.
(A) 28 MPa (B) 41 MPa (C) 59 MPa (D) 71 MPa (E) 98 MPa

118. (English) A steel rod just fits into position. What stress is induced if $\Delta T = +80°$ F? $E = 30 \times 10^6$ psi, $\alpha = 6.5 \times 10^{-6}° F^{-1}$.
(A) 7600 psi (B) 9800 psi (C) 11,600 psi (D) 13,200 psi (E) 15,600 psi

---

119. The maximum vertical shearing stress due to VQ/Ib in a beam with a rectangular cross section
(A) occurs where the normal stress is a maximum.
(B) occurs at the neutral axis.
(C) occurs at the outermost fiber.
(D) is linearly dependent on the distance from the neutral axis.
(E) depends on the moment applied to the section.

120. The ratio of the maximum normal stress to the maximum vertical shearing stress in a cantilever beam of length L with rectangular cross section $h \times b$ (h is the height) and loaded at the end is given by
(A) L/h (B) $2\ L/h$ (C) $4\ L/h$ (D) $2\ L/b$ (E) $4\ L/b$

121. Find the maximum vertical shear in the beam if $w = P/L$.
(A) $P/4$
(B) $P/2$
(C) $3\ P/4$
(D) P
(E) $5\ P/4$

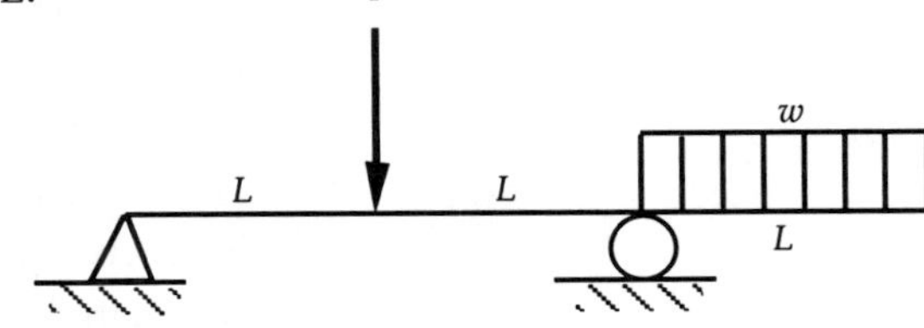

122. The maximum bending moment in the beam of Prob. 121 is
(A) $PL/4$ (B) $13PL/8$ (C) $PL/2$ (D) $5PL/8$ (E) $3PL/4$

Select either of the following two problems:

123. (metric) A torque of 400 N·m acts on a 4-cm-dia solid shaft. The maximum shearing stress is
(A) 3.2 MPa (B) 4.8 MPa (C) 5.2 MPa (D) 6.0 MPa (E) 6.4 MPa

123. (English) A torque of 4000 in-lb acts on a 4-in-dia solid shaft. The maximum shearing stress is
(A) 320 psi (B) 640 psi (C) 1280 psi (D) 2560 psi (E) 5120 psi

---

Select either of the following two problems:

124. (metric) A man climbs to the top of a 20-m-high, 6-cm-dia steel flag pole. His maximum weight (with a factor of safety of 2) should be at most ($E = 210 \times 10^6$ kPa)

(A) 1126 N (B) 982 N (C) 824 N (D) 630 N (E) 412 N

124. (English) A man climbs to the top of a 60-ft-high, 2-in-dia steel flag pole. His maximum weight (with a factor of safety of 2) should be at most ($E = 30 \times 10^6$ psi)

(A) 186 lb (B) 152 lb (C) 112 lb (D) 85 lb (E) 56 lb

Select either of the following two problems:

125. (metric) A 4-cm-dia, 20-cm-long cantilevered shaft is subjected to a torque of 500 N·m and a vertical load of 2000 N at the free end. What is the maximum shearing stress in the shaft?

(A) 50.9 MPa (B) 48.2 MPa (C) 41.1 MPa (D) 39.8 MPa (E) 28.2 MPa

125. (English) A 2-in-dia, 10-in-long cantilevered shaft is subjected to a torque of 5000 in-lb and a vertical load of 200 lb at the free end. What is the maximum shearing stress in the shaft?

(A) 3420 psi (B) 3180 psi (C) 2560 psi (D) 1870 psi (E) 1120 psi

---

126. The vertical shear diagram and bending moment diagram for a simply supported beam with a uniform load are, respectively,
(A) rectangle, triangle
(B) parabola, triangle
(C) triangle, rectangle
(D) rectangle, parabola
(E) triangle, parabola

127. Flow through a sudden contraction in a pipe
(A) is usually laminar.
(B) results in a significant energy loss.
(C) can be analyzed using Bernoulli's equation.
(D) has a relatively small loss coefficient.
(E) does not tend to separate.

128. To determine the flow rate using a square-edged orifice, the pressure must be measured upstream of the orifice and
(A) at the orifice.
(B) just upstream of the orifice.
(C) just downstream of the orifice at the vena contracta.
(D) at least 10 diameters downstream of the orifice.
(E) at least 20 diameters downstream of the orifice.

129. Freon flows through an expansion valve. Select the best statement.
(A) The temperature increases.
(B) The internal energy remains constant.
(C) The pressure remains constant.
(D) The enthalpy remains constant.
(E) The entropy decreases.

Select either of the following two problems:

130. (metric) The force P required per mèter of gate width (into the paper) is
(A) 29 kN
(B) 33 kN
(C) 35 kN
(D) 102 kN
(E) 980 kN

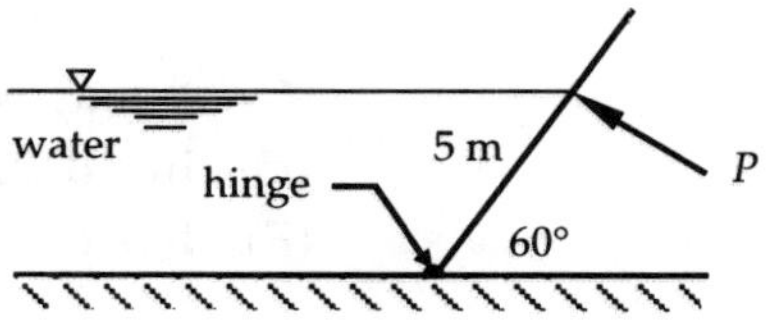

130. (English) The force P required per foot of gate width (into the paper) is
(A) 1260 lb
(B) 1870 lb
(C) 2000 lb
(D) 3000 lb
(E) 6720 lb

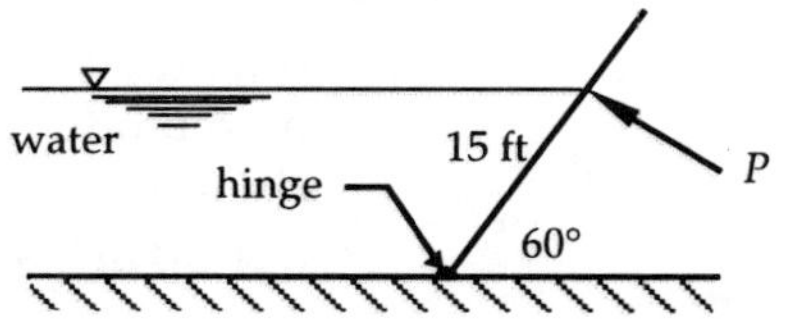

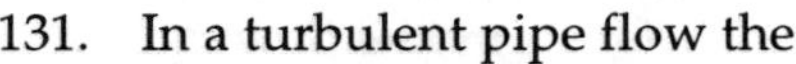

131. In a turbulent pipe flow the
(A) shear forces dominate inertial forces.
(B) inertial forces dominate shear forces.
(C) pressure forces dominate inertial forces.
(D) body forces dominate inertial forces.
(E) inertial forces are negligible.

132. Which of the following ratios of velocity V, diameter D, power $\dot{W}$, and density ρ is dimensionless?

(A) $\dfrac{\dot{W}D}{\rho V^2}$ (B) $\dfrac{\dot{W}}{\rho V^2 D^2}$ (C) $\dfrac{\dot{W}\rho}{V^3 D^2}$ (D) $\dfrac{V\dot{W}}{\rho D^3}$ (E) $\dfrac{\dot{W}}{\rho V^3 D^2}$

133. Select the false statement for the Bernoulli equation.
(A) It can be used for an unsteady flow.
(B) It can be used along a streamline.
(C) It can be used in an inertial coordinate system.
(D) It can be used in an inviscid flow.
(E) It can be used in a low speed air flow.

134. The velocity head in a section of a pipe is given by
(A) the difference between the *EGL* and the *HGL* at the section.
(B) the elevation of the *EGL*.
(C) the elevation of the *HGL*.
(D) Bernoulli's equation applied from the inlet to the section.
(E) a momentum balance on the fluid between the inlet and the section.

135. The pressure loss due to friction in a horizontal section of constant diameter pipe is usually determined using the
   (A) continuity equation.
   (B) Chezy-Manning equation.
   (C) Bernoulli's equation.
   (D) Navier-Stokes equation.
   (E) Darcy equation.

136. Select the correct statement concerning a supersonic flow.
   (A) It occurs after a converging nozzle for sufficiently low receiver pressures.
   (B) In a converging nozzle the velocity increases.
   (C) It may occur in a converging-diverging nozzle attached to a reservoir.
   (D) In a diverging nozzle the temperature decreases.
   (E) In a converging nozzle the enthalpy decreases.

<u>**Select either of the following two problems:**</u>

137. (metric) Flow over a 27-m-high dam is to be studied in a lab with a 3-m-high model. If the river has a flow rate of 74 m^3/s, the model flow rate should be
   (A) 30 m^3/s (B) 3 m^3/s (C) 0.3 m^3/s D) 9 m^3/s (E) 0.9 m^3/s

137. (English) Flow over a 90-ft-high dam is to be studied in a lab with a 10-ft-high model. If the river has a flow rate of 740 cfs, the model flow rate should be
   (A) 300 cfs (B) 30 cfs (C) 3 cfs (D) 90 cfs (E) 9 cfs

Questions 138-140 relate to the following figure.

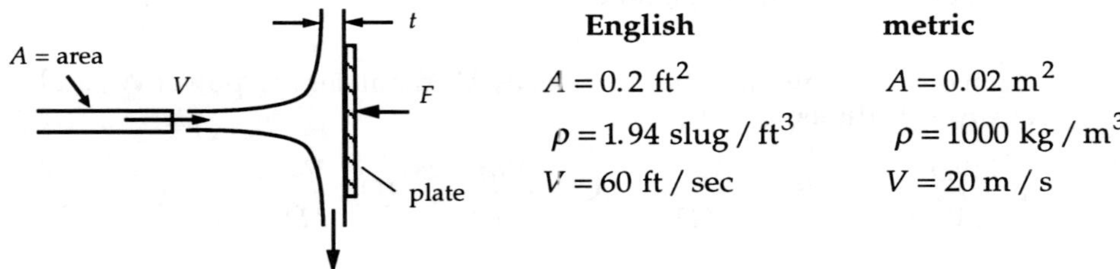

<u>**Select either of the following two problems:**</u>

138. (metric) At a plate radius of 20 cm t is about
   (A) 8 mm (B) 16 mm (C) 30 mm (D) 52 mm (E) 85 mm

138. (English) At a plate radius of 1 ft t is about
   (A) 0.192" (B) 0.384" (C) 0.72" (D) 1.2" (E) 2.16"

Select either of the following two problems:

139. (metric) For a stationary plate the force F is
(A) 4000 N (B) 6000 N (C) 8000 N (D) 10 000 N (E) 12 000 N

139. (English) For a stationary plate the force F is
(A) 1000 lb (B) 1200 lb (C) 1400 lb (D) 1600 lb (E) 1800 lb

Select either of the following two problems:

140. (metric) If the single plate is to move at 10 m/s to the right the force F is
(A) 1000 N (B) 2000 N (C) 3000 N (D) 4000 N (E) 5000 N

140. (English) If the single plate is to move at 30 ft/sec to the right the force F is
(A) 300 lb (B) 350 lb (C) 400 lb (D) 450 lb (E) 500 lb

STOP!

If you finish before 4 hours is up, you may return to any of the problems in the morning session. Do not look at the afternoon session yet. Take a 1 hour break and then continue with the afternoon session.

Afternoon Session Answer Sheet

Mathematics

1. A B C D E
2. A B C D E
3. A B C D E
4. A B C D E
5. A B C D E
6. A B C D E
7. A B C D E
8. A B C D E
9. A B C D E
10. A B C D E
11. A B C D E
12. A B C D E
13. A B C D E
14. A B C D E
15. A B C D E
16. A B C D E
17. A B C D E
18. A B C D E
19. A B C D E
20. A B C D E

score________

Mechanics

21. A B C D E
22. A B C D E
23. A B C D E
24. A B C D E
25. A B C D E
26. A B C D E
27. A B C D E
28. A B C D E
29. A B C D E
30. A B C D E
31. A B C D E
32. A B C D E
33. A B C D E
34. A B C D E
35. A B C D E
36. A B C D E
37. A B C D E
38. A B C D E
39. A B C D E
40. A B C D E

score________

Electrical Circuits

41. A B C D E
42. A B C D E
43. A B C D E
44. A B C D E
45. A B C D E
46. A B C D E
47. A B C D E
48. A B C D E
49. A B C D E
50. A B C D E

score________

Economics

51. A B C D E
52. A B C D E
53. A B C D E
54. A B C D E
55. A B C D E
56. A B C D E
57. A B C D E
58. A B C D E
59. A B C D E
60. A B C D E

score________

Thermodynamics/ Fluid Mechanics

61. A B C D E
62. A B C D E
63. A B C D E
64. A B C D E
65. A B C D E
66. A B C D E
67. A B C D E
68. A B C D E
69. A B C D E
70. A B C D E

score________

AFTERNOON SESSION

4 hours maximum

Each of the 70 problems is followed by five answers. You must answer all 70 questions. Select the response that is best and circle the corresponding letter on the answer sheet.

Questions 1-5:

Four vectors are given in Cartesian coordinates as

$$\mathbf{A} = 3\mathbf{i} + 2\mathbf{j} + 2\mathbf{k}, \quad \mathbf{B} = \mathbf{j} - \mathbf{k}, \quad \mathbf{C} = 2\mathbf{i} + 2\mathbf{j} + \mathbf{k}, \quad \mathbf{D} = x^2\mathbf{i} + xy\mathbf{j} + z^2\mathbf{k}$$

1. The component of **A** in the direction of **C** is
   (A) 1 (B) 2 (C) 3 (D) 4 (E) 5

2. The angle between vectors **A** and **B** is
   (A) 0° (B) 45° (C) 90° (D) 135° (E) 180°

3. The vector product (cross product) **B** × **C** is
   (A) $5\mathbf{i} - 4\mathbf{j} - 2\mathbf{k}$
   (B) $3\mathbf{i} - 4\mathbf{j} - 2\mathbf{k}$
   (C) $3\mathbf{i} + 4\mathbf{j} - 2\mathbf{k}$
   (D) $5\mathbf{i} - 4\mathbf{j} + 2\mathbf{k}$
   (E) $3\mathbf{i} - 2\mathbf{j} - 2\mathbf{k}$

4. The divergence of **D** at the point (1, 0, –1) is
   (A) $2\mathbf{i} + \mathbf{j} + 2\mathbf{k}$ (B) $2\mathbf{i} + 2\mathbf{k}$ (C) 0 (D) 1 (E) 2

5. The area of a parallelogram with sides **B** and **C** is
   (A) 4.12 (B) 3.24 (C) 3 (D) 2.08 (E) 1.0

Questions 6-10:

The motion of an undamped spring acted upon by a force $F(t) = 16 \sin \omega t$ is described by

$$\ddot{y} + 25y = 16 \sin \omega t$$

where y is the displacement and t is the time. The initial conditions are $y(0) = 0$ and $\dot{y}(0) = 8$.

6. The general solution of the homogeneous equation is
   (A) $c_1 e^{5t} + c_2 e^{-5t}$ (B) $c_1 \sin 5t$ (C) $c_1 \cos 5t$
   (D) $c_1 \cos \omega t + c_2 \sin \omega t$ (E) $c_1 \cos 5t + c_2 \sin 5t$

7. If $\omega = 3$ the particular solution is
(A) $c_3 \sin 3t$ (B) $\sin 3t$ (C) $c_1 \sin 3t + c_2 \cos 5t$
(D) $c_1 \cos 5t + c_2 \sin 5t$ (E) e^{-3t}

8. Resonance occurs if ω is
(A) 1 (B) 2 (C) 3 (D) 4 (E) 5

9. If $\omega = 3$ the solution that satisfies the initial conditions is
(A) $\sin 5t + \cos 5t + \sin 3t$ (B) $\cos 5t + \sin 3t$ (C) $\sin 5t - \sin 3t$
(D) $\sin 5t + \sin 3t$ (E) $\sin 3t - \sin 5t$

10. If a damping term $8\dot{y}$ were added to the original differential equation with $\omega = 5$ the steady-state solution would have an amplitude of
(A) 0.5 (B) 0.4 (C) 0.3 (D) 0.2 (E) 0.1

Questions 11-15:

Given the parabola and the straight line as shown.

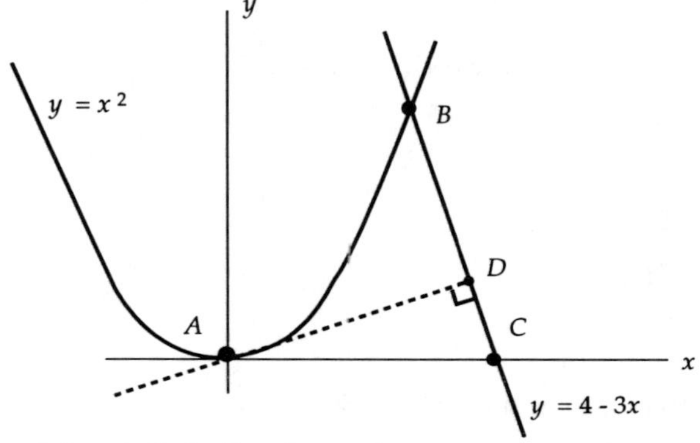

11. The slope of the straight line is
(A) –4 (B) –3 (C) 4/3 (D) 3 (E) 4

12. Find the intersection point B of the two curves.
(A) (2, 1) (B) (1, 1) (C) (1, 2) (D) (2, 3) (E) (2, 4)

13. Find the y-coordinate of the centroid of the area between the two curves and the x-axis.
(A) 0.256 (B) 0.289 (C) 0.311 (D) 0.332 (E) 0.451

14. What is the equation of the line passing through A and D? $AD \perp BC$.
(A) $y = 3x$ (B) $y = 4x$ (C) $4y = 3x$ (D) $3y = x$ (E) $4y = x$

15. The parabola is rotated about the y-axis. What volume is contained up to the surface $y = 2$?
(A) π (B) 2π (C) 3π (D) 4π (E) 5π

Questions 16-20:

Given the set of equations represented by $[a_{ij}][x_j] = [r_i]$:

$$2x_1 + 3x_2 = 12$$
$$2x_1 + x_2 = 2$$

16. What is the determinant $|a_{ij}|$?

(A) –5 (B) –4 (C) –2 (D) 0 (E) 4

17. What is the adjoint matrix $[a_{ij}]^{+}$?

(A) $\begin{bmatrix} 1 & -3 \\ -2 & 2 \end{bmatrix}$ (B) $\begin{bmatrix} 1 & -1 \\ -3 & 2 \end{bmatrix}$ (C) $\begin{bmatrix} 1 & 1 \\ 3 & 2 \end{bmatrix}$ (D) $\begin{bmatrix} 1 & 3 \\ 1 & 2 \end{bmatrix}$ (E) $\begin{bmatrix} 1 & -3 \\ 1 & 2 \end{bmatrix}$

18. What is the inverse matrix $[a_{ij}]^{-1}$?

(A) $\begin{bmatrix} -1 & -3 \\ -1 & -2 \end{bmatrix}$ (B) $\frac{1}{4}\begin{bmatrix} -1 & 3 \\ 2 & -2 \end{bmatrix}$ (C) $\frac{1}{2}\begin{bmatrix} 1 & -1 \\ -3 & 2 \end{bmatrix}$

(D) $\frac{1}{5}\begin{bmatrix} 1 & -3 \\ -1 & 2 \end{bmatrix}$ (E) $\frac{1}{4}\begin{bmatrix} -1 & 1 \\ 3 & -2 \end{bmatrix}$

19. When using Cramer's rule, x_2 can be found by evaluating a determinant $|b_{ij}|$ and then dividing by the determinant $|a_{ij}|$. What is $|b_{ij}|$ for x_2?

(A) 20 (B) 4 (C) 0 (D) –4 (E) –20

20. What are the eigenvalues (characteristic values) associated with the matrix $[a_{ij}]$?

(A) 2, 6 (B) 5, 0 (C) 0, 4 (D) 4, –1 (E) 1, 2

Questions 21-25:

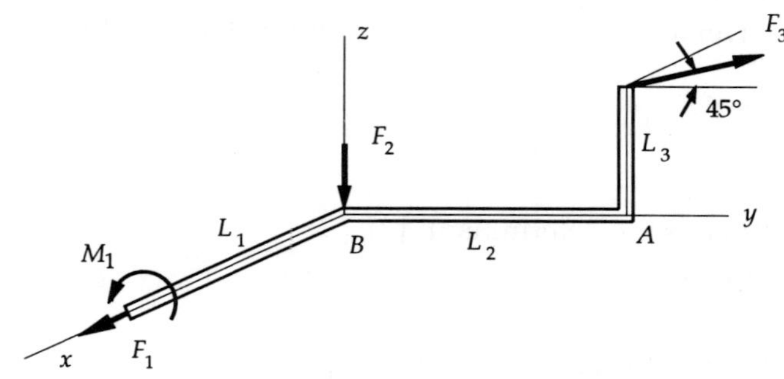

L_1 = 20 cm (7.87 in)
L_2 = 15 cm (5.91 in)
L_3 = 10 cm (3.94 in)
F_1 = 60 N (13.5 lb)
F_2 = 20 N (4.5 lb)
F_3 = 40 N (9.0 lb) —has no z-component
M_1 = 10 N·m (88.5 in-lb)

21. The system of forces is
(A) planar. (B) concurrent. (C) parallel. (D) 2-D. (E) 3-D.

22. If the rigid assembly is held in equilibrium with a support at A, a force at A must have magnitude
(A) 47 N (10.6 lb) (B) 52 N (11.7 lb) (C) 58 N (13.0 lb)
(D) 61 N (13.7 lb) (E) 65 N (14.6 lb)

23. If the rigid assembly is held in equilibrium with a support at A, a moment at A must have magnitude
(A) 10 N·m (88.5 in-lb) (B) 14 N·m (123 in-lb) (C) 20 N·m (177 in-lb)
(D) 26 N·m (230 in-lb) (E) 42 N·m (372 in-lb)

24. If F_1 were moved to point B, it would be replaced by a
(A) force only.
(B) force and a clockwise moment.
(C) clockwise moment only.
(D) force and a counterclockwise moment.
(E) counterclockwise moment only.

25. If F_1 and M_1 were moved to point A, the resulting moment would have magnitude
(A) 9.4 N·m (83 in-lb)
(B) 11.2 N·m (99 in-lb)
(C) 13.4 N·m (118 in-lb)
(D) 15.2 N·m (135 in-lb)
(E) 19.3 N·m (171 in-lb)

Questions 26-30:

L_1 = 30cm (11.8 in) $\omega_1 = 10$ rad/s

L_2 = 50 cm (19.7 in) $\alpha_1 = 0$

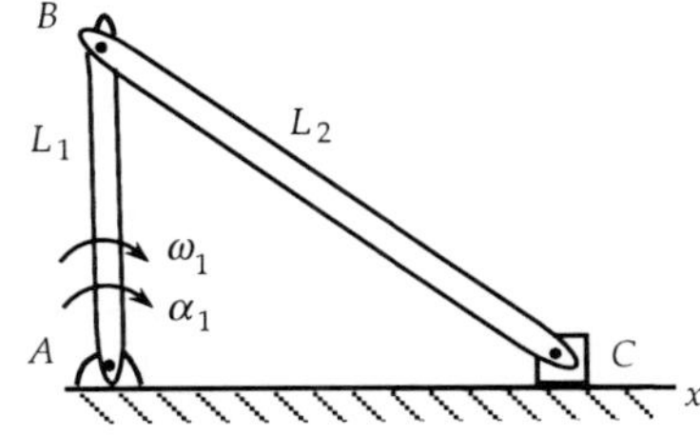

26. The angular velocity of link BC is
(A) 0 (D) 15 rad/s
(B) 5 rad/s (E) 20 rad/s
(C) 10 rad/s

27. The velocity of the slider at C is
(A) 1 m/s (3.28 ft/sec)
(B) 2 m/s (6.56 ft/sec)
(C) 3 m/s (9.83 ft/sec)
(D) 4 m/s (13.1 ft/sec)
(E) 5 m/s (16.4 ft/sec)

28. The angular acceleration of link BC is
(A) 0 (B) 25 rad/s^2 (C) 50 rad/s^2 (D) 75 rad/s^2 (E) 100 rad/s^2

29. The acceleration of C is
(A) 15 m/s^2 (49.2 ft/sec^2))
(B) 17.5 m/s^2 (57.4 ft/sec^2)
(C) 20 m/s^2 (65.6 ft/sec^2)
(D) 22.5 m/s^2 (73.8 ft/sec^2)
(E) 25 m/s^2 (82.0 ft/sec^2)

30. If a bead slides down link AB at a constant velocity of 2 m/s (6.56 ft/sec), what is the magnitude of its acceleration when it is half-way between A and B ?
(A) 24 m/s^2 (78.7 ft/sec^2)
(B) 30 m/s^2 (98.4 ft/sec^2)
(C) 36 m/s^2 (118 ft/sec^2)
(D) 40 m/s^2 (131 ft/sec^2)
(E) 43 m/s^2 (141 ft/sec^2)

Questions 31-35:

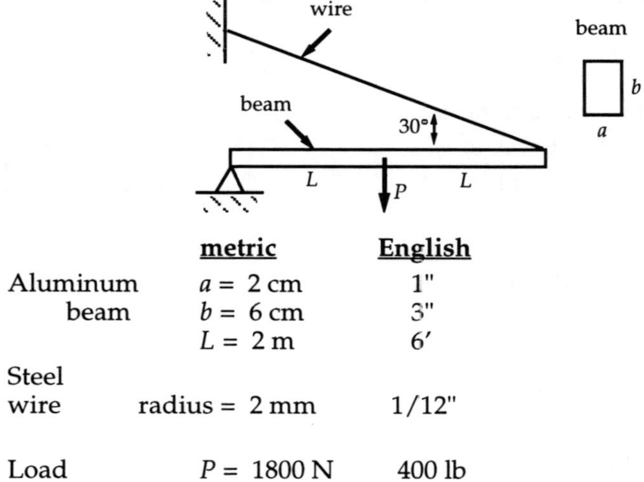

	metric	English
Aluminum beam	a = 2 cm	1"
	b = 6 cm	3"
	L = 2 m	6′
Steel wire	radius = 2 mm	1/12"
Load	P = 1800 N	400 lb

31. Assuming the weightless beam to be rigid, the right end will drop a distance of

metric	(A) 0.4 mm	(B) 2.6 mm	(C) 6.3 mm	(D) 17.3 mm	(E) 47 mm
English	(A) 0.02 in	(B) 0.1 in	(C) 0.2 in	(D) 0.48 in	(E) 1.3 in

32. The maximum tensile stress in the weightless beam is

metric	(A) 24 MPa	(B) 56 MPa	(C) 92 MPa	(D) 120 MPa	(E) 150 MPa
English	(A) 1200 psi	(B) 2300 psi	(C) 4700 psi	(D) 6200 psi	(E) 9600 psi

33. The maximum shear stress in the weightless beam, due to the vertical shear, is

metric	(A) 0	(B) 240 kPa	(C) 520 kPa	(D) 890 kPa	(E) 1.12 MPa
English	(A) 0	(B) 20 psi	(C) 50 psi	(D) 80 psi	(E) 100 psi

34. The maximum shear stress in the wire is

metric	(A) 0	(B) 24 MPa	(C) 48 MPa	(D) 64 MPa	(E) 72 MPa
English	(A) 0	(B) 3000 psi	(C) 6100 psi	(D) 8200 psi	(E) 9200 psi

35. The bending moment diagram resembles a

(A) circle (B) rectangle (C) ellipse (D) parabola (E) triangle

Questions 36-40:

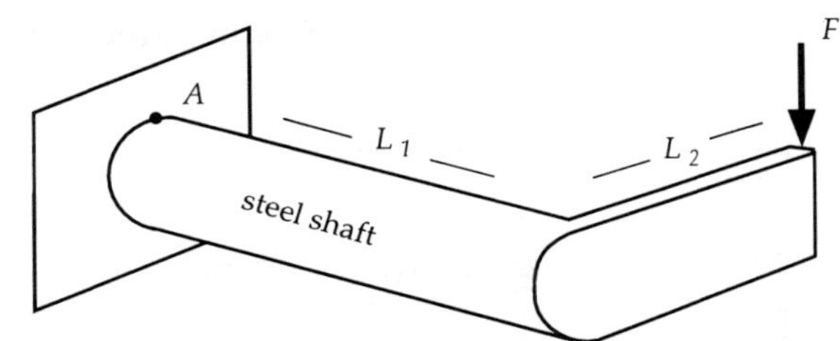

rectanglular beam — 2 cm x 5 cm (0.787" x 1.97")
shaft — 5 cm (1.97") diameter

$F = 2000$ N (450 lb)
$L_1 = 2$ m (6.56 ft)
$L_2 = 1$ m (3.28 ft)
$E = 210 \times 10^9$ Pa$\left(30 \times 10^6 \text{ psi}\right)$

36. The maximum normal stress in the rectangular beam is
(A) 120 MPa (17,400 psi)
(B) 180 MPa (26,000 psi)
(C) 240 MPa (34,800 psi)
(D) 360 MPa (52,200 psi)
(E) 480 MPa (69,000 psi)

37. The maximum vertical shearing stress in the rectangular beam is
(A) 1200 kPa (174 psi)
(B) 1800 kPa (261 psi)
(C) 2400 kPa (348 psi)
(D) 3000 kPa (435 psi)
(E) 3600 kPa (522 psi)

38. The stress situation at pt. A is described by which element? (Point A is at the top.)

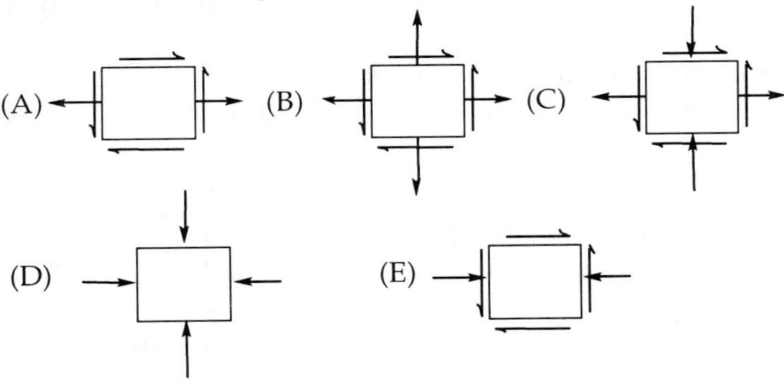

39. The maximum normal stress in the shaft is
(A) 163 MPa (23,640 psi)
(B) 182 MPa (26,400 psi)
(C) 247 MPa (35,800 psi)
(D) 326 MPa (47,300 psi)
(E) 345 MPa (50,000 psi)

40. The maximum deflection of the shaft is
(A) 3.9 cm (1.54 in)
(B) 5.1 cm (2 in)
(C) 6.6 cm (2.6 in)
(D) 8.3 cm (2.85 in)
(E) 10.5 cm (4.13 in)

Questions 41-42:

41. The DC (average) current through the $+j$ 40 ohm inductor is

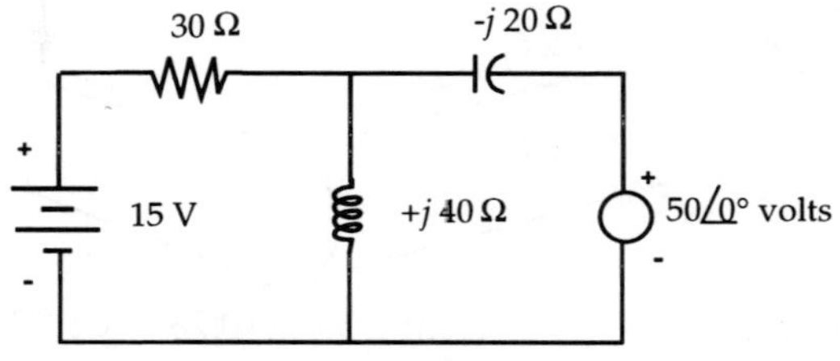

(A) 0.3 A (B) 0.5 A (C) 0.8 A (D) 1.2 A (E) 1.5 A

42. The rms current through the $+j$ 40 ohm inductor is
(A) 0.5 A (B) 0.93 A (C) 1.58 A (D) 2.11 A (E) 2.84 A

Questions 43-45:

43. Using the source voltage 100 $\angle 0°$ rms as the reference, what is the voltage across the $+j$ 35 ohm inductor?

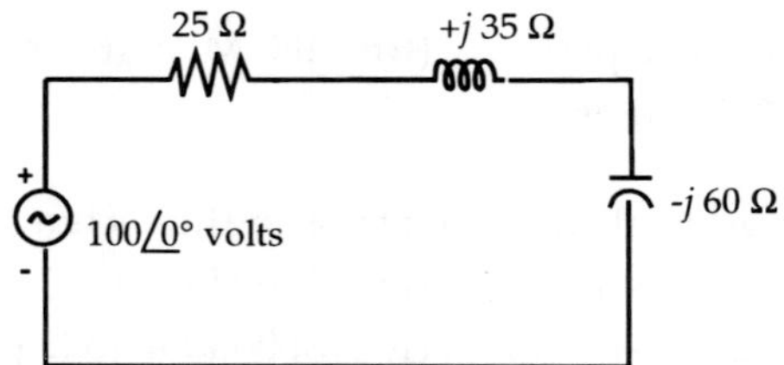

(A) $35\angle 90°$ (B) $100\angle 90°$ (C) $170\angle -45°$ (D) $2.83\angle 45°$ (E) $99.0\angle 135°$

44. What is the total average real power delivered to the network?
(A) 280 W (B) 200 W (C) 170 W (D) 120 W (E) 70.8 W

45. What is the power factor for this network?
(A) 0.38 (B) 0.42 (C) 0.58 (D) 0.71 (E) 0.81

Questions 46-47:

46. The current through the capacitor is
(A) 0.21 A (D) 2.86 A
(B) 0.57 A (E) 4.85 A
(C) 1.0 A

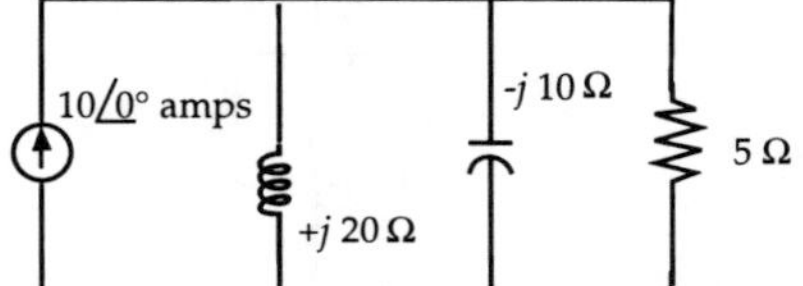

47. The voltage across the 5-ohm resistor is
(A) 0.50 V (B) 1.61 V (C) 2.06 V (D) 10.0 V (E) 48.5 V

Questions 48-50:

48. The voltage V_{An} in magnitude and phase is

(A) $120\angle 0^\circ$ (D) $208\angle -120^\circ$

(B) $120\angle 30^\circ$ (E) $360\angle -30^\circ$

(C) $120\angle -30^\circ$

49. The current, in magnitude and phase, in the line CC' is

(A) $2.4\angle 36.9^\circ$ (D) $4.16\angle 67^\circ$

(B) $2.4\angle 53.1^\circ$ (E) $4.16\angle 113.1^\circ$

(C) $2.4\angle 173.1^\circ$

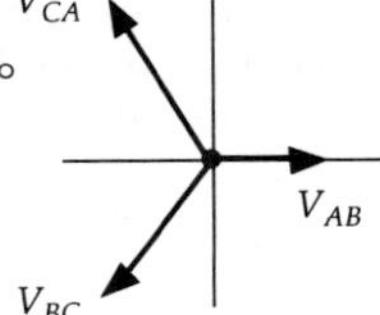

50. The total real power delivered to the wye connected load is

(A) 518 W (B) 691 W (C) 1557 W (D) 2077 W (E) 3124 W

Questions 51-53:

The City of Sprawl operates a public transportation system. Because the system is a government entity, it pays no taxes and can procure capital by selling bonds which pay 6% annual interest on par value.

51. A fleet of vehicles is needed for the city's new dial-a-ride service. Two models are being considered. Cost estimates are:

	Guppy	Shark
First cost	$25,500	$42,000
Annual maintenance	$ 1200	$ 1800
Life (years)	3	6
Salvage value	$ 900	$ 1500

What is the annual advantage of selecting the Guppy?

(A) –$16,500 (B) –$850 (C) –$600 (D) –$330 (E) $10,250

52. If interest on the bonds were paid semi-annually rather than annually, what would be the effective annual rate?

(A) 6.1% (B) 6.25% (C) 16% (D) 106% (E) 402%

53. A Sprawl Transportation Authority bond has par value of $5000 and term of 10 years. The bond pays 5% nominal annual interest on par value. Estimate the selling price of the bond if the market interest rate is 6%.

(A) $4632 (B) $5000 (C) $7500 (D) $8000 (E) $5386

Questions 54-55:

A new bus route was added last week, but the average number of passengers on that bus has been only 20; the capacity is 60. The fare is $0.50 and the cost of operating the bus is $0.05 per passenger plus $25 per trip. The marketing director suggests offering a $0.25 fare to people who have never before taken the bus.

54. Assume that the director's suggestion is implemented, and that 15 additional passengers per trip are gained. The total profit per trip will be

(A) –$26.75 (B) –$13 (C) –$3.75 (D) $10 (E) $13.75

55. Assume that the director's suggestion is implemented, and that 15 additional passengers per trip are gained, but that all passengers claim the $0.25 fare. The total profit per trip will be

(A) –$25 (B) –$23 (C) –$18 (D) $7 (E) $8.75

Questions 56-60:

A corporation evaluates all capital investments using a 20% annual rate of return before taxes. The corporation must purchase a new tangent scanner. The following estimates pertain to the two models available:

	Scanx	Holo-Scan
First cost	$90,000	$170,000
Life (years)	5	5
Salvage value	$15,000	$ 50,000
Annual cost	$44,000	$ 70,000
Scanner generated income (annual)	$100,000	$160,000

56. Using general straight line depreciation, compute the book value of the Scanx at the end of two years.

(A) $15,000 (B) $30,000 (C) $45,000 (D) $60,000 (E) $75,000

57. Suppose the Scanx is depreciated by the general double declining balance method. At the end of which year will book value be reduced to salvage value?

(A) 1 (B) 2 (C) 3 (D) 4 (E) 5

58. If income is disregarded, an annual cost comparison indicates that the preferred model is

(A) Scanx by about $22,000/yr.
(B) Scanx by about $26,000/yr.
(C) Scanx by about $48,000/yr.
(D) Holo-Scan by about $2700/yr.
(E) Holo-Scan by about $53,000/yr.

59. The present worth of costs and income for the two models indicates that Holo-Scan is worth about how much more than Scanx?

(A) $80,000 (B) $65,000 (C) $45,000 (D) $36,000 (E) $7600

60. The scanner is served by high-voltage electric lines. Total cost of the lines is $F + tI + L/t$ where t is the thickness of insulation on the lines, F is a fixed cost independent of insulation, tI is the cost of insulation, and L/t is the cost of leakage through the insulation. The thickness of the insulation that will give the lowest cost is

(A) $\sqrt{L/I}$ (B) $F + L/I$ (C) $\sqrt{I/L}$ (D) $[(I+F)/L]^2$ (E) $[L/(I+F)]^2$

Questions 61-65: (use the tables on the next page)

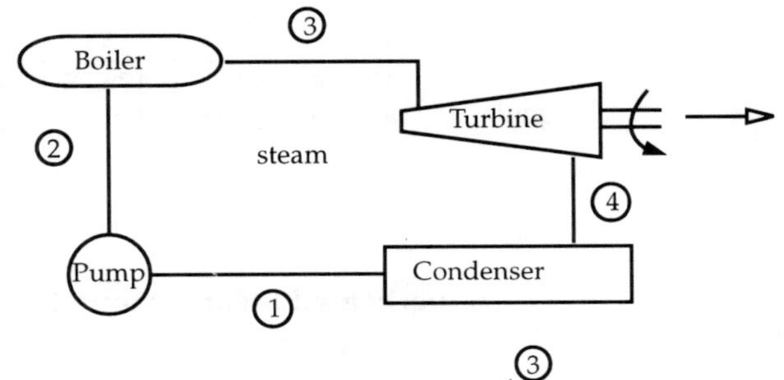

		metric	English
P(high)	=	6 MPa	400 psia
P(low)	=	10 kPa	2 psia
T_3	=	600° C	1000° F
$\dot{m}$	=	5 kg/s	10 lb/sec
ρ (water)	=	1000 kg/m^3	62.4 lb/ft^3

T, P_{high}, P_{low}, 1, 2, 3, 4, s

61. The quality of the steam leaving the turbine is

metric (A) 0.83 (B) 0.85 (C) 0.87 (D) 0.89 (E) 0.91

English (A) 0.87 (B) 0.89 (C) 0.91 (D) 0.93 (E) 0.95

62. The maximum output for the turbine is closest to

metric (A) 4 MW (B) 5 MW (C) 6 MW (D) 7 MW (E) 8 MW

English (A) 4000 Hp (B) 5000 Hp (C) 6000 Hp (D) 7000 Hp (E) 8000 Hp

63. The minimum pump power input is

metric (A) 20 kW (B) 30 kW (C) 50 kW (D) 75 kW (E) 150 kW

English (A) 10 Hp (B) 17 Hp (C) 25 Hp (D) 40 Hp (E) 80 Hp

64. The energy requirement by the boiler is

metric (A) 14.1 MJ/s (B) 17.3 MJ/s (C) 22.3 MJ/s (D) 31.0 MJ/s (E) 39.2 MJ/s

English (A) 11,300 BTU/s (B) 14,300 BTU/s (C) 18,100 BTU/s (D) 24,800 BTU/s (E) 32,300 BTU/s

65. The efficiency of this Rankine cycle is

metric (A) 25% (B) 30% (C) 35% (D) 40% (E) 45%

English (A) 20% (B) 25% (C) 30% (D) 35% (E) 40%

Saturated H_2O — Pressure Table (Metric and English)

P, MPa	T, °C	v_f	v_g	u_f	u_g	h_f	h_{fg}	h_g	s_f	s_{fg}	s_g
0.01	45.8	.00101	14.67	191.8	2437.9	191.8	2392.8	2584.6	.6491	7.5019	8.151
6	275.6	.00132	.0324	1205.4	2589.7	1213.3	1571.0	2784.3	3.0273	2.8627	5.890

P, psia	T, °F	v_f	v_g	u_f	u_g	h_f	h_{fg}	h_g	s_f	s_{fg}	s_g
2	126.04	.01623	173.75	94.02	1051.8	94.02	1022.1	1116.1	.17499	1.7448	1.9198
400	444.70	.01934	1.162	422.8	1119.5	424.2	781.2	1205.5	.6218	.8638	1.4856

Superheated Steam (Metric)
P = 6 MPa

	v, m^3/kg	u, kJ/kg	h, kJ/kg	s, kJ/kg·K
600°C	.06525	3266.9	3658.4	7.1699

Superheated Steam (English)
P = 400 psia

	v, ft^3/lbm	u, BTU/lbm	h, BTU/lbm	s, BTU/lbm·F
1000°F	2.136	1365.5	1523.6	1.7632

Questions 66-70:

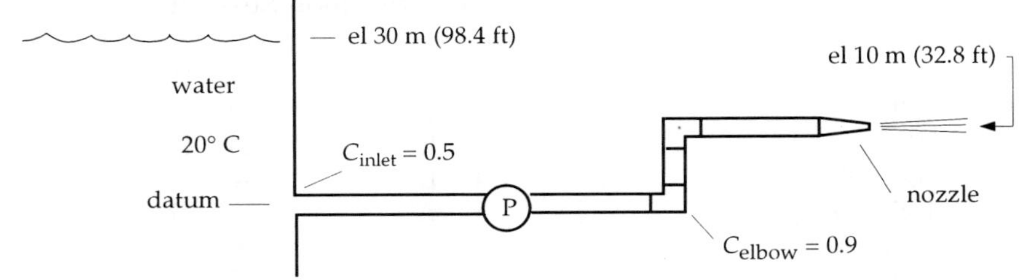

pipe material	= cast iron
pipe diameter	= 5 cm (1.97 in)
pipe length	= 200 m (656 ft)
nozzle exit dia	= 2 cm (0.79 in)
density of water	= 1000 kg/m^3 (1.94 slug/ft^3)
flow rate	= 0.02 m^3/s (0.706 ft^3/sec)

66. The velocity in the pipe is
(A) 6 m/s (20 ft/sec) (B) 8 m/s (26 ft/sec) (C)10 m/s (33 ft/sec)
(D) 12 m/s (40 ft/sec) (E) 14 m/s (46 ft/sec)

67. The pressure just upstream of the nozzle is
(A) 520 kPa (75 psi) (B) 720 kPa (104 psi) (C) 930 kPa (135 psi)
(D) 1220 kPa (182 psi) (E) 1980 kPa (284 psi)

68. What is the maximum distance the pump can be located from the reservoir inlet to avoid cavitation?
(A) 4 m (13 ft) (B) 6 m (20 ft) (C) 10 m (33 ft)
(D) 16 m (52 ft) (E) 24 m (79 ft)

69. Using the pump energy requirement of 157 Hp, find the maximum pressure rise across the pump. (1 Hp = 746 W.)
(A) 2100 kPa (300 psi) (B) 4300 kPa (620 psi) (C) 5900 kPa (850 psi)
(D) 8100 kPa (1200 psi) (E) 9200 kPa (1330 psi)

70. What force is needed to hold the nozzle onto the pipe?
(A) 1200 N (270 lb)
(B) 2820 N (626 lb)
(C) 4600 N (1030 lb)
(D) 6200 N (1400 lb)
(E) 7800 N (1750 lb)

STOP!

•You are finished with the Diagnostic Exam!•

If you finish before 4 hours is up, you may return to any of the problems in the afternoon session. You may not return to the morning session.

Answers to the Practice Examination

The Morning Session

1. B	51. D	101. E	11. A	61. B	111. C
2. D	52. E	102. D	12. D	62. C	112. A
3. E	53. A	103. C	13. E	63. D	113. D
4. C	54. C	104. B	14. B	64. D	114. B
5. B	55. D	105. D	15. C	65. D	115. D
6. C	56. B	106. A	16. C	66. E	116. A
7. B	57. E	107. A	17. B	67. B	117. D
8. A	58. C	108. C	18. E	68. B	118. E
9. D	59. C	109. E	19. A	69. A	119. B
10. C	60. E	110. E	20. A	70. A	120. C

1. B
2. D
3. E
4. C
5. B
6. C
7. B
8. A
9. D
10. C
11. A
12. D
13. E
14. B
15. C
16. C
17. B
18. E
19. A
20. A
21. A
22. B
23. B
24. D
25. C
26. E
27. B
28. C
29. C
30. E
31. E
32. D
33. A
34. B
35. C
36. B
37. D
38. B
39. E
40. B
41. C
42. C
43. B
44. D
45. C
46. B
47. D
48. B
49. C
50. B
51. D
52. E
53. A
54. C
55. D
56. B
57. E
58. C
59. C
60. E
61. B
62. C
63. D
64. D
65. D
66. E
67. B
68. B
69. A
70. A
71. C
72. B
73. B
74. C
75. B
76. A
77. A
78. B
79. C
80. B
81. D
82. B
83. E
84. A
85. B
86. D
87. E
88. E
89. B
90. B
91. D
92. E
93. E
94. A
95. C
96. D
97. B
98. D
99. C
100. E
101. E
102. D
103. C
104. B
105. D
106. A
107. A
108. C
109. E
110. E
111. C
112. A
113. D
114. B
115. D
116. A
117. D
118. E
119. B
120. C
121. D
122. C
123. A
124. C
125. A
126. E
127. B
128. C
129. D
130. C
131. B
132. E
133. A
134. A
135. E
136. C
137. C
138. B
139. C
140. B

The Afternoon Session

1. D
2. C
3. E
4. D
5. A
6. E
7. B
8. E
9. D
10. B
11. B
12. B
13. C
14. D
15. B
16. B
17. A
18. B
19. E
20. D
21. E
22. A
23. B
24. A
25. C
26. A
27. C
28. D
29. D
30. E
31. C
32. E
33. E
34. E
35. E
36. C
37. D
38. A
39. E
40. D
41. B
42. C
43. E
44. B
45. D
46. E
47. E
48. C
49. C
50. A
51. D
52. A
53. A
54. B
55. C
56. D
57. D
58. C
59. D
60. A
61. C
62. D
63. B
64. B
65. D
66. C
67. E
68. C
69. C
70. B

Practice Exam Solutions—Morning

1. **B** The point (1, 4) must satisfy the equation. Only (B) does this.
   $m=(1-4)/(4-1)=-1.$ $y-4=-1(x-1)$ or $x+y=5.$

2. **D** $(x-3)^2+(y-3)^2=16.$ $r^2=16.$ $\therefore r=4.$

3. **E** $0+9+0-0-0-12=-3.$

4. **C** $\dfrac{\cos 0}{1}=1.$ (We differentiated both numerator and denominator.)

5. **B** The curves intersect at the origin and at (1, 1). Using a horizontal strip.
   $\text{Area}=\int(x_2-x_1)dy=\int_0^1\left(\sqrt{y}-y^2\right)dy=\frac{2}{3}(1)-\frac{1}{3}=\frac{1}{3}.$

6. **C** $i=e^{\pi i/2}.$ $\therefore i^{1/3}=e^{\pi i/6}=\cos\frac{\pi}{6}+i\sin\frac{\pi}{6}=\frac{\sqrt{3}}{2}+\frac{1}{2}i.$

7. **B** $2x-4+2y\dfrac{dy}{dx}+4\dfrac{dy}{dx}=0.$ $\therefore\dfrac{dy}{dx}=\dfrac{2-x}{y+2}=0.$ $\therefore x=2.$

8. **A** $\dfrac{dy}{dt}=-4y.$ $\dfrac{dy}{y}=-4dt.$ $\therefore \ln y=-4t+\ln C.$ $\therefore\dfrac{y}{C}=e^{-4t}.$
   $y(0)=2.$ $\therefore C=2.$ $\therefore y(t)=2e^{-4t}.$

9. **D** $y_p=A.$ $4A=20.$ $\therefore A=5.$ $\therefore y_p=5.$ Note: $y_h(\infty)=0.$

10. **C** $x_1-x_3=2$ and $x_3=0.$ $\therefore x_1=2$ and $x_2=1.$

11. **A** $\mathbf{Ax}=\lambda\mathbf{x}$ or $(\mathbf{A}-\lambda\mathbf{I})\mathbf{x}=\mathbf{0}$. This is true if $|\mathbf{A}-\lambda\mathbf{I}|=\mathbf{0}$.

12. **D** $\mathbf{A}\cdot\mathbf{i}_B=(10\mathbf{i}-2\mathbf{j}+8\mathbf{k})\cdot\dfrac{2\mathbf{i}-6\mathbf{j}+3\mathbf{k}}{\sqrt{4+36+9}}=\frac{1}{7}(20+12+24)=8.$

13. **E** $\nabla\cdot\mathbf{u}=\dfrac{\partial}{\partial x}(xy)+\dfrac{\partial}{\partial y}\left(2y^2\right)+\dfrac{\partial}{\partial z}(-yz)=y+4y-y=4(1)=4.$

14. **B** $$x_c=\frac{\int\left(\frac{y}{2}\right)y\,dx}{\int y\,dx}=\frac{\frac{1}{2}\int_0^2 x^4dx}{\int_0^2 x^2dx}=\frac{1}{2}\times\frac{32/5}{8/3}=\frac{6}{5}=1.2.$$
    Note: A horizontal strip was used.

15. **C** $V=\int_0^4\pi x^2dy=\pi\int_0^4 y\,dy=\pi\dfrac{4^2}{2}=8\pi.$

16. **C** $P(\text{red})P(\text{black})=\dfrac{40}{100}\times\dfrac{25}{100}=0.4\times0.25=0.1.$ $0.1\times100=10.$

17. **B** $\sigma = \left[\dfrac{\left(5-4^2\right)+(2-4)^2+(6-4)^2+(3-4)^2}{3}\right]^{1/2}$

$= \left(\dfrac{1+4+4+1}{3}\right)^{1/2} = \sqrt{\dfrac{10}{3}} = 1.826.$

18. **E** $16^2 x + 16y + 16^0 z = 739.$

Let $x = 2.$ $16^2 \times 2 = 512.$ $739 - 512 = 227.$

Let $y = 14 = E.$ $16 \times 14 = 224.$ $739 - 736 = 3.$

Let $z = 3.$ $\therefore (2E3)_{16} = (739)_{10}.$

19. A

20. A

21. **D** See section 3.2 of the book.

25. C Beyond the region of linear stress-strain response, plastic deformation begins, and strain is no longer a linear function of stress.

26. **E** The onset of plastic flow or the point at which nonlinear stress-strain behavior begins is often difficult to ascertain. Thus, as a convention, the stress corresponding to the point of intersection of a straight line parallel to the linear position of the stress-strain curve, and at a total strain of 0.002, is defined as the 0.2% yield stress.

27. **B** The occurrence of upper- and lower-yield points is a phenomenon associated with the barrier to dislocation motion posed by impurity or solute atoms. In mild steel, the carbon atoms produce such a barrier. A higher stress (upper-yield point) is needed to "break loose" the dislocations from these barriers; once freed they can move at a lower stress (lower yield-point).

28. C At an elevated temperature, and at a constant load even much lower than the load necessary for yielding, metals and alloys fail due to the nucleation and growth of voids.

29. C The ductility of carbon steel is very sensitive to temperature, and brittle fracture occurs at a temperature below the ductile-brittle transition temperature. This temperature increases as the carbon content increases.

30. **E** The reason fatigue failure is so treacherous is because it can occur well below the yield stress of a metal or alloy. Failure of an airplane wing section due to repeated vibration is a drastic example.

31. **E** The presence of grain boundaries strengthens steel. Also small non-elongated grains give rise to a more isotropic deformation behavior. Grain refinement often is a technique to increase strength of steel.

32. **D** Annealing of cold worked metals or alloys produces recrystallization and grain growth. Associated with this process is the loss of cold-worked hardness or strength.

33. **A** See the Fe-C phase diagram. At 1100° C, a 0.18% C sample is 100 austenite (γ-phase). On cooling first the γ + Fe_3C phase field is

crossed, and then at temperatures below 723° C the sample is in the ferrite + pearlite phase field.

34. **B** Note the difference between the terms ending in *tic* and *toid* (eutectic, eutectoid, peritectic, peritectoid, etc.). Eutec*tic* is the reaction between two liquids to form a solid. Eutec*toid* is the reaction between two solids to form a third solid.

35. **C** If you know that Cl and I are monovalent and form HCl and HI, then At, in the same group (VIIA) will also form HAt, not H_2At. Alternatively, you could examine the electron configuration and discover that At is one electron short of Rn (86), and is therefore monovalent.

36. **B** Look at the periodic table. Sr is the furthest to the left, therefore the least electronegative, therefore the least likely to form a negative ion.

37. **D** Negative ions migrate towards the positive anode. Br^- is the only negative ion present (H_2O^- is not a stable ion).

38. **B** See Avogadro's Law: You therefore know that 3.7 l of N_2 contains the same number of molecules (or moles) as 3.7 l of He, i.e. 3.7/22.4.

39. **E** If you know that the oxidation number of K is +1 and O is –2, then Cl must be +7 to give a net of 0 (+1 – 2(4) + 7 = 0).

40. **B** From the collision theory of reaction rate it is intuitive that more collisions result in a faster rate and fewer result in a slower rate.

41. **C** Since HCl is written over the arrows, it is most likely a catalyst. Note that it does not appear in the reactants or products. Catalysts do not normally alter the equilibrium constant.

42. **C** Since the moles of gas is the same on both sides of the equilibrium, the volume of gas is unchanged. Therefore, changing *P* will not affect the equilibrium. Since heat is input on the left, increasing *T* will shift the equilibrium to the right (Le Chatelier principle).

43. **B** It is sufficient to recognize that one molecule of propene (3 carbon atoms) gives one molecule of CH_3CO_2H (2 carbon atoms) and one of CO_2 (1 carbon atom). Therefore one mole of propene gives one mole of CO_2 = 22.4 l.

44. **D** First you must balance the equation:

$$2\ Fe + 3\ CuSO_4 = Fe_2(SO4)_3\ + 3Cu$$

Then it is clear that 2 gram-atoms of Fe gives 3 gram-atoms of Cu and one gram-atom of Fe gives 1.5 gram-atom or 1.5(63.5) = 95 g of Cu.

45. **C** The valence electronic configuration is that of the outermost shell of an atom, i.e. the 3 level in sulfur.

46. **B** If you know that Mg is in the same group as Cu and Be (group IIA) and therefore divalent, and Cl is in group VIIA (F, Br, I) and therefore monovalent, then the answer clearly is $MgCl_2$. You could also see from the periodic table that Mg (atomic number = 12) contains 2 more electrons than the noble gas neon (atomic number = 10) and is therefore divalent. Cl (atomic number = 17) has one less electron than the noble gas At (atomic number = 18) and is therefore monovalent.

47. **D** Ethane, a non-polar organic molecule, is held together by covalent bonds, as is the norm for compounds of carbon and hydrogen.

48. **B** Carbon, a solid, does not enter into the equilibrium constant expression since its concentration is a constant. Cl_2 enters as $(Cl_2)^2$ since there are 2 moles of Cl_2 formed from every mole of CCl_4.

49. **C** $P = 200(P/A)_{10}^{6} = \$1472.$

50. **B** By inspection.

51. **D** $F = 12{,}000(F/P)_{3}^{12} - 4000 = \$12{,}860.$

52. **E** $D = (20{,}000 - 2{,}000)/5 = 3600.\ \ V_3 = 20{,}000 - 3\times 3600 = \$9200.$

53. **A** $A = (10{,}000 - 1{,}500)(A/F)_{10}^{4} = \$708.$

54. **C** $P = 1600(P/A)_{\infty}^{6} + 1400(P/A)_{5}^{6} = \$32{,}600.$

55. **D** $1000 = 300(P/A)_{4}^{i}.\ \ (P/A)_{4}^{i} = 3.33.\ \ i = 7.7\%.\ \ \therefore i_n = 52\times 7.7 = 400\%.$

56. **B** $P = 12{,}000 + 500(P/A)_{3}^{10} + 1000(P/G)_{3}^{10} = \$15{,}570.$

57. **E** $2P = (F/P)_{5}^{i}P.\ \ (F/P)_{5}^{i} = (1+i)^5 = 2.\ \ \therefore i = 15\%.$

58. **C** $i = \frac{12\%}{12} = 1\%.\ \ i_e = 1.01^{12} - 1 = 0.127.\ \ \therefore i_e = 12.7\%.$

59. **C** $P = 100{,}000(P/A)_{\infty}^{8} = \$1{,}250{,}000.$

60. **E** Reduce the circuit using equivalent resistances. $I = \frac{V}{R} = \frac{12}{4} = 3\text{ amp}.$

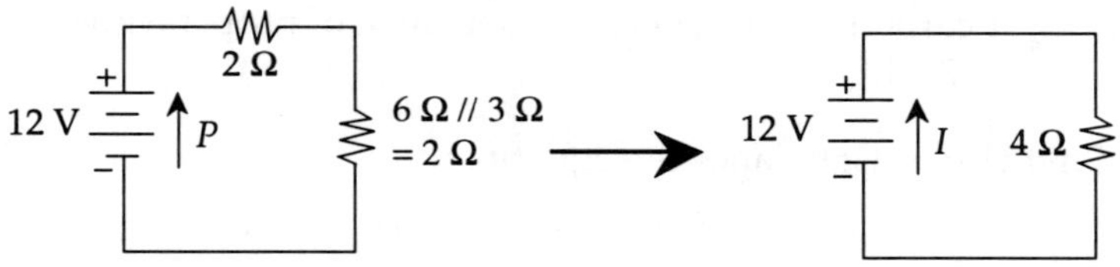

12 V, 2 Ω, P, 6 Ω // 3 Ω = 2 Ω → 12 V, I, 4 Ω

61. **B** $9 = 18I.\ \ \therefore I = 1/2.\ \ \therefore V = IR = \frac{1}{2}(5) = 2.5V.$

62. **C** $8 + 4 = 12\,\Omega.\ \ \frac{1}{\frac{1}{6}+\frac{1}{12}} = \frac{72}{18} = 4\,\Omega.\ \ \therefore 4 + 4 = 8\,\Omega.$

63. **D** First find the voltage across the 5Ω resistor. $V_{ab} = \frac{4}{12} \times 12 = 4$ volts

12 V, 8 Ω, a, b, 5 Ω // 20 Ω = 4 Ω

$$P_5 = \frac{V_{ab}^2}{R} = \frac{4^2}{5} = 3.2 \text{ W.}$$

64. **D** The rms source voltage is $V_s = \frac{V_{max}}{\sqrt{2}} = \frac{100}{1.414} = 70.7$ V. The voltage across the 20Ω resistor is $V_{20} = \frac{20}{20+5} V_s = \frac{4}{5} \times 70.7 = 56.6$ V.

65. **D** In order for V_S and I_S to be in phase, the circuit must be resonant. This requires that the inductor and capacitor have equal values of susceptance.

$$Y_L + Y_C = 0, \quad \frac{1}{j20} + \frac{1}{-jX_C} = 0, \quad -j\frac{1}{20} + j\frac{1}{X_C} = 0, \quad X_C = 20\ \Omega.$$

66. **E** Note that the voltage across the 10Ω resistor does not change as X_C changes. It remains at $V_S = 100\angle 0°$ V. $\therefore P_{10} = \frac{V_S^2}{R} = \frac{100^2}{10} = 1000$ W.

67. **B** Compute the rms value of the square wave. Plot $v^2(t)$ for one period:

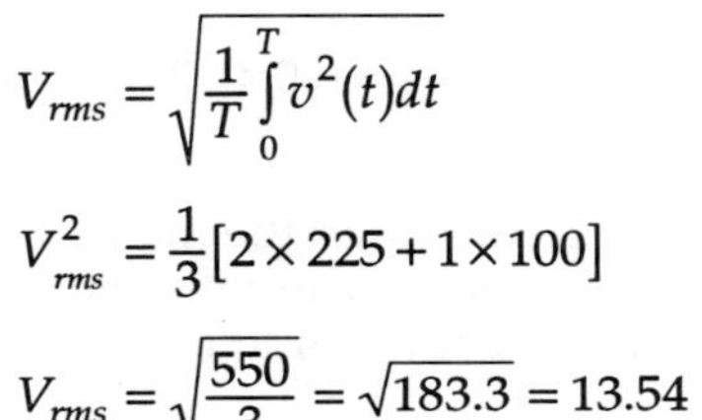

$$V_{rms} = \sqrt{\frac{1}{T}\int_0^T v^2(t)dt}$$

$$V_{rms}^2 = \frac{1}{3}[2 \times 225 + 1 \times 100]$$

$$V_{rms} = \sqrt{\frac{550}{3}} = \sqrt{183.3} = 13.54 \text{ V}$$

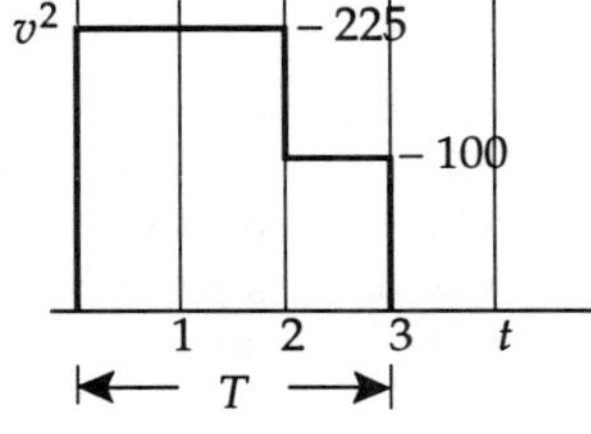

68. **B** The rms source voltage is $V_S = \frac{V_{max}}{\sqrt{2}} = \frac{141.4}{\sqrt{2}} = 100$

$$|V_C| = |I||X_C| \le 50$$

$$|I|\,50 \le 50, \quad |I| \le 1, \quad I = \frac{100}{R - j50}$$

$$|I| = \frac{100}{\sqrt{R^2 + 2500}} \le 1, \quad 10000 \le R^2 + 2500, \quad R^2 \ge 7500. \quad \therefore R \ge 86.6\,\Omega.$$

69. **A** $p(t) = v(t)i(t) = 70.7\sin(377t) \times 5.1\sin(377t + 0.16)$
$p(0.001) = 70.7\sin(0.377) \times 5.1\sin(0.377 + 0.16) = 67.9$ W.

70. **A** $I(t) = I_\infty + (I_0 - I_\infty)e^{-\frac{t}{T}}$. The current through the inductor cannot change instantly. $\therefore I_0 = 0$. The inductor has zero resistance.

$$\therefore I_\infty = \frac{15}{75} = 0.2, \; T = \frac{L}{R} = \frac{0.15 \times 10^{-3}}{75} = 2 \times 10^{-6} \text{ and}$$

$$I\left(10^{-6}\right) = 0.2 + (0 - 0.2)e^{-\frac{10^{-6}}{2\times 10^{-6}}} = 0.079 \text{ A.}$$

71. **C** $B = \dfrac{\mu NI}{\ell} = \dfrac{4\pi \times 10^{-9} N2.5}{0.1} = 0.05$ Tesla. Solve for N:

$N = \dfrac{5 \times 10^{-3}}{\pi \times 10^{-6}} = 1591$ turns.

72. **B** $\mathbf{F} = q\mathbf{E},\ \ E = \dfrac{V}{d} \rightarrow F = \dfrac{qV}{d} = \dfrac{1.6 \times 10^{-19} \times 2500}{0.01} = 4 \times 10^{-14}$ N, toward the positive plate.

73. **B** The E field created by one line of charge is $E = \dfrac{\rho_L}{2\pi\varepsilon r} \dfrac{\text{volts}}{\text{meter}} \left(\dfrac{\text{newtons}}{\text{coulomb}}\right)$.

The force on a dq in the other line of charge is $df = dqE = \dfrac{\rho_L dq}{2\pi\varepsilon r}$. Sum the differential forces to get the total force on a 1 meter length. Use $dq = \rho_L dx$.

$$F = \int_0^1 df = \int_0^1 \frac{\rho_L dx \rho_L}{2\pi\varepsilon(3)} = \frac{\rho_L{}^2(1)}{2\pi\varepsilon(3)} = \frac{\left(5.6 \times 10^{-6}\right)^2}{2\pi \times 3 \times 8.85 \times 10^{-12}} = 0.19 \text{ N}.$$

74. **C** Adiabatic $\Rightarrow Q = 0$. Paddle wheel $\Rightarrow W \neq 0$. $\therefore -W = \Delta U$.
Note: $W \neq \int PdV$ since the process is not in quasiequilibrium.

75. **B** $x = \dfrac{m_g}{m_f + m_g} = \dfrac{\dfrac{V}{v_g}}{\dfrac{V}{v_f} + \dfrac{V}{v_g}} = \dfrac{\dfrac{1}{1.673}}{\dfrac{1}{0.00104} + \dfrac{1}{1.673}} = 0.000621$ (using metric units).

The mass of liquid is much, much greater than the mass of vapor. So $x = .0006$ is probably the answer. Check by using the saturated pressure table. We also use mass = Volume/specific volume.

76. **A** Rigid $\Rightarrow W = 0$ (reversible work only).

77. **A** In a rigid vessel, the volume remains constant. If we move straight up on the T-v diagram, x increases. (Temperature increases as heat is added.) Note: it is important to know that the diagram is skewed quite markedly to the left.

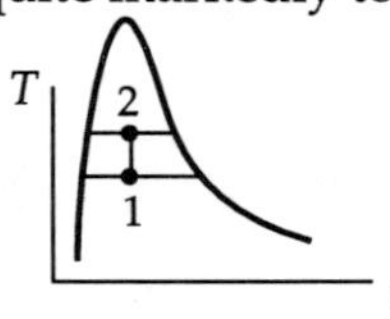

78. **B** The heat transfer is zero since for an ideal (reversible) isentropic process $ds = \dfrac{\delta Q}{T} = 0$.

79. **C** $P_1V_1 = mRT_1$. $(2P_1)(2V_1) = mRT_2$. $\therefore T_2 = \dfrac{4P_1V_1}{mR} = 4T_1$.

80. **B** All processes in a Carnot cycle are reversible.

81. **D** $\dot{W} = \dot{m}\Delta h = 90 \times 4.18(100 - 20) = 30096$ kW or 30 MW

$\dot{W} = \dot{m}\Delta h = 3000 \times 1.00(170 - 70) = 300{,}000$ BTU/hr

82. **C** Work done by a system is positive. Work done on a system is negative.
$Q - W = \Delta U$. $Q = 20 + (-40) = -20$ J

$Q = 2 + (-4000/778) = 3.14$ BTU

83. **E** $\Delta U = mc_v \Delta T = 0$ since $\Delta T = 0$ (isothermal).

84. **A** If the process is reversible $\Delta S_{net} = 0$.

85. **B** δW and δQ are inexact differentials. That's why we don't write dW and dQ.

86. **D** $\phi_2 - \phi_1 = w_{max}$.

87. **E** $Q = \Delta h = 390 - 150 = 240$ kJ/kg. See Fig. 7.6.
$Q = \Delta h = 103 - 0 = 103$ BTU/lbm. See Fig. 7.6E.

88. **E** $\mathbf{M} = \mathbf{F} \times \mathbf{r} = (10\mathbf{i} - 20\mathbf{j} + 40\mathbf{k}) \times (2\mathbf{i} + \mathbf{j} + \mathbf{k}) = (-40 - 20)\mathbf{i} + \_\mathbf{j} + \_\mathbf{k}$.

$\therefore M_x = 60$ N·m.

89. **B** Moments about B: $\sum M_B$: $12F_A = 600 \times 3 + 600 \times 8$. $\therefore F_A = 550$ N.
Note: The force that replaces the triangular distribution acts through the centroid of the triangle; i.e., 4 m from A.

90. **B** Sketch the force triangle.
$\sum M_C$: $1200 \times 3 = F_{AB} \times 4$. $\therefore F_{AB} = 900$ N.

91. **D**
$$\left.\begin{aligned} \frac{4}{5}F_{AB} &= \frac{1}{2}F_{AC} \\ \frac{3}{5}F_{AB} + 0.866F_{AC} &= 1000 \end{aligned}\right\} F_{AB} = 504 \text{ N.}$$

Forces act at point A as shown.

92. **A** Moments about D: $\sum M_D$: $1200 \times 0.866L = F_A \times 3L$. $\therefore F_A = 346$ N.

93. **E** Cut links GF, BF, and BC.

$\sum F_A$: $F_{AG} \times 0.866 = 346$. $\therefore F_{AG} = 400$.

$\sum F_G$: $0.866 \times 400 = 0.866F_{GB}$. $\therefore F_{GB} = 400$.

$\sum F_B$: $0.866 \times 400 = 0.866F_{BF}$. $\therefore F_{BF} = 400$.

94. **A** If the component of the weight acting down the plane exceeds the friction force μF, motion results.

$\sum F_n$: $N = 0.866$ W.

$\sum F_t$: $F = 0.5W = \mu N = \mu \times 0.866W$. $\therefore \mu = \frac{0.5}{0.866} = 0.577.$

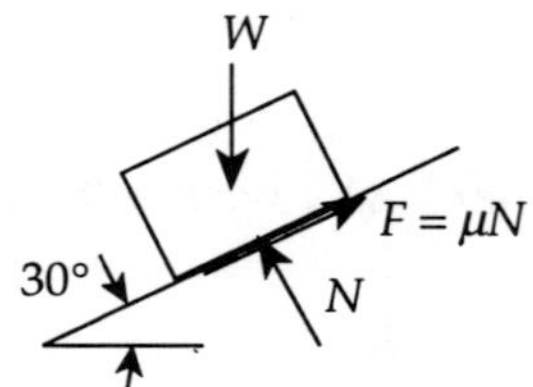

95. **C** The top right pulley:

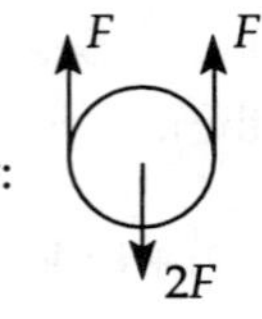

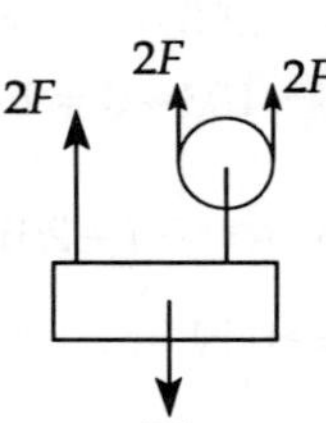

$W = T + 2T$. $2F = T$. $\therefore F = \frac{T}{2} = \frac{W/3}{2} = \frac{W}{6}$.

96. **D** $(2\mathbf{i} - 3\mathbf{j} + \mathbf{k}) \cdot (\mathbf{i} + 2\mathbf{j} - \mathbf{k}) = 2 - 6 - 1 = -5$.

97. **B** The maximum moment occurs at the wall:
$M_{\max} = 1000 \times 4 + 800 \times 2 = 5600$ N · m.

98. **D** $F = 100e^{0.5(2\pi)} = 2314$ N.

99. **C** $\bar{y} = \frac{12(5) + 20(1)}{12 + 20} = 2.5.$

100. **E** Divide the area into a vertical rectangle 2 x 8 and two other rectangles each 4 x 2. Then $I_x = \frac{8 \times 2^3}{3} + \frac{2 \times 8^3}{3} = 362.7$.

101. **E** $320 = \frac{b \times 4^3}{3}$. $\therefore b = 15$. $I_x = I_c + Ad^2 = \frac{15 \times 4^3}{12} + 60 \times 6^2 = 2240.$

102. **D** $V = \sqrt{2gh}$. $h = \frac{40^2}{2 \times 9.8} = 81.6$ m.

103. **C** $g = \frac{V^2}{R}$. $9.8 = \frac{V^2}{6\,400\,000}$. $\therefore V = 7920$ m/s.

104. **B** $v = \frac{ds}{dt}$. $\Delta s = \int v dt = \int_0^6 \frac{20}{6} t dt = \frac{10}{3} \times \frac{6^2}{2} = 60$ m.

105. D $\omega = \omega_0 + \alpha t$ $\quad$ $\omega^2 = \omega_0^2 + 2\alpha(\theta - \theta_0)$

$0 = 20 - 10t$ $\quad$ $0 = 20^2 - 2(10)\theta$ $\quad$ When $t = 4$ sec, $\theta = -20$ rad.

$\therefore t = 2$ when $\omega = 0$. $\quad$ $\therefore \theta = 20$ rad at $t = 2$ sec. $\quad$ $\therefore \theta_{total} = 40$ rad, or 6.37 rev.

106. A $v = v_0 + at$ $\quad$ $v^2 = v_0^2 + 2a(s - s_0)$ $\quad$ $s = s_0 + v_0 t + \frac{1}{2}at^2$

$0 = -10 + 2t$ $\quad$ $0 = 10^2 + 2(2)s$ $\quad$ $= \frac{1}{2}\times 2\times 10^2 = 100$ m

$\therefore t = 5$ when $v = 0$. $\quad$ $\therefore s = -25$ m at $t = 5$ sec. $\quad$ $\therefore s_{total} = 100 + 25 = 125$ m.

107. A $\left.\begin{aligned} 30 &= v_0 t\times 0.866 \\ 10 &= v_0 t\times 0.5 - \frac{1}{2}\times 9.8t^2 \end{aligned}\right\}$ $\quad$ $v_0 t = 34.64$ $\quad$ $10 = 34.64\times 0.5 - \frac{1}{2}\times 9.8t^2$

$\therefore t = 1.222$ sec. $\quad$ $\therefore v_0 = 28.3$ m/s.

108. C $\mathbf{v}_C = \mathbf{v}_B + \mathbf{v}_{C/B}$. $v_B = r\omega_{AB} = 0.5\omega_{AB}$. $v_{C/B} = 0.5\omega_{BC}$.
The velocity polygon with $\mathbf{v}_B \perp \mathbf{r}_{AB}$ and $\mathbf{v}_{C/B} \perp \mathbf{r}_{BC}$ is an equilateral triangle with all sides equal.
$\therefore v_C = 20 = 0.5\omega_{AB}$. $\quad$ $\therefore \omega_{AB} = 40$ rad/s.

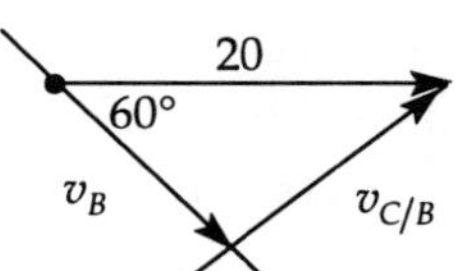

109. E $\mathbf{a}_C = \mathbf{a}_B + \mathbf{a}_{C/B}$. $(a_B)_n = r\omega_{AB}^2$. $(a_B)_t = r\alpha_{AB}$.
$(a_{C/B})_n = r\omega_{CB}^2$. $(a_{C/B})_t = r\alpha_{CB}$. From symmetry of the velocity polygon of No. 108, $\omega_{CB} = \omega_{AB} = 40$ rad/s.
Draw the acceleration polygon:
$(0.5\alpha_{AB})\times 0.5 = (0.5\times 40^2)\times 0.866$. $\quad$ $\therefore \alpha_{AB} = 2771$ rad/s^2.

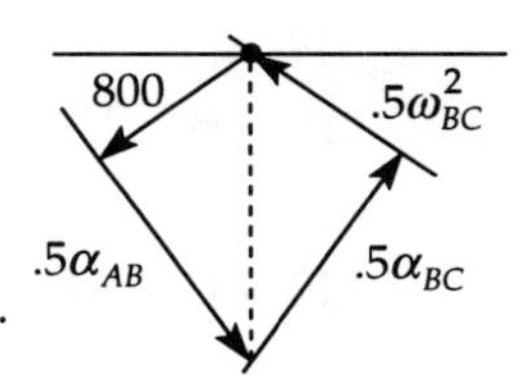

110. E $a_{Coriolis} = 2v\omega = 2\times 4\times \frac{5\times 2\pi}{60} = 4.19$ m/s^2.

111. C $\left.\begin{aligned} \sum M &= I\alpha. \quad 0.4T = \frac{1}{2}\times 50\times 0.4^2(a/0.4) \\ \sum F &= ma. \quad 20\times 9.8 - T = 20a \end{aligned}\right\}$ $\quad$ $\therefore T = 109$ N.

112. A At the extreme position the velocity is zero and the acceleration is maximum. Draw a free-body at the 30° position. Forces in the tangential direction provide:

$$\sum F_t = ma_t$$

$$mg\cos 60° = ma_t. \quad \therefore a_t = 9.8\times 0.5 = 4.9 \text{ m/s}^2.$$

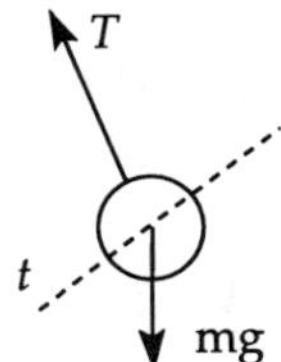

113. D $\Delta PE + \Delta KE = 0$. $\quad -(20\times 9.8)\times 2 + \frac{1}{2}\times 20\times (0.4\omega)^2 + \frac{1}{2}\left(\frac{1}{2}\times 50\times 0.4^2\right)\omega^2 = 0$.

$\therefore \omega = 10.43$ rad/s.

114. **B** $\frac{1}{2}mv^2 = \frac{1}{2}Kx^2$. $\frac{1}{2} \times 50 \times 40^2 = \frac{1}{2} \times 20\,000x^2$. $\therefore x = 2$ m.

115. **D** $e = -\frac{\text{separation velocity}}{\text{approach velocity}}$. $0.8 = \frac{v' \sin\theta}{v \sin 30°} = \frac{v \cos 30° \sin\theta}{v \sin 30° \cos\theta}$.

$\tan\theta = 0.8\frac{\sin 30°}{\cos 30°} = 0.462$. $\therefore \theta = 24.8°$.

Note: $v \cos 30° = v' \cos\theta$ in the tangential direction.

116. **A** $\delta = \frac{PL}{AE} = \frac{40\,000 \times 20}{\pi \times 0.01^2 \times 210 \times 10^9} = 0.0121$ m or 1.21 cm.

$= \frac{10{,}000 \times 60 \times 12}{\pi \times 0.4^2 \times 30 \times 10^6} = 0.477$ in.

117. **D** $\sigma_{max} = \frac{6+0}{2} + \left[(6-0)^2/4 + 4^2\right] = 8$ kPa.

118. **E** $\delta_t = \alpha(T - T_0)L = \frac{PL}{AE} = \sigma\frac{L}{E}$.

$11.7 \times 10^{-6} \times 40 = \frac{\sigma}{210 \times 10^9}$. $\therefore \sigma = 98.3 \times 10^6$ Pa

$6.5 \times 10^{-6} \times 80 = \frac{\sigma}{30 \times 10^6}$. $\therefore \sigma = 15{,}600$ psi

119. **B** The vertical shear stress varies parabolically over the area.

120. **C** $M = PL$ and $V = P$. $\frac{\sigma}{\tau} = \frac{My/I}{VQ/Ib} = \frac{Myb}{VQ} = \frac{PL(h/2)b}{P(hb/2)(h/4)} = 4\frac{L}{h}$.

121. **D** Find the reactive forces. (Draw the shear diagram. The maximum shear is P.) $R_{left} \times 2L = PL - P\frac{L}{2}$. $\therefore R_{left} = P/4$ and $R_{right} = 7P/4$.
$\therefore V_{max} = P$ at right support.

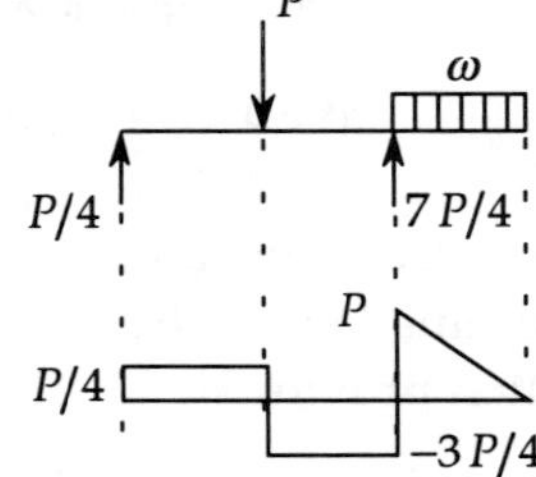

122. **C** The moment is the area under the shear diagram.
$M_{max} = PL/2$ over right support.

123. **A** $\tau = \frac{Tr}{J} = \frac{40 \times 0.02}{\pi \times 0.02^4/2} = 3.18 \times 10^6$ Pa

$= \frac{4000 \times 2}{\pi \times 2^4/2} = 318$ psi

124. **E** $P_{cr} = \dfrac{\pi^2 EI}{k^2L^2} = \dfrac{\pi^2 \times 210\times10^9 \times \pi \times 0.03^4/4}{2^2\times 20^2} = 824\text{ N.} \quad \therefore P_{actual} = 412\text{ N.}$

$= \dfrac{\pi^2 \times 30\times10^6 \times \pi \times 1^4/4}{2^2\times(60\times12)^2} = 112\text{ lb.} \quad \therefore P_{actual} = 56\text{ lb.}$

125. **A** $\sigma = \dfrac{My}{I} = \dfrac{2000\times0.2\times0.02}{\pi\times0.02^4/4} = 63.7\times10^6\text{ Pa.} \quad \tau = \dfrac{Tr}{J} = \dfrac{500\times0.02}{\pi\times0.02^4/2} = 39.8\times10^6\text{ Pa.}$

$= \dfrac{200\times10\times1}{\pi\times1^4/4} = 2546\text{ psi.} \qquad = \dfrac{5000\times1}{\pi\times1^4/2} = 3183\text{ psi.}$

$\therefore \tau_{max} = 50.9$ MPa and 3428 psi using Mohr's circle (or the equations).

126. **E**

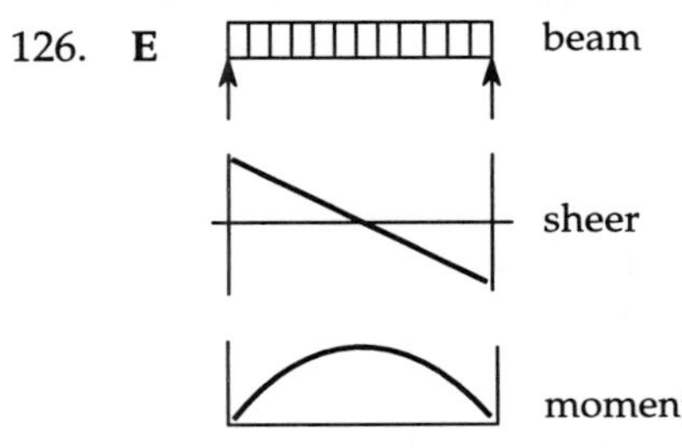

127. **B** The flow separates upstream and downstream of the contraction with relatively large losses.

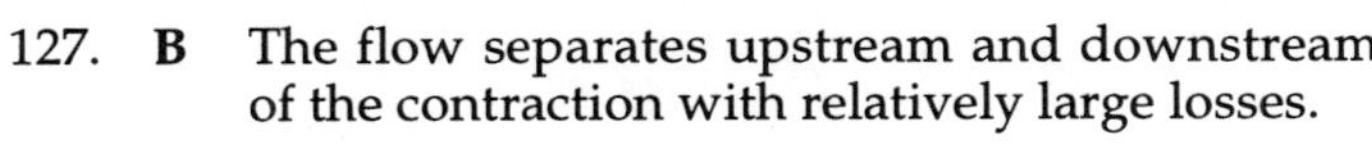

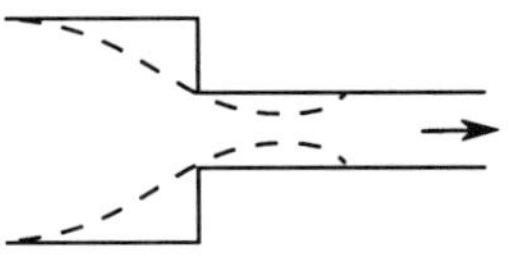

128. **C** In an orifice, the minimum pressure (maximum velocity) occurs at the vena contracta. The loss coefficient is based on this minimum pressure.

129. **D** The energy equation for a control volume, with no work or heat transfer demands that $\Delta h = 0$.

130. **C** The pressure distribution is triangular ($p = 0$ at the surface and $p = \gamma h$ at the hinge.) The force of the water on the gate acts through the centroid of the triangular distribution; i.e., 5/3 m from the hinge. Moments about the hinge gives:

$F = \gamma h_c A = 9800\times(5\times0.866/2)\times(5\times1) = 106\,000\text{ N.} \quad \therefore P = F/3 = 35.3\text{ kN.}$

$= 62.4\times(15\times0.866/2)\times(15\times1) = 6080\text{ lb.} \qquad = 2020\text{ lb.}$

131. **B** Shear forces result from viscosity which dominates in a laminar flow. The turbulent stress due to $\overline{u'v'}$ is an inertial effect.

132. **E** $[V] = L/T,\ [D] = L,\ [\dot{W}] = [\text{force}\times\text{velocity}] = M\dfrac{L}{T^2}\times\dfrac{L}{T} = ML^2/T^3,$

$[\rho] = M/L^3$. To cancel T, we must have $\dot{W}/V^3$. To cancel M, we divide by ρ: $\dot{W}/V^3\rho$. To cancel L we have $\dot{W}/V^3\rho D^2$.

133. **A** Bernoulli's equation is only applicable in a steady, viscous, incompressible, flow along a streamline in an inertial coordinate system.

It must be modified to accommodate other effects, such as compressible viscous flow.

134. **A** EGL has elevation $\frac{V^2}{2g}+\frac{p}{\gamma}+z$. The HGL has elevation $\frac{p}{\gamma}+z$.

135. **E** Darcy's equation is $h_L = f\frac{L}{D}\frac{V^2}{2g}$ and is $\frac{\Delta p}{\gamma}$ for a horizontal section.

136. **C** To obtain supersonic flow from a reservoir the subsonic velocity must increase in a converging section until $m = 1$ at the throat and then continue to increase in the diverging section. As velocity increases temperature (and enthalpy) decrease. Velocity decreases in a converging section if $M > 1$.

137. **C** For free surface flows (dams, ships, open channels, etc.) the Froude number governs. Hence:

$$(\mathrm{Fr})_m = (\mathrm{Fr})_p. \quad \frac{V_m^2}{L_m g_m} = \frac{V_p^2}{L_p g_p}. \quad \therefore \frac{V_m}{V_p} = \sqrt{\frac{L_m}{L_p}} = \frac{1}{3}.$$

$$Q_m^* = Q_p^*. \quad \frac{Q_m}{V_m L_m^2} = \frac{Q_p}{V_p L_p^2}. \quad \therefore Q_m = Q_p \frac{V_m}{V_p}\frac{L_m^2}{L_p^2} = 80 \times \frac{1}{3} \times \left(\frac{1}{9}\right)^2 = 0.329 \text{ m}^3/\text{s}.$$

$$= 800 \times \frac{1}{3} \times \left(\frac{1}{9}\right)^2 = 3.29 \text{ cfs}.$$

138. **B** $A_1V_1 = A_2V_2$. $\quad 0.02 \times 20 = 2\pi \times 0.2t \times 20. \quad \therefore t = 0.0159$ m or 16 mm.

$0.2 \times 60 = 2\pi \times t \times 60. \quad \therefore t = 0.0318$ ft or 0.382 in.

Note: $V_2 = V$ since Bernoulli's eq. demands that if $p_1 = p_2$.

139. **C** Momentum: $-F = \rho AV(V_{2x} - V_{1x})$. $\quad F = 1000 \times 0.02 \times 20(20-0) = 8000$ N.

$= 1.94 \times 0.2 \times 60(60-0) = 1368$ lb.

140. **B** Momentum: $-F = \rho A v_r (v_{r2} - v_{r1})_x$. $\quad F = 1000 \times 0.02 \times 10(10-0) = 2000$ N.

(v_r is relative speed) $\quad = 1.94 \times 0.2 \times 30(30-0) = 349$ lb.

Note: With a single blade we must also use relative speed for $\dot{m} = \rho AV$ since not all of the fluid leaving the jet has its momentum changed.

Practice Exam Solutions—Afternoon

1. **D** $\mathbf{A}\cdot\mathbf{i}_C = (3\mathbf{i}+2\mathbf{j}+2\mathbf{k})\cdot\dfrac{2\mathbf{i}+2\mathbf{j}+\mathbf{k}}{\sqrt{4+4+1}} = \frac{1}{3}(6+4+2) = 4.$

2. **C** $\mathrm{A}\cdot\mathrm{B} = AB\cos\theta.\quad (3\mathrm{i}+2\mathrm{j}+2\mathrm{k})\cdot(\mathrm{j}-\mathrm{k}) = \sqrt{9+4+4}\sqrt{1+1}\cos\theta.$

   $\therefore 0 = \cos\theta.\quad \therefore \theta = 90°.$

3. **E** $\mathrm{B}\cdot\mathrm{C} = \begin{vmatrix} 0 & 1 & -1 \\ 2 & 2 & 1 \\ \mathbf{i} & \mathbf{j} & \mathbf{k} \end{vmatrix} = (1+2)\mathrm{i}+(-2-0)\mathrm{j}+(0-2)\mathrm{k} = 3\mathrm{i}-2\mathrm{j}-2\mathrm{k}.$

4. **D** $\nabla\cdot\mathbf{D} = \left(\dfrac{\partial}{\partial x}\mathbf{i}+\dfrac{\partial}{\partial y}\mathbf{j}+\dfrac{\partial}{\partial z}\mathbf{k}\right)\cdot\left(x^2\mathbf{i}+xy\mathbf{j}+z^2\mathbf{k}\right) = 2x+x+2z = 3-2 = 1.$

5. **A** $\nabla\times\mathbf{D} = \begin{vmatrix} \dfrac{\partial}{\partial x} & \dfrac{\partial}{\partial y} & \dfrac{\partial}{\partial z} \\ x^2 & xy & z^2 \\ \mathbf{i} & \mathbf{j} & \mathbf{k} \end{vmatrix} = \left(\dfrac{\partial z^2}{\partial y}-\dfrac{\partial xy}{\partial z}\right)\mathbf{i}+\left(\dfrac{\partial x^2}{\partial z}-\dfrac{\partial z^2}{\partial x}\right)\mathbf{j}+\left(\dfrac{\partial xy}{\partial x}-\dfrac{\partial x^2}{\partial y}\right)\mathbf{k} = y\mathbf{k} = 0\,.$

6. **E** $\ddot{y}+25y = 0.\quad r^2+25 = 0.\quad \therefore r = \pm 5i.\quad \therefore y_h(t) = c_1e^{5it}+c_2e^{-5it} = A\cos 5t + B\sin 5t.$

7. **B** $y_p = A\sin 3t.\quad -9A+25A = 16.\quad \therefore A = 1.\quad \therefore y_p(t) = \sin 3t.$

8. **E** The natural frequency is the coefficient of t in the homogeneous solution: $\omega_n = 5$ rad/s. (See No. 6).

9. **D** $y(t) = y_h(t)+y_p(t) = A\cos 5t + B\sin 5t + \sin 3t.$

   $y(0) = 0 = A.$

   $\dot{y}(0) = 8 = 5B+3.\quad \therefore B = 1.\quad \therefore y(t) = \sin 5t + \sin 3t.$

10. **B** Steady state is the particular solution. (Damping always drives the homogeneous solution to zero for large time.) Note: Substitute into the differential equation.

    $y_p = A\sin 5t + B\cos 5t.$

    $-25A\sin 5t - 25B\cos 5t + 40A\cos 5t - 40B\sin 5t + 25A\sin 5t + 25B\cos 5t = 16\sin 5t.$

    $$\left.\begin{aligned} \sin 5t:&\quad -40B = 16.\quad \therefore B = -0.4 \\ \cos 5t:&\quad 40A = 0.\quad \therefore A = 0 \end{aligned}\right\} \therefore y_p(t) = y_{\text{steady-state}} = -0.4\cos 5t.\quad \text{Amp} = 0.4.$$

11. **B** $y = mx+b = -3x+4.\quad \therefore m = -3.$

12. **B** $y = 4 - 3x = x^2$. $x^2 + 3x - 4 = 0$. $x = \frac{-3 \pm \sqrt{9+16}}{2} = \frac{-3 \pm 5}{2} = 1$ or -4.

$\therefore$ Use $x = 1$. $\therefore y = 1^2 = 1$.

13. **C** $\bar{y} = \frac{\int y(x_2 - x_1)dy}{\int (x_2 - x_1)dy} = \frac{\int_0^1 y\left(\frac{4-y}{3} - \sqrt{y}\right)dy}{\int_0^1 \left(\frac{4-y}{3} - \sqrt{y}\right)dy} = \frac{\frac{2}{3} - \frac{1}{9} - \frac{2}{5}}{\frac{4}{3} - \frac{1}{6} - \frac{2}{3}} = 0.311.$

14. **D** $y = mx + b = \frac{1}{3}x + 0$. $\therefore x = 3y$.

15. **B** Use a horizontal disk of thickness dy. $V = \int_0^2 \pi x^2 dy = \pi \int_0^2 y dy = 2\pi$.

16. **B** $\begin{vmatrix} 2 & 3 \\ 2 & 1 \end{vmatrix} = 2 - 6 = -4$.

17. **A** $\begin{bmatrix} 2 & 3 \\ 2 & 1 \end{bmatrix}^+ = \text{adj}\,\mathbf{A} = \begin{bmatrix} 1 & -3 \\ -2 & 2 \end{bmatrix}$. The adjoint is the transpose of the matrix whose elements are the respective cofactors of the elements of the given matrix.

18. **B** $\mathbf{A}^{-1} = \frac{\text{adj}\,\mathbf{A}}{|\mathbf{A}|} = -\frac{1}{4}\begin{bmatrix} 1 & -3 \\ -2 & 2 \end{bmatrix} = \frac{1}{4}\begin{bmatrix} -1 & 3 \\ 2 & -2 \end{bmatrix}$.

19. E $|b_{ij}| = \begin{vmatrix} 2 & 12 \\ 2 & 2 \end{vmatrix} = 4 - 24 = -20.$

20. **D** $\begin{vmatrix} 2-\lambda & 3 \\ 2 & 1-\lambda \end{vmatrix} = (2-\lambda)(1-\lambda) - 6 = \lambda^2 - 3\lambda - 4 = 0 = (\lambda - 4)(\lambda + 1).$

$\therefore \lambda = 4, -1.$

21. E

22. **A** $\sum \mathrm{F} = 0$. $\mathrm{F}_A - 20\mathbf{k} + 60\mathbf{i} - 0.707(40)\mathbf{i} + 0.707(40)\mathbf{j} = 0$.

$\therefore \mathrm{F}_A = -31.7\mathbf{i} - 28.3\mathbf{j} + 20\mathbf{k}$. $F_A = 47$ N.

23. **B** $\sum \mathrm{M}_A = 0$.

$\mathrm{M}_A + 10\mathbf{i} + (20 \times 0.15)\mathbf{i} + (60 \times 0.15)\mathbf{k} - (28.3 \times 0.1)\mathbf{i} - (28.3 \times 0.1)\mathbf{j} = 0$.

$\therefore \mathrm{M}_A = -10.2\mathbf{i} - 2.83\mathbf{j} + 9\mathbf{k}$. $M_A = \sqrt{10.2^2 + 2.83^2 + 9^2} = 13.9$ N·m.

24. **A**

25. **C** $\mathbf{M} = 10\mathbf{i} + (60 \times 0.15)\mathbf{k} = 10\mathbf{i} + 9\mathbf{k}$. $M = \sqrt{100 + 81} = 13.45$ N·m.

26. **A** $\mathbf{v}_C = \mathbf{v}_B + \mathbf{v}_{C/B}$. A velocity polygon shows that $\mathbf{v}_B = \mathbf{v}_C$ so that $\omega_{BC} = 0$.

27. **C** $v_C = v_B = r\omega = 0.3 \times 10 = 3 \text{ m/s}$.

28. **D** Draw an acceleration polygon. $\mathbf{a}_C = \mathbf{a}_B + \mathbf{a}_{C/B}$. $\mathbf{a}_B$ is in y-direction, $\mathbf{a}_C$ is in x-direction, and $\mathbf{a}_{C/B}$ is $\perp \mathbf{r}_{BC}$. Using $a_B = 30$, $a_{C/B} = 30 \times \frac{5}{4} = 37.5 = 0.5\alpha_{BC}$. $\therefore \alpha_{BC} = 75 \text{ rad/s}^2$.

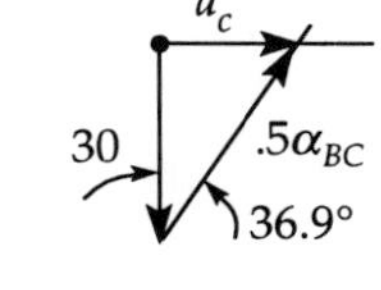

29. **D** From the acceleration polygon above, $a_C = 30 \times \frac{3}{4} = 22.5 \text{ rad/s}^2$.

30. **E**

$$\left.\begin{aligned} a_{\text{Coriolis}} &= 2\omega v = 2 \times 10 \times 2 = 40 \text{ m/s}^2. \\ a_{\text{normal}} &= r\omega^2 = 0.15 \times 10^2 = 15 \text{ m/s}^2. \end{aligned}\right\} \quad a = \sqrt{1600 + 225} = 42.7 \text{ m/s}^2.$$

Note: The Coriolis acceleration is normal to the direction of **v**, normal to the arm.

31. **C** $T \sin 30° \times 2L = PL$. $\therefore T = 1800$ N. $\delta = \dfrac{1800 \times 4/0.866}{\pi \times 0.002^2 \times 210 \times 10^9} = 3.15 \times 10^{-3}$ m.
Note: The end drops straight down, but the wire stretches in the direction of the wire. $\therefore$ The end drops $3.15/\sin 30° = 6.3$ mm.

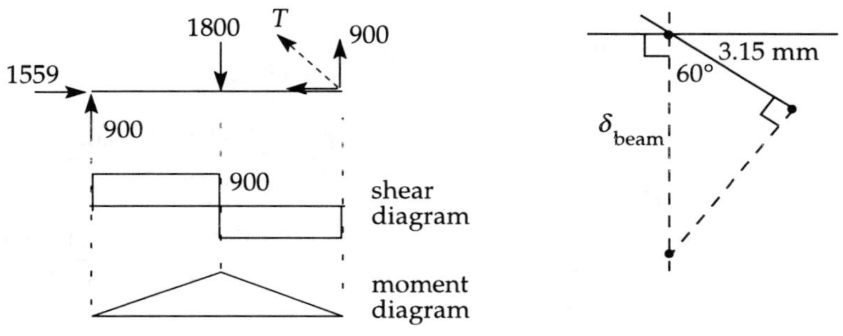

32. **E** The maximum tensile stress occurs at the outer fiber on the bottom of the beam where the moment is largest (i.e., under P).

$$\sigma_{\max} = \frac{My}{I} = \frac{(P/2 \times L)b/2}{ab^3/12} = \frac{3PL}{ab^2} = \frac{3 \times 1800 \times 2}{0.02 \times 0.06^2} = 150 \times 10^6 \text{ Pa.}$$

$$= \frac{3 \times 400 \times 72}{1 \times 3^2} = 9600 \text{ psi.}$$

33. **E**

$$\tau = \frac{VQ}{Ib} = \frac{(P/2)(ab/2)(b/4)}{(ab^3/12)a} = \frac{3P}{4ab} = \frac{3 \times 1800}{4 \times 0.02 \times 0.06} = 1.125 \times 10^6 \text{ Pa.}$$

$$= \frac{3 \times 400}{4 \times 1 \times 3} = 100 \text{ psi.}$$

34. **E** $\tau_{\max} = \dfrac{\sigma}{2} = \dfrac{P/A}{2} = \dfrac{1800}{2\pi \times 0.002^2} = 71.6 \times 10^6$ Pa. (Think of Mohr's circle.)

35. **E** The moment diagram results from integrating the shear diagram.

36. C $M_{max} = FL_2 = 2000 \times 1 = 2000\ \text{N} \cdot \text{m}.$

$$= 450 \times 3.28 = 1476\ \text{ft} \cdot \text{lb}.$$

$$\sigma = \frac{My}{I} = \frac{2000 \times 0.025}{0.02 \times 0.05^3/12} = 240 \times 10^6\ \text{Pa}.$$

$$= \frac{(1476 \times 12) \times 1.97/2}{0.787 \times 1.97^3/12} = 34{,}790\ \text{psi}.$$

37. D $\frac{VQ}{Ib} = \frac{2000 \times (0.02 \times 0.025) \times 0.0125}{\left(0.02 \times 0.05^3/12\right) \times 0.02} = 3 \times 10^6\ \text{Pa}.$

$$= \frac{450 \times (0.787 \times 0.985) \times 0.4925}{\left(0.787 \times 1.97^3/12\right) \times 0.787} = 435\ \text{psi}.$$

38. A The force F provides a torque of 2000 N·m and a force of 2000 N acting on the end of the steel shaft. This is best described by (A) since the top fibers experience tension. There is no normal stress in the circumferential direction, ruling out (B), (C) and (D).

39. E $\sigma = \frac{My}{I} = \frac{(2000 \times 2) \times 0.025}{\pi \times 0.025^4/4} = 326 \times 10^6\ \text{Pa}.$

$$= \frac{(450 \times 6.56 \times 12) \times 0.985}{\pi \times 0.985^4/4} = 47{,}200\ \text{psi}.$$

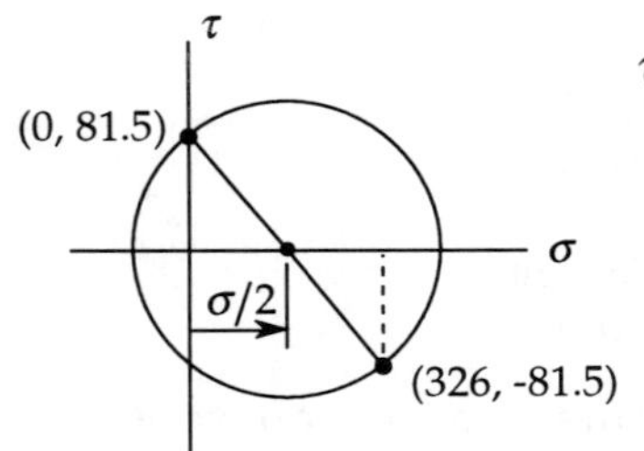

$$\tau = \frac{Tr}{J} = \frac{(2000 \times 1) \times 0.025}{\pi \times 0.025^4/2} = 81.5 \times 10^6\ \text{Pa}.$$

$$= \frac{(450 \times 3.28 \times 12) \times 0.985}{\pi \times 0.985^4/2} = 11{,}800\ \text{psi}.$$

$$\therefore \sigma_{max} = 163 + \left[163^2 + 81.5^2\right]^{1/2} = 345\ \text{MPa}.$$

$$\therefore \sigma_{max} = 23{,}600 + \left[23{,}600^2 + 11{,}800^2\right]^{1/2} = 50{,}000\ \text{psi}.$$

40. D $\delta = \frac{PL^3}{3EI} = \frac{2000 \times 2^3}{3 \times 210 \times 10^9 \times \pi \times 0.025^4/4} = 0.0828\ \text{m}.$

$$= \frac{450 \times (6.25 \times 12)^3}{3 \times 30 \times 10^6 \times \pi \times 0.985^4/4} = 2.85\ \text{in}.$$

41. B Sketch the DC circuit, inductor = short, capacitor = open:

15 V, 30 Ω, I_{AC}, Open

$$I_{DC} = \frac{15}{30} = 0.5\ \text{amp}$$

42. A The rms current through the inductor is composed of I_{DC} and I_{AC}: $I_{rms} = \sqrt{I_{DC}^2 + I_{AC}^2}$, $I_{DC} = 0.5$ A, from Problem 41. Sketch the AC circuit (the battery is a short):

$$I_{AC} = \frac{50\angle 0}{-j20 + \dfrac{(30)(+j40)}{30 + j40}} = \frac{50}{800 + j600} = 0.05\angle -36.9^\circ$$

$$I_{rms} = \sqrt{(0.5)^2 + (0.05)^2} = 0.502 \text{ A}$$

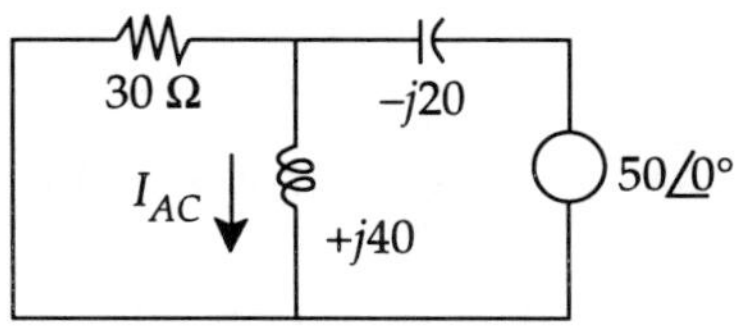

43. E Find the current, then the voltage across the inductor is $V_L = Iz$.

$$Iz = \frac{100\angle 0^\circ}{25 + j35 - j60} = 2.83\angle 45^\circ \text{ A}$$

$$V_L = (2.83\angle 45^\circ)(j35) = (2.83\angle 45^\circ)(35\angle 90^\circ) = 99\angle 135^\circ \text{ volts}$$

44. B $P = I^2R = (2.83)^2 25 = 200$ W.

$I = 2.83\angle 45^\circ$

θ

45. D power factor = $\cos\theta$. $pf = \cos 45^\circ = 0.707 \approx 0.71$.

$V = 100\angle 0^\circ$

46. E Find the voltage across the parallel combination $V = Iz_{eq}$ where $I = 10\angle 0^\circ$, $\dfrac{1}{z_{eq}} = \dfrac{1}{z_1} + \dfrac{1}{z_2} + \dfrac{1}{z_3} = \dfrac{1}{j20} + \dfrac{1}{-j10} + \dfrac{1}{5} = 0.2 + j0.05$,
$z_{eq} = 4.71 - j1.18$

$$V = (10\angle 0^\circ)(4.71 - j1.18) = 47.1 - j11.8 = 48.5\angle -14.04^\circ \text{ V}$$

$$I_c = \frac{V}{z_c} = \frac{48.5\angle -14.04^\circ}{-j10} = 4.85\angle 76^\circ \text{ A}$$

47. E From Problem 46, $V = 48.5\angle -14^\circ$ volts,

48. C Note that the phase sequence is V_{AM}, V_{BM}, V_{CM}, or ABC. Also: $V_{AB} = V_{AM} + V_{MB}$. In the problem statement V_{AB} is shown at 0°, thus V_{AM} will be at −30°. Since $V_{ph} = \dfrac{V_{line}}{\sqrt{3}}$, $V_{AM} = \dfrac{208}{\sqrt{3}}\angle -30^\circ = 120\angle -30^\circ$ V.

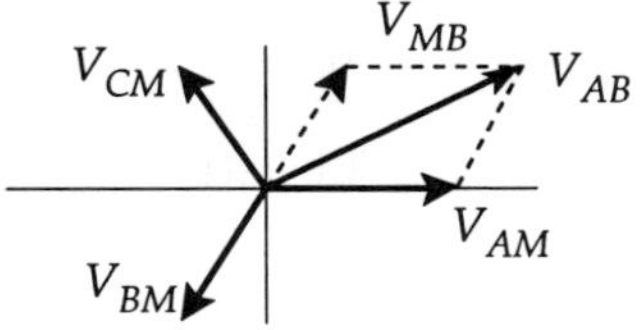

49. C From Problem 48: $V_{CM} = 120\angle 120^\circ$.

$$I_{CC'} = \frac{V_{CM}}{30 - j40} = \frac{120\angle 120^\circ}{5\angle -53.1^\circ} = 2.4\angle 173.1^\circ \text{ amp.}$$

50. **A** Since this is a balanced 3ϕ load, $P_T = 3P_{ph}$.

$$P_T = 3\left(I_{ph}\right)^2 R_{ph} = 3(2.4)^2 30 = 518.4 \text{ W}.$$

51. **D**

$$\text{ANEV} = \left[-25{,}500(A/P)_3^6 - 1200 + 900(A/F)_3^6\right] - \left[-42{,}000(A/P)_6^6 - 1800 + 1500(A/F)_6^6\right] = -\$330$$

52. **A** $i = \frac{6\%}{2} = 3\%;\quad i_e = (1.03)^2 - 1 = 0.061;\quad i_e = 6.1\%.$

53. **A** $0.05 \times 5000 = \$250$

$$P = 250(P/A)_{10}^6 + 5000(P/F)_{10}^6 = \$4632.$$

54. **B** $P = 20 \times (0.50 - 0.05) + 15(0.25 - 0.05) - 25 = -\$13.$

55. **C** $P = (20 + 15) \times (0.25 - 0.05) - 25 = -\$18.$

56. **D** $D = \frac{(90{,}000 - 15{,}000)}{5} = 15{,}000;\quad V_2 = 90{,}000 - 2 \times 15{,}000 = \$60{,}000.$

57. **D**

$$D_1 = \frac{90{,}000 \times 2}{5} = 36{,}000.\quad V_1 = 90{,}000 - 36{,}000 = 54{,}000$$

$$D_2 = \frac{54{,}000 \times 2}{5} = 21{,}600.\quad V_2 = 54{,}000 - 21{,}600 = 32{,}400$$

$$D_3 = \frac{32{,}400 \times 2}{5} = 12{,}960.\quad V_3 = 32{,}400 - 12{,}960 = 19{,}440$$

$$D_4 = \frac{19{,}440 \times 2}{5} = 7{,}776.\quad V_4 = 19{,}440 - 7{,}776 = 11{,}664 < 15{,}000$$

58. **C**

$$\text{ANEV} = -(170{,}000 - 90{,}000)(A/P)_5^{20} - (70{,}000 - 44{,}000) + (50{,}000 - 15{,}000)(A/F)_5^{20} = -\$48{,}000$$

59. **D**

$$\text{PNEV} = (160{,}000 - 100{,}000)(P/A)_5^{20} - (170{,}000 - 90{,}000) - (70{,}000 - 44{,}000)(P/A)_5^{20} + (50{,}000 - 15{,}000)(P/F)_5^{20} = \$36{,}000.$$

60. **A** $dC/dt = I - \left(L/t^2\right) = 0;\quad t = \sqrt{L/I}.$

61. **C** $s_3 = s_4 = 7.1677 = 0.6491 + x_4(7.5019).\quad \therefore x_4 = 0.869.$

$= 1.7632 = 0.17499 + x_4(1.7448).\quad \therefore x_4 = 0.910.$

62. **D** $\dot{W}_T = \dot{m}(h_3 - h_4) = 5(3658.4 - 2271) = 6940 \text{ kW}.$

$= 10(1523.6 - 1024) = 5000 \text{ BTU/sec} \text{ or } 7070 \text{ Hp}.$

$h_4 = 191.8 + 0.869 \times 2392.8 = 2271$ where x_4 is from Problem 61.

$= 94.02 + 0.91 \times 1022.1 = 1024$ where x_4 is from Problem 61.

63. **B** Use the energy equation (thermo or fluids):

$$\dot{Q} - \dot{W} = \dot{m}\left[\frac{p_2 - p_1}{\rho} + \frac{V_2^2 - V_1^2}{2} + g(z_2 - z_1)\right].$$

$$\dot{W}_P = \dot{m}\frac{p_2 - p_1}{\rho} = 5\frac{6000 - 10}{1000} = 30 \text{ kW}.$$

$$= 10\frac{(400 - 2)144}{62.4} = 9170 \text{ ft} \cdot \text{lb/sec} \text{ or } 16.7 \text{ Hp}.$$

64. **B** $\dot{Q}_B = \dot{m}(h_3 - h_2) = 5(3658.4 - 191.8) = 17\,330 \text{ kJ/s}.$

$= 10(1523.6 - 94.02) = 14,300 \text{ BTU/sec}.$

65. **D** $\eta = \dfrac{\dot{W}_T}{\dot{Q}_B} = \dfrac{6940}{17330} = 0.400.$

$$= \frac{5000}{14,300} = 0.350.$$

66. **C** $V = \dfrac{Q}{A} = \dfrac{0.02}{\pi \times 0.025^2} = 10.2 \text{ m/s}.$

$$= \frac{0.706}{\pi \times 0.985^2/144} = 33.4 \text{ fps}.$$

67. **E** $V_{\text{exit}} = V\dfrac{A}{A_{\text{exit}}} = 10.2\dfrac{5^2}{2^2} = 63.8 \text{ m/s}.$

$$= 33.4\frac{1.97^2}{0.79^2} = 208 \text{ fps}.$$

Bernoulli:

$$\frac{10.2^2}{2 \times 9.8} + \frac{p}{9800} = \frac{63.8^2}{2 \times 9.8} + \frac{0}{9800}. \quad \therefore p = 198.3 \times 10^4 \text{ Pa or } 1983 \text{ kPa}.$$

$$\frac{33.4^2}{2 \times 32.2} + \frac{p}{62.4} = \frac{208^2}{2 \times 32.2} + \frac{0}{62.4}. \quad \therefore p = 40,840 \text{ psf or } 284 \text{ psi}.$$

68. **C** Energy (surface to inlet): Use

$$\text{Re} = \frac{VD}{\nu} = \frac{10.2\times 0.05}{10^{-6}} = 5.1\times 10^5. \quad \frac{e}{D} = \frac{0.26}{50} = 0.0052. \quad \therefore f = 0.03.$$

$$30 = \frac{10.2^2}{2\times 9.8} + \frac{-100\,000}{9800} + \left(0.03\frac{L}{0.05} + 0.5\right)\frac{10.2^2}{2\times 9.8}. \quad \therefore L = 10.1\text{ m}.$$

$$\text{Re} = \frac{33.4\times 1.97/12}{10^{-5}} = 5.5\times 10^5. \quad \frac{e}{D} = \frac{0.00085}{1.97/12} = 0.0052. \quad \therefore f = 0.03.$$

$$98.4 = \frac{33.4^2}{2\times 32.2} + \frac{-14.7\times 144}{62.4} + \left(0.03\frac{L}{1.97/12} + 0.5\right)\frac{33.4^2}{2\times 32.2}. \quad \therefore L = 33.6\text{ ft}.$$

69. **C** $\dot{W}_P = \dot{m}\frac{\Delta p}{\rho} = 0.02\times 1000\frac{\Delta p}{1000} = 157\times 0.746. \quad \therefore \Delta p = 5860\text{ kPa}.$

$$= 0.706\times 1.94\frac{\Delta p}{1.94} = 157\times 550. \quad \therefore \Delta p = 122{,}300\text{ psf or } 850\text{ psi}.$$

70. **B** $\sum F = \rho Q(V_2 - V_1). \qquad p_1A_1 - F = \rho Q(V_2 - V_1).$

$$1983\,000\times\pi\times 0.025^2 - F = 1000\times 0.02(63.8 - 10.2). \quad \therefore F = 2820\text{ N}.$$

$$284\times\pi\times 0.985^2 - F = 1.94\times 0.706(208 - 33.4). \quad \therefore F = 626\text{ lb}.$$

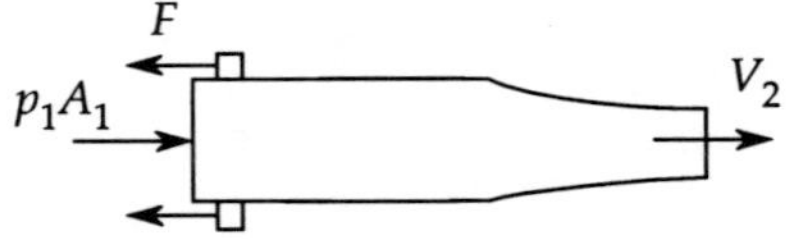

Equation Summaries

These summaries are useful for study

The following pages are Equation Summary Sheets of the FE/EIT subjects which rely most on equations. They are intended to be used for quick overviews, handy problem solving—and as part of a special strategy for preparing for the exam in its new format.

The selected equations are presented in the same format and nomenclature as is to be found in the NCEES Reference Handbook, a newsprint booklet which is given to all exam applicants. Be prepared! You may find some of the nomenclature to be different from what you are used to using, as the NCEES has apparently obtained some of it from older or obscure texts.

They can help you anticipate and prepare for NCEES obstacles

The current edition of the NCEES Reference Handbook has many extraneous equations which will in all likelihood be of little aid in solving exam problems. As discussed earlier, an excellent exam preparation strategy is to identify the equations you find to be most useful, then highlight them and become acquainted with their position in the Handbook, so that you may quickly access them in the 'clean' Handbook during the exam itself. This strategy should maximize your ability to perform as well as possible, given the present exam format.

Calculators are a good way to access equations during the exam

The Reference Handbook provided at the end of this book (and with your application kit) is not allowed into the exam. A clean one will be issued instead. You may pre-program your calculator to solve many of the problems on the FE exam, e.g., problems involving matrices, permutations, standard deviations, Mohr's circle problems, etc. Or you may choose to buy the HP-48G, which has hundreds of equations and constants preprogrammed. (We offer this superior calculator at a significant discount.) Not all states allow the use of the HP-48G, so be sure you check with your state board.

Mathematics

—Selected Equations from the NCEES Reference Handbook—

Straight Line:

$y = mx + b$ (slope - intercept form)

$y - y_1 = m(x - x_1)$ (point - slope form)

$m = \dfrac{y_2 - y_1}{x_2 - x_1}$ (slope)

$m_1 = -\dfrac{1}{m_2}$ (two perpendicular lines)

Quadratic Equation: $ax^2 + bx + c = 0$ roots $= \dfrac{-b \pm \sqrt{b^2 - 4ac}}{2a}$

Conic Sections:

	General Form	$h = k = 0$
Parabola:	$(y-k)^2 = 2p(x-h)$	$y^2 = 2px$ Focus: $(p/2, 0)$ Directrix: $x = -p/2$
Ellipse:	$\dfrac{(x-h)^2}{a^2} + \dfrac{(y-k)^2}{b^2} = 1$	$\dfrac{x^2}{a^2} + \dfrac{y^2}{b^2} = 1$ Focus: $\left(\sqrt{a^2 - b^2},\ 0\right)$
Hyperbola:	$\dfrac{(x-h)^2}{a^2} - \dfrac{(y-k)^2}{b^2} = 1$	$\dfrac{x^2}{a^2} - \dfrac{y^2}{b^2} = 1$ Focus: $\left(\sqrt{a^2 + b^2},\ 0\right)$
Circle:	$(x-h)^2 + (y-k)^2 = r^2$	$x^2 + y^2 = r^2$

Logarithms:

$\ln x = 2.3026 \log x$ $\quad \log xy = \log x + \log y$ $\quad \log x/y = \log x - \log y$

$\log_b b^n = n$ $\quad \log_b b = 1$

$\log x^c = c \log x$ $\quad \log 1 = 0$

If $b^c = x$, then $\log_b x = c$

Trigonometry:

$\sin\theta = y / r$ $\quad \cos\theta = x / r$

$\tan\theta = y / x$ $\quad \cot\theta = x / y$

$\csc\theta = r / y$ $\quad \sec\theta = r / x$

Law of Sines: $\dfrac{a}{\sin A} = \dfrac{b}{\sin B} = \dfrac{c}{\sin C}$

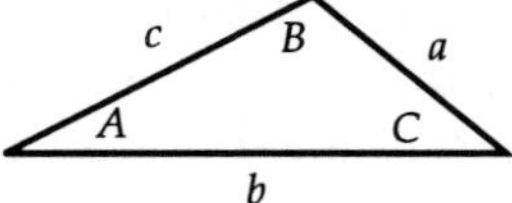

Law of Cosines: $a^2 = b^2 + c^2 - 2bc\cos A$

Identities:

$\tan\theta = \sin\theta / \cos\theta$

$\sin^2\theta + \cos^2\theta = 1$

$\sin(\alpha + \beta) = \sin\alpha\cos\beta + \cos\alpha\sin\beta$

$\cos(\alpha + \beta) = \cos\alpha\cos\beta - \sin\alpha\sin\beta$

$\sin 2\alpha = 2\sin\alpha\cos\alpha$

$\cos 2\alpha = \cos^2\alpha - \sin^2\alpha$

$= 2\cos^2\alpha - 1$

$= 1 - 2\sin^2\alpha$

Complex Numbers:

$i = \sqrt{-1}$ $\quad x + iy = re^{i\theta}$ $\quad \cos\theta = \dfrac{e^{i\theta} + e^{-i\theta}}{2}$

$r = \sqrt{x^2 + y^2}$ $\quad e^{i\theta} = \cos\theta + i\sin\theta$ $\quad \sin\theta = \dfrac{e^{i\theta} - e^{-i\theta}}{2i}$

$$(x + iy)^n = r^n(\cos n\theta + i\sin n\theta)$$

Matrices:

Transpose: $\mathbf{B} = \mathbf{A}^T$ if $b_{ji} = a_{ij}$

Inverse: $\mathbf{A}^{-1} = \dfrac{\text{adj}(\mathbf{A})}{|\mathbf{A}|}$

Adjoint: $\text{adj}(\mathbf{A}) =$ matrix formed by replacing $\mathbf{A}^T$ elements with their cofactors

Cofactor: cofactor = minor $\times (-1)^{h+k}$ where h = column, k = row

Minor: minor = determinant that remains after the common row and column are struck out

Vectors:

$$\mathbf{A}\cdot\mathbf{B} = a_x b_x + a_y b_y + a_z b_z = |\mathbf{A}||\mathbf{B}|\cos\theta = \mathbf{B}\cdot\mathbf{A}$$

$$\mathbf{A}\times\mathbf{B} = \begin{vmatrix} \mathbf{i} & \mathbf{j} & \mathbf{k} \\ a_x & a_y & a_z \\ b_x & b_y & b_z \end{vmatrix} = |\mathbf{A}||\mathbf{B}|\mathbf{n}\sin\theta = -\mathbf{B}\times\mathbf{A}$$

where $\mathbf{n}$ is $\perp$ plane of $\mathbf{A}$ and $\mathbf{B}$

$\mathbf{i}\cdot\mathbf{i} = \mathbf{j}\cdot\mathbf{j} = \mathbf{k}\cdot\mathbf{k} = 1 \qquad \mathbf{i}\times\mathbf{j} = \mathbf{k}, \quad \mathbf{j}\times\mathbf{k} = \mathbf{i}, \quad \mathbf{k}\times\mathbf{i} = \mathbf{j}$

Taylor Series: $f(x) = f(a) + \dfrac{f'(a)}{1!}(x-a) + \dfrac{f''(a)}{2!}(x-a)^2 + \cdots$

Maclaurin Series: a Taylor series with $a = 0$

Probability and Statistics:

$P(n,r) = \dfrac{n!}{(n-r)!}$ (permutation of n things taken r at a time)

$C(n,r) = \dfrac{P(n,r)}{r!} = \dfrac{n!}{r!(n-r)!}$ (combination of n things taken r at a time)

$\bar{x} = \dfrac{x_1 + x_2 + \cdots + x_n}{n}$ (arithmetic mean)

$\sigma^2 = \dfrac{\sum(x_i - \bar{x})^2}{n-1}$ (variance)

$\sigma = \sqrt{\text{variance}}$ (sample standard deviation)

$$\text{median} = \begin{cases} \text{middle value if odd number of items} \\ \tfrac{1}{2}(\text{sum of middle two values}) \text{ if even number of items} \end{cases}$$

mode = value that occurs most often

Calculus: $f'(x)=0\begin{cases}\text{maximum} & \text{if } f''(x)<0\\ \text{minimum} & \text{if } f''(x)>0\end{cases}$

L'Hospital's Rule: $\lim_{x\to a}\frac{f(x)}{g(x)}=\lim_{x\to a}\frac{f'(x)}{g'(x)}$ if $\frac{f(a)}{g(a)}=\frac{0}{0}$ or $\frac{\infty}{\infty}$

$$\frac{d}{dx}(uv)=u\frac{dv}{dx}+v\frac{du}{dx}$$

$$\frac{d}{dx}(\ln u)=\frac{1}{u}\frac{du}{dx}$$

$$\frac{d}{dx}(\sin u)=\cos u\frac{du}{dx}$$

$$\frac{d}{dx}\left(\frac{u}{v}\right)=\frac{v\,du/dx-u\,dv/dx}{v^2}$$

$$\frac{d}{dx}\left(e^u\right)=e^u\frac{du}{dx}$$

$$\frac{d}{dx}(\cos u)=-\sin u\frac{du}{dx}$$

$$\frac{d}{dx}\left(u^n\right)=nu^{n-1}\frac{du}{dx}$$

$$\int x^n dx=\frac{x^{n+1}}{n+1}\qquad n\neq -1$$

$$\int \sin x\,dx=-\cos x$$

$$\int \sin^2 x\,dx=\frac{x}{2}-\frac{\sin 2x}{4}$$

$$\int\frac{dx}{ax+b}=\frac{1}{a}\ln|ax+b|$$

$$\int \cos x\,dx=\sin x$$

$$\int \cos^2 x\,dx=\frac{x}{2}+\frac{\sin 2x}{4}$$

$$\int e^{ax}dx=\frac{1}{a}e^{ax}$$

Differential Equations: $y''+2ay'+by=f(x)$ (linear, 2nd order, constant coefficient, nonhomogeneous)

Homogeneous solution:

$$y_h(x)=C_1e^{r_1x}+C_2e^{r_2x}\quad \text{if } r_1\neq r_2\qquad \text{where } r^2+2ar+b=0$$

$$=(C_1+C_2x)e^{r_1x}\quad \text{if } r_1=r_2$$

$$=e^{-ax}(C_1\cos\beta x+C_2\sin\beta x)\quad \text{if } a^2<b.\qquad \beta=\sqrt{b-a^2}$$

Particular solution:

$$y_p=B \qquad \text{if } f(x)=A$$

$$=Be^{\alpha x}\qquad \text{if } f(x)=Ae^{\alpha x}$$

$$=B_1\sin\omega x+B_2\cos\omega x\qquad \text{if } f(x)=A_1\sin\omega x+A_2\cos\omega x$$

General solution: $y(x)=y_h(x)+y_p(x)$

Mechanics of Materials

—Selected equations from the NCEES Reference Handbook—

Definitions:

$$\sigma = \varepsilon E$$
$$\tau = \gamma G$$
$$E = 2G(1+\nu)$$
$$\nu = -\frac{\varepsilon_{lateral}}{\varepsilon_{longitudinal}}$$

E = modulus of elasticity

G = shear modulus

σ and τ = normal and shear stress

ε and γ = normal and shear strain

ν = Poisson's ratio

Uniaxial Loading:

$$\left.\begin{array}{l}\sigma = \dfrac{P}{A} \\ \varepsilon = \dfrac{\delta}{L}\end{array}\right\} \quad \delta = \frac{PL}{AE}$$

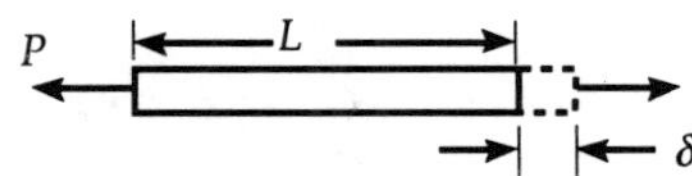

Thermal Deformation:

$$\delta_t = \alpha L(T - T_o)$$

α = coefficient of thermal expansion

Thin-walled Pressure Vessel:

$$\sigma_t = \frac{pD}{2t} \quad \text{hoop (circumferential) stress}$$

$$\sigma_a = \frac{pD}{4t} \quad \text{axial (longitudinal) stress}$$

t = cylinder thickness

D = cylinder diameter

p = pressure

Stress and Strain:

Stress Condition | Mohr's Circle | Maximum and Minimum Stresses

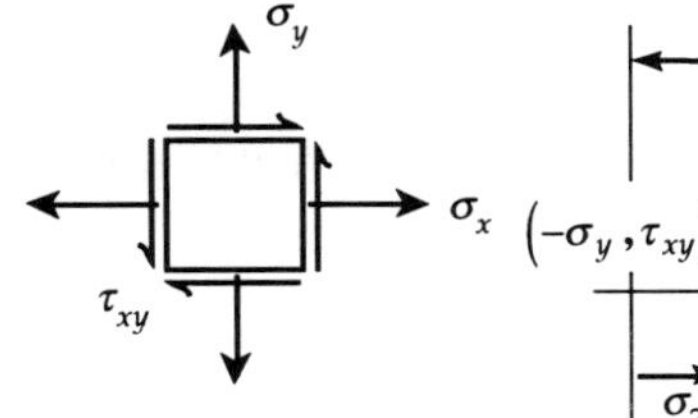

$$\sigma_1 = \sigma_{max} = \frac{\sigma_x + \sigma_y}{2} + \left[\left(\sigma_x - \sigma_y\right)^2/4 + \tau_{xy}^2\right]^{1/2}$$

$$\sigma_2 = \sigma_{min} = \frac{\sigma_x + \sigma_y}{2} - \left[\left(\sigma_x - \sigma_y\right)^2/4 + \tau_{xy}^2\right]^{1/2}$$

$$\tau_{max} = \frac{\sigma_1 - \sigma_2}{2} = \text{radius of Mohr's circle}$$

3-D Strain:

$$\varepsilon_x = \frac{1}{E}\left[\sigma_x - \nu\left(\sigma_y + \sigma_z\right)\right]$$

$$\gamma_{xy} = \frac{\tau_{xy}}{G}$$

Torsion:

$$\tau = \frac{Tr}{J} \quad \text{(shear stress)}$$

$$\phi = \frac{TL}{JG} \quad \text{(angle of twist)}$$

J = polar moment of inertia $= \pi r^4/2$ for a circle

T

L

Beams:

$$V = \frac{dM}{dx}$$

V = vertical shear force, M = bending moment

$$\sigma = -\frac{My}{I}$$

I = centroidal moment of inertia
$= bh^3/12$ for a rectangle
$= \pi r^4/4$ for a circle

y = distance from neutral axis

$$\tau = \frac{VQ}{Ib}$$

Q = moment of area between y-position and top or bottom

$$EIy'' = M$$

differential equation of deflection curve

Columns:

$$P_{cr} = \frac{\pi^2 EI}{k^2 L^2} \qquad k = \begin{cases} 1 & \text{ends pinned} \\ 0.5 & \text{ends fixed} \\ 0.7 & \text{one pinned, one fixed} \\ 2 & \text{one fixed, one free} \end{cases}$$

Dynamics

—Selected equations from the NCEES Reference Handbook—

Kinematics (motion only)

Tangential and Normal Components:

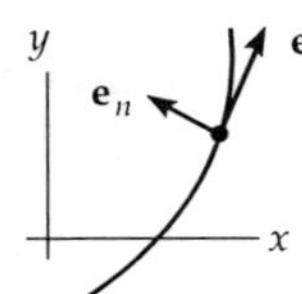

$$\mathbf{a} = \frac{dv_t}{dt}\mathbf{e_t} + \frac{v_t^2}{\rho}\mathbf{e_n}$$

$$\mathbf{v} = v_t \mathbf{e_t}$$

ρ = radius of curvature

Plane Circular Motion:

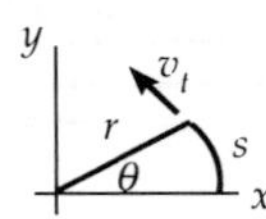

$$\mathbf{e_r} = -\mathbf{e_n} \qquad \mathbf{e}_\theta = \mathbf{e_t}$$

$$\omega = \dot{\theta} = \frac{v_t}{r} \qquad a_t = r\alpha$$

$$\alpha = \dot{\omega} = \ddot{\theta} = \frac{a_t}{r} \qquad a_n = \frac{v_t^2}{r} = r\omega^2$$

$$v_t = r\omega \qquad s = r\theta$$

Straight Line Motion:

$$s = s_o + v_o t + a_o t^2/2$$

$$v = v_o + a_o t$$

$$v^2 = v_o^2 + 2a_o(s - s_o)$$

Projectile Motion:

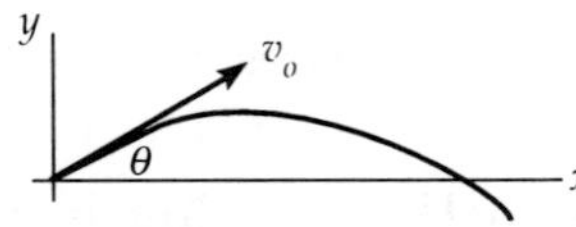

$$a_x = 0, \qquad a_y = -g$$

$$v_x = v_o \cos\theta \qquad x = v_o t \cos\theta$$

$$v_y = v_o \sin\theta - gt \qquad y = v_o t \sin\theta - \tfrac{1}{2}gt^2$$

Kinematics (forces and motion)

$$\sum \mathbf{F} = \frac{d}{dt}(m\mathbf{v}), \qquad \sum F_t = ma_t = m\frac{dv_t}{dt}, \qquad \sum F_n = ma_n = m\frac{v_t^2}{\rho}$$

Impulse and Momentum: $m[v_x(t) - v_x(0)] = \int_0^t F_x(t)dt$ or change in momentum = impulse

Work and Energy: $PE_1 + KE_1 + W_{1\to 2} = PE_2 + KE_2$ where

$$KE = \tfrac{1}{2}mv^2$$

$$PE = mgh \quad \text{(gravity)}$$

$$= \tfrac{1}{2}kx^2 \quad \text{(spring)}$$

$W_{1\to 2}$ = friction force work

Impact: $m_1v_1 + m_2v_2 = m_1v_1' + m_2v_2'$

$$e = -\frac{v_{1n}' - v_{2n}'}{v_{1n} - v_{2n}} = \begin{cases} 1 & \text{elastic} \\ 0 & \text{plastic} \end{cases}$$

v_1, v_2 = velocities before impact

v_1', v_2' = velocities after impact

Rotation: $I_o\alpha = \sum M_o$ where $I_o = \int(x^2 + y^2)dm$, rotation about O.

constant M:

$$\alpha = \frac{M}{I}$$

$$\omega = \omega_o + \frac{M}{I}t$$

$$\theta = \theta_o + \omega_o t + \frac{M}{2I}t^2$$

work and energy: $I_o\frac{\omega^2}{2} - I_o\frac{\omega_o^2}{2} = \int_{\theta_o}^{\theta} M d\theta$

Banking of Curves: $\tan\theta = \frac{v^2}{rg}$ where r = radius of curvature

θ = angle between surface and horizontal

Electric Circuits

—Selected Equations from the NCEES Reference Handbook—

Electrostatics:

$F_2 = \dfrac{Q_1 Q_2}{4\pi\varepsilon r^2}$ (force on charge 2 due to charge 1) ε = permittivity — $\mathrm{C^2/N \cdot m^2 = F/m}$ $= 8.85 \times 10^{-12}$ for air or free space

$E = \dfrac{Q}{4\pi\varepsilon r^2}$ (electric field intensity due to point charge Q — C)

$E_L = \dfrac{\rho_L}{2\pi\varepsilon r}$ (radial field due to line charge ρ_L — C/m)

$E_s = \dfrac{\rho_s}{2\varepsilon}$ (plane field due to sheet charge ρ_s — $\mathrm{C/m^2}$)

$Q = \oint \varepsilon \mathbf{E} \cdot d\mathbf{A}$ (enclosed charge — C)

$E = \dfrac{V}{d}$ (electric field between plates with potential difference V separated by the distance d)

$H = \dfrac{I}{2\pi r}$ (magnetic field strength due to current in long wire)

$B = \mu H$ (magnetic flux density)

$\mathbf{F} = I\mathbf{L} \times \mathbf{B}$ (force on conductor) $\mathbf{L}$ = length vector of conductor

DC Circuits:

Resistors: R

$V = IR$ (Ohm's law) $R_T = R_1 + R_2 + \cdots$ (series)

$P = VI = \dfrac{V^2}{R} = I^2 R$ (power) $R_T = \left[\frac{1}{R_1} + \frac{1}{R_2} + \cdots\right]^{-1}$ (parallel)

Capacitors: C, $+v(t)-$, $i(t)$

$i = C\dfrac{dv}{dt}$ $\quad$ $\underset{\text{stored}}{\text{energy}} = \frac{1}{2}Cv^2$ $\quad$ $C_{eq} = C_1 + C_2 + \cdots$ (parallel)

$v = \dfrac{1}{C}\int i\,dt$ $\quad$ $C_{eq} = \left[\frac{1}{C_1} + \frac{1}{C_2} + \cdots\right]^{-1}$ (series)

Inductors: L, $+v(t)-$

$i = \dfrac{1}{L}\int v\,dt$ $\quad$ $\underset{\text{stored}}{\text{energy}} = \frac{1}{2}Li^2$ $\quad$ $L_{eq} = L_1 + L_2 + \cdots$ (series)

$v = L\dfrac{di}{dt}$ $\quad$ $L_{eq} = \left[\frac{1}{L_1} + \frac{1}{L_2} + \cdots\right]^{-1}$ (parallel)

Kirchhoff Voltage Law (KVL): $\sum V_{rises} = \sum V_{drops} = 0$

Kirchhoff Current Law (KCL): $\sum I_{in} = \sum I_{out}$

Thevenin equivalent circuit:

V_{eq}, R_{eq}, a, b

$R_{eq} = \dfrac{V_{eq}}{I_{sc}}$ $\quad$ I_{sc} = short circuit current

V_{eq} = open circuit voltage

RC Transients:

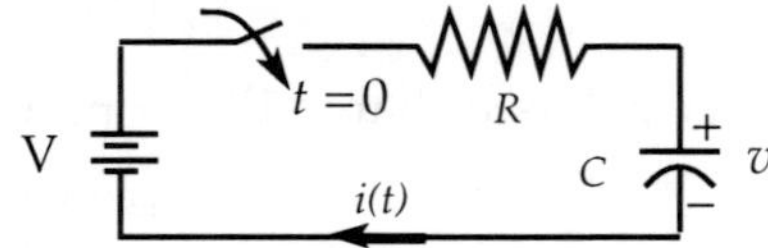

$$v_C(t) = v_C(0)e^{-t/RC} + V\left(1 - e^{-t/RC}\right)$$

$$i(t) = \{[V - v_C(0)]/R\}e^{-t/RC}$$

RL Transients:

V, t = 0, R, L, v_L, i(t)

$$v_L(t) = -i(0)Re^{-Rt/L} + Ve^{-Rt/L}$$

$$i(t) = i(0)e^{-Rt/L} + \frac{V}{R}\left(1 - e^{-Rt/L}\right)$$

Operational Amplifiers:

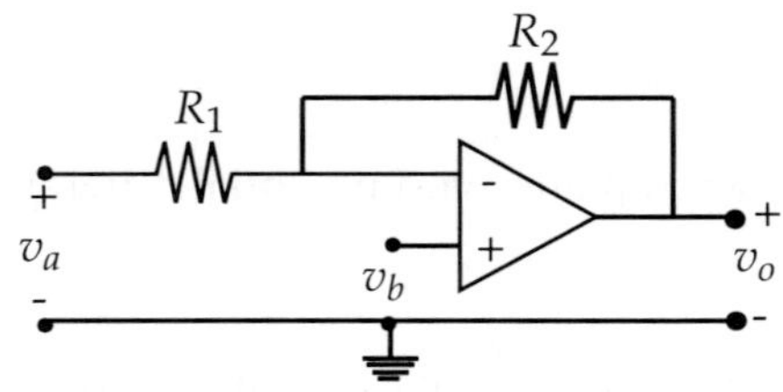

$$v_o = -\frac{R_2}{R_1}v_a + \left(1 + \frac{R_2}{R_1}\right)v_b$$

inverting if $v_b = 0$

non - inverting if $v_a = 0$

AC Circuits: (single phase)

$$f = \frac{1}{T} = \frac{\omega}{2\pi}$$

f = frequency (Hz)
T = period (sec)
ω = angular frequency (rad / s)

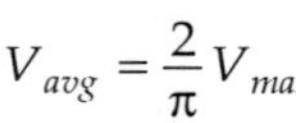

$V_{avg} = \frac{2}{\pi}V_{max}$ (full-wave rectified sine wave)

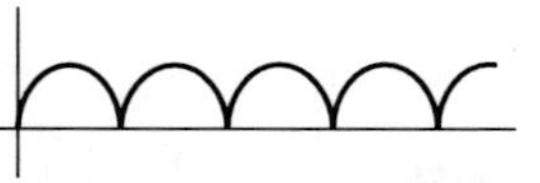

$V_{avg} = \frac{1}{\pi}V_{max}$ (half-wave rectified sine wave)

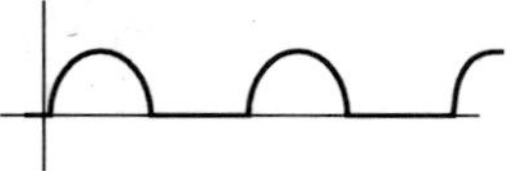

$V_{rms} = \frac{1}{\sqrt{2}}V_{max}$ (full-wave rectified sine wave)

$V_{rms} = \frac{1}{2}V_{max}$ (half-wave rectified sine wave)

Resistor: $Z = R$ Z = Impedance

Capacitor: $Z = -\frac{j}{\omega C} = -jX$ X = Reactance

Inductor: $Z = j\omega L = jX$

$V = IZ$

$P = \frac{1}{2}V_{max}I_{max}\cos\theta = V_{rms}I_{rms}\cos\theta$ (real power) ($\theta = 0$ for resistors)

$Q = \frac{1}{2}V_{max}I_{max}\sin\theta = V_{rms}I_{rms}\sin\theta$ (reactive power)

$\text{p.f.} = \cos\theta$ (power factor)

Resonance: $f = \frac{1}{2\pi\sqrt{LC}}$ (resonant frequency for series and parallel circuits)

Fluid Mechanics

—Selected Equations from the NCEES Reference Handbook—

Properties:

$\rho = \frac{m}{V}$ (density)

$\gamma = \rho g$ (specific weight)

$\nu = \frac{\mu}{\rho}$ (kinematic viscosity)

$\tau_n = -p$ (normal stress)

$\tau_t = \mu \frac{dv}{dy}$ (tangential stress)

$\mu =$ dynamic viscosity

Statics:

$p_2 - p_1 = -\gamma h$ (*h* is vertical upward)

$F_{\text{buoyant}} = \gamma V_{\text{displaced}}$ (Archimedes' principle)

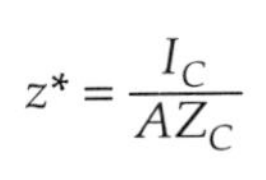

$F = \gamma h_C A$

$z^* = \frac{I_C}{A Z_C}$

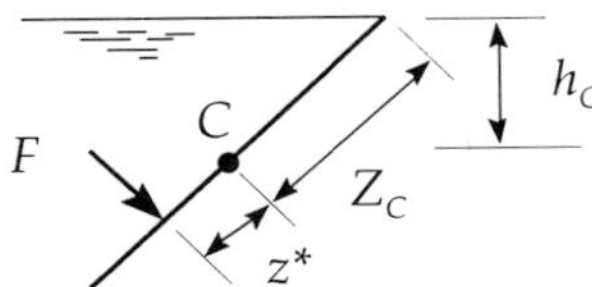

One-Dimensional Flows:

$A_1 V_1 = A_2 V_2$ (continuity equation)

$Q = AV$ (flow rate)

$\dot{m} = \rho A V$ (mass flow rate)

$$-\frac{\dot{W}_S}{\gamma Q} + \frac{p_1}{\gamma} + \frac{V_1^2}{2g} + z_1 = \frac{p_2}{\gamma} + \frac{V_2^2}{2g} + z_2 + h_f$$ (Energy Equation — if $h_f = \dot{W}_s = 0$, then Bernoulli Eq.)

$$h_f = f \frac{L}{D} \frac{V^2}{2g}$$ (Darcy's Equation — find *f* on Moody Diagram)

$$\text{Re} = \frac{VD\rho}{\mu}$$ (Reynolds Number)

$$h_{f,\text{ fitting}} = C \frac{V^2}{2g}$$ (minor losses — *C* is loss coefficient)

$$\sum \mathbf{F} = \rho Q(\mathbf{V}_2 - \mathbf{V}_1)$$ (Momentum equation)

Perfect Gas:

$p = \rho R T$ (perfect gas law)

$c = \sqrt{kRT}$ (speed of sound)

$\text{M} = \frac{V}{c}$ (Mach number)

Similitude:

If viscous effects dominate (internal flows) then Reynolds numbers on prototype and model are equated:

$$(\text{Re})_p = (\text{Re})_m \quad \text{or} \quad \left(\frac{V\ell\rho}{\mu}\right)_p = \left(\frac{V\ell\rho}{\mu}\right)_m$$

If gravity dominates (dams, weirs, ships) then Froude numbers are equated:

$$(\text{Fr})_p = (\text{Fr})_m \quad \text{or} \quad \left(\frac{V^2}{\ell g}\right)_p = \left(\frac{V^2}{\ell g}\right)_m$$

Open Channel:

$$Q = \frac{C}{n} A R^{2/3} S^{1/2} \quad \text{where} \quad R = \frac{A}{P_{wetted}}$$

$$C = \begin{cases} 1.0 & \text{metric} \\ 1.49 & \text{english} \end{cases}$$

Thermodynamics

—Selected Equations from the NCEES Reference Handbook—

Properties:

P	(absolute pressure, kPa or lbf/in^2)
$v = \dfrac{V}{m}$	(specific volume, m^3/kg or ft^3/lbm)
u	(internal energy, kJ/kg or Btu/lbm)
$h = u + Pv$	(enthalpy, kJ/kg or Btu/lbm)
s	(entropy, kJ/kg·K or Btu/lbm-°R)
C_p	(constant pressure specific heat, kJ/kg·K or Btu/lbm-°R)
C_v	(constant volume specific heat, kJ/kg·K or Btu/lbm-°R)
$x = \dfrac{m_v}{m_{total}}$	(quality)

Two phase system:

$v = v_f + xv_{fg}$ where $v_{fg} = v_g - v_f$ $\quad v_f$ = saturated liquid value

$h = h_f + xh_{fg}$ $\quad v_g$ = saturated vapor value

Ideal gas:

$Pv = RT, \quad PV = mRT$ where $R = \dfrac{\overline{R}}{M}, \quad \overline{R} = 8.314\dfrac{\text{kJ}}{\text{kmol}\cdot\text{K}}$ or $1545\dfrac{\text{ft - lbf}}{\text{lbmol - °R}}$

$\Delta u = C_v \Delta T, \qquad \Delta h = C_p \Delta T$

$$\Delta s = C_p \ln\frac{T_2}{T_1} - R\ln\frac{P_2}{P_1} = C_v \ln\frac{T_2}{T_1} + R\ln\frac{v_2}{v_1}$$

$$\frac{T_2}{T_1} = \left(\frac{P_2}{P_1}\right)^{\frac{k-1}{k}} = \left(\frac{v_1}{v_2}\right)^{k-1}, \qquad \left.\begin{aligned} P_2 v_2^k &= P_1 v_1^k \\ k &= C_p/C_v \end{aligned}\right\} \text{(constant entropy process)}$$

First law (system):

$q - w = \Delta u$ where $w = \int P dv$

$$= RT\ln\frac{v_2}{v_1} = RT\ln\frac{P_1}{P_2}$$ (isothermal process with ideal gas)

First law (control volume):

$h_i + V_i^2/2 = h_e + V_e^2/2$	(nozzles, diffusers)	i = inlet
$h_i = h_e + w$	(turbine, compressor)	e = exit
$h_i = h_e$	(throttling device, valve)	
$h_i + q = h_e$	(boilers, condensers, evaporators)	

Cycles: $\eta = \frac{W}{Q_H} = \frac{Q_H - Q_L}{Q_H}$ (efficiency) $\quad$ $COP = \frac{Q_H}{W}$ (heat pump)

$= 1 - \frac{T_L}{T_H}$ (Carnot cycle) $\quad$ $= \frac{Q_L}{W}$ (refrigerator)

Second Law: No engine can produce work while transferring heat with a single reservoir. (Kelvin-Planck)

No refrigerator can operate without a work input. (Clausius)

$\Delta S \geq \int \frac{\delta Q}{T}$ $\quad$ $\Delta S = \frac{Q}{T}$ (reservoir or T = const)

$\Delta S_{total} = \Delta S_{surr} + \Delta S_{system} \geq 0$ $\quad$ $\Delta S = C_p \ln \frac{T_2}{T_1}$ (solid or liquid)

Heat Transfer: $q = -kA\frac{dT}{dx}$ (conduction) $\quad$ k = conductivity

$= -kA\frac{T_2 - T_1}{L}$ (through a wall) $\quad$ $R = \frac{L}{kA}$ (resistance factor)

$q = hA(T_1 - T_2)$ (convection) $\quad$ $R = \frac{1}{hA}$ (resistance factor)

$= \varepsilon\sigma A\left(T_1^4 - T_2^4\right)F_{12}$ (radiation) $\quad$ h = convection coefficient

$\varepsilon = 1$ for black body (emissivity)

$\sigma = 5.67 \times 10^{-8} \frac{\text{W}}{\text{m}^2 \cdot \text{K}^4}$ (Stefan - Boltzmann constant)

$F_{12} = 1$ if one body encloses the other (shape factor)

English and SI Units

The following tables present the SI (Systems International) units and the conversion of English units to SI units, along with some of the more common conversion factors.

SI Prefixes

Multiplication Factor	*Prefix*	*Symbol*
10^{15}	peta	P
10^{12}	tera	T
10^{9}	giga	G
10^{6}	mega	M
10^{3}	kilo	k
10^{-1}	deci	d
10^{-2}	centi	c
10^{-3}	mili	m
10^{-6}	micro	μ
10^{-9}	nano	n
10^{-12}	pico	p
10^{-15}	femto	f

SI Base Units

Quanitiy	*Name*	*Symbol*
length	meter	m
mass	kilogram	kg
time	second	s
electric current	ampere	A
temperature	kelvin	K
amount of substance	mole	mol
luminous intensity	candela	cd

SI Derived Units

Quantity	*Name*	*Symbol*	*In Terms of Other Units*
area	square meter		m^2
volume	cubic meter		m^3
velocity	meter per second		m/s
acceleration	meter per second squared		m/s^2
density	kilogram per cubic meter		kg/m^3
specific volume	cubic meter per kilogram		m^3/kg
frequency	hertz	Hz	s^{-1}
force	newton	N	$m \cdot kg/s^2$
pressure, stress	pascal	Pa	$kg/(m \cdot s^2)$
energy, work, heat	joule	J	N · m
power	watt	W	J/s
electric charge	coulomb	C	A s
electric potential	volt	V	W/A
capacitance	farad	F	C/V
electric resistance	ohm	Ω	V/A
conductance	siemens	S	A/V
magnetic flux	weber	Wb	V · s
inductance	henry	H	Wb/A
viscosity	pascal second		Pa · s
moment (torque)	meter newton		N · m
heat flux	watt per square meter		W/m^2
entropy	joule per kelvin		J/K
specific heat	joule per kilogram-kelvin		J/(kg · K)
conductivity	watt per meter-kelvin		W/(m · K)

Conversion Factors to SI Units

English	*SI*	*SI Symbol*	*To Convert from English to SI Multiply by*
Area			
square inch	square centimeter	cm^2	6.452
square foot	square meter	m^2	0.09290
acre	hectare	ha	0.4047
Length			
inch	centimeter	cm	2.54
foot	meter	m	0.3048
mile	kilometer	km	1.6093
Volume			
cubic inch	cubic centimeter	cm^3	16.387
cubic foot	cubic meter	m^3	0.02832
gallon	cubic meter	m^3	0.003785
gallon	liter	L	3.785
Mass			
pound mass	kilogram	kg	0.4536
slug	kilogram	kg	14.59
Force			
pound	newton	N	4.448
kip (1000 lb)	newton	N	4448
Density			
pound/cubic foot	kilogram/cubic meter	kg/m^3	16.02
pound/cubic foot	grams/liter	g/L	16.02
Work, Energy, Heat			
foot-pound	joule	J	1.356
BTU	joule	J	1055
BTU	kilowatt-hour	kWh	0.000293
therm	kilowatt-hour	kWh	29.3

Conversion Factors to SI Units (continued)

English	*SI*	*SI Symbol*	*To Convert from English to SI Multiply by*
	Power, Heat, Rate		
horsepower	watt	W	745.7
foot pound/sec	watt	W	1.356
BTU/hour	watt	W	0.2931
BTU/hour-ft^2-°F	watt/meter squared-°C	$W/m^2 \cdot °C$	5.678
tons of refrig.	kilowatts	kW	3.517
	Pressure		
pound/square inch	kilopascal	kPa	6.895
pound/square foot	kilopascal	kPa	0.04788
inches of H_20	kilopascal	kPa	0.2486
inches of Hg	kilopascal	kPa	3.374
one atmosphere	kilopascal	kPa	101.3
	Temperature		
Fahrenheit	Celsius	°C	5/9 (°F-32)
Fahrenheit	kelvin	K	5/9 (°F+460)
	Velocity		
foot/second	meter/second	m/s	0.3048
mile/hour	meter/second	m/s	0.4470
mile/hour	kilometer/hour	km/h	1.609
	Acceleration		
foot/second squared	meter/second squared	m/s^2	0.3048
	Torque		
pound-foot	newton-meter	$N \cdot m$	1.356
pound-inch	newton-meter	$N \cdot m$	0.1130
	Viscosity, Kinematic Viscosity		
pound-sec/square foot	newton-sec/square meter	$N \cdot s/m^2$	47.88
square foot/second	square meter/second	m^2/s	0.09290
	Flow Rate		
cubic foot/minute	cubic meter/second	m^3/s	0.0004719
cubic foot/minute	liter/second	L/s	0.4719
	Frequency		
cycles/second	hertz	Hz	1.00

Conversion Factors

Length

1 cm	= 0.3937 in
1 m	= 3.281 ft
1 yd	= 3 ft
1 mi	= 5280 ft
1 mi	= 1760 yd
1 km	= 3281 ft

Area

1 cm^2	= 0.155 in^2
1 m^2	= 10.76 ft^2
1 ha	= 10^4 m^2
1 acre	= 100 m^2
1 acre	= 4047 m^2
1 acre	= 43560 ft^2

Volume

1 ft^3	= 28.32 L
1 L	= 0.03531 ft^3
1 L	= 0.2642 gal
1 m^3	= 264.2 gal
1 ft^3	= 7.481 gal
1 m^3	= 35.31 ft^3
1 acre-ft	= 43,560 ft^3
1 m^3	= 1000 L

Velocity

1 m/s	= 3.281 ft/s
1 mph	= 1.467 ft/s
1 mph	= 0.8684 knot
1 knot	= 1.688 ft/s
1 km/h	= 0.2778 m/s
1 km/h	= 0.6214 mph

Force

1 lb	= 4.448 x 10^5 dyne
1 lb	= 32.17 pdl
1 lb	= 0.4536 kg
1 N	= 10^5 dyne
1 N	= 0.2248 lb
1 kip	= 1000 lb

Mass

1 oz	= 28.35 g
1 lb	= 0.4536 kg
1 kg	= 2.205 lb
1 slug	= 14.59 kg
1 slug	= 32.17 lb

Work and Heat

1 BTU	= 778.2 ft-lb
1 BTU	= 1055 J
1 Cal	= 3.088 ft-lb
1 J	= 10^7 ergs
1 kJ	= 0.9478 ft-lb
1 BTU	= 0.2929W · hr
1 ton	= 12,000 BTU/hr
1 kWh	= 3414 BTU
1 quad	= 10^{15} BTU
1 therm	= 10^5 BTU

Power

1 Hp	= 550 ft-lb/s
1 HP	= 33,000 ft-lb/min
1 Hp	= 0.7067 BTU/s
1 Hp	= 2545 BTU/hr
1 Hp	= 745.7 W
1 W	= 3.414 BTU/hr
1 kW	= 1.341 Hp

Volume Flow Rate

1 cfm	= 7.481 gal/min
1 cfm	= 0.4719 L/s
1 m^3/s	= 35.31 ft^3/s
1 m^3/s	= 2119 cfm
1 gal/min	= 0.1337 cfm

Torque

1 N · m	= 10^7 dyne · cm
1 N · m	= 0.7376 lb-ft
1 N · m	= 10 197 g · cm
1 lb-ft	= 1.356 N · m

Viscosity

1 lb-s/ft^2	= 478 poise
1 poise	= 1 g/cm · s
1 N · s/m^2	= 0.02089 lb-s/ft^2

Pressure

1 atm	= 14.7 psi
1 atm	= 29.92 in Hg
1 atm	= 33.93 ft H_20
1 atm	= 1.013 bar
1 atm	= 1.033 kg/cm^2
1 atm	= 101.3 kPa
1 psi	= 2.036 in Hg
1 psi	= 6.895 kPa
1 psi	= 68 950 dyne/cm^2
1 ft H_20	= 0.4331 psi
1 kPa	= 0.145 psi

Index

CATALOG of titles by GREAT LAKES PRESS

Editor and Principal Author: Dr. Merle C. Potter, Ph.D., P.E.

NEW 5th!

FUNDAMENTALS OF ENGINEERING, 5th edition. Retail: $49.95

By far the most effective review for the FE/EIT exam. Written by *seven Ph.D. teaching professors.* Contains everything you need to pass the exam—no more, no less—for one low price. All new info-rich, easy-to-use text layout and organization. (Caution: other books contain massive quantities of information *that are not on the exam.*) •Includes hardcover reference book which reviews all 11 engineering subjects tested; with many tables and charts; 1100 problems and full solutions; two 8-hour practice exams (pre-test & readiness test); NCEES equation summaries and free NCEES Reference Handbook. 600 pages; 8.5"x11".

PROFESSIONAL CIVIL ENGINEERING EXAM REVIEW

1st edition. Retail: $69.95

A highly-effective review for the Principles & Practice of Engineering (PE) Civil Exam. Also an excellent desk reference for civil engineers. •The **only** review written by ***Ph.D. professors***—ten experts in their fields. •Includes reviews of all 15 major subject areas tested; hundreds of excellent exam-simulating practice problems with full solutions; equation summaries. The only review with all you need in one volume. 650 pages; hardcover.

GRE TIME•SAVER™, 2nd edition. Retail: $17.95

The most authoritative study guide for the GRE. All authors are *Ph.D. teaching professors* from major universities. Praised by the national Psychology honor society, Psi Chi. •Study same subjects as in other reviews...in *half* the time! •Includes diagnostic test and evaluation to identify your strengths and weaknesses, two practice tests and score-analysis charts, hundreds of problems, full solutions, complete reviews of all 3 subjects, and 3000-word vocabulary list with quick-phrases for easy memory. Hardcover, 510 pages.

GMAT TIME•SAVER™, 2nd edition. Retail: $17.95

The most authoritative study guide for the GMAT Test—authors are *Ph.D. teaching professors* from major universities. Now with fax-in writing evaluation offer lets you maximize the new writing portion of the GMAT. •Study the same material as in other reviews...in *half* the time! •Includes a diagnostic test and evaluation to identify your skills, two practice tests and score-analysis charts, hundreds of problems and full solutions, complete reviews of all subjects. 397 pages.

GRE ENGINEERING REVIEW, 1st edition. Retail: $19.95

The only full review for the difficult GRE Engineering Subject Test. •The *only* prep that reviews all the subjects covered in the exam—and *fully solves all problems.* •The *only* prep written by qualified *Ph.D. professors* (nine experts). •Includes a diagnostic test, two practice tests and score-analysis charts, complete reviews of all subjects, and a self-assessment scale. 297 pages; hardcover.

ORDER FORM

Name (print): ____________________

Street Address: ____________________

____

____

Day ph.: ____________________

Credit Card: VISA MC (circle one)

Number: ____________________ Exp.: __________

Name of Title	Price	Quan.	Multi-Discount!	Subtotal
Civil PE Exam Review	$69.95 x		− Take 20% off each book on orders of 3 or more!	=
Fundamentals of Engr	$49.95 x		−	=
GRE Time•Saver	$17.95 x		−	=
GMAT Time•Saver	$17.95 x		−	=
GRE/Engr Review	$19.95 x		−	=
Shipping	—		x $3 per item	=
			(Mich. residents add 6% sales tax.) TOTAL	$

Mail this order form, credit info., check or money order to:

Great Lakes Press POB 172 Grover, MO 63040

Ph. 1-800-837-0201/fax 1-314-273-6086

Your Suggestions and Comments help us continuously improve this review.